6

AF760849

Biofuel Crops

Production, Physiology and Genetics

FSC
www.fsc.org
MIX
Paper from
responsible sources
FSC® C013604

Biofuel Crops

Production, Physiology and Genetics

Edited by

Bharat P. Singh

Fort Valley State University
Fort Valley, Georgia
USA

www.cabi.org

CABI is a trading name of CAB International

CABI
Nosworthy Way
Wallingford
Oxfordshire OX10 8DE
UK

Tel: +44 (0)1491 832111
Fax: +44 (0)1491 833508
E-mail: info@cabi.org
Website: www.cabi.org

CABI
38 Chauncey Street
Suite 1002
Boston, MA 02111
USA

Tel: +1 800 552 3083 (toll free)
Tel: +1 (0)617 395 4051
E-mail: cabi-nao@cabi.org

© CAB International 2013. All rights reserved. No part of this publication may be reproduced in any form or by any means, electronically, mechanically, by photocopying, recording or otherwise, without the prior permission of the copyright owners.

A catalogue record for this book is available from the British Library, London, UK.

Library of Congress Cataloging-in-Publication Data

Biofuel crops : production, physiology, and genetics / edited by Bharat P. Singh, Fort Valley State University, Fort Valley, Georgia, USA.
pages cm
Includes bibliographical references and index.
ISBN 978-1-84593-885-7 (hbk)
1. Energy crops. 2. Energy crops--Breeding. 3. Energy crops--Genetics. 4. Biomass energy. I. Singh, Bharat P.

SB288.B57 2013
631--dc23

2013008421

ISBN: 978 1 84593 885 7

Commissioning editor: Claire Parfitt
Editorial assistant: Emma McCann
Production editor: Tracy Head

Typeset by SPi, Pondicherry, India.
Printed and bound in the UK by CPI Group (UK) Ltd, Croydon, CR0 4YY.

Contents

Contributors

Kossonou Guillaume Anzoua, Field Science Center for Northern Biosphere, Hokkaido University, Sapporo 060-0811, Japan. E-mail: koss@fsc.hokudai.ac.jp

Surinder S. Banga, ICAR National Professor, Punjab Agricultural University, Ludhiana-141004, India. E-mail: nppbg@pau.edu

Gary Stephan Bañuelos, Research Scientist, USDA-ARS, Water Management Research Unit, 9611 S. Riverbend Ave., Parlier, CA 93648, USA. E-mail: gary.banuelos@ars.usda.gov

Natalie S. Betts, ARC Centre of Excellence in Plant Cell Walls, School of Agriculture, Food and Wine, University of Adelaide, Urrbrae 5064, Australia. E-mail: natalie.betts@adelaide.edu.au

Sujata Bhargava, University of Pune, Pune, India. E-mail: sujata@unipune.ac.in

T.K. Biswas, Murray-Darling Basin Authority, GPO Box 1801, Canberra, ACT 2601, Australia. E-mail: tapas.biswas@mdba.gov.au

Serge Braconnier, Centre de Coopération Internationale en Recherche Agronomique pour le Développement (CIRAD), Montpellier, France. E-mail: serge.braconnier@cirad.fr

Rachel A. Burton, ARC Centre of Excellence in Plant Cell Walls, School of Agriculture, Food and Wine, University of Adelaide, Urrbrae 5064, Australia. E-mail: rachel.burton@adelaide.edu.au

Caitlin S. Byrt, ARC Centre of Excellence in Plant Cell Walls, School of Agriculture, Food and Wine, University of Adelaide, Urrbrae 5064, Australia. E-mail: caitlin.byrt@adelaide.edu.au

Isabel S. Carvalho, IBB/CGB – Faculty of Sciences and Technology, University of Algarve Campus de Gambelas, Ed. 8, 8005-139 Faro, Portugal. E-mail: icarva@ualg.pt

M. Czakó, Department of Biological Sciences, University of Southern Carolina, 700 Sumter St, Columbia, SC 29208, USA. E-mail: Czako-mihaly@sc.edu

Karaj Singh Dhillon, Professor (Retd), Department of Soil Science, Punjab Agricultural University, Ludhiana 141 004, Punjab, India. E-mail: dhillon_karaj@yahoo.com

Ismail Dweikat, University of Nebraska, Lincoln, Nebraska, USA. E-mail: idweikat@unlnotes.unl.edu

C. Eynck, Visiting Scientist, Agriculture and Agri-Food Canada, Plant Breeder, Linnaeus Plant Sciences Inc., 107 Science Place, Saskatoon, Saskatchewan, Canada S7N 0X2. E-mail: christina.eynck@agr.gc.ca

N.K. Fageria, National Rice and Bean Research Center of Embrapa, Caixa Postal 179, Santo Antônio de Goiás, Goiás State, CEP 75375-000, Brazil. E-mail: fageria@cnpaf.embrapa.br

K.C. Falk, Research Scientist, Agriculture and Agri-Food Canada, 107 Science Place, Saskatoon, Saskatchewan S7N 0X2, Canada. E-mail: kevin.falk@agr.gc.ca

Naser Farrokhi, Department of Agronomy and Plant Breeding, Faculty of Agriculture, Shahrood University of Technology, Shahrood, Iran. E-mail: nfarrokh@nigeb.ac.ir

John H. Fike, Crop and Soil Environmental Sciences, Virginia Polytechnic Institute and State University, Blacksburg, VA 24061, USA. E-mail: jfike@vt.edu

Leslie H. Groom, USDA Forest Service, Southern Research Station, Pineville, LA 71360, USA. E-mail: lgroom@fs.fed.us

Anna Hale, USDA-ARS Sugarcane Research Unit, 5883 USDA Rd, Houma, LA 70360, USA. E-mail: Anna.Hale@ars.usda.gov

S. Hemaiswarya, Department of Biotechnology, Indian Institute of Technology Madras, Chennai-600 036, India. Present address: IBB/CGB – Faculty of Sciences and Technology, University of Algarve, Campus de Gambelas, Ed. 8, 8005-139 Faro, Portugal. E-mail: iswaryahema@gmail.com

J.G. Isebrands, Environmental Forestry Consultants, POB 54, New London, WI 54961l, USA. E-mail: efcllc@athenet.net

Kurt H. Johnsen, USDA Forest Service, Southern Research Station, Southern Institute of Forest Ecosystems Biology, Research Triangle Park, NC 27709, USA. E-mail: kjohnsen@fs.fed.us

Eric J. Jokela, School of Forest Resources and Conservation, University of Florida, Gainesville, FL 32611, USA. E-mail: ejokela@ufl.edu

Masazumi Kayama, Kyushu Research Centre, Forestry and Forest Products Research Institute, Kumamoto, Japan. Current address: Japan International Research Center for Agricultural Sciences (JIRCAS) 1-1, Ohwashi, Tsukuba, Ibaraki 305-8686, Japan. E-mail: kayama@affrc.go.jp

Bill Kovarik, Box 6932, School of Communication, Radford University, Radford, VA 24142, USA. E-mail: wkovarik@radford.edu

C. Ganesh Kumar, Chemical Biology Laboratory, Indian Institute of Chemical Technology (IICT), Uppal Road, Hyderabad 500607, India. E-mail: cgkumar@iict.res.in; cgkumar5@gmail.com

Christopher Q. Lan, Department of Chemical and Biological Engineering, University of Ottawa, 161 Louis Pasteur Private, Ottawa, Ontario, Canada, K1N 6N5. E-mail: Christopher.Lan@uottawa.ca

Ratikanta Maiti, Advisor, Vibha Seeds, Hyderabad-500081, India and Ex-Professor, Universidad de las Americas-Puebla, Santa Catarina Martir, Cholula, CP 72 820, Puebla, Mexico. E-mail: ratikanta.maiti@gmail.com

L. Márton, Department of Biological Sciences, University of Southern Carolina, 700 Sumter St, Columbia, SC 29208, USA. E-mail: drlaszlomarton@gmail.com

Steven E. McKeand, Department of Forestry and Environmental Resources, North Carolina State University, Raleigh, NC 27695, USA. E-mail: Steve_McKeand@ncsu.edu

P.J. Minogue, North Florida Research and Education Center, University of Florida, 155 Research Road, Quincy, FL 32351-5677, USA. E-mail: pminogue@ufl.edu

L.A.C. Moraes, National Soybean Research Center of Embrapa, Caixa Postal 231, Londrina, Paraná State, CEP 86001-970, Brazil. E-mail: larissa.moraes@cnpso.embrapa.br

A. Moreira, National Soybean Research Center of Embrapa, Caixa Postal 231, Londrina, Paraná State, CEP 86001-970, Brazil. E-mail: Adonis@cnpso.embrapa.br

K. Muthukumarappan, Professor, Department of Agricultural and Biosystems Engineering, South Dakota State University, Brookings, South Dakota, USA. E-mail: muthukum@sdstate.edu

C. Dana Nelson, USDA Forest Service, Southern Research Station, Southern Institute of Forest Genetics, Saucier, MS 39574, USA. E-mail: dananelson@fs.fed.us

Eric Obeng, 1005 State University Drive, Fort Valley State University, Fort Valley, GA 31030, USA. E-mail: obenge@fvsu.edu

David J. Parrish, Crop and Soil Environmental Sciences, Virginia Polytechnic Institute and State University, Blacksburg, VA 24061, USA. E-mail: dparrish@vt.edu

J.V. Patil, Directorate of Sorghum Research (DSR), Rajendranagar, Hyderabad-500030, India. E-mail: jvp@sorghum.res.in

Gary F. Peter, School of Forest Resources and Conservation, University of Florida, Gainesville, FL 32611, USA. E-mail: gfpeter@ufl.edu

Somashekhar Punnuri, 1005 State University Drive, Fort Valley State University, Fort Valley, GA 31030, USA. E-mail: punnuris@fvsu.edu

R. Raja, Aquatic Energy LLC, Lake Charles, LA 70629, USA. Present address: IBB/CGB – Faculty of Sciences and Technology, University of Algarve, Campus de Gambelas, Ed. 8, 8005-139 Faro, Portugal. E-mail: rraja307@yahoo.co.in

R. Ravikumar, Aquatic Energy LLC, Lake Charles, LA 70629, USA. E-mail: drravialg@yahoo.com

Belum V.S. Reddy, International Crops Research Institute for the Semi-Arid Tropics (ICRISAT), Patancheru-502324, India. E-mail: b.reddy@cgiar.org

Donald L. Rockwood, Professor Emeritus, School of Forest Resources and Conservation, University of Florida, 118 Newins-Ziegler Hall, Room 2, Bldg 182, Gainesville, FL 32611-0410, USA. E-mail: dlrock@ufl.edu

Robert B. Rummer, USDA Forest Service, Southern Research Station, Auburn, AL 36849, USA. E-mail: rrummer@fs.fed.us

Pratik Satya, Senior Scientist, Central Research Institute for Jute and Allied Fibres, Barrackpore-700120, Kolkata, India. E-mail: pscrijaf@gmail.com

Peer M. Schenk, Algae Biotechnology Laboratory, School of Agriculture and Food Sciences, The University of Queensland, Brisbane, Queensland 4072, Australia. E-mail: p.schenk@uq.edu.au

Holger Schuhmann, Algae Biotechnology Laboratory, School of Agriculture and Food Sciences, The University of Queensland, Brisbane, Queensland 4072, Australia. E-mail: h.schuhmann@uq.edu.au

Ralph E.H. Sims, Professor of Sustainable Energy, Director, Centre for Energy Research, School of Engineering and Technology, Massey University, Private Bag 11222, Palmerston North, New Zealand. E-mail: R.E.Sims@massey.ac.nz

Bharat P. Singh, 1005 State University Drive, Fort Valley State University, Fort Valley, GA 31030, USA. E-mail: singhb@fvsu.edu

Hari P. Singh, 1005 State University Drive, Fort Valley State University, Fort Valley, GA 31030, USA. E-mail: singhh@fvsu.edu

P. Srinivasa Rao, International Crops Research Institute for the Semi-Arid Tropics (ICRISAT), Patancheru-502324, India. E-mail: p.srinivasarao@cgiar.org; psrao72@gmail.com

A.V. Umakanth, Directorate of Sorghum Research (DSR), Rajendranagar, Hyderabad 500030, India. E-mail: umakanth@sorghum.res.in

Ryan P. Viator, USDA-ARS Sugarcane Research Unit, 5883 USDA Rd, Houma, LA 70360, USA. E-mail: Ryan.Viator@ars.usda.gov

Suhas P. Wani, Resilient Dryland Systems, International Crops Research Institute for the Semi-Arid Tropics (ICRISAT), Patancheru-502324, Andhra Pradesh, India. E-mail: s.wani@cgiar.org

Martin Weih, Department of Crop Production Ecology, Swedish University of Agricultural Sciences, PO Box 7043, SE-750 07, Uppsala, Sweden. E-mail: martin.weih@slu.se

C.M.J. Williams, Formerly, Senior Scientist, Sustainable Systems, South Australian Research and Development Institute, GPO Box 397, Adelaide, South Australia, 5001, Australia. Now retired. Present address: 2 Grand Central Avenue, Hallett Cove, South Australia, 5158, Australia. E-mail: cmjwill@bigpond.com

Toshihiko Yamada, Field Science Center for Northern Biosphere, Hokkaido University, Sapporo 060-0811, Japan. E-mail: yamada@fsc.hokudai.ac.jp

Preface

Liquid biofuels have been around for more than a century. Inventors of automotive engines had envisioned farm-grown energy sources to play an important role in supplying fuel to run these vehicles. Biofuels have met their extinction several times due to the price advantage enjoyed by fossil fuels, but always managed to resurface after an interval during times of petroleum shortage. Farm survival and rural development has been the main impetus behind promoting biofuels. However, during the last three or four decades environmental benefits and energy security have also emerged as reasons for the use of biofuel thereby providing it a more stable market base. Global biofuel production has grown from 16 billion l in 2000 to more than 100 billion l in 2010. Currently, biofuels account for approximately 3% of total world road transport fuel, but they are projected to meet up to 27% of the world demand by 2050. Under this scenario, biofuels may help reduce CO_2 emission by 2.1 gigatonnes (Gt) per year.

To be sure, critics of biofuel abound. There has been fierce debate concerning the wisdom of using food crops for biofuel and holding it responsible for grain price rises of recent years. The greenhouse gas remediation value of biofuel has also been questioned with the argument that the inputs of fossil fuels required in tillage, fertilizers, transportation and processing essentially negates the perceived CO_2 sequestration advantage. Regardless of the validity of these claims and the data base from which these conclusions were derived, these criticisms had the positive impacts of pushing the biofuel industry to chart a future course based on more sustainable and non-food source lignocellulosic feedstocks that indisputably provide CO_2 remediation benefits and to develop processes essential for their efficient conversion to biofuel. In addition, much research and development emphasis has been placed on making the biofuel fully compatible with the existing distribution systems and to make the volumetric energy content of biofuels equivalent to fossil fuels.

Through this book, I have attempted to disseminate my own knowledge and experience together with those of other eminent scientists in the biofuel field concerning best management practices for the production of the important biofuel crops. Physiological and genetic characteristics of these crops are also fully discussed because of their value for future improvements in their adaptability, productivity and management. Hopefully, all segments of people involved with biofuel – researchers, professionals, farmers, industry and public energy policy makers – will find this book useful. Keeping in mind that many plant science and bioprocessing disciplines now offer courses in biofuel crops and graduate degree programmes in bioenergy are in the planning or initial offering stages, especial efforts have been made to make the content beneficial for teachers and students engaged in teaching or learning about biofuels. The book begins with a review of the history of biofuel narrated by a noted historian on this subject. To make it broad based and suitable for textbook adoption, separate chapters have been devoted to emerging technologies for biofuel production, cell wall structure and destructing approaches, cellulosic biofuel crops role in phytoremediation, and feedstock pretreatment methods. Detailed discussion on the physiological and genetic research challenges and opportunities for the improvement of biofuel crops at large also constitutes a chapter. The book comprehensively covers taxonomy, genetics, breeding, physiology and biochemistry of algae and different technologies of their conversion to biofuel in view of the future potential of algal biofuel. I was fortunate to be able to assemble experts from different parts of the world who have worked on the crops described and willing to take time out of their busy schedules to present an in-depth discussion on topics assigned to them. I am grateful for their contribution. I am also obliged to my associate, Eric Obeng, who has been immensely helpful during the compilation and editing of this book.

Bharat Singh
Fort Valley, Georgia, USA
September 2012

1 Biofuels in History

Bill Kovarik
School of Communication, Radford University, Virginia, USA

1.1 Introduction

Biofuels were humanity's first liquid fuels. They include vegetable oils, animal fats, ethanol from crops, and methanol and turpentine from wood – all of these pre-date the general use of petroleum for illumination, cooking, heating and transportation (Bailey, 1975). During the early stages of the industrial revolution, biofuels powered the first lamps and internal combustion engines. The shift from biofuel to petroleum products like kerosene and gasoline as the primary fuel source took place in the 1860s for oil lamps, and in the early 20th century for automotive fuels. The shift was gradual in Europe but rapid in the USA due to differences in tax policies. Biofuels have continued to be used, straight or in blends with petroleum products, as fuels for diesel and spark ignition engines throughout the 19th and 20th centuries. Biofuels have been especially useful as additives for gasoline in order to safely improve fuel octane.

During the 20th century, biofuels were employed in most countries, especially during wartime emergencies or when there was concern about national energy self-sufficiency. Although biofuels such as ethanol and vegetable oils tend to be somewhat more expensive than petroleum in global markets, the issue of expense is not merely a short term market problem but also a long-term issue involving environment, agriculture and national security trade-offs.

Biofuels have been used in some form over four broad historical epochs:

1. Until the mid-19th century, as primary lamp and cooking fuels until the advent of petroleum and other competing energy forms such as gas and electric lighting.
2. Around the early 20th century, for internal combustion engines, when fuel quality and petroleum depletion were serious concerns.
3. In the mid- to late-20th century, for internal combustion engines, when international oil politics, especially the Arab oil embargoes of the 1970s, spurred national energy security investments.
4. In the early 21st century, for internal combustion engines, when global concerns about climate change, biodiversity and sustainability framed the larger context of energy research and policy debates.

Research into biofuels has progressed steadily over more than a century concentrated in the fields of agriculture, automotive and chemical engineering. Although the research was often overshadowed by political debates, scientists have understood that their work has the potential to shape these policies over the long run. Biofuels research can be characterized into: (i) optimizing production of agricultural, forest and aquatic biomass resources; (ii) improving

©CAB International 2013. *Biofuel Crops: Production, Physiology and Genetics* (ed. B.P. Singh)

the conversion of traditional starch and sugar crops (first generation) and cellulosic materials (second generation) via microbial and thermal methods; and (iii) testing of biofuels for engine compatibility and emissions modelling (Bente, 1984). Novel approaches for more direct low-energy production of renewable alkenes have been suggested as characterizing a third or next generation for biofuels (Kovarik, 2009). Other horizons include use of relatively limitless marine biomass systems and direct production of hydrogen from photosynthetic processes.

The practical aims of biofuels researchers have historically been: to assist in rural economic development; to increase national energy and economic security; and to provide cleaner ('reformulated') gasoline by replacing octane-enhancing additives such as tetraethyl lead and benzene, which are public health hazards.

On a somewhat more idealistic level, scientists have searched for methods to use solar energy in liquid fuel form in order to live within energy 'income' as opposed to squandering fossil fuel 'capital'. Many of the great scientists and engineers of the 20th century expressed enthusiastic support for biofuels, including Rudolph Diesel, Henry Ford, Alexander Graham Bell, Harry Ricardo and Charles F. Kettering (Kovarik, 1998).

Biofuels such as ethanol and vegetable oils may have serious drawbacks, especially in situations where they may compete with food or food crops. It is significant that similar concerns have historically guided biofuels development, especially in Europe during World War II and in Asia in the 1950s. Another serious but more recent concern involves the loss of biodiversity when tropical forests are sacrificed for biofuels plantations in Asia, Africa and Latin America.

Like any technology, biofuels hold both promise and peril, and it is a mark of their increasing significance that the debate over biofuels has become more contentious in the early decades of the 21st century. Hope exists that policy initiatives facilitating the social construction of this technology can curb the worst potential abuses. This hope is particularly animated by a high level of scientific concern that renewable energy technologies need to be deployed on a broad scale in order to shift away from fossil energy dependence, promote rural development and, in the process, stave off the worst impacts of climate change over the coming decades and centuries (Gomez *et al.*, 2008).

1.2 Biofuels for Illumination in the 19th and Early 20th Centuries

Forms of liquid energy derived from renewable plant material were well known and widely used for many thousands of years, and by the early 1700s lamps fuelled by vegetable oils and fats lit up the major streets in European and American cities (Crosby, 2006). New kinds of lamps, such as Ami Argand's 1783 carburetted design, contributed to better street and home lighting.

Among the biofuels available in the 1800s were whale oil (US$1.30–2.50 per gallon at the time), lard oil (US$0.90 per gallon), and camphene (at US$0.50 per gallon). Camphene was a somewhat irregular blend of ethanol, turpentine and camphor oil (Williamson and Daum, 1959). In the 1830s, camphene 'easily took the lead as the illuminant' because it was 'a decided improvement on other oils then in use,' especially lard oils, according to a lamp manufacturer's *History of Light* (Welsbach Gas Co., 1909). By 1860, thousands of distilleries produced tens of millions of gallons of alcohol per year for lighting in the USA and Europe (Brachvogel *et al.*, 1907).

Most existing histories of energy mistakenly depict a direct transition from whale oil to kerosene in the mid-19th century when, in fact, a large and diverse market for lamps and fuels was in place long before petroleum was introduced. Although whale oil was one of the significant fuel industries in the early to mid-19th century, it was fairly small in comparison to others. Whale oil use peaked in the USA at around 15 to 18 million gallons in 1847 (Starbuck, 1878), but the market for camphene was far larger in the USA at around 90 million gallons.

In 1862, the US Congress imposed a tax of US$2.08 per gallon on alcohol as part of the Internal Revenue Act to pay for the US Civil War. The tax was meant to apply to beverage alcohol, but without any specific exemption, it also applied to camphene as well. The tax crushed competition from biofuels and provided an important boost for the petroleum industry.

'The imposition of the internal-revenue tax on distilled spirits ... increased the cost of this "burning fluid" beyond the possibility of using it in competition with kerosene', said Rufus F. Herrick, an engineer with the Edison Electric Testing Laboratory who wrote one of the first books on the use of alcohol fuel (Herrick, 1907). The tax 'had the effect of upsetting [the distilleries] and in some cases destroying them,' said US Internal Revenue Service commissioner David A. Wells in 1872 (US Senate, 1907). The effect was 'disastrous to great industries', according to a chemical engineer of the era (Tweedy, 1917).

1.2.1 German biofuels programmes 1890–1916

The competition between petroleum and biofuel ethanol was quite different in Europe, where there were no tax barriers holding back industrial uses of ethanol. With few domestic oil reserves, German and French governments wanted to develop alternatives to petroleum. In addition, European governments were attempting to strike a balance between conservative agrarians who wanted strong farm markets and the progressive urbanites who wanted cheap food. Both groups could be served by creating rural value-added industries using surplus farm products.

Germany created the world's first large-scale biofuels industry in the decades before World War I as a way to promote rural development and national self-sufficiency. German Kaiser Wilhelm found he could 'satisfy the discontented agrarians' by encouraging the use of alcohol fuel made from potatoes (London Times, 1902). The German programme involved tariffs on imported oil, farm distillery construction, promotion of ethanol-fuelled appliances, and research into ethanol-fuelled trucks, automobiles and locomotives. Beginning in 1899, the programme was administered by the Centrale fur Spiritus Verwerthung (Office of Alcohol Sales). In 1903, the Reichstag approved a tariff on oil to expand the farm ethanol production infrastructure. Potato alcohol was seen as the 'final solution of the oil problem and the means by which the grasp of the great [Standard oil] monopoly will be broken' (New York Times, 1903). A network of small farm 'Materialbrennereien' distilleries was put in place. Estimates of its size vary. By one 1906 account, some 72,000 distilleries operated, of which 57,000 were small farm 'Materialbrennereien' stills producing a total of 27 million gallons (New York Times, 1906b). Another account, from 1914, put the number at 6000 distilleries producing 66 million gallons of alcohol per year (Nathan, 1928).

The German government also promoted ethanol-fuelled household appliances such as 'spirit' lamps, water heaters, laundry irons and cooking stoves. By one estimate, some 95,000 alcohol-fuelled stoves and 37,000 spirit lamps were made in Germany by 1902 (Tweedy, 1917).

The nationalistic aims of the alcohol fuel movement were clear. 'To Kaiser William II, it seems, we are indebted for the great, new industry,' said a *New York Times* magazine writer in 1906. 'Not that he discovered the fuel, but that he forced its use on Germany. The Kaiser was enraged at the Oil Trust of his country, and offered prizes to his subjects and cash assistance ... to adapt [alcohol] to use in the industries' (New York Times, 1906c).

Germany's 'Materialbrennereien' programme was an early approach to household and small-scale energy systems that would be replaced in coming decades with electricity. Attempts to export the programme met with interest but not as much government-level enthusiasm. An exhibit developed in Germany around 1900 was devoted to alcohol-powered automobiles, farm machinery (Fig. 1.1) and a wide variety of lamps, stoves, heaters, laundry irons, hair curlers, coffee roasters and every conceivable household appliance and agricultural engine powered by alcohol (Automobile-Club de France, 1902). The exhibit travelled to France, Italy and Spain between 1901 and 1904, and was then sent to the USA and displayed in Norfolk, Virginia and Baltimore, Maryland in 1907 and 1908 (Lucke and Woodward, 1907).

Fig. 1.1. Ethanol-fuelled farm tractor, Deutz company, Germany, *c.*1900.

1.2.2 USA repeals tax on biofuels in 1906

The German and French push for an agriculturally based fuel generated a great deal of interest in the USA, leading at first to an 1896 congressional investigation and a decade of debate over the folly of having taxed industrial alcohol off the market (US Congress, 1897). By 1905 it was common for Americans to read about German potato alcohol winning markets for farmers in the battle against the Standard Oil monopoly (New York Times, 1905).

The US oil industry was held in disdain in the early 20th century for its monopolistic and underhanded business practices, and it was not difficult for the US tax on fuel alcohol to be easily repealed in May 1906. US President Theodore Roosevelt supported the repeal and said: 'Standard Oil Company has, largely by unfair or unlawful methods, crushed out home competition. It is highly desirable that an element of competition should be introduced by ... putting alcohol used in the arts and manufactures upon the free list' (Washington Post, 1906a).

Thousands of news articles expressed the hope that, like their counterparts in Germany and France, American farmers could recapture some of the markets being lost to the new automotive industry. If horses were being replaced by horseless carriages, perhaps farmers could at least grow the fuel to supply them. However, this proved more difficult than expected due to remaining regulatory, market and cultural barriers. Regulatory barriers included the high cost of meeting regulations for 'denaturing' (rendering alcohol non-potable). Market barriers included the low price of kerosene and gasoline, especially in regions where serious competition emerged; and cultural barriers included state-by-state (and later, national) prohibition of beverage alcohol. In theory, the distilleries could have switched to fuel alcohol, and indeed some people (like Henry Ford) argued that they should (Washington Post, 1916b). However, in practice, prohibitionist politics and the potential for law breaking worked against the conversion of beverage distilleries to fuel production in the USA in the 1912–1932 timeframe.

Strong interest in biofuels continued during this period. Enthusiastic endorsements by scientists and engineers of all kinds were commonplace. Typical was a statement by inventor

Alexander Graham Bell, who wrote in *National Geographic* in 1917 that ethanol 'makes a beautiful, clean and efficient fuel ... that can be manufactured from corn [maize] stalks, and in fact from almost any vegetable matter capable of fermentation ... We need never fear the exhaustion of our present fuel supplies so long as we can produce an annual crop of alcohol to any extent desired' (Bell, 1917). It was a 'universal assumption,' said *Scientific American* in 1920, 'that (ethyl) alcohol in some form will be a constituent of the motor fuel of the future' (Scientific American, 1920).

1.2.3 British interest in biofuels 1907–1930s

The apparent scarcity of oil resources was another major reason for interest in biofuels at the beginning of the 20th century, and a 1907 British commission noted that 'a famine in petrol appears to be inevitable' (Motor Union, 1907). Surveying the possible substitutes, the commission said: 'Of all the liquid fuels which have been considered by the Committee, the one holding out the greatest promise is alcohol,' since it was renewable and production could be 'unlimited in any amount'. The optimism about biofuels was so widespread that even a 1915 book for youngsters, entitled *Modern Inventions*, had a chapter entitled 'Alcohol Motors and the Fuel of the Future' sandwiched amid the zeppelins and submarines (Johnson, 1915).

Although British foreign policy focused mainly on securing supplies of petroleum from the Middle East, an Alcohol Motor Fuel Committee was created in 1914 as part of the defence research effort (London Times, 1914). The committee was charged with considering sources of supply, methods of manufacture and costs of production for alcohol fuel (Fox, 1924). The commission concluded in 1921 that the cost of alcohol in comparison to petroleum made alcohol a likely fuel only in tropical and remote areas of the world where sugarcane, cassava, Jerusalem artichokes and other crops would give a high yield per acre (London Times, 1921).

One committee member was Harry Ricardo, who at the time was one of world's leading engine designers. In his pioneering 1923 book, *The High Speed Internal Combustion Engine*, Ricardo said:

> It is a matter of absolute necessity to find an alternative fuel. Fortunately, such a fuel is in sight in the form of alcohol; this is a vegetable product whose consumption involves no drain on the world's storage and which, in tropical countries at all events, can ultimately be produced in quantities sufficient to meet the world's demand, at all events at the present rate of consumption. By the use of a fuel derived from vegetation, mankind is adapting the sun's heat to the development of motive power, as it becomes available from day to day; by using mineral fuels, he is consuming a legacy – and a limited legacy at that – of heat stored away many thousands of years ago. In the one case he is, as it were, living within his income, in the other he is squandering his capital. It is perfectly well known that alcohol is an excellent fuel, and there is little doubt but that sufficient supplies could be produced within the tropical regions of the British empire.

1.2.4 French biofuels programmes 1900–1930s

The French ethanol fuel programme was supported by the Ministry of Agriculture before World War I, and French biofuel production rose from 2.7 million gallons in 1900 to 5.7 million gallons in 1903 and 8.3 million in 1905. Its main purpose was to help support French sugarbeet markets and curtail the rising surplus of many other crops. Another concern was the increase in oil imports from Russia and the USA, along with the lack of domestic oil reserves. However, without a contentious agrarian movement, the French did not embark on a large-scale distillery building programme like the Germans.

After World War I, a French committee recommended that a 'national fuel' of 40–50% ethanol with gasoline be created, and on 28 February 1923, Article Six was passed requiring gasoline importers to buy alcohol for 10% blends from the State Alcohol Service (Fox, 1924). Administering the alcohol law was the

Comptoir des Ventes du Carburant National. The mandatory blending level was revised to 25% alcohol in 1928. Several brands were marketed: 'Carburant Poids Lourds' (truck fuel), 'Tourisme' and 'Supercarburant'. By 1931 blending stocks had passed 87 million l and the Alcohol Service acquired even more alcohol from the struggling wine industry. Biofuels use peaked in 1935 at 406 million l, accounting for over 7% of all fuel use, and declined to 194 million l by 1937 due to poor harvests (Egloff, 1939).

The French, British and German biofuels laws and research had a worldwide impact. Engineers in Asia and Latin America who studied in European universities took home ideas about national self-sufficiency, fuel improvement and agrarian support that would form the basis of biofuels programmes in their own countries during the 1930s and into the 1970s (New York Times, 1931).

1.3 Biofuels for Internal Combustion Engines

At least a dozen inventors tried to develop some form of internal combustion engine between the 17th century and the early 19th century, according to historian Lyle Cummins (Cummins, 1989). The first authentic internal combustion engine using volatile liquid fuel, a carburettor and a spark-ignition piston engine was developed by US engineer Samuel Morey at the surprisingly early date of 1826. Morey's engine ran on ethyl alcohol and turpentine (camphene) and powered a small boat at 8 miles per hour up the Connecticut River. Morey remained relatively unknown because was never able to attract financing; only one prototype engine was ever built (Farell, 1915; Goodwin and Duryea, 1931; Hardenberg, 1992).

A more successful figure was German inventor Nicholas August Otto. In 1860, Otto used ethyl alcohol as a fuel in an early engine because it was widely available for spirit lamps throughout Europe. He devised a carburettor, which, like Morey's, heated the alcohol to help it vaporize when the engine was started. But a January 1861 patent application with the Kingdom of Prussia was turned down, probably because the principle of heated alcohol carburetion was already being widely used in spirit lamps (Cummins, 1989). It is interesting to note that Otto's financing came from Eugen Langen, who owned a sugar refining company that probably had links to the alcohol lamp fuel markets of Europe. Of course, the Otto and Langen Company went on to success in the 1870s by producing stationary piston engines, which were usually powered by coal gas. The four-stroke 'Otto-cycle' engine, for the automobile, was developed in the 1880s and was fuelled primarily with gasoline, which was a cheap by-product of the kerosene refining process at the time. Still, like most early engines, it was adaptable to a variety of fuels such as alcohol or benzene.

German inventor Rudolph Diesel also designed his compression ignition 'diesel' engine for heavy fuels from oil, but he also found that groundnut, castor and palm oils worked quite well. 'One cannot predict what part these (vegetable) oils will play in the colonies in the future,' he wrote in 1912. 'In any case, they make it certain that motor-power can still be produced from the heat of the sun, which is always available for agricultural purposes, even when all our natural stores of solid and liquid fuels are exhausted' (Knothe, 2001).

American automotive engineers also favoured the use of biofuels for a variety of reasons. In 1906, one automotive industry representative of the Detroit Board of Commerce told a US Senate committee that alcohol was 'preferable' to gasoline because it was safer, 'absolutely clean and sanitary' and because 'artificial shortages' could not raise the price in the future. In fact, the biggest problem for auto makers, the representative said, was not so much cost as the question of long term supply (US Senate, 1907).

Among American engineers was Henry Ford, who was well informed about the German alcohol fuel programme and hoped a similar programme could contribute to rural prosperity in the USA (Wik, 1963). In 1906, when the alcohol tax was repealed, Ford said that carburettors on his Model T cars would be designed to use either gasoline or alcohol (Washington Post, 1906b). When World War I threatened to create a gasoline shortage, 'he announced in 1915 that ... the new Fordson tractor would be designed to burn alcohol as well as gasoline; thus the supply of fuel would be unlimited' (Wik, 1963). In 1925, Ford told a *New York Times* reporter that ethyl alcohol was 'the fuel of the future', which 'is going to come from fruit like that sumac out by the road, or from apples, weeds, sawdust – almost anything. There is fuel in every bit of vegetable matter that can be fermented. There's enough alcohol in one year's yield of an acre of potatoes to drive the machinery necessary to cultivate the fields for a hundred years' (New York Times, 1925). Throughout his life, Ford hoped that these kinds of developments would bring on 'the greatest era of prosperity and happiness we have ever known' (New York Times, 1938).

1.3.1 Farm Chemurgy in the USA, 1930s

Henry Ford's ideas about alcohol fuels inspired a broader campaign for industrial uses for farm crops to help fight the Great Depression. The programme was called 'Farm Chemurgy' – literally meaning chemical work but actually aimed at industrializing agriculture through scientific research. At the time, agricultural research in the government and universities tended to be aimed at food production and not at opening new industrial markets for farm products, although some pioneering scientists, such as George Washington Carver, had already seen the need. Politically, Chemurgy was a populist Republican alternative to Democratic President Franklin Delano Roosevelt's supply-constraining agricultural policies. While the Chemurgy movement had practical goals and a scientific vision for agriculture, it was also occasionally derailed by self-described saviours of the farmer and bitter political infighting (Wright, 1995).

Practical projects backed by the Chemurgy movement included the use of soybeans for plastics, Southern pine for paper pulp and guayule for rubber. But the signature and most controversial project was the use of ethanol as an octane booster in gasoline. In 1936, following a series of conferences sponsored by Ford in Detroit, many felt that the time had come to compete directly with the oil industry. An experimental alcohol manufacturing and blending programme began in Atchison, Kansas. By 1937 motorists from Indiana to South Dakota were being urged to use its product, 'Agrol'. Two types were available: Agrol 5, with 5–7% alcohol and Agrol 10, with 12.5–17.5% alcohol. 'Try a tankfull – you'll be thankful,' the Agrol brochures said (National Agrol Co., 1938). The blend was sold with much initial enthusiasm at 2000 service stations (Fig. 1.2). However, Agrol plant managers complained of sabotage, market manipulation and bitter infighting from the oil industry. By 1939, the Atchison Agrol plant closed its doors. The experiment had failed, but it was not the end of the story. The Atchison plant and others would be rebuilt a few years later to make fuels and chemicals, and especially synthetic rubber, during World War II (Bernton *et al.*, 2010).

Fig. 1.2. Agrol was a US ethanol–gasoline blended fuel that came in 5, 10 and 15% blends of ethanol in gasoline. It was used extensively in the Midwest from 1937–1939 but phased out due to heavy competition from the oil industry.

1.3.2 The octane paradox and fuel additives, 1920s–present

Early in the 20th century, engineers who were thinking about the relative advantages of ethanol and gasoline turned up an interesting paradox. They found that ethanol could be used in both high and low compression engines, but gasoline could only be used in low compression engines. If gasoline was used in high compression engines, the engine would knock. If ethanol was used in a low compression engine, fuel consumption would go up. But if the fuel was matched to the engine – that is, ethanol in high compression and gasoline in lower compression engines – both would have about the same fuel consumption and both would run smoothly.

Between 1907 and 1909, the US Navy, the US Department of Agriculture and the US Geological Survey (USGS) performed over 2000 engine and fuel tests and reached some of these conclusions (Lucke and Woodward, 1907; Strong, 1909). The Edison Electric Testing Laboratory and Columbia University found the same thing. 'A gallon of alcohol will develop substantially the same power in an internal combustion engine as a gallon of gasoline ... owing to the superior efficiency of operation,' said the Edison report (New York Times, 1906c).

Efficiency, at the time, is what would later become known as the 'octane' rating. Gasoline's octane rating was around 55 in the post-World War I era, while ethanol had about double the octane rating, at 110. A high octane fuel could be used in a more efficient high-compression engine. As a result, auto racers preferred high compression engines fuelled by ethanol and benzene, despite the somewhat higher cost of fuel, while regular motorists used heavier, lower-powered engines fuelled by cheaper gasoline. The cost differentials generally amounted to about one-third more for ethanol, although this varied considerably from country to country and time to time. Frequently enough, ethanol was cheaper than gasoline, especially in rural areas or in the tropics.

In the post-World War I era, demand for fuel was accelerating while the quality of gasoline declined as lower quality oil reserves were brought into the market. Geologists estimated that only 20 or 30 years' worth of oil were left in the USA and a 'gasoline famine' was likely (Smith, 1920). The USGS estimated US oil reserves at 7 billion barrels while consumption was at 330 million barrels per year and rapidly increasing (Pratt, 1945). Automotive engineers worried about 'a calamity, seriously disorganizing an indispensable system of transportation' (Scientific American, 1919).

One technological path would involve building low-compression engines that were more tolerant of low-grade fuels, but far less efficient. A piece in *Scientific American* (1919) said 'The burden falls upon the engine, it must adapt itself to less volatile fuel, and it must be made to burn the fuel with less waste.... Automotive engineers must turn their thoughts away from questions of speed and weight ... and comfort and endurance and focus on averting the calamity.'

At this technological crossroads, many European and American engineers disagreed with the idea of downgrading the engine and worked on various solutions to the problem. In 1919, Charles F. Kettering, then vice president of research for General Motors, urged engineers to avoid compromising engine design (as *Scientific American* had suggested). Instead, they should improve the fuel and raise compression. Kettering (1919) opined: 'Geologists tell us that at our present rate of consumption the domestic supply of crude oil will be exhausted in less than 15 years. If we could sufficiently raise the compression of our motors ... we could double the mileage and thereby lengthen this period to 30 years.'

To raise engine compression, Kettering suggested two ways to improve fuel: 'high percentage' and 'low percentage' gasoline additives. Blending gasoline with 20% ethanol or up to 40% benzene was an example of the high percentage solution. The low percentage solution, on the other hand, could be some small volume additive easily blended at the petroleum refinery. In 1921, Kettering and associates discovered the anti-knock effect of tetraethyl lead. They began marketing it in 1923 through their 'Ethyl' partnership with Standard Oil Co, despite vehement public objections from health experts at the time and throughout subsequent decades. A mass of 3 g of tetraethyl lead would raise fuel octane value of a gallon of gasoline by five points, and since it was a relatively cheap way to improve fuel, leaded gasoline came to dominate world fuel markets. However, leaded gasoline was finally forced off the markets by international public health agencies around the turn of the 21st century (DePasquale, 2011).

What is not well understood is that Kettering's strategy, at the time, continued to involve both the low percentage solution (leaded gasoline) and the high percentage solution (biofuels blends) over the long term. According to a 1936 du Pont legal study: 'An important special motive for this (tetraethyl lead) research was General Motors' desire to fortify itself against the exhaustion or prohibitive cost of the gasoline supply, which was then believed to be impending in about twenty-five years; the thought being that the high compression motors which should by that time have been brought into general use if knocking could be overcome could more advantageously be switched to [ethyl] alcohol' (Wescott, 1936).

British researchers, lacking commercial ties to Standard Oil, began to consider fuels from the standpoint of the highest useful compression ratio that could be achieved without engine knocking. In Britain, Harry Ricardo observed that knock was reduced when fuels had high levels of benzene, methanol or ethanol. Ethanol had a 7.5 value, as opposed to commercial gasoline then available at 4.5 to 6. He concluded that the low burning rate of alcohol lessens the tendency to knock, and that, using toluene as the reference point at a 100 anti-knock value, ethanol had a 130 rating – the highest of any other fuel (Ricardo, 1921).

Ricardo never abandoned alcohol as a way to improve a fuel's highest useful compression ratio. In 1921 he patented racing fuels RD1 and RD2 (for Ricardo Discol) that contained methanol and ethanol, acetone and small amounts of water. These were widely used on race tracks throughout Europe and the USA in the 1920s and 1930s, but were regarded as a 'pleasant foible' rather like the smell of castor oil around the race track (Pleeth, 1949). Still, his advocacy of ethanol for general use was challenged in the 1920s by technical problems with alcohol production, such as the need for better azeotropic processing and competition from tetraethyl lead.

As the technical problems cleared up, and ethanol blending could be more easily accomplished in the 1930s, Ricardo worked with National Distillers Co. and Cleveland Oil Co. on an alcohol fuel blend called 'Discol', which soon became very popular on a commercial level (Society of Automotive Engineers, 1992). The formula was said to have 'monopolized' racing fuels. Advertisements in the *London Times* boasted that 'Racing Motors Run on Cleveland * Ricardo * Discol' (Cleveland Discol, 1935). Hundreds of other advertisements and articles about Cleveland Discol (Fig. 1.3) are found in the British newspapers and magazines from the 1930s until 1968. The brand was historically the second-longest of any commercial alcohol fuel blending programme in the world, after Brazil's programme, which started in the 1930s.

In the long run, the most effective solutions to the octane paradox turned out to be improvements to gasoline refining, especially the Houdry catalytic process (in place in the 1930s) and the Haensel/Universal catalytic reforming process (in place in the late 1940s to early 1950s). These brought gasoline octane ratings up to the low 80s by the 21st century. And yet, by the 21st century, the standard gasoline pool from a refinery still averaged about 84 octane, and some additive or additional severe reforming process would still be needed for octane to reach the minimum standard of 87 or higher.

Additive choices today are limited by economic and public health concerns. Severe reforming at the refinery is expensive and increases the amount of carcinogenic benzene, toluene and xylene, known as 'air toxics'. Other petroleum-based additives, such as methyl tertiary butyl ether (MTBE), present severe water pollution problems (Navarro, 2009). For this reason, many countries – even those with large oil reserves and petroleum refinery capacity – continue a long historical tradition of using agriculturally derived ethanol blends in gasoline.

Fig. 1.3. Cleveland Discol ethanol–gasoline fuel was developed by Harry Ricardo as a racing fuel in the 1920s but marketed in the UK through 1964.

1.4 Worldwide Experience with Biofuels 1920s–1930s

Most industrial and developing countries have a long history of producing biofuels, especially blending ethanol from sugar or starch crops into gasoline. However, ethanol from paper processes, gases and liquids from wood pyrolysis units and vegetable oils for diesel engines were also common. So, too, were the familiar concerns about biofuels that surface today, such as questions about using food for fuel, or about developing biofuels on an industrial scale for export versus developing industry for import substitution. In addition, many of the same politically motivated objections to biofuels typical today are also found in the historical record, including technical problems with fuel blending or claims that only the marketplace, and not government policies, can or should animate energy choices.

1.4.1 Brazil and Philippines develop new markets for sugarcane 1900s–1930s

Following the agrarian movements in Germany, France and the USA, biofuels were most economically attractive in countries that produced sugarcane in the early 20th century. With the high cost of gasoline imports and readily available sugarcane processing equipment, it was natural that developing nations would develop biofuels. Two that were especially active were Brazil and the Philippines.

The Brazilian programme is usually said to have started around 1919 when the governor of the north-eastern state of Pernambuco ordered official vehicles to operate on ethanol, and by 1921 distilleries in the state produced 2.2 million gallons of ethanol. A year later, the Brazilian Congress of Coal and other National Fuels recommended forming alcohol cooperative societies equipped with fermenting, distilling and denaturing plants, with tank wagons for distribution. They also recommended organization of alcohol marketing agencies throughout Brazil, government vehicle use of ethanol and reduced taxes for ethanol (Fox, 1924).

The first major alcohol fuel plant was built in Recife (Pernambuco state) in June 1927. Brands of alcohol and gasoline blended fuel included Azuline and Motorin, which were said to be popular throughout the sugarcane-growing regions of Brazil (Boletim Enfoque, 2000). As alcohol production grew and proved profitable, Brazil contemplated a national fuel programme. In 1931, Brazilian engineer Eduardo do Sabino de Oliviera said that he 'had already perfected a fuel consisting of alcohol mixed with other chemicals which he is satisfied will replace gasoline' (New York Times, 1931). Difficulties included redesigning the carburettor and additives to help with cold starting. Sabino later said that the Brazilian alcohol programme of the 1930s was inspired by the French biofuels programme of the 1920s, since he and colleagues studied there as engineering students. Sabino and colleagues, in turn, helped initiate the Pro-Alcool programme a generation later.

By 1931 a Brazilian law required gasoline importers to buy alcohol in volumes of 5% of their imports under the supervision of the Minister of Agriculture. At the time, gasoline cost about US$0.41 per gallon while alcohol was less than half the cost (New York Times, 1931). The number of Brazilian distilleries producing fuel-grade ethanol increased from just one in 1933 to 31 by 1939 and to 54 by 1945. Fuel alcohol production rose from 100,000 l in 1933 to 51.5 million l in 1937, or about 7% of the nation's fuel consumption.

To help keep track of research, promote biofuels and lend technical assistance, Brazil's Instituto do Assucar e do Alcool was established in 1933. Sales tax exemptions for blends and reductions on taxes of high compression motors for pure alcohol use were also instituted. Much of the interest in alcohol fuels came from sugarcane planters, who often used the pure alcohol engines. But blends of various proportions were also marketed, including a 90% alcohol 10% gasoline blend, a 70% alcohol and 30% di-ethyl ether blend, and a 12% alcohol 88% gasoline blend.

In the Philippines, alcohol was apparently first used as an engine fuel around 1914 at the Calamba Sugar Estate, an American-operated sugar and coconut plantation. Some technical problems, especially cold starting, were noted. The Philippine Motor Alcohol Corp. was incorporated in August 1922 in Manila. A variety of fuel types were tested, and by 1931 'Gasonol' (spelled with an 'n') blends of 20% ethanol and 5% kerosene were being used on a commercial scale (Fox, 1924).

Unlike Brazil and France, where ethanol blending with gasoline was mandatory, the Philippine policy was to use sugarcane ethanol as a pure fuel in autos, buses, trucks and railway locomotives. Studebaker, McCormack, General Motors and International Harvester sold pure alcohol-fuelled cars and trucks, advertising them as 'more economical ... (and) free from carbon'.

Three large bus companies, including Manila's Batangas Transportation Co., were running their buses on 100% ethanol, while buses and trucks on Negros and Panay also used pure alcohol as a common fuel (Fig. 1.4). No compulsory blending or tax advantages were given to alcohol fuels in the Philippines, but one US Commerce Department official commented, 'The sugar interests have felt reasonably well satisfied'. Ethanol fuel use reached 90 million l in 1939.

1.4.2 Other biofuels programmes 1930s

Brazil and the Philippines were not unique. At least 30 industrial nations had some kind of tax incentive or mandatory ethanol blending programme in place by 1932 (Fulmer *et al.*, 1932). Most were either in sugarcane-growing tropical regions, where alcohol could be produced cheaply, or in Europe, where octane-boosting additives were needed for high-compression automotive engines.

Cuba, for instance, produced about 20 million litres per year in 1922 for a blend of 80% gasoline and 20% alcohol called 'Espiritu,' and the blend amounted to half the gasoline sold. Standard Oil president Walter Teagle at one point noted with alarm that in Cuba,

Fig. 1.4. Advertisement from a Manila newspaper around 1937. Studebaker, McCormack, GM and International Harvester sold pure alcohol-fuelled cars and trucks in the Philippines in the 1930s, advertising them as 'more economical ... (and) free from carbon'. The Philippine government did not give ethanol tax advantages or make alcohol–gasoline blending mandatory, as in Brazil, France and other countries; it was produced cheaply enough to compete directly with gasoline and diesel fuels.

'Industrial alcohol is in very substantial competition with gasoline'. The price of alcohol in raw state was from US$0.21 to US$0.22 per gallon while gasoline sold for US$0.355 to US$0.36 per gallon (Wall Street Journal, 1923).

In Panama, import taxes on gasoline favoured locally produced ethanol, but a 'price war' kept distilleries from expanding their markets in the early 1930s. And in Puerto Rico, an Arecibo-based company sold 'Alco-Motor' at a price of US$0.25 per gallon – US$0.10 cheaper than gasoline for about 2 years, until the price of its feedstock of molasses suddenly went up and the price of gasoline suddenly dropped (Fox, 1924).

In Europe, scepticism about leaded gasoline was one reason for the widespread adoption of alcohol blending in gasoline from the late 1920s to the 1940s. 'These lead halides, being comparatively soluble, are obviously toxic,' said Myer Coplans in the *British Medical Journal*. 'Legislation and regulation upon this subject are urgently called for' (Coplans, 1928).

In the Czech Republic (then Czechoslovakia), 'Dynakol,' a mixture of about 50% alcohol, 20% benzene and 30% gasoline was typical. Some 66 million l, or 12% of the total fuel supply, were sold in 1936. The government subsidized low cost power alcohol with a tax on beverage alcohol. Hungary had 'Moltaco', a blend of 20% ethanol that was made compulsory by royal decree in 1929. Poland had a state alcohol monopoly that was financed by a 5-year advance purchase by Standard Oil Company. About 1500 small alcohol plants serving farm communities were included in a network centred around the Kutno Chemical Works. In Sweden, ethanol production was confined to paper mills, and a 25% ethanol and 75% gasoline blend called 'Lattbentyl' was standard (Fulmer *et al.*, 1932; United Nations, 1952).

1.4.3 War emergency programmes 1930s–1940s

During World War I and World War II, the French, English and Americans were said to have 'floated to victory on a wave of oil'. However, German self-sufficiency in alcohol fuel helped extend World War I. 'Every motor car in the empire was adapted to run on alcohol,' according to Tweedy. 'It is possible that Germany would have been beaten [by 1917] if production of alcohol had not formed an important part of the agricultural economy' (Tweedy, 1917). By the 1940s, Germany once again tried to avoid oil shortages, but at that point coal-based fuels were more feasible on an industrial basis. In 1942, the peak year for Germany's synthetic fuel production, about 1.7 billion l of fuel per year came from coal (which, we must note, often involved concentration camp labour and crimes against humanity). About 267 million l of fuel ethanol, mostly from potatoes, were also produced. All told, 54% of the pre-war German fuel production was derived from non-petroleum sources, of which only 8% was ethanol from renewable sources (Egloff, 1942).

Although the allied armies tended to have plenty of oil for the war effort, the scarcity of oil resources elsewhere tended to promote substitution and innovation. Among the many substitute biofuels in Europe during World War II were the wood gas generators typical from France to Finland called 'gasogens'. The units heated wood to release combustible gases, and although nearly three times more efficient than wood-to-ethanol systems, the gases created high levels of engine wear (Reed, 1975).

The war forced innovation in other ways as well. In China and India, where food was scarce, inedible molasses from sugarcane was usually turned into alcohol for fuel. India's Utter Pradesh province passed a 20% law mandating alcohol blending, but all fuels were scarce during the war. In China, 'Benzolite', a mixture of 55% alcohol, 40% benzene and 5% kerosene, was widely sold in the 1930s. When war broke out, molasses distillers especially were turned towards ethanol production and often provided the only form of fuel available. A spokesman for the Chia Yee Solvent Works noted at a United Nations conference on power alcohol in 1952 that 'the shortage of gasoline was so acute it became impossible for civilians to get any amount of gas'. At that time the use of alcohol was no longer a question of costs or efficiency, but of necessity (United Nations, 1952). There is at least one report of a diesel-powered bus that ran on a number of vegetable oils, including those extracted from groundnuts, tea leaves, poppy seeds, tung, cotton seed and cabbage seed. Many American soldiers in China in World War II remember the potent, potable alcohol (Fig. 1.5) that doubled as a fuel for their jeeps and generators (Bernton *et al.*, 2010). Joseph Needham, a British scientist on a mission to China, was a 'strong believer' in power alcohol and found his ethanol-powered truck had far more power than wood-fuelled gasogens (Winchester, 2008).

The Brazilian experience was typical of sugarcane countries in Latin America. There, ethanol production increased from 51 to 77 million l per year between 1937 and 1944. Mandatory blending levels rose as high as 50% at the height of the war, when submarine attacks ravaged world oil-tanker fleets. As the war ended, cheap imported oil was once more readily available and alcohol blends were marketed sporadically, mostly to offset sugar surplus (Pischinger and Pinto, 1979). Blending continued intermittently through the 1950s as an outlet for sugar surpluses and began again in 1975 with the National Alcohol Program (PNA).

Fig. 1.5. The National Resources Commission Power Alcohol Works near Neichiang, China. (The tower houses, rectifying columns and the wood and bamboo vats in the foreground are for the fermentation of molasses.)

After the war in India, about 8 million l of alcohol were used in 1946, increasing to 9 million at the peak use in 1948. Another 20 million l were used in blends, out of about 1 billion l of gasoline used in 1951. Indian leaders were conscious of possible 'food or fuel' conflicts and prohibited the use of grains and root crops as feedstocks, but also felt that the power alcohol industry had to be protected from petroleum interests. The 1948 'Indian Alcohol Act' mandated 20% blending where feasible, but it was not widely adopted (United Nations, 1952).

During World War II, Philippine ethanol production reached a standstill, but climbed back to 30 million l by 1950. A 4-year plan was then in place to produce 120 million l, or 20% of the nation's fuel supply, but it was abandoned as new sources of cheap oil became available. According to the Philippine delegate to a 1952 United Nations power alcohol conference: 'The use of blended motor fuel was abandoned, for the simple reason that the gasoline interests fought hard to kill it. After such a very sad experience, we fully realize that proper legislation similar to that in India should be adopted in the Philippines.'

The main problem, though, was the increasing availability of cheap oil from the Middle East. By the 1950s most alternative fuels programmes had been abandoned as far too costly in comparison. Only in the UK did blends of alcohol and gasoline continue to be sold by the Cleveland Discol company through the 1960s.

1.5 Biofuels and the Global Energy Crisis 1970s–2000s

In the 25 years after World War II, global oil consumption grew by five and a half times, and the world became dependent on cheap oil from the Middle East. Discussions about raising prices preoccupied meetings of the Organization of Petroleum Exporting Countries (OPEC) for years, but in 1973, a Middle Eastern war conflated tight oil supplies into an international energy crisis. It began on 6 October 1973, when Egypt, Saudi Arabia and other Arab countries launched an attack on Israel in an attempt to regain land lost in the 1967 war. By 17 October 1973, as the attack faltered and US military aid flowed to the Israelis, Arab oil ministers gathered in Kuwait agreed to institute a total oil embargo against the USA and other countries friendly to Israel (Salameh, 2004). They would drop production by 5% per month until their demands were met. The price of oil quadrupled, from US$4.50 to US$22.60 per barrel. The shortages created long queues for gasoline and sparked panic buying across the USA and Europe. Although the embargo ended in March 1974, the US gross national product (GNP) plunged 6% and unemployment doubled to 9% by 1975.

The first energy crisis also spurred a widespread search for alternative energy sources, especially the 'Pro-Alcool' National Alcohol Program in Brazil in 1975, involving mandatory blending of about 20% ethanol in gasoline. Around the same time, states in the American Midwest, particularly Nebraska, began researching the potential of ethanol from maize (corn) in a blend with gasoline (Fig. 1.6).

The second energy crisis took place when Iranian dissent grew into an October 1978 strike in the nation's oil refineries, shutting down 5% of world oil exports. This in turn grew into a violent revolution that overthrew the pro-western government of the Shah in January 1979. Once again, panic buying led to price increases, this time up to US$34.50 a barrel.

The Brazilian ethanol programme began growing quickly, but a proposed US ethanol programme became mired in controversy and opposition from the oil industry. The Brazilian programme was seen as an outgrowth of sugarcane growing traditions and was part of the economic movement towards import substitution and industrialization. It was also able to enlist the full support of Brazil's automotive industry.

In the USA, the oil industry insisted that ethanol was an inferior fuel and that it caused insurmountable technical problems when blended with gasoline. Support for biofuels came mostly from a farm movement that saw ethanol in the light of farm prosperity and independence from the oil industry. Although the US auto industry was more inclined to back the oil industry, early proof that ethanol blending caused only minor engine problems that were

Fig. 1.6. The Earle Coryell gasoline company marketed several hundred thousand gallons of 'Corn Alcohol Gasoline Blend' in Nebraska in 1933. (Photo courtesy of the Nebraska State Historical Society.)

easily solved came from both the Brazilian Pro-Alcool programme and the Nebraska state Corn Products Utilization Committee, later known as the Gasohol Committee, which initiated a million miles of road tests on ethanol blends.

'Automotive companies were denouncing the idea of promoting biofuels on the grounds that they couldn't be taken seriously,' said Scott Sklar, then an aide to Sen. Jacob Javits and today a renewable energy expert with the Stella Group. 'Brazil's move to subsidize their vast sugar industry and work with automotive engineers to make ethanol-tolerant cars created immense tension in Washington,' Sklar said. 'Biofuels advocates were able to point to the "Brazilian experience" every time the US oil and auto industry said that it couldn't be done or that it would never be successful' (Kovarik, 2006).

In 1980, in the waning months of his administration, President Jimmy Carter created a fledgling US ethanol programme by signing a bill giving a US$0.54 per gallon ethanol tax incentive and, at the same time, excluding Latin American ethanol from the US market. These two pieces of legislation protected the US ethanol industry through its infancy and up to the mid-1990s, when it grew to over 1 billion gallons per year of capacity.

1.5.1 Lead gas phase-out and oxygenate blending, 1980s–2000s

The most important reason for the success of the maize-based ethanol programme in the USA was the dilemma facing the oil industry after it lost its most effective (and polluting) octane additives. Tetraethyl lead (TEL), the ingredient in leaded gasoline, was removed from US fuel in the late 1970s (and internationally by 2012) for engineering and public health reasons.

To replace TEL in the 1980s, most unleaded gasoline was made using a petroleum refining process called 'severe reforming', which boosted the levels of benzene, toluene and xylene (aromatic/BTX) compounds from 25% to as high as 40% of the fuel. High levels of these carcinogenic compounds in fuels and in automotive exhaust worried environmental policy makers. 'Cars with catalytic converters are 99 percent cleaner than they were in 1970,' said C. Boyden Gray, a legal advisor to President George H.W. Bush in 1990. 'But gasoline has gotten dirtier. It doesn't have lead (TEL) but it is far dirtier for aromatics (BTX)' (Lyman, 1990).

The Bush administration worked with Congress to create the 1990 Clean Air Act and empower the Environmental Protection Agency to order the oil industry to remove BTX 'air toxics' from fuel. In many cases, the fuels would contain a class of octane boosting compounds called 'oxygenates' that were far more benign than BTX or TEL. These included ethanol, methanol and other alcohols. However, as the proposal wound its way through Congress, another oxygenate made from petroleum was included in the Clean Air Act. This was MTBE (methyl tertiary butyl ether), which was made from natural gas and butane, both of which the industry possessed in abundance. So by the time regulations were written for reformulated gasoline, both ethanol and MTBE were to be used as octane boosting additives in gasoline.

The problem with MTBE, however, was that it fouled the water while cleaning the air. MTBE could be detected as a harsh chemical odour and taste in concentrations as low as 2 parts per billion. Minor leaks of the hydroscopic gasoline additive into a town's water supply would make it undrinkable. By 2004, some 17 states had banned MTBE and clean-up costs for over 1100 water systems were estimated at US$24 billion (Environmental Working Group, 2004). One result of the MTBE fiasco was that in the absence of competition from the oil industry, ethanol production from maize ramped up very quickly, from 2 billion gallons per year in 2002, doubling by 2005, doubling again to 8 billion gallons by 2008 and rising to 14 billion gallons by 2012 (Renewable Fuels Association, 2012).

1.5.2 Food OR fuel? Or food AND fuel?

The rapid expansion of the American maize ethanol programme proved the resilience and flexibility of renewable energy systems, but it raised serious questions about the extent to which food crops can or should be used for fuel. Many of the questions had been raised in previous years, for example by Lester Brown in a 1980 Worldwatch Institute paper, where he noted that the competition took place in an indirect form, putting pressure on crop prices, land use and infrastructure financing (Brown, 1980).

In 2007 Jean Zigler, special rapporteur for food rights at the UN Food and Agriculture Organization, said biofuel from food crops were a 'crime against humanity' and called for a 5-year moratorium until cellulosic biofuels processes could be developed (Ferrett, 2007; Kleiner, 2007). Later that year, other FAO officials said biofuels technologies held both promise and peril, and that it was up to the world community to put food and human needs first, local development including fuels second, and fuel exports third (Aguilar, 2007).

The food or fuel issue is more complex than it usually appears. The ethanol industry points out that maize ethanol production uses a type of maize fed to livestock. It is rock-hard and not edible by humans. The ethanol industry removes the maize starch and adds yeast in the ethanol process. In the end, the leftover distillers' grains have 90% of the protein that was originally on its way to livestock feeding pens. In the short run, then, maize ethanol is not a threat to international food supplies (Renewable Fuels Association, 2011).

But the mid- to long-range problem is a far more serious concern, and the possibility that the volatile energy market will become bound up with the price of food is sobering to say the least. 'If oil goes to $150 per barrel or more, the price of grain will follow it upward as it becomes ever more profitable to convert grain into oil substitutes,' said Lester Brown in a 2011 article in *Foreign Policy*. 'And it's not just a US phenomenon: Brazil, which distills ethanol from sugar cane, ranks second in production after the United States, while the European Union's goal of getting 10 percent of its transport energy from renewables, mostly biofuels, by 2020 is also diverting land from food crops' (Brown, 2011). It is the same reason that people have been hoping for a cellulose ethanol industry – or a similar advanced non-food biofuels industry – for almost a century.

1.6 Cellulosic Biomass: Non-food Biofuels

When the 1952 United Nations conference on biofuels was organized in Lucknow, India, the speaker who was chosen to open the conference was Kanaiyalal Manekial Munshi, the governor of Uttar Pradesh. Munshi was a journalist, a literary scholar and former agricultural minister who had launched a popular tree-planting celebration called Van Mahotsav only 2 years before. With characteristic eloquence, he explained the problem that would bedevil biofuels researchers for the next seven decades.

'In India – with a scarcity of food – grains and root crops cannot be utilized for the production of power alcohol,' Munshi said. 'If our farm waste products or other cellulosic wastes like bagasse or wood chips ... are utilized, it would greatly increase our capacity for substitute motor fuels and provide an additional stabilizing factor for our farm economy. Research, therefore, needs to be concentrated on such substitute raw materials' (Munshi, 1952).

As Munshi would go on to acknowledge, cellulose ethanol technology was relatively old; research had begun over a century beforehand. In 1819, French chemist Henri Braconnot discovered that a sulfuric acid treatment could convert straw, wood or cotton into glucose. And another French chemist, Anselme Payen, isolated and purified cellulose in 1838. As polymer chemistry came to be better understood, a variety of inventors found ways to use it in billiard balls, shirt collars and camera film in the 1870s and 1880s (Klemm *et al.*, 2005). Cellulose seemed even more useful when, around 1900, scientists analysed the composition of cellulose and understood that it could be broken down into glucose molecules and converted into a wide variety of chemicals and fuels. By World War I, two commercial cellulose-to-ethanol plants using acid hydrolysis had been built in Georgetown, North Carolina and Fullerton, Louisiana in the USA (Bente, 1984).

Around the early 20th century, news articles about cellulose as a feedstock for biofuels were common, reflecting not only advances in science but also hope for alternative biofuels in the face of a possible oil shortage. The *Washington Post*, for instance, said: 'One of the most important of recent discoveries ... is that ethyl alcohol ... is so remarkably cheap that it can be obtained from ordinary sawdust (and) seems destined to solve the problem of motor fuel ... Wood, then, in place of petroleum, is to be for (auto drivers) the future source of supply. The problem, in fact, has been fully worked out.' The article noted that the University of Wisconsin was using a hydrochloric acid process to convert wood into sugar, which is then fermented into alcohol (Washington Post, 1916a). The same process was also tried on novel sources of cellulose. *Scientific American* and many other journals of the era reported that in 1918, the Pasteur Institute had been able to produce about 10 gallons of fuel ethanol per ton of seaweed (Scientific American, 1918).

The urgency of the biofuels problem picked up in the 1920s with widespread estimates of a serious pending oil shortage (New York Times, 1920). One of the most influential chemists of the era, Harold Hibbert of Yale University, said in 1921 that oil shortages posed a serious challenge for public policy and for science. 'Does the average citizen understand what this means?' he asked. 'In from 10 to 20 years this country will be dependent entirely upon outside sources for a supply of liquid fuels ... paying out vast sums yearly in order to obtain supplies of crude oil from Mexico, Russia and Persia.' Chemists might be able to solve the problem, Hibbert said, by making ethanol from abundant cellulose waste – materials such as seaweed, sawdust, corn stalks and wheat straw. 'It is believed that the chemist is capable of solving this difficult problem. ... (and) it would seem that cellulose in one form or another is capable of filling that role' (Hibbert, 1921). Harold Hibbert helped form the American Chemical Society's cellulose division the next year and became its first chair.

As the years went by, the biofuels potential of cellulose – the most abundant organic material on earth – was a recurrent theme in scientific and popular literature. Henry Ford, for example, told the *New York Times* in 1925 that the 'fuel of the future' would come not from oilfields but rather fields full of weeds (New York Times, 1925). In 1927 the British Fuel Research Board reported advances in cellulose conversion (New York Times, 1927), and in 1928, the *Washington Post* noted that research into cellulose fuels could help with farm relief (Washington Post, 1929).

1.6.1 Research in the 1930s and 1940s

Early attempts to hydrolyse cellulose through the varieties of acid-based processes proved difficult and expensive, but in Germany in the 1930s Heinrich Scholler developed a process that used weak acid to percolate through wood chips to hydrolyse cellulose and remove wood sugars at the same time. The Scholler process doubled yields, and about 50 parts of sugar were obtained for every 100 parts of wood (Bente, 1984). Three Scholler plants were built in the 1930s in Germany and one in Switzerland, while a US version of the process was tested and modified by the Madison, Wisconsin Forest Product Laboratories. During World War II, this modified Madison Wood Sugar Process was used to build a plant in Springfield, Oregon, in the USA, and in several locations in Russia, to make ethanol for chemicals such as synthetic rubber.

One of the best known scientists working in this area at the time was Ernst Berl, a chemical professor at the Technical University at Darmstadt who went to work at Carnegie Mellon University. A Jewish scientist who fled the Nazis, Berl made contributions to cellulose research by examining the pressurizing process for reducing cellulose from all kinds of plant materials to either liquid or solid fuels. This work 'assures mankind of an illimitable supply of the prime movers of the wheels of civilization for all time, after natural deposits have been exhausted', wrote the *New York Times* science correspondent William L. Laurence (Laurence, 1940). It was an 'astonishing announcement', said *Time* magazine; Berl had made fuel from grass, leaves, seaweed, sawdust, scrap lumber, maize, maize stalks and cotton (Time, 1940). A few years later, at another American Chemical Society meeting, Berl once again caught the world's attention. Associated Press science writer Howard Blakeslee said that a way had been found to supply the world with 'gasoline from plants' (Blakeslee, 1944). Future farmers, Berl said, might have simple installations to make their own fuels to run their tractors and heat their homes. And the *New York Times* science editor said: 'The process means that no nation need import oil or coal if it has land enough to grow carbohydrates' (Kaempffert, 1944).

Meanwhile, as Berl's research amazed people on the home front, science was also advancing in the forests of south-eastern Asia during World War II. It seemed that US and British troops found that their equipment was not standing up to 'jungle rot'. Soldiers' cotton uniforms would disintegrate into rags after only a few weeks in the tropical environment. Researchers were sent across the Pacific collecting strains of fungus to be tested in the US Army's quartermaster labs in Natick, Massachusetts (Voosen, 2011). Elwyn T. Reese, a chemist with the Army, studied a greenish yellow mould and was able to isolate a stable variety. While the original idea had been to eliminate the fungus, Reese and others in the lab realized that the enzyme from the fungus was turning cellulose cotton uniforms into glucose, not by the old acid hydrolysis process, but rather through an enzyme hydrolysis process never really considered before (Reese, 1956). The fungus was eventually named *Trichoderma reesei* in his honour.

Although the research showed that the fungus could be useful for fuels and chemicals from cellulose in the 1950s, cheap oil from the Middle East made the research moot. Scientists would have to wait another generation before the economics of cellulose biofuels became interesting again.

1.6.2 Cellulose biofuels after the Arab oil embargo

When the 1974 Arab oil embargo raised the price of oil to the point where cellulosic ethanol was interesting again, Reese's protégés were among the first wave of scientists to describe petroleum alternatives to Congressional hearings in Washington DC. Cellulosic biomass could be put into operation 'on a very large scale' by 1980 at a cost of US$0.35 per gallon, said Natick scientist Leo Spano in a committee hearing in 1974 (Steiger, 1974).

'In the laboratory, filled with test tubes and incubators, I felt apart from the world,' Spano said later. 'But it was there that I realized that a tiny enzyme could change the world ... the compounds could eat up our poisonous wastes and convert them to useful substances. Just think of all the waste cellulose ... sewage, wood pulp, corn cobs ... it can all be used to better mankind' (Bernton *et al.*, 2010). Spano's optimism notwithstanding, cellulosic biofuels proved to be an enormously complex area of biochemical engineering. Researchers in hundreds of university and government labs have taken decades to create an industry that is nearing commercial status. They have been isolating, characterizing and testing the complex chemical structures of plants, and working on cascading systems of enzyme reactions, and measuring their progress against the roller coaster of oil prices.

One of the scientists intrigued with cellulosic hydrolysis was Patrick Foody, who founded Iogen Corp. in 1974. The company now has a commercial scale enzyme biorefinery in Saskatchewan, Canada. Other research efforts on the enzyme process in the 1970s and 1980s took place at Rutgers University, Virginia Polytechnic Institute and State University and the University of California at Berkeley in the USA.

The idea of using fungus to convert cellulose was interesting but not surprising for those who had been paying attention to the progress of biochemistry. Science fiction writer Isaac Asimov found it fascinating, and in a 1986 non-fiction article, noted that cellulose hydrolysis was an attractive option for research. 'Cellulose is self-renewing if we are careful to conserve our forests,' Asimov said, 'so the fuel we get from it could last indefinitely, whereas oil from the ground must be completely used up eventually' (Asimov, 1986).

1.7 Conclusion

The history of biofuels research and policy is extensive enough to fill an encyclopaedia, and this chapter has provided only a brief overview of major historical highlights.

The original fuels, biofuels from renewable resources, were pushed into niche markets by low-cost petroleum in the late 19th and early 20th centuries. Yet when faced with emergency fuel shortages or agricultural surpluses, most countries opted to protect biofuels markets through tax policies or mandatory blending.

Biofuels have been used in some form over four epochs: (i) as a lamp fuel from prehistory to the mid-19th century; (ii) as an internal combustion engine fuel from the early 20th century; (iii) as replacement for petroleum during oil shortages of the 1970s; and (iv) as a safe octane booster for gasoline substituting for leaded gasoline (tetra ethyl lead) and MTBE (methyl tertiary butyl ether) in the 20th and 21st centuries.

In recent years, serious questions about the impacts of biofuels on climate change, food rights, biodiversity and sustainability have framed the larger context of energy research and policy debates. These questions are needed to inform the social construction of optimal sustainable energy systems; and they are needed because technologies do not simply emerge from intrinsic properties and launch out on a predetermined path. Optimal systems development requires vision and, given the current environmental crises, it is absolutely vital that scientists have a sense of the history and motives of those who worked before them in this area.

Scientists and engineers from all continents– Nicholas Otto, Rudolph Diesel, Henry Ford, Harry Ricardo, Eduardo do Sabino de Oliviera, Kanaiyalal Munshi, among others – have seen biofuels as a path towards reconciliation between rich and poor, urban and rural interests, and industrial and environmental orientations for technology. Ricardo's idea that biofuels presents a way to live within solar 'income' and save fossil fuel 'capital' is one that should be better known. Another is Munshi's thought that food should not be used for fuel in countries like India, but that using cellulosic wastes could be a stabilizing force for agriculture.

The development of sustainable systems is beyond its infancy, but it may still be in its childhood. All too often vested interests manage to put renewable energy into a subordinate position, as 'alternatives' or 'substitutes'. All too often, concerns that are legitimately part of the social construction process are made to seem like insurmountable obstacles.

And yet, continued research on novel crops, on cellulosic biofuels, on third-generation crops and on other ideas that are only beginning to form, is producing remarkable results. These are all the more remarkable for having been developed under budgetary pressures and the unfortunate undermining of scientific research worldwide.

The pressing complex of global energy, environment and agriculture issues would seem to require the best possible efforts in the biosciences. Even if some governments have shirked these responsibilities, certainly most individual scientists have not. It is a matter of course, as well as pride, to be among those whose vision far exceeds their budget, or even their lifespan. While our toils may now be obscure, scientists who are on the verge of unlocking renewable energy sources today will certainly earn the gratitude of generations to come.

References

Aguilar, E. (2007) UN head calls for more biofuels research. Science Development Network, 15 November 2007. Available at: http://www.scidev.net/en/climate-change-and-energy/news/un-head-calls-for-more-biofuels-research.html (accessed 23 April 2012).

Asimov, I. (1986) The Most Common. *American Way*, 27 May 1986.

Automobile-Club de France (1902) *Congress des Applications de L'Alcool Denature, 16 au 23 Dec. Paris.* Automobile-Club de France, Paris.

Bailey, D.M. (1975) *A Catalogue of Lamps in the British Museum.* British Museum, London.

Bell, A.G. (1917) Prizes for the inventor: some of the problems awaiting solution. *National Geographic*, February 1917, p. 131.

Bente, P. (1984) *International Bio-Energy Handbook.* The Bio-Energy Council, Washington, DC.

Bernton, H., Kovarik, B. and Sklar, S. (2010) *The Forbidden Fuel: A History of Power Alcohol.* Bison Press, Lincoln, Nebraska.

Blakeslee, H. (1944) Way found to supply world with gasoline from plants. *Washington Post*, 9 April 1944, p. B3.

Boletim Enfoque (2000) Em 1927, O primeiro grande empreendimento brasileiro em alcool combustivel. *Adone Vamous, Boletim Enfoque*, June 2000, No. 7. Sao Paulo, Brazil.

Brachvogel, J., Thatcher, C. and Marcker, M. (1907) *Industrial Alcohol, Its Manufacture and Uses.* Munn and Co., New York.

Brown, L. (1980) *Food or Fuel: New Competition for the World's Cropland.* Worldwatch Institute, Washington, DC.

Brown, L. (2011) The new geopolitics of food. *Foreign Policy*, April–May, 2011.

Cleveland Discol (1935) The British alcohol motor spirit. *London Times*, 29 November 1935, p. 13.

Coplans, M. (1928) Tetra-ethyl lead. *British Medical Journal*, 14 January 1928, p. 74.

Crosby, A.W. (2006) *Children of the Sun.* W.W. Norton, New York.

Cummins, L. (1989) *Internal Fire.* Society of Automotive Engineers, Warrendale, Pennsylvania.

DePasquale, R. (2011) Leaded gasoline to be eliminated. *Associated Press*, 27 October 2011.

Egloff, G. (1939) *Motor Fuel Economy of Europe.* American Petroleum Institute, Washington, DC.

Egloff, G. (1942) Substitute fuels as a war economy. *Journal of the Institute of Petroleum* 28, 115–132.

Environmental Working Group (2004) Like Oil and Water: MTBE Contamination By State, Environmental Working Group paper. Available at: http://www.ewg.org/node/19752 (accessed 23 April 2012).

Farell, G. (1915) *Capt. Samuel Morey Who Built a Steamboat Fourteen Years Before Fulton.* Standard Book Co., Manchester, New Hampshire.

Ferrett, G. (2007) Biofuels: crime against humanity. *BBC News*, 27 October 2007.

Fox, H.S. (1924) Alcohol Motor Fuels. In: *Supplementary Report to World Trade in Gasoline.* Bureau of Domestic and Foreign Commerce, US Dept of Commerce Monograph, Trade Promotion Series No. 20, Government Printing Office, Washington, DC.

Fulmer, E.I., Hixon, R.M., Christensen, L.M. and Coover, W.F. (1932) A preliminary survey of the use of alcohol as a motor fuel in various countries. *Conference to Discuss the Utilization of Agricultural Surpluses.* Department of Chemistry, Iowa State University Archives, Ames, Iowa, 10 December 1932.

Gomez, L.D., Steele-King, C.G. and McQueen-Mason, S.J. (2008) Sustainable liquid biofuels from biomass: the writing's on the walls. *New Phytologist* 178, 473–485.

Goodwin, K. and Duryea, C.E. (1931) *Captain Samuel Morey: The Edison of His Day*. The Vermonter Press, White River Junction, Vermont.
Hardenberg, H. (1992) *Samuel Morey and His Atmospheric Engine*. Society of Automotive Engineers, Warrendale, Pennsylvania, SP922.
Herrick, R.F. (1907) *Denatured or Industrial Alcohol*. John Wiley and Sons, New York.
Hibbert, H. (1921) The role of the chemist in relation to the future supply of liquid fuel. *Journal of Industrial and Chemical Engineering* 13, 841.
Johnson, V.E. (1915) Alcohol motors and the fuel of the future. In: *Modern Inventions*. TC and EC Jack, Ltd, London, pp. 286–293.
Kaempffert, W. (1944) Science in review. *New York Times*, 9 April 1944, p. E9.
Kettering, C.F. (1919) Studying the knocks. *Scientific American*, 11 October 1919, p. 364.
Kleiner, K. (2007) The Backlash Against Biofuels. Nature Climate Change. Available at: http://www.nature.com/climate/2008/0801 /full/climate.2007.71.html (accessed 23 April 2012).
Klemm, K., Heublein, B.l., Fink, H.-P. and Bohn, A. (2005) Cellulose: fascinating biopolymer and sustainable raw material. *Angewandte Chemie* 44, 3358–3393.
Knothe, G. (2001) Historical perspectives on vegetable oil-based diesel fuel. *Inform* 12, 1103–1107.
Kovarik, B. (1998) Henry Ford, Charles Kettering and the fuel of the Future. *Journal of the Society of Automotive Historians*. Available at: http://www.radford.edu/~wkovarik/papers/fuel.html (accessed 23 March 2012).
Kovarik, B. (2006) Looking South: the world ethanol industry is booming – thanks to the Brazilian example. *Com Sciencia*, Brazil, Fall 2006.
Kovarik, B. (2009) Biofuels: the sequel. *SE Journal,* Summer 2009, pp. 28. Available at: http://www.sej.org/publications/sejournal-su09/biofuels-the-sequel (accessed 23 April 2012).
Laurence, W. (1940) Makes coal or oil of grass in hour. *New York Times*, 13 September 1940, p. 19.
London Times (1902) The Kaiser's new scheme. Reprinted from *New York Times*, 24 April 1902, p. 9.
London Times (1914) Alcohol motor fuel. *London Times*, 22 January 1914, p. 10.
London Times (1921) Fuel from waste. *London Times*, 24 December 1921, p. 14.
Lucke, C.E. and Woodward, S.M. (1907) *The Use of Alcohol and Gasoline in Farm Engines*. USDA Farmers Bulletin No. 277. US Government Printing Office, Washington, DC.
Lyman, F. (1990) The gassing of America. *Washington Post*, 13 April 1990, p. C5.
Motor Union (1907) *Report of the Fuels Committee of the Motor Union of Great Britain and Ireland*, London, 1907.
Munshi, K.M. (1952) Inaugural Address: The production and use of power alcohol in Asia and the Far East. *Report of a Seminar Held at Lucknow, India*, 23 October 1952. Organized by the Technical Assistance Administration and the Economic Commission for Asia and the Far East, United Nations, New York.
Nathan, Col S.F. (1928) Alcohol for power purposes. In: *The Transactions of the World Power Congress*, London, 24 September–6 October 1928.
National Agrol Co. (1938) Try a Tankful – You'll be Thankful. Sales brochure, William J. Hale collection, Michigan State University Archives, Ann Arbor, Michigan.
Navarro, M. (2009) City awarded $105 million in Exxon Mobil lawsuit. *New York Times*, 20 October 2009.
New York Times (1903) A blow at American oil. *New York Times*, 15 January, 1903, p. 9.
New York Times (1905) Potato alcohol vs Standard Oil. *New York Times*, 18 June 1905, p. SM3.
New York Times (1906a) Future of alcohol in the industries. *New York Times*, 5 August 1906, p. 4.
New York Times (1906b) Free alcohol distilleries. *New York Times*, 13 September 1906.
New York Times (1906c) Launching of a great industry: the making of cheap alcohol. *New York Times*, 25 November 1906, Section III, p. 3.
New York Times (1920) Asserts Americans face oil shortage. *New York Times*, 3 May 1920, p. 22.
New York Times (1925) Ford predicts fuel from vegetation. *New York Times*, 20 September 1925, p. 24.
New York Times (1927) Use vegetable waste to make motor fuel. *New York Times*, 28 November 1927, p. 2.
New York Times (1931) Brazil seeks to cut gasoline payments. *New York Times*, 11 January 1931, p. 60.
New York Times (1938) Ford predicts new era of prosperity. *New York Times*, 4 June 1938, p. 2.
Pischinger, G. and Pinto, N.L.M. (1979) *Experiences with the Utilization of Ethanol*. Third International Alcohol fuels symposium, Asilomar, California.
Pleeth, S.J.W. (1949) *Alcohol: A Fuel for Internal Combustion Engines*. Chapman and Hall, London.
Pratt, W.B. (1945) Our oil and natural gas reserves. In: Fanning, L.M (ed.) *Our Oil Resources*. McGraw Hill Book Co. Inc., New York.
Reed, T.B. (1975) *Use of alcohols and other synthetic fuels in Europe from 1930–1950. Impact of methanol fuel on urban air pollution*. American Institute of Chemical Engineers, Boston, Massachusetts, 7 September 1975, paper 31-C.

Reese, E.T. (1956) Enzymatic hydrolysis of cellulose. *Applied Microbiology* 4, 39–45.
Renewable Fuels Association (2012) US Ethanol Industry Overview. Renewable Fuels Association. Available at: http://www.ethanolrfa.org/pages/statistics (accessed 23 March 2012).
Ricardo, H.R. (1921) The influence of various fuels on engine performance. *Automobile Engineer*, February, 1921.
Ricardo, H.R. (1923) *The High Speed Internal Combustion Engine*. Blackie and Son, Ltd, London.
Salameh, M.G. (2004) Oil crises, historical perspective. In: Cleveland, C.J. (ed.) *Encyclopedia of Energy*. Elsevier Academic Press, Boston, Massachusetts.
Scientific American (1918) Seaweed as a source of alcohol. *Scientific American*, 9 November 1918, p. 371.
Scientific American (1919) The declining supply of motor fuel. *Scientific American*, 8 March 1919, p. 220.
Scientific American (1920) Recent patents on mixed fuels. *Scientific American*,11 December 1920, p. 593.
Smith, G.O. (1920) Where the world gets oil and where will our children get it when American wells cease to flow? *National Geographic*, February 1920, p. 202.
Society of Automotive Engineers (1992) *The Ricardo Story: The Autobiography of Sir Harry Ricardo, Pioneer of Engine Research*. SAE Historical Series, Society of Automotive Engineers, Warrendale, Pennsylvania.
Starbuck, A. (1878) *History of the American Whale Fishery from its Earliest Inception to the Year 1876*. US Government Printing Office, Washington, DC.
Steiger, P. (1974) Army lab finds gas substitutes. *Washington Post*, 5 May 1974, p. K5.
Strong, R.M. (1909) *Commercial Deductions from Comparisons of Gasoline and Alcohol Tests on Internal Combustion Engines*. Dept. of the Interior, US Geological Survey, Bulletin 392. US Government Printing Office, Washington, DC.
Time (1940) Recipe for fuel. *Time Magazine*, 23 September 1940.
Tweedy, R.N. (1917) *Industrial Alcohol*. Plunkett House, Dublin, Ireland.
US Congress (1897) *Report of the Joint Select Committee on Alcohol in the Manufactures and Arts*. US Government Printing Office, Washington, DC.
US Senate (1907) Free Alcohol Law, Senate Finance Committee Hearings on HR 24816, Feb. 1907, Doc. No. 362, 320 pp.
United Nations (1952) The production and use of power alcohol in Asia and the Far East. *Report of a Seminar Held at Lucknow, India*, 23 October 1952. Organized by the Technical Assistance Administration and the Economic Commission for Asia and the Far East, United Nations, New York.
Voosen, P. (2011) No eureka moments in long US campaign to crack cellulosic code. *New York Times*, 13 July 2011.
Wall Street Journal (1923) Teagle states gasoline is 36 cents in Havana. *Wall Street Journal*, 15 March 1923.
Washington Post (1906a) President flays the Oil Trust. *Washington Post*, 5 May 1906, p. 1.
Washington Post (1906b) With the automobilists. *Washington Post*, 22 May 1906, p. 8.
Washington Post (1916a) How we eat wood and wear it. *Washington Post*, 17 December 1916, p. MT5.
Washington Post (1916b) Sees use for breweries; Ford would have plants make denatured alcohol for autos. *Washington Post*, 14 November 1916, p. 2.
Washington Post (1929) Farm relief through chemistry. *Washington Post*, 12 September 1929, p. 66.
Welsbach Gas Co. (1909) *History of Light*. Welsbach Publishing, Philadelphia, Pennsylvania. Available at: Smithsonian collection of Advertising, Museum of American History, Washington, DC.
Wescott, N.P. (1936) *Origins and Early History of the Tetraethyl Lead Business*. Du Pont Corp. Report No. D-1013, Longwood ms group 10, Series A, 418-426, GM Anti-Trust Suit. Hagley Museum and Library, Wilmington, Delaware.
Wik, R.M. (1963) Henry Ford's science and technology for rural America. *Technology and Culture*, Summer 1963, pp. 247–258.
Williamson, H.F. and Daum, A.R. (1959) *The American Petroleum Industry*. Northwestern University Press, Evanston, Illinois, pp. 1859–1899.
Winchester, S. (2008) *The Man who loved China*. Harper Collins, New York.
Wright, D.E. (1995) Agricultural editors Wheeler McMillen and Clifford V. Gregory and the farm chemurgic movement. *Agricultural History* 69, 272–287.

2 Status, Innovations and Challenges of Next Generation Biofuel Technologies

Ralph E.H. Sims
School of Engineering and Technology, Massey University, New Zealand

2.1 Introduction

The current debate over biofuels produced from food crops has pinned a lot of hope on advanced biofuels. These are produced from agriculture and forest residues and from non-food energy crops. This chapter examines the current state of the art, the production costs, the supporting policies and the challenges for advanced biofuel technologies and practices.

It is well understood that biofuels produced primarily from grains, sugar crops and oil seeds are limited in their ability to substitute for petroleum products as well as reduce greenhouse gas (GHG) emissions and provide energy security and economic growth. The sustainable production of the relevant biomass feedstocks is under review, as is competition for land and water used for food and fibre production. A possible exception that appears to meet many of the acceptable sustainability criteria is ethanol produced from sugarcane. Even in this case, vulnerability of rural communities in developing countries (mainly in Africa and South America) due to large scale land-use change from traditional farming and forestry to sugarcane plantations affecting livelihood and ecosystem services has given reasons for close scrutiny.

The cumulative impact of these concerns has increased the interest in developing biofuels produced from non-food biomass. Such lignocellulosic materials include cereal straw, sugarcane bagasse, forest residues, and purpose-grown vegetative grass crops such as switchgrass and miscanthus, together with short rotation forests such as poplar or eucalyptus. The use of these feedstocks could avoid many of the 'food versus fuel' concerns currently facing first-generation biofuels and in the longer term they offer greater GHG emission reductions and potentially lower costs. Major constraints remain on the production of advanced biofuels and their commercial deployment although significant progress continues to be made to overcome the technical and economic challenges.

2.2 Current Biofuel Production Technologies

2.2.1 Ethanol production

The traditional biological conversion routes for bioethanol production are well established (Fig. 2.1). The main raw materials needing to be extracted are sucrose or starch. For sucrose from sugarcane or sugarbeet crops, the juices are first mechanically pressed from the cooked biomass followed by fractionation. The sucrose

©CAB International 2013. *Biofuel Crops: Production, Physiology and Genetics* (ed. B.P. Singh)

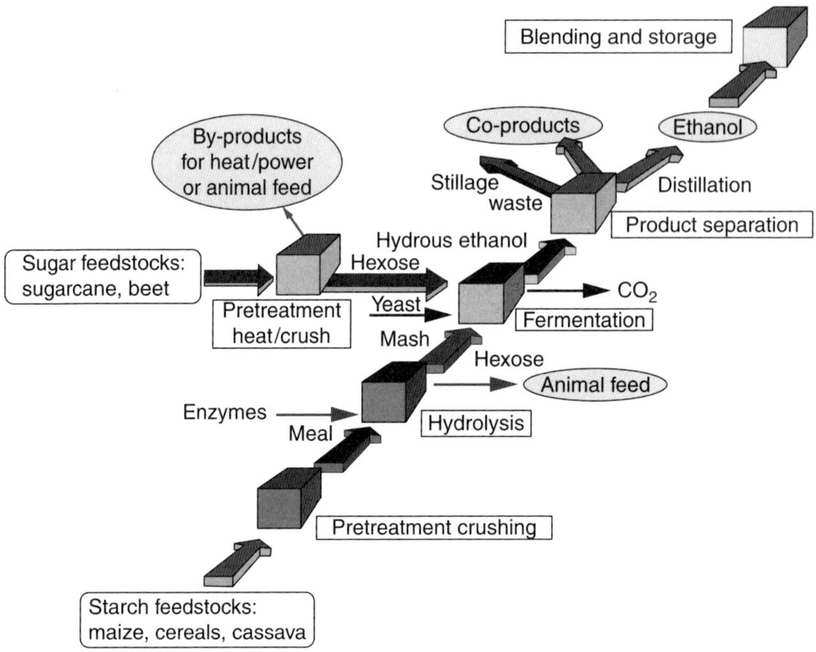

Fig. 2.1. Conversion routes for sugar or starch feedstocks to produce ethanol and co-products such as animal feeds or heat and power generation from combustion of the residues.

is metabolized by yeast cells fermenting the hexoses and the ethanol is then recovered by distillation. Starch crops must first be hydrolysed into glucose before the yeast cells can convert the carbohydrates into ethanol. Pretreatment of cereal crop feedstocks consists of milling the grain followed by liquefaction and fractionation. Acidic or enzymatic hydrolysis then occurs prior to fermentation of the resulting hexoses.

Although highly efficient, the starch grain-based route consumes more energy (and thus potentially emits more CO_2 into the atmosphere depending on the energy sources used) than the sucrose-based route. From the fermentation process onwards, both routes are almost identical. Overall either sugar or starch feedstock utilization use mature technologies to which few significant improvements have been made in recent years.

Although mature, the development of these conversion routes continues with step-by-step improvements and from time to time new technology solutions do emerge. Generally, research and development (R&D) now focuses on the optimization of energy integration and finding value-added solutions for the co-products, previously considered to be 'wastes' in existing production facilities, which have therefore become multi-product 'biorefineries'.

2.2.2 Biodiesel production

The basic inter-esterification process for biodiesel manufacture at normal pressure and ambient temperature (Fig. 2.2) can easily be reproduced at both small and large scales, although the quality of the resulting fuel can vary as unwanted reactions and chemical substances can develop and contaminate the fuel. International standards are now in place to ensure stringent fuel specifications are met at the commercial production scale. As a basis for quality assurance, the development of the European biodiesel standard EN 14214 led to biodiesel becoming accepted as a reliable fuel by many diesel engine and fuel-injection equipment manufacturers.

Quality assurance of the product is imperative, just as it is for diesel and other petroleum

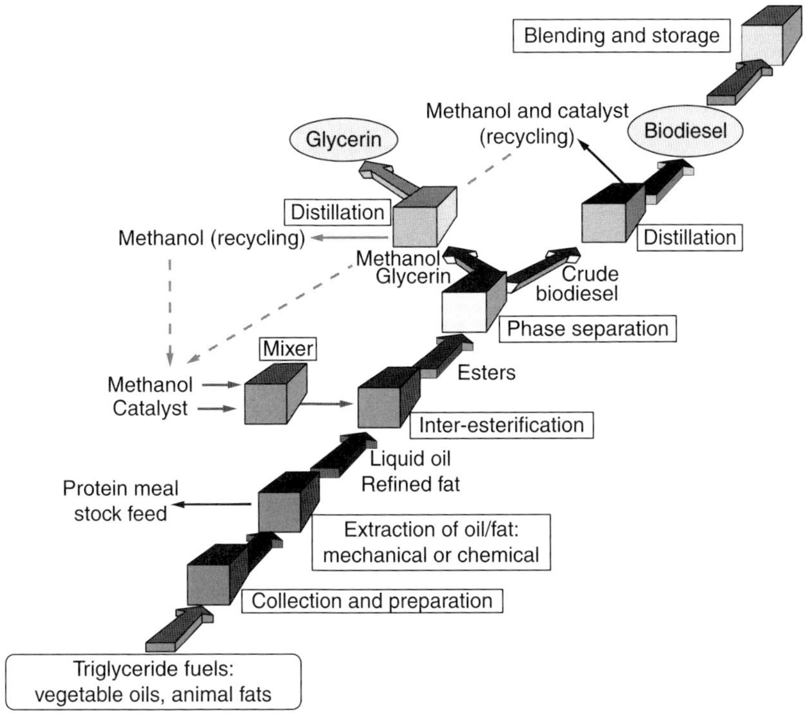

Fig. 2.2. The inter-esterification process to convert oils extracted from vegetable seeds and animal fats (triglycerides) to biodiesel esters also has the co-products of glycerin and, from some vegetable oils, high protein meals for animal feeds are also derived.

fuels for which universally accepted standards have long been in place. The difference with biodiesel is that it can be manufactured by numerous small producers so that if sold (rather than used in engines owned by the producer), maintaining quality is very difficult because frequent testing of batches would be necessary to ensure consistency.

2.3 Sustainability Issues

Biofuels may have positive or negative environmental, social and economic impacts on soil, water and air quality, agrobiodiversity, landscape and food security. Opportunities also exist to gain the co-benefits of local employment and rural development. Internationally accepted sustainability criteria and certification schemes that are implemented locally can help to ensure that the production and use of biofuels have positive impacts overall (IEA, 2011b).

Domestic mandates can ensure that markets for biofuels exist over the long term, but they can also distort markets for related energy and agricultural products. The co-existence of mandates with other policy instruments such as subsidies, tariffs, import quotas, export taxes and non-tariff barriers have not always resulted in effective deployment and efficient production of biofuels. They can even restrict some of the opportunities that biofuels present (Sims, 2008).

The current negative image of biofuels presented by some critics, provoked in part by a rather complex set of national public support schemes and concerns over sustainable land use, is threatening the fulfilment of their promise. If increased biofuel deployment is the goal, then an orderly and defined schedule for the elimination of fossil fuel subsidies can be a critical strategy. Such a strategy can be undertaken in parallel with the gradual implementation of mandates for sustainable biofuels. These measures can provide the necessary conditions to reduce risks and to attract investments to develop and expand sustainable production.

Several approaches to reach these goals are ongoing, including multilateral, regional and bilateral negotiations, as well as unilateral actions. Public and private instruments, such as quality standards, product specifications and certification, may also prove useful approaches when addressing the various technical and sustainability issues (Fritsche *et al.*, 2012). In addition, the development of a global scheme for sustainable production, combined with technical and financial support to facilitate compliance, will ensure that sustainability and trade agendas are complementary.

2.4 Biofuels Trade

Many countries have a competitive advantage when producing biofuels in excess of their current demand so that these can then be exported to other countries unable to meet their biofuel demand from domestic sources (Fig. 2.3). Therefore, increased international trade in biofuels holds promise, possibly in parallel with trade in carbon emission reduction units. Trade in biofuels surplus to local requirements can open up new markets and stimulate the investment needed to promote the full potential for sustainable biofuel production in many impoverished countries.

Currently most biofuels are exported from Argentina, Brazil, the USA, Indonesia and Malaysia to Europe with bioethanol also traded between Brazil and the USA in both directions. For example, when the sugar commodity price is high and sugarcane yields are low due to adverse weather conditions in Brazil, ethanol becomes in short supply to meet the mandated blend levels necessitating imports from the USA to Brazil (Fig. 2.4).

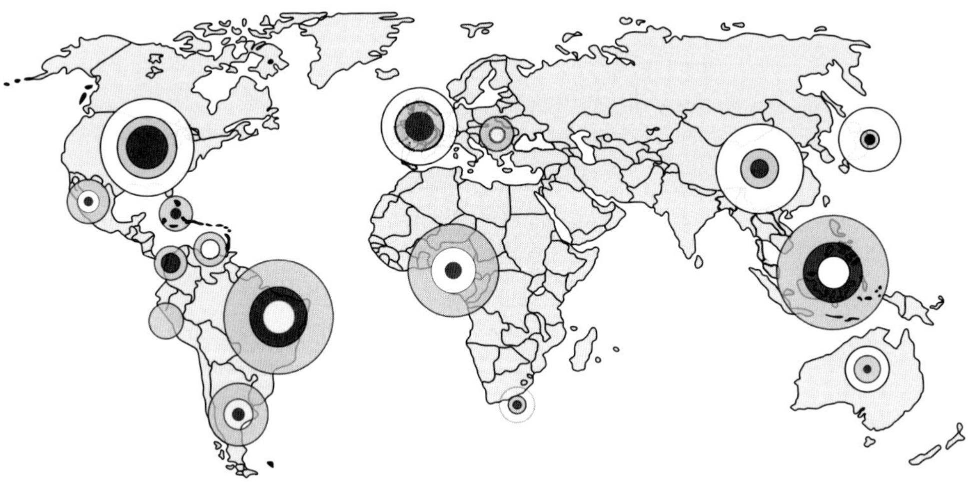

- Feedstock potential based on land available for devotion to first generation biofuel feedstocks.
- Theoretical biofuel demand, assessed to be ~30% of liquid transport fuel consumption in 2006.
- Biofuel production capacity in place at year end 2006.
- Feedstock potential exceeds biofuel demand and surplus production capacity – so export.
- Capacity less than biofuel demand so investment in infrastructure warranted to encourage export potential.
- Feedstock constrained and capacity less than demand – so import.

Source: New Energy Finance http://www.newenergyfinance.com

Fig. 2.3. Indication of regional biofuel feedstock potentials, theoretical production demands, production plant capacities installed at end of 2006, and hence the export potential or import requirements for each selected world region (Sims, 2008; available at: http://www.renewableenergyworld.com/rea/news/article/2008/07/reaching-consensus-on-sustainable-biofuels-52692). Note: areas of circles approximately depict comparative scales between regions and countries.

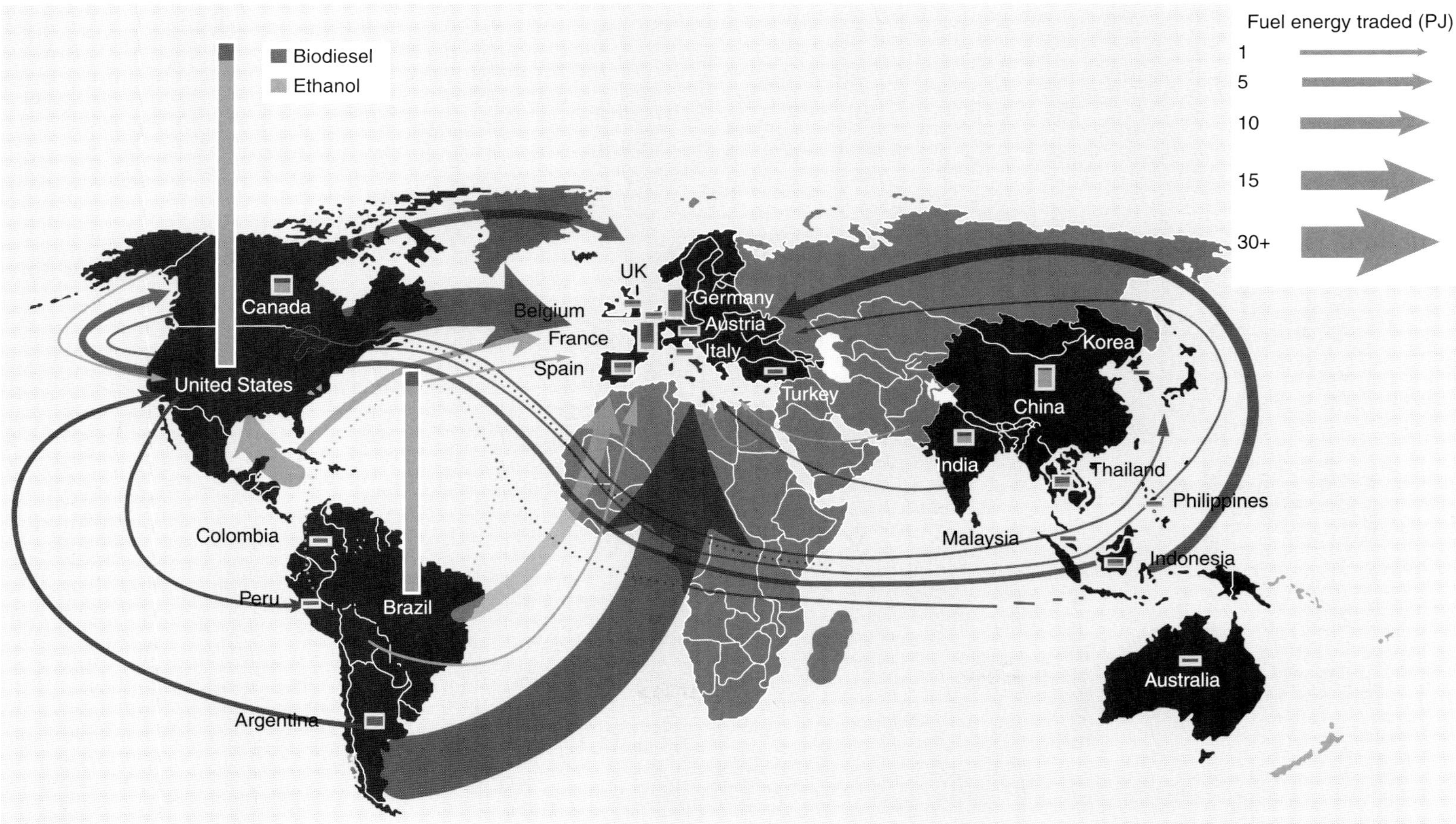

Fig. 2.4. Ethanol (light grey bars) and biodiesel (dark grey bars) comparative production levels and the main international biofuel trade flows in 2009 (Chum *et al.*, 2011). Note: the vertical bars showing annual biofuel production levels by country are approximately to scale with, as a measure, the USA producing around 41 billion l of ethanol and 2 billion l of biodiesel.

Future biofuel markets could be characterized by a diverse set of supplying and consuming regions. From the somewhat concentrated, current supply and demand of biofuels, a future international market could evolve into a truly global market, supplied by many producers, resulting in stable and reliable biofuel sources. This balancing role of an open market and free trade is a crucial precondition for developing biofuel production capacities worldwide (Sims, 2008).

2.5 Biofuel Markets

The production of first-generation biofuels, such as sugarcane ethanol in Brazil, maize ethanol in the USA, oilseed rape biodiesel in Germany and palm oil biodiesel in Malaysia, is characterized by mature commercial markets and well understood technologies. The global demand for liquid biofuels more than quadrupled from 2000 to 2010 and future targets and investment plans suggest strong growth will continue into the near future.

Over the last decade, global biofuel production has steadily increased to reach around 3% of total road transport fuel demand (IEA, 2011a). This growth was largely driven by fluctuating oil prices and policies aimed at improving energy security by reducing dependence on oil imports, reducing GHG emissions of the transport sector, and supporting the agricultural sector that produces the biomass feedstocks.

The main drivers behind the policies in OECD (Convention on the Organization for Economic Cooperation and Development) countries that have encouraged this growth are reduction of dependency on oil imports, energy supply security, support for agricultural industries and rural development of local communities and the potential for GHG mitigation. Recent fluctuating oil prices and future supply constraints have highlighted the need for non-petroleum alternatives. Several non-OECD countries have developed their own biofuel industries to produce fuels for local use, as well as for export, in order to aid their economic development. Many others are considering replicating this model encouraged by supportive policy actions of their national governments (REN21, 2012) and aid agencies.

2.6 Production Costs of First-generation Biofuels

The current costs of producing biofuels are wide ranging and vary with type, feedstock and location (Fig. 2.5). Sugarcane ethanol produced in Brazil and other countries has relatively low life cycle GHG characteristics as well as the lowest production costs due to relatively high biomass yields. This perennial C_4 plant with high photosynthetic efficiency can produce sufficient biomass to be converted to over 6000 l bioethanol ha^{-1} harvested. In addition, the bagasse co-product can be combusted to produce heat and power at the processing plant. Any electricity generated that is surplus to on-site requirements can also be sold to the grid to provide an additional revenue stream.

Typically, ethanol production costs vary between US\$0.25 and US\$0.75 l^{-1} (US\$10–30 GJ^{-1} or around US\$0.45–1.35 l^{-1} gasoline equivalent) and biodiesel costs vary between US\$0.40 and US\$1.60 l^{-1} (US\$10–40 GJ^{-1} or around US\$0.45–1.80 l^{-1} diesel equivalent). The higher cost ranges for maize ethanol and rapeseed oil biodiesel explain why various federal and state financial support mechanisms are required to enable biofuels to compete with petroleum products when oil prices are below approximately US\$100 per barrel.

2.7 Biofuels and Greenhouse Gases

Life cycle assessments (LCA) of biofuels tend to only consider carbon dioxide emissions, and usually exclude the direct and indirect impacts of land use change (Fig. 2.6). Future LCA analyses need to be able to show that positive GHG balances result from biofuel production even when fertilizer and agri-chemical inputs, land use changes, and direct and indirect energy inputs into both biomass production and biofuel processing, are all included. Eventually, having a long-term, internationally accepted, sustainable land use strategy for all agricultural

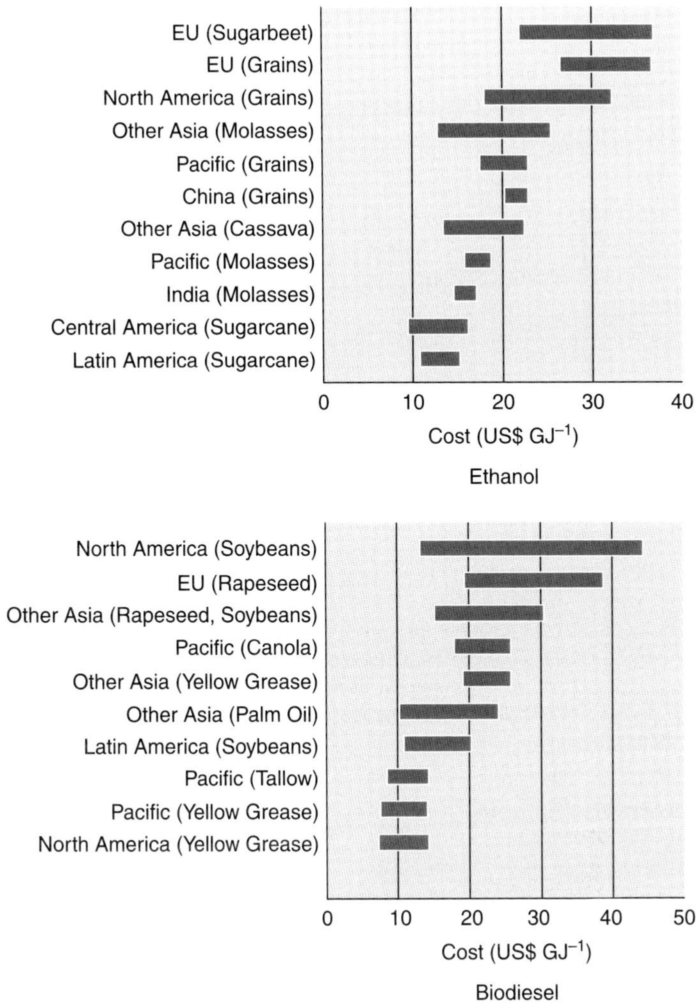

Fig. 2.5. Estimated biofuel production cost ranges for ethanol and biodiesel in 2009 from various feedstocks, exclusive of support subsidies and financial incentives (Chum *et al.*, 2011).

and forestry products, not just for biomass for bioenergy and biofuels, would be an ideal protocol to follow. Meanwhile, product-focused sustainability certification schemes will continue to monitor feedstocks used specifically for biofuels.

In the International Energy Agency's BLUE Map scenario that defines one possible pathway to achieve atmospheric GHG stabilization levels to give no more than a 2°C maximum global temperature rise (IEA, 2010), around 65 EJ of biomass will be needed for transport biofuels in 2050 and a further 80 EJ for heat and power generation (IEA, 2012). The share of transport fuels from biofuels, including for aviation, would need to reach 25%, or perhaps higher if a range of demand-side efficiency measures and/or the uptake of electric vehicles could effectively reduce the total liquid fuel demand growth in the transport sector.

In the long term, combining the biochemical production routes for biofuels with carbon dioxide capture and storage (CCS) and/or combining the thermo-chemical routes with incorporating the solid biochar produced into the soil to increase its carbon content (IBI, 2012),

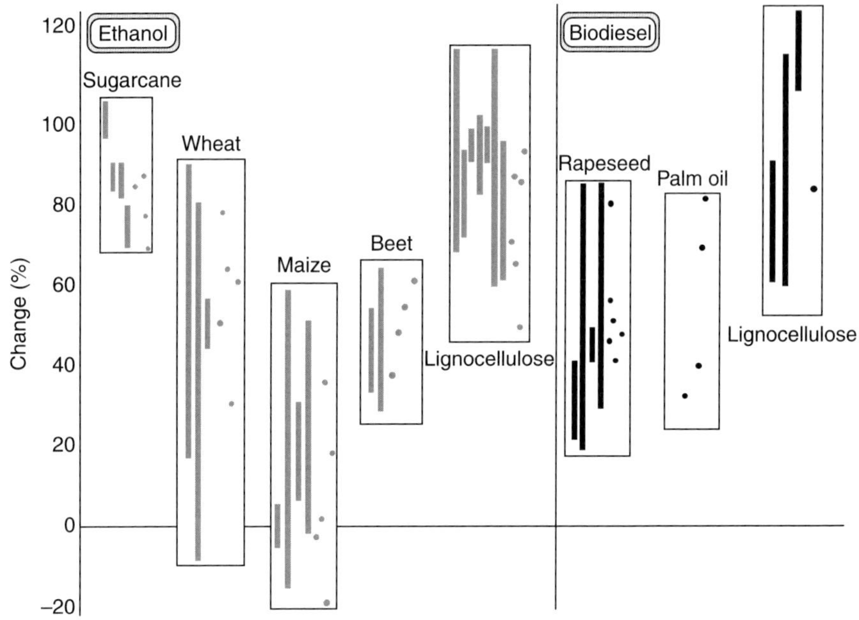

Fig. 2.6. Well-to-wheel emission changes for a range of first- and second-generation biofuels (excluding land use change) compared with emissions from gasoline or mineral diesel (OECD, 2008). Note: based on IEA and UNEP analysis of 60 published life cycle analysis studies giving either ranges (shown by bars) or specific data points (shown by dots).

could lead to net negative annual GHG emissions. Reducing the atmospheric concentration of carbon dioxide by such chemical or physical means will need to be achieved before the end of this century if atmospheric GHG levels are to be stabilized at around 450 ppm (parts per million) and global mean temperature rise is to be restricted to below 2°C as agreed by all nations attending the 15th Conference of Parties of the United Nations Framework Convention on Climate Change held in Copenhagen in 2010.

2.8 Advanced Biofuels

At present the share of global biofuels provided by advanced processes remains very small, most being ethanol production from the enzymatic hydrolysis of lignocellulose (Fig. 2.7).

Interest is now growing in producing 'drop-in fuels' (such as iso-butanol) that can be easily blended with existing fuels for use in markets where conventional biofuels have technical limitations. Revised international aviation fuel standards now officially allow commercial airlines to blend conventional jet fuel with up to 50% of approved biofuels in order to consolidate the future of aviation biofuels (REN21, 2012). KLM Royal Dutch Airlines became the first airline to operate a commercial passenger flight using a low-blend of biofuels and several other airlines are following suit, driven by both rising oil prices and GHG emission restrictions over European air space under the EU emissions trading scheme (Euractiv, 2012).

Advanced biofuels are relatively immature (although R&D has been ongoing for at least four decades). Hence the potential for reducing costs and increasing production efficiency levels is possible as more experience is gained at the commercial scale. Several demonstration plants are now operating, mainly in Europe and the USA. A database of demonstration plants is provided by Task 39, Biofuels, of the IEA Bioenergy implementing agreement (full details can be found at: http://demoplants.bioenergy2020.eu). Depending partly on future

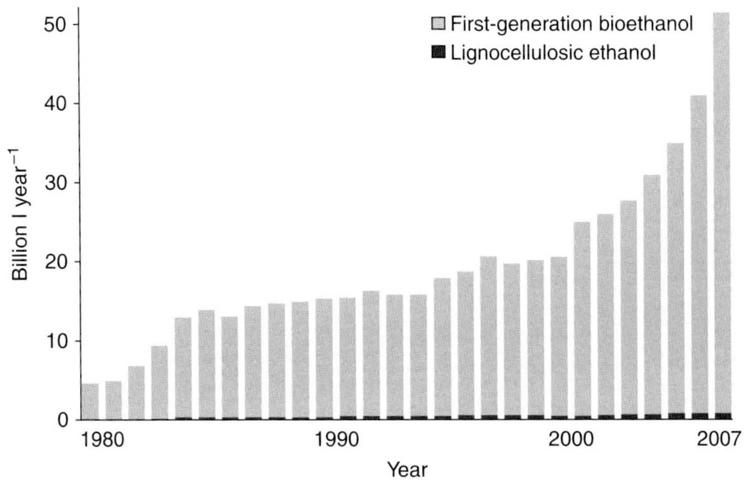

Fig. 2.7. Growth in world ethanol production using lignocellulosic biomass feedstocks has remained very slow compared with production from first-generation ethanol produced from food crops (Mabee and Saddler, 2007).

oil prices, these biofuels are therefore likely to become a part of the solution to the challenge of shifting the transport sector towards more sustainable energy sources at some stage in the medium term. However, major technical and economic hurdles are still to be faced before advanced biofuels can be widely deployed at current oil prices.

To address these issues, significant investment in research, development and demonstration (RD&D) funding by both public and private sources is occurring in order to develop pilot and demonstration facilities. However, much more investment is likely to be required in the near future if rapid commercial deployment of these technologies is to eventuate.

The range of first-generation biofuels, particularly sugarcane ethanol, will play a continuing role in meeting future biofuel demands. The transition to an integrated first- and second-generation biofuel landscape is therefore most likely to encompass the next one or two decades. The infrastructure and experiences gained from deploying and using first-generation biofuels can be transferred to support and guide advanced biofuel deployment. Once these technologies are fully commercialized, it is likely they will be favoured over many first-generation alternatives if policies can be designed to also reward national objectives such as improved environmental performance or security of supply.

2.9 Future Feedstock Supply

Low-cost crop and forest residues, wood process wastes and the organic fraction of municipal solid wastes (MSW) can all be used as lignocellulosic feedstocks for advanced biofuels. Where these materials are available, it should be possible to produce biofuels with virtually no additional land requirements or impacts on food and fibre crop production. The one proviso is to ensure removal of the chosen volumes of biomass from a given land area does not result in a significant lowering of soil nutrients. Where residues and organic waste feedstocks are in limited supply, the growing and harvesting of energy crops such as vegetative grasses or short rotation forests will be necessary to provide supplementary supplies. Soil nutrient levels are likely to decline without the addition of chemical fertilizers or the recycling of nutrients, for example from addition of ash after combustion.

Crop yields could increase significantly over time since breeding research (including genetic modification) is at an early stage compared with the breeding of various species and varieties of food crops. New varieties of energy crops may lead to increased yields, reduced water demand and lower dependency on agri-chemical inputs. In some regions where low intensity farming is currently practised, improved management of existing crops grown on arable land could result in higher yields per hectare and/or shorter maturity periods. Increased productivity could free up land so as to enable energy crops to be grown without the need for increased deforestation or any reduction in present food and fibre supplies.

2.10 Supply Chain Issues

Feedstock costs are an important component of the total biofuel costs, so the aim should be to reduce feedstock delivery and storage costs. This issue is often inadequately taken into account and supply logistics could become more important as competition for biomass feedstocks develops. To service a commercial-scale, advanced biofuel processing plant, the lignocellulosic feedstock will need to be harvested or collected, then pretreated, transported, stored and delivered in large volumes at a desired quality (without soil contamination and at the correct moisture content), in order to keep the plant running all year round and hence maximize the return on investment.

Biomass suppliers need to be contracted in advance to provide minimum volumes of guaranteed quality, ideally over a period of several years, in order to reduce the project investment risks. The failure to secure the feedstock in sufficient quantities has proved to be a major challenge for some bioenergy plants, which have either closed down or the project abandoned prior to commissioning.

The fibrous nature of lignocellulosic biomass and its low energy density (particularly with a high moisture content) make it difficult and costly to collect, store, handle and transport. The present mechanical handling methods available are largely derived from agricultural and forestry systems. They tend to be costly and could be inefficient for handling the large quantities of biomass feedstocks needed to supply large-scale, advanced biofuel production plants.

Biomass process residues that are removed from the growing area at the time of harvesting the primary product can accumulate at the processing plant (for example bagasse, rice husks, sawdust and bark from pulp logs). They are therefore considered to be free-on-site or can even have a negative value if the costs of disposal are avoided by their utilization as feedstocks. Other biomass residues need collection from the forest or field after harvest of the main product, which can be a costly activity.

Crop and forest residues can be collected, stored and delivered at relatively low cost under some circumstances, but dedicated energy crop feedstocks are likely to be significantly more costly to grow, harvest and deliver. Important reductions in the delivered costs of these feedstocks need to be achieved to help lower the cost of biofuel production. The choice of collection system can have a considerable impact on the cost of the biomass delivered to the processing plant gate (Fig. 2.8). In this example, for a transport distance over 80 km of varying road types, the total delivered cost of forest residues was less than US\$2 GJ^{-1} for system A (based on transport to a centralized chipper with no intermediate storage), and was around half the delivered cost of system E (based on a mobile, in-forest chipper/forwarder system).

Most forms of biomass and bioenergy carriers tend to have a relatively low energy density per unit of volume (e.g. $GJ\ m^{-3}$) or mass (e.g. $MJ\ kg^{-1}$) compared with fossil fuels with the same energy equivalent content. For example, ethanol has an energy content of ~21 $MJ\ l^{-1}$ whereas gasoline is ~34 $MJ\ l^{-1}$; air-dried woody biomass is ~12–15 $GJ\ t^{-1}$ and sub-bituminous coal ~20–25 $GJ\ t^{-1}$ (low heat values). This usually makes handling, storage and transport more costly per unit of energy carried in the biomass. The variable moisture content of biomass adds a further complexity (Box 2.1).

Some forest and crop residues may be less cost competitive because the biomass resource is dispersed over large areas, which leads to higher collection and transport costs. Where road transport cannot be avoided (for example, where rail is not feasible), environmental and social negative impacts from numerous vehicle

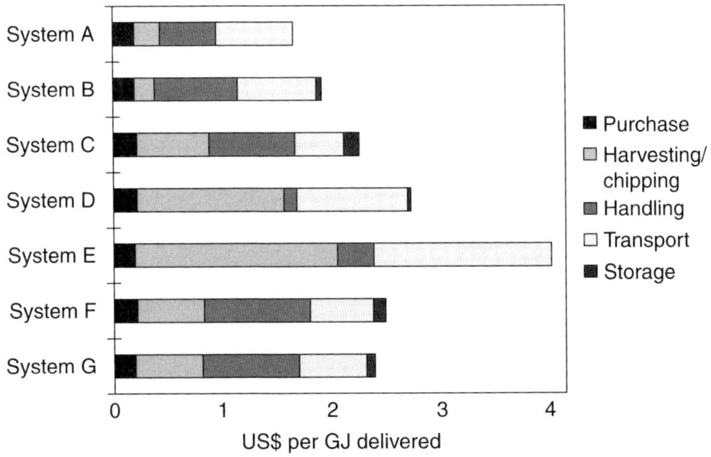

Fig. 2.8. Delivered costs of forest residues over 80 km distance can vary with the choice of harvesting, handling, transport and storage system together with the purchase price for the biomass feedstock (IEA, 2007).

Box 2.1. Moisture content of biomass

Take, for example, a truck carting wood chips produced from the branches and tops of a freshly harvested plantation forest 35 km to a bioenergy plant. At the time of harvest, trees typically contain at least 50% water by weight. So if a full load of fresh wood chips in a truck weighs 26 t, 13 t is the 'dry matter' of the biomass and 13 t is the water in the biomass. On combustion, the load of fuelwood would have an available energy content of around 200 GJ. If the truck is now fully loaded with the same volume of chipped biomass, but it is only chipped after the branches have been left in the forest for a few weeks to naturally air dry down to say 20% moisture content (wet basis), then the load would now weigh 16.3 t, of which the dry matter remains at 13 t but the moisture present in the biomass now only weighs 3.3 t. Being drier, the fuelwood will burn more efficiently. The energy content of the load will be higher at around 280 GJ but it is also lighter to transport.

If the truck had a 26 t maximum payload capacity, then the original load size would be limited by the weight of the wet, freshly harvested chips and not by the volume of the load. If the load of air-dry wood chips carried could have been increased in volume so as to meet the 26 t maximum payload (assuming the truck design allowed for a bigger size load, perhaps by adding higher extension sides), then the load would now contain 21.8 t of dry biomass and 5.2 t moisture, thus giving an overall energy content of 370 GJ. Hence the transport cost per GJ of available energy would be a lot less when the truck is operating at full load capacity and less water is being transported (Fig. 2.9).

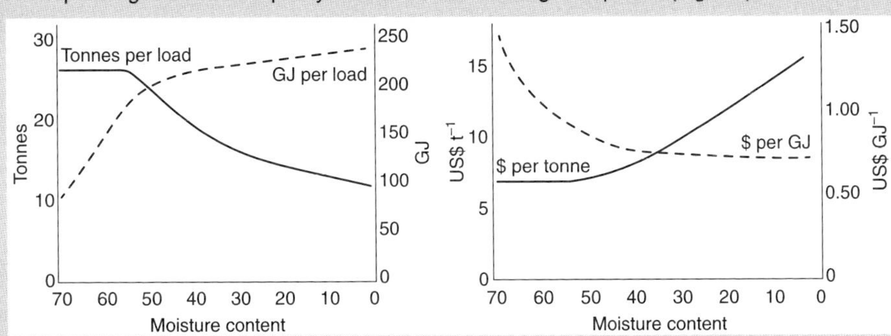

Fig. 2.9. Delivered biomass energy costs (US$ GJ^{-1}) depend on optimizing the moisture content. Note: analysis is based on a 26 t maximum payload truck travelling over a cartage distance of 35 km at a charge of US$0.42 t^{-1} km^{-1}.

Table 2.1. Typical scale of operation for commercial-scale advanced biofuel processing plants using energy crop or forest residues as lignocellulosic feedstocks.

Scale of plant	Plant capacity ranges, and assumed annual hours of operation	Biomass fuel required (oven dry t $year^{-1}$)	Truck vehicle movements for delivery to the plant	Land area required to produce the biomass[a] (% of total land within the given radius)
Small pilot	15,000–25,000 l $year^{-1}$ 2,000 h	40–60	3–5 $year^{-1}$	1–3% within 1 km radius
Demonstration	40,000–500,000 l $year^{-1}$ 3,000 h	100–1,200	10–140 $year^{-1}$	5–10% within 2 km radius
Pre-commercial	1–4 Ml $year^{-1}$ 4,000 h	2,000–10,000	25–100 $month^{-1}$	1–3% within 10 km radius
Commercial	25–50 Ml $year^{-1}$ 5,000 h	60,000–120,000	10–20 day^{-1}	5–10% within 20 km radius
Large commercial	150–250 Ml $year^{-1}$ 7,000 h	350,000–600,000	100–200 day^{-1} and night	1–2% within 100 km radius

[a]The land area requirement would be reduced where crop and forest residue feedstocks are available to provide some of the biomass required by the plant.

movements can be inevitable due to the low energy density of many solid and liquid forms of biomass making many trips necessary to supply the bioenergy plant (Table 2.1).

The forest and sugar industries have largely overcome the large-scale transport problems of bulky feedstocks after many years of experience. Sugarcane processing plants, for example, typically handle around 300,000 t of billets during the 6–7-month harvesting season and therefore efficient road or mini-rail transport systems have evolved.

2.11 Advanced Biofuel Conversion Processes

Several technological conversion routes exist for producing advanced liquid or gaseous biofuels from biomass (Table 2.2). None has yet reached the fully commercial stage; hence no clear technology leader or pathway has emerged. The biorefinery concept, usually based on either thermo- or biochemical routes, is where biofuels are produced from single or multi-feedstocks along with one or more co-products, as well as possibly heat and power produced for use on site and/or for export. The concept of producing small quantities of high value products (e.g. chemicals) and larger quantities of low value products (e.g. biofuels) theoretically maximizes returns from the biomass feedstock by improving economic performance in the same way that oil refineries produce multi-products from crude oil today.

Several of these advanced biofuel routes are under evaluation in research laboratories and pilot-plants. Some biofuel products are similar to those produced from the two main conversion routes for lignocellulosic feedstocks (some examples of which are highlighted in Table 2.2). Others can produce drop-in fuels that can be easily blended with existing petroleum fuels.

2.11.1 Biochemical route

After comminution of the biomass feedstock and pretreatment, ethanol can be produced by the hydrolysis of lignocellulosic raw materials, followed by the fermentation of the extracted sugars then distillation and formulation to give the final fuel product (Fig. 2.10). Fermentation of glucose sugars is a mature commercial technology, but the hydrolysis of agricultural residues and woody biomass and fermentation of pentose sugars still need further development.

Table 2.2. Classification of a range of advanced biofuels and conversion processes.

Biofuel group	Specific biofuel or process	Biomass feedstock	Production process
Bioethanol	Cellulosic ethanol	Lignocellulosic materials	[a]Advanced enzymatic hydrolysis and fermentation
Synthetic biofuels	Biomass-to-liquids (BTL) Fischer-Tropsch diesel (FT) Synthetic diesel Biomethanol Heavier alcohols (butanol and mixed) Dimethyl ether (DME) P-series (ethanol + MTHF, etc.)	Lignocellulosic materials	[b]Gasification and synthesis
Biodiesel hybrid combining first-generation with advanced processed biofuels	NExBTL H-Bio Green pyrolysis diesel Algal oil	Vegetable oils and animal fats Lignocellulosic materials Algae	Hydrogenation (refining) [b]Pyrolysis Cultivation
Methane	Biosynthetic natural gas (SNG)	Lignocellulosic materials	[b]Gasification and synthesis
Biohydrogen	Hydrogen	Lignocellulosic materials	[b]Gasification and synthesis or [a]biological processes

[a]Major biological routes; [b]major thermo-chemical routes. Some of the other routes are described in Section 2.13.
MTHF, methyltetrahydrofuran.

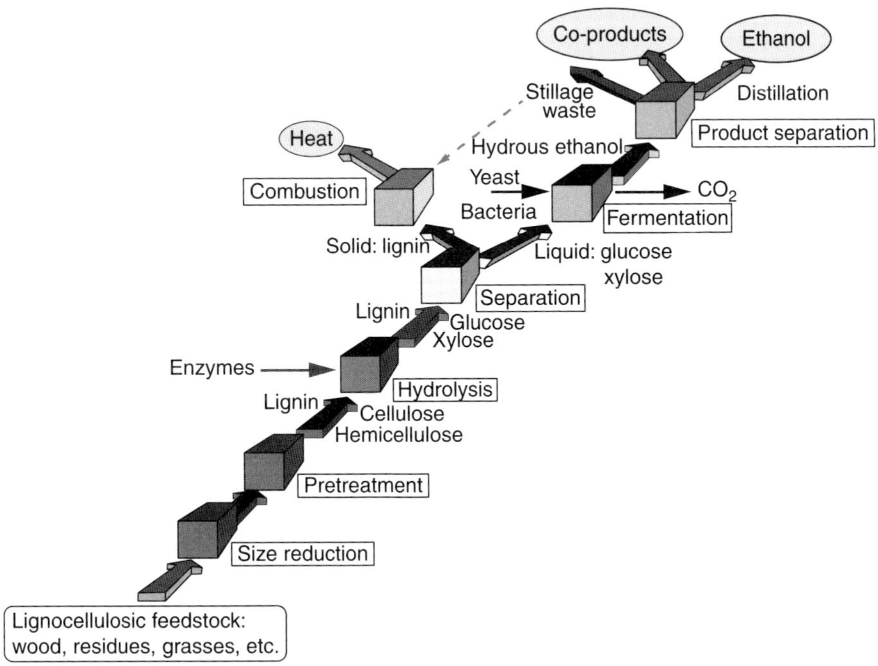

Fig. 2.10. Biochemical pathway for conversion of lignocellulosic feedstocks into ethanol and co-products.

2.11.2 Thermo-chemical route

Indirect liquefaction methods require the biomass to be first gasified and the product gas cleaned and processed to form synthesis gas (syngas) (mainly CO and H_2) (Fig. 2.11).This gaseous mixture can then be used in a commercial chemical process to synthesize a range of liquid biofuels including methanol, Fischer-Tropsch diesel, DME (dimethyl ether) or as gaseous methane or hydrogen fuels.

2.11.3 Comparisons of routes

There is currently no clear commercial or technical advantage between the biochemical and thermo-chemical pathways (Sims *et al.*, 2008). Both sets of technologies remain unproven at the fully commercial scale, are under continual development and evaluation, and have significant technical and environmental barriers yet to be overcome. For the biochemical route, much remains to be done in terms of: improving feedstock characteristics; reducing the costs by perfecting the pretreatment; improving the performance of enzymes and lowering their production costs; and improving integration of the overall process. The potential advantage of the biochemical route is that cost reductions have proved reasonably successful to date, so it could possibly provide cheaper biofuels than via the thermo-chemical route. As a broad generalization, there are fewer technical hurdles to the thermo-chemical route since much of the technology is already proven, so there is perhaps less opportunity for improvements and hence cost reductions.

Improving efficiency is vital to the extent that it reduces the final product cost and improves environmental performance, but it should not be a goal in itself. One key difference of the biochemical process route is that the lignin component is a residue of the enzymatic hydrolysis process and hence is available to be used separately, for example for heat and power generation. In the BTL thermo-chemical process it is converted into synthesis gas along with the cellulose and hemicellulose biomass components. Both processes can potentially convert 1 dry t of biomass (~20 GJ t^{-1}) to around

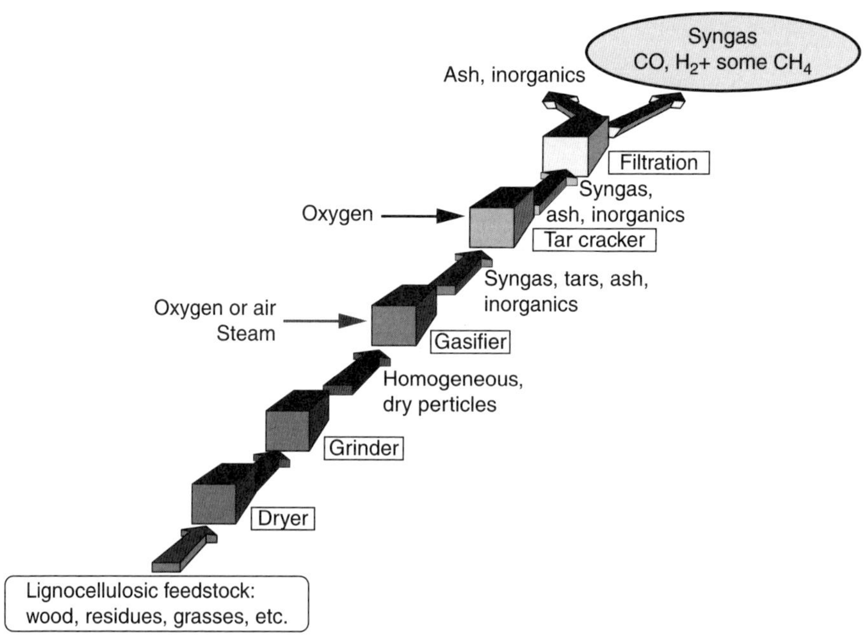

Fig. 2.11. Thermo-chemical pathway for conversion of lignocellulosic feedstocks into synthesis gas and co-products.

6.5 GJ t^{-1} of biofuel energy carrier giving an overall biomass to biofuel conversion efficiency of around 35%. Although this efficiency appears relatively low, overall efficiencies of the biochemical process can be improved when surplus heat, power and co-product generation are included within the total system.

Although both routes have similar potential yields in energy terms, different volumetric biofuel yields (in terms of litres per tonne of feedstock) occur in practice (Table 2.3). Major variations between the various processes under development, together with variations between biofuel yields from different feedstocks, gives a complex picture with wide ranges quoted in the literature. Typically enzyme hydrolysis could be expected to produce up to 300 l ethanol per dry t biomass whereas the biomass-to-liquids (BTL) route could yield up to 200 l synthetic diesel t^{-1}. The similar overall yield in energy terms (around 6.5 GJ t^{-1} biofuels at the top of the range) is because synthetic diesel has a higher energy density by volume than ethanol.

The biochemical routes produce mainly ethanol whereas the thermo-chemical routes can also be used to produce a range of longer-chain hydrocarbons from the synthesis gas. These include biofuels better suited for aviation and marine purposes. Only time will tell which conversion route will be preferred, but whereas there may be alternative drive systems becoming available for light duty vehicles in future (including hybrids, electric plug-ins and fuel cells), such alternatives for aeroplanes, boats and heavy trucks are less likely and liquid fuels will continue to dominate.

2.12 Production Costs

The full production costs associated with advanced biofuel pathways remain uncertain and are treated with a high degree of commercial propriety. Comparisons between the biochemical and thermo-chemical routes have proven to be very contentious within the industry, with the lack of any real published cost data being a major limitation for comparative analysis. Production costs at the commercial scale have been estimated to be in the range of US$0.80–1.00 l^{-1} of gasoline equivalent (lge) for ethanol and at least US$1 l^{-1} of diesel equivalent for synthetic diesel. This range broadly relates to gasoline or diesel wholesale prices (measured in US$ lge^{-1}) when the crude oil price is between US$100 and US$130 bbl^{-1} (Fig. 2.12). The present widely fluctuating oil and gas prices therefore make investment in advanced biofuels a high risk venture. By 2030, production costs could drop to a lower range between US$0.50 and US$0.75 l^{-1} equivalent. Algal-based biofuels remain at very high costs of production in spite of considerable R&D investment over several decades.

Success in the commercial development and deployment of advanced biofuel technologies will require significant progress in a number of areas if the technological and cost barriers they currently face are to be overcome (Sims *et al.*, 2008). Areas that need attention include:

- improved understanding of feedstocks, reduction in costs of delivered biomass to the processing plant and further

Table 2.3. Indicative biofuel yield ranges per dry tonne of feedstock from biochemical and thermo-chemical process routes (Sims *et al.*, 2008).

	Biofuel yield (l dry t^{-1})		Energy content (MJ l^{-1})	Energy yields (GJ t^{-1} dry matter)	
Process	Low	High	Low heat value	Low	High
Biochemical					
Enzymatic hydrolysis ethanol	110	300	21.1	2.3	6.3
Thermo-chemical					
Syngas to Fischer Tropsch diesel	75	200	34.4	2.6	6.9
Syngas to ethanol	120	160	21.1	2.5	3.4

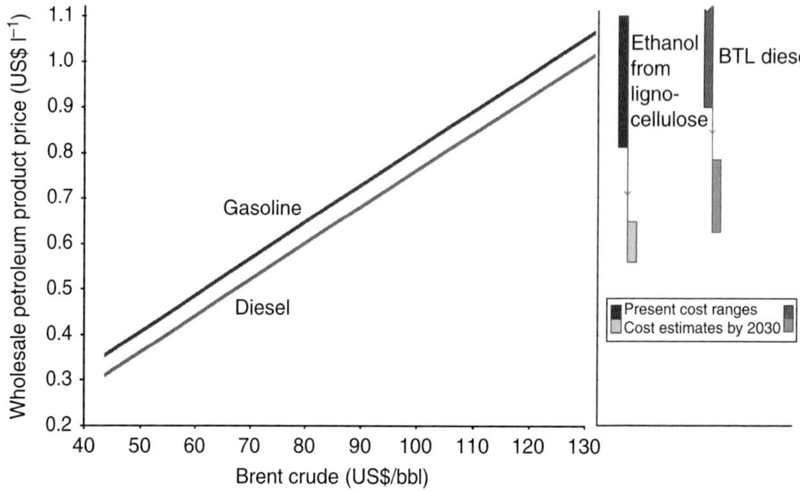

Fig. 2.12. Production cost range assessments for advanced biochemical and thermo-chemical (BTL) biofuels (US$ l^{-1} gasoline equivalent) compared with wholesale petroleum fuel prices correlated with the crude oil price over a 16-month period, and 2030 biofuel cost projections assuming significant investment in research, development and demonstration (RD&D) (Sims *et al.*, 2008).

development of energy crops to improve productivity while reducing inputs;
- technology improvements for the biochemical route in terms of feedstock pretreatment, enzymes and improvement in process efficiency and overall cost reductions;
- technology improvements for the thermochemical route in terms of feedstock quality, improved gasification process and gas clean-up, and improvements in efficiency and overall cost reductions; and
- development of co-products and improved integration of the process in biorefineries.

Market assessments of the biofuels and their co-products need to take into account all the benefits, costs and co-benefits including rural development, employment, energy security and carbon sequestration potential, if a thorough assessment of their deployment is to be made.

2.13 Other Liquid and Gaseous Biofuels

Some other routes for producing advanced biofuels are briefly outlined below (see also Table 2.2). Full details of many of these technologies are provided in the literature to be found on the website of Task 39, Biofuels of the IEA Bioenergy Implementing Agreement (http://www.task39.org/Home.aspx) and in the IEA Biofuels Roadmap (IEA, 2011b).

2.13.1 Hydrothermal processing

Vegetable oils and fats can undergo thermal depolymerization or hydrothermal processing to yield a diesel biofuel that is more similar to fossil-derived diesel than is conventional biodiesel. The process involves biomass reacting in water at high temperatures and pressure to form oils and residual solids. After the reaction the organics are separated from water and a distillate component suitable for combustion in compression ignition (diesel) engines is produced.

2.13.2 Hydrodrogenation

An advanced diesel referred to as HVO (hydrogenated vegetable oil) can be produced from fatty acids by well understood hydro-processing technologies, but the product is a synthetic diesel fuel containing no sulfur and having a cetane number of 90 to 100. Biomass feedstocks for this fuel are similar to those for conventional biodiesel and therefore subject to the same sustainability concerns. One of the advantages of this 'green diesel' drop-in fuel is that it does not contain oxygen and can therefore be blended with other hydrocarbons (such as diesel and kerosene) in all proportions. The benefits that HVOs have over biodiesel include good performance at low temperatures, no storage stability problems and no susceptibility to microbial attack. HVO processing projects are in place in Singapore and Finland (Bacovsky *et al.*, 2010).

2.13.3 Algal biofuels

The overall concept for producing biofuels from oil-containing algal feedstocks involves similar process steps to those used for other biofuels. The final synthetic liquid products are a diesel-type fuel of high quality and synthetic aviation fuels. Claimed advantages for algae biofuels include: potentially high productivity as well as the use of non-food feedstocks; use of otherwise non-productive, non-arable land; possible utilization of fresh, brackish or saline water sources and wastewater; production of valuable co-products together with the biofuels; and the potential for recycling CO_2 and other nutrient waste streams (Darzins *et al.*, 2010).

2.13.4 Biomethanol

This basic alcohol can be produced by the thermo-chemical conversion of lignocellulosic feedstocks after gasification. The cleaned synthesis gas mixture is condensed into liquid forms by adding a catalyst.

2.13.5 Butanol

This drop-in fuel is produced by the bacterial fermentation of similar feedstocks to those presently used for ethanol production (sugars, carbohydrates, lignocellulose or algae). It can be distributed through existing pipelines and used directly in spark-ignition Otto cycle internal combustion engines as it has properties similar to gasoline, including a higher energy density (~29 MJ l^{-1}) than either ethanol or methanol. Several demonstration plants are in place.

2.13.6 Dimethyl ether

Dimethyl ether (DME) can be produced from either the catalytic dehydration of methanol or from the syngas produced after gasifying biomass. DME is a basic ether that can be used to produce a range of organic compounds or used in fuel applications as a substitute for liquefied petroleum gas (LPG) propane. It therefore also has good potential for use as a cooking fuel. A locally produced, clean burning, DME fuel could become an acceptable substitute for the use of traditional biomass in developing countries and hence avoid the health issues from smoke inhalation, the daily drudgery of fuelwood collection, mainly by women, and hence reduce the related deforestation and sustainability issues.

2.13.7 Biogas

Biogas is produced by anaerobic bacteria through the digestion of organic waste materials. Dedicated green energy crops such as maize and grass are also suitable feedstocks as they can be stored as silage to provide year-round supplies. Anaerobic digestion also occurs naturally in landfills to produce landfill gas. Biogas is mainly used to provide electricity and heat, but it can also be upgraded to biomethane (by removal of hydrogen sulfide and carbon dioxide) and injected into the natural gas pipeline network or used in vehicle engines designed to run on compressed natural gas (CNG).

2.13.8 Biosynthetic natural gas (SNG)

This methane-rich gas is produced by the thermo-chemical conversion of any form of biomass, usually woody biomass, into a combination of synthetic gases including hydrogen, carbon monoxide and methane. The mixture is cleaned and then converted into 'bioSNG' methane through a catalytic process. Water and carbon dioxide are removed prior to use or injection into gas pipelines.

2.13.9 Hydrogen

This gas can be produced via many routes including the gasification of biomass, at the large scale by employing an efficient integrated gasification combined cycle system. Most hydrogen produced commercially is a by-product of industrial processes or produced from natural gas by the steam reforming of methane. Biomass could be used as the feedstock for this process but the scale of plant needed to be economic would require large volumes of biomass to be consumed annually.

2.14 Policy Considerations

To integrate and better coordinate biofuels policy frameworks requires:

- coordinating national and international actions among the key sectors involved in biofuel development and use including agriculture, energy, environment and transport;
- negotiating a schedule to gradually eliminate the tariff and non-tariff barriers to trade;
- agreeing on internationally compatible fuel quality technical standards, while recognizing that several countries are already engaged in efforts to harmonize these standards;
- providing transparency in blending and other regulatory requirements at national and sub-national levels;
- reviewing policies in agriculture, energy and other sectors that could contribute to the inefficient production of biomass feedstock or biofuels leading to possible market distortion; and
- adopting local, bilateral, regional and/or other frameworks for biofuels trade agreements with the objective of collaborating with existing frameworks to achieve convergence towards a comprehensive, international land use improvement agreement.

Assessing and monitoring the benefits and impacts of biofuels trade, use and production, requires agreeing on sustainability principles and criteria that include effective, mutually agreed and attainable systems using means such as:

- certification, consistent with World Trade Organization (WTO) rules;
- recognizing that several key international efforts are already underway both in governmental and non-governmental contexts and that an iterative review of such criteria should be undertaken in order to continually raise the standards through advances in knowledge from research and through experience gained in the field;
- harmonizing life cycle analysis methodologies for biofuels, including GHG life cycle accounting methodologies, recognizing that efforts both at the international and national levels are already under way;
- mapping and continued assessment of suitability of degraded and marginal land; and
- evaluation and mapping of carbon stocks, areas rich in biodiversity, and other high conservation value areas to avoid land use change from competition.

Addressing negative indirect effects of biofuels trade, use and production calls for:

- continued global research to identify and quantify links between biofuels and land use change;
- mechanisms to promote biofuels that do not have negative land use change impacts;
- mechanisms that mitigate these negative impacts but do not unduly increase transaction costs for producers; and
- social safeguards at the national level that ensure that vulnerable people, particularly small landholders, are not further disadvantaged through food and energy price increases and other potential negative economic side effects.

Rewarding positive impacts and investments, including through carbon management, will require the use of existing and innovative tools to ensure that markets reward environmental and social performance, including carbon sequestration. Ensuring that biofuels production is accomplished by shared benefits, rights and rules of law is essential.

There is a link between bioenergy and rural development for improving rural incomes and abating poverty and thus providing a basis for increased investment and more efficient and sustainable agriculture. To create significant rural and social development benefits from biofuel projects will need to overcome difficulties in integrating smallholders into markets, tendencies to concentrate buying power within supply chains, and a lack of financial markets for small-producers. Many business models exist that equitably share benefits throughout the supply chain, especially at the farmer level. National policies, bilateral agreements, foreign assistance and international financial institutions should give preferential treatment to these types of production systems where feasible, as well as to projects that encourage development of small-scale production and regional biofuels markets. Trade, energy security, health, transport and climate change policies should be developed in association with each other. Some benefits from biofuels use do not have an associated income stream. Therefore even sustainable trade will not necessarily flow to the best performers. Underfunded benefits fall into the categories of rural and social development, ecosystem services including biological carbon fixation and water resource management, and better practices that might reduce crop yields but restore ecosystem health, such as conservation agriculture.

Promoting an informed and continuous dialogue engaging all relevant stakeholders, particularly biomass producers, is key to ensuring equitable distribution of benefits resulting from biofuel projects, and to addressing other elements of sustainability. To be effective, these dialogues should be translated into the allocation of public and private budgets to meet agreements on priorities for RD&D portfolios and for specific projects.

While countries could consider other climate-related initiatives besides biofuels, the goals of public and private RD&D investments related to biofuel trade, use and production should include (but not be limited to):

- producing cost-effective advanced biofuels;
- enabling sustainability lessons learned from first-generation biofuels to be used for future biofuel production;
- increasing conversion technology performance;
- maximizing climate change mitigation opportunities;
- evaluating the costs and benefits of increasing soil carbon content;
- increasing crop productivity; and
- improving ecosystem health through better management techniques, mechanization, water management, ongoing plant breeding and selection, and precision farming techniques that reduce losses when applying fertilizers and agri-chemicals.

Capacity-building programmes are needed for farmers, foresters and small and medium-sized enterprises that are active in biofuels. Education should include soil improvement techniques and water management technologies. Capacity building is also needed for the development of effective technology innovation systems involving research and education, extension, industrial capacity to participate in joint ventures with supportive government agencies and an engaged civil society.

In developing countries, there is a need for a clear commitment for national climate change policies that promote biofuels and for related overseas aid to be additional to that provided for sustainable development. This is best achieved by climate change policies that drive direct foreign investment by engaging the energy sector to ensure it is in harmony with trade policies and sustainability requirements. Guided by a consensus of national stakeholders in recipient countries, overseas aid connected with biofuels development should be partnered with development and UN agencies and the private sector to help in reducing the overall transaction costs of sustainable development schemes.

2.15 Conclusions

The uptake of first-generation biofuels, currently produced in many countries mainly from food crop feedstocks, has increased rapidly over the past few years in response to concerns about energy supply security, rising oil prices and climate change. The wide production and deployment in terms of local production, blending, distribution, engine compatibility, development of flex-fuel vehicles, international trade, etc. has been successful in general. This experience should ease the transition to advanced biofuels over the coming years.

'Good' biofuels have low GHG emissions, are produced sustainably, can be cost-effective without government support mechanisms, provide useful and valuable co-products and, if carefully managed with due regard given to sustainable land use, can support the drive for sustainable development in many developing countries.

Crop and forest residues and organic wastes are considered to be produced more sustainably when properly managed and soil nutrient levels are carefully maintained. The production of high energy yielding, non-food energy crops grown specifically for feedstocks can have direct and indirect land use change impacts. Non-competing land use opportunities may exist such as crop production on marginal or degraded lands, although productivity is likely to be low. Improving the overall biomass feedstock supply chain is needed in order to provide consistent quality feedstock delivered all year round to the conversion plant gate.

Several large-scale demonstration projects are operating or under development in the USA, Europe and elsewhere. However, even with generous government subsidies, the commercial risks remain high, especially with the recent widely fluctuating oil prices and global financial investment uncertainties. Full commercialization of either the biochemical or thermo-chemical conversion routes for producing advanced biofuels appears to remain some years away. The 'best technology pathway' cannot be easily determined between the biochemical and thermo-chemical routes, which are at a similar stage of maturity, both having been the subject of R&D for four decades or longer. Technologies, enzymes and processes have improved during this period, but no major breakthroughs have been apparent to drive down the costs of production.

Even if oil prices remain around the US$80–120 per barrel range, advanced biofuels will probably not become fully commercially viable against gasoline and diesel, nor enter the market for several years to come, without significant additional government support. Higher oil prices may improve their competitiveness although increased fertilizer, construction and transport costs would also result.

Due to improved understanding of GHG emissions as a result of detailed life cycle analyses and related direct and indirect land use change issues, the perceived environmental benefits of advanced biofuels continue to show promise, though further analysis will be needed once fully commercial plants become operational and accurate data can be obtained.

Policies designed to reward environmental performance and the sustainable production of biofuels, as well as to encourage the provision of a more abundant and geographically extensive feedstock supply, could see advanced biofuels begin to displace some of the first-generation alternatives in the medium to longer term. This transition will benefit from the biofuel deployment experiences learned to date, but it may take substantial time before large volumes of commercial-scale, drop-in biofuels become available, particularly for aviation and marine applications.

Acknowledgements

Prof. Jack Saddler and Dr Warren Mabee were co-authors with Michael Taylor and the author for the IEA report 'From 1st to 2nd generation biofuel technologies' from which much of the material in this chapter has been drawn. The major contributions they have made to the technical and scientific knowledge on advanced biofuels is well recognized. Prof. Saddler is also closely linked with Task 39 Biofuels of the IEA Bioenergy implementing agreement that provides a considerable source of literature on advanced biofuels on its website and maintains a database of the numerous demonstration plants operating around the world. My co-authors of the Sustainable Biofuels Consensus, also drawn on for this chapter, are also gratefully acknowledged, especially Gustavo Best, Jane Earley, André Faaij, Uwe Fritsche, Annette Hester, Suzanne Hunt, Tetsunari Iida, Francis Johnson, Geraldine Kutas, John Mathews, Plinio Nastari, Charlotte Opal, Martina Otto, Sergio Trindade, Jeffrey Tschirley, Simonetta Zarrilli and particularly Peter Read, recently deceased.

References

Bacovsky, D., Dallos, M. and Worgetter, M. (2010) Status of 2nd generation biofuel demonstration facilities in June 2010. Task 39 Biofuels, IEA Bioenergy. Available at: http://www.task39.org/Publications.aspx (accessed 5 April 2012).

Chum, H., Faaij, A., Moreira, J., Berndes, G., Dhamija, P. *et al.* (2011) Bioenergy. In: *IPCC Special Report on Renewable Energy Sources and Climate Change Mitigation*. Intergovernmental Panel on Climate Change, Cambridge University Press, Cambridge.

Darzins, A., Pienkos, P. and Edye, L. (2010) Current status and potential for algal biofuels production. Task 39, Biofuels, IEA Bioenergy. Available at: http://www.task39.org/Publications.aspx (accessed 5 April 2012).

Euractiv (2012) Airlines set to win carbon credits from biofuel flights, Euractiv, 13 February. Available at: http://www.euractiv.com/transport/airlines-set-win-carbon-credits-biofuel-flights-news-510800 (accessed 5 April 2012).

Fritsche, U., van Steen, H., Petersen, J.-E., Marelli, L., Andersson, K. *et al.* (2012) *Proceedings of workshop on 'Biofuels and indirect land use change'*, IP/A/ENVI/WS/2011-18, PE 475.090. February 2012. p. 75. Available at: http://www.europarl.europa.eu/studies (accessed 5 April 2012).

IBI (2012) International biochar initiative. Available at: http://www.biochar-international.org (accessed 5 April 2012).

IEA (2007) Bioenergy project development and biomass supply – good practice guidelines. International Energy Agency, OECD/IEA Paris. Available at: http://www.iea.org/publications/free_new_Desc.asp?PUBS_ID=1933 (accessed 5 April 2012).

IEA (2010) Energy technology perspectives 2010. International Energy Agency, OECD/IEA Paris. Available at: http://www.iea.org/publications/free_new_Desc.asp?PUBS_ID=2100 (accessed 5 April 2012).

IEA (2011a) World energy outlook 2011, International Energy Agency, OECD/IEA Paris. Available at: http://www.iea.org (accessed 5 April 2012).

IEA (2011b) Technology roadmap – biofuels for transport. International Energy Agency, OECD/IEA Paris. Available at: http://www.iea.org/publications/free_new_Desc.asp?PUBS_ID=2389 (accessed 5 April 2012).

IEA (2012) Technology roadmap – bioenergy for heat and power. International Energy Agency, OECD/IEA Paris. Available at: http://www.iea.org (accessed 5 April 2012).

Mabee, W. and Saddler, J. (2007) Deployment of 2nd-generation biofuels. *Technology Learning and Deployment Workshop, IEA*, Paris 11–12 June. Forest Products Biotechnology, University of British Columbia, Canada.

OECD (2008) Economic assessment of biofuel support policies. Organisation for Economic Co-operation and Development, OECD, Paris. Available at: http://www.oecd.org/tad/bioenergy (accessed 5 April 2012).

REN21 (2012) Renewables 2012, global status report, Renewable Energy Policy Network for the 21st Century, REN21 Secretariat, Paris, 116 pp. Available at: http://www.ren21.net (accessed 5 April 2012).

Sims, R.E.H. (2008) Reaching consensus on sustainable biofuels. Renewable Energy World, 9 July. Available at: http://www.renewableenergyworld.com/rea/news/article/2008/07/reaching-consensus-on-sustainable-biofuels-52692 (accessed 5 April 2012).

Sims, R.E.H., Taylor, M., Saddler, J. and Mabee, W. (2008) From 1st- to 2nd- generation biofuel technologies. International Energy Agency, IEA/OECD Paris. pp.184. Available at: http://www.iea.org/publications/free_new_Desc.asp?PUBS_ID=2079 (accessed 5 April 2012).

3 Microalgae Taxonomy and Breeding

S. Hemaiswarya,[1,2] R. Raja,[2]* R. Ravikumar[3] and Isabel S. Carvalho[2]
[1]*Department of Biotechnology, Indian Institute of Technology Madras, India;* [2]*Faculty of Sciences and Technology, Food Sciences Laboratory, IBB/CGB, University of Algarve Campus de Gambelas, Portugal;* [3]*Aquatic Energy LLC, Louisiana, USA*

3.1 Introduction

Microalgae are a diverse group of organisms that form the basic component of many ecosystems. The systemic classification of algae is primarily based on their pigment composition and is divided into nine classes. The largest groups are *Chlorophyceae* (green algae), *Phaeophyceae* (brown algae), *Pyrrophyceae* (dinoflagellates), *Chrysophyceae* (golden brown algae), *Bacillariophyceae* (diatoms) and *Rhodophyceae* (red algae). It has been estimated that between 22,000 and 26,000 species exist (Norton *et al.*, 1996), of which only a few species have been identified to be useful for commercial application, such as *Spirulina, Chlorella, Haematococcus, Dunaliella, Botryococcus, Phaeodactylum* and *Porphyridium*. Several other species that are also cultivated commercially for the hatcheries in the aquaculture field include *Chaetoceros, Crypthecodinium, Isochrysis, Nannochloris, Nitzschia, Schizochytrium, Tetraselmis* and *Skeletonema* (Raja *et al.*, 2008). Commercial use of microalgae as sources of specific chemicals began with *D. salina* for the production of β-carotene in the 1970s (Borowitzka and Borowitzka, 1988; Raja *et al.*, 2007a,c) followed by the use of *Haematococcus pluvialis* Flotow as a source of astaxanthin (Lorenz and Cysewski, 2000) and *Crypthecodinium cohnii* Seligo for long-chain polyunsaturated fatty acid (PUFA) and docosahexaenoic acid (DHA) (Kyle *et al.*, 1998). Microalgae have the capacity to assimilate nutrients from different sources, making them a better option for wastewater treatment (Raja *et al.*, 2004) and they have also shown to be effective both under *in vitro* and *in vivo* conditions as anticancer agents (Raja, 2003; Raja *et al.*, 2007b). Recent developments in bioprocess engineering and increased understanding of algal physiology have paved the way for current initiatives to mass culture microalgae for bioenergy applications.

The lipid content of some microalga makes them a potential source of biofuel. Lipids are traditionally subdivided in two main classes, polar (phospholipids and glycolipids) and neutral lipids (mono-, di-, triglycerides, waxes and isoprenoid-like lipids). Microalgal fatty acids and lipids perform cell membrane-related functions and are used as energy storage boxes and in metabolic process. Microalgae also synthesize a group of essential fatty acids that are not synthesized by animals (Pohl and Zurheide, 1979). The lipids in these algae are diverse: some algae possess the ability to synthesize medium-chain fatty acids (e.g. C10, C12 and C14), whereas others produce very long-chain fatty acids (>C20). Generally, algae synthesize fatty acids having

* Corresponding author: rraja307@yahoo.co.in

©CAB International 2013. *Biofuel Crops: Production, Physiology and Genetics* (ed. B.P. Singh)

chain lengths that range from C16 to C18 and contain palmitic acid (C16:1), stearic acid (C18:0), oleic acid (C18:1) and linoleic acid (C18:2) (Hu *et al.*, 2008; Francisco *et al.*, 2009). Thus, the C16 and C18 series content are often used to evaluate the oil productivity from algae (Xu *et al.*, 2001; Khozin-Goldberg and Cohen, 2006; Converti *et al.*, 2009). This composition is similar to diesel fuel, which mainly consists of C16 and C18 compounds (Hu *et al.*, 2008). Important fuel properties of biodiesel that are influenced by the fatty acid profile are cetane number, heat of combustion, cold flow, oxidative stability, viscosity, lubricity and ultimately exhaust emissions (Knothe, 2005). Generally, cetane number, heat of combustion and viscosity increase with increasing chain length and decrease with increasing degree of unsaturation, which means that long chain length (C16–18) and low degree of unsaturation are preferable for biodiesel (Francisco *et al.*, 2009). Also, long chain fatty acids (C16 and C18) give these oils unique hydrating and stabilizing properties, which improve the quality of biodiesel product (Knothe, 2005).

Microalgae are classified into two prokaryotic divisions (*Cyanophyta* and *Prochlorophyta*) and nine eukaryotic divisions (*Glaucophyta, Rhodophyta, Heterokontophyta, Haptophyta, Cryptophyta, Dinophyta, Euglenophyta, Chlorarachniophyta* and *Chlorophyta*). However, according to Khan *et al.* (2009), the most important groups of algae in terms of abundance are: diatoms, green algae, blue-green algae and golden algae. There is potential for further exploitation of these organisms for production of value added products and biofuels (Mutanda *et al.*, 2011).

3.2 Microalgae Taxonomy Emphasizing Lipid Content

Algae produce varied classes of fatty acids such as C14–C22, saturated and cis-unsaturated fatty acids (Pohl and Zurheide, 1979; Alvarez and Zarco, 1989). Eukaryotic algae contain generally more unsaturated fatty acids. The major saturated fatty acid in algae is invariably palmitate, while oleate is much less abundant than in higher plants. Palmitoleate is the dominant monoene. Some strains of algae can produce lipids up to 20–50% of dry cell weight under specific growth conditions.

A number of studies have shown that taxonomic positions of microalgae alter based on the fatty acid compositions (Morris and Culkin, 1976; Pohl and Zurheide, 1979). The composition of specific fatty acids or their ratios have been used as indicators for the identification of bacteria (chemotaxonomic investigations) (Shaw, 1974). Species specificity differences among fatty acid classes, their phylogenic significance have been shown by various reports (Holton *et al.*, 1963; Ackman *et al.*, 1968; Kenyon and Stainer, 1970; Kenyon, 1972; Ben-Amotz *et al.*, 1985, 1987). Numerous algal species and strains belonging to *Chlorophyceae* have been reported to contain, based on conventional solvent extraction and gravimetric analyses, high levels (25–50% of dry weight) of neutral lipids, suggesting that this class of microalgae may represent a large pool of organisms that could be useful for lipid/oil feedstock production (Hu *et al.*, 2006, 2008). In a study by Mourente *et al.* (1990), it was shown that *Nannochloropsis* species presented higher levels of 20:4n-6 and 20:5n-3 than *Chlorophyceae. Nannochloris* and *Chlorella* species (*Chlorophyceae*) showed high proportions of short-chained polyunsaturated fatty acids, with 16:2n-6, 16:3n-3, 18:2n-6 and 18:3n-3 as major fatty acids, but low levels of highly unsaturated fatty acids such as 20:5n-3 or 22:6n-3 (Table 3.1). It was hypothesized that there is no common pathway of fatty acid biosynthesis for all microalgae (Alvarez and Zarco, 1989).

The fatty acids of 34 strains of unicellular blue-green algae provisionally assigned to the genera *Synechococcus, Aphanocapsa, Gloeocapsa, Microcystis* and *Chlorogloea* by Stanier *et al.* (1971) have been chemically characterized. The strains analysed can be divided into a series of compositional groups based upon the highest degree of unsaturation of the major cellular fatty acids. Twenty strains fall into the group characterized by one trienoic fatty acid isomer (α-linolenic acid), and seven strains fall into a group characterized by another trienoic acid isomer (γ-linolenic acid) (Kenyon, 1972). These groups in many cases correlate well with groupings based upon other phenotypic characters of the strains, e.g. deoxyribonucleic acid base composition. The assignment of a strain to a compositional group is not

Table 3.1. Microalgal lipid production based on dry weight (Mourente *et al.*, 1990).

Microalgae	Source	% dry matter
Eustigmatophyceae		
Nannochloropsis oculata	Marine	11.02
Nannochloropsis salina	Marine	10.30
Nannochloropsis gaditana	Marine	16.11
Nannochloropsis sp.	Marine	15.30
Chlorophyceae		
Nannochloris atomus	Marine	8.55
Nannochloris maculata	Marine	6.37
Nannochloris sp.	Marine	8.77
Chlorella autotrophica	Marine	5.50
Chlorella stigmatophora	Marine	7.85
Dunaliella salina	Marine	6.55
Botryococcus braunii	Freshwater	30–40
Prasinophyceae		
Tetraselmis suecica	Marine	8.25
Haptophyceae		
Isochrysis galbana	Marine	9.25

altered when the strain is grown under a variety of different culture conditions (Kenyon, 1972).

Triacylglycerol profiling of microalgal strains (*Botryococcus braunii, Nannochloropsis gaditana, Neochloris oleoabundans, Phaeodactylum tricornutum, Porphyridium aerugineum* and *Scenedesmus obliquus*) were directly analysed by ultra-high pressure liquid chromatography–mass spectrometry (MacDougall *et al.*, 2011). It was evident from the chromatograms that only *B. braunii* and *S. obliquus* produced significant amounts of triacylglycerol (TAGs) under the growth conditions employed. In contrast, *N. oleoabundans* yielded a high abundance of polar lipids, as well as phosphocholine peaks, but relatively small amounts of TAGs. The TAG profiles varied significantly between species, with *B. braunii* and *S. obliquus* exhibiting the broadest range of TAGs. In particular, over 70 different TAGs were identified in *B. braunii*, with the TAGs 18:1/18:1/18:1, 18:1/18:1/18:3, 28:1/18:1/18:1 and 28:2/18:1/18:1 being the most abundant. Other TAGs containing long-chain fatty acids, such as 30:1, 30:2 and 36:2, were also detected in *B. braunii*. Although relatively rare, these fatty acids are known to be present at significant abundance in *B. braunii* (Banerjee *et al.*, 2002). Further, both *B. braunii* and *S. obliquus* also have a high abundance of TAGs containing oleic acid (18:1), which is favourable for biodiesel (Durrett *et al.*, 2008).

Viso and Marty (1993) determined profiles of fatty acids (FAs) of nine different marine microalgal groups and were able to define species-specific lipid compositions. In this study, a total of 2076 microalgal strains from the algal culture collection centre of Göttingen University (SAG) were included. Overall, 76 different fatty acids and ten other lipophilic substances were identified and quantified. Lang *et al.* (2011) found that the fatty acid distribution patterns reflected phylogenetic relationships at the level of phyla and classes. In contrast, at lower taxonomic levels, e.g. between closely related species and even among multiple isolates of the same species, the variation was in the fatty acid content.

3.3 Collection and Selection of New Microalgal Strains for Higher Lipids and Other Traits

Microalgae can be purified from a wide range of environments using a variety of techniques, including physical extraction from crude water samples (e.g. micromanipulation), dilution to resolve individual cells, antibiotic selection and enrichment cultures using specific selection pressures (e.g. photoautotrophic conditions). Individual algal strains can be purified based on traditional colony selection or high-throughput fluorescence-activated cell sorting (FACS) approaches before the resulting axenic cultures are cryopreserved for storage to prevent genetic drift. Screening of algae is often based on optimal growth or the production of specific metabolites and the use of response surface modelling and principle component analysis to identify these conditions and the key variables controlling those (Larkum *et al.*, 2012). It is important to select the strain that grows well in the particular geographical location and environment, as it easier to manipulate acclimatized strains. Most companies have recognized the importance of utilizing the highest performing algal strains. Whereas some favour screening locally harvested natural strains others prefer to genetically modify strains (Lundquist *et al.*, 2010). At least one company, Kuehnle Agrosystems Inc. (Honolulu, Hawaii, USA), has specialized into being a 'seed company' focused on strain selection, optimization and certified pure inoculums supply for customers and specific sites. This company uses over 40 selection characteristics to screen natural strains, to include oil yield and composition, protein composition, productivity level at the desired temperature, pH, salinity etc. (Kuehnle, 2010). These strains have to be screened for biofuel production, accumulation of desired storage compounds (e.g. high lipid strains), high-value products (e.g. unsaturated fatty acids), salt and temperature tolerance, resistance to predation and ease of harvesting.

3.4 Maintenance of Freshwater and Marine Water Microalgae Strains without Contamination

Algae isolated from natural environments are unlikely to be optimally adapted to the pond environment. Therefore genetic selection may significantly improve productivities of the algal mass cultures. For example, when a single limiting factor exists such as the nutrient source, genetic selection will offer a greater advantage for growth. If two organisms differ through a hereditable genetic difference to utilize the same limiting substance, then the one that is better in capturing the limiting nutrient will prevail in that environment. Eventually competition for one limiting factor will result in a single type of dominant organism. There are many factors that determine the species dominance, including: (i) resource–growth rate relationships for the different algae; (ii) variable environmental conditions; (iii) inhibition of one organism by another through excreted substances (allelopathy); and (iv) loss of growth due to predation or sinking. Free fatty acids, cell wall degradation products, exometabolites produced by algae and cyanobacteria, such as cyanobacterin (Gleason *et al.*, 1983; Gleason and Paulson, 1984; Gleason and Baxa, 1986) and fischerellin (Gross *et al.*, 1991; Hagmann and Jüttner, 1996) could also have allelopathic potential. Simple mathematical competition models can be designed for two organisms that compete in the pond. The above factors can be included as terms in the model. The effect of single or specific variables or combination variables on the growth of organisms in a laboratory scale can be made. The key issue is difficulty in simulation of the outdoor environment in the laboratory conditions or extrapolation of data from lab to mass-culture condition.

Maintaining a specific, genetically selected mono-algal culture of a specific inoculated strain in outdoor ponds is necessary to exhibit the high lipid productivities. The problem of species dominance and competition in outdoor mass cultures can be controlled by variations of biota, light, temperature, pH, oxygen and nutrient supplies. For example, high ammonia or pH will inhibit most zooplankton infestations, or selecting a growth environment for specific microalgae species such as a very high alkalinity selects for *Spirulina* and high salinity for *Dunaliella*, the two major species. These techniques are expensive and result in severe reduction of algal productivities. Severe contamination in *Chlorella* production could be solved by high density inoculations and semi-batch operations, but this resulted in reduced overall productivity. Both *Dunaliella* and *Chlorella* are dominant in their optimal environments. Contamination can be better managed in closed photobioreactors, but upon continuous cultivation both open and closed systems become more susceptible. Careful strain isolation and characterization, cultivation parameters of individual microalgal candidates and anticipation of the expected contaminants in the region can reduce the contamination issues.

3.5 Breeding Strategies to Improve Strains

Microalgae have significant advantages over land plants for breeding, which include: (i) life cycles of hours or days instead of seasonal cycles; (ii) unicellular nature greatly assists in the miniaturization of breeding systems, which reduces cost; (iii) replicate both sexually and asexually, which accelerates the generation of genetic diversity compared to prokaryotic organisms; and (iv) selection, screening and UV and chemical mutagenesis can be more easily applied (Larkum *et al.*, 2012). Though many microalgae can accumulate more than half their dry weight biomass as lipids including triacylglycerol and high value nutraceutical compounds, unimproved algae are unlikely to possess all of the traits necessary for economic production of biofuels (Wijffels and Barbosa, 2010). Robust molecular biology tools are required for strain optimization (Radokovits *et al.*, 2010) and further application of the strain in industrial production. Adaptive evolution is essential for species survival and reproduction. But both genetically and non-genetically modified organisms (GMOs and non-GMOs) can be screened for efficient phenotypes. The lack of transformation techniques is currently a major limitation. An alternative strategy is the isolation and breeding of highly efficient non-GMO strains. Such an approach can involve high-throughput screening of libraries after chemical treatment or UV mutagenesis, which avoids the regulatory problems of using GMO strains in outdoor production systems (Larkum *et al.*, 2012).

The green alga *Chlamydomonas reinhardtii* is currently the eukaryotic model alga of choice, because of its genetic tractability (Sager, 1955), array of molecular tools (Harris, 2001; Ledford *et al.*, 2004) and comprehensively annotated nuclear genome sequence (Merchant *et al.*, 2007). Genome sequences of several algae of ecological, economic and/or phylogenetic significance have been determined recently (Tirichine and Bowler, 2011). Transformation methods have been developed for several algal species, most notably the model diatom *Phaeodactylum tricornutum* (Apt *et al.*, 1996; Zaslavskaia *et al.*, 2000); however, the molecular genetic tools available for these species are generally quite limited in comparison with *Chlamydomonas*. An efficient transformation method was developed for *Nannochloropsis* sp., a fast-growing, unicellular alga capable of accumulating large amounts of oil by generated knockouts of the genes encoding nitrate reductase and nitrite reductase, resulting in strains that were unable to grow on nitrate and nitrate/nitrite, respectively (Kilian *et al.*, 2011).

3.5.1 Development of selectable markers

Efficient isolation of genetic transformants is greatly facilitated by the use of selection markers, including antibiotic resistance and/or fluorescent/biochemical markers. Antibiotic resistance genes can be inserted as a selectable marker. Marker systems that take advantage of the ability to genetically complement auxotrophic and metabolism mutants have also been achieved (Debuchy *et al.*, 1989; Kindle *et al.*, 1989). The major concern is that the environmental implication of widespread antibiotic use could exacerbate current problems with increased antibiotic-resistant microbes. Several different antibiotic resistance genes have been used successfully for microalgal transformant selection, including bleomycin (Apt *et al.*, 1996), spectinomycin (Doetsch *et al.*, 2001), streptomycin (Doetsch *et al.*, 2001), G418 (Dunahay *et al.*, 1995), hygromycin (Berthold *et al.*, 2002) etc. Other markers that have been used include luciferase (Fuhrmann *et al.*, 2004), β-glucuronidase (El-Sheekh, 1999), β-galactosidase (Gan *et al.*, 2003) and green fluorescent protein (GFP) (Fuhrmann *et al.*, 1999).

Alternatively, markers with functional diversity in genes and pathways are important in determining lipid production and related traits. The genomes of *C. reinhardtii* (Merchant *et al.*, 2007), *Chlorella variabilis* (Blanc *et al.*, 2010), *Micromonas* (Worden *et al.*, 2009), *Ostreococcus tauri* and *Ostrecoccus lucimarinus* (Palenik *et al.*, 2007), the red algae *Cyanidioschyzon merola* (Nozaki *et al.*, 2007) and the diatoms *Thalassiosira pseudonana* (Armbrust *et al.*, 2004) and *Phaeodactylum tricornutum* (Bowler *et al.*, 2008) can serve as a base for novel phylogenetic markers or functional markers for strain typing of microalgae. These markers, once validated and tested over a broad range of isolates, might enable genotyping-based strategy for strain characterization. This would be better than the phenotype-based selection strategies and would enhance favoured strain selection (Wang *et al.*, 2012). Fatty acid and lipid composition have also been used as biomarkers to distinguish closely related microalgae at the species and the generic levels (Volkman *et al.*, 1998; Leblond *et al.*, 2005).

3.5.2 Genetic engineering methods to improve lipid production

Metabolic pathways are integrated through complex mechanisms that direct them to synthesize protein, carbohydrates, nucleic acids, lipids and hydrogen. In order to engineer these metabolic pathways for optimized lipid production in microalgae, it is necessary to completely understand the biosynthesis and degradation of precursors, intermediates, metabolic end-products and their regulatory networks. Microalgae are capable of synthesizing a range of biofuels. Lipids and carbohydrates represent the main energy storage molecules of algae, therefore bioengineering techniques should be adopted to switch the microalgal metabolism towards more lipid production. The strategies to engineer the metabolic pathway to optimize lipid production include over-expressing fatty acid biosynthetic enzymes, increasing the availability of precursor molecules such as acetyl-CoA, down-regulate fatty acid catabolism by inhibiting β-oxidation or lipase hydrolysis, alter saturation profiles by regulation of desaturases and optimizing fatty acid length with thioesterases. Genetic engineering of microalgae has also been attempted to improve lipid productivity. In 1995, one of the key enzymes for lipid biosynthesis, encoded by the acetyl-CoA carboxylase (ACCase) gene, was identified. Targeting the gene, transgenic *Cyclotella cryptica* and *Navicula saprophila* were developed, but neither showed increased lipid content thereby necessitating the need to explore in microalgal metabolism, the genetic engineering approaches and the feasibility of the modified strains in mass culture production (Roessler and Ohlrogge, 1993; Roessler *et al.*, 1994; Dunahay *et al.*, 1995; Li *et al.*, 2010).

More recently, in a *Chlamydomonas reinhardtii* mutant with inactivated ADP-glucose pyrophosphorylase, an enzyme involved in starch synthesis, the triacylglycerol content exhibited a tenfold increase, from 2 to 20.5%, under nitrogen starvation (Li *et al.*, 2010). But the fatty acid and TAG profiles of microalgae are variable and around 80 mutants with altered FAS activity were observed in the large scale mutant screening of the *C. reinhardtii* insertional library (Beer *et al.*, 2009). Inactivation of ADP-glucose pyrophosphoryl in a *Chlamydomonas* starchless mutant led to a tenfold increase in TAG, suggesting that shunting of photosynthetic carbon partitioning from starch to TAG synthesis may represent a more effective strategy than direct manipulation of the lipid synthesis pathway to overproduce TAG. However, this also slowed down growth, resulting in lower overall lipid yield than that of the wild type. So far, there have been few success stories in improving oil productivity by engineering individual genes or pathways in oleaginous microalgae (Li *et al.*, 2010).

Many microalgae initiate TAG storage during the day and deplete those stores at night to support cellular ATP demands and/or cell division. Consequently, inhibition of β-oxidation would prevent the loss of TAG during the night, but most likely at the cost of reduced growth. This strategy, therefore, may not be beneficial for microalgae grown in outdoor open ponds, but it may be a valid strategy to increase lipid production in microalgae grown in photobioreactors with exogenous carbon sources and/or continuous light (Radakovits *et al.*, 2010).

The use of regulatory factors such as transcription factors (TFs) to control the abundance or activity of multiple enzymes relevant to the production of desired products has provoked widespread interest. Over-expression of lipid synthesis genes may still be beneficial if they can be controlled by an inducible promoter that can be activated once the microalgal cells have grown to a high density and have entered stationary phase. Examples of inducible promoters in algae include copper-responsive elements in *C. reinhardtii* (Quinn and Merchant, 1995) and a nitrate-responsive promoter in diatoms (Poulsen and Kroger, 2005).

Applying physiological stresses such as nutrient-depletion to channel metabolic flux to lipid biosynthesis, is the most mature and most widely employed technique. Although there is a lack of success in lipid overproduction using transformed microalgal strains, the knowledge obtained in studies on lipid pathways and genetically transformed organisms for enhanced lipid synthesis will help in future for commercial application of these strains or developing a new one devoid of drawbacks.

3.6 Conclusion

Microalgae are extraordinarily diverse with complex evolutionary relationships. Natural variability in chemical composition has been exploited commercially to produce stable, high-quality products. Microalgae rich in lipids are used for production of lipid-derived biofuel and biodiesel in particular. Generally lipid levels and composition, including fatty acid profiles, vary according to taxonomic entity, season, location and growth conditions in microalgae. Breeding of microalgal strains rich in lipid through mutagenesis, transformation or using robust molecular biology tools are required for strain optimization for effective biofuel production. Alternatively, genetically modified microalgae can offer varied advantages, which is achieved by engineering the metabolic pathway to flux the metabolites towards the desired lipid production or altering the regulatory factors such as transcription factors for over-expression of lipid synthesis genes. Collection, cultivation, selection and breeding of naturally occurring strains with high lipid productivity or genetically modified microalgae with desired phenotype is required for both increase in biomass as well as lipid production.

Acknowledgements

The authors are grateful to Dr Velusubramani, Senior Scientist, Refining Technology, Beyond Petroleum Pvt Ltd, Neperville, USA and Prof. R. Manivasakan, Indian Institute of Technology Madras, Chennai, India for their critical review of the manuscript.

References

Ackman, R.G., Tocher, C.S. and McLachlan, J. (1968) Marine phytoplankter fatty acids. *Journal of Fisheries Research Board of Canada* 25, 1603–1620.

Alvarez, M. and Zarco, J. (1989) Lipids in microalgae: a review. *Biochemistry. Grasasy Aceites* 40, 118–145.

Apt, K.E., Grossman, A.R. and Kroth-Pancic, P.G. (1996) Stable nuclear transformation of the diatom *Phaeodactylum tricornutum. Molecular and General Genetics* 252, 572–579.

Armbrust, E., Berges, J.A., Bowler, C., Green, B.R., Martinez, D. *et al.* (2004) The genome of the diatom, *Thalassiosira pseudonana*: ecology, evolution, and metabolism. *Science* 306, 79–86.

Banerjee, A., Sharma, R., Chisti, Y. and Banerjee, U.C. (2002) *Botryococcus braunii*: a renewable source of hydrocarbons and other chemicals. *Critical Reviews in Biotechnology* 22, 245–279.

Beer, L.L., Boyd, E.S., Peters, J.W. and Posewitz, M.C. (2009) Engineering algae for biohydrogen and biofuel production. *Current Opinion in Biotechnology* 20, 264–271.

Ben-Amotz, A., Tornabene, T.G. and Thomas, W.H. (1985) Chemical profiles of selected species of microalgae with emphasis on lipids. *Journal of Phycology* 21, 72–81.

Ben-Amotz, A., Fishler, R. and Schneller, A. (1987) Chemical composition of dietary species of marine unicellular algae and rotifers with emphasis on fatty acids. *Marine Biology* 95, 31–36.

Berthold, P., Schmitt, R. and Mages, W. (2002) An engineered *Streptomyces hygroscopicus* aph 7 gene mediates dominant resistance against hygromycin B in *Chlamydomonas reinhardtii. Protist* 153, 401–412.

Blanc, G., Duncan, G., Agarkova, I., Borodovsky, M., Gurnon, J. *et al.* (2010) The *Chlorella variabilis* NC64A genome reveals adaptation to photosymbiosis, coevolution with viruses, and cryptic sex. *Plant Cell Online* 22, 2943.

Borowitzka, M.A. and Borowitzka, L.J. (1988) *Dunaliella*. In: Borowitzka, M.A. and Borowitzka, L.J. (eds) *Microalgal Biotechnology*. Cambridge University Press, New York, pp. 27–58.

Bowler, C., Allen, A.E., Badger, J.H., Grimwood, J., Jabbari, K. *et al.* (2008) The *Phaeodactylum* genome reveals the evolutionary history of diatom genomes. *Nature* 456, 23–244.

Converti, A., Casazza, A., Ortiz, Y., Perego, P. and Del Borghi, M. (2009) Effect of temperature and nitrogen concentration on the growth and lipid content of *Nannochloropsis oculata* and *Chlorella vulgaris* for biodiesel production. *Chemical Engineering and Proceeding: Process Intensification* 48, 1146–1151.

Debuchy, R., Purton, S. and Rochaix, J.-D. (1989) The argininosuccinate lyase gene of *Chlamydomonas reinhardtii*: an important tool for nuclear transformation and for correlating the genetic and molecular maps of the ARG7 locus. *EMBO Journal* 8, 2803–2809.

Doetsch, N.A., Favreau, M.R., Kuscuoglu, N., Thompson, M.D. and Hallick, R.B. (2001) Chloroplast transformation in *Euglena gracilis*: splicing of a group III twintron transcribed from a transgenic psbK operon. *Current Genetics* 39, 49–60.

Dunahay, T.G., Jarvis, E.E. and Roessler, P.G. (1995) Genetic transformation of the diatoms *Cyclotella cryptica* and *Navicula saprophila*. *Journal of Phycology* 31, 1004–1011.
Durrett, T.P., Benning, C. and Ohlrogge, J. (2008) Plant triacylglycerols as feedstocks for the production of biofuels. *The Plant Journal* 54, 593–607.
El-Sheekh, M.M. (1999) Stable transformation of the intact cells of *Chlorella kessleri* with high velocity microprojectiles. *Plant Biology* 42, 209–216.
Francisco, E., Neves, D., Lopes, E. and Franco, T. (2009) Microalgae as feedstock for biodiesel production: carbon dioxide sequestration, lipid production and biofuel quality. *Journal of Chemical Technology and Biotechnology* 85, 395–403.
Fuhrmann, M., Oertel, W. and Hegemann, P. (1999) A synthetic gene coding for the green fluorescent protein (GFP) is a versatile reporter in *Chlamydomonas reinhardtii*. *The Plant Journal* 19, 353–362.
Fuhrmann, M., Hausherr, A., Ferbitz, L., Schödl, T., Heitzer, M. *et al.* (2004) Monitoring dynamic expression of nuclear genes in *Chlamydomonas reinhardtii* by using a synthetic luciferase reporter gene. *Plant Molecular Biology* 55, 869–881.
Gan, S.Y., Qin, S., Othman, R.Y., Yu, D. and Phang, S.M. (2003) Transient expression of lacZ in particle bombarded *Gracilaria changii* (Gracilariales, Rhodophyta). *Journal of Applied Phycology* 15, 345–349.
Gleason, F.K. and Baxa, C.A. (1986) Activity of the natural algicide, cyanobacterin, on eukaryotic microorganisms. *FEMS Microbiology Letters* 33, 85–88.
Gleason, F.K. and Paulson, J.L. (1984) Site of action of the natural algicide, cyanobacterin, in the blue-green alga, *Synechococcus* sp. *Archives in Microbiology* 138, 273–277.
Gleason, F.K., Michurski, S. and Wood, J.M. (1983) Biosynthesis of the antibiotic cyanobacterin. *Federation Proceedings* 42, 2209.
Gross, E., Wolk, P.C. and Jüttner, F. (1991) Fischerellin, a new allelochemical from the freshwater cyanobacterium *Fischerella muscicola*. *Journal of Phycology* 27, 686–692.
Hagmann, L. and Jüttner, F. (1996) Fischerellin A, a novel Photosystem-II-inhibiting allelochemical of the cyanobacterium *Fischerella muscicola* with antifungal and herbicidal activity. *Tetrahedron Letters* 37, 6539–6542.
Harris, E.H. (2001) Chlamydomonas as a model organism. *Annual Reviews in Plant Physiology and Plant Molecular Biology* 52, 363–406.
Holton, R.W., Blecker, H.H. and Stevens, T.S. (1963) Fatty acids in blue-green algae: possible relation to phylogenetic position. *Science* 160, 545–547.
Hu, Q., Zhang, C. and Sommerfeld, M. (2006) Biodiesel from algae: lessons learned over the past 60 years and future perspectives. *Annual Meeting of the Phycological Society of America*, 7–12 July, Juneau, Alaska, pp. 40–41.
Hu, Q., Sommerfeld, M., Jarvis, E., Ghirardi, M., Posewitz, M. *et al.* (2008) Microalgal triacylglycerols as feedstocks for biofuel production: perspectives and advances. *The Plant Journal* 54, 621–639.
Kenyon, C.N. (1972) Fatty acid composition of unicellular strains of blue-green algae. *Journal of Bacteriology* 109, 827–834.
Kenyon, C.N. and Stanier, R.Y. (1970) Possible evolutionary significance of polyunsaturated fatty acids in blue-green algae. *Nature* 227, 1164–1166.
Khan, S.A., Rashmi, Hussain, M.Z., Prasad, S. and Banerjee, U.C. (2009) Prospects of biodiesel production from microalgae in India. *Renewable and Sustainable Energy Reviews* 13, 2361–2372.
Khozin-Goldberg, I. and Cohen, Z. (2006) The effect of phosphate starvation on the lipid and fatty acid composition of the fresh water eustigmatophyte *Monodus subterraneus*. *Phytochemistry* 67, 696–701.
Kilian, O., Benemann, C.S.E., Niyogi, K.K. and Vick, B. (2011) High-efficiency homologous recombination in the oil-producing alga *Nannochloropsis* sp. *Proceedings of National Academy of Sciences* 108, 21265–21269.
Kindle, K.L., Schnell, R.A., Fernandez, E. and Lefebvre, P. (1989) Stable nuclear transformation of *Chlamydomonas* using the *Chlamydomonas* gene for nitrate reductase. *Journal of Cell Biology* 109, 2589–2601.
Knothe, G. (2005) Dependence of biodiesel fuel properties on the structure of fatty acid alkyl esters. *Fuel Processing Technology* 86, 1059–1070.
Kuehnle, A. (2010) President of Kuehnle Agrosystems, presentation at Asia-Pacific Clean Energy Summit and Expo, September 2010.
Kyle, D.J., Reeb, S.E. and Sicoiie, V.J. (1998) Dinoflagellate biomass, methods for its production and compositions containing the same. *USA Patent Number 5711983*.
Lang, I., Hodac, L., Friedl, T. and Feussner, I. (2011) Fatty acid profiles and their distribution patterns in microalgae: a comprehensive analysis of more than 2000 strains from the SAG culture collection. *BMC Plant Biology* 11, 124.

Larkum, A.W.D., Ross, I.L., Kruse, O. and Hankamer, B. (2012) Selection, breeding and engineering of microalgae for bioenergy and biofuel production. *Trends in Biotechnology* 30, 1–8.
Leblond, J.D., Dahmen, J.L., Seipelt, R.L., Elrod-Erickson, M.J., Kincaid, R. *et al.* (2005) Lipid composition of chlorarachniophytes (Chlorarachniophyceae) from the genera *Bigelowiella, Gymnochlora,* and *Lotharella. Journal of Phycology* 41, 311–321.
Ledford, H.K., Dent, R.M. and Niyogi, K.K. (2004) Chlamydomonas genomics: new approaches for a classical model system. In: Leister, D. (ed.) *Plant Functional Genomics*. Food Products, Binghamton, New York, pp. 153–202.
Li, Y., Han, D., Hu, G., Dauvillee, D., Sommerfeld, M., Ball, S. and Hu, Q. (2010) *Chlamydomonas* starchless mutant defective in ADP-glucose pyrophosphorylase hyper-accumulates triacylglycerol. *Metabolic Engineering* 12, 387–391.
Lorenz, T.R. and Cysewski, G.R. (2000) Commercial potential for *Haematococcus* microalgae as a natural source of astaxanthin. *Trends in Biotechnology* 18, 160–167.
Lundquist, T.J., Woertz, I.C., Quinn, N.W.T. and Benemann, J.R. (2010) *A Realistic Technology and Engineering Assessment of Algae Biofuel Production*. Energy Biosciences Institute, University of California, Berkeley, California.
MacDougall, K.M., McNichol, J., McGinn, P.J., O'Leary, S.J. and Melanson, J.E. (2011) Triacylglycerol profiling of microalgae strains for biofuel feedstock by liquid chromatography-high-resolution mass spectrometry. *Analytical and Bioanalytical Chemistry* 40, 2609–2616.
Merchant, S.S., Prochnik, S.E., Vallon, O., Harris, E.H., Karpowicz, S.J. *et al.* (2007) The *Chlamydomonas* genome reveals the evolution of key animal and plant functions. *Science* 318, 245–250.
Morris, R.J. and Culkin, F. (1976) Marine lipids: analytical technique and fatty acid ester analyses. *Oceanography and Marine Biology Annual Review* 14, 391–433.
Mourente, G., Lubián, L.M. and Odriozola, J.M. (1990) Total fatty acid composition as a taxonomic index of some marine microalgae used as food in marine aquaculture. *Hydrobiologia* 203, 147–154.
Mutanda, T., Ramesh, D., Karthikeyan, S., Kumari, S., Anadraj, A. and Bux, F. (2011) Bioprospecting for hyper-lipid producing microalgal strains for sustainable biofuel production. *Bioresource Technology* 102, 57–70.
Norton, T.A., Melkonian, M. and Andersen, R.A. (1996) Algal biodiversity. *Phycologia* 35, 308–326.
Nozaki, H., Takano, H., Misumi, O., Terasawa, K., Matsuzaki, M. *et al.* (2007) A 100%-complete sequence reveals unusually simple genomic features in the hot-spring red alga *Cyanidioschyzon merolae. BMC Biology* 5, 28.
Palenik, B., Grimwood, J., Aerts, A., Rouzé, P., Salamov, A. *et al.* (2007) The tiny eukaryote *Ostreococcus* provides genomic insights into the paradox of plankton speciation. *Proceedings of National Academy of Sciences* 104, 7705–7710.
Pohl, P. and Zurheide, F. (1979) Fatty acids and lipids of marine algae and the control of their biosynthesis by environmental factors. In: Hoppe, H.A., Levring, T. and Tanaka, Y. (eds) *Marine Algae in Pharmaceutical Science*. Walter de Gruyter, Berlin-New York, pp. 473–523.
Poulsen, N. and Kroger, N. (2005) A new molecular tool for transgenic diatoms. *FEBS Journal* 272, 3413–3423.
Quinn, J.M. and Merchant, S. (1995) Two copper-responsive elements associated with the *Chlamydomonas* Cyc6 gene function as targets for transcriptional activators. *Plant Cell* 7, 623–638.
Radakovits, R., Jinkerson, R.E., Darzins, A. and Posewitz, M.C. (2010) Genetic engineering of algae for enhanced biofuel production. *Eukaryotic cell* 9, 486–501.
Raja, R. (2003) Studies on *Dunaliella salina* (Dunal) Teod. with special reference to its anticancer properties. PhD thesis, University of Madras, Chennai, India.
Raja, R., Anbazhagan, C., Ganesan, V. and Rengasamy, R. (2004) Efficacy of *Dunaliella salina* (Volvocales, Chlorophyta) in salt refinery effluent treatment. *Asian Journal of Chemistry* 16, 1081–1088.
Raja, R., Hemaiswarya, S., Balasubramanyam, D. and Rengasamy, R. (2007a) PCR identification of *Dunaliella salina* (Volvocales, Chlorophyceae) and its growth characteristics. *Microbiological Research* 162, 168–176.
Raja, R., Hemaiswarya, S., Balasubramanyam, D. and Rengasamy, R. (2007b) Protective effect of *Dunaliella salina* (Volvocales, Chlorophyta) against experimentally induced fibrosarcoma on wistar rats. *Microbiological Research* 162, 177–184.
Raja, R., Hemaiswarya, S. and Rengasamy, R. (2007c) Exploitation of *Dunaliella* for β-carotene production. *Applied Microbiology and Biotechnology* 74, 517–523.
Raja, R., Hemaiswarya, S., Ashokkumar, N., Sridhar, S. and Rengasamy, R. (2008) A perspective on the biotechnological potential of microalgae. *Critical Reviews in Microbiology* 34, 77–88.

Roessler, P. and Ohlrogge, J. (1993) Cloning and characterization of the gene that encodes acetylcoenzyme A carboxylase in the alga, *Cyclotella cryptica. Journal of Biological Chemistry* 268, 19254–19259.
Roessler, P.G., Bleibaum, J.L., Thompson, G.A. and Ohlrogge, J.B. (1994) Characteristics of the gene that encodes acetyl CoA carboxylase in the diatom *Cyclotella cryptica. Annals of the New York Academy of Sciences* 721, 250–256.
Sager, R. (1955) Inheritance in the green alga, *Chlamydomonas reinhardi. Genetics* 40, 476–489.
Shaw, N. (1974) Lipid composition as the guide to the classification of bacteria. In: Perlam, D. (ed.) *Advances in Applied Microbiology*. Academic Press, New York, pp. 63–105.
Stanier, R.Y., Kunisawa, R., Mandel, M. and Cohen-Bazire, G. (1971) The purification and properties of unicellular blue-green algae (order Chroococcales). *Bacteriological Reviews* 35, 171–205.
Tirichine, L. and Bowler, C. (2011) Decoding algal genomes: tracing back the history of photosynthetic life on Earth. *The Plant Journal* 66, 45–57.
Viso, A.C. and Marty, J.C. (1993) Fatty acids from 28 marine microalgae. *Phytochemistry* 34, 1521–1533.
Volkman, J.K., Barrett, S.M., Blackburn, S.I., Mansour, M.P., Sikes, E.L. *et al.* (1998) Microalgal biomarkers: a review of recent research developments. *Organic Geochemistry* 29, 1163–1179.
Wang, D., Lu, Y., Huang, H. and Xu, J. (2012) Establishing oleaginous microalgae research models for consolidated bioprocesing of solar energy. *Advances in Biochemical Engineering/Biotechnology* 1–16.
Wijffels, R.H. and Barbosa, M.J. (2010) An outlook on microalgal biofuels. *Science* 329, 796–799.
Worden, A.Z., Lee, J.H., Mock, T., Rouzé, P., Simmons, M.P. *et al.* (2009) Green evolution and dynamic adaptations revealed by genomes of the marine picoeukaryotes *Micromonas. Science* 324, 268–272.
Xu, N., Zhang, X., Fan, X., Han, L., Zeng, C. *et al.* (2001) Effects of nitrogen source and concentration on growth rate and fatty acid composition of *Ellipsoidion* sp. Eustigmatophyta. *Journal of Applied Phycology* 13, 463–469.
Zaslavskaia, L.A., Lippmeier, J.C., Kroth, P.G., Grossman, A.R. and Apt, K.E. (2000) Transformation of the diatom, *Phaeodactylum tricornutum* (Bacillariophyceae) with a variety of selectable markers and reporter genes. *Journal of Phycology* 36, 379–386.

4 Physiology, Biochemistry and Genetics of Microalgal Growth and Lipid Production

Holger Schuhmann and Peer M. Schenk
School of Agriculture and Food Sciences, The University of Queensland, Australia

4.1 Introduction

Algae are an attractive alternative to oil crops for the production of biodiesel. Their production may not require arable land, and – if not already grown in aquaculture systems – algae growth facilities can be operated with wastewater or, for marine strains, seawater and brackish water. Macroalgae have only very recently attracted attention as potential biodiesel feedstock, and available data are very sparse. Many species have a potential to accumulate high amounts of triacylglycerol (TAG) within the cell. Their biggest advantages, however, are their fast growth rate and higher light conversion rate as compared to land plants, therefore much less area is needed to produce a comparable amount of TAG (Chisti, 2007; Schenk *et al.*, 2008). The term 'microalga' itself is a very broad and diffuse description for a wide range of organisms and more a practical rather than a phylogenetic term. In its widest definition it means any unicellular organism capable of photosynthesis. A somewhat narrower definition excludes prokaryotic cyanobacteria, i.e. microalgae are unicellular, eukaryotic, photosynthetic cells. This still includes green, red, brown (diatoms) and golden algae, as well as haptophytes and eustigmatophytes (Fig. 4.1). A specific microalgae species (e.g. a green alga) might be much more closely related to a multicellular organism (e.g. a tree) than to another microalga (e.g. a diatom). This of course has consequences of how far we can transfer our much better knowledge of lipid biochemistry, physiology and genetics in multicellular organisms to microalgae. On the other hand, it provides the researcher with a plethora of different pathways and lipids from which to choose.

4.2 Classification of Lipids

Like every other organism, microalgae produce a huge variety of different lipids for a wide range of purposes, including membrane formation and as storage molecules. In general, a lipid is defined as a small, hydrophobic or amphipathic molecule, usually derived from the condensation of thioesters (fatty acids, glycerolipids, glycerophospholipids, sphingolipids, saccharolipids, polyketides) or isoprene units (sterols and prenol lipids) (Fahy *et al.*, 2009).

Chemically, biodiesel is a mixture of methyl esters of fatty acids (FA). FAs are carboxylic acids with a long aliphatic chain, ranging from 4 to 28 carbon atoms. They are one of the many types of lipids present in every organism (Table 4.1) and a building block of many other lipids. In saturated FAs, the aliphatic chain only contains carbon–carbon single

©CAB International 2013. *Biofuel Crops: Production, Physiology and Genetics* (ed. B.P. Singh)

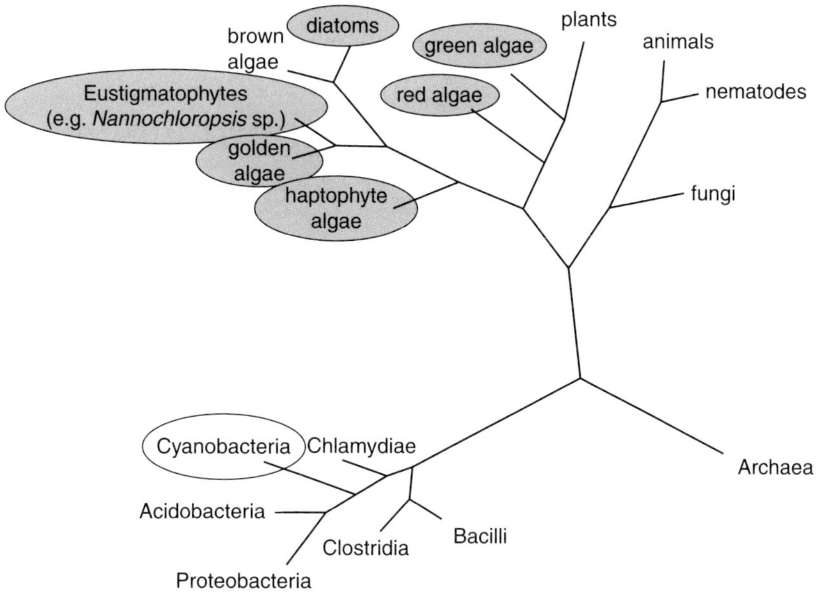

Fig. 4.1. Schematic representation of the tree of life. Green algae, red algae, diatoms, golden algae, eustigmatophytes and haptophyte algae are microalgae in a stricter sense (shaded in grey). A broader definition of microalgae includes prokaryotic cyanobacteria (encircled). (Adapted from Schuhmann *et al.*, 2012.)

bonds, whereas unsaturated FAs contain one or several (in the case of polyunsaturated FA) carbon–carbon double bonds. Several nomenclature systems can be found in the literature to describe and categorize FAs (Table 4.2), with each system having advantages and drawbacks with regards to unambiguousness and conciseness. A double bond in the aliphatic chain inhibits rotation around the carbon–carbon bond, resulting in a more rigid molecule. This double bond can be either in *cis*- (Z) or in *trans*- (E) configuration (Table 4.1). The *cis*-conformation (which is predominant in most naturally occurring unsaturated FAs) induces a bend in the chain, therefore molecules with *cis*-configured unsaturated FAs as building blocks cannot be packed as densely as molecules based on saturated FAs in the cell, resulting in lower melting temperatures and higher fluidity. On the other hand, a double bond in the *trans*-configuration does not have such an impact on the properties of the FA and is less common in naturally occurring FAs.

As mentioned above, FAs are building blocks of many lipid molecules in the cell. From a biofuel perspective, TAGs are the most important lipids, since the general protocol for biodiesel production is based on the transesterification of TAG with methanol. This process yields FA-methyl esters, i.e. biodiesel and glycerol as a by-product. In TAGs, all hydroxyl groups of glycerol are esterified with FAs, resulting in an overall apolar molecule, which can be densely packed without interfering with the osmotic status of the cell. As a result, TAGs are the main storage molecules in lipid-storing cells. Other glycerol-based lipids, such as glycosylglycerols and glycerophospholipids, contain polar head groups and are main constituents of membranes of the cell (Table 4.1) (Ohlrogge and Browse, 1995). FAs are also building blocks of sphingolipids and saccharolipids. In sphingolipids, the backbone of the molecule is not glycerol, but a long-chain unsaturated aliphatic amino-alcohol, which forms an amide bond with an FA (Table 4.1). The backbone of saccharolipids, on the other hand, is a sugar molecule, which is esterified with FAs (Table 4.1). Glucosamin, a main constituent of the bacterial cell wall in gram-negative bacteria, is a well examined example of saccharolipid (Osborn *et al.*, 1964).

Table 4.1. Types of lipids.

Lipid	Example
Fatty acid	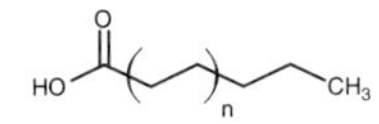 Saturated fatty acid
	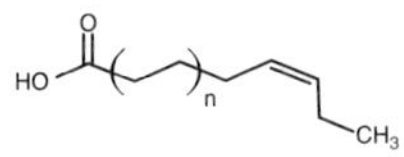 Unsaturated fatty acid, Z configuration
	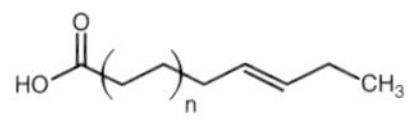 Unsaturated fatty acid, E configuration
Triacylglycerol	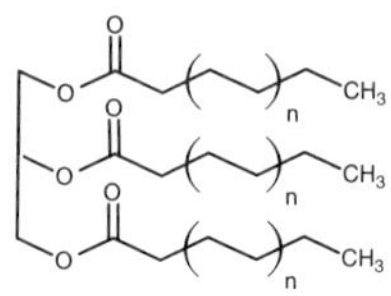
Glycerophospholipids	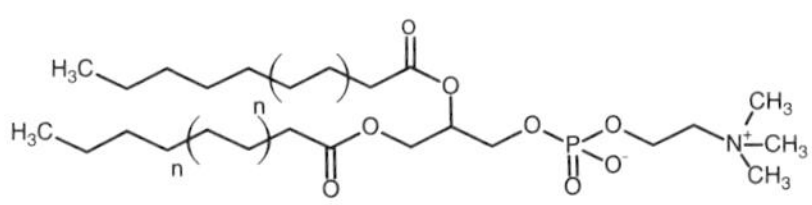 Phosphatidylcholine
Glycosylglycerols	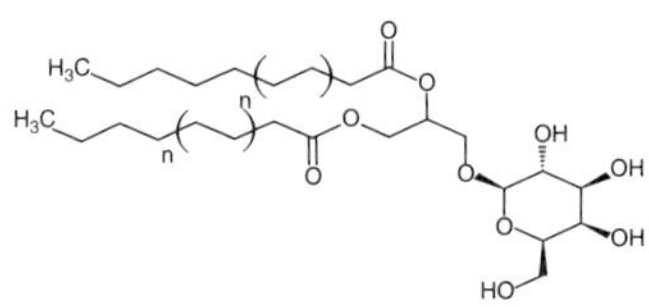 Monogalactosyldiacylglycerol
Sphingolipids	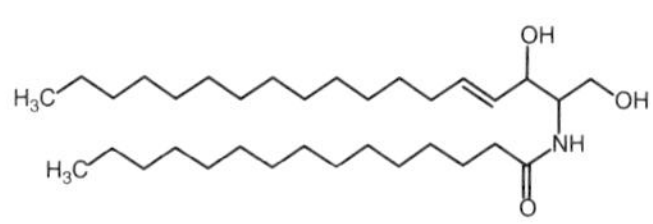 Ceramide
Saccharolipids	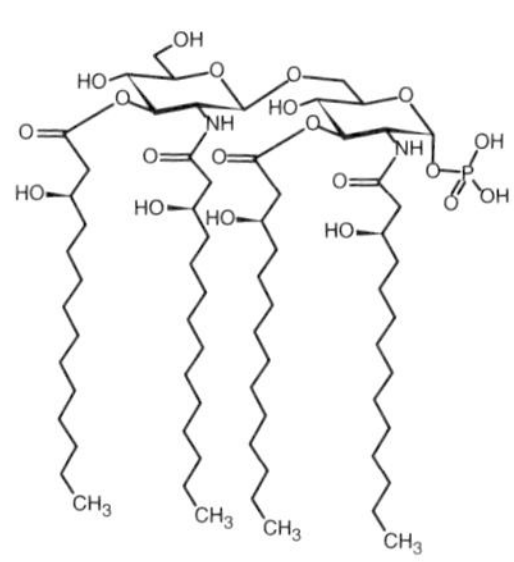 Lipid A-disaccharide-1-phosphate
Sterols	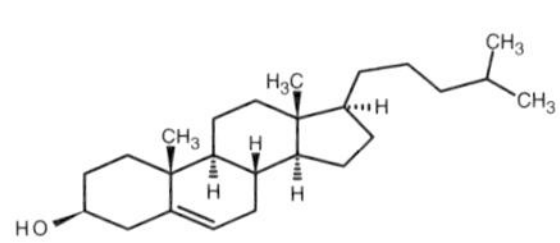 Cholesterol
Prenol lipids	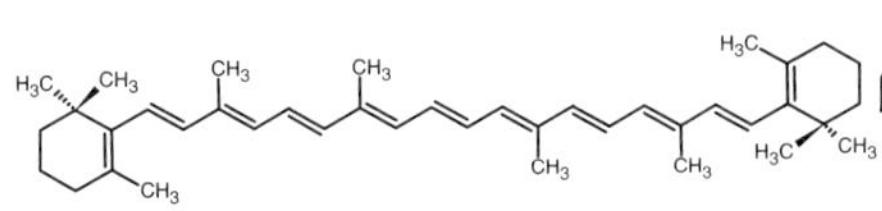 β-carotene

Table 4.2. Nomenclature for the description of fatty acids.

Trivial name	For example palmitic acid, linoleic acid, or oleic acid. Usually, this name is based on distinctive characteristics of the FA (e.g. smell) or on a typical source where this FA can be found in nature (e.g. palm oil or olive oil). The trivial names are historical not systematic and can be ambiguous.
Systematic IUPAC name	This system unambiguously describes one specific molecule, including number, position (counting from the carboxylic end) and configuration of the double bonds. For example, (Z,Z)-9,12-octadecadienoic acid is a carboxylic acid with 18 carbon atoms and two carbon–carbon double bonds in cis(Z)-configuration between the 9th and 10th, and the 12th and 13th carbon atom, respectively (linoleic acid). Although technically clear and unambiguous, it is rather lengthy and may be cumbersome.
Omega-*x*, ω-*x*, or n-*x*	This nomenclature is usually used in physiological and nutritional contexts and summarizes FA with the same biosynthetic pathway; *x* indicates the first double bond counting from the non-carboxylic end. For example, linoleic acid ((Z,Z)-9,12-octadecadienoic acid) and arachidonic acid ((Z,Z,Z,Z)-5,8,11,14-eicosatetraenoic acid) are both ω-6 FA.
C*x*:*y* nomenclature	Also described as 'lipid nomenclature'; *x* indicates the number of carbon atoms, *y* the number of double bonds. Since no information about position and configuration of the double bonds is included, this nomenclature system is ambiguous for unsaturated FA. However, it is very short.

Two other important classes of lipids, sterol (e.g. cholesterol) and prenol lipids (e.g. β-carotene, Table 4.1), do not contain FA. Many secondary metabolites from algae, such as β-carotene, lutein and astaxanthin, belong to this lipid class and are produced on a commercial scale (Guedes *et al.*, 2011). Most of these products are used in animal feed or human food additives, but not many sterol or prenol lipids have attracted a lot of attention from the biofuel research community. As a notable exception, botryococcenes were suggested as a feedstock for biofuel production (Banerjee *et al.*, 2002). These hydrocarbons are long chain (up to 37 carbon atoms) isoprenoids, which are excreted by *Botryococcus braunii* and can be converted to biofuels using standard petrochemistry. However, due to unsolved culturing problems with this species, no large-scale or commercial production scheme has been developed so far.

4.3 Biochemistry of Lipid Accumulation

4.3.1 Pathways of lipid biosynthesis

Compared to the vast amount of data about lipid metabolism in higher plants that is available in the literature, our knowledge about the algal biosynthesis, storage and degradation mechanisms is still sparse. At least for green algae (chlorophytes), which are the closest relatives to higher plants in the taxonomically heterogeneous group of algae, it is reasonable to expect that most pathways resemble those in model plants such as *Arabidopsis thaliana*. However, caution has to be exercised (Khozin-Goldberg and Cohen, 2011). Especially with organisms containing plastids resulting from a secondary endocytobiosis event, e.g. diatoms, lipid metabolism might differ in some details from the pathways observed in organisms from the 'green' line. On the other hand, with an increasing amount of genomic, biochemical and transcriptomic data from microalgae available, it was possible to identify algal counterparts of plant genes involved in lipid metabolism (Khozin-Goldberg and Cohen, 2011). These data indicate that the overall pathways are shared between all or most photosynthetic organisms, and it is reasonable to quickly summarize the knowledge gained from the analysis of higher plants (Fig. 4.2).

4.3.2 Building blocks and energy for lipid synthesis

The energy and building blocks necessary for the biosynthesis of FA are derived from the capture of light and the fixation of CO_2, two

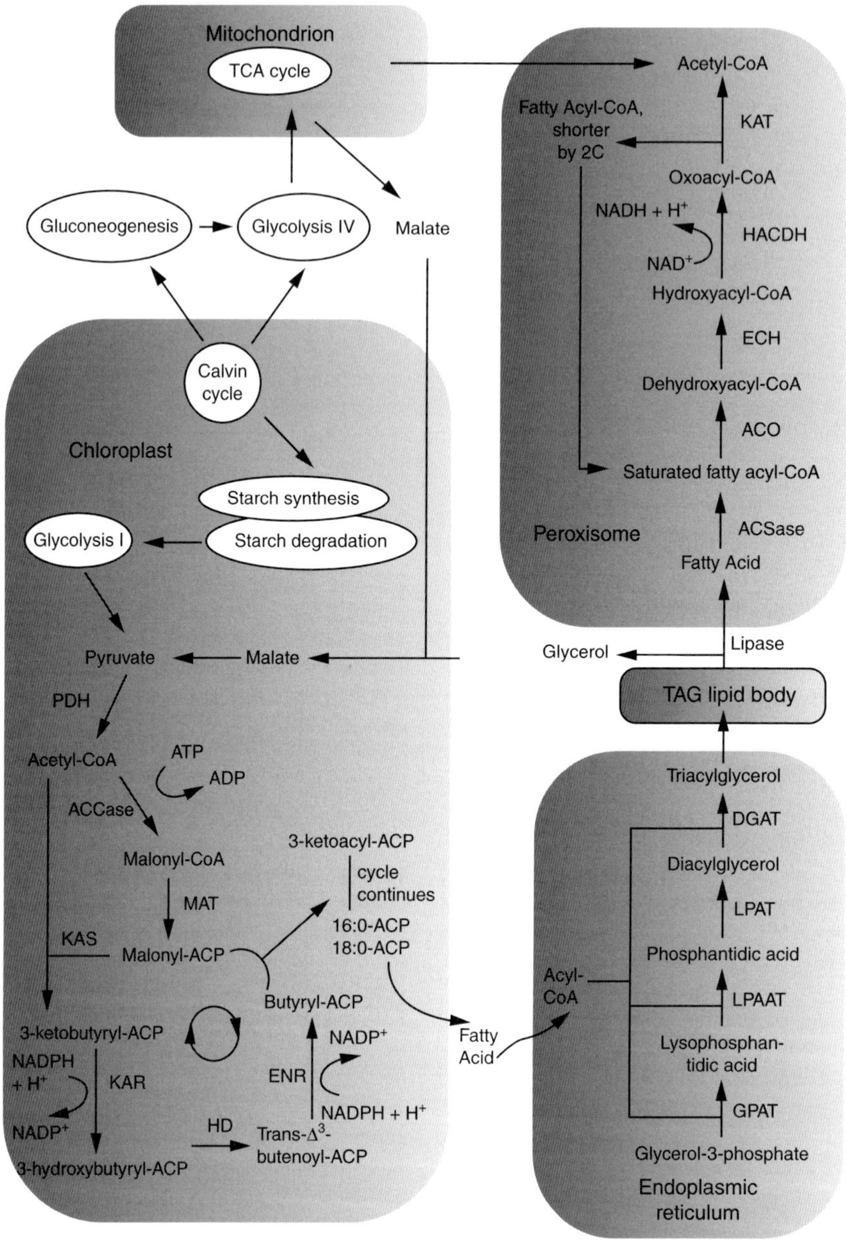

Fig. 4.2. Overview of lipid metabolism pathways in the cell. Included are fatty acid biosynthesis in the chloroplast, TAG formation in the endoplasmic reticulum and fatty acid β-oxidation in the peroxisome. Pathways providing building blocks (white ellipsoids) are indicated as well. A detailed description of the pathways is presented in the text. ACCase, acetyl-CoA carboxylase; ACO, acyl-CoA oxidase; ACP, acyl carrier protein; ACSase, acyl-CoA synthetase; DGAT, diacylglycerol acyltransferase; ECH, enoyl-CoA hydratase; ENR, enoyl-ACP reductase; GPAT, glycerol-3-phosphate acyltransferase; HACDH, hydroxyacyl-CoA dehydrogenase; HD, 3-hydroxyacyl-ACP dehydratase; KAR, 3-ketoacyl-ACP reductase; KAS, 3-ketoacyl-ACP synthase; KAT, ketoacyl-CoA thiolase; LPAAT, lyso-phosphantidic acid acyltransferase; LPAT, lyso-phosphantidylcholine acyltransferase; MAT, malonyl-CoA:ACP transacylase; PDH, pyruvate dehydrogenase complex; TAG, triacylglycerol; TCA, tricarboxylic acid. (Adapted from Schuhmann *et al.*, 2012.)

processes that shaped the atmosphere and surface of our planet and enable life as we know it today. The molecular machinery responsible for the conversion of solar to chemical energy is embedded in thylakoid membranes of cyanobacteria, higher plants and algae. After absorption of an electromagnetic quantum by pigments (the exact nature of this pigment depends on the organism), the energy of the photon is utilized to split water, thereby creating reduction potential (an electron on a range of electron carrier proteins), protons and oxygen as a waste product. The pumping of electrons along proteins of the electron transport chain results in the creation of an electrochemical potential gradient of protons, which in turn is responsible for the generation of adenosine triphosphate (ATP) by ATP synthase. Finally, the photosynthetic process produces a reduced 'hydrogen carrier' molecule, nicotin amide adenine dinucleotide phosphate (NADPH, see Barber (2009) for a comprehensive review of photosynthesis).

Both ATP and NADPH are necessary for the fixation of CO_2 in the Calvin-Benson cycle, also called light-independent reaction or 'dark reaction' (although the latter term is misleading, since CO_2 fixation only occurs when the organism is exposed to light; Raven, 2010). The key enzyme of this cycle is Ribulose-1,5-bisphosphate carboxylase oxygenase (RuBisCO), probably the most abundant enzyme on Earth. It catalyses the incorporation of one molecule CO_2 into ribulose-1,5-bisphosphate, which is the point where the inorganic carbon enters the cell's metabolism and becomes available for all anabolic reactions. The product of the Calvin-Benson cycle is glyceraldehyde-3-phospate (G3P) at the expense of ATP and NADPH, which are needed to regenerate ribulose-1,5-bisphosphate. Subsequently, G3P can be converted into a variety of different carbon hydrates, e.g. glucose, starch and sucrose.

The biosynthesis of FA in the plastid requires acetyl-CoA as a major building block. In photosynthetically active cells of *Arabidopsis thaliana*, carbon atoms fixed in the Calvin-Benson cycle are quickly incorporated into FA (Bao *et al.*, 2000), but the exact metabolic route these atoms take is still under debate (Rawsthorne, 2002). Most likely, acetyl-CoA is derived from the decarboxylation of pyruvate by a plastidial pyruvate dehydrogenase complex. The pyruvate necessary for this reaction could be obtained from a variety of sources, including cytosolic glycolysis and subsequent import of metabolites into the plastid, direct glycolysis in the chloroplast or import of malate from the cytosol (Rawsthorne, 2002).

Regardless of the exact route the carbon atoms take from their first introduction into the metabolism of the cell, photosynthesis and Calvin-Benson cycle provide the cell with all the building blocks necessary for the biosynthesis of FA: ATP, NADPH and acetyl-CoA.

4.3.3 Fatty acid biosynthesis pathway

As mentioned above, most of our knowledge about lipid biosynthesis pathways is gathered from studying higher land plants. Since there is no indication so far that these pathways are totally different in any microalgae species, it is reasonable to believe that the overall pathway of FA synthesis is more or less the same for every photosynthetically active eukaryote. However, this of course does not rule out that some details might differ from the scheme presented here.

De novo FA biosynthesis starts with the irreversible carboxylation of acetyl-CoA to form malonyl-CoA at the expense of one molecule of ATP, being the committed step of the pathway. This reaction is catalysed by acetyl-CoA carboxylase (ACCase), a biotin-dependent multifunctional enzyme, comprising a biotin carrier function, biotin carboxylase activity and carboxyl transferase activity in one protein or protein complex, respectively. ACCase proteins exist in two different forms. The prokaryotic form, present in prokaryotes and most chloroplasts from organisms of the *Viridiplantae* line, is assembled from four individual polypeptide chains (biotin carrier protein, biotin carboxylase, and carboxyltransferase alpha and beta), and each polypeptide is encoded by an individual gene. In plants, the first three proteins are encoded in the nuclear genome, whereas the gene for carboxyltransferase beta is in the plastidial genome. Apart from the heteromeric prokaryotic form, plants also posses the homomeric so-called eukaryotic form, which is best characterized from mammalian and fungal mitochondria and cytosol. This protein comprises all enzymatic functions on one single polypeptide chain. Interestingly, although chloroplasts are of prokaryotic origins according to the endosymbiont theory, some microalgae and higher plants (e.g. grasses) seem to exclusively employ a single-chain ACCase for the biosynthesis of FA (Sasaki and Nagano, 2004). Since ACCase is the key enzyme for shuttling acetyl-CoA into lipid biosynthesis, it is not surprising that its gene transcription, assembly and activity is highly regulated (Sasaki and Nagano, 2004). Under light exposure, a photosynthetically active cell accumulates energy and building block precursors as described above, and FA synthesis is possible. Indeed, ACCase activity also increases when the cell is exposed to light, possibly regulated by a thioredoxin-mediated redox mechanism (reviewed in Geigenberger *et al.*, 2005).

After carboxylation of acetyl-CoA, the product malonyl-CoA is then transferred to an acyl carrier protein (ACP) by malonyl-CoA:ACP transacylase. This molecule is condensed with another molecule acetyl-CoA by 3-ketoacyl-ACP synthetase, releasing CO_2 during the reaction and thus rendering this process irreversible. The product of this step is a 3-ketoacyl-ACP molecule, which in turn is reduced by 3-ketoacyl-ACP reductase using NADPH produced during photosynthesis. Subsequently, the product is dehydrated by 3-hydroxyacyl-ACP dehydratase, and reduced once more, this time by enoyl-ACP reductase. The product of these reaction steps is butyryl-ACP, or more general, an acyl-ACP molecule that has been elongated by two carbon atoms. This acyl-ACP can now enter another reaction cycle with the condensation of another molecule of malonyl-ACP. The growing acyl-ACP molecule continues to cycle through the condensation, reduction and dehydration steps, until it reaches a chain length of 16 or 18 carbon atoms. It is then either released from the carrier protein by an acyl-ACP thioesterase yielding palmitic acid (16:0) or stearic acid (18:0), respectively, or is desaturated by stearoyl desaturase and subsequently released. As an alternative to the release and the production of free FA, the acyl moiety can be transferred to glycerol-3-phosphate. It can then be subjected to additional modification steps, including further elongation and desaturation, and stays in the chloroplast as membrane or storage lipid.

FAs that were released by a thiolase are transported outside the chloroplast, where they are again transferred to CoA by acyl-CoA synthetase. The acyl-CoAs then enter the endoplasmic reticulum (ER) for incorporation into glycerol lipids and further modifications.

4.3.4 Glycerolipid biosynthesis and FA modification

The acyl moieties of most of the acyl-CoAs are then transferred to glycerol-3-phosphate (G3P), yielding phospatidic acid (PA), which can be further modified to other glycerolipids, e.g. phosphatidylcholine (PC), phosphatidylethanolamine, and phosphatidylinositol. This pool of polar lipids is highly dynamic. Head groups can be exchanged and modified and acyl moieties can be exchanged by the action of acyltransferases (Snyder *et al.*, 2009). Furthermore, the acyl residues themselves can be modified. Endoplasmic elongases produce saturated or mono-unsaturated very long chain (VLC) FA (Napier and Graham, 2010), whereas alternating actions of desaturases and elongases on PC produce very long-chain polyunsaturated FA (VLC-PUFA) residues. These modified FA can then be incorporated into other lipids by acyltransferases (Abbadi *et al.*, 2004). It is noteworthy, however, that most higher plants do not produce VL PUFA, and the main source for these nutritionally valuable FA are marine microalgae and marine organisms higher in the food chain (Khozin-Goldberg *et al.*, 2011). Other modifications that can be introduced into the FA moieties of PC include hydroxylation and epoxidation, and these create a range of unusual FAs (Shanklin and Cahoon, 1998; Napier, 2007).

From a biofuel perspective, the most interesting lipid in a cell is TAG, since cells accumulate it under certain conditions (see section 4.4 below), and it can be extracted from the cell quite easily. Several pathways exist how a fatty acid synthesized in the chloroplast, or modified in the ER, can be incorporated into TAGs, the most straightforward being the Kennedy pathway (Ohlrogge and Browse, 1995). Here, two acyl-CoAs are subsequently transferred to G3P to yield PA, which is then dephosphorylated to diacylglycerol (DAG). In an enzymatic reaction unique to TAG biosynthesis, acyl-CoA:diacylglycerol acyltransferase (DGAT) catalyses the esterification of the third hydroxy group of the glycerol moiety. This step is also the committing and rate-limiting step of TAG biosynthesis, irreversibly leading to the final product (Lung and Weselake, 2006).

Polyunsaturated, elongated, hydroxylated, or otherwise modified FA can enter TAG in several different pathways. First, acyl-CoA:PC acyltransferases catalyse the exchange of FA that were modified as a part of PC between this molecule and acyl-CoA, so they enter the acyl-CoA pool and become available for TAG biosynthesis via the Kennedy pathway. Second, the whole DAG portion PC can be made available for TAG synthesis, since the formation of PC from DAG (resulting from dephosphorylation of PA) and activated choline is highly reversible. Therefore, PC containing modified FA can lose its polar head group, and become a substrate for DGAT. Alternatively, FA from PC can become incorporated into TAG via an acyl-CoA independent pathway. Phospholipid:diacylglycerol acyltransferases are able to catalyse the direct transfer of an acyl group from PC to DAG, resulting in TAG (Dahlqvist *et al.*, 2000).

4.3.5 TAG storage and mobilization

TAGs enable the cell to store energy and lipid building blocks without altering the osmolytic properties of the cytoplasm. Since they lack a polar head group, neutral lipids such as TAGs are unable to integrate into lipid bilayers and are stored in lipid particles, also called lipid bodies (lb), oil bodies or lipid droplets. These particles are comprised of a neutral lipid core covered by a phospholipid monolayer and proteins. Being dynamic structures, they can grow due to incorporation of new TAGs synthesized by enzymes present on the lb surface. For example, proteins from the TAG biosynthesis pathway, which were found associated with lipid particles, include glycerol-3-phosphate acyltransferase, phospholipid:diacylglycerol acyltransferase, 3-ketoacyl-CoA synthase and FA desaturases (Nguyen *et al.*, 2011). TAG storage in lb, as well as their biogenesis and importance in the cell, are described in more detail in Section 4.4, below.

The first step in metabolism of lipid from lb is the hydrolysis of the ester bonds by tri-, di-, and monoacylglycerol lipases, yielding FA and glycerol. During this process, some coating proteins on the particle surface might play a critical role in providing docking stations for the lipases (Athenstaedt and Daum, 2006; Matos and Pham-Thi, 2009).

In plants, the glycerol produced by the lipases is phosphorylated to G3P (Eastmond, 2004), which is converted to dihydroxyacetone phosphate (DHAP) by G3P dehydrogenase (Quettier *et al.*, 2008). Although the glycerol moiety of TAG contains only about 5% of the carbon content of storage oils in seedlings, this fraction can be decisive for proper growth of the organism under certain conditions (Eastmond, 2004).

However, most of the energy and carbon atoms stored in TAG is contained within the FA molecules. These are transported for further processing into the peroxisome, either as acyl-CoA, or as free acids (Graham, 2008). Which import pathway is involved may depend on the species, as well as the FA chain length (Graham, 2008). In the peroxisome, FAs are activated for degradation as thioesters by several acyl-CoA synthetases under the expense of ATP and then enter the peroxisomal beta-oxidation pathway (Goepfert and Poirier, 2007).

The core of this pathway is a series of four reactions, which shorten the acyl-CoA molecule by two carbon atoms and are repeated in a circle until the FA moiety is entirely consumed. The first step is the oxidation of acyl-CoA to (2E)-enoyl-CoA by a group of acyl-CoA-oxidases (ACX). ACX use flavin adenine dinucleotide (FAD) as a co-factor and differ in the preferences of the substrate chain length (Goepfert and Poirier, 2007, and references therein).

Then, the (2E)-enoyl-CoA molecule is hydrated to (3S)-hydroxyacyl-CoA, catalysed by a multifunctional protein (MFP). MFP also catalyses the next step, the oxidation to 3-ketoacyl-CoA under the expense of oxidized nicotin amide adenine dinucleotide (NAD^+).

In the final step, thiolase catalyses the nucleophilic attack of another molecule of CoA on ketoacyl-CoA, yielding acetyl-CoA and an acyl-CoA that is two molecules shorter than the original one that entered the reaction cycle. Acetyl-CoA can then enter cellular metabolism, while the remaining acyl-CoA is subjected to another round of peroxisomal beta-oxidation.

The metabolism of unsaturated FA requires additional steps. First, the process is essentially the same as described for saturated molecules until a 2,5-dienoyl intermediate product is formed. The subsequent actions of Δ^2,Δ^3 isomerase, $\Delta^{3,5},\Delta^{2,4}$ isomerase, 2,4-dienoyl-CoA reductase, and again Δ^2,Δ^3 isomerase transform this intermediate product into (2E)-enoyl-CoA, which can then be processed in the normal beta-oxidation pathway.

Although our knowledge about lipid metabolism in photosynthetic organisms is mainly based on studies in higher plants, especially the model organism *A. thaliana*, it is reasonable to assume that the overall picture is also true for microalgae (Fig. 4.2). However, details might differ. For example, it was found that the green alga *Chlamydomonas reinhardtii* is able to produce and store TAG directly in its chloroplasts, although no gene encoding a protein for a chloroplastidial TAG biosynthesis machinery has been identified so far (Fan *et al.*, 2011). This illustrates that microalgal TAG production might show much more variation of the standard pathways.

Also, almost all higher plants are unable to synthesize VLC-PUFA, whereas many microalgae, especially from a marine environment, are primary producers of these FA (Venegas-Caleron *et al.*, 2010). Another difference between higher land plants and some microalgae species is the presence of additional membranes, especially in the chloroplast. The chloroplast of diatoms, for example, is believed to originate from a secondary endosymbiotic event, where a non-photosynthetic cell incorporated a red algae-type eukaryote with subsequent dissolution of the prey cell's nucleus (Agrawal and Striepen, 2010). This resulted in a plastid that is surrounded by four membrane layers, as opposed to two membranes in land plants. Partitioning of TAG biosynthesis, with some processes localized in the chloroplast and others in the ER, requires additional transmembrane transport mechanisms in these organisms, or different allocation of biosynthetic pathways (compare Fan *et al.*, 2011).

4.4 Physiology of Microalgal Lipid Accumulation

4.4.1 Polar lipids and membranes

As in every other organism, glycerol lipids fulfil two major roles in the algal cell. Charged, polar lipids such as phospho- and glycolipids form the membranes within and around a cell, thereby creating different compartments. Thus, they separate molecules, reactions and charges, which enables the existence of gradients (e.g. the proton gradient created during photosynthesis). Without membranes consisting of lipid bilayers, life would not be possible.

During growth phase, algal cells increase their size and eventually divide into daughter cells (asexual reproduction). Since an increase of a cell's volume (i.e. size) also means an increase of surfaces, membranes have to grow as well, and polar lipids are constantly synthesized to provide membrane building blocks. A growing cell channels most of its resources for FA and glycerol lipid biosynthesis into synthesis of membrane building blocks (Thompson, 1996), especially chloroplast membranes. For example, in a growing green algal cell, chloroplast membrane lipids can make up to 70% of all structural lipids (Thompson, 1996).

Cells can alter the physical properties of their membranes and adjust them to the prevalent environmental conditions. The melting temperature, and therefore the fluidity of a membrane, is controlled by the ratio of saturated versus unsaturated FA in the glycerol lipids forming the lipid bilayer. The double bond of an unsaturated FA induces a 'kink' in the carbon chain, preventing neighbouring lipids from close packaging. This, in turn, increases the fluidity of the membrane and ensures its biological activity and integrity under, for example, lower temperature conditions.

4.4.2 The benefits of accumulating triacylglycerols

Of more importance for the biodiesel researcher, however, are neutral lipids, especially TAGs. TAG contains more FA moieties per molecule than polar lipids (three versus two), which can be trans-esterified into biodiesel. Therefore, understanding the physiology of lipid algal metabolism is essential for selecting and breeding high performance algal lipid producers. Whereas polar membrane lipids are preferably synthesized under conditions permitting cell and organelle growth and division, many algae species are able to rearrange the lipid metabolism when the situation becomes more unfavourable (Hu *et al.*, 2008). This is the case when the environment becomes depleted of (at least) one essential nutrient, e.g. N, P, or Si for diatoms, but a carbon source (e.g. CO_2) and light energy is still available in abundance. Under these conditions, some algae strains can accumulate TAG to up to 77% of their total dry weight (Bigogno *et al.*, 2002), whereas in other species (i.e. *C. reinhardtii*) starch is the dominant storage compound. The physiological advantages of lipid accumulation for the cell are not fully understood yet, and may depend on the species and the exact conditions. A main function of TAG is probably the storage of energy. Since future energy availability is unpredictable for the individual organism, the one that utilizes the available energy better as compared to others can have an evolutionary advantage. When conditions become more favourable for cell division, an organism with high energy reserves is able to immediately restart cell growth and cell division. Therefore, many algae species accumulate either TAG or starch under nutrient limitation.

However, compared to starch, TAG provides an opportunity to store membrane building blocks (e.g. glycerol and FA) as well. For example, TAG can be synthesized even under non-limiting conditions in the light, and are degraded in favour of polar lipid synthesis during darkness (Thompson, 1996). Similar to accumulating energy reserves during non-favourable conditions, it might be beneficial for the organism to accumulate a reservoir of membrane building blocks in order to quickly restart membrane biosynthesis when conditions improve. This view is supported by the presence of LC-PUFA moieties in TAGs of some algae species, e.g. *Parietochloris incisa* (Bigogno *et al.*, 2002). After re-establishing growth conditions, TAG stored in lipid bodies was metabolized, and FAs were recycled in the biosynthesis of new membranes, especially chloroplastidial (Khozin-Goldberg *et al.*, 2005).

Additional benefits might arise from the shuttling of metabolites such as ATP and NADPH into TAG biosynthesis (Thompson, 1996). When sufficient nutrients are available, a large fraction of ATP and NADPH produced by photosynthesis is consumed for cell growth, i.e. biosynthesis of membranes, proteins, nucleic acids and others. When nutrients such as N become limited this route is blocked. However, since light-driven photosynthesis is mainly controlled by the availability of light, it cannot be easily switched off by the cell, and the pool of ATP and NADPH precursors (i.e. ADP and $NADP^+$) can become depleted. For example, $NADP^+$ is the final electron acceptor of the photosynthetic electron transport chain. If $NADP^+$ levels are low, but photosynthesis is still running, reduced intermediates of the electron transport chain accumulate. This, in combination with the presence of photosynthesis-derived oxygen, leads to the formation of reactive oxygen species (ROS), which damage cell walls, proteins and nucleic acids.

To avoid this potentially dangerous situation, NADPH is consumed by the biosynthesis of FA. According to Fig. 4.2, the synthesis of one molecule of octadecanoic acid (stearic acid, C18:0) requires 16 molecules of NADPH in the condensation cycle alone, plus additional molecules are required for the fixation of carbon dioxide and the conversion into acetyl-CoA. $NADP^+$ produced during FA biosynthesis becomes then available again as an electron acceptor in photosynthesis (Hu *et al.*, 2008).

Additionally, the production of TAG can be coupled with the biosynthesis of pigments such as astaxanthin (Zhekisheva *et al.*, 2002). These hydrophobic molecules are readily soluble in TAG containing lipid bodies (see below), and provide shading and thereby protection.

4.4.3 Lipid bodies – storage of triacylglycerols

TAGs are usually stored in specialized structures within the cytoplasm, called lipid bodies (lb), also known as lipid particles, lipid droplets, or oil bodies. Although there are still some uncertainties about their exact origin, it is now widely believed that cytoplasmic lb are formed by vesiculation from the ER (Murphy, 2001; Athenstaedt and Daum, 2006), which is also the site of TAG biosynthesis (see 'Biochemistry of lipid accumulation'). Electron microscopy pictures obtained from starved *C. reinhardtii* cells show that lb maintain contact with the ER membrane, and also might form contact with the outer envelope of the chloroplastidial membrane. These findings suggest a model of ER–chloroplast communication, where FA synthesized in the latter can travel towards the ER for further processing (Goodson *et al.*, 2011).

Compared to organelles, they are relatively simple structures, with a core formed almost exclusively by TAGs, and surrounded by a monolayer of polar lipids. In the seeds of plants, this outer layer may itself be coated by oleosins, small (15–24 kDa) proteins that protect the lb from degradation and from fusing with each other (Murphy, 2001). However, no oleosin orthologue was found by searching for orthologous genes in the genome of the green alga *C. reinhardtii*, or by analysis of the algal lb proteome (Moellering and Benning, 2010). On the other hand, the same study also revealed the presence of a previously unidentified protein in the lb of green algae, name MLDP. This protein could fulfil a similar role to oleosins, since a reduction in *MLDP* gene expression increased the lb size in *C. reinhardtii*. (Moellering and Benning, 2010). Besides MLDP, lb also contain enzymes of lipid metabolism, i.e. TAG biosynthesis and lipid degradation (Moellering and Benning, 2010; Nguyen *et al.*, 2011). Together with the observation that externally applied FAs are predominantly incorporated into TAG in lipid bodies, but later appear in plastidial membranes (Khozin *et al.*, 1997), this suggests a rather dynamic nature of algal lb. More than just being carbon storage facilities within the cell, they are also a pool of polar lipid building blocks that can be exploited during times of increased cell growth and/or membrane turnover.

Lipid bodies have also been described in the chloroplasts of algae and higher plants (Murphy, 2001; Fan *et al.*, 2011; Goodson *et al.*, 2011), where they are also called plastoglobuli. Similar to cytoplasmic lb, they accumulate and grow under conditions where cell division (and therefore membrane growth) is limited, but energy and a carbon source is readily available (Fan *et al.*, 2011; Goodson *et al.*, 2011).

4.5 Environmental Factors Inducing TAG Accumulation

As described above, TAG accumulation usually constitutes a response to a stress condition that reduces cell growth and/or threatens the integrity of the photosynthetic apparatus. The following conditions proved to be efficient in inducing this stress in at least some microalgae strains with potential as biodiesel feedstock.

4.5.1 Nitrogen starvation

Many microalgae accumulate TAG when the growth medium becomes depleted for one or more nutrients. Of all nutrients, nitrogen availability seems to have the highest impact on lipid metabolism, and many algae species accumulate TAG as a response to low N levels (Hu *et al.*, 2008; Schenk *et al.*, 2008). Nitrogen uptake is indispensible for *de novo* amino acid biosynthesis, and therefore for protein biosynthesis and ultimately cell growth. Under nitrogen-depleted conditions, cells still might be able to synthesize new proteins by recycling their proteome, but cell growth and division is impossible. Consequently, all the hazards associated with photosynthesis under limiting nutrient conditions as described in the previous section might trigger TAG accumulation in the microalgal cell. However, reduced nitrogen availability results in additional problems. Nitrogen assimilation pathways rely on reducing agents such as NAD(P)H (Fernandez and Galvan, 2008). When assimilation is not possible due to low nitrogen concentrations, the respective pathways are down-regulated (Hockin *et al.*, 2012), and the cell requires alternative electron sinks to keep the photosynthetic process running. Again, this can be provided by increased FA biosynthesis and, eventually, TAG accumulation.

Since so many microalgae have been described as TAG accumulators under N deficiency, it is far beyond the limit of this chapter to provide an exhaustive list of reports. However, some more recent examples will be provided (see Griffiths and Harrison (2009) for a more detailed list). Green algae seem to be the best examined microalgae with regards to nitrogen starvation and TAG content for biodiesel production. Illman *et al.* (2000) compared the productivity of several algae species from the *Chlorella* genus (*C. vulgaris, C. emersonii, C. protothecoides, C. sorokiniana, C. minutissima*). All species showed an increased lipid content (based on percentage of dry weight) when grown in low-nitrogen media, with the biggest difference in *C. emersonii*, where lipid levels increased from 29% (high N regime) to 63% (low N regime). Additionally, overall dry weight of cell mass was also higher in N-depleted cultures, therefore the increase in lipid levels was the result of true lipid accumulation and resulted in higher overall yields (Illman *et al.*, 2000). However, the exact nature of the lipid species, i.e. TAG, phospholipid, or galactolipid, was not addressed. For *C. vulgaris*, this was analysed in a different study (Stephenson *et al.*, 2010). Here, when grown in nitrogen-containing medium, lipid content was 14% of cell dry mass, but TAG contributed only 3% to the total lipid content. However, when cultured under nitrogen-free conditions, lipid content increased to 19% of total dry mass, with TAG levels of more than 50% (Stephenson *et al.*, 2010). Recently, the changes in the proteome and the transcriptome upon nitrogen starvation was analysed for the same species, providing insights into the cell's response on a molecular level (Guarnieri *et al.*, 2011). FA and TAG biosynthetic pathways were up-regulated under N-depletion conditions, as expected for TAG accumulation conditions. Notably, this up-regulation was much more pronounced for the TAG pathway, compared to FA biosynthesis, and an additional down-regulation for TAG metabolizing enzymes was shown (Guarnieri *et al.*, 2011).

Besides microalgae from the *Chlorella* genus, other green algae have attracted attention, too.

For example, the freshwater alga *Neochloris oleoabundans* was analysed for its potential as biodiesel production feedstock (Li *et al.*, 2008). Here, it was interesting to note that the cellular lipid content was highest with the lowest nitrogen content in the medium, however, the highest *productivity* per culture volume was obtained with a slightly higher nitrogen concentration (Li *et al.*, 2008).

Microalgae from the *Dunaliella* genus have been commercially exploited for their carotenoids for many years. With increasing interest in TAG production for biodiesel, they were also analysed for their lipid accumulation potential, since they combine rapid growth with a high tolerance towards environmental conditions, e.g. salinity. As only one example, the influence of several nutrients, and the lack thereof, on the neutral lipid level of *Dunaliella tertiolecta* was investigated (Chen *et al.*, 2011). Nitrogen starvation showed the highest impact on TAG levels within a cell, however, cultures under N-repleted conditions yielded the highest amount of TAG, due to higher biomass production under normal growth conditions. Therefore, the authors suggest to grow *D. tertiolecta* under normal conditions until a high biomass level is reached, and then subject the cells to N-depleted conditions. A reduction of cobalt and iron led to an intracellular lipid accumulation, although not as pronounced as the lack of a nitrogen source (Chen *et al.*, 2011).

The relationship of the accumulation of the two major storage compounds, TAG and starch, was examined in *Pseudochlorococcum* sp. (Li *et al.*, 2011). On the onset of nitrogen starvation, starch is the major storage compound produced by the cells. However, as the starvation phase continues, starch levels decrease in favour of TAG levels, which becomes the major cellular component after prolonged N-depleted periods. These results indicate that TAG might be more suitable as a long-term storage product, compared to starch. Additional benefits (e.g. protection of the photosynthetic apparatus, see above) make it attractive for the cell to convert starch to lipids (Li *et al.*, 2011).

The model organism *C. reinhardtii* has a special role in microalgal lipid research. Its fully sequenced genome (Merchant *et al.*, 2007) and a wide range of molecular biological tools applicable to it, make this species probably the microalga with the best examined lipid metabolism. Although naturally being a starch accumulator under nitrogen starvation conditions, lb can also be observed in addition to starch granules (Moellering and Benning, 2010). Recently, the changes in the transcriptome of the alga, as a response to nitrogen limitation, was analysed (Miller *et al.*, 2010). Indeed, genes corresponding to enzymes of the FA and TAG biosynthetic pathways were upregulated. Contrarily, genes encoding proteins of the photosynthetic apparatus and carbon fixation were down-regulated. This is consistent with the hypothesis that microalgal cells might experience stress associated with hyperactive photosynthesis during nitrogen starvation. However, since photosynthesis and carbon fixation provide the building blocks for TAG formation, these pathways might be a promising point for genetic engineering in order to increase lipid yield for biofuel production.

Since *C. reinhardtii* produces both starch and TAG as a response to nitrogen starvation, several studies tried to elucidate a potential competition between both biosynthetic pathways. Analysis of starchless mutants showed an increased TAG content, suggesting that starch accumulation indeed uses resources otherwise available for lipid synthesis (Wang *et al.*, 2009; Li, Y. *et al.*, 2010; Work *et al.*, 2010). However, careful analysis of the genetic backgrounds of the respective algae strains used in these works, as well as complementation studies, rejected the hypothesis of a simple competition between starch and TAG (Siaut *et al.*, 2011). Future work will have to show if it is possible to shuttle precursors from other anabolic pathways into TAG biosynthesis.

Not only green algae accumulate lipids when deprived of a nitrogen source. Although originally examined for its potential as a supply for valuable LC-PUFA, *Phaeodactylum tricornutum* also gained interest as a biodiesel feedstock. Similar to green algae, the fraction of lipids on the dry weight increases under high light and low nitrogen conditions (Qian and Borowitzka, 1993) at the cost of proteins and carbohydrates. This is mainly due to an accumulation of TAG, whereas the amount of galactolipids is reduced when facing low N concentrations (Alonso *et al.*, 2000). Contrary to the lipid species composition, the FA composition was reported to be fairly stable. However, PUFA slightly decreased, whereas the amount of 16:0 (more suitable for biodiesel production) and of 16:1 FA increased (Alonso *et al.*, 2000). Similarly, the freshwater diatom *Stephanodiscus minulutus* showed increased TAG levels under N-limited growth (Lynn *et al.*, 2000).

The eustigmatophyte *Nannochloropsis* sp. can accumulate up to 55% of its dry weight as lipids when exposed to low nitrogen concentrations, the majority of it being TAG (Suen *et al.*, 1987). These findings were generally confirmed by later studies (Converti *et al.*, 2009; Pal *et al.*, 2011). Additionally, FA analysis showed that under these conditions, *Nannochloropsis* contains high amounts of 16:0 oleic acid, making it a very suitable feedstock for biodiesel production (Converti *et al.*, 2009; Pal *et al.*, 2011).

In summary, nitrogen starvation has been proven to be a trigger for TAG accumulation in many microalgae species, over the whole range of the tree of life. However, some algae do not show lipid accumulation when deprived of nitrogen. Also, the exact conditions, i.e. the onset of nitrogen starvation, light regime, salinity, etc., have to be evaluated for every species, or even for every strain.

4.5.2 Phosphorus starvation

Upon entering phosphorus starvation, microalgal cells rearrange their polar membrane lipids in order to save valuable phosphorus resources. Phospholipids in the membranes can be largely (although not totally) substituted by lipids containing sulfur or nitrogen in their polar head groups, e.g. betaine lipids and sulfoquinovosyldiacylglycerol (Van Mooy *et al.*, 2009). Compared to nitrogen starvation, the lack of phosphorus seems to have less potential to trigger TAG accumulation in microalgae. This might be a consequence of phosphorus nearly always being present as a derivative of phosphate, i.e. phosphorus metabolism requires no reduction/oxidation steps. Therefore, there is also no need for reducing agents produced by photosynthesis, and lowered phosphate availability does not generate a surplus of these agents as does reduced nitrogen metabolism.

However, some examples from the literature are available where low phosphorus conditions induce lipid accumulation in microalgae cells. For green algae from the *Scenedesmus* genus, two research groups reported an increase in lipid content under phosphate-depleted conditions. Mandal and Mallick (2009) report lipid accumulation of up to 23% of dry cell mass when cells were transferred to phosphate-free media, although the strain showed an even more pronounced response to nitrogen starvation. The *Scenedesmus* strains used by Li *et al.* (2010a) even achieved lipid levels of 55% when cultivated at total phosphate concentrations of 0.1 mg l^{-1}. However, as with other starvation experiments, both groups noted a decrease in cell growth, which led to reduced overall lipid productivity (Mandal and Mallick, 2009; Li, X. *et al.*, 2010). Another chlorophyte, the extremophile *Chlamydomonas acidophila*, was analysed with respect to its FA content under P-limited growth conditions. Although TAG levels were not determined in this study, an increase of saturated and mono-unsaturated FA with decreasing cellular phosphorus levels indicate an accumulation of this lipid species as a result of P starvation stress, especially when additional carbon was available in the form of CO_2 (Spijkerman and Wacker, 2011).

A few reports exist where algae other than green algae accumulate lipids when deprived of phosphorus. *Phaeodactylum tricornutum*, *Chaetocerus* sp., *Isochrysis galbana* and *Pavlova lutheri* all showed increased cellular lipid content with decreasing phosphorus availability. Again, nutrient limitation also resulted in a change of the FA composition, with saturated and mono-unsaturated FA levels rising at the cost of PUFA levels (Reitan *et al.*, 1994). For *P. tricornutum* and *Chaetocerus muelleri*, the increase in overall lipid was confirmed using Fourier-transform infrared spectroscopy (FTIR), in addition to gas chromatography (Liang *et al.*, 2006). Again, starved cells showed higher levels of saturated and mono-unsaturated FA, and decreased PUFA levels. However, in this study, phosphorus starvation was not induced, but cultures were allowed to age. Knowledge of the starting conditions, and of nitrogen/phosphorus assimilation rates from the literature, led the authors to the conclusion that older cultures suffer from phosphorus starvation stress (Liang *et al.*, 2006).

Another diatom, the freshwater alga *S. minulutus*, also accumulates TAG under phosphorus limitation (Lynn *et al.*, 2000). Whereas most algae species show a much stronger response to nitrogen limitation than to phosphorus limitation, when analysed simultaneously, this species reacted much more strongly to the lack of the latter. This not only resulted in higher cellular lipid content, but also in higher lipid yields per culture volume as compared to nutrient repleted conditions (Lynn *et al.*, 2000).

Monodus subterraneus, a freshwater eustigmatophyte, is the last example presented here of a microalga that accumulates TAG when starved of phosphorus (Khozin-Goldberg and Cohen, 2006). Again, phosphorus deprivation caused the stimulated cells to synthesize more TAG. In accordance with the other studies presented here, the FA profile also changed from PUFA-rich to saturated/mono-unsaturated FA rich (Khozin-Goldberg and Cohen, 2006).

In summary, phosphorus starvation does not seem to be the most prominent trigger for TAG accumulation in microalgae cells, however, some species show a response, especially at a later stage when P reserves may be used up. It is intriguing to see that the majority of reports presented above deal with freshwater species. Whether this is a consequence of their physiology, or just due to the limited number of overall reports, cannot be answered here.

4.5.3 Silicon starvation

Most diatoms have a hard external shell mainly made of silicon dioxide (SiO_2). As the cell grows and divides, the outer shell also has to be expanded, and divided, requiring a constant supply of silicate. A low supply of Si will lead to retarded growth, and diatoms face the hazards of overly active photosynthesis as outlined above. Therefore, it does not come as a surprise that some diatoms were shown to increase the activity of their lipid biosynthesis apparatus under Si-depleted conditions. The freshwater diatom *S. minulutus* has been already mentioned before in this chapter. Besides accumulating TAG when deprived of nitrogen and phosphorus, Si starvation also leads to an increase in cellular TAG content (Lynn *et al.*, 2000). In fact, the latter condition triggered the strongest response considering total lipid levels. However, although cellular TAG content was much higher than under control or nitrogen-depleted conditions, phospholipids were the main contributor to the overall increase in lipid content (Lynn *et al.*, 2000). Silicon deficiency also leads to lipid accumulation in another diatom, *Cyclotella cryptica*. Here, enzymatic activity of carboxyl-CoA-carboxylase rose up to fourfold in extracts from Si-depleted cells, indicating the rearrangement of the cell's metabolism towards increased FA and lipid synthesis (Roessler, 1988). Increased FA biosynthesis activity was observed in yet another diatom, *Skeletonema marinoi*, although total lipid levels were not analysed in this study (Ribalet *et al.*, 2009). Although most microalgae produce more saturated or mono-unsaturated FA when entering stress conditions, *S. marinoi* had higher levels of PUFA under Si depletion. However, as only FA content was analysed, it remains unclear if these FA were sequestered within TAG, other lipids, or existing as free FA in the cell.

Although Si depletion holds the potential to trigger TAG accumulation, its applicability is naturally limited to diatoms. Furthermore, the number of reports is much less for Si-depletion-induced lipid production, compared to reports of experimental setups using nitrogen starvation. Still, it might prove to be a valuable tool in lipid production from diatoms, and provide valuable insight into the physiology of TAG accumulation in these algae.

Apart from nitrogen, phosphorus and silicon limitation, several other nutrients have been described to influence TAG accumulation in microalgae, for example sulfur starvation led to TAG production in *C. reinhardtii* (Timmins *et al.*, 2009).

4.5.4 Salinity

Microalgae are well known for their ability to thrive in a wide range of salinities, from fresh water to salt lakes. Algae from the *Dunaliella* genus, for example, tolerate salinities of up to several M NaCl (Azachi *et al.*, 2002). Therefore, it is not surprising that algae developed a variety of mechanisms to cope with osmotic stress. Although the physiological consequences are not fully understood yet, some strains show lipid accumulation, and/or FA profile alteration when experiencing increased salt concentrations. For example, an osmotic shock induces the transcription of a microsomal ketoacyl-CoA synthase, an enzyme which catalyses the elongation of C16:0 to LCFA (Azachi *et al.*, 2002). Although it is unknown if the increase in enzyme activity was accompanied by increased neutral lipid levels, it shows that the salinity of the environment influences lipid metabolism. An increase in TAG levels was demonstrated for one strain of the green alga *B. braunii* (Zhila *et al.*, 2011). When the NaCl concentration in the growth medium was increased from 0 to 0.7 M, TAG content rose from 4.5 to 31% of the total dry weight. Additionally, as seen under starvation conditions, the FA profile changed, with more saturated FA being present under high salinity.

In *Nannochloropsis* sp., the effect of increased NaCl conditions was shown to be dependent on the nutrient regime (Pal *et al.*, 2011). Under nitrogen-replete conditions, higher NaCl levels resulted in increased TAG content. However, when cells were grown under nitrogen-deplete conditions to induce TAG accumulation, increasing the salinity failed to increase TAG levels further. Indeed, high salinities had a negative effect on lipid yield (Pal *et al.*, 2011). Contrarily to the reports cited above, two studies found decreasing lipid contents with increased salinities for the diatoms *Nitzschia laevis* (Chen *et al.*, 2008) and *Chaetoceros calcitrans* (Raghavan *et al.*, 2008). Furthermore, in *N. laevis* the reduced lipid content is concomitant with an increase in polar lipids and PUFAs (Chen *et al.*, 2008).

Taken together, research indicates that the response of an algal strain to altered salinities depends on the particular strain, and it is hard to predict the actual behaviour. This could be the reason why it does not seem to be very common to induce lipid accumulation by applying osmotic stress to the algal cell, contrary to e.g. nutrient starvation.

4.5.5 Temperature

The impact of temperature on the lipid composition, especially the relative levels of different FA in plants and algae, has been thoroughly investigated (see Guschina and Harwood, 2009 for examples). The length of the FA moiety of a membrane lipid, and its degree of unsaturation, has a pronounced effect on the overall fluidity of the membrane, with unsaturated species making the membrane more 'liquid'. Therefore, decreasing temperatures usually result in increased incorporation of unsaturated or polyunsaturated FA into membranes. However, reports about the accumulation of neutral lipids as a response to temperature change (other than determining optimal growth conditions) are sparse.

As an example, Converti *et al.* (2009) examined the lipid content of *C. vulgaris* and *N. oculata*. In the former, an increase in temperature (30°C compared to 25°C) resulted in a lower growth rate, lower lipid content, and consequently in a lower lipid productivity (measured per culture volume and time). For *N. oculata*, on the other hand, results were different. Although growth was also affected by elevated temperatures (25°C), this was compensated for by an increased lipid content. These two effects almost completely annihilated each other, and cells grown at optimal (20°C) and elevated temperature showed almost equal lipid productivity (Converti *et al.*, 2009).

The haptophyte *P. lutheri* was reported to accumulate TAG under elevated temperatures (25°C compared to 15°C) (Tatsuzawa and Takizawa, 1995). At the higher of these temperatures, total cellular lipids consisted of 40% TAG, whereas at 15°C betaine was the predominant lipid species. In accordance with other reports, the FA composition also changed towards more saturated FAs with increasing temperature, and a decreasing proportion of LC-PUFA. Although temperature is an important growth parameter, altering it does not seem to be a widely used trigger to induce TAG accumulation in microalgae. However, it affects biomass production, potentially the efficiency of other triggers (e.g. nitrogen starvation), and therefore influences overall lipid productivity.

4.5.6 Light

As mentioned above, the accumulation of TAG can protect cells from excess light, either by serving as an electron sink, or as a solvent for shading pigments in lipid bodies. Light conditions can change quite rapidly for a microalgae cell. Apart from the diurnal changes also experienced by land plants, some algae might have to cope with tides, i.e. the presence and absence of several metres of a light-absorbing water column. Furthermore, pelagic microalgae are exposed to the effect of waves, and might be driven from the sunlight-flooded surface to deeper layers (and vice versa) within minutes. It is not surprising that microalgae therefore developed a range of light-protection/adaptation mechanisms, including TAG accumulation.

For example, *Haematococcus pluvialis*, a monocellular, freshwater-dwelling green alga, is well known for its ability to accumulate high amounts of astaxanthin. When subjected to light stress (300 μmol photons m^{-2} s^{-1}), this alga also accumulated high amounts of TAG in cytoplasmic lb (Damiani *et al.*, 2010), increasing the TAG content from 9.2% dry weight under low-light to 19.8% under high-light regime. Additionally, the fraction of saturated FA increased. Furthermore, it is interesting to note that a combination of light stress and nitrogen depletion did not result in an additional increase of TAG content, but cells showed a similar lipid content and FA composition as cells subjected to high-light stress alone (Damiani *et al.*, 2010). A shift in lipid composition as a result of light stress was also observed for the red alga *Tichocarpus crinitus* (Khotimchenko and Yakovleva, 2005). When shifted to high-light conditions, the fraction of TAG on the total lipids increased at the cost of polar lipids. However, since the total lipid yield decreased, the total TAG yield per gram dry weight only improved slightly.

The eustigmatophyte *Nannochloropsis* sp. also showed TAG accumulation as a response to increased light intensity (500 and 700 μmol photons m^{-2} s^{-1}, respectively; Pal *et al.*, 2011; Su *et al.*, 2011). Although in both studies TAG accumulation was triggered by a transfer of the cells to nitrogen-depleted media, higher light irradiation increased the TAG content even more. This is consistent with a model that microalgae cells synthesize TAG as a way to sequester photosynthesis products. Under nitrogen-depleted and therefore non-growth conditions, even low light intensities might be too high to support normal cell metabolism only, which is even more pronounced when the cell experiences higher light intensities – resulting in increased TAG accumulation. The effect of light on *Nannochloropsis* was also demonstrated in an inverse experimental setup (Fisher *et al.*, 1998). Cells grown under high-light conditions, containing a lot of lb, metabolized these after transfer to lower light intensities (Fisher *et al.*, 1998).

Higher light intensities also resulted in a different lipid composition in the diatom *Thalassiosira pseudonana* (Brown *et al.*, 1996). Cells grown at 100 μmol photons m^{-2} s^{-1} in continuous light accumulated TAG, and, as observed for *H. pluvialis*, contained more saturated FA than cells grown under lower light and light/dark cycles (Brown *et al.*, 1996; Damiani *et al.*, 2010).

In summary, many microalgal species across the phylogenetic tree show increased TAG production when experiencing high light conditions. This supports the hypothesis that TAG accumulation serves as a mechanism to protect the cell from an overly active photosynthetic apparatus, as outlined above. The exact definition of 'high light', however, depends on the microalgae species, and strain, and even on the culturing history of the strain. Furthermore, cell density influences how much light an individual cell receives. In very dense cultures, under strong illumination, cells at the surface might suffer from light stress, but shade deeper cells from light. As a result, the culture on average, although strongly illuminated, does not experience high light stress.

When looking at reports about lipid productivity for a particular species, under a particular stress condition, comparability becomes a major problem. The most obvious difference between reports is the actual lipid species that is analysed – total lipid content, TAG (or other lipid class) content, or FA. Then, usually the lipid content (regardless of the lipid species) is reported as an amount (mass) per dry cell weight or per cell. This, however, does not necessarily reflect the usefulness of an algae species as biodiesel feedstock, since high cellular lipid content might be accompanied by low cell density in the culture. The only value the biofuel researcher can be interested in is therefore the yield of TAG per culture volume (or area) and time, i.e. the lipid productivity. However, even the productivity for a given species can (and usually does) differ between two reports. This is due to different culturing regimes, i.e. light, temperature, nutrients, type of culture (batch, pond, shaking), CO_2/air bubbling, etc. The strain or isolate of a given microalgae species can also be different between two laboratories. It is therefore necessary to carefully compare different culturing conditions when a decision has to be made about which strain to grow for TAG production, sometimes resulting in additional experiments comparing strains under identical conditions.

4.6 Genetics of Microalgal Lipid Accumulation

The potential of every cell is defined by its genome. Here, all the enzymes are encoded for every biochemical reaction of which the cell is capable, together with information when and how strongly this reaction will take place. It is therefore not surprising that huge efforts are made to sequence whole microalgal genomes, or to identify key genes responsible for lipid production in microalgae cells. This section will provide a short overview about sequencing projects, and will show examples of important lipid metabolism genes discovered in microalgal genomes. Furthermore, we will discuss the option of creating genetically modified organisms (GMO) for the production of biofuels.

The first algal nuclear genome to be sequenced was the genome of *Cyanidioschyzon merolae* in 2004 (Matsuzaki *et al.*, 2004), a small monocellular red alga with a genome size of approximately only 16.5 mega base pairs (Mbp), encoding 5300 genes. This was shortly followed by the first fully sequenced and annotated nuclear genome sequence of a diatom, *T. pseudonana*, in the same year (Armbrust *et al.*, 2004). In 2012, there are currently ten complete algal genomes publicly available, the green algae and model organism *C. reinhardtii* (Merchant *et al.*, 2007), *Chlorella variabilis* NC64 (Blanc *et al.*, 2010), *Micromonas pusilla* (Worden *et al.*, 2009), *Micromonas* sp. RCC299 (Worden *et al.*, 2009), *Ostreococcus lucimarinus* (Palenik *et al.*, 2007), *Ostreococcus tauri* (Derelle *et al.*, 2006) and *Volvox carteri* (Prochnik *et al.*, 2010), the diatoms *P. tricornutum* (Bowler *et al.*, 2008) and *Thalassiosira pseudonana* (Armbrust *et al.*, 2004) and the red alga *Cyanidioschyzon merolae* (Matsuzaki *et al.*, 2004) (Table 4.3). Apart from nuclear genomes, algae also possess genomes for their

Table 4.3. Completed algal genomes.

Organism	Genome database	Size	Reference
Green algae			
Chlamydomonas reinhardtii	Phytozome (http://www.phytozome.net/chlamy.php)	112 Mb 17,114 genes	Merchant *et al.* (2007)
Chlorella variabilis NC64A	Joint Genome Institute (http://genome.jgi-psf.org/ChlNC64A_1/ChlNC64A_1.home.html)	64 Mb 9,791 genes	Blanc *et al.* (2010)
Micromonas pusilla	Joint Genome Institute (http://genome.jgi-psf.org/MicpuC3/MicpuC3.home.html)	21.9 Mb 10,575 genes	Worden *et al.* (2009)
Micromonas sp. RCC299	Joint Genome Institute (http://genome.jgi-psf.org/MicpuN3/MicpuN3.home.html)	20.9 Mb 10,056 genes	Worden *et al.* (2009)
Ostreococcus lucimarinus	Joint Genome Institute (http://genome.jgi-psf.org/Ost9901_3/Ost9901_3.home.html)	13.2 Mb 7,651 genes	Palenik *et al.* (2007)
Ostreococcus tauri	Joint Genome Institute (http://genome.jgi-psf.org/Ostta4/Ostta4.home.html)	12.6 Mb 7,892 genes	Derelle *et al.* (2006)
Volvox carteri	Phytozome (http://www.phytozome.net/volvox.php)	138 Mb 14,491 genes	Prochnik *et al.* (2010)
Heterokont algae, diatoms and others			
Phaeodactylum tricornutum	Joint Genome Institute (http://genome.jgi-psf.org/Phatr2/Phatr2.home.html)	27.4 Mb 10,402 genes	Bowler *et al.* (2008)
Thalassiosira pseudonana	Joint Genome Institute (http://genome.jgi-psf.org/Thaps3/Thaps3.home.html)	32.4 Mb 11,776 genes	Armbrust *et al.* (2004)
Red algae			
Cyanidioschyzon merolae	Cyanidioschyzon merolae Genome Project (http://merolae.biol.s.u-tokyo.ac.jp)	16.5 Mb 5,331 genes	Matsuzaki *et al.* (2004)

chloroplasts and mitochondria, with a multitude of sequencing projects for these in progress or already complete (see Lu *et al.* (2011) for an overview). With knowledge obtained from much better analysed model organisms such as *A. thaliana*, it is possible to deduct and analyse biochemical pathways in these organisms, e.g. FA and TAG biosynthesis, and peroxisomal beta-oxidation. For example, it was possible to identify the genes necessary for glycerol lipid biosynthesis in *C. reinhardtii* (Riekhof *et al.*, 2005). Additionally, carbon acquisition pathways and carbon hydrate metabolism, which compete with or feed into lipid metabolism, could be identified analysing the genome sequence of the diatom *P. tricornutum* (Bowler *et al.*, 2008). Recent advances in sequencing technologies ('next generation sequencing', see Metzker, 2010 for a review) greatly decreased the costs as well as the time needed for whole genome sequencing, so many more algal genome sequencing projects can be expected to be complete very soon. So far, analyses of the sequenced genomes indicate that, at least for green and red algae, the biosynthetic pathways for the production of FA and TAG seem to be very similar to those present in higher plants (Khozin-Goldberg and Cohen, 2011).

Unfortunately, the genome sequence only tells us what the cell is potentially able to do, not what it 'prepares' to do, or actually is doing in a certain situation. Analysing the transcriptome and the proteome, respectively, can give us insight into how a cell reacts to an external stimulus, such as excessive light, or the lack of nutrients. This again can lead to the identification of key regulators in pathways, enzymes that control levels of metabolites within a cell. Of course, from a biodiesel perspective, conditions that induce the accumulation of TAG are the most interesting. However, given the value of data they can create, large scale transcriptome analyses in microalgae seem to be surprisingly sparse. Until now (2012), there have only been three studies addressing the response of microalgae to nitrogen starvation using this approach. This is probably due to the limited number of fully sequenced algal genomes, as this information greatly facilitates transcriptomics data analysis. This, however, might change in the future. When the model organism *C. reinhardtii* is subjected to nitrogen deprivation, genes encoding enzymes of the beta-oxidation pathway, which are responsible for degradation of FA, and genes associated with TAG biosynthesis are up-regulated (Miller *et al.*, 2010). This is conclusive with the observation that *C. reinhardtii* accumulates lipid bodies under N starvation. However, the abundance of transcripts for genes from the FA biosynthesis pathway was only slightly higher than under control conditions, leading the authors to suspect intensive membrane recycling for TAG accumulation instead of FA *de novo* synthesis. This is surprising, as suppressed cell growth and division should, as outlined in the 'physiology' section of this chapter, result in increased FA biosynthesis. It could be possible, however, that under normal growth conditions FA production is not controlled at the level of enzyme concentration, but by metabolite availability and partitioning. It was found in the same study that transcript levels of a gene encoding for a subunit of a pyruvate dehydrogenase complex were up-regulated under nitrogen-depleted conditions. This enzyme complex converts pyruvate to acetyl-CoA, thereby shuttling metabolites into the FA biosynthesis pathway. Thus, FA production could be increased without up-regulation of the genes encoding the respective biosynthetic machinery.

This hypothesis is supported by a recent study of *C. reinhardtii* under nitrogen limitation, using semi-quantitative reverse transcriptase PCR (Msanne *et al.*, 2012). In this study, the authors confirmed up-regulation of DGAT genes for the production of TAG. However, transcript abundance of genes for proteins from the FA biosynthetic machinery decreased during nitrogen starvation, despite the accumulation of TAG during the observation time (Msanne *et al.*, 2012).

In a second transcriptomics study, the transcriptome of *Dunaliella tertiolecta* was analysed under nitrogen-repleted, nitrogen-depleted and high salinity conditions in order to identify genes for the biosynthesis of storage products, i.e. lipids and starch (Risman-Yazdi *et al.*, 2011). The authors identified and reconstructed pathways of FA, TAG and starch biosynthesis, beta-oxidation and TAG- and starch degradation by comparing the obtained sequence data to the fully sequenced genomes of *C. reinhardtii* and *V. carteri* and metabolic pathway databases. Here, the authors did not rely on a fully sequenced and annotated genome, but used transcriptome data for gene identification instead (Rismani-Yazdi *et al*., 2011).

A third study, investigating the response of the previously unsequenced *Chlorella vulgaris* to nitrogen deprivation, combined transcriptomics with analysis of the changes in protein abundance (Guarnieri *et al*., 2011). Since no complete genome sequence is available for *C. vulgaris*, the authors first created a guiding scaffold from the transcriptome under normal growth conditions. Gene assembly and identification then enabled the analysis of the proteomics data, leading to the identification of differentially expressed proteins under normal growth and nitrogen-depleted conditions. Under the latter conditions, the abundance of proteins for FA biosynthesis increased up to twofold. Proteins for TAG biosynthesis, however, showed an even more pronounced up-regulation, especially DGAT, which is responsible for catalysing the committed step in TAG synthesis, which was up-regulated more than 130-fold (Guarnieri *et al*., 2011). Apart from transcriptomics projects in algae focusing on lipid metabolism and TAG accumulation, several other studies are available that are of potential interest to the biofuels researcher. These include studies on the response of green algae to light stress and day/night cycles, sulfur deprivation, and of diatoms to Si deprivation, and have been reviewed by Schuhmann *et al*. (2012).

Most of our knowledge about lipid metabolism genes from microalgae is derived from high-throughput projects. The list of individual lipid metabolism genes from microalgae that have been characterized functionally is still relatively short. Our knowledge can be best described as fragmented. A recent review provides an overview of genes identified and characterized (Khozin-Goldberg and Cohen, 2011). Elongases and desaturases, necessary for the production of long-chain polyunsaturated FA, seem to be in the focus of research (Khozin-Goldberg and Cohen, 2011). This is probably due to the possibility to express these algae-derived genes in transgenic food plants. By this method, these plants could be enabled to produce beneficial omega-3 and omega-6 FA for human consumption. However, several genes interesting from a biodiesel perspective have been characterized as well. These include several acyltransferases (AT), e.g. acyl-CoA:glycerol-3-phosphate AT, acyl-CoA:lysophosphatidic AT, acyl-CoA:diacylglycerol AT (DGAT) and others (see Khozin-Goldberg and Cohen, 2011 for references).

As a more recent addition to this list, a DGAT from *P. tricornutum* was identified and characterized (Guiheneuf *et al*., 2011). This gene, which encodes an enzyme of the DGAT 1 type (see Lung and Weselake, 2006), is expressed as two different splicing variants. Both forms are up-regulated under nitrogen deprivation, but only the shorter one exhibited DGAT enzyme activity by producing TAG in a yeast complementation assay. Additionally, the fact that both cloned splicing variants differed from the hypothetical transcript predicted from genomic data (Guiheneuf *et al*., 2011), underlines the need of laborious characterization of individual genes, even in the '-omics' era. Another gene that was not covered by the aforementioned review encodes a lipid storage protein in VLC-PUFA *H. pluvialis*. The protein, which seems to be a homologue of the MLDP identified in *C. reinhardtii* (Moellering and Benning, 2010), accumulates under high-light and nitrogen deprivation conditions, i.e. when the alga produces high amounts of lipid bodies. It seems reasonable to assume that the *Haematococcus* protein fulfils a similar role in stabilizing lipid bodies, as its *Chlamydomonas* orthologue.

One major incentive to research the genetic potential of microalgae for the production of TAG is, of course, the prospect to either modify strains by genetic manipulation, or to select for strains using marker-assisted breeding. During the last decade, genetic manipulation of microalgae has made immense progress. Of course, this is mainly due to the development of *C. reinhardtii* as a model organism (Specht *et al*., 2010), but methods developed for this alga have been successfully applied to and improved for other species. *Chlamydomonas reinhardtii, D. salina, Chlorella* sp., *P. tricornutum, Thalassiosira* sp. and many more can now be – more or less – routinely genetically transformed, either as nuclear, chloroplastidial or transient transformants (Hallmann, 2007; Hannon *et al.*, 2010).

The most popular method for microalgal transformation is bombardment with DNA-coated micro-particles (Zaslavskaia *et al.*, 2000). Other options include electroporation (Maruyama *et al.*, 1994), Si carbide whiskers (e.g. Dunahay, 1993), agitation with glass beads in the presence of DNA (e.g. Feng *et al.*, 2009), protoplast transformation (e.g. Hawkins and Nakamura, 1999) and transformation mediated by the bacterium *Agrobacterium tumefaciens* (e.g. Cha *et al.*, 2011; see Hannon *et al.* (2010) for a tabular review and additional references). Together with these methods, a wide range of selection markers were developed for use in microalgae to detect successful transformation (Table 4.4). However, transformation methods have to be developed for every species individually. For example, recombinant gene expression can be low due to codon bias of the respective species, or to poor performance of the chosen promoter. Although the widely used *Cauliflower mosaic virus* 35S promoter (CaMV35S) is successfully used in plant biotechnology and exhibits a strong expression pattern in a wide range of green organisms (Corrado and Karali, 2009), it may be advisable to use an endogenous promoter derived from the target organism directly. A popular promoter for chloroplastidal expression of recombinant genes in *C. reinhardtii* is the promoter of the *psbA* gene (Manuell *et al.*, 2007). An alternative for the same purpose is the combination of the 16S RNA promoter with a 5′ untranslated region (UTR) from the *atpA* gene (Rasala *et al.*, 2011). Inducible promoters are also available for *C. reinhardtii*, e.g. the promoter of the *cyt6* gene, which is induced under copper limitation (Quinn and Merchant, 1995). In diatoms such as *T. pseudonana* and *P. tricornutum*, constitutive promoters from *fcpA* and *fcpB* genes (Zaslavskaia *et al.*, 2000; Poulsen *et al.*, 2006), as well as a nitrate-responsive inducible promoter from the nitrate reductase gene (Poulsen *et al.*, 2006), have been successfully used for recombinant gene expression. Furthermore, promoters from two violaxanthin/chlorophyll a-binding protein (VCP) genes were successfully employed in the transformation of *Nannochloropsis* sp., giving molecular biologists even more tools for the genetic manipulation of microalgae (Kilian *et al.*, 2011).

Table 4.4. Selection markers for algae transformation.

Marker	Reference
Bleomycin/Zeocin/Phleomycin	Apt *et al.* (1996); Lumbreras *et al.* (1998); Fischer *et al.* (1999); Fuhrmann *et al.* (1999); Zaslavskaia *et al.* (2000); Walker *et al.* (2005); Kilian *et al.* (2011)
Spectinomycin	Cerutti *et al.* (1997); Doetsch *et al.* (2001)
Streptomycin	Doetsch *et al.* (2001)
Paromomycin	Sizova *et al.* (2001); Jakobiak *et al.* (2004)
Nourseothricin	Zaslavskaia *et al.* (2000); Poulsen *et al.* (2006)
G418	Dunahay *et al.* (1995); Zaslavskaja *et al.* (2000); Poulsen *et al.* (2006)
Hygromycin	Berthold *et al.* (2002); Anila *et al.* (2011)
Chloramphenicol	Tang *et al.* (1995); Geng *et al.* (2003)
Kanamycin	Tan *et al.* (2005)
GFP	Fuhrmann *et al.* (1999); Zaslavskaia *et al.* (2000); Ender *et al.* (2002); Franklin *et al.* (2002); Poulsen and Kroger (2005); Anila *et al.* (2011)
Beta-glucuronidase	Chow and Tung (1999); El-Sheekh (1999); Falciatore *et al.* (1999); Chen *et al.* (2001); Tan *et al.* (2005)
Luciferase	Jarvis and Brown (1991); Falciatore *et al.* (1999); Fuhrmann *et al.* (2004)
Beta-galactosidase	Gan *et al.* (2003); Jiang *et al.* (2003)
Norflurazon	Steinbrenner and Sandmann (2006); Huang *et al.* (2008)
BASTA/Phosphinothricin/Glufosinate	Tan *et al.* (2005); Feng *et al.* (2009)

However, one has to keep in mind that the development of transformation strategies, selection markers and promoter/terminator sequences for these organisms is a recent development. Furthermore, as indicated above, all these parameters have to be adjusted to every species, or even every strain. It is therefore not surprising that efforts of genetic engineering of microalgae in order to increase lipid yield are still limited and success is sparse. Since there were several detailed reviews on genetic metabolic engineering of microalgae published recently (Courchesne *et al.*, 2009; Hannon *et al.*, 2010; Radakovits *et al.*, 2010; Zeng *et al.*, 2011; Schuhmann *et al.*, 2012), in this chapter we discuss the main principles and only provide selected examples.

Principally, genetic engineering can address one or more pathways related to lipid metabolism in order to increase lipid yield: (i) improve light harvesting and carbon fixation, which provide energy and building blocks; (ii) increase FA biosynthesis rates; (iii) increase TAG biosynthesis; (iv) manipulate lipid storage in order to alleviate harvesting or decrease TAG susceptibility for degradation; (v) decrease TAG and FA degradation; and (vi) down-regulate competing pathways such as starch biosynthesis. An example for improving light harvesting characteristics are light harvesting complex mutants of *C. reinhardtii* (Polle *et al.*, 2003, Mussgnug *et al.*, 2007) which showed reduced antenna sizes. Although this reduction leads to a *decrease* of the energy uptake of an individual cell, it also means that the light can penetrate much deeper into the algal culture. In a culture of wild-type cells, most of the irradiance is absorbed by cells close to the surface. However, these cells then actually experience an excess of energy and have to employ protection mechanisms such as chlorophyll fluorescence and non-photochemical quenching, thereby 'wasting' energy. In culture of mutant cells, on the other hand, less energy is wasted and therefore more is available for biosynthetic processes in the whole culture (Polle *et al.*, 2003; Mussgnug *et al.*, 2007).

Manipulating FA and TAG biosynthesis by over-expression of key proteins such as ACCase or DGAT seems to be the most obvious strategy, and was successfully employed in vascular plants, yeasts and fungi (Courchesne *et al.*, 2009). However, reports about experiments in microalgae are surprisingly sparse, probably due to a lack of identified genes (see above). ACCase from the diatom *Cyclotella cryptica* was over-expressed in the same organism, and additionally in another diatom, *Navicula sapuvila* (Dunahay *et al.*, 1996). However, no increased lipid content could be reported. In another diatom, *P. tricornutum*, a thioesterase was over-expressed that is usually up-regulated under nitrogen-depleted conditions (Gong *et al.*, 2011). This thioesterase was suggested to be involved in the release of FA from the biosynthetic machinery in the chloroplast. Its over-expression did not alter the FA profile of the diatom, but increased total FA content by 72%, although it remains unclear if these FA were present as free FA or sequestered as TAG in the organism (Gong *et al.*, 2011).

The storage of TAG was successfully altered by suppressing the production of a major lipid droplet protein (MLDP) in *C. reinhardtii* (Moellering and Benning, 2010). In mutant strains where MLDP levels were reduced, lipid droplet size increased. Although total TAG contents did not increase, this might have potential impacts on TAG harvesting and stability.

So far, TAG and FA degradation pathways such as beta-oxidation have not been targeted by genetic engineering in microalgae, probably due to a lack of gene sequence data. However, there are multiple reports addressing the relationship of lipid biosynthesis and competing starch biosynthesis in microalgae, mostly *C. reinhardtii*. Several reports claim increased TAG accumulation in mutant strains with deficiencies in starch biosynthesis (Wang *et al.*, 2009; Li, Y. *et al.*, 2010; Work *et al.*, 2010). However, careful analysis of the genetic backgrounds of the strains used showed that decreased starch and increased TAG content might not necessarily be linked to each other (Siaut *et al.*, 2011), which was also demonstrated using over-expression lines showing increased starch and TAG content (Work *et al.*, 2010). In *Chlorella pyrenoidosa*, however, increased TAG accumulation could be connected to reduced starch biosynthesis (Ramazanov and Ramazanov, 2006).

Apart from addressing a single pathway such as FA biosynthesis or TAG degradation, it might be feasible to modify several pathways simultaneously through the over-expression/silencing of transcription factors (Courchesne *et al.*, 2009). However, not much knowledge is available about transcription factors from microalgae. In summary, genetic engineering of microalgae for increased TAG accumulation is still in its infancy, and achievements are very limited.

4.7 Concluding Remarks

In the previous sections, we have reviewed the biochemistry, physiology and genetics of algal lipid accumulation. It has to be noted that our knowledge in these three fields still has to be described as rudimentary at best. Of course, this is at least partly due to the arbitrary definition of the word 'microalgae', a definition that groups together species that are located on very distant branches on the tree of life, such as the green alga *C. reinhardtii* and the diatom *P. tricornutum*. However, several fully sequenced genomes of microalgae are now available, and even much more fragmental genetic data are present for non-sequenced organisms. Therefore, it is feasible to expect that our knowledge about lipid metabolism in algae will be refined during the next years.

Until now, establishing TAG production in microalgae is still largely based on 'trial and error', where lipid accumulation conditions have to be developed for each strain individually. So far, it is impossible to predict if one strain, species, or even phylum will yield high or low lipid production strains. It is also necessary to keep in mind what defines a 'suitable' strain for biodiesel production:

- *Yield per time unit*: for commercial purposes, the overall yield per time unit is decisive, not the maximum achievable TAG content of each cell. A strain with medium TAG content, but fast growth (e.g. *Nannochloropsis oculata*) will likely be preferred over a strain that accumulates large amounts of lipids, but has very slow growth (e.g. *B. braunii*). The ideal strain therefore would combine high TAG content with rapid growth.
- *Robustness*: contamination by low performance strains and species decreases the yield. Strains that thrive under a wide range of conditions and exhibit rapid growth, should be preferred.
- *Suitable FA and lipid composition*: saturated, short-to-medium chain FA sequestered in TAG are preferred for biodiesel production. Of course, this changes if the focus is, for example, on the production of high-value health products such as omega-3 FA.
- *Harvestability and dewatering*: harvesting and dewatering seem to be two of the main obstacles for profitable large-scale culture of microalgae for biofuel production. As unicellular organism living dispersed in liquid environments, microalgae tend to be difficult to harvest economically on a large scale. Strains that can be induced to support harvesting by e.g. clumping/autoflocculation therefore will be easier to grow than others.
- *Oil extraction and performance in downstream processing*: the easier TAG can be isolated from the rest of the cell content, the better is the performance of the strain in postharvesting processing. For example, the rigid silica shell of some diatoms might be incompatible with some isolation methods, due to inacceptable abrasion. Similarly, pretreatment of some microalgae with strong cell walls (e.g. *Nannochloropsis* sp.) may be required to extract maximum TAG. Harvesting and TAG isolation might be greatly facilitated by strains that excrete lipids into the growth media, which then can be easily separated from the aqueous phase due to their hydrophobicity. However, according to our knowledge, the feasibility of this approach has never been demonstrated so far.

Fortunately, the variability of microalgae seems to be vast, and we have only targeted a tiny fraction of all species and strains on Earth. Together with improved knowledge about microalgae biochemistry, physiology and genetics, this will enable us to turn microalgae into a reliable, economical and ecologically sustainable crop for biodiesel production.

References

Abbadi, A., Domergue, F., Bauer, J., Napier, J.A., Welti, *et al.* (2004) Biosynthesis of very-long-chain polyunsaturated fatty acids in transgenic oilseeds: constraints on their accumulation. *Plant Cell* 16, 2734–2748.

Agrawal, S. and Striepen, B. (2010) More membranes, more proteins: complex protein import mechanisms into secondary plastids. *Protist* 161, 672–687.

Alonso, D.L., Belarbi, E.H., Fernandez-Sevilla, J.M., Rodriguez-Ruiz, J. and Grima, E.M. (2000) Acyl lipid composition variation related to culture age and nitrogen concentration in continuous culture of the microalga *Phaeodactylum tricornutum*. *Phytochemistry* 54, 461–471.

Anila, N., Chandrashekar, A., Ravishankar, G.A. and Sarada, R. (2011) Establishment of *Agrobacterium tumefaciens*-mediated genetic transformation in *Dunaliella bardawil*. *European Journal of Phycology* 46, 36–44.

Apt, K.E., KrothPancic, P.G. and Grossman, A.R. (1996) Stable nuclear transformation of the diatom *Phaeodactylum tricornutum*. *Molecular and General Genetics* 252, 572–579.

Armbrust, E.V., Berges, J.A., Bowler, C., Green, B.R., Martinez, D. *et al.* (2004) The genome of the diatom *Thalassiosira pseudonana*: ecology, evolution, and metabolism. *Science* 306, 79–86.

Athenstaedt, K. and Daum, G. (2006) The life cycle of neutral lipids: synthesis, storage and degradation. *Cellular and Molecular Life Sciences* 63, 1355–1369.

Azachi, M., Sadka, A., Fisher, M., Goldshlag, P., Gokhman, I. *et al.* (2002) Salt induction of fatty acid elongase and membrane lipid modifications in the extreme halotolerant alga *Dunaliella salina*. *Plant Physiology* 129, 1320–1329.

Banerjee, U.C., Banerjee, A., Sharma, R. and Chisti, Y. (2002) *Botryococcus braunii*: a renewable source of hydrocarbons and other chemicals. *Critical Reviews in Biotechnology* 22, 245–279.

Bao, X., Focke, M., Pollard, M. and Ohlrogge, J. (2000) Understanding *in vivo* carbon precursor supply for fatty acid synthesis in leaf tissue. *Plant Journal* 22, 39–50.

Barber, J. (2009) Photosynthetic energy conversion: natural and artificial. *Chemical Society Reviews* 38, 185–196.

Berthold, P., Schmitt, R. and Mages, W. (2002) An engineered *Streptomyces hygroscopicus* aph 7′ gene mediates dominant resistance against hygromycin B in *Chlamydomonas reinhardtii*. *Protist* 153, 401–412.

Bigogno, C., Khozin-Goldberg, I., Boussiba, S., Vonshak, A. and Cohen, Z. (2002) Lipid and fatty acid composition of the green oleaginous alga *Parietochloris incisa*, the richest plant source of arachidonic acid. *Phytochemistry* 60, 497–503.

Blanc, G., Duncan, G., Agarkova, I., Borodovsky, M., Gurnon, J. *et al.* (2010) The *Chlorella variabilis* NC64A genome reveals adaptation to photosymbiosis, coevolution with viruses, and cryptic sex. *Plant Cell* 22, 2943–2955.

Bowler, C., Allen, A.E., Badger, J.H., Grimwood, J., Jabbari, K. *et al.* (2008) The *Phaeodactylum* genome reveals the evolutionary history of diatom genomes. *Nature* 456, 239–244.

Brown, M.R., Dunstan, G.A., Norwood, S.J. and Miller, K.A. (1996) Effects of harvest stage and light on the biochemical composition of the diatom *Thalassiosira pseudonana*. *Journal of Phycology* 32, 64–73.

Cerutti, H., Johnson, A.M., Gillham, N.W. and Boynton, J.E. (1997) A eubacterial gene conferring spectinomycin resistance on *Chlamydomonas reinhardtii*: Integration into the nuclear genome and gene expression. *Genetics* 145, 97–110.

Cha, T.S., Chen, C.F., Yee, W., Aziz, A. and Loh, S.H. (2011) Cinnamic acid, coumarin and vanillin: alternative phenolic compounds for efficient *Agrobacterium*-mediated transformation of the unicellular green alga, *Nannochloropsis* sp. *Journal of Microbiological Methods* 84, 430–434.

Chen, G.Q., Jiang, Y. and Chen, F. (2008) Salt-induced alterations in lipid composition of diatom *Nitzschia laevis* (Bacillariophyceae) under heterotrophic culture condition. *Journal of Phycology* 44, 1309–1314.

Chen, M., Tang, H.Y., Ma, H.Z., Holland, T.C., Ng, K.Y.S. *et al.* (2011) Effect of nutrients on growth and lipid accumulation in the green algae *Dunaliella tertiolecta*. *Bioresource Technology* 102, 1649–1655.

Chen, Y., Wang, Y.Q., Sun, Y.R., Zhang, L.M. and Li, W.B. (2001) Highly efficient expression of rabbit neutrophil peptide-1 gene in *Chlorella ellipsoidea* cells. *Current Genetics* 39, 365–370.

Chisti, Y. (2007) Biodiesel from microalgae. *Biotechnology Advances* 25, 294–306.

Chow, K.C. and Tung, W.L. (1999) Electrotransformation of *Chlorella vulgaris*. *Plant Cell Reports* 18, 778–780.

Converti, A., Casazza, A.A., Ortiz, E.Y., Perego, P. and Del Borghi, M. (2009) Effect of temperature and nitrogen concentration on the growth and lipid content of *Nannochloropsis oculata* and *Chlorella vulgaris* for biodiesel production. *Chemical Engineering and Processing* 48, 1146–1151.

Corrado, G. and Karali, M. (2009) Inducible gene expression systems and plant biotechnology. *Biotechnology Advances* 27, 733–743.
Courchesne, N.M.D., Parisien, A., Wang, B. and Lan, C.Q. (2009) Enhancement of lipid production using biochemical, genetic and transcription factor engineering approaches. *Journal of Biotechnology* 141, 31–41.
Dahlqvist, A., Stahl, U., Lenman, M., Banas, A., Lee, M. *et al.* (2000) Phospholipid: diacylglycerol acyltransferase: an enzyme that catalyzes the acyl-CoA-independent formation of triacylglycerol in yeast and plants. *Proceedings of the National Academy of Sciences of the United States of America* 97, 6487–6492.
Damiani, M.C., Popovich, C.A., Constenla, D. and Leonardi, P.I. (2010) Lipid analysis in *Haematococcus pluvialis* to assess its potential use as a biodiesel feedstock. *Bioresource Technology* 101, 3801–3807.
Derelle, E., Ferraz, C., Rombauts, S., Rouze, P., Worden, A.Z. *et al.* (2006) Genome analysis of the smallest free-living eukaryote *Ostreococcus tauri* unveils many unique features. *Proceedings of the National Academy of Sciences of the United States of America* 103, 11647–11652.
Doetsch, N.A., Favreau, M.R., Kuscuoglu, N., Thompson, M.D. and Hallick, R.B. (2001) Chloroplast transformation in *Euglena gracilis*: splicing of a group III twintron transcribed from a transgenic psbK operon. *Current Genetics* 39, 49–60.
Dunahay, T.G. (1993) Transformation of *Chlamydomonas reinhardtii* with silicon-carbide whiskers. *Biotechniques* 15, 452.
Dunahay, T.G., Jarvis, E.E. and Roessler, P.G. (1995) Genetic transformation of the diatoms *Cyclotella cryptica* and *Navicula saprophila*. *Journal of Phycology* 31, 1004–1012.
Dunahay, T.G., Jarvis, E.E., Dais, S.S. and Roessler, P.G. (1996) Manipulation of microalgal lipid production using genetic engineering. *Applied Biochemistry and Biotechnology* 57, 223–231.
Eastmond, P.J. (2004) Glycerol-insensitive *Arabidopsis* mutants: *gli1* seedlings lack glycerol kinase, accumulate glycerol and are more resistant to abiotic stress. *Plant Journal* 37, 617–625.
El-Sheekh, M.M. (1999) Stable transformation of the intact cells of *Chlorella kessleri* with high velocity microprojectiles. *Biologia Plantarum* 42, 209–216.
Ender, F., Godl, K., Wenzl, S. and Sumper, M. (2002) Evidence for autocatalytic cross-linking of hydroxyproline-rich glycoproteins during extracellular matrix assembly in volvox. *Plant Cell* 14, 1147–1160.
Fahy, E., Subramaniam, S., Murphy, R.C., Nishijima, M., Raetz, C.R. *et al.* (2009) Update of the LIPID MAPS comprehensive classification system for lipids. *Journal of Lipid Research* 50(Suppl.), S9–14.
Falciatore, A., Casotti, R., Leblanc, C., Abrescia, C. and Bowler, C. (1999) Transformation of nonselectable reporter genes in marine diatoms. *Marine Biotechnology* 1, 239–251.
Fan, J.L., Andre, C. and Xu, C.C. (2011) A chloroplast pathway for the de novo biosynthesis of triacylglycerol in *Chlamydomonas reinhardtii*. *Febs Letters* 585, 1985–1991.
Feng, S.Y., Xue, L.X., Liu, H.T. and Lu, P.J. (2009) Improvement of efficiency of genetic transformation for *Dunaliella salina* by glass beads method. *Molecular Biology Reports* 36, 1433–1439.
Fernandez, E. and Galvan, A. (2008) Nitrate assimilation in *Chlamydomonas*. *Eukaryotic Cell* 7, 555–559.
Fischer, H., Robl, I., Sumper, M. and Kroger, N. (1999) Targeting and covalent modification of cell wall and membrane proteins heterologously expressed in the diatom *Cylindrotheca fusiformis* (Bacillariophyceae). *Journal of Phycology* 35, 113–120.
Fisher, T., Berner, T., Iluz, D. and Dubinsky, Z. (1998) The kinetics of the photoacclimation response of *Nannochloropsis* sp. (Eustigmatophyceae): a study of changes in ultrastructure and PSU density. *Journal of Phycology* 34, 818–824.
Franklin, S., Ngo, B., Efuet, E. and Mayfield, S.P. (2002) Development of a GFP reporter gene for *Chlamydomonas reinhardtii* chloroplast. *Plant Journal* 30, 733–744.
Fuhrmann, M., Oertel, W. and Hegemann, P. (1999) A synthetic gene coding for the green fluorescent protein (GFP) is a versatile reporter in *Chlamydomonas reinhardtii*. *Plant Journal* 19, 353–361.
Fuhrmann, M., Hausherr, A., Ferbitz, L., Schodl, T., Heitzer, M. and Hegemann, P. (2004) Monitoring dynamic expression of nuclear genes in *Chlamydomonas reinhardtii* by using a synthetic luciferase reporter gene. *Plant Molecular Biology* 55, 869–881.
Gan, S.Y., Qin, S., Othman, R.Y. and Phang, S.M. (2003) Transient expression of lacZ in particle bombarded *Gracilaria changii* (Gracilariales, Rhodophyta). *Journal of Applied Phycology* 15, 345–349.
Geigenberger, P., Kolbe, A. and Tiessen, A. (2005) Redox regulation of carbon storage and partitioning in response to light and sugars. *Journal of Experimental Botany* 56, 1469–1479.
Geng, D.G., Wang, Y.Q., Wang, P., Li, W.B. and Sun, Y.R. (2003) Stable expression of hepatitis B surface antigen gene in *Dunaliella salina* (Chlorophyta). *Journal of Applied Phycology* 15, 451–456.

Goepfert, S. and Poirier, Y. (2007) Beta-oxidation in fatty acid degradation and beyond. *Current Opinion in Plant Biology* 10, 245–251.
Gong, Y.M., Guo, X.J., Wan, X., Liang, Z. and Jiang, M. (2011) Characterization of a novel thioesterase (PtTE) from *Phaeodactylum tricornutum*. *Journal of Basic Microbiology* 51, 666–672.
Goodson, C., Roth, R., Wang, Z.T. and Goodenough, U. (2011) Structural correlates of cytoplasmic and chloroplast lipid body synthesis in *Chlamydomonas reinhardtii* and stimulation of lipid body production with acetate boost. *Eukaryotic Cell* 10, 1592–1606.
Graham, I.A. (2008) Seed storage oil mobilization. *Annual Review of Plant Biology* 59, 115–142.
Griffiths, M.J. and Harrison, S.T.L. (2009) Lipid productivity as a key characteristic for choosing algal species for biodiesel production. *Journal of Applied Phycology* 21, 493–507.
Guarnieri, M.T., Nag, A., Smolinski, S.L., Darzins, A., Seibert, M. *et al.* (2011) Examination of triacylglycerol biosynthetic pathways via *de novo* transcriptomic and proteomic analyses in an unsequenced microalga. *PLoS ONE* 6, 10.
Guedes, A.C., Amaro, H.M. and Malcata, F.X. (2011) Microalgae as sources of carotenoids. *Marine Drugs* 9, 625–644.
Guiheneuf, F., Leu, S., Zarka, A., Khozin-Goldberg, I., Khalilov, I. *et al.* (2011) Cloning and molecular characterization of a novel acyl-CoA:diacylglycerol acyltransferase 1-like gene (PtDGAT1) from the diatom *Phaeodactylum tricornutum*. *FEBS Journal* 278, 3651–3666.
Guschina, I.A. and Harwood, J.L. (2009) Algal lipids and effect of the environment on their biochemistry. *Lipids in Aquatic Ecosystems* 1, 24.
Hallmann, A. (2007) Algal transgenics and biotechnology. *Transgenic Plant Journal* 1, 81–98.
Hannon, M., Gimpel, J., Tran, M., Rasala, B. and Mayfield, S. (2010) Biofuels from algae: challenges and potential. *Biofuels* 1, 763–784.
Hawkins, R.L. and Nakamura, M. (1999) Expression of human growth hormone by the eukaryotic alga, *Chlorella*. *Current Microbiology* 38, 335–341.
Hockin, N.L., Mock, T., Mulholland, F., Kopriva, S. and Malin, G. (2012) The response of diatom central carbon metabolism to nitrogen starvation is different from that of green algae and higher plants. *Plant Physiology* 158, 299–312.
Hu, Q., Sommerfeld, M., Jarvis, E., Ghirardi, M., Posewitz, M. *et al.* (2008) Microalgal triacylglycerols as feedstocks for biofuel production: perspectives and advances. *Plant Journal* 54, 621–639.
Huang, J.C., Liu, J., Li, Y.T. and Chen, F. (2008) Isolation and characterization of the phytoene desaturase gene as a potential selective marker for genetic engineering of the astaxanthin-producing green alga *Chlorella zofingiensis* (Chlorophyta). *Journal of Phycology* 44, 684–690.
Illman, A.M., Scragg, A.H. and Shales, S.W. (2000) Increase in *Chlorella* strains calorific values when grown in low nitrogen medium. *Enzyme and Microbial Technology* 27, 631–635.
Jakobiak, T., Mages, W., Scharf, B., Babinger, P., Stark, K. *et al.* (2004) The bacterial paromomycin resistance gene, *aphH*, as a dominant selectable marker in *Volvox carteri*. *Protist* 155, 381–393.
Jarvis, E.E. and Brown, L.M. (1991) Transient expression of firefly luciferase in protoplasts of the green-alga *Chlorella ellipsoidea*. *Current Genetics* 19, 317–321.
Jiang, P., Qin, S. and Tseng, C.K. (2003) Expression of the *lacZ* reporter gene in sporophytes of the seaweed *Laminaria japonica* (Phaeophyceae) by gametophyte-targeted transformation. *Plant Cell Reports* 21, 1211–1216.
Khotimchenko, S.V. and Yakovleva, I.M. (2005) Lipid composition of the red alga *Tichocarpus crinitus* exposed to different levels of photon irradiance. *Phytochemistry* 66, 73–79.
Khozin, I., Adlerstein, D., Bigongo, C., Heimer, Y.M. and Cohen, C. (1997) Elucidation of the biosynthesis of eicosapentaenoic acid in the microalga *Porphyridium cruentum*. 2. Studies with radiolabeled precursors. *Plant Physiology* 114, 223–230.
Khozin-Goldberg, I. and Cohen, Z. (2006) The effect of phosphate starvation on the lipid and fatty acid composition of the fresh water eustigmatophyte *Monodus subterraneus*. *Phytochemistry* 67, 696–701.
Khozin-Goldberg, I. and Cohen, Z. (2011) Unraveling algal lipid metabolism: Recent advances in gene identification. *Biochimie* 93, 91–100.
Khozin-Goldberg, I., Shrestha, P. and Cohen, Z. (2005) Mobilization of arachidonyl moieties from triacylglycerols into chloroplastic lipids following recovery from nitrogen starvation of the microalga *Parietochloris incisa*. *Biochimica et Biophysica Acta – Molecular and Cell Biology of Lipids* 1738, 63–71.
Khozin-Goldberg, I., Iskandarov, U. and Cohen, Z. (2011) LC-PUFA from photosynthetic microalgae: occurrence, biosynthesis, and prospects in biotechnology. *Applied Microbiology and Biotechnology* 91, 905–915.

Kilian, O., Benemann, C.S.E., Niyogi, K.K. and Vick, B. (2011) High-efficiency homologous recombination in the oil-producing alga *Nannochloropsis* sp. *Proceedings of the National Academy of Sciences of the United States of America* 108, 21265–21269.
Li, X., Hu, H.Y., Gan, K. and Sun, X.Y. (2010) Effects of different nitrogen and phosphorus concentrations on the growth, nutrient uptake, and lipid accumulation of a freshwater microalga *Scenedesmus* sp. *Bioresource Technology* 101, 5494–5500.
Li, Y., Horsman, M., Wang, B., Wu, N. and Lan, C.Q. (2008) Effects of nitrogen sources on cell growth and lipid accumulation of green alga *Neochloris oleoabundans*. *Applied Microbiology and Biotechnology* 81, 629–636.
Li, Y., Han, D., Hu, G., Dauvillee, D., Sommerfeld, M. *et al.* (2010) *Chlamydomonas* starchless mutant defective in ADP-glucose pyrophosphorylase hyper-accumulates triacylglycerol. *Metabolic Engineering* 12, 387–391.
Li, Y., Han, D., Sommerfeld, M. and Hu, Q. (2011) Photosynthetic carbon partitioning and lipid production in the oleaginous microalga *Pseudochlorococcum* sp. (Chlorophyceae) under nitrogen-limited conditions. *Bioresource Technology* 102, 123–129.
Liang, Y., Beardall, J. and Heraud, P. (2006) Changes in growth, chlorophyll fluorescence and fatty acid composition with culture age in batch cultures of *Phaeodactylum tricornutum* and *Chaetoceros muelleri* (Bacillariophyceae). *Botanica Marina* 49, 165–173.
Lu, J., Sheahan, C. and Fu, P.C. (2011) Metabolic engineering of algae for fourth generation biofuels production. *Energy and Environmental Science* 4, 2451–2466.
Lumbreras, V., Stevens, D.R. and Purton, S. (1998) Efficient foreign gene expression in *Chlamydomonas reinhardtii* mediated by an endogenous intron. *Plant Journal* 14, 441–447.
Lung, S.C. and Weselake, R.J. (2006) Diacylglycerol acyltransferase: a key mediator of plant triacylglycerol synthesis. *Lipids* 41, 1073–1088.
Lynn, S.G., Kilham, S.S., Kreeger, D.A. and Interlandi, S.J. (2000) Effect of nutrient availability on the biochemical and elemental stoichiometry in the freshwater diatom *Stephanodiscus minutulus* (Bacillariophyceae). *Journal of Phycology* 36, 510–522.
Mandal, S. and Mallick, N. (2009) Microalga *Scenedesmus obliquus* as a potential source for biodiesel production. *Applied Microbiology and Biotechnology* 84, 281–291.
Manuell, A.L., Beligni, M.V., Elder, J.H., Siefker, D.T., Tran, M. *et al.* (2007) Robust expression of a bioactive mammalian protein in *Chlamydomonas* chloroplast. *Plant Biotechnology Journal* 5, 402–412.
Maruyama, M., Horakova, I., Honda, H., Xing, X.H., Shiragami, N. *et al.* (1994) Introduction of foreign DNA into *Chlorella saccharophila* by Electroporation. *Biotechnology Techniques* 8, 821–826.
Matos, A.R. and Pham-Thi, A.T. (2009) Lipid deacylating enzymes in plants: old activities, new genes. *Plant Physiology and Biochemistry* 47, 491–503.
Matsuzaki, M., Misumi, O., Shin-I, T., Maruyama, S., Takahara, M. *et al.* (2004) Genome sequence of the ultrasmall unicellular red alga *Cyanidioschyzon merolae* 10D. *Nature* 428, 653–657.
Merchant, S.S., Prochnik, S.E., Vallon, O., Harris, E.H., Karpowicz, S.J. *et al.* (2007) The *Chlamydomonas* genome reveals the evolution of key animal and plant functions. *Science* 318, 245–251.
Metzker, M.L. (2010) Next generation technologies: basics and applications. *Environmental and Molecular Mutagenesis* 51, 691–691.
Miller, R., Wu, G.X., Deshpande, R.R., Vieler, A., Gartner, K. *et al.* (2010) Changes in transcript abundance in *Chlamydomonas reinhardtii* following nitrogen deprivation predict diversion of metabolism. *Plant Physiology* 154, 1737–1752.
Moellering, E.R. and Benning, C. (2010) RNA interference silencing of a major lipid droplet protein affects lipid droplet size in *Chlamydomonas reinhardtii*. *Eukaryotic Cell* 9, 97–106.
Msanne, J., Xu, D., Konda, A.R., Casas-Mollano, J.A., Awada, T. *et al.* (2012) Metabolic and gene expression changes triggered by nitrogen deprivation in the photoautotrophically grown microalgae *Chlamydomonas reinhardtii* and *Coccomyxa* sp. C-169. *Phytochemistry* 75, 50–59.
Murphy, D.J. (2001) The biogenesis and functions of lipid bodies in animals, plants and microorganisms. *Progress in Lipid Research* 40, 325–438.
Mussgnug, J.H., Thomas-Hall, S., Rupprecht, J., Foo, A., Klassen, V. *et al.* (2007) Engineering photosynthetic light capture: impacts on improved solar energy to biomass conversion. *Plant Biotechnology Journal* 5, 802–814.
Napier, J.A. (2007) The production of unusual fatty acids in transgenic plants. *Annual Review of Plant Biology* 58, 295–319.
Napier, J.A. and Graham, I.A. (2010) Tailoring plant lipid composition: designer oilseeds come of age. *Current Opinion in Plant Biology* 13, 330–337.

Nguyen, H.M., Baudet, M., Cuine, S., Adriano, J.M., Barthe, D. *et al.* (2011) Proteomic profiling of oil bodies isolated from the unicellular green microalga *Chlamydomonas reinhardtii*: with focus on proteins involved in lipid metabolism. *Proteomics* 11, 4266–4273.
Ohlrogge, J. and Browse, J. (1995) Lipid biosynthesis. *Plant Cell* 7, 957–970.
Osborn, M.J., Zeleznick, L.D., Horecker, B.L., Rothfield, L. and Rosen, S.M. (1964) Lipopolysaccharide of gram-negative cell wall – biosynthesis of complex heteropolysaccharide occurs by successive addition of specific sugar residues. *Science* 145, 783–789.
Pal, D., Khozin-Goldberg, I., Cohen, Z. and Boussiba, S. (2011) The effect of light, salinity, and nitrogen availability on lipid production by *Nannochloropsis* sp. *Applied Microbiology and Biotechnology* 90, 1429–1441.
Palenik, B., Grimwood, J., Aerts, A., Rouze, P., Salamov, A. *et al.* (2007) The tiny eukaryote *Ostreococcus* provides genomic insights into the paradox of plankton speciation. *Proceedings of the National Academy of Sciences of the United States of America* 104, 7705–7710.
Polle, J.E., Kanakagiri, S.D. and Melis, A. (2003) tla1, a DNA insertional transformant of the green alga *Chlamydomonas reinhardtii* with a truncated light-harvesting chlorophyll antenna size. *Planta* 217, 49–59.
Poulsen, N. and Kroger, N. (2005) A new molecular tool for transgenic diatoms – control of mRNA and protein biosynthesis by an inducible promoter-terminator cassette. *FEBS Journal* 272, 3413–3423.
Poulsen, N., Chesley, P.M. and Kroger, N. (2006) Molecular genetic manipulation of the diatom *Thalassiosira pseudonana* (Bacillariophyceae). *Journal of Phycology* 42, 1059–1065.
Prochnik, S.E., Umen, J., Nedelcu, A.M., Hallmann, A., Miller, S.M. *et al.* (2010) Genomic analysis of organismal complexity in the multicellular green alga *Volvox carteri*. *Science* 329, 223–226.
Qian, K.X. and Borowitzka, M.A. (1993) Light and nitrogen deficiency effects on the growth and composition of *Phaeodactylum tricornutum*. *Applied Biochemistry and Biotechnology* 38, 93–103.
Quettier, A.L., Shaw, E. and Eastmond, P.J. (2008) SUGAR-DEPENDENT6 encodes a mitochondrial flavin adenine dinucleotide-dependent glycerol-3-P dehydrogenase, which is required for glycerol catabolism and postgerminative seedling growth in *Arabidopsis*. *Plant Physiology* 148, 519–528.
Quinn, J.M. and Merchant, S. (1995) Two copper-responsive elements associated with the *Chlamydomonas* Cyc6 gene-function as targets for transcriptional activators. *Plant Cell* 7, 623–638.
Radakovits, R., Jinkerson, R.E., Darzins, A. and Posewitz, M.C. (2010) Genetic engineering of algae for enhanced biofuel production. *Eukaryotic Cell* 9, 486–501.
Raghavan, G., Haridevi, C.K. and Gopinathan, C.P. (2008) Growth and proximate composition of the *Chaetoceros calcitrans* f. *pumilus* under different temperature, salinity and carbon dioxide levels. *Aquaculture Research* 39, 1053–1058.
Ramazanov, Z. and Ramazanov, A. (2006) Isolation and characterization of a starchless mutant of *Chlorella pyrenoidosa* STL-PI with a high growth rate, and high protein and polyunsaturated fatty acid content. *Phycological Research* 54, 255–259.
Rasala, B.A., Muto, M., Sullivan, J. and Mayfield, S.P. (2011) Improved heterologous protein expression in the chloroplast of *Chlamydomonas reinhardtii* through promoter and 5′ untranslated region optimization. *Plant Biotechnology Journal* 9, 674–683.
Raven, J.A. (2010) Inorganic carbon acquisition by eukaryotic algae: four current questions. *Photosynthesis Research* 106, 123–134.
Rawsthorne, S. (2002) Carbon flux and fatty acid synthesis in plants. *Progress in Lipid Research* 41, 182–196.
Reitan, K.I., Rainuzzo, J.R. and Olsen, Y. (1994) Effect of nutrient limitation on fatty-acid and lipid-content of marine microalgae. *Journal of Phycology* 30, 972–979.
Ribalet, F., Vidoudez, C., Cassin, D., Pohnert, G., Ianora, A. *et al.* (2009) High plasticity in the production of diatom-derived polyunsaturated aldehydes under nutrient limitation: physiological and ecological implications. *Protist* 160, 444–451.
Riekhof, W.R., Sears, B.B. and Benning, C. (2005) Annotation of genes involved in glycerolipid biosynthesis in *Chlamydomonas reinhardtii*: discovery of the betaine lipid synthase BTA1(Cr). *Eukaryotic Cell* 4, 242–252.
Rismani-Yazdi, H., Haznedaroglu, B.Z., Bibby, K. and Peccia, J. (2011) Transcriptome sequencing and annotation of the microalgae *Dunaliella tertiolecta*: pathway description and gene discovery for production of next-generation biofuels. *BMC Genomics* 12, 148. Available at: http://www.biomedcentral.com/content/pdf/ 1471-2164-12-148.pdf (accessed 25 March 2012).
Roessler, P.G. (1988) Changes in the activities of various lipid and carbohydrate biosynthetic-enzymes in the diatom *Cyclotella cryptica* in response to silicon deficiency. *Archives of Biochemistry and Biophysics* 267, 521–528.

Sasaki, Y. and Nagano, Y. (2004) Plant acetyl-CoA carboxylase: structure, biosynthesis, regulation, and gene manipulation for plant breeding. *Bioscience Biotechnology and Biochemistry* 68, 1175–1184.
Schenk, P.M., Thomas-Hall, S.R., Stephens, E., Marx, U.C., Mussgnug, J.H. *et al.* (2008) Second generation biofuels: high-efficiency microalgae for biodiesel production. *Bioenergy Research* 1, 20–43.
Schuhmann, H., Lim, K. and Schenk, P.M. (2012) Perspectives on metabolic engineering for increased lipid contents in microalgae. *Biofuels* 3, 71–86.
Shanklin, J. and Cahoon, E.B. (1998) Desaturation and related modifications of fatty acids. *Annual Review of Plant Physiology and Plant Molecular Biology* 49, 611–641.
Siaut, M., Cuine, S., Cagnon, C., Fessler, B., Nguyen, M. *et al.* (2011) Oil accumulation in the model green alga *Chlamydomonas reinhardtii*: characterization, variability between common laboratory strains and relationship with starch reserves. *BMC Biotechnology* 11, 7.
Sizova, I., Fuhrmann, M. and Hegemann, P. (2001) A *Streptomyces rimosus* aphVIII gene coding for a new type phosphotransferase provides stable antibiotic resistance to *Chlamydomonas reinhardtii*. *Gene* 277, 221–229.
Snyder, C.L., Yurchenko, O.P., Siloto, R.M.P., Chen, X., Liu, Q. *et al.* (2009) Acyltransferase action in the modification of seed oil biosynthesis. *New Biotechnology* 26, 11–16.
Specht, E., Miyake-Stoner, S. and Mayfield, S. (2010) Microalgae come of age as a platform for recombinant protein production. *Biotechnology Letters* 32, 1373–1383.
Spijkerman, E. and Wacker, A. (2011) Interactions between P-limitation and different C conditions on the fatty acid composition of an extremophile microalga. *Extremophiles* 15, 597–609.
Steinbrenner, J. and Sandmann, G. (2006) Transformation of the green alga *Haematococcus pluvialis* with a phytoene desaturase for accelerated astaxanthin biosynthesis. *Applied and Environmental Microbiology* 72, 7477–7484.
Stephenson, A., Dennis, J., Howe, C., Scott, S. and Smith, A. (2010) Influence of nitrogen-limitation regime on the production by *Chlorella vulgaris* of lipids for biodiesel feedstocks. *Biofuels* 1, 47–58.
Su, C.H., Chien, L.J., Gomes, J., Lin, Y.S., Yu, Y.K. *et al.* (2011) Factors affecting lipid accumulation by *Nannochloropsis oculata* in a two-stage cultivation process. *Journal of Applied Phycology* 23, 903–908.
Suen, Y., Hubbard, J.S., Holzer, G. and Tornabene, T.G. (1987) Total lipid production of the green-alga *Nannochloropsis* sp Qii under different nitrogen regimes. *Journal of Phycology* 23, 289–296.
Tan, C.P., Qin, S., Zhang, Q., Jiang, P. and Zhao, F.Q. (2005) Establishment of a micro-particle bombardment transformation system for *Dunaliella saliva*. *Journal of Microbiology* 43, 361–365.
Tang, D.K.H., Qiao, S.Y. and Wu, M. (1995) Insertion mutagenesis of *Chlamydomonas reinhardtii* by electroporation and heterologous DNA. *Biochemistry and Molecular Biology International* 36, 1025–1035.
Tatsuzawa, H. and Takizawa, E. (1995) Changes in lipid and fatty-acid composition of *Pavlova lutheri*. *Phytochemistry* 40, 397–400.
Thompson, G.A. (1996). Lipids and membrane function in green algae. *Biochimica et Biophysica Acta – Lipids and Lipid Metabolism* 1302, 17–45.
Timmins, M., Zhou, W., Lim, L., Thomas-Hall, S.R., Hankamer, B. *et al.* (2009) The metabolome of *Chlamydomonas reinhardtii* following induction of anaerobic H_2 production by sulphur deprivation. *Journal of Biological Chemistry* 284, 23415–23425.
Van Mooy, B.A.S., Fredricks, H.F., Pedler, B.E., Dyhrman, S.T., Karl, D.M. *et al.* (2009) Phytoplankton in the ocean use non-phosphorus lipids in response to phosphorus scarcity. *Nature* 458, 69–72.
Venegas-Caleron, M., Sayanova, O. and Napier, J.A. (2010) An alternative to fish oils: metabolic engineering of oil-seed crops to produce omega-3 long chain polyunsaturated fatty acids. *Progress in Lipid Research* 49, 108–119.
Walker, T.L., Becker, D.K., Dale, J.L. and Collet, C. (2005) Towards the development of a nuclear transformation system for *Dunaliella tertiolecta*. *Journal of Applied Phycology* 17, 363–368.
Wang, Z.T., Ullrich, N., Joo, S., Waffenschmidt, S. and Goodenough, U. (2009) Algal lipid bodies: stress induction, purification, and biochemical characterization in wild-type and starchless *Chlamydomonas reinhardtii*. *Eukaryotic Cell* 8, 1856–1868.
Worden, A.Z., Lee, J.H., Mock, T., Rouze, P., Simmons, M.P. *et al.* (2009) Green evolution and dynamic adaptations revealed by genomes of the marine Picoeukaryotes Micromonas. *Science* 324, 268–272.
Work, V.H., Radakovits, R., Jinkerson, R.E., Meuser, J.E., Elliott, L.G. *et al.* (2010) Increased lipid accumulation in the *Chlamydomonas reinhardtii* sta7-10 starchless isoamylase mutant and increased carbohydrate synthesis in complemented strains. *Eukaryotic Cell* 9, 1251–1261.

Zaslavskaia, L.A., Lippmeier, J.C., Kroth, P.G., Grossman, A.R. and Apt, K.E. (2000) Transformation of the diatom *Phaeodactylum tricornutum* (Bacillariophyceae) with a variety of selectable marker and reporter genes. *Journal of Phycology* 36, 379–386.

Zeng, X.H., Danquah, M.K., Chen, X.D. and Lu, Y.H. (2011) Microalgae bioengineering: from CO_2 fixation to biofuel production. *Renewable and Sustainable Energy Reviews* 15, 3252–3260.

Zhekisheva, M., Boussiba, S., Khozin-Goldberg, I., Zarka, A. and Cohen, Z. (2002) Accumulation of oleic acid in *Haematococcus pluvialis* (Chlorophyceae) under nitrogen starvation or high light is correlated with that of astaxanthin esters. *Journal of Phycology* 38, 325–331.

Zhila, N.O., Kalacheva, G.S. and Volova, T.G. (2011) Effect of salinity on the biochemical composition of the alga *Botryococcus braunii* Kutz IPPAS H-252. *Journal of Applied Phycology* 23, 47–52.

5 Cultivation of Microalgae for Biofuel Production

Christopher Q. Lan
Department of Chemical and Biological Engineering, University of Ottawa, Canada

5.1 Introduction

Microalgae, similar to plants, possess the ability to convert solar energy into chemical energy through photosynthesis. They also possess several characteristics that make them preferred advanced biofuel feedstock that can provide a meaningful alternative to fossil fuel. Plants invest a substantial portion of their energy in building lignocellulosic structures of roots and stems for anchorage, which are difficult to deconstruct while microalgae are supported in water and store their energy in readily accessible forms. Land plants usually take a full growing season to produce one crop while the exponential growth rate of microalgae doubles their biomass in as short a period as 3.5 h (Spolaore *et al.*, 2006). Moreover, microalgae can be produced throughout the year, thus the resulting oil yield far exceeds well known terrestrial oil crops. For example, microalgae producing 30% by weight oil yield 58,700 l ha^{-1} compared to 5950 l ha^{-1} from oil palm, 2689 l ha^{-1} from coconut, 1190 l ha^{-1} from rapeseed, 1892 l ha^{-1} from jatropha and 446 l ha^{-1} from soybean (Singh and Gu, 2010).

Historically, microalgae have been produced and applied industrially in different areas such as foods and animal feeds (Knuckey *et al.*, 2006), nutraceutical products (Shi *et al.*, 2002), pigments (Del Campo *et al.*, 2007) and wastewater treatment (Delanoue *et al.*, 1992; Travieso *et al.*, 2006; Lan and Wang, 2010; Wang and Lan, 2011a). Their potentials in CO_2 biofixation have also been recognized (Huntley and Redalje, 2007; Wang *et al.*, 2008). Nevertheless, the relatively high production cost of microalgal biomass in comparison to agricultural and forestry biomasses has been one of the major challenges in its commercial production.

Extensive efforts have been dedicated to selecting or engineering energy microalga strains that could accumulate large quantities of high energy content and easy-to-convert storage materials such as neutral lipids (e.g. triglycerides and hydrocarbons) and starch or small molecular weight extracellular metabolites such as hydrogen as biofuels or precursors of biofuels.

Physiological studies aiming at understanding the interaction between microalgal cells and the environment, including the effects of both nutritional and operational conditions, have provided valuable information for the development and optimization of microalgal cultivation processes (Li *et al.*, 2008; Wang and Lan, 2011b).

Developing cost-effective and high-efficiency microalgal cultivation systems has been challenging for biochemical engineers. An efficient cultivation system must be able to maximize light capture, optimize light distribution and allow convenient manipulation of

©CAB International 2013. *Biofuel Crops: Production, Physiology and Genetics* (ed. B.P. Singh)

operational parameters in such a way to encourage optimal light utilization for cell growth. In addition, all these objectives must be achieved in a cost-effective manner.

Downstream processes, including biomass harvest and downstream processing (i.e. biorefinery), are one of the most challenging aspects of algal biofuel production. Due to light limitation, biomass concentration of algal culture is typically lower than 0.5% (w/v). Consequently, a large volume of culture must be processed to harvest the biomass needed to feed a commercial production line designed for mass biofuel production. In addition, the energy content of algal biomass is much lower than that of most low-grade fossil fuels (e.g. heavy oil contained in oil sands), making biorefinery much more complex than the upgrading of these low-grade fossil fuels.

Ultimately, the feasibility of algal biofuel as a major renewable alternative to fossil fuels relies on the life cycle energy balance and economics. Minimizing energy consumption and maximizing the cost-effectiveness of microalgal biofuels remains the number one goal of studies in this field.

5.2 Microalgae for Biofuel Production

More than 40,000 species of microalgae and cyanobacteria have been identified to date and they can be classified into ten divisions according to their photosynthetic pigment composition, biochemical constituents, ultrastructure and life cycle. Among these microalgae, six classes are the most interesting to CO_2 biomitigation and biofuel production: diatoms (class *Bacillariophyceae*), green algae (class *Chlorophyceae*), golden-brown algae (class *Chrysophyceae*), golden algae (class *Prymnesiophyceae*), eustigmatophytes (class *Eustigmatophyceae*) and blue-green algae or cyanobacteria (class *Cyanophyceae*) (Sheehan, 1998). An ideal energy algal strain should have one or more of the following traits: (i) fast growing; (ii) capable of accumulating large cell contents of high energy content storage materials such as neutral lipids and carbohydrates; (iii) easy to harvest; (iv) adapted to local climate; (v) capable of growth in wastewaters, seawater, or brackish water; and (vi) capable of producing high-value by-products.

5.3 Nutrition of Microalgae

The growth medium must provide sufficient nutrients for microalgal growth. Carbon (C), nitrogen (N), phosphorus (P), oxygen (O), hydrogen (H) and sulfur (S) are the most important elements constituting algal cells. Other essential elements include iron, magnesium, trace elements and, in some cases, silicon (Rebolloso-Fuentes *et al.*, 2001). Cell growth and survival also requires the consumption of bioenergy for cellular metabolism. While oxygen and hydrogen are readily available from water, other elements need to be supplied in the medium for algae to grow. It is important to develop balanced media for optimal microalgal cultivation and CO_2 fixation (Mandalam and Palsson, 1998).

5.3.1 Energy

Microalgae and cyanobacteria utilize light (photosynthetic growth) or organic compounds such as sugars (heterotrophic growth) as energy for cell growth and cell maintenance. Although heterotrophic cultivation of microalgae has been employed for the production of speciality fuels such as jet fuels for airplanes, they do not fulfil the mandate of biofuel production, i.e. converting solar energy to fuels that are suitable for different utilities. Indeed, the energy source of heterotrophic growth microalgae and cyanobacteria, i.e. organic compounds such as sugars, has first to be produced by other primary producers one way or the other. Therefore, we will focus on photosynthetic growth of microalgae and cyanobacteria that capture solar energy in this chapter.

5.3.2 Carbon

Microalgae and cyanobacteria can utilize either organic compounds such as sugars, lipids and proteins or inorganic carbon compounds such as CO_2 and carbonates as carbon sources. According to the carbon and energy sources utilized by a microalgal strain, microalgae could be divided into three groups: (i) photoautotrophic microalgae, which utilize inorganic carbon compounds (e.g. CO_2 and carbonate) as the carbon source and light as the energy source; (ii) heterotrophic microalgae, which utilize organic compounds as the carbon source and energy source as well; and (iii) mixotrophic microalgae, which could utilize both organic and inorganic compounds as a carbon source. Most algal groups are photoautotrophs, i.e. depending entirely upon photosynthesis for cell growth and cell maintenance using sunlight as the source of energy and CO_2 as the source of carbon. There are some colourless heterotrophic species that depend on organic carbon sources assimilated from the external environment either by taking up dissolved substances (osmotrophy) or by engulfing bacteria and other microorganisms (phagotrophy) as the carbon source and energy source. A large group of algae can utilize both inorganic and organic carbon sources and is referred to as mixotrophs. Some mixotrophs are primarily photosynthetic and only use organic energy sources occasionally. Others meet most of their nutritional demand by phagotrophy but may use some of the products of photosynthesis from sequestered prey chloroplasts. Of particular importance to biofuel production are phototrophs and mixtrophs that are primarily photosynthetic because, as mentioned previously, heterotrophic growth of microalgae does not fulfil the mandate of converting solar energy to chemical energy stored in biomass.

Photoautotrophic microalgae can fix CO_2 from different sources, including: (i) CO_2 from the atmosphere; (ii) CO_2 from industrial exhaust gases (e.g. flue gas and flaring gas); and (iii) fixed CO_2 in the form of soluble carbonates (e.g. $NaHCO_3$ and Na_2CO_3).

Traditionally, microalgae are cultivated in closed systems or open ponds, which are aerated or exposed to air to allow microalgae to capture CO_2 for cell growth. Since the atmosphere contains only 0.03–0.06% CO_2, mass transfer of CO_2 could limit the growth of microalgae (Chelf *et al.*, 1993). Industrial exhaust gases such as flue gas contain up to 15% CO_2, providing a CO_2-rich source for microalgal cultivation and a potentially more efficient route for CO_2 biofixation. The third route is to fix CO_2 by chemical reaction to produce carbonates (e.g. Na_2CO_3) and use the latter as the carbon source for microalgal cultivation.

Flue gases from power plants are responsible for more than 7% of the total world CO_2 emissions (Sakai *et al.*, 1995). Carbon dioxide in flue gas is available at little or no cost. As estimated by the IPCC criteria, the CO_2 concentration of flue gas is up to 15% (Maeda *et al.*, 1995). Therefore, it would be beneficial if microalgae are tolerant to elevated CO_2 levels should they be used for CO_2 fixation from flue gases.

Adoption of carbonate-utilizing strains for CO_2 fixation could be advantageous in many aspects: (i) CO_2 released at night time from industrial facilities could be converted to carbonate salts and stored for conversion in daytime; (ii) since only a limited number of microalgal species thrive in media containing high concentrations of carbonate salts, species control (i.e. preventing wild-type microalgal species from contaminating the cultivation system) is relatively simple; and (iii) most of these species have high pH optima (in the range of 9.0 to 11), further simplifying species control (Ginzburg, 1993).

5.3.3 Nitrogen

Nitrate, ammonia and urea (or combinations of them) are the most common nitrogen sources for microalgae (Lourenco *et al.*, 1998). Ammonium is the chemical form of nitrogen most readily taken up and assimilated by phytoplankton. Unlike nitrate, it does not require reduction prior to being assimilated into amino acids. However, research indicated that ammonium at high concentration has toxic effects on microalgal growth (Lourenco *et al.*, 2002), and ammonia may escape into the atmosphere, raising valid environmental and economic concerns.

A cheap source of nitrogen is wastewaters or secondary wastewaters, which contain large quantities of different forms of nitrogen sources. However, the use of wastewater for microalgal cultivation may cause contamination problems and/or complicate downstream processing and therefore should be used with precautions.

It would be necessary for the microalgae to have a high calorific value if they are used for biofuel production. Studies have shown that the calorific values of microalgae biomass could be enhanced by cultivation in nitrogen-limiting conditions. For instance, the calorific value of *Chlorella vulgaris* biomass, which was 18 KJ g^{-1} when growing in nitrogen-sufficient medium, increased to 23 KJ g^{-1} in low nitrogen medium (Illman *et al.*, 2000). To put this in perspective, the caloric value of diesel is 43 KJ g^{-1}.

From the energetic point of view, lipids are the most desirable component of microalgal cells, followed by carbohydrates. Cells with high lipid content have elevated calorific value and produce higher yields of oil when processed via, for instance, biomass liquefaction (Tornabene *et al.*, 1983; Ginzburg, 1993; Illman *et al.*, 2000). Nitrogen limitation was observed to lead to the increase of the lipid content in some *Chlorella* strains such as *C. emersonii* (63%), *C. minutissima* (56%), *C. vulgaris* (57.9%), *C. luteoviridis* (28.8%), *C. capsulata* (11.4%) and *C. pyrenoidosa* (29.2%) (Illman *et al.*, 2000). An oil-rich microalgal species, *Neochloris oleoabundans*, was reported to accumulate 35–54% lipids of its cell dry weight under nitrogen-deficient conditions (Tornabene *et al.*, 1983). It was also observed by Yamaberi *et al.* (1998) that the volume of triglycerides accumulated in *Nannochloris* sp. cells could be 2.2 times higher than in the cells in nitrogen-sufficient cultures.

However, there is a dilemma in the fact that high-lipid and high-caloric cells are usually produced in stress states, which is associated with reduced cell division (Ratledge, 2002). Biomass yield and overall lipid/energy productivity will therefore be compromised as a result. It is obvious that it is important to find the balance between producing high caloric value cells and maintaining high biomass productivity by optimizing nitrogen in the medium for the growth of microalgae.

It was observed that algal cells could utilize intracellular nitrogen reservoirs such as chlorophylls as a nitrogen source for cell growth, and tripling of biomass concentration was reported after the nitrogen source in the medium was depleted. Cell growth was accompanied by the decrease of chlorophyll content of cells (Li *et al.*, 2008).

5.3.4 Phosphorus

Phosphorus is preferentially assimilated as inorganic phosphates in the form of $H_2PO_4^-$ and HPO_4 (Gauthier and Turpin, 1997; Martinez *et al.*, 1999; Knuckey *et al.*, 2006). It has been pointed out that phosphates may form complexes or precipitations with some metal ions and not all the added phosphorus is bioavailable (Yun *et al.*, 1997). Commonly used sources of phosphorus include sodium phosphate and potassium phosphate salts. Wastewaters, especially high strength wastewaters such as that from piggery, poultry and dairy farms or extracts of manure are rich sources of phosphorus.

5.3.5 Other elements

Sulfur, iron, magnesium and other elements are also indispensable for the growth of microalgae. Sulfur is an essential component of cysteine and methionine. In the absence of sulfur, protein biosynthesis is impeded and the photosynthetic system PSII repair cycle is blocked (Zhang *et al.*, 2002). Magnesium is required for nitrogenase activity using a creatine phosphate/kinase/ATP generating system as one of its roles in cell metabolism. Iron is involved in electron flow from H_2O to $NADP^+$ (Roden and Zachara, 1996). Some trace metals play key roles in (non-cyclic) photosynthetic electron transport (Raven *et al.*, 1999). For instance, manganese is essential for O_2 evolution and calcium has an important role in the thylakoid lumen in facilitating H_2O dehydrogenation and O_2 evolution.

5.4 Environmental Factors Affecting Microalga Growth

5.4.1 Light intensity, light spectral quality and light/dark cycle

Light is the energy for cell growth in photoautotrophic microalga cultivation. Light intensity plays an important role in microalgal photosynthesis. While light limitation could occur at low light intensity, light saturation and photoinhibition might also occur when light intensity goes beyond a critical level. Photo-inhibition can be reversible or irreversible, depending on the light stress and the length of time the microalgae are exposed to the stress. Light inhibition should be avoided as much as possible.

Light spectral quality is another important factor to consider in microalga cultivation. While sunlight covers a wide spectral range, only light within the range of 400 to 700 nm is photosynthetically active radiation (PAR), which accounts for approximately 50% of sunlight (Suh and Lee, 2003). There is therefore a natural barrier for enhancing the photosynthetic efficiency and the actual photosynthetic efficiency is even lower due to losses such as light reflection and cellular respiration.

Another important aspect is the effects of the natural rhythm of the light/dark cycle, which have remarkable effects on the overall efficiency of solar energy capturing. Due to the lack of light energy, microalgae would have to conduct respiration during the night time to support metabolism for the maintenance of cell viability. As a result, up to 42% of biomass produced during the daytime could be lost throughout the subsequent night (Jacob-Lopes *et al.*, 2009). Respiration during night time should be minimized.

A different type of light/dark cycle is commonly encountered by microalgae in a controlled cultivation system even during the daytime. Due to the light gradient inside a photobioreactor (PBR), algae are exposed to different microenvironments, from the light region near the surface to the dark zone in the centre of a solar tube, in a cyclic manner. As a result, algae are subjected to certain light/dark cycles. Barbosa *et al.* (2003) demonstrated that long light/dark cycle time might have negative effects on cell growth and light utilization of algae but the negative effect was eliminated when the cycle time was sufficiently short (e.g. 10 s or less).

Some parameters that could affect light distribution include light scattering by algal cells in solution and absorption by intracellular pigments. In high density microalgal cultures, mutual shading between different cells becomes the main factor influencing light distribution among cells. In an under-agitated culture, an undesirable scenario could occur where some cells (i.e. those on the light-receiving surface) are exposed to an overdose of light whereas some other cells (i.e. these at regions distant from the surface) receive less or zero light. Limiting the length of light path and improving mixing are the most commonly adopted strategies to improve light distribution.

5.4.2 Dissolved carbon dioxide and oxygen concentration

Cell growth of microalgae involves three competing cellular processes: photosynthesis, photorespiration and (dark) respiration. Photosynthesis of microalgae utilizes light energy to fix CO_2 and releases O_2 as a by-product. However, at high dissolved oxygen (dO) concentration, ribulose-1,5-bisphosphate carboxylase oxygenase (Rubisco), the prime carboxylating enzyme that furnishes CO_2 for the Calvin cycle, can also utilize O_2 for photorespiration, which consumes O_2 to produce CO_2.

Furthermore, a high concentration of dO is toxic to microalgal cells. Upon exposure to strong sunlight radiation, for instance, reactive oxygen species (ROS) such as oxygen radicals may develop, which may damage cytoplasmic membranes and other cellular components. High temperatures and photon flux density (PFD), in combination with CO_2 limitation, may intensify the physiological inhibitory effects of dO. It was reported that accumulation of photosynthetically generated oxygen could lead to severe inhibition of algal growth, especially in closed systems such as tubular PBRs in which oxygen accumulates during the loop cycle and is removed only at the degasser (Weissman *et al.*, 1988).

CO_2 is the carbon source for microalgae in autophototrophic culture and could be the limiting factor if the CO_2 concentration is low in the feed gas (e.g. when air is used as the CO_2 source) or when mixing is not sufficient. On the other hand, a high concentration of dissolved CO_2 (dCO_2) would result in a low culture pH, which could be inhibitory to some microalgal cells. Therefore, maintaining the dCO_2 in the optimal range is important.

5.4.3 Temperature

In general, the cultivation temperature should ideally be as close as possible to the temperature at which the microalgae were collected (polar regions <10°C temperate regions 10–25°C and tropical regions >25°C). Some species of microalgae tolerate temperatures above 30°C or even 42°C.

A few thermotolerant strains have been selected. For instance, several unicellular green algal strains, identified as species of *Chlorella*, were isolated from hot springs in Japan (Sakai *et al.*, 1995). These strains grew at temperatures up to 42°C and in air containing more than 40% CO_2. Their tolerance to both high temperature and high CO_2 content makes them potentially the appropriate microbial cellular reactors for bioCO_2 mitigation from flue gas. The tolerance of microalgae to relatively high temperature is very important in reducing cooling costs. Using these thermotolerant strains may also simplify species-control because the temperature optima of most microalgal species locate in the range of 20 to 30°C. On the other hand, selection of energy algae that grow well under relatively low temperature conditions would greatly enlarge the production window and reduce the heating costs when cultivating microalgae in cold regions.

5.4.4 pH

The optimal pH of most cultured algal species is in the range of 7 to 9. However, some species have pH optima in more acid or basic ranges. For instance, the optimal pH of the cyanobacterium *Spirulina platensis* is in the range of 8.0 to 10.0 (Vieira Costa *et al.*, 2004). It is crucial to maintain culture pH in the optimal range as complete culture collapse may occur due to the disruption of cellular processes by extreme pH.

In the case of high-density algal culture in controlled systems using air enriched with CO_2 (pure CO_2 or high CO_2 flue gases), the concentration of dCO_2 may be the dominant factor that determines the culture pH (Sanchez *et al.*, 2003). In such a scenario, dCO_2 is the result of the balance between the mass transfer of CO_2 from gas phase to liquid phase and the consumption of CO_2 by cells. Therefore, the pH control mechanism should be integrated with the aeration system, as adding basic solution to the culture, which is a standard practice in conventional bioreactors, would not be able to control culture pH effectively.

5.4.5 Mixing

Microalgae live in their natural habitats at a density of approximately 10^3 cells ml^{-1} and at distances of more than 1000 μm between cells. However, in high-cell-density microalgal cultures such as in PBRs, the cell density could be as high as 10^9 cells ml^{-1}. The high cell density in such environments may result in drastically reduced transmission of light and increased rates of dCO_2 consumption and dO accumulation, as well as quick increase of culture temperature. Mixing is therefore an important feature to consider. More specifically, mixing of microalgal cultures is necessary to: (i) prevent sedimentation of algal cells; (ii) ensure that all cells of the population have uniform average exposure to light and nutrients; (iii) facilitate heat transfer and avoid thermal stratification; and (iv) improve gas exchange between the culture medium and the air phase.

Depending on the scale and the choice of cultivation system, mixing could be accomplished by aeration, pumping, mechanical agitation (e.g. rotation wheels, static mixer), or a combination of these means. It should be noticed that not all algal species can tolerate vigorous mixing. Mechanical agitation and bubble break-up often lead to hydrodynamic stress, resulting in restriction to algal growth and metabolic activity (Suh and Lee, 2003; Vunjak-Novakovic *et al.*, 2005).

5.4.6 Sterility (species control)

It is well accepted that a certain amount of impurity in microalga cultures must be tolerated when the processes are designed for low-value objectives such as biofuel production and CO_2 sequestration. Fortunately, for autotrophic microalgal farming facilities, contamination of heterotrophic microorganisms is usually not of significant concern due to the lack of organic carbon sources in the system. However, the control of exotic and invasive algal species and predators is critical for stable continuous operations and also for stable quality of products. Species control could be particularly difficult for cultivation of relatively slow-growing microalgal species.

5.5 Microalgae Cultivation Systems

Currently, most industrial microalgal cultivation systems are open ponds. These systems are preferred for their low capital and operational costs but are limited by the lack of control on operational conditions and hence low biomass productivity. Furthermore, difficulties in species control tend to limit open ponds to algal cultivation in highly selective environments such as high salinity and extreme pH conditions that exclude contaminating organisms (Suh and Lee, 2003). A large variety of closed PBRs have been developed in the last four decades and a few commercial demonstration tubular closed PBRs have been installed around the world. Closed PBRs can support large algal productivity but are impeded by the high capital and operational costs. To harness the benefits of both open and closed systems, a concept was proposed to integrate both open ponds and closed PBRs to a hybrid system, with the closed system for cultivation of microalgae in the early stage mainly dedicated to biomass production and the open ponds employed for the later stage when nutrient starvation is applied to promote the accumulation of carbon storage materials such as algal oil.

5.5.1 Open pond system

Figure 5.1. shows the picture of an open air pond (left) and a raceway open pond system composed of a series of ponds with varied sizes. The major disadvantage of an open air pond is there is no agitation, leading to poor mass transfer and heat transfer. In addition, algal growth is limited to the surface layer where light can penetrate. Using shallow ponds (depth of 0.3 m or less) could to a certain extent alleviate this problem.

The raceway pond is the most common open pond system. It is made of a closed loop recirculation channel that is typically about 0.3 m deep (Fig. 5.1). Mixing and circulation are achieved with a paddlewheel and the flow is guided around bends by baffles placed in the flow channel. Raceway channels are built in concrete or compacted earth, and may be lined with white plastic (Chisti, 2007).

Production of microalgal lipids for biodiesel production was extensively evaluated in raceway ponds in the Algal Species Program (ASP) sponsored by the US Department of Energy (Sheehan, 1998). Nevertheless, the world's largest raceway pond system is currently the one owned by Earthrise Nutritionals (http://www.earthrise.com) for production of food-grade *Spirulina* as human food. Long 15.2 m

Fig. 5.1. Typical open algae cultivation systems: open air ponds (left) and raceway ponds (right).

paddlewheels are used for mixing and CO_2 is bubbled into the ponds to facilitate algal growth.

In both open air ponds and raceways, the only means of cooling is water evaporation. Temperature fluctuates within a diurnal cycle and seasonally. Evaporative water loss and CO_2 escape (when CO_2 is used to bubble the ponds) are more significant than in PBRs. Terry and Raymond (1985) provide a detailed discussion on raceways and other open pond systems.

5.5.2 Closed photobioreactors

A variety of different PBRs have been developed in the last four decades, which can be generally classified into three categories according to the configuration of their solar energy capture units: vertical columns, tubular and flat panel PBRs. In general, vertical columns such as bubble column and air-lift PBRs offer excellent gas exchange features but limited surface area to volume ratios. For this reason, they are commonly considered to be less efficient than tubular or flat panel PBRs.

Tubular PBRs

Tubular PBR (TPBR) is one of the most popular configurations of PBRs (Travieso *et al.*, 2001). As depicted in Fig. 5.2, a horizontal TPBR system is composed of a tubular solar array for sunshine harvest, a biomass harvest unit for biomass separation and a gas exchange column for gas exchange and cooling as well. A pump is used to circulate medium between the solar panel and the gas exchanger. Algal biomass is harvested before entering the gas exchanger, fresh air (or air enriched with CO_2) and fresh medium are fed into the column. The gas exchange column could be a bubble column or an air-lift column.

Horizontal TPBRs provide large surface area to volume ratio and a better angle for incident light compared to vertical columns, allowing for more efficient light harvesting. However, this also causes the generation of a large amount of heat, requiring sometimes expensive temperature control systems (Richmond, 1987). They also have rather poor mass transfer features, making oxygen accumulation and CO_2 depletion a major concern. All these factors make it very difficult to operate horizontal TPBRs at a large scale. In one case, for instance, a commercial horizontal TPBR facility (Fig. 5.3) failed to perform up to expectation and had to be abandoned (Mirón *et al.*, 1999).

To overcome the disadvantages of both the vertical columns and the horizontal TPBRs, a hybrid between a horizontal and vertical TPBR, the helical PBR as shown in Fig. 5.4, was devised. Watanabe (2004) studied this reactor design for cultivation of *Spirulina platensis* and pilot plants in the UK and Australia called Biocoil were set up using this model. One disadvantage of the helical TPBR is that a portion of the irradiation is lost to the middle basal region of the reactor at high incident light angles, such as midday sunshine. Watanabe (2004) proposed a modified cone-shaped TPBR in order to remedy this problem.

Another TPBR design improving upon the horizontal TPBR and at the same time avoiding the complex structure of the aforementioned helical TPBR is the near-horizontal TPBR

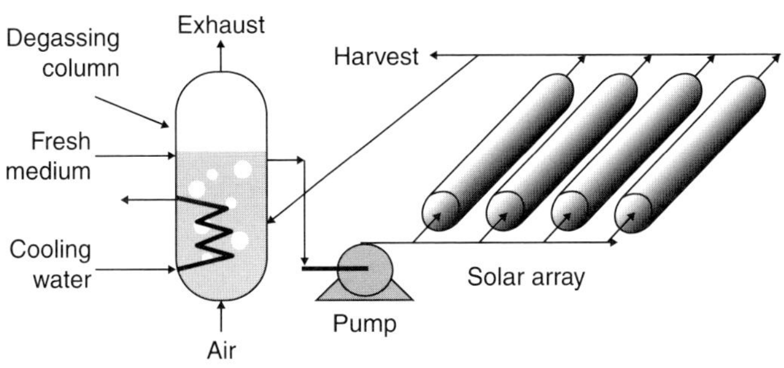

Fig. 5.2. Schematic diagram of a horizontal tubular photobioreactor (Wen and Johnson, 2009).

Fig. 5.3. A commercial horizontal tubular bioreactor facility (Mirón *et al.*, 1999). This failed facility was located in Cartagena, Spain, and it was owned by Photobioreactors Ltd.

(Fig. 5.5). This type of reactor consists of a series of thin tubes connected at the bottom by a manifold that supplies compressed gas, and at the top by a degasser. Ugwu *et al.* (2002) investigated the effect of column inclination on the gas transfer characteristics of a near-horizontal TPBR and observed increased gas transfer coefficient and gas holdup and decreased mixing time with increasing angles, which at the extreme, would lead to a vertical

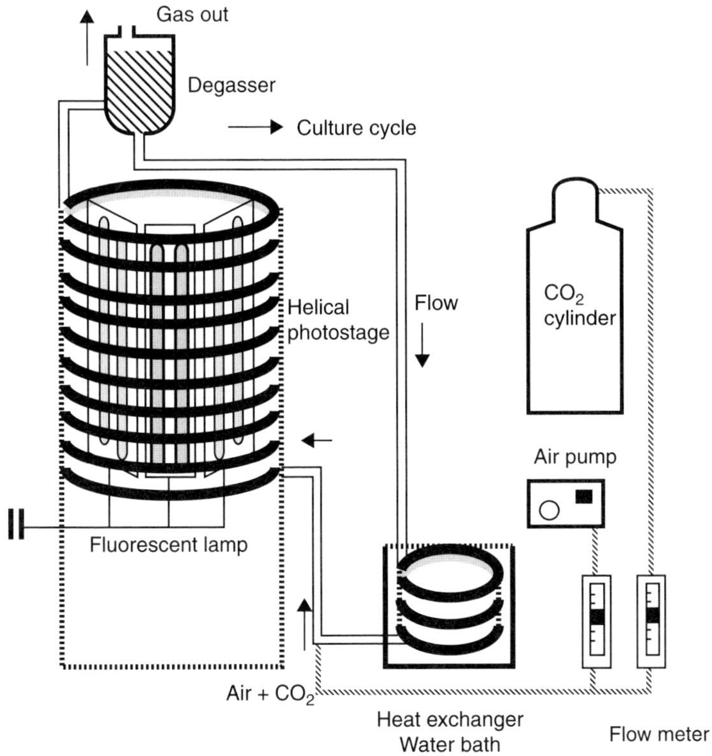

Fig. 5.4. Schematic diagram of a helical photobioreactor (Osada *et al.*, 2004).

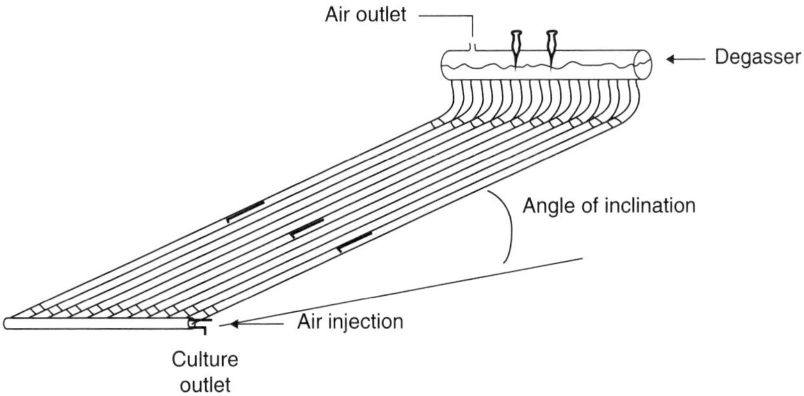

Fig. 5.5. Schematic diagram of a near-horizontal tubular PBR (Tredici *et al.*, 1998).

column. However, it was suggested that 45° was the best choice, because maintaining the column at angles greater than this would cause increases in cost for structural supports.

The mass transfer problem associated with TPBRs could be partially addressed by enriching the air stream with CO_2 to avoid CO_2 limitation. However, significant losses of CO_2 may occur, causing the cost increase and reduction of CO_2 fixation efficiency. This problem could be alleviated by injection at multiple sites within the bioreactor.

Furthermore, Ugwu *et al.* (2002) reported the use of static mixers (baffles) to improve the gas holdup and oxygen mass transfer coefficient (k_La). They noted that increasing the number of static mixers inside the column increased the gas transfer coefficient and gas holdup time. With four static mixers in the riser tube, cultures of *Chlorella sorokiniana* reached productivities of 1.47 g l^{-1} day^{-1}, which was 15–70% higher than the tubes without static mixers, depending on the time of day and amount of illumination. Nevertheless, the static mixers were also found to increase the mixing time, therefore a compromise had to be made for optimal results.

Flat panel photobioreactors

Flat panel photobioreactors (FP-PBRs) are closed PBRs with a narrow light path and are characterized by large illuminated surface to volume ratio. These PBRs can be oriented into the direct path of light to obtain maximum exposure to solar energy.

The first FP-PBR was described in the 1950s (Burlew, 1953). Since then, many different types of FP-PBRs have been developed. As shown in Fig. 5.6, these PBRs can be classified into two categories, pump-driven and airlift FP-PBR, according to the means of mixing. Pump-driven FP-PBR depends on the flow of liquid created by pumping to generate the necessary turbulence for mixing while airlift FP-PBR depends on compressed air to deliver the power of mixing. It was reported that a 17 l airlift FP-PBR was capable of supporting *Haematococcus pluvialis* NIES-144 growth with a maximum cell density of 4.1 × 10^5 cell ml^{-1} and specific growth rate of 0.52 day^{-1}. A similar level of performance was obtained from a 90 l air-lift FP-PBR of the same structure with a cell density of 40 × 10^4 cell ml^{-1} and a slightly smaller specific growth rate of 0.39 day^{-1} (Issarapayup *et al.*, 2009).

Conventional FP-PBRs suffer from deficiencies in culture flow control and the engineering problems encountered in the construction of suitably cost-effective panels (Pirt *et al.*, 1983). To remedy these problems, Tredici and his co-workers (Tredici *et al.*, 1991; Tredici and Materassi, 1992) proposed a vertical alveolar panel PBR (VAP) made of 1.6 cm thick rigid alveolar Plexiglas sheets that had a transparency of about 95% to the visible fraction of solar radiation. The VAP had a surface-to-volume ratio of 80 m^2 m^{-3} and a culture thickness of approximately 12.5 mm. Net biomass productivity of up to 16 g m^{-2} day^{-1} were achieved at a biomass concentration of 28 g m^{-2} when a VAP having a 5 m^2 panel was used for outdoor cultivation of *Anabaena azollae*. In addition to the relatively inexpensive manufacturing costs,

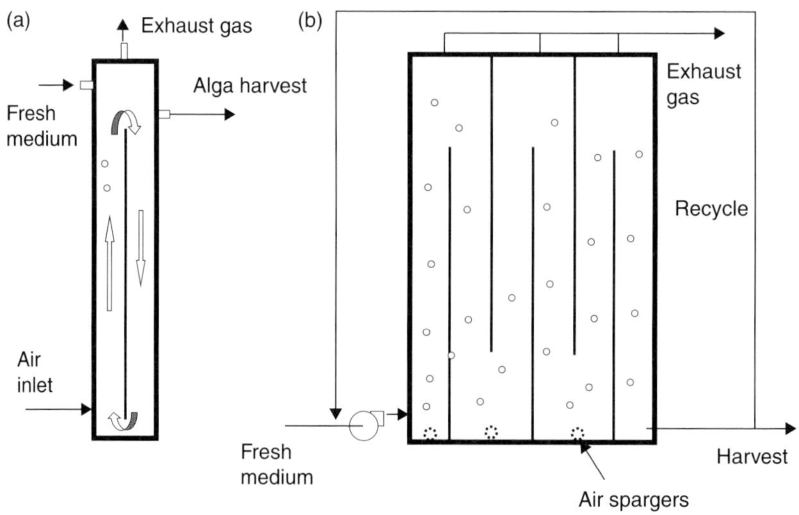

Fig. 5.6. Schematic diagram of (a) an air-lift FP-PBR and (b) a pump-driven FP-PBR.

the VAP offered advantages such as large surface-to-volume ratio, flexibility and good mixing and gas exchange features. However, it was also pointed out that temperature control and light inhibition were two major challenges in designing such a PBR.

The thickness of plate is very important in the design of a FP-PBR because it determines the surface area/volume ratio and the length of light path. A small thickness will allow for better diffusion and distribution of light. In general, the shorter the light path, or thickness, the higher the optimal cell density (OCD) and biomass productivity (Hu and Richmond, 1996; Zou and Richmond, 1999). However, it is worth noting that thin PBRs are usually more expensive to make, more difficult to clean, and more readily subjected to light inhibition and temperature fluctuation.

The angle of a FP-PBR relative to the ground is referred to as the tilt angle of the reactor. Throughout the year, the optimal tilt of the reactor that allows maximal incident light will change due to the position of the sun. Hu *et al.* (1998) studied the effect of the tilt angle on the biomass productivity of *S. platensis*. It was shown that, in the summer when the sun was high, the smaller tilt angles of 10° and 30° resulted in higher biomass production. In the winter, when the sun is lower, the larger tilt angles resulted in higher biomass production with 60° being the optimal. They pointed out that, as a general rule, the optimal angle for year round biomass production is equal to the geographic latitude of the location and that increasing the tilt angle for the winter increases biomass production. On a different notion, Zhang *et al.* (2002) showed that a west–east facing bioreactor had a 1.4-fold higher slope of productivity versus irradiation curve than a south–north facing reactor at equal solar irradiation.

Photobioreactor design is evolving rapidly to meet the needs of industrial production. A large variety of different PBRs have been developed in the past three decades and FP-PBR and TPBR are the two most promising configurations for industrial processing. Building on these two basic designs, developments aiming to improve light capturing and distribution (e.g. internal illumination and spectral shifting), mass transfer (e.g. MPBR) and other aspects of PBR operation are expected to elevate the overall PBR efficiency to a new level in the near future.

5.5.3 Hybrid systems

So far, industrial-scale cultivation has been limited to open pond technology and was successful only for a handful of species (Borowitzka, 1999) and cultivation in open ponds of microalgae for biofuel production has proved unsustainable. On the other hand, while closed-system PBRs are well recognized for their excellent ability to control sterility and, in so doing, to permit continuous cultivation of a large variety of species, their application to industrial production has been limited by small scale, generally less than 1000 l. Furthermore, capital costs of PBRs have typically been substantially greater than for open ponds.

To address the apparent technical dilemma, the concept of a hybrid system coupling PBRs and open ponds was proposed (Huntley and Redalje, 2007). In this system, closed PBRs are used in the early stages, which require maintaining constant conditions that favour continuous cell division and prevent contamination of the culture by other organisms. Open ponds are used for the later stages when cell growth is completed or close to completion and cells are ready to be exposed to nutrient limitation or other environmental stresses that lead, as rapidly as possible, to accumulation of secondary metabolites, e.g. lipids or starch. The danger of contamination is minimized by two factors: (i) high cell density of the target algae; and (ii) nutrient limitation (most commonly nitrogen limitation) or physiological stress (most commonly high salinity).

5.6 Harvesting, Cell Disruption and Drying

Harvesting is an expensive and problematic part of the industrial production of microalgal biomass due to the low cell density achievable with microalgal cultures, which is typically in the range of 0.3–0.5 g dry cell weight (DCW) l^{-1}

and with exceptional cases reaching 5 g DCW l^{-1}. On the other hand, the optimal material for industrial conversion is a cell sludge containing at least 300–400 g DCW l^{-1}. This implies that the effluent algal suspension needs to be concentrated 100–1000 times (Ginzburg, 1993), which could be an energy-intensive process.

There is no single harvest method that is suited to every case and therefore selecting the right technologies and optimizing the harvesting process are important (Sanchez *et al.*, 2002). A reasonable approach is to harvest biomass in two steps (Ginzburg, 1993). The first step is bulk harvesting, which is a large-scale operation on the entire bulk, with the aim of separating biomass from the bulk algal suspension, leading to a concentration factor of 100 to 800 times, depending on the initial biomass concentration and the technologies used for bulk harvesting. The second step is thickening, which will further concentrate the slurry by an additional 10 to 30 times. Flocculation and flotation are widely used for the first step of bulk separation. The mechanism of flocculation is to neutralize or reduce the negative charge on the microalgal cellular surface in order to aggregate cells in suspension, which can be achieved by adding flocculants such as multivalent cations and cationic polymers (Sanchez Miron *et al.*, 2002). Another mechanism of flocculation (Knuckey *et al.*, 2006) entails adjustment of culture pH to between 10 and 10.6 using NaOH to neutralize the negative charges on the cell surface, followed by the addition of non-ionic polymer Magnafloc LT-25 to a final concentration of 0.5 mg l^{-1}. The ensuing flocculate is harvested, and neutralized to give a final concentration factor of between 200- and 800-fold. This process has been successfully applied to harvesting cells of many species with efficiencies >80%. Another method, flotation, depends on trapping cells by dispersed micro-air bubbles. It is very attractive as it does not require any addition of chemicals, resulting in a very clean sludge. However, flotation engineering could be challenging, especially at large scale.

The second step, i.e. thickening, consists of thickening the sludge 10–30 times. Existing technologies such as centrifugation and filtration are the most commonly used for this purpose. Centrifugation is an efficient but energy intensive method (Sanchez Miron *et al.*, 2002). Its efficiency depends on the settling characteristics of the cells, the residence time of the cell slurry and the settling depth. Settling depth can be kept small by the design of the centrifuge. The residence time of the slurry in the centrifuge can be controlled by the flow rate. Centrifugal recovery of the biomass is feasible for high-value products, since it can process large volumes relatively rapidly and the biomass remains fully contained during recovery (Heasman *et al.*, 2000). Filtration, which operates under pressure or vacuum, is a preferred method for harvesting relatively large filamentous microalgae such as *S. platensis*. For the small cells of microalgae such as *Chlorella* and *Dunaliella*, conventional filtration has not proven practical while microfiltration has been suggested to be a possible alternative technology (Hooper *et al.*, 1998; Hung and Liu, 2006). However, current large-scale microalgal biomass production facilities do not generally use membrane filtration because of the cost concern, to which membrane replacement and pumping are the major contributors. Generally speaking, however, microfiltration can be more cost-effective than centrifugation (Sanchez Miron *et al.*, 2002).

A strategy for saving biomass harvesting costs is to select easy-to-harvest microalgal strains for CO_2 fixation and biomass production. Since separating processes for microalgae is in general an energy-intensive process (~1.0 MJ kg^{-1} biomass) (Ginzburg, 1993), use of filamentous or colony-forming algae (Dote *et al.*, 1994) could lead to valuable energy and cost savings.

Biomass drying before further extraction and/or thermochemical processing is another step that needs to be taken into consideration. Sun drying is probably the cheapest drying method that has been employed for the processing of microalgal biomass (Millamena *et al.*, 1990; Prakash *et al.*, 1997). However, this method takes a long drying time, requires a large drying surface, and risks the loss of some bioreactive products. Low-pressure shelf drying is another low-cost drying technology that has been investigated (Prakash *et al.*, 1997). It is nevertheless ineffective, requires long processing time and risks decomposition of bioproducts. More efficient but more costly drying technologies that have been investigated for drying microalgae include drum drying (Prakash *et al.*, 1997), spray drying (Leach *et al.*, 1998; Desmorieux and Decaen, 2005), fluidized bed drying (Leach *et al.*, 1998), freeze drying (Millamena *et al.*, 1990) and refractance window dehydration technology (Nindo and Tang, 2007). It is important to find the balance between the drying efficiency and cost-effectiveness.

Depending on the products and biorefinery process to be adopted, microalgal cells may need to be physically disrupted before drying or downstream processing. Both ball mills and high-pressure homogenizers have been used successfully to disrupt microalgal cells of *Haematococcus* at a commercial scale. Use of solvents and enzymes might help with cellular disruption and product recovery. The energy consumption associated with the cell disruption process must be minimized to maximize the net energy outcome when the objective is to produce microalgal biomass. In some cases, cell autolysis of algal cells under well controlled conditions could be employed to facilitate cell disruption.

5.7 Commercial Production of Microalgal Biomass and Biofuels

Commercial large-scale culture of microalgae started in the early 1960s with the culture of *Chlorella* by Nihon Chlorella (Taipei, Taiwan). In 1976, a group of US visionary pioneers founded the progenitor of Earthrise (California, USA), Proteus Corporation, to develop *Spirulina* blue-green algae as a food resource. Earthrise now operates the world's largest *Spirulina* farm on a 108 acre site with 30 5000 m^2 raceway ponds to supply over 20 countries with *Spirulina* powder, capsules or tablets as food sources. By 1980, there were 46 large-scale factories in Asia producing more than 12,000 kg of microalgae (mainly *Chlorella*) per annum. The commercial facilities for production of *Dunaliella salina* as a source of β-carotene were established by Western Biotechnology (Hutt Lagoon, Australia) and Cognis Nutrition and Health (Whyalla, Australia) in 1986. Several plants producing *Haematococcus pluvialis* as a source of astaxanthin have also been established in the USA, India and China. It was estimated that the microalgal biomass market produced more than 7000 t DCW year^{-1} in 2006 (Spolaore *et al.*, 2006) for applications as human nutrition, animal feed, aquaculture and cosmetics.

Algae were initially examined as a potential replacement fuel source for fossil fuels in the 1970s as a result of the OPEC oil embargo that started in the autumn of 1973. Subsequent studies, continued through the 1980s and heightened in the last decade, are enabling the commercial production of microalgae to shift from aquaculture, fine chemicals and health food to fuel production. It is interesting to note that, even though the two major projects sponsored by the US government and the Japanese government concluded that algal oil was not economically feasible, the private sector has moved forward in building commercial facilities to produce biodiesel using algal oils. A privately funded US$20 million programme successfully operated for several years a 2 ha commercial-scale modular production system coupling PBRs with open ponds in a two-stage process to produce *Haematococcus pluvialis* for biodiesel production (Nindo and Tang, 2007). The annual averaged rate of achieved microbial oil production was equivalent to 420 GJ ha^{-1} year^{-1} with the maximum production rate achieved being equivalent to 1014 GJ ha^{-1} year^{-1}. It was predicted that a rate of 3200 GJ ha^{-1} year^{-1} is feasible using *Chlorella* under conditions that prevail in the existing production system, a rate possible to replace reliance on current fossil fuel usage equivalent to about 300 EJ year^{-1} and eliminate fossil fuel emissions of CO_2 of about 6.5 gigatons of carbon (GtC) per year using only 7.3% of the surplus arable land projected to be available by 2050 (Huntley and Redalje, 2007). To put this in perspective, Canada's total energy consumption in 2004 was 8.5 EJ (Natural Resources Canada, 2006).

Replacing all the transport fuel consumed in the USA with biodiesel requires 0.53 billion m^3 of biodiesel annually at the current rate of consumption. Chisti (2007) calculated that the land needed to meet 50% of the US transport fuel demands is 4.5 Mha or 3% of the total US cropping area. The calculation was based on the assumption of biomass productivity at a rather realistic cell oil content of 30% (by weight in dry biomass), which led to an oil productivity of 58,700 l ha^{-1} $year^{-1}$.

While the tremendous potential of microalgae to satisfy the world demand for biofuels has been demonstrated without doubt, one must be aware that microalgal biofuel production is at present an energy-intensive process and the ultimate success of it depends on the success of reducing the energy intensity of this process to maximize the net energy output. The cost to produce a kilogram of microalgal biomass with an average oil content of 30% was estimated to be between US$2.95 and US$3.80, assuming CO_2 to be available and free (Chisti, 2007). Considering that the energy content of microalgal biomass is much lower than crude oil and the price of the latter oscillated around US$0.67 l^{-1} ($80 per barrel) in 2011, there are still significant challenges to be addressed before alga-based biofuels become competitive to fossil fuels.

5.8 Conclusion

Microalgal biotechnology in areas such as strain selection and engineering for photobioreactor design and process optimization have established beyond doubt the potential of microalgae to convert solar to biomass energy as a sustainable and renewable replacement of fossil fuels. Nevertheless, the primary challenge remains the high energy intensity and low cost-effectiveness of the production processes.

References

Barbosa, M.J., Janssen, M., Ham, N., Tramper, J. and Wijffels, R.H. (2003) Microalgae cultivation in air-lift reactors: modeling biomass yield and growth rate as a function of mixing frequency. *Biotechnology and Bioengineering* 82, 170–179.

Borowitzka, M.A. (1999) Commercial production of microalgae: ponds, tanks, tubes and fermenters. *Journal of Biotechnology* 70, 313–321.

Burlew, J. (1953) *Algal Culture from Laboratory to Pilot Plant*. Carnegie Institution of Washington, Publication No. 600, Washington, DC, 357 pp.

Chelf, P., Brown, L.M. and Wyman, C.E. (1993) Aquatic biomass resources and carbon dioxide trapping. *Biomass and Bioenergy* 4, 175–183.

Chisti, Y. (2007) Biodiesel from microalgae. *Biotechnology Advances* 25, 294–306.

Delanoue, J., Laliberte, G. and Proulx, D. (1992) Algae and waste-water. *Journal of Applied Phycology* 4, 247–254.

Del Campo, J.A., García-González, M. and Guerrero, M.G. (2007) Outdoor cultivation of microalgae for carotenoid production: current state and perspectives. *Applied Microbiology and Biotechnology* 74, 1163–1174.

Desmorieux, H. and Decaen, N. (2005) Convective drying of spirulina in thin layer. *Journal of Food Engineering* 66, 497–503.

Dote, Y., Sawayama, S., Inoue, S., Minowa, T. and Yokoyama, S. (1994) Recovery of liquid fuel from hydrocarbon-rich microalgae by thermochemical liquefaction. *Fuel* 73, 1855–1857.

Gauthier, D.A. and Turpin, D.H. (1997) Interactions between inorganic phosphate (P(i)) assimilation, photosynthesis and respiration in the P(i)-limited green alga *Selenastrum minutum*. *Plant, Cell and Environment* 20, 12–24.

Ginzburg, B.Z. (1993) Liquid fuel (oil) from halophilic algae: a renewable source of non-polluting energy. *Renewable Energy* 3, 249–252.

Heasman, M., Diemar, J., O'Connor, W., Sushames, T. and Foulkes, L. (2000) Development of extended shelf-life microalgae concentrate diets harvested by centrifugation for bivalve molluscs – a summary. *Aquaculture Research* 31, 637–659.

Hooper, L.A., Hollein, H.C. and Slater, C.S. (1998) Microfiltration of *Streptomyces rimosus*: cell harvesting process studies. *Separation Science and Technology* 33, 1747–1765.

Hu, Q. and Richmond, A. (1996) Productivity and photosynthetic efficiency of *Spirulina platensis* as affected by light intensity, algal density and rate of mixing in a flat plate photobioreactor. *Journal of Applied Phycology* 8, 139–145.

Hu, Q., Faiman, D. and Richmond, A. (1998) Optimal tilt angles of enclosed reactors for growing photoautotrophic microorganisms outdoors. *Journal of Fermentation and Bioengineering* 85, 230–236.

Hung, M.T. and Liu, J.C. (2006) Microfiltration for separation of green algae from water. *Colloids and Surfaces B: Biointerfaces* 51, 157–164.

Huntley, M.E. and Redalje, D.G. (2007) CO_2 mitigation and renewable oil from photosynthetic microbes: a new appraisal. *Mitigation and Adaptation Strategies for Global Change* 12, 573–608.

Illman, A.M., Scragg, A.H. and Shales, S.W. (2000) Increase in *Chlorella* strains calorific values when grown in low nitrogen medium. *Enzyme and Microbial Technology* 27, 631–635.

Issarapayup, K., Powtongsook, S. and Pavasant, P. (2009) Flat panel airlift photobioreactors for cultivation of vegetative cells of microalga *Haematococcus pluvialis*. *Journal of Biotechnology* 142, 227–232.

Jacob-Lopes, E., Scoparo, C.H.G., Lacerda, L.M.C.F. and Franco, T.T. (2009) Effect of light cycles (night/day) on CO_2 fixation and biomass production by microalgae in photobioreactors. *Chemical Engineering and Processing: Process Intensification* 48, 306–310.

Knuckey, R.M., Brown, M.R., Robert, R. and Frampton, D.M.F. (2006) Production of microalgal concentrates by flocculation and their assessment as aquaculture feeds. *Aquacultural Engineering* 35, 300–313.

Lan, C.Q. and Wang, B. (2010) *Microalgae for Biofuel Production and CO_2 Sequestration*. Nova Science Publishers, Hauppauge, New York, 180 pp.

Leach, G., Oliveira, G. and Morais, R. (1998) Spray-drying of *Dunaliella salina* to produce a ß-carotene rich powder. *Journal of Industrial Microbiology and Biotechnology* 20, 82–85.

Li, Y.Q., Horsman, M., Wang, B., Wu, N. and Lan, C.Q. (2008) Effects of nitrogen sources on cell growth and lipid accumulation of green alga *Neochloris oleoabundans*. *Applied Microbiology and Biotechnology* 81, 629–636.

Lourenco, S.O., Barbarino, E., Lanfer Marquez, U.M. and Aidar, E. (1998) Distribution of intracellular nitrogen in marine microalgae: basis for the calculation of specific nitrogen-to-protein conversion factors. *Journal of Phycology* 34, 798–811.

Lourenco, S.O., Barbarino, E., Mancini-Filho, J., Schinke, K.P. and Aidar, E. (2002) Effects of different nitrogen sources on the growth and biochemical profile of 10 marine microalgae in batch culture: an evaluation for aquaculture. *Phycologia* 41, 158–168.

Maeda, K., Owada, M., Kimura, N., Omata, K. and Karube, I. (1995) CO_2 fixation from the flue gas on coal-fired thermal power plant by microalgae. *Energy Conversion and Management* 36, 717–720.

Mandalam, R.K. and Palsson, B. (1998) Elemental balancing of biomass and medium composition enhances growth capacity in high-density *Chlorella vulgaris* cultures. *Biotechnology and Bioengineering* 59, 605–611.

Martinez, M.E., Jimenez, J.M. and El Yousfi, F. (1999) Influence of phosphorus concentration and temperature on growth and phosphorus uptake by the microalga *Scenedesmus obliquus*. *Bioresource Technology* 67, 233–240.

Millamena, O.M., Aujero, E.J. and Borlongan, I.G. (1990) Techniques on algae harvesting and preservation for use in culture and as larval food. *Aquacultural Engineering* 9, 295–304.

Mirón, A.S., Gómez, A.C., Camacho, F.G. and Grima, E.M. (1999) Comparative evaluation of compact photobioreactors for large-scale monoculture of microalgae. *Journal of Biotechnology* 70, 249–270.

Natural Resources Canada (2006) *Energy Use Data Handbook, 1990 and 1998 to 2004*. Energy Publications, Ottawa, Ontario, Canada, 81 pp.

Nindo, C.I. and Tang, J. (2007) Refractance window dehydration technology: a novel contact drying method. *Drying Technology* 25, 37–48.

Osada, M., Sato, T., Watanabe, M., Adschiri, T. and Arai, K. (2004) Low-temperature catalytic gasification of lignin and cellulose with a ruthenium catalyst in supercritical water. *Energy and Fuels* 18, 327–333.

Prakash, J., Pushparaj, B., Carlozzi, P., Torzillo, G., Montaini, E. *et al.* (1997) Microalgal biomass drying by a simple solar device. *International Journal of Solar Energy* 18, 303–311.

Ratledge, C. (2002) Regulation of lipid accumulation in oleaginous micro-organisms. *Biochemical Society Transactions* 30, 1047–1050.

Raven, J.A., Evans, M.C.W. and Korb, R.E. (1999) The role of trace metals in photosynthetic electron transport in O_2-evolving organisms. *Photosynthesis Research* 60, 111–149.

Rebolloso-Fuentes, M.M., Navarro-Perez, A., Garcia-Camacho, F., Ramos-Miras, J.J. and Guil-Guerrero, J.L. (2001) Biomass nutrient profiles of the microalga *Nannochloropsis*. *Journal of Agricultural and Food Chemistry* 49, 2966–2972.

Richmond, A. (1987) The challenge confronting industrial microagriculture: high photosynthetic efficiency in large-scale reactors. *Hydrobiologia* 151/152, 117–121.

Roden, E.E. and Zachara, J.M. (1996) Microbial reduction of crystalline iron (III) oxides: influence of oxide surface area and potential for cell growth. *Environmental Science and Technology* 30, 1618–1628.

Sakai, N., Sakamoto, Y., Kishimoto, N., Chihara, M. and Karube, I. (1995) *Chlorella* strains from hot springs tolerant to high temperature and high CO_2. *Energy Conversion and Management* 36, 693–696.
Sanchez, J.L.G., Berenguel, M., Rodriguez, F., Sevilla, J.M.F., Alias, C.B. *et al.* (2003) Minimization of carbon losses in pilot-scale outdoor photobioreactors by model-based predictive control. *Biotechnology and Bioengineering* 84, 533–543.
Sanchez Miron, A., Garcia, M.C., Garcia Camacho, F., Molina Grima, E. and Chisti, Y. (2002) Growth and biochemical characterization of microalgal biomass produced in bubble column and airlift photobioreactors: studies in fed-batch culture. *Enzyme and Microbial Technology* 31, 1015–1023.
Sheehan, J. (1998) A look back at the US Department of Energy's aquous species program: biodiesel from Algae; NREL/TP-580-24190. The National Renewable Energy Laboratory, US Department of Energy.
Shi, X.M., Jiang, Y. and Chen, F. (2002) High-yield production of lutein by the green microalga *Chlorella protothecoides* in heterotrophic fed-batch culture. *Biotechnology Progress* 18, 723–727.
Singh, J. and Gu, S. (2010) Commercialization potential of microalgae for biofuel production. *Renewable and Sustainable Energy Reviews* 14, 2596–2610.
Spolaore, P., Joannis-Cassan, C., Duran, E. and Isambert, A. (2006) Commercial applications of microalgae. *Journal of Bioscience and Bioengineering* 101, 87–96.
Suh, I.S. and Lee, C.G. (2003) Photobioreactor engineering: design and performance. *Biotechnology and Bioprocess Engineering* 8, 313–321.
Terry, K.L. and Raymond, L.P. (1985) System design for the autotrophic production of microalgae. *Enzyme and Microbial Technology* 7, 474–487.
Tornabene, T.G., Holzer, G., Lien, S. and Burris, N. (1983) Lipid composition of the nitrogen starved green alga *Neochloris oleoabundans*. *Enzyme and Microbial Technology* 5, 435–440.
Travieso, L., Hall, D.O., Rao, K.K., Benitez, F., Sanchez, E. *et al.* (2001) A helical tubular photobioreactor producing *Spirulina* in a semicontinuous mode. *International Biodeterioration and Biodegradation* 47, 151–155.
Travieso, L., Benitez, F., Sanchez, E., Borja, R., Martin, A. *et al.* (2006) Batch mixed culture of *Chlorella vulgaris* using settled and diluted piggery waste. *Ecological Engineering* 28, 158–165.
Tredici, M.R. and Materassi, R. (1992) From open ponds to vertical alveolar panels: the Italian experience in the development of reactors for the mass cultivation of phototrophic microorganisms. *Journal of Applied Phycology* 4, 221–231.
Tredici, M.R., Carlozzi, P., Zittelli, G.C. and Materassi, R. (1991) A vertical alveolar panel (VAP) for outdoor mass cultivation of microalgae and cyanobacteria. *Bioresource Technology* 38, 153–159.
Tredici, M., Zittelli, G. and Benemann, J. (1998) Photobioreactor for biological hydrogen production: preliminary cost analysis. In: Zaborsky, O., Benemann, J., Matsunaga, T., Miyake, J. and San Pietro, A. (eds) *BioHydrogen*. Plenum Press, New York, pp. 391–402.
Ugwu, C.U., Ogbonna, J.C. and Tanaka, H. (2002) Improvement of mass transfer characteristics and productivities of inclined tubular photobioreactors by installation of internal static mixers. *Applied Microbiology and Biotechnology* 58, 600–607.
Vieira Costa, J.A., Colla, L.M. and Duarte Filho, P.F. (2004) Improving *Spirulina platensis* biomass yield using a fed-batch process. *Bioresource Technology* 92, 237–241.
Vunjak-Novakovic, G., Kim, Y., Wu, X., Berzin, I. and Merchuk, J.C. (2005) Air-lift bioreactors for algal growth on flue gas: mathematical modeling and pilot-plant studies. *Industrial and Engineering Chemistry Research* 44, 6154–6163.
Wang, B. and Lan, C.Q. (2011a) Biomass production and nitrogen and phosphorus removal by the green alga *Neochloris oleoabundans* in simulated wastewater and secondary municipal wastewater effluent. *Bioresource Technology* 102, 5639–5644.
Wang, B. and Lan, C.Q. (2011b) Optimising the lipid production of the green alga *Neochloris oleoabundans* using box-behnken experimental design. *Canadian Journal of Chemical Engineering* 89, 932–939.
Wang, B., Li, Y.Q., Wu, N. and Lan, C.Q. (2008) CO_2 bio-mitigation using microalgae. *Applied Microbiology and Biotechnology* 79, 707–718.
Watanabe, Y. (2004) Design and experimental practice of photobioreactor incorporating microalgae for efficient photosynthetic CO_2 fixation performance. *Studies in Surface Science and Catalysis* 153, 445–452.
Weissman, J.C., Goebel, R.P. and Benemann, J.R. (1988) Photobioreactor design: mixing, carbon utilization, and oxygen accumulation. *Biotechnology and Bioengineering* 31, 336–344.
Wen, Z. and Johnson, M.B. (2009) Microalgae as a Feedstock for Biofuel Production. Available at: http://pubs.ext.vt.edu/442/442-886/442-886.pdf (accessed 23 April 2012).

Yamaberi, K., Takagi, M. and Yoshida, T. (1998) Nitrogen depletion for intracellular triglyceride accumulation to enhance liquefaction yield of marine microalgal cells into a fuel oil. *Journal of Marine Biotechnology* 6, 44–48.

Yun, Y.S., Lee, S.B., Park, J.M., Lee, C.I. and Yang, J.W. (1997) Carbon dioxide fixation by algal cultivation using wastewater nutrients. *Journal of Chemical Technology and Biotechnology* 69, 451–455.

Zhang, L., Happe, T. and Melis, A. (2002) Biochemical and morphological characterization of sulfur-deprived and H2-producing *Chlamydomonas reinhardtii* (green alga). *Planta* 214, 552–561.

Zou, N. and Richmond, A. (1999) Effect of light-path length in outdoor flat plate reactors on output rate of cell mass of EPA in *Nannochloropsis* sp. *Journal of Biotechnology* 70, 351–356.

6 Physiology and Genetics of Biofuel Crop Yield

Bharat P. Singh
Fort Valley State University, Georgia, USA

6.1 Introduction

The quantity of economic biomass accumulated by a plant depends on the: (i) amount of solar radiation intercepted by its leaves; (ii) efficiency of leaves to utilize the solar energy to produce carbohydrate; (iii) effectiveness of plant to convert photosynthate into new biomass; (iv) cost of maintenance as proportion of total metabolites produced; (v) sufficient sink capacity in the organ of economic value of accommodate metabolic products; and (vi) plant's choice to translocate metabolites to organ of economic interest. It so happens that the interest of the plant to produce grains for regeneration coincides with human's desire to harvest grain for human and animal consumption, and of late for biofuel production. But this human and plant concurrence of interests may not exist in the case of perennial grass crops. The continued survival of perennial grasses lies in extended longevity by strengthened root systems and range expansion through production of prolific amounts of seeds. Both of these plant's interests may run counter to the human's interest. Perennial grasses are grown for their vegetative biomass, which is harvested for forage often several times during one growing season depriving the plant of the opportunity to translocate maximum amounts of metabolites for storage in roots for initiating next year's growth. Many perennial grasses can also become noxious weeds in row crop production, so the harvest scheduling intentionally prevents these grasses from going into seed. In sweet sorghum, an annual crop, human and plant interests also appear to be in conflict. It would appear logical for the sweet sorghum plant to accumulate sugar in the stem as a ready source for translocation to seeds during grain filling. On the contrary, to get maximum yield of sugar, the plant is harvested in the soft dough stage to prevent this translocation. While it is true that several sweet sorghum cultivars have been developed, especially for growers in developing countries, that simultaneously produce reasonable grain yields, the act of diverting a significant portion of stem sugar into a separate harvestable commodity still interferes with the original plant intent of accumulating sugar in the stem for the sole purpose of supporting grain formation and filling.

Biofuel can be produced from both grain and vegetative parts of the plant. When grain is the raw material, the increase in biofuel yield can be attained by improving harvest per unit land area as well as by augmenting starch content or oil content of the grain. For crops such as maize and sorghum, either grain alone or in combination with stover can be of interest for conversion to biofuel. Both the total amount of biomass production ability and biomass amenability to deconstruction for conversion to biofuel are traits desirable in potential cellulosic

©CAB International 2013. *Biofuel Crops: Production, Physiology and Genetics* (ed. B.P. Singh)

biofuel crops. While selecting crops that show promise as cellulosic biofuel feedstock, a crop with high dry matter production ability that allocates greater proportions of assimilate to the vegetative parts over reproductive parts, thus, would be well suited for biofuel feedstock. In this respect, perennial grass such as *Miscanthus* is ideal as it sends less than 2% of the total photosynthate to the reproductive portion (Heaton *et al.*, 2008). Cellulose, hemicelluloses and lignin are key components of plant cell walls. Cellulose and hemicelluloses are of interest for ethanol and other alcohols, out of which cellulose is more amenable to deconstruction. Thus, high biomass producing plant species with high cell wall cellulose obviously are preferred biofuel feedstocks. Lignin lowers biofuel conversion efficiency. It is, however, important to plant structural integrity. Thus, although low lignin brown midrib (BMR) sorghum possesses greater energy conversion efficiency, questions linger on its usefulness as biofuel feedstock because these properties may also cause reduced dry matter yield, plant height and tiller number, and greater susceptibility to lodging (McCollum *et al.*, 2005). Identifying more efficient cell-wall-degrading enzymes and expressing them in recombinant transgenic plants may provide a way to overcome recalcitrance of lignified cell walls to hydrolysis.

Yield potential is commonly used to express the full economic yielding capability of a crop. The connotation includes the conditionality of ideal environment for growth, i.e. the crop is grown in an adapted climate, has been supplied with adequate nutrients and water, and weed, insects and diseases have been effectively controlled. The underlying assumption is that there are genes for yield whose expression depends on the existence of a favourable environment (Evans and Fischer, 1999). The farmer always strives to attain the full yield potential of crops to maximize return. However, one of the main criteria in selection of new dedicated biofuel crops is that they have the ability to grow on marginal lands and achieve adequate growth with less demand for external inputs. Thus, it is obvious that in case of the yield from biofuel crops the expectation is not to achieve full yield potential, but rather maximum yield under less than ideal growth conditions. Marginal lands have been classified such because it is not feasible to turn them into prime cropland. Many conditions, such as eroded top soil, slope, soil texture and structure, and toxicity and salinity may present impediment to growing crops with high nutrient and timely moisture demands on these soils. However, there are annual crops such as sorghum and millets that can grow on less than fertile soils, and adjust to the timing and amount of rainfall. Perennial grasses such as elephantgrass, switchgrass and *Miscanthus* are suited for marginal lands because of their long growing season and capacity to produce high biomass volume utilizing C_4 photosynthesis, high water use efficiency, and ability to sequester nutrients in rhizomes at the end of the growing season (Tilman *et al.*, 2006).

6.2 Yield Optimization

6.2.1 Approaches

Improvement in plant components that contribute to grain formation and filling can increase total grain yield in cereals. A number of plant characteristics, including plant height, number of tillers, time of anthesis, number of panicles per plant, panicle length, number of grains per panicle and size of grain contribute to the final yield. Both classical and molecular genetics have provided information regarding the genes that control several important yield controlling traits. In sorghum, six genes attributed to maturity (Rooney *et al.*, 2000) and four responsible for dwarfism (Hadley, 1957) have been identified by classical genetics while controlling regimes of the genome for plant height (Rami *et al.*, 1998), maturity (Lin *et al.*, 1995), number of tillers (Paterson *et al.*, 1995), seed weight (Pereira and Lee, 1995), panicle characteristics (Pereira and Lee, 1995) and stay green (Tao *et al.*, 2000) have been located using molecular markers. Hart *et al.* (2001) mapped a minimum of 27 quantitative trait loci (QTLs) in sorghum that control variation in nine morphological traits. Depending on the environment, four QTLs explained from 48.9 to 86.3% of the additive genetic variance in the number of basal tillers with head and seven QTLs were responsible for from 47.9 to 85.9% of variation in panicle width. Increase in starch concentration can also enhance starch yield. Grain starch content up to 135% of control has been reported following the constitutive expression of a regulatory variant of bacterial ADP-glucose pyrophosphorylase (Stark *et al.*, 1992).

Improvement in subsidiary traits can contribute toward increasing or stabilizing yield. For example, increasing photosynthesis, enhancing nitrogen assimilation and carbohydrate metabolism, and improving tolerance to stress are separate traits, but the composite of their individual contribution would reflect in the final yield. Hormones also effect plant growth and development cumulating into yield of biomass crops. Sakamoto and Matsuoka (2004) reported that modifying plant architecture improved photosynthesis and dry matter accumulation.

Biofuel yield enhancement in cellulosic crops can be achieved either by improving the overall biomass accumulation, or by increasing the polysaccharide content of the cell wall. Carpita and McCann (2008) have reported annotation and assembly into gene family of about 1200 cell wall-related genes involved in substrate generation, polysaccharide synthesis, membrane trafficking, assembling and turnover, secondary wall formation and signalling. Research on the pathway of cellulose synthesis is progressing and enzymes involved in the biosynthesis are being characterized (Bolwell, 2000; Persson *et al.*, 2005). Understanding of hemicelluloses' biosynthesis pathway, however, is farther behind (Sticklen, 2008).

Plant yield is a complex trait influenced significantly by both genetics and environment. Conventional breeding has remained the main route for breeding for yield and the gains made have been impressive, e.g. it has contributed to almost ninefold increase in grain yield of maize since the introduction of hybrid vigour cultivars (Dhugga, 2007). Conventional breeding aims at exploiting the natural diversity in a population. This exercise, however, can be aided by well-mapped populations of recombinant inbred lines (RILs), which permits fast gene identification by association mapping or QTLs of genes of value (Yu and Buckler, 2006).

Conventional breeding in self-pollinated crops for yield has relied on classical Mendelian segregation and selection from population segregates with the highest yield. Further generations are inbred to achieve homozygosity. On the other hand, in cross-pollinated crops, the main mode of yield improvement has been to exploit hybrid vigour. Conventional breeding has been effective for crops with normal ploidy ($2n$). Although breeders have relied on phenotypic markers for a long time, the advent of molecular markers has been a boon to efficient and economic breeding.

Without any other option, breeders had to rely for a long time on conventional methods for breeding new varieties of polyploid crops. Among biofuel crops, breeding has been particularly difficult for sugarcane because of polyploidy (generally an octaploid). Cultivated sugarcane varieties originated from the cross between *Saccharum officinarum* ($2n = 80$) and *Saccharum spontaneum* ($2n = 40$–128) and subsequently backcrossed twice (known as nobilization) with *S. officinarum* as the recurrent parent. Nobilization increased the genome complexity. Modern sugarcane cultivars have originated from first nobilized hybrids or their progenies. Because of ploidy, each allele may occur in multiple copies and replacing poor alleles with desirable ones is unreliable and much more time consuming than with diploid species. The introduction of recessive alleles is particularly difficult because they would not express themselves before each original allele is replaced by the introduced version. The probability of finding such a combination is infinitesimal. If a trait is quantitatively inherited where multiple genes contribute modestly and moreover if they complement each other for the expression, the probability of such recombination is also extremely small. It generally takes 12–15 years to complete evaluation in order to release a new sugarcane cultivar.

Genetic engineering has shown promise in improving the yield of biofuel crops through either increased yield or improved conversion efficiency. Eriksson *et al.* (2000) found that transgenic hybrid poplar grew faster and accumulated greater amount of biomass than control due to increased rate of gibberellin biosynthesis. A 20% increase in plant biomass in rice was achieved by improving the expression of ADP-glucose pyrophosphorylase using an endosperm-specific promoter (Smidansky *et al.*, 2003).

How this biomass increase came about is not clear. BSES Limited, an organization owned by Australian sugarcane growers and millers, intends to modify sucrose accumulation with RNAi constructs containing gene fragments from a common plant. The statement released by the organization informs that the purpose of breeding is to alter sucrose transport, carbohydrate metabolism or osmotic tolerance. The details have been declared confidential. The sugarcane genotypes developed from this effort are currently in field trials in Australia (DIR 095, 2009). Genetic screening for yield-related quality trait mutants can be useful in yield improvement. Robertson's Mutator (McCarty *et al.*, 2005), Activator (Bai *et al.*, 2007), RescueMu (Fernandes *et al.*, 2004) and TILLING (Targeting Induced Local Lesions IN Genomes) (Weil, 2007) are among means of generating rich allelic series.

Incorporation of classic GM traits such as insect and herbicide resistance into promising biofuel crops should be expected in the not too distant future. Dow AgroSciences (Indianapolis, Indiana, USA) has now perfected the skill to stack six Bt genes in maize in order to override insects' ability to develop resistance to one particular gene. The company has also developed a maize variety ('Enlist') that is resistant to 2-4 D and can thus be grown where 2-4 D is sprayed to control glyphosate-resistant weeds (personal communication). RNA silencing may also prove to be an effective tool in breeding for virus, bacterial and fungal pathogen resistant biofuel crops (Eamens *et al.*, 2008).

6.2.2 Future directions

As the number of researchers engaged in important biofuel crops increases, the research will expand to new areas that may be challenging but show long-term promise. A few examples are presented here.

There exists the possibility of creating two versions of annual grain crop species, one for high grain yield and another for high biomass. Crops such as maize and sorghum have wide natural diversity in plant size and architecture, and selection of genotypes directed at biomass yield offers promise. Mutations, such as *brachytics2* in maize and *dw3* in sorghum, that add extra layers of cells inside the stem show a possible avenue for adding extra biomass to these crops (Multani *et al.*, 2003). Thus, a high grain-yielding cultivar of maize for grain can be grown on fertile soil and another cultivar with high biomass production ability can be produced on marginal soils. Usually, grain production is more affected by the constraints of fertility and moisture than biomass because plant has a longer time period to assimilate vegetative dry matter than reproductive dry matter.

It is perceived that if the terpene content of pine can be significantly increased, it could provide abundant raw material for ethanol considering the large pine hectarage in different parts of the world and concentrations in regions such as the south-eastern USA. Terpenes can be easily converted to ethanol. The average percentages of terpenes in loblolly pine have been reported to be 2.3% in heartwood, 0.77% in inner sapwood and 0.35% in outer sapwood (Thompson *et al.*, 2006). Concentrated efforts are underway to increase pine terpene content. Semiz *et al.* (2007) reported the existence of significant genetic variation in terpene content in Scots pine grown across Finland and Turkey. Terpene synthase genes hold the key to the production of terpenoids. Up-regulation of terpene synthase has been achieved in *Arabidopsis* through genetic modification (Aharoni *et al.*, 2003). The terpene synthase gene has also been inserted into tobacco (Lücker *et al.*, 2004). Scientists from several institutions in the USA (University of Florida, University of California, Berkeley, National Renewable Energy Laboratory) aim to increase the current 3–5% terpene content of the existing loblolly pine cultivars to new ones that accumulate more than 20% terpene (personal communication).

There is an ongoing effort to change the C_3 mode of CO_2 fixation into C_4 mode with the assumption that it would result in increased photosynthate production (Farago, 2007). This projection is based on findings that the quantum yield of C_4 plants is higher at above 30°C (gap diminished with temperature decrease or rise in CO_2 partial pressure) and they possess a greater rate of CO_2 assimilation per unit leaf nitrogen (difference reduced as leaf area index and/or canopy nitrogen increased) (Evans and Von Caemmerer, 2000). However, the task of such transformation is formidable. The C_4

pathway depends on a delicately balanced conductance rate across the bundle sheath to maintain a high level of inside CO_2 but yet not so low as to prevent escape or diffusion of oxygen out of it. Evans and Von Caemmerer (2000) suggest that engineering of novel Rubisco into C_3 plants could be a more achievable goal. Sage and Zhu (2011) are hopeful that the induction of new C_4 model species, *Setaria viridis*, will help in developing better understanding of the C_4 mechanism through discovery of new genes involved in this type of photosynthesis, making C_4 engineering feasible in the next few decades. The success in engineering of major enzymes of the C_4 metabolic cycle into C_3 plants has been encouraging, but knowledge of genes controlling Kranz anatomy is still lacking.

Recently, research interest has been aroused in introducing a perennial character into cereal grains (Peterson, 2008). This approach may particularly be of value in developing cereals for biomass suitable as biofuel feedstock. Success in this line of research will add to the time period of vegetative growth during early season when cold soils either prevent or retard the seed germination while above-ground temperature is adequate to support photosynthesis. Perennial crops rely on root-stored carbohydrate to begin growth and support till above-ground growth becomes self-sustaining. Perennials have another advantage over annuals in that while annuals have to rebuild their roots every year, perennials contribute annually a portion of the dry matter for the growth and maintenance of the roots. In a comparison in southern England, it was found that *Miscanthus*, a perennial, appropriated approximately 40% of the total biomass to the root over a 3-year period against 14–26% annually by the annual maize (Beale and Long, 1995; Bonifas and Lindquist, 2006). Annuals not only have to share a larger portion of the total accumulation with the roots, but this translocation also entails expense in dry matter, which has to be converted to energy to carry out the transport.

Results of the research by Hu *et al.* (2003) suggest that the progenitors of annual cereals were of a perennial nature. The annual growth habit is an adaptation mechanism subsequently developed for adjustment to environmental changes. The annual cultivated and wild perennial species currently co-exist in many grass genera. Researchers at the Land Institute, Kansas (USA), reported shoot emergence from the rhizomes of perennial sorghum at least 1 month in advance of the shoots from seeds planted normally around 15 May (Cox *et al.*, 2006). Shoots from the rhizome also grew faster compared to those from seeds, which could be due to a readily available supply of food from the underground rhizome. Piper and Kulakow (1994) obtained 62% higher above-ground biomass dry weight in the progeny of annual × perennial sorghum compared to the annual parent. The molecular mapping of F_2 from the cross between cultivated rice (*Oryza sativa*) and a perennial wild relative (*Oryza lonistaminata*) revealed that two dominant-complementary genes (*Rhz2* and *Rhz3*) controlled rhizome trait (Hu *et al.*, 2003). *Rhz3* was located between markers OSR16 (1.3 centimorgans (cM)) and OSR 13 (8.1 cM) on chromosomes 4 and *Rhz2* between RM119 (2.2 cM) and RM273 (7.4 cM) on chromosome 3. Comparative mapping indicated that the two genes closely corresponded to two QTLs controlling rhizome trait in the perennial wild relative of sorghum, *Sorghum propinquum*. DNA markers closely linked to *Rhz2* and *Rhz3* genes will facilitate perennial cereals development. Perennial cereal production systems utilizing a low resource demand crop like sorghum could be of special interest for cultivation on eroded soils and fragile ecosystems.

The increase in sugar percentage of stem juice should leave more of the photosynthate in a form easy to convert into ethanol. Efforts to increase stem sugar content, however, have proved difficult in sugarcane because of the large number of genes involved in sugar content determination, each with only modest effect. In addition, sugarcane being polyploid, multiple copies of each gene occur in the cell, making it very difficult to accumulate ‘sugary alleles’ in one genotype. This may explain why sugar content in sugarcane cultivars has remained unchanged for the past several decades. Research is in progress for breaking the homeostasis barrier for sugar sink. It will be very important to the enhancement of stem sugar content in both sugarcane and sweet sorghum.

To bypass the complex sugar synthesis and regulation mechanisms, an alternative approach to incorporate an isomeric form of sucrose that does not participate in plant growth and maintenance metabolism process is gaining favour. It has been possible to double the stem sugar content in sugarcane by introducing a gene for sucrose isomerase enzyme (Wu and Birch, 2007). The modified plant produced normal amounts of sucrose but also an equal amount of isomaltulose. As isomaltulose is not utilized by the plant, it is shifted to storage tissues. Several regulatory elements (promoters) from sugarcane and maize are being tested for their control on the functioning of sucrose isomerase enzyme. Different combinations of regulatory elements are also being tried to obtain maximum isomaltulose without detrimental effect on the plant (Tammisola, 2010). The performance evaluation of a number of isomaltulose sugarcane lines under field conditions to isolate the best among them for commercialization is currently underway in Australia.

Cold tolerance research should find priority in future biofuel crop development as many of the promising species are of tropical origin and adaptation to cold could be requisite to their successful establishment in temperate climates. Gene expression profiles of sugarcane in low and normal temperatures have revealed the existence of 20 cold-responsive genes (Nogueira *et al.*, 2003). Elucidation of the functions of these genes will tell if their adjustment will lead to improved cold tolerance. Readers are referred to the energycane section of Chapter 8, this volume, for a detailed account of ongoing cold tolerance efforts in *Saccharum* spp. Genetic engineering should help introgression of cold-tolerance genes from alien species. Hairgrass (*Deschampsia antarctica*) survives as sole grass species in the Antarctic peninsula. It can withstand temperatures down to −30°C in winter and occasionally to −15°C during the growing season as *IRIP* (ice recrystallization inhibition protein) genes prevent ice recrystallization inhibiting small ice crystals from recrystallizing into lethal aggregations (John *et al.*, 2009). The gene *DaIRIP4* from the IRIP family expressed in *Arabidopsis* made it also tolerant to freezing. Thus these genes are a potential valuable resource for improving a number of important cold-sensitive grass species such as rice, wheat, sugarcane and elephantgrass.

6.3 Biomass Accumulation

The following equation describes the basis for primary productivity (P_p) in crop species (Long *et al.*, 2006):

$$P_p = S_t\, \varepsilon_i\, \varepsilon_c / k \qquad (6.1)$$

where S_t = total solar radiation across the growing season; ε_i = crop photosynthetically active radiation (PAR) interception efficiency; ε_c = crop efficiency of converting solar energy to biomass energy; k = energy content of plant mass (≈17.5 MJ g^{-1} for vegetative biomass).

Improvements in biomass accumulation by field crops until now have been largely achieved through ε_i enhancement. Current crop cultivars have been improved for rapid ground cover and lodging resistance to enable leaves to be fully exposed to incoming solar radiation for the complete life span. Beadle and Long (1985) have pointed out that now most improved field crop cultivars have attained $\varepsilon_i > 0.9$, leaving little scope for further improvement in this value. Based on the above reality, enhanced photosynthesis is the remaining channel for future improvement in crop biomass productivity.

Current estimates of the maximum ε_c over the growing season for C_3 and C_4 species are around 0.024 and 0.034, respectively, with brief period peaks of approximately 0.035 for C_3 and 0.043 for C_4 species (Monteith, 1977; Beadle and Long, 1985, 1995). Long *et al.* (2006) state that the theoretical maxima for ε_c in C_3 and C_4 plants are approximately 0.051 and 0.060, respectively, leaving the possibility open for improving biomass production through photosynthesis. At present, the leaves become light-saturated at peak periods at much lower light intensities and the rest of the energy is wasted. In a leaf canopy with a suitable architecture so as to disperse the sunlight to leaves below, the energy otherwise wasted could then be utilized in photosynthesis. Ort and Long (2003) have observed that approximately 700 μmol m^{-2} s^{-1} light energy is intercepted during the maximum full sunlight by a leaf positioned at 75° from horizontal, an amount sufficient to saturate the photosynthesis of the top leaves of C_3 plants but leaving 1300 μmol m^{-2} s^{-1} of direct light to reach lower leaves for added photosynthesis. A collateral benefit of steeper leaf angles in the upper canopy is reduced heat load during midday, leading to increase in water use efficiency and decrease in the risk of overheating (King, 1997). The deficiency in the biosynthesis of the plant hormone brassinosteroid has been shown to result in the formation of very erect leaves in rice enabling plants to capture light efficiently at high density and increase biomass and grain yield (Morinaka *et al.*, 2006; Sakamoto *et al.*, 2006).

High light intensity at midday without dissipation can cause damage to photosystem II (PSII). Epoxidated xanthophylls are produced by plants to dissipate the increase in thermal energy and protect PSII. However, such events decrease the quantum yield of PSII and uptake of CO_2 (Long *et al.*, 2006). A time period elapses between the onset of photoprotection and its relaxation at the cost of photosynthesis and ε_c gain. If relaxation lag period could be shortened, it would result in an increase in ε_c. Long *et al.* (1994) observed that there was a close relationship between the rate of photoprotection recovery and xanthophyll cycle capacity. Wang *et al.* (2002) reported that 'super high yield' rice cultivars recovered much faster from photoprotection than lower yielding cultivars. A breakthrough in ε_c gain, of course, will be possible if the need for photoprotection were to be completely eliminated. The findings of Falkowski and Dubindky (1981) show such promise. They identified algae associated with corals with the ability to withstand 1.5 times full sunlight and function at maximum photosynthetic efficiency without any need for photoprotection.

Crop species with the inherent ability for accumulating biomass actively for a longer period under given climatic conditions are able to accumulate greater amount of biomass than another species with similar canopy but shorter life span. In the Midwest USA, maize and *Miscanthus* × *giganteus* developed a similar canopy with $\varepsilon_i > 0.9$ but the latter accumulated 59% more biomass (Dohleman and Long, 2009). This was due to the greater cold tolerance of *M.* × *giganteus*, which allowed it to initiate vegetative growth earlier in the season and develop closed canopy 4 weeks before maize and maintain it 4 weeks after maize senesced. The extended growing season allowed *M.* × *giganteus* to cumulatively intercept approximately 61% more radiation than maize.

Interesting contrasts in the photosynthetic behaviour from the above and similar studies have emerged, which points to the opportunities in both *Miscanthus* and maize for improvement in photosynthate accumulation. Beale *et al.* (1996) had previously shown that the rate of photosynthesis in *M.* × *giganteus* was inhibited at high light intensities as evidenced by CO_2 uptake to incident photon flux. In the study of Dohleman and Long (2009) also, CO_2 uptake of *M.* × *giganteus* during summer months was significantly lower than maize. But on the other hand, not only had *M.* × *giganteus* a longer growth period than maize, it showed greater CO_2 uptake than maize during the early and late periods of the maize growing season when the temperatures had moderated. Naidu *et al.* (2003) reported that at 10°C, light saturated photosynthetic rate of *M.* × *giganteus* was 10 μmol m^{-2} s^{-1} compared to 1.5 μmol m^{-2} s^{-1} for maize. Thus, opportunities for improvement in photosynthesis exist in both, for maize at low temperatures and for *Miscanthus* at high irradiance.

Cold tolerance of C_4 crops is of special importance in developing cellulosic biofuel feedstocks in temperate climate zones, as several important perennial grass species being targeted for adoption behave more like maize than *Miscanthus*. It will not be sufficient to improve cold tolerance to improve survival for expanding the production range; in order to get full benefit the adapted crops also should behave like *Miscanthus* for ability to actively grow for longer duration in the year to maximize biomass output. Although a number of C_4 species also can survive cold weather, they do not possess the unique low temperature photosynthesizing ability of *Miscanthus*.

Thus, the cold tolerance and cold photosynthesis traits need separate examination. Matsuoka *et al.* (2001) reported that metabolic control of light-saturated C_4 photosynthesis in cold-tolerant *Flaveria* and *Amaranthus* may reside in the photosynthetic carbon metabolism enzymes pyruvate orthophosphate dikinase

(PPDK), Rubisco (ribulose 1,5-biphosphate carboxylase), and phosphophenolpyruvate dikinase. Pitterman and Sage (2000) suggest that while PPDK plays the main role, other C_4 enzymes such as Rubisco may also influence in low temperature photosynthesis. Naidu *et al.* (2003) did not observe any change in PPDK amount in either maize or *M.* × *giganteus* at cold temperatures (14°C/11°C day/night) compared to warm temperature (25°C/20°C day/night), but at low temperature large subunits of PPDK decreased >50% and Rubisco >30% in maize while no such decrease happened in *M.* × *giganteus*.

In sorghum, the biomass accumulation period can be substantially increased by selecting photoperiod-sensitive cultivars for production in temperate climates. The species comprises both photoperiod-sensitive and photoperiod-neutral types. Photoperiod-sensitive (PS) cultivars need day length to decrease to less than 12 h and 20 min to initiate floral development. As the day length required for flowering arrives too late in the growing season of PS cultivars at most latitudes of the continental USA, the growing season is essentially devoted to vegetative growth. According to McCollum *et al.* (2005), in Texas (USA), high biomass PS sorghum yielded 26–43% more than photoperiod-insensitive cultivars. The delayed maturity of high biomass PS sorghum also postpones the decline in feedstock quality that comes with drying and allows prolonged harvest. Rooney *et al.* (2000) noted that the task of breeding PS high biomass sorghum is somewhat facilitated by the fact that they can be even derived by the cross of two parental photoperiod-insensitive parents.

Cellulosic biomass yield comes mostly from the plant vegetative portion comprising leaves and stems and meristematic tissues such as apical meristem and vascular cambium which generate these vegetative tissues. When a plant enters into the reproductive phase from the vegetative phase, the vegetative growth either greatly slows down or stops. Thus, a logical way to increase the cellulosic biomass would be to prolong the vegetative growth period by postponing flower initiation. Over-expression of the flowering time gene, *FLOWERING LOCUS C* (*FLC*) from *Arabidopsis* to tobacco resulted in significant delay in flowering and simultaneous increase in biomass yield (Salehi *et al.*, 2005). Melzer *et al.* (2008) reported that simultaneous mutations of *Arabidopsis* flowering-time genes, *SUPPRESSOR OF OVEREXPRESSION OF CONSTANS 1* (*SOC1*) and *FRUITFULL* (*FUL*) or *FLOWERING LOCUS T* (*FT*) and *FUL*, prolonged the vascular cambium activity and accelerated vegetative growth and dry matter gain. In tree crops, wood (also known as secondary xylem) constitutes the bulk of biomass, which, like herbaceous crops, is also differentiated from vascular cambium. Schrader *et al.* (2004) have profiled vascular cambium transcriptome of poplar and identified a number of potential regulators of which several are putative orthologues of *Arabidopsis* apical stem identity genes.

Traditionally, breeding has been targeted to shorten the stature of crop varieties to prevent lodging. Heights of tall grass species were also reduced to increase the leafiness for improved forage quality and palatability. In biofuel, taller plants should have the advantage of providing a greater sink for photosynthate accumulation. Plant hormones play an important role in regulating plant height. Gibberellin (GA) is well known for its cell elongation ability. By altering GA biosynthesis or signalling, it therefore should be feasible to engineer taller biofuel crops by increasing cell elongation. Over-expression of GA20-oxidase in transgenic poplar enhanced GA production, causing increase in plant height, xylem fibre length and stem dry weight (Eriksson *et al.*, 2000). Over-expression of GA receptor GA INSENSITIVE DWARF 1 (GID1) degraded GA signalling suppressor DELLA and transduction of GA signal (Hirano *et al.*, 2008) resulting in growth and stem elongation in aspen (Mauriat and Moritz, 2009).

6.3.1 C_4 photosynthesis and biomass accumulation

Since C_4 grass species feature prominently in the list of promising biofuel feedstocks, a review of their desirable features that bestows upon them the capability to produce high amounts of biomass is presented. On first appearances, C_4 plants seem to have a disadvantage over C_3 plants because they spend typically 2 ATP more per CO_2 molecule assimilated. But this disadvantage is more than compensated by the successful protection it provides to C_4 plants from photorespiration to which C_3 plants lose about 30% of carbohydrate formed during photosynthesis. At the heart of the discrepancy in photorespiration between C_3 and C_4 lies the behaviour of the opportunistic enzyme Rubisco, a key enzyme in photosynthesis. It partakes in both carboxylation and oxygenation. Carboxylation and oxygenation behaviour of Rubisco depends on the relative concentration of CO_2 and O_2 in its vicinity. Rubisco oxygenation is the cause of photorespiration in C_3 plants while C_4 plants can suppress this reaction by 80% or more (Kanai and Edwards, 1999) by maintaining high tissue CO_2 concentration. The C_4 mechanism is basically a CO_2 pumping mechanism added on to the C_3 CO_2 fixation process, which raises the concentration of CO_2 around Rubisco to about 10 times the current atmospheric CO_2 levels (Hatch, 1987). This is feasible because of the 'Kranz' anatomy of C_4 leaves. In this arrangement, Rubisco is isolated in the inner bundle sheath surrounded by mesophyll tissues. The only place for atmospheric gases to occupy in the leaf is the intercellular spaces within mesophyll tissues and CO_2 is subsequently transferred through chemical reaction to the bundle sheath for assimilation into carbohydrate. This two-phase travel mode separates CO_2 movement in the leaf from other atmospheric gases, but the cost of this additional travel is around 2 ATP.

The advantage of C_4 photosynthesis is more during warm days because as the solubility of CO_2 relative to O_2 declines with temperature, photorespiration in proportion to photosynthesis increases in C_3 plants but no such change occurs in C_4 plants. But below-leaf light saturation levels and at leaf temperatures at or below 25°C photosynthesis is linearly associated with radiative flux, and because C_3 plants have lower energy requirement for carbon assimilation, they are more efficient than C_4 plants (Long *et al.*, 1993). In light-saturated leaves even below 25°C, the added cost of CO_2 assimilation is more than compensated by lack of photorespiration for C_4 to prove more efficient than C_3 plants (Long *et al.*, 2006). Total biomass accumulation is dependent upon photosynthesis during the entire growing season; temperatures in tropical climates and during summer months in temperate climates when most of plant growth takes place are significantly above minimum threshold, and irradiance also considerably exceeds leaf light saturation. Thus, Rubisco operates at all times at CO_2 saturation and does not get the opportunity to engage in oxygenation. Since the rate of photosynthesis increases with temperature, the catalytic activity of Rubisco is from two- to fivefold higher in C_4 plants than C_3 plants in warmer climates (Seemann *et al.*, 1984; Sage, 2002). As an outcome, at daytime temperatures above 25–30°C, C_4 plants may show radiation use efficiency that is more than 50% above C_3 species (Long, 1999). Similarly, C_4 species possess 1.3 to 4 times greater nitrogen use efficiency and water use efficiency than C_3 species at warmer temperatures because of elevated photosynthetic capacity (Sage and Pearcy, 2000; Ghannoum *et al.*, 2011). Among the secondary benefits of high Rubisco efficiency in C_4 species is that they require from 50 to 80% less Rubisco at a given photosynthetic rate than C_3 species (Sage and Percy, 1987, 2000). This frees up N reserves for use in other purposes.

6.4 Mineral Nutrition

The meaning of the terms 'high input', 'low input' and 'no input' in reference to mineral nutrition is often ambiguous. Nature is dynamic, and be it from rain or from the microbial and earthworm activities in the soil, nutrient balance and organic matter content is constantly in a state of flux. So, 'no input' only implies the lack of external inputs and in no way minimizes the role nature plays in soil management. It is difficult to produce crops without external input because plants' nutrient needs for optimal growth are usually greater than the amount soil is capable of supplying from its reserve and nature's replenishment year after year. Therefore, it is a common practice to supplement nutrients to compensate for the soil deficiency. 'Low input' is the first order nutrient management alternative. Actually, the term 'low input' is a misnomer; it should be correctly characterized as 'efficient input'. The true intent in low input systems is to apply nutrients in the form of fertilizer only to the amount that cannot be supplemented through other innovative alterations in the crop management system and ensure that the placement of fertilizer is precise to the root uptake zone. Some of the soil inherent improvements can be accomplished through increased organic matter and nitrogen supplementation through legumes, adoption of minimum tillage to prevent nutrient loss from erosion and runoff and improved genetic ability of the crops for soil mining ability as a result of an extended root system, mycorrhizal association and enzyme systems that enhance nutrient concentration in soil solutions for uptake. The term 'high input' implies fertilizer applied to meet all perceived crop needs but also accompanies connotation of an excess left over.

There are two basic components to the mineral nutrient use by a plant. First, the plant has to extract the minerals from the soil and second to assimilate them into various compounds needed for physiological functions, and growth and development. The root is the plant organ of accumulation and its character determines how well the plant can accumulate nutrients from various soil types and under different conditions. Root characteristics can be broadly separately into two groups: (i) ability to explore the soil and reach to where different nutrients are located; and (ii) ability to extract the nutrients from soil and transport them to other plant parts as needed. To grow under low input conditions, a plant may need a large root system with long lateral roots to find patches in soil rich in mineral nutrients. There is a critical difference in the seasonal root growth pattern of low-input efficient annual versus perennial crops. Annual crops, in order to fully utilize the nutrients collected, should have the ability to remobilize the nutrients assimilated into the root to sinks of economic value during the terminal growth period. On the other hand, storage organs in the roots of perennial crops support plants' early growth in the coming season, thus, to grow efficiently under low input requires translocation of nutrients from senescing top to the root for an adequate reservoir.

6.4.1 Nitrogen nutrition

Nitrogen is required in the largest quantity by the plant but its availability is usually the most limited of all the essential elements. Paucity of N is ironic considering the fact that about 78% of air is made of nitrogen. The amount of N taken up by plants is often greater than the amount replenished through the natural process. The upper soil horizon contains most of N in the form of organic fractions. Although N content of organic matter averages approximately 50 g N kg^{-1} (w/w), the organic matter content of the soil plough layer ranges between 0.2 and 4 g N kg^{-1} (Barber, 1995). The gap between available soil N and the crop need must be met by the external input to achieve satisfactory growth and yield (Fageria *et al.*, 2006).

There has been a sevenfold increase in N-fertilizer use during the past four decades, but it has been accompanied by only a twofold increase in agricultural food production (Hirel *et al.*, 2007). It is estimated that only 25 to 50% of the applied N is taken up by plants (Peterson and Frye, 1989). The rest, except for a small portion immobilized by microbial processes in the soil, is not only lost but also becomes a matter of environmental concern. The left-over fertilizer N can vaporize as nitrous oxide, a potent greenhouse gas, emit toxic ammonia into the atmosphere, leach to groundwater to create a hazard to the drinking water supply, or run off to freshwater and marine ecosystems causing eutrophication. The Haber-Bosch process of N-fertilizer manufacturing utilizes fossil fuel for feedstocks and with the rising cost of fossil fuel, farmers have experienced steep increases in N costs and a resulting squeeze on profitability.

Biofuel crops provide a new opportunity in farming based on plant species or cultivars with recognized ability to produce high feedstock yield at low N input. Annual crops such as sorghum (Anderson *et al.*, 1995), perennial grass crops such as *Miscanthus* (Lewandovski *et al.*, 2000) and tree crops such as willow (Heller *et al.*, 2004) and crop varieties selected for N-efficiency can ensure farm profitability but reduce the crop demand for N-fertilizer. Symbiotic N-fixation is estimated to contribute approximately half the amount of N annually applied as N fertilizer (Smil, 2006) and is the historically established way of increasing soil N content. Integrating legumes to the biofuel crop production systems could further reduce the need for external N input.

As N in the soil moves with ease by mass flow, root size and spatial distribution is not considered limiting for N acquisition when soil N is adequate (Sinclair and Vadez, 2002). Under conditions of low N supply or moisture stress, however, N supply to plants is constrained as mass flow may not meet the demands of the growing plant. In these situations, root size and distribution assumes critical significance for the utilization of spatially distributed N. Several researchers (Wang, Y. *et al.*, 2003; Chun *et al.*, 2005a; Tian *et al.*, 2005) report axial and lateral root elongation enhanced uptake at low N supply. However, there is a minimum N threshold below which lateral root growth is inhibited and only external N input can restore this growth (Chun *et al.*, 2005b; Guo *et al.*, 2005).

Strong interaction has been shown between N acquisition and N utilization under low N input (Moll *et al.*, 1982; Bertin and Gallais, 2001). Ismande and Touraine (1994) found N accumulation to be regulated by shoot demand. In a model proposed by Cooper and Clarkson (1989), balance in the movement of N upwards via xylem to the top and downward through phloem to the root determines the rate of N uptake. Under high demand most of the N translocated via xylem to the top is utilized for shoot growth and a small amount is re-translocated to the root, signalling roots to increase uptake. On the other hand, when N is translocated to shoots in excess, it is translocated back to the root, which serves as a signal to decrease the uptake.

Plant roots absorb N as NO_3^- and NH_4^+. Nitrogen is mainly taken up as NO_3^- in aerated soils and as NH_4^+ in submerged or poorly aerated or cold soils. In agricultural soils, NO_3^- content ranges from 7 to 140 mg kg^{-1} (Wiren *et al.*, 1997) while NH_4^+ is available in much smaller amounts. NO_3^- assimilation requires up to 20 ATP mol^{-1} of energy while NH_4^+ assimilation needs only 5 ATP mol^{-1} (Salsac *et al.*, 1987).

Nitrate uptake is carried out by two nitrate transporters (NRT) systems, the high affinity transport system (HATS) and the low affinity transport system (LATS), both encoded by multiple genes (Glass *et al.*, 2002; Orsel *et al.*, 2002). The expression of HATS and LATS is either constitutive to roots or induced by nitrate concentration (Forde, 2000). High affinity and low affinity ammonium transporters are responsible for ammonium uptake by the root system. Multiple gene families encode these transporters (von Wirén *et al.*, 2001; Glass *et al.*, 2002). Ammonium is the only form in which nitrogen is assimilated into plant tissues (Harrison *et al.*, 2000). Nitrate is reduced in two steps, first into nitrite and then to ammonium catalysed by nitrate reductase and nitrite reductase, respectively. Ammonium is converted into amino acid glutamine by enzyme couple glutamine synthetase (GS)/glutamate synthase for assimilation into the plant (Miflin and Lea, 1976; Andrews *et al.*, 2004). Alternatively, ammonium is assimilated into asparagine by asparagine synthetase (Hirel and Lea, 2002). Nitrogen is assimilated into glutamine during the day and into asparagine in the dark (Lillo, 2004). The GS activity significantly influences plant N status irrespective of soil N conditions (Hirel *et al.*, 2005; Kichey *et al.*, 2006). Gallais and Hirel (2004) showed a strong positive correlation between GS activity and kernel yield in maize. The QTLs for GS have been found to coincide with QTLs for yield (Hirel *et al.*, 2001). A QTL for thousand kernel weight was coincident with *Gln1-4* locus and two QTLs for thousand kernel weight and yield were coincident with *Gln1-3*. Over-expression of *Gln1-3* in transgenic plants was also accompanied with an increase in the number of kernels (Below, 2002). A host of additional enzymes participate in further transformation of basic assimilates for different plant functions and organ formation. A number of studies (Groat *et al.*, 1984; Jessen *et al.*, 1987; Degenhart *et al.*, 1992) have shown that it is possible to select for increased or decreased activities of several enzymes involved in the synthesis of N into different compounds.

Plant N composition is mostly in the form of proteins and nucleic acids (deoxyribonucleic acid (DNA) and ribonucleic acid (RNA)). These compounds make up the major portion of protoplasm, and N shortage results in their scarcity and the resultant cell division inhibition and growth reduction. Nitrogen is a main constituent of chlorophyll and thereby plays a major role in photosynthesis (Murata and Matsushima, 1975; Swank *et al.*, 1982). Close to 50% of the total soluble leaf protein of C_3 species (Mae *et al.*, 1983) and up to 20% of C_4

species (Sage *et al.*, 1987) is in the form of a single enzyme Rubisco, a catalyst for CO_2 fixation. In grain crops, balance between N allocation to maintain photosynthesis and its reallocation for seed formation is of importance. On the other hand, keeping the leaf active as long as possible during the growing season would be a priority in order to continue photosynthesis for maximum biomass formation where vegetative dry weight is of concern. The 'stay green' trait has been studied extensively for prolonged leaf photosynthetic activity (Ma and Dwyer, 1998; Borrell *et al.*, 2001). In both sorghum and maize, 'stay green' plants had prolonged photosynthesis and simultaneous N uptake. This resulted in more biomass accumulation in sorghum and higher yield in maize. Extrapolating, 'stay green' should have a distinct advantage in obtaining high biomass yield in cellulosic sorghum while it would be of questionable value for sweet sorghum where maximum sugar development takes place when the leaves are green in the normal course. Subedi and Ma (2005) observed that 'stay green' phenotype expression in maize was only under adequate N supply, but more extensive studies under different N levels and on different crops is needed to fully comprehend where this trait can be useful.

Nitrogen has a significant effect on plant vegetative and reproductive development. Vegetative components affected by soil N include number of leaves and their rate of appearance, leaf area, and node and tiller number (Snyder and Bunce, 1983; Vos and Biemond, 1992; Sagon *et al.*, 1993; Lemaire *et al.*, 2007). Nitrogen influences vegetative growth of grasses by affecting cell division rather than final cell size (Gastal and Nelson, 1994). Response to N-deficiency is variable among crop species; in potato, leaf area is reduced while in maize the leaf area is almost unchanged but photosynthesis and radiation use efficiency is reduced (Vos and van de Putten, 1998; Vos *et al.*, 2005). In cereals, the reproductive parameter most affected by the N deficiency is the grain number (Mae, 1997; Martre *et al.*, 2003).

Nitrogen translocation is based on organ needs at different growth periods. More N is allocated to the roots at early growth so that the plant can establish a sizeable root system capable of scrounging nutrients from a large volume of soil throughout growth and development (Li *et al.*, 2001; Niu *et al.*, 2007). Nitrogen is translocated from older leaves to younger leaves and from vegetative organs to reproductive organs. Singletary and Below (1990) observed that remobilization from the stem was the source of a substantial portion of N for young developing maize kernels. Translocation of N from vegetative tissues accounted for more than 75% of grain N in oat and wheat (Cataldo *et al.*, 1975; Dalling *et al.*, 1976).

For efficient management of N inputs, it is important to know the growth span when it is needed most by the plant. For example, Fageria *et al.* (2006) report that lowland rice demand for N is at the peak between 20 and 100 days and for maize from 20 to 90 days after germination. Uptake of 110 kg N ha^{-1} in rice produced 5580 kg grain and 7740 kg shoot dry weight, and 210 kg N ha^{-1} in maize yielded 8150 kg grain and 13,700 kg dry shoot matter. The plant N status before anthesis is important for kernel set, number of kernels and translocation of carbon assimilates to kernels in maize (Below, 1987; Neumann *et al.*, 2002). In spring wheat and rice, N deficiency around anthesis results in reduction in grain number, the degree of reduction dependent upon the duration and intensity (Mae, 1997; Demotes-Mainard *et al.*, 1999).

Nitrogen use efficiency (NUE) is achieved by the integrated improvement in the plant N uptake, assimilation and remobilization capability (Good *et al.*, 2004). Plant species and genotypes within species differ in their ability for N uptake, utilization and conversion into economic entity under low input. Lemaire *et al.* (1996) observed that N uptake by maize and sorghum were similar at high N input, but under N limitation, sorghum plants acquired significantly more N than maize. A more developed and branched root system in sorghum possibly permitted it to scrounge for N from a greater soil mass than maize under N shortage (Hirel *et al.*, 2007). Anderson *et al.* (1995) reported similar ethanol yield in sorghum with less than 50% total N compared to maize. The existence of genotypic variation for NUE has been recognized for a long time (Smith, 1934; Springfield and Salter, 1934). Svecnjak and Rengel (2005) observed differences among spring canola genotypes in NUE, which resulted in greater biomass with parallel performance at high and low N inputs. In winter wheat, Le Gouis *et al.* (2000) noted significant genetic variability for grain yield at low N, which was accompanied by significant genotype × N level interaction. Tian *et al.* (2006) observed that N-efficient maize genotypes developed more roots compared to inefficient genotypes utilizing the same amount of N. As NUE is genetically determined, it obviously could be modified through breeding. Yanagisawa *et al.* (2004) reported enhanced plant growth and nitrogen content by the overexpression of DOF1, a transcription factor involved in the activation of genes encoding enzymes associated with organic acid metabolism. This would suggest that manipulating the level of expression of regulatory proteins may provide a route for improving NUE in plants. Rauh *et al.* (2002) and Loudet *et al.* (2003) have identified a number of loci associated with total, mineral and organic N content and NUE in *Arabidopsis*. It should pave the way for studying the role of these different N forms in other crop species.

Most of the breeding of crops for productivity has been carried out at adequate N levels. However, Below (2002) found cultivars that performed well at high N were also superior at low N. Improved performance, thus, appeared to be due to comprehensive improvement in physiological traits related to yield consistent at all N levels. On the other hand, Presterl *et al.* (2003) and Gallais and Coque (2005) are of the opinion that selection for low-input should be carried out at N-deficiency conditions. Lian *et al.* (2005) observed in rice that most of the QTLs related to shoot and root seedling growth were different under high and low N conditions.

Joffre (1990) observed that the difference in the life cycle between annual and perennial grass species affected soil N availability and plant uptake. In south-western Spain, yearly accumulation of N in soils under perennial grasses was found to be greater than under annual grasses (103 kg N ha^{-1} versus 83 kg N ha^{-1}, respectively). The greater efficiency of perennials to utilize soil N contributed to their ability to commence growth with the onset of the rainy season while a significant portion of soil N leached while seeds of annual grasses germinated and established. Internal N cycling plays an important role in early season growth of perennial grasses. The deciduous and evergreen grasses, however, differ in their mode of remobilization (Thornton and Millard, 1993; Bausenwein *et al.*, 2001a). Thornton and Millard (1993) quantified the internal cycling of N in deciduous purple moorgrass. New shoot growth in spring was supported by remobilization of N from both roots and basal internodes. The contribution of N from roots was greater than the basal internodes. The remobilization of N mainly occurred before the full N uptake by roots was established. The amount of N remobilized to new shoots was independent of soil N supply. However, the same research team (Bausenwein *et al.* (2001b) and Thornton and Bausenwein (2000)) later confirmed that 40–50% of plant N in purple moorgrass can be derived from current uptake as early as March, suggesting that both sources of early season N supply existed simultaneously. Bausenwein *et al.* (2001b) studied the N translocation pattern in evergreen temperate grasses, red fescue and colonial bentgrass. Contrary to deciduous purple moorgrass, the root system did not contribute to the remobilization of N for new growth in red fescue or colonial bentgrass. The N cycling to the new above-ground growth was supplied by overwintering tillers. Remobilized N met 70 and 82% of red fescue and colonial bentgrass needs, respectively, in early spring declining to 34 and 45%, respectively, by mid-June. So it can be summarized that previously stored N is the major source of N to next season's perennial early growth, the reserve N can be stored in different plant parts and both N uptake and remobilization can function simultaneously depending on the plant species and the prevailing environmental conditions favouring soil N mineralization. Trees have a considerable influence on the nutrient status of the soil (Ritter *et al.*, 2003) and can alter soil properties relatively rapidly (Reich *et al.*, 2005). Meiresonne *et al.* (2007) noted that nutrient cycling in poplar plantations was very efficient and negligible loss occurred from leaching. Approximately 80% of total N input came from litterfall. Berthelot *et al.* (2000) observed similar N input from litterfall in coppiced poplar.

Several N_2-fixing diazotrophs, *Gluconacetobacter diazotrophicus, Herbaspirillum* spp. and *Burkholderia* spp., have been isolated from the interior of sugarcane plants (Boddey *et al.*, 2003). These diazotrophs are called endophytic because of their preference for the internal tissues of the plant. It has been shown that some Brazilian sugarcane varieties are able to derive substantial bacterial nitrogen fixation benefits from such an association (Urquiaga *et al.*, 1992). However, it is not yet clear which endophyte is responsible for N gain and the site or sites within the sugarcane plant where N_2-fixation mostly takes place. Maximum N benefit is derived when molybdenum is adequately present in the soil and moisture is not limiting. However, until the phenomenon of N_2-fixing diazotrophs is deciphered, the transfer of such benefit to other crop plants is not feasible.

6.4.2 Phosphorus nutrition

Phosphorus is another major element needed by plants in large quantity. Soils often contain abundant amounts of P but in a form unavailable for plant uptake and as a result P deficiency is widely prevalent (Tiessen, 2008). The P concentration in the plant varies from 0.05 to 0.50% on a dry weight basis (Vance, 2010). The average free P in the soil solution is in the range of from 1 to 5 μM and P gradient from soil solution to plant cell increases by 2000-fold or more (Bieleski, 1973; Schachtman *et al.*, 1998). Phosphorus is present in the soil as Ca-phosphates, adsorbed phosphates, occluded phosphates and organic phosphates (Mengel, 1985). The proportion of different fractions depends on the soil type and organic matter content. The form in which P is present in the soil dilute solution varies with pH:

$$H_3PO_4 \underset{+H^+}{\overset{-H^+}{\longleftrightarrow}} H_2PO_4^- \underset{+H^+}{\overset{-H^+}{\longleftrightarrow}} HPO_4^{2-} \underset{+H^+}{\overset{-H^+}{\longleftrightarrow}} PO_4^{3-} \qquad (6.2)$$

The pH of most soils range from 5 to 8 and in this range the amounts of H_3PO_4 and PO_4^{3-} are negligible and soil phosphate is predominantly present as $H_2PO_4^-$ and HPO_4^{2-}. At pH 6, 94% of P in soil solution is $H_2PO_4^-$ but drops to 60% at pH 7 (Stevenson and Cole, 1999).

On average not more than 20% of P applied as fertilizer is actually taken up by the plant (Friesen *et al.*, 1997). Fertilizer P availability is particularly problematic in acid soils because it gets fixed by active Al and Fe present in high amounts in these soils. The uptake of P by plants depends on the roots' ability and size to absorb P from soil dilute solutions and mine soil mass. In some soils P is present in insoluble form and plants possessing the metabolic ability to convert it into soluble forms are better equipped for survival. Loneragan (1978) reported that plant species differ in the rate at which they absorb P from soil solutions. In soils with a low concentration of P in solution but a high buffering capacity, plants with greater ability to absorb P at lower concentrations can survive better as they continue to absorb P as depleted solution is replenished. However, roots need different attributes in soils where the quantity of P is limited and the buffering capacity is also low. In such a situation, plants that can absorb P before soil is exhausted would be in a better position to access available P.

Organic acids and CO_2 released from roots dissolve insoluble P and make it available to the plant (Loneragan, 1978). For example, pigeon pea is capable of adapting to acidic P-deficient soils, as it secrets piscidic, malonic and oxalic acids to solubilize soil P and also protects itself from Al toxicity by chelation (Otani *et al.*, 1996). Phosphatases present on the roots or excreted to the soil have been shown to hydrolyse soil organic phosphate compounds. The phosphatase activity increases with P deficiency in some plants, resulting in increased soluble P availability and absorption (Bieleski, 1973). Considering the relatively immobile nature of P and the fact that its highest concentration is present in the top soil layer, high root concentration at the top is important for improved P uptake. Nye *et al.* (1965) found that large root surface area was important to P absorption. Cluster roots of white lupin with a dense mat of root hair are capable of absorbing more P because of the ability to dissolve soil P by secreting large quantities of citric acid and then absorbing it through a large surface area (Hocking *et al.*, 1997). Several researchers

(Hermans *et al.*, 2006; Jain *et al.*, 2007) report a relationship between active carbon assimilation and P-deficiency-induced increase in lateral root and leaf hair formation. Summing up, roots of P-efficient plant species possess a combination of desirable mechanisms that can be categorized into two groups: (i) morphological, comprising increased adventitious root formation, enhanced lateral root proliferation, shallow horizontal basal root growth angle, and augmented root hair density and length; and (ii) metabolic, consisting of high activity of phosphate (Pi) transporters, high activity of internal phosphatase, and exudation of organic acids and phosphatases into the rhizosphere (Raghothama, 1999; Lynch and Brown, 2001; Vance *et al.*, 2003).

Sufficient variations exist within species for P utilization efficiency (Gahoonia and Nielsen, 2004; Wang *et al.*, 2010). Fageria (2001) observed that kg grain/kg P accumulated was higher in cereals compared to legumes; the values equalling to 360, 361, 251 and 78 for upland rice, maize, common bean and soybean, respectively. Upland rice accumulated about 11 kg P ha^{-1} to produce 4340 kg grain and 6040 kg straw while maize accumulated 23 kg P ha^{-1} to produce 8150 kg grain and 13,670 kg straw. Similarly, 4 kg ha^{-1} P in common bean was proportioned to 1670 kg grain and 1770 kg straw and 17 kg P ha^{-1} in soybean was distributed between 1320 kg grain and 2900 kg straw.

Phosphorus is required for the synthesis of phospholipids, nucleotides, ATP, glycophosphates and other phosphate esters and its deficiency results in slowing of cell division and growth (Fageria *et al.*, 2006). The deficiency of P affects both below- and above-ground plant growth (Lynch and Brown, 2001; Hernândez *et al.*, 2007). The results of P shortage are reduced photosynthetic ability, delayed leaf development, stunted growth, higher root to shoot ratio and impaired flower development.

Several reports suggest that breeding targeted to root morphology, architecture and root exudates appears to be the best strategy to achieve P-efficiency (Wang *et al.*, 2004; Liu *et al.*, 2008; Cheng *et al.*, 2010). It has been shown that a plant's ability to survive in P-deficient soils is regulated by many suites of genes regulated in a coordinated fashion (Vance, 2010). P-deficiency tolerance QTLs have been reported in a number of crop species including rice, wheat, common bean, soybean, barley and maize (Vance, 2010). Wissuwa *et al.* (2002) located a low P tolerance QTL in rice designated *phosphorus uptake1* (*Pup1*) on the long arm of chromosome 12 from a mapping population derived from a cross between the tolerant landrace 'Kasalath' and intolerant cultivar 'Nipponbare'. Presence of *Pup1* has been found in more than 50% of rice genotypes adapted to P-deficiency (Chin *et al.*, 2010). Liao *et al.* (2004) found 16 QTLs controlling root traits in common bean and those related to low P mainly situated on linkage groups B2 and B9. One QTL alone on linkage group B9 accounted for 61% of the total phenotypic variation. Chen *et al.* (2008) noted a QTL for P efficiency and topsoil root dry weight at bin 1.06 in maize. In *Arabidopsis*, a *low-phosphate root* (*LPR1*) has been identified in the 2.5 Mb region at the top of chromosome 1 (Reymond *et al.*, 2006). The root meristem and root cap contained the most transcripts for *LPR1*. Furthermore, *LPR1* locus encoded a multicopper oxidase enzyme (Svistoonoff *et al.*, 2007). *LPR1* interacts with *phosphate deficiency 2* (*PDR2*) to mediate the developmental response of root meristem to P availability and the root cap plays a crucial role in sensing of P in the immediate soil vicinity (Ticconi *et al.*, 2009).

Sucrose and non-coding small RNAs (miRNAs) have been reported to be involved in transmission of P-deficiency signal in plants (Chou *et al.*, 2006; Hammond and White, 2008; Liu *et al.*, 2010). Sucrose enhances the expression of P-starvation-induced genes. Liu *et al.* (2005) found dark reduced gene expression to non-detectable levels within a few hours but the gene function was restored back with light. Hummel *et al.* (2009) has described a sucrose non-fermenting 1 kinase:calcineurin B-like protein kinase (SNF1:CIPK) pathway as transduction system for sugar signalling, but it awaits confirmation whether this pathway actually regulates sugar signalling during P-deficiency. Chou *et al.* (2006) showed the induction of miR399 under P stress in *Arabidopsis*, while Bari *et al.* (2006) noted rapid decline in miR399 transcript as P was added and complete absence when P was adequate.

It has been further shown that miR399 travels in the phloem, carrying the signal for phosphate homeostasis (Pant *et al.*, 2008).

Colonization of the plant root by mycorrhizal fungi is helpful in P uptake under deficient conditions (Clark and Zeto, 2000). The fungus prefers to develop an association with some species such as maize over others such as *Arabidopsis* (Paszkowski and Boller, 2002). The fungal response to different genotypes within a species is also variable (Kaeppler *et al.*, 2000) and the ability of genotypes to benefit from this association differs (Sawers, 2009). The symbiosis between the root and the fungus begins with the exudation of flavonoids and strigolactones from root (Steinkellner *et al.*, 2007) followed by establishment of a novel intracellular apparatus in root epidermal cells by the fungus before infection (Genre *et al.*, 2005). The exchange of metabolites between plant and fungus is carried out by membrane transport systems, some of the transporters being mycorrhizal specific (Glassop *et al.*, 2005; Nagy *et al.*, 2006).

6.5 Water Use

In the case of biofuel crops, water use basically relates to efficiency in use of stored soil water and ability to survive under periodic drought conditions. Irrigation minimally enters into the equation because of the emphasis on minimal inputs in biofuel crop production systems. Drought is different from most abiotic stresses such as acidity and salinity in that while the latter are static in character, occurrence of drought, severity and timing during crop growth stage varies from one growing season to another. The transient nature of drought gives rise to complex genotype × year and genotype × timing of stress interactions. In addition, even though soil may appear homogeneous, usually intrinsic variations exist within a relatively short area for soil texture and plant available water.

Plants with the ability to maintain leaf water potential (drought avoidance) and/or adjust osmotic potential to sustain cell hydration and leaf turgor (drought tolerance) can continue with minimal interruption of cellular processes under drought (Blum, 2006). A deep root system permits plants to avoid drought by thoroughly utilizing water in the soil profile. Osmotic adjustment (OA) is the most common response of plants to drought by reducing cellular water potential through accumulating a variety of organic and inorganic solutes in cells (Blum, 2005; Sankar *et al.*, 2008; Ashraf, 2010). These manoeuvres permit plants to sustain leaf gas exchange, cellular membrane and protein functions, and chloroplast volume and functions (Blum, 2006).

A number of additional morphological and physiological characteristics have been associated with drought tolerance in plants (Innes *et al.*, 1984; Blum, 1988; Loss and Siddique, 1994; Morgan 1995; Mitra, 2001). Morphological traits of value include early ground cover and good capacity for stem reserves and mobilization, delayed leaf-rolling and stay-green while physiological functions of importance consist of capability to accumulate abscisic acid (ABA) and increase cytokinin production under drought.

Selecting biofuel feedstocks for semi-arid production regions is particularly challenging. Crops with C_4 photosynthetic pathways are more suited for these environments as according to Samson and Knof (1994) they produce 30% more dry matter per unit of water than C_3 crops. The crops that can grow successfully in these climates also have the ability to override drought situations. Sorghum, a member of the C_4 group of plants, is often cited as an ideal crop for drought-prone areas. However, it should be noted that sorghum is very sensitive to dehydration with lethal leaf water potential around 3 MPa (Santamaria *et al.*, 1990) while some forage grasses (also C_4) dehydrate at water potential as high as −13 MPa (Wilson *et al.*, 1980); so sorghum's drought resistance is dependent upon its ability of not allowing leaf water potential to reach the danger zone. Habyarimana *et al.* (2004) found that sorghum dry matter yield did not decrease with lowering of plant density in accordance with soil water availability because the plant could compensate by higher leaf weight, greater number of tillers per plant and an extensive root system 1.5 to 2.5 m deep and up to 1 m away from the stem to avail soil water to the fullest extent. Wright *et al.* (1983) also attributed the higher yield of dryland sorghum hybrid cultivar DK 57 compared to the lesser adapted hybrid RS671 to the former's longer roots allowing extraction of soil moisture from deeper soil profiles to sustain an optimum level of leaf water potential. Thus, they suggested that sorghum's ability to withstand drought was due to drought avoidance rather than drought tolerance. Xin *et al.* (2009) identified high transpiration efficiency (TE) genotypes (total biomass produced per unit water transpired) in sorghum and suggested that enhancing TE may be an effective approach to increase sorghum yield in arid and semi-arid regions under no or limited irrigation. Sorghum also escapes wilting during drought by becoming dormant (Smith and Frederiksen, 2000).

Several hundred stress-related genes have been identified in different plant species (Seki *et al.*, 2003; Zhang *et al.*, 2004). Umezawa *et al.* (2006) compiled a list of genes encoding osmolyte metabolism, protective proteins, reactive oxygen species (ROS) scavenging proteins, ion transport, ABA biosynthesis, stomatal regulation and ABA catabolism that have been used successfully for genetic engineering in attempts to enhance drought tolerance. The osmolytes whose increased accumulation has been achieved include proline, glycine betaine, polyamines, mannitol, trehalose, fructan and galactinol. Introduction of genes over-expressing the late-embryogenesis-abundant (LEA) class of proteins have been considered important to detoxification and protection to cells during dehydration. They are considered effective in protein aggregation induced by desiccation and function as chaperone-like protective molecules (Bartels and Sunkar, 2005; Goyal *et al.*, 2005). In addition to structural proteins, regulatory proteins such as transcriptional factors (TFs) consisting of the AP2/ERF family, basic leucine-zipper (bZIP) protein, MYB/MYC, zinc-finger protein and others, and signalling factors consisting of protein kinases and others that controlled a broad range of stress responses also exist.

The cellular response to abiotic responses is often similar, e.g. drought, salinity and freezing all exhibit dehydration. Quantitative trait loci (QTL) associated with water-use efficiency and drought tolerance, frost resistance (Tsarouhas *et al.*, 2004) and growth (Rönnberg-Wästljung *et al.*, 2005; Weih *et al.*, 2006) in willow, salt tolerance in rice (Lin *et al.*, 2004) and drought tolerance in cotton (Saranga *et al.*, 2004) have been reported. In maize, QTLs for leaf elongation rate in response to soil moisture, temperature and evaporative demand have been identified (Reymond *et al.*, 2003). The combined effect of these QTLs was able to predict 74% of leaf elongation rate variability. While most of the identified genes are functional genes, several are also regulatory governing a subset of stress-related genes. Umezawa *et al.* (2006) are of the opinion that genetic engineering of regulatory genes provide the most promising strategy for inducing drought tolerance.

From the point of view of yield, it is not enough that a plant survives drought episodes but it must also be accompanied by the competence to harvest available water maximally and utilize it efficiently for maximum dry matter production. It brings roots to the forefront of drought tolerance development. Bengough *et al.* (2006) have shown that roots possess a high degree of developmental plasticity in response to water and nutrients. Understanding of the role that root plasticity plays in root elongation under varying soil moisture conditions and assessment of genotype × environment interactions is essential to improving root architecture for stress conditions. Several root constitutive traits of consequence in case of drought, such as root thickness, the ability to penetrate compacted soil, root length and mass, are present in certain adapted species even in the absence of moisture-stress conditions (Pathan *et al.*, 2004). Several QTLs for root traits have been identified in rice (Price *et al.*, 2002) and maize (Tuberosa *et al.*, 2002). Following an effort to transfer four QTL alleles identified for promoting deep root in 'Azucena', an upland rice cultivar adapted to rainfed conditions, to 'IR64', another rice cultivar with shallow root system using marker assisted selection (MAS), phenotypic evaluations indicated that MAS was successful for three QTL regions (Courtois *et al.*, 2003). In a second study, the same 'Azucena' donor parent provided the four QTLs to another rice cultivar 'Kalinga III', but out of four introgressed segments, only one on chromosome 9 was effective in increasing root length in the backcross population (Steele *et al.*, 2009). Thus, it was obvious that QTL allele expression was limited by the genetic background of the recipient genotype. Genetic engineering is being actively tried to increase drought tolerance in sugarcane. A gene responsible for trehalose production, a sugar that protects cells from damage, has been introduced from a mushroom species into sugarcane in China (Zhang *et al.*, 2006). The GM plants exhibited high accumulation of trehalose and normal growth. In field trials, these plants tolerated drought better and recovered faster thereafter than non-GM controls. Tuberosa and Salvi (2006) point to the need of seeking drought-tolerance genes in wild relatives of crops. A case in point is of wild barley (*Hordeum spontaneum*), which has turned out to be a valuable source of alleles for drought tolerance for cultivated barley (*Hordeum vulgare*). Out of a total of 81 QTLs identified for agronomic traits under different moisture conditions in a backcross population between *H. vulgare* and *H. spontaneum*, 43 beneficial alleles were contributed by *H. spontaneum* (Talamè *et al.*, 2004).

6.6 Salinity Tolerance

Saline soils are a category of marginal lands that provide opportunity to expand the hectarage of biofuel crops. A rise of soluble salts (usually NaCl) in soil and water causes salinity. Primary salinity is common in arid and semi-arid areas because there is not enough water to move salt originated from natural processes down through the soil so that it is removed from the plant root zone (Brown,1978). Secondary salinity can arise when land is cleared of deep-rooted native vegetation and replanted with shallow-rooted plants causing a rise in the water table carrying dissolved salts towards the top. Some degree of salt or Na^+ accumulation is common on irrigated lands (Bernstein and Hayward, 1958). Irrigation with high salt water in countries like India and Pakistan is a major contributor to salinity. Flowers and Yeo (1995) report approximately 20% of irrigated agricultural land is facing salinity problems. According to Ashraf (1994), more than 50% of arable land is expected to encounter salinity problems by 2050. Some of the saline irrigated lands have been reclaimed by expensive drainage to wash out the salt and alleviate waterlogging.

Halophytes are a group of plant species, comprising probably 2% of total plant species, that are adapted to grow on saline soils; the remaining plant species are called glycophytes and are damaged by salt in the soil. Salt, once inside the cell, can cause ionic stresses. Metabolic processes including protein synthesis are inhibited by Na^+ (and Cl^-). Na^+ can accumulate to toxic levels in older leaves and become lethal, reducing the leaf area available for photosynthesis. The most logical way to improve the ability of glycophytes for normal growth on saline soils is to incorporate traits that make halophytes tolerant of salt (Tester and Davenport, 2003; Flowers, 2004). Halophytes have acquired salt tolerance by adopting two escape mechanisms: (i) to limit uptake of Na^+ by roots thus avoiding its transport into the plant; and (ii) to store Na^+ in the vacuole, thus separating it from cytoplasm, the functioning portion of the cell.

Many plant species belonging to the glycophytes have some level of salt tolerance and genetic differences within species for salt tolerance also exist (Munns *et al.*, 2006; James *et al.*, 2008). Barley is considered a moderately tolerant forage crop and a highly tolerant grain crop (Francois and Mass, 1999). Durum wheat lacks the D genome of bread wheat and tends to accumulate more Na^+ than bread wheat. Recently two salt-tolerance genes, *Nax1* and *Nax2*, were located during germplasm screening for salt tolerance of durum wheat in Australia introduced inadvertently into it during an earlier breeding programme from a wheat ancestor, *Triticum monococcum* (Huang *et al.*, 2006; Byrt *et al.*, 2007). *Nax1* helps reduce Na^+ concentration in leaf blades by retaining Na^+ in the leaf sheath while *Nax2* removes Na^+ from root xylem preventing upward movement. Genotypes with low Na^+ transport have been reported in rice (Yeo *et al.*, 1988). Wang, W. *et al.* (2003) have grouped genes that are induced under saline conditions and the encoded protein into three categories: (i) signalling and transcriptional control; (ii) protection of membranes and proteins; and (iii) transport water and ion. Chinnusamy *et al.* (2005) have compiled an excellent review of the physiology and genetics of salt tolerance in crops.

The saline soil can be profitably utilized for growing economically important high biomass salt-tolerant trees and shrubs such as *Eucalyptus* and salt brush. Sugarcane germplasm exhibits natural variation to salt conditions and some genotypes possessing somewhat better ability to avoid intake of Cl^- ions or of transferring them to older leaves have been identified (Wahid *et al.*, 1997; Huwyzeh *et al.*, 2008). The issue of producing crops on saline soils has to be considered in two ways: (i) identifying crops that can withstand and can be produced on saline soils; and (ii) identifying specific differences and improving genetic potential of crops currently produced on those soils.

Aside from conventional breeding, genetic engineering shows promise for improving salt tolerance. Single gene manipulations to achieve over-expression of salt overly sensitive gene, *SOS1* in the model plant *Arabidopsis* (Shi *et al.*, 2003) and NHX type ion transporter, *NHX1* in tomato and *Brassica* (Zhang and Blumwald, 2001; Zhang *et al.*, 2001) have resulted in significant increases in salt tolerance without any growth abnormalities. Tolerance genes are being imported from known salt-tolerant species into cultivated plants using genetic engineering. The salt tolerance in the seashore plant seepweed (*Sueada salsa*) is due to its ability to transport Na^+ ion through bordering membrane into the vacuole and is governed by the vacuolar Na^+/H^+ antiporter gene, *SsNHX1*. In the model plant *Arabidopsis*, over-expression of this gene resulted in enhancement in salt as well as cold tolerance and it resulted from accumulation of Na^+ taken up by roots into stem and leaf vacuoles. Two genes from another salt-tolerant species, mangrove plant *Bruguiera gymnorhiza*, also provided salt tolerance to *Arabidopsis*, with one of them up to 150 mM NaCl (Ezawa and Tada, 2009). Existence of a pump that sends salt back out of the cell has been reported in a moss, *Pyscomitrella patens*. Furthermore, the gene *PpENA1* that codes for a Na^+ pumping ATPase protein has been introduced into rice and barley and its expression noted in the GM plants (Jacobs *et al.*, 2007). The gene if exclusively expressed in roots would be a preferred option because root is the main site for the leakage of Na^+ into the plant and by doing so other plant metabolites will be left unaffected.

6.7 Acidity Tolerance

Soils with pH below 5.5 are designated acidic. Approximately 27% of Africa's tropical land, 38% of Asia's tropical land and 68% of America's tropical land have been classified acidic (Martinez and Estrella, 1999). There are several factors that limit plant growth on acid soils. These include soluble toxic levels of Al, Mn and Fe and decreased soluble levels of several elements, P being the most important among them. The effect of soil acidity on phosphorus availability has already been deliberated under Section 6.4.2.

Aluminium toxicity is the major hurdle to crop production on acid soils (McLean, 1976). There are several excellent reviews on Al toxicity and plant tolerance mechanisms (Kochian and Jones, 1997; Ma *et al.*, 2001; Matsumoto *et al.*, 2001; Barcelo and Poschenrieder, 2002; Garvin and Carver, 2003). The important features of Al toxicity are: (i) rapid inhibition of plant root growth restricting water and mineral nutrient uptake; (ii) site of Al toxicity is localized to the root apex; (iii) interference with signal transduction pathways, in particular Ca^{2+} homeostasis and signalling; and (iv) induction of reactive oxygen species (ROS) as well as peroxidative damage to membranes (Kochian *et al.*, 2004).

Plant species differ significantly in their tolerance to soil acidity. Species such as tea, azalea, pineapple, rye, cranberry, weeping lovegrass, bermudagrass, buckwheat, rubber, blueberry and Norway spruce are adapted to acid soils. One of the reasons for sugarcane thriving so well in Hawaii is its tolerance to low Ca and high Al. Variation for Al tolerance within species has been found in important crop species such as rice, wheat, lucerne, snap bean, soybean, cotton, maize, sunflower, pea and sweet potato. Because of the wide existence of acid soils in tropical areas of Central and South America, CIAT (International Centre for Tropical Agriculture) research has been aimed at breeding Al-tolerant cultivars in most promising pasture and forage species and as a result a number of acid soil-tolerant cultivars in several grass species are already available. This bodes well for biofuel feedstocks since a large proportion of advanced biofuels will depend upon the high cellulosic production capability of grass species.

Evidence suggests that acid-tolerant plants possess an Al exclusion mechanism in their root apex. It has been shown that Al-tolerant wheat genotypes accumulate three- to eightfold less Al in the root apex than Al-sensitive genotypes (Kochian *et al.*, 2004). This is supposedly due to the continuous release of organic acid (OA) from the root apex that chelates with Al in the rhizosphere, preventing Al entry into the root (Delhaize *et al.*, 1993; Ryan *et al.*, 1995). Wheat roots responded to Al very rapidly, within minutes, and the response was localized within the first few millimetres of the tip of the root apex. Because the exudation was continuous, the apex of the root was protected as the root continued to grow and penetrate into new regions of the acid soil. However, crop species differ in the localization of exudates as well as rapidity of response to Al exposure. As against the OA release in the terminal few millimetres in wheat, it can extend to as far back as 3 cm or more in maize and sorghum. Similarly, while in wheat OA release is extremely rapid and remains stable over time, in rye there is a noticeable lag period for the start and the rate of exudation accelerates over the first 12 to 24 h of Al exposure (Li *et al.*, 2000). Ma and associates (Ma *et al.*, 2001; Ma and Furukawa, 2003) propose the existence of two different mechanisms of Al tolerance in crop plants. In case of wheat, an already expressed OA transporter is present and responds rapidly to Al exposure and in such cases gene activation is not involved. On the contrary, increased rate of OA exudation with longer Al exposure as in rye and triticale is indicative of a triggering gene responsible for OA exudation.

Malate, citrate and oxalate have been identified as Al-chelating OA in different plant species; some species may possess a single OA while others possess more than one (Kochian *et al.*, 2004). Citrate is most prevalent in crop genotypes expressing Al tolerance. The biofuel crop rape has been reported to release both citrate and malate (Zheng *et al.*, 1998). Citrate^{3-}, being a tricarboxylic acid anion, chelates much more strongly with Al and thus is a more effective detoxifier than dicarboxylic malate^{2-} anion (formation constant for Al: citrate and Al: malate equal to 9.6 and 5.7, respectively). As there are a number of OA present in the cytoplasm but Al exposure activates only exudation of one or two OAs, it presents the likelihood that Al activation is to a specific plasma membrane transporter (Ryan *et al.*, 1995; Osawa and Matsumoto, 2001). Sasaki *et al.* (2004), while studying a pair of wheat lines differing at a single Al tolerance locus, were successful in identifying a gene, named *ALMT1*, that encodes the root tip Al-activated malate transporter. Expression of *ALMT1* also increased Al tolerance in tobacco suspension cells and in transgenic rice seedlings. In sorghum, inheritance of Al tolerance appears to be a qualitative trait with one locus explaining most of the differences (Kochian *et al.*, 2004). However, the Alt_{SB} locus of sorghum does not occupy the homologous chromosomal location of Al-tolerance genes belonging to *Triticeae* and differ in tolerance response, which in this case is inducible and takes days to fully manifest (Magalhaes, 2002). Thus, it is inferred that the Al tolerance pathway of sorghum is different from wheat and its relatives.

In buckwheat, in addition to oxalate exudation from the root tip to exclude Al, an additional protective mechanism of Al chelation with oxalate in the leaves and citrate in the xylem also exists (Ma *et al.*, 1998; Ma and Hiradate, 2000). The evidence of more than one mechanism for Al tolerance is buttressed by the evidence that Al tolerance in rice, the most tolerant among small grain cereals, is a quantitative trait with multiple contributing loci (Ma *et al.*, 2002). Overexpression of citrate synthase and malate dehydrogenase (OA metabolism enzymes) in transgenic tobacco, *Arabidopsis* and lucerne have been shown, in several studies, to increase root OA content and exudation and resultant enhanced Al tolerance (de la Fuente *et al.*, 1997; Koyama *et al.*, 2000; Tesfaye *et al.*, 2001).

Manganese toxicity is another threat to crops on acid soils. It differs from Al toxicity in that while damage by Al toxicity is usually to the root, Mn toxicity symptoms are expressed in the shoot. Affected plants have stunted growth, chlorosis, and necrotic lesions in the leaves. It is believed to be due to Mn-induced oxidative stress and genotypic variation in crop ability to withstand this stress has been reported (Horst *et al.*, 1999). Several transporters (CAX2, ECA1 and ShMTP1) with the ability to sequester potentially toxic levels of Mn in the vacuole have been identified (Hirschi *et al.*, 2000; Wu *et al.*, 2002; Delhaize *et al.*, 2003).

6.8 Concluding Remarks

The physiology and genetics of starch grain crops used for first-generation biofuels was well studied because of their importance as food and feed. The advanced biofuels will rely greatly on cellulosic biomass and grass species will figure prominently among new feedstock species. Some of the grass species will be the same as otherwise used for animal forage and hence have been improved for characters such as disease and pest resistance, digestibility and forage yield. These will, indeed, be useful in developing grass biofuel feedstocks. However, the production systems, harvesting methods, timing and post-harvest treatments, and biomass quality requirements between grass for biofuel feedstock and for forage will obviously have several differences due to different end purposes. The search for 'energy genes' should be helped by the existence of approximately 10,000 wild grass species and vast array of traits they harbour. Although wide differences in chromosome number exist among members of *Poaceae*, the species are closely related and variation in their gene content is not great. High variation of ploidy level among the species is mainly responsible for the chromosome number differences. Thus, the likelihood of transfer of the achievements obtained in one grass species to another species holds promise especially with the aid of genetic engineering. Wild relatives of promising grasses should be able to supply novel germplasm for growth traits and tolerances to adverse growing conditions.

The physiology and genetics of neo-domesticated species such as *Jatropha*, hitherto largely unknown, will require a special concentrated effort and it may take considerable time to exploit fully their biofuel potential. First of all, realistic appraisal of the yield potential and sustainability in different climates and soils of the new species must be carried out in multiple-year large field trails and simultaneously ecological consequences of large scale production assessed and invasiveness ascertained. Preliminary success in improving the plant can be rather rapid based mainly on visual and phenotypic selection, but thereafter it will require study of the physiology of different metabolic processes in the plant and development of suitable strategies utilizing the modern genetic tools in combination with conventional breeding to bring about subsequent improvements. Being unexploited, however, novel species should possess wide genetic diversity for different traits.

To avoid infringement upon lands customarily devoted to field crops, biofuel crops will have to rely to a great extent for production on soils that are limiting to plant growth by either acidity or salinity and may have to be planted in areas with limited rainfall or occasional drought. Fortunately, a number of grass species are adapted to growing in less than optimal conditions. Traditional breeding and mutant selection has been applied in the past to develop crop resistance to salinity and acidity but results have not been very encouraging. Applications of molecular approaches to breeding save time and increase confidence levels of success, and therefore have received much attention, but release of crop varieties from such efforts is still awaited. As knowledge of salinity, stress and water tolerance increases in other crops and methods of improvements are devised, important biofuel feedstocks will surely be modified accordingly.

References

Aharoni, A., Giri, A.P., Deuerlein, S., Griepink, F., de Kogel, W. *et al.* (2003) Terpenoid metabolism in wild-type and transgenic *Arabidopsis* plants. *The Plant Cell* 15, 2866–2884.

Anderson, I.C., Buxton, D.B., Hallam, A. and Hunter, E. (1995) Biomass production and ethanol yield potential from sweet sorghum. *Leopold Center for Sustainable Agriculture, Iowa State University, Ames, Iowa* 4, 97–101.

Andrews, M., Lea, P.J., Raven, J.A. and Lindsey, K. (2004) Can genetic manipulation of plant nitrogen assimilation enzymes result in increased crop yield and greater N-use efficiency? *Annals of Applied Biology* 145, 25–40.

Ashraf, M. (1994) Breeding for salinity tolerance in plants. *Critical Reviews in Plant Science* 13, 17–42.
Ashraf, M. (2010) Inducing drought tolerance in plants: recent advances. *Biotechnology Advances* 28, 169–183.
Bai, L., Singh, M., Pitt, L., Sweeney, M. and Brutnell, T.P. (2007) Generating novel allelic variation through *Activator* insertional mutagenesis in maize. *Genetics* 175, 981–992.
Barber, S.A. (1995) *Soil Nutrient Bioavailability: A Mechanistic Approach*, 2nd edn. John Wiley, New York, 414 pp.
Barcelo, J. and Poschenrieder, C. (2002) Fast root growth responses, root exudates, and internal detoxification as clues to the mechanisms of aluminum toxicity and resistance: a review. *Environmental and Experimental Botany* 48, 75–92.
Bari, R., Pant, B.D., Stitt, M. and Scheible, W.R. (2006) Pho2, microRNA399, and PHR1 define a phosphate-signaling pathway in plants. *Plant Physiology* 141, 988–999.
Bartels, D. and Sunkar, R. (2005) Drought and salt tolerance in plants. *Critical Reviews in Plant Sciences* 24, 23–58.
Bausenwein, U., Millard, P. and Raven, J.A. (2001a) Remobilized old-leaf nitrogen predominates for spring growth in two temperate grasses. *New Phytologist* 152, 283–290.
Bausenwein, U., Millard, P., Thornton, B. and Raven, J.A. (2001b) Seasonal nitrogen storage and remobilization in the forb *Rumex acetosa*. *Functional Ecology* 15, 370–377.
Beadle, C.L. and Long, S.P. (1985) Photosynthesis – is it limiting to biomass production. *Biomass* 8, 119–168.
Beale, C.V. and Long, S.P. (1995) Can perennial C4 grasses attain high efficiencies of radiant energy conversion in cool climates? *Plant, Cell and Environment* 18, 641–650.
Beale, C.V., Bint, D.A. and Long, S.P. (1996) Leaf photosynthesis in the C4-grass *Miscanthus* x *giganteus*, growing in the cool temperate climate of southern England. *Journal of Experimental Botany* 47, 267–273.
Below, F.E. (1987) Growth and productivity of maize under nitrogen stress. In: Edmeades, G.O., Bänzinger, M., Mickelson, H.R. and Peña-Valdivia, C.B. (eds) *Developing Drought- and Low N-Tolerant Maize*. CIMMYT, El Batán, Mexico, pp. 243–240.
Below, F.E. (2002) Nitrogen metabolism and crop productivity. In: Pessarakli, M. (ed.) *Handbook of Plant and Crop Physiology*, 2nd edn. Marcel Dekker, New York, pp. 85–406.
Bengough, A.G., Bransby, M.F., Hans, J., McKenna, S.J., Tim, J. *et al.* (2006) Root responses to soil physical conditions, growth dynamics form field to cell. *Journal of Experimental Botany* 57, 437–447.
Bernstein, L. and Hayward, H.E. (1958) Physiology of salt tolerance. *Annual Review of Plant Physiology* 9, 25–46.
Berthelot, A., Ranger, J. and Gelhaye, D. (2000) Nutrient uptake and immobilization in a short-rotation coppice stand of hybrid poplars in north-west France. *Forest Ecology and Management* 128, 167–179.
Bertin, P. and Gallais, A. (2001) Physiological and genetic basis of nitrogen use efficiency. II. QTL detection and coincidences. *Maydica* 46, 53–68.
Bieleski, R.L. (1973) Phosphate pools, phosphate transport, and phosphate availability. *Annual Review of Plant Physiology* 24, 225–252.
Blum, A. (1988) *Plant Breeding for Stress Environments*. CRC Press, Boca Raton, Florida.
Blum, A. (2005) Drought resistance, water-use efficiency and yield potential – are they compatible, dissonant or mutually exclusive? *Australian Journal of Agricultural Research* 56, 1159–1168.
Blum, A. (2006) Drought adaptation in cereal crops: a prologue. In: Ribaut, J.M. (ed.) *Drought Adaptation in Cereals*. Food Products Press, New York, pp. 3–15.
Boddey, R.M., Urquiaga, S., Alves, B.J.R. and Reis, V. (2003) Endophytic nitrogen fixation in sugarcane; present knowledge and future applications. *Plant and Soil* 252, 139–149.
Bolwell, G.P. (2000) Biosynthesis of plant cell wall polysaccharides. *Trends in Glycoscience and Glycotechnology* 12, 143–160.
Bonifas, K.D. and Lindquist, J.L. (2006) Predicting biomass partitioning to root versus shoot in corn and velvetleaf (*Abutilon theophrasti*). *Weed Science* 54, 133–137.
Borrell, A.K., Hammer, G.L. and Van Oosterom, E. (2001) Stay-green: a consequence of the balance between supply and demand for nitrogen during grain filling. *Annals of Applied Biology* 138, 91–95.
Brown, J.C. (1978) Physiology of salt tolerance to alkaline soils. In: Jung, G.A. (ed.) *Crop Tolerance to Suboptimal Land Conditions*. ASA, CSSA and SSSA, Madison, Wisconsin, pp. 257–276.
Byrt, C.S., Platten, J.D., Spielmeyer, W., James, R.A., Lagudah, E.S. *et al.* (2007) HKT1;5-like transporters linked to Na+ exclusion loci in wheat Nax2 and Kna1[1[OA]]. *Plant Physiology* 143, 1918–1928.

Carpita, N.C. and McCann, M. (2008) Maize and sorghum: genetic resources for bioenergy grasses. *Trends in Plant Science* 13, 415–420.
Cataldo, D.A., Schrader, L.E., Peterson, D.M. and Smith, D. (1975) Factors affecting seed protein concentration in oats: I. Metabolism and distribution of N and carbohydrate in two cultivars that differ in groat protein concentration. *Crop Science* 15, 19–23.
Chen, J., Xu, L., Cai, Y. and Xu, J. (2008) QTL mapping of phosphorus efficiency and relative biologic characteristics in maize (*Zea mays* L.) at two sites. *Plant and Soil* 313, 251–266.
Cheng, F., Tu, P., Yan, X., Wang, X. and Liao, H. (2010) Phosphorus nutrition characters for new soybean germplasms with phosphorus efficiency on acid red soils. *Plant Nutrition and Fertilizer Science* 16, 71–78.
Chin, J.H., Haefele, S.M., Gamuyao, R., Ismail, A., Wissuwa, M. *et al.* (2010) Development and application of gene based markers for the major rice QTL phosphorus uptake 1. *Theoretical and Applied Genetics* 120, 1073–1086.
Chinnusamy, V., Jagendorf, A. and Zhu, J.K. (2005) Understanding and improving salt tolerance in plants. *Crop Science* 45, 437–448.
Chou, T.J., Aung, K., Lin, S.L., Wu, C.C., Chiang, S.F. and Su, C.L. (2006) Regulation of phosphate homeostatsis by microRNA in *Arabidopsis*. *Plant Cell* 18, 412–421.
Chun, L., Chen, F., Zhang, F. and Mi, G.H. (2005a) Root growth, nitrogen uptake and yield formation of hybrid maize with different N efficiency. *Plant Nutrition and Fertilizer Science* 11, 615–619.
Chun, L., Mi, G.H., Li, J., Chen, F. and Zhang, F. (2005b) Genetic analysis of maize roots characteristics in response to low nitrogen stress. *Plant and Soil* 276, 369–382.
Clark, R.B. and Zeto, S.K. (2000) Mineral acquisition by arbuscular mycorrhizal plants. *Journal of Plant Nutrition* 23, 867–902.
Cooper, H.D. and Clarkson, D.T. (1989) Cycling of amino nitrogen and other nutrients between shoot and root in cereals: a possible mechanism integrating shoot and root in the regulation of nutrient uptake. *Journal of Experimental Botany* 40, 753–762.
Courtois, B., Shen, L., Petalcorin, W., Carandang, S., Mauleon, R. *et al.* (2003) Locating QTLs controlling constitutive root traits in the rice population IAC 165 × Co39. *Euphytica* 134, 335–345.
Cox, T.S., Glover, J.D., Van tassel, D.L., Cox, C.M. and DeHaan, L.R. (2006) Prospects for developing perennial grain crops. *BioScience* 56, 649–659.
Dalling, M.J., Boland, G. and Wilson, J.H. (1976) Relation between acid proteinase activity and redistribution of nitrogen during grain development in wheat. *Australian Journal of Plant Physiology* 3, 721–730.
Degenhart, N.R., Barnes, D.K. and Vance, C.P. (1992) Divergent selection for nodule aspartate aminotransferase and asparagines synthetase activities in alfalfa. *Crop Science* 32, 313–317.
de la Fuente, J.M., Ramirez-Rodriguez, V., Cabrera-Ponce, J.L. and Herrera-Estrella, L. (1997) Aluminum tolerance in transgenic plants by alteration of citrate synthesis. *Science* 276, 1566–1568.
Delhaize, E., Craig, S., Beaton, C.D., Bennet, R.J., Jagadish, V.C. *et al.* (1993) Aluminum tolerance in wheat (*Triticum aestivum* L.): I. Uptake and distribution of aluminum in root apices. *Plant Physiology* 103, 685–693.
Delhaize, E., Kataoka, T., Hebb, D.M., White, R.G. and Ryan, P.R. (2003) Gene encoding proteins of the cation diffusion facilitator family that confer manganese tolerance. *Plant Cell* 15, 1131–1142.
Demotes-Mainard, S.D., Jeufrroy, M.H. and Robin, S. (1999) Spike and dry matter accumulation before anthesis in wheat as affected by nitrogen fertilizer: relationship to kernel per spike. *Field Crops Research* 64, 249–259.
Dhugga, K. (2007) Maize biomass yield and composition for biofuels. *Crop Science* 7, 2211–2227.
DIR 095, Australia Government (2009) Limited and controlled release of sugarcane genetically modified for altered plant growth, enhanced drought tolerance, enhanced nitrogen use efficiency, altered sucrose accumulation and improved cellulosic ethanol production from sugarcane biomass/BSES Ltd. *Full Risk Assessment and Risk Management Plan.* Office of Gene Technology Regulation, Department of Health and Aging, Australian Government 2009, 92.
Dohleman, F.G. and Long, S.P. (2009) More productive than maize in the Midwest – how does *Miscanthus* do it? *Plant Physiology* 150, 2104–2115.
Eamens, A., Wang, M.B., Smith, N.A. and Waterhouse, P.M. (2008) RNA silencing in plants: yesterday, today and tomorrow. *Plant Physiology* 147, 456–468.
Eriksson, M.E., Israelsson, M., Olsson, O. and Moritiz, T. (2000) Increased gibberellin biosynthesis in transgenic trees promotes growth, biomass production and xylem fiber length. *Nature Biotechnology* 18, 784–788.

Evans, J.R. and Von Caemmerer, S. (2000) Would C_4 rice produce more biomass than C_3 rice? In: Sheehy, J.E., Mitchell, P.L. and Hardy, B. (eds) *Redesigning Rice Photosynthesis to Increase Yield.* International Rice Institute, Los Baños, the Philippines, pp. 53–72.
Evans, L.T. and Fischer, R.A. (1999) Yield potential: its definition, measurement, and significance. *Crop Science* 39, 1544–1551.
Ezawa, S. and Tada, Y. (2009) Identification of salt tolerance genes from the mangrove plant *Bruguiera gymnorhiza* using Agrobacterium functional screening. *Plant Science* 176, 222–228.
Fageria, N.K. (2001) Response of upland rice, dry bean, corn and soybean to base saturation in cerrado soil. *Revista Brasileria de Engenharia Agricola e Ambiental* 5, 416–424.
Fageria, N.K., Baligar, V.C. and Clark, R.B. (2006) *Physiology of Crop Production*, 2nd edn. Food Products Press, Binghamton, New York, 345 pp.
Falkowski, P.G. and Dubindky, Z. (1981) Light shade adaption of *Stylophora pistillata*, a hermatypic coral from the Gulf of Eilat. *Nature* 172–174.
Farago, J. (2007) Biotechnological tools to improve ethanol production from plant biomass. *Nova Biotechnologica* 7, 63–68.
Fernandes, J., Dong, Q., Schneider, B., Morrow, D.J., Nan, G.N. *et al.* (2004) Genome-wide mutagenesis of *Zea mays* L. using ResueMu transposons. *Genome Biology* 5, R82. Available at: http://genomebiology.com/2004/5/10/R82 (accessed 28 January 2012).
Flowers, T.J. (2004) Improving crop salt tolerance. *Journal of Experimental Botany* 55, 307–319.
Flowers, T.J. and Yeo, A.R. (1995) Breeding for salinity resistance in crop plants. Where next? *Australian Journal of Plant Physiology* 22, 875–884.
Forde, B.G. (2000) Nitrate transporters in plants: structure, function and regulation. *Biochimica et Biophysica Acta* 1465, 219–235.
Francois, L.E. and Maas, E.V. (1999) Crop response and management of salt-affected soils. In: Pessarakli, M. (ed.) *Handbook of Plant and Crop Stress*, 2nd edn. Marcel-Dekker, New York, pp. 169–201.
Friesen, D.K., Rao, I.M., Thomas, R.J., Oberson, A. and Sanz, J.I. (1997) Phosphorus acquisition and cycling in crop and pasture systems in low fertility tropical soils. In: Ando, T., Fujita, K., Mae, T., Tsumoto, H., Mori, S. and Sekiya, J. (eds) *Plant Nutrition for Sustainable Food Production and Environment.* Kluwer, Dordrecht, the Netherlands, pp. 493–498.
Gahoonia, T.S. and Nielsen, N.E. (2004) Root traits as tools for creating phosphorus efficient crop varieties. *Plant and Soil* 260, 47–57.
Gallais, A. and Coque, M. (2005) Genetic variation and selection for nitrogen use efficiency in maize: a synthesis. *Maydica* 50, 531–537.
Gallais, A. and Hirel, B. (2004) An approach of the genetics of nitrogen use efficiency in maize. *Journal of Experimental Botany* 55, 295–306.
Garvin, D.F. and Carver, B.F. (2003) Role of genotype in tolerance to acidity and aluminum toxicity. In: Rengel, Z. (ed.) *Handbook of Soil Acidity*. Marcel Dekker, New York.
Gastal, F. and Nelson, C.J. (1994) Nitrogen use within the growing leaf blade of tall fescue. *Plant Physiology* 105, 191–197.
Genre, A., Chabaud, M., Timmers, T., Bonfante, P. and Barker, D.G. (2005) Arbuscular mycorrhizal fungi elicit a novel intracellular apparatus in *Medicago truncatula* root epidermal cells before infection. *Plant Cell* 17, 3489–3499.
Ghannoum, O., Evans, J.R. and von Caemmerer, S. (2011) Nitrogen and water use efficiency in C_4 plants. In: Raghavendra, A.S. and Sage, R.F. (eds) *C_4 Photosynthesis and Related CO_2 Concentrating Mechanisms.* Springer, Dordrecht, the Netherlands, pp. 129–146.
Glass, A.D.M., Britto, D.T., Kaiser, B.N., Kinghorn, J.R., Kronzucker, H.J. *et al.* (2002) The regulation of nitrate and ammonium transport systems in plants. *Journal of Experimental Botany* 53, 855–864.
Glassop, D., Smith, S.E. and Smith, F.W. (2005) Cereal phosphate transporters associated with the mycorrhizal pathway of phosphate uptake into roots. *Planta* 222, 688–698.
Good, A.G., Shrawat, A.K. and Muench, D.G. (2004) Can less yield more? Is reducing nutrient input into the environment compatible with maintaining crop production. *Trends in Plant Science* 9, 597–605.
Goyal, K., Walton, L.J. and Tunnacliffe, A. (2005) LEA proteins prevent protein aggregation due to water stress. *Biochemical Journal* 388, 151–157.
Groat, R.C., Vance, C.P. and Barnes, D.K. (1984) Host plant nodule enzymes associated with selection for increased N_2 fixation in alfalfa. *Crop Science* 24, 895–898.
Guo, Y., Mi, G.H., Chen, F. and Zhang, F. (2005) Effect of NO_3^- supply in later root growth in maize plants. *Journal of Plant Physiology and Molecular Biology* 31, 90–96.

Habyarimana, E., Bonardi, P., Laureti, D., Di Bari, V., Cosentino, C. *et al.* (2004) Multilocational evaluation of biomass sorghum hybrids under two stand densities and variable water supply in Italy. *Industrial Crops and Products* 20, 3–9.

Hadley, H.H. (1957) An analysis of variation in height in sorghum. *Agronomy Journal* 49, 144–147.

Hammond, J.P. and White, P.J. (2008) How do plants respond to nutrient shortage by biomass accumulation? *Trends in Plant Science* 11, 610–617.

Harrison, J., Brugière, N., Phillipson, B., Ferrario-Mery, S., Becker, T. *et al.* (2000) Manipulating the pathway of ammonium assimilation through genetic engineering and breeding: consequences to plant physiology and plant development. *Plant and Soil* 221, 81–93.

Hart, G., Schertz, K., Peng, Y. and Syed, N. (2001) Genetic mapping of *Sorghum bicolor* (L.) Moench QTLs that control variation in tillering and other morphological character. *Theoretical and Applied Genetics* 103, 1232–1242.

Hatch, M.D. (1987) C-4 photosynthesis – a unique blend of modified biochemistry, anatomy and ultrastructure. *Biochimica et Biophysica Acta* 895, 81–106.

Heaton, E.A., Dohleman, F.G. and Long, S.P. (2008) Meeting US biofuel goals with less land: the potential of *Miscanthus*. *Global Change Biology* 14, 2000–2014.

Heller, M.C., Keoleian, G.A., Mann, M.K. and Volk, T.A. (2004) Life cycle energy and environmental benefits of generating electricity from willow. *Renewable Energy* 29, 1023–1042.

Hermans, C., Hammond, J.P., White, P.J. and Verbruggen, N. (2006) How do plants respond to nutrients shortage by biomass allocation? *Trends in Plant Science* 11, 610–617.

Hernåndez, G., Ramirez, M., Valdės-Lopez, O., Tesfaye, M., Graham, M.A. *et al.* (2007) Phosphorus stress in common bean: root transcript and metabolic responses. *Plant Physiology* 144, 752–767.

Hirano, K., Ueguchi-Tanaka, M. and Matsuoka, M. (2008) GID1-mediated gibberellin signaling in plants. *Trends in Plant Science* 13, 192–199.

Hirel, B. and Lea, P.J. (2002) The biochemistry, molecular biology and genetic manipulation of primary ammonia assimilation. In: Foyer, H. and Noctor, G. (eds) *Photosynthetic Nitrogen Assimilation and Associated Carbon and Respiratory Metabolism*. Kluwer, Dordrecht, the Netherlands, pp. 71–92.

Hirel, B., Berlin, P., Quillere, I., Bourdoncle, W., Attagnant, C. *et al.* (2001) Towards a better understanding of the genetic and physiological basis for nitrogen use efficiency in maize. *Plant Physiology* 125, 1258–1270.

Hirel, B., Andrieu, B., Valadier, M.H., Renard, S., Quillerè, I. *et al.* (2005) Physiology of maize. II. Identification of physiological markers representative of the nitrogen status of maize (*Zea mays* L.) leaves, during grain filling. *Physiologia Plantarum* 124, 178–188.

Hirel, B., Gouis, J.L., Ney, B. and Gallais, A. (2007) The challenge of improving nitrogen use efficiency in crop plants: towards a more central role for genetic variability and quantitative genetics within integrated approaches. *Journal of Experimental Botany* 58, 2369–2387.

Hirschi, K.D., Korenkov, V.D., Wilganowski, N.L. and Wagner, G.J. (2000) Expression of Arabidopsis CAX2 in tobacco. Altered metal accumulation and increased manganese tolerance. *Plant Physiology* 124, 125–133.

Hocking, P.J., Keerthisinghe, G., Smith, F.W. and Randall, P.J. (1997) Comparison of the ability of different crop species to access poorly-available soil phosphorus. In: Ando, T., Fujita, K., Mae, T., Tsumoto, H., Mori, S. and Sekiya, J. (eds) *Plant Nutrition for Sustainable Food Production and Environment*. Kluwer, Dordrecht, the Netherlands, pp. 305–308.

Horst, W., Fecht, M., Naumann, A., Wissemeier, A. and Maier, P. (1999) Physiology of manganese toxicity and tolerance in *Vigna unguiculata* (L.) Walp. *Journal of Plant Nutrition* 162, 263–274.

Hu, F.Y., Tao, D.Y., Sacks, E., Fu, B.Y., Xu, P. *et al.* (2003) Convergent evolution of perenniality in rice and sorghum. *Proceeding of the National Academy of Sciences USA* 100, 4050–4054.

Huang, S., Spielmeyer, W., Lagudah, E.S., James, R.A., Platten, J.D. *et al.* (2006) A sodium transporter (HKT7) is a candidate for NaX1, a gene for salt tolerance in durum wheat. *Plant Physiology* 142, 1718–1727.

Hummel, M., Rahamani, F., Smeekens, S. and Hanson, J. (2009) Sucrose mediated translational control. *Annals of Botany (London)* 104, 1–7.

Huwyzeh, M.S., Maibody, S.A. and Arzani, A. (2008) Evaluation of salt tolerance of sugarcane (*Saccharum officinarum* L.) genotypes based on the ability to regulate ion uptake and transport at early stage of growth. *Journal of Science and Technology of Agriculture and Natural Resources* 11, 55–66.

Innes, P., Blackwell, R.D. and Quarrie, S.A. (1984) Some effects of genetic variation in drought-induced abscisic acid accumulation on the yield and water-use of spring wheat. *Journal of Agricultural Science (Cambridge)* 102, 341–351.

Ismande, J. and Touraine, B. (1994) N demand and the regulation of nitrate uptake. *Plant Physiology* 105, 3–7.

Jacobs, A., Lunde, C., Bacic, A., Tester, M. and Roessner, U. (2007) The impact of constitutive heterologous expression of a moss Na^+ transporter on the metabolomes of rice and barley. *Metabolomics* 3, 307–317.

Jain, A., Poling, M.D., Karthikeyan, A.S., Blakeslee, J.J., Peer, W.A. *et al.* (2007) differential effects of sucrose and auxin on localized phosphate deficiency-induced modulation of different traits of root system architecture in Arabidopsis. *Plant Physiology* 144, 232–247.

James, R.A., von Caemmerer, S., Condon, A.G., Zwart, A.B. and Munns, R. (2008) Genetic variation in tolerance to the osmotic stress component of salinity stress in durum wheat. *Functional Plant Biology* 35, 111–123.

Jessen, D.L., Barnes, D.K., Vance, C.P. and Heichel, G.H. (1987) Variation for activity of nodule nitrogen and carbon assimilating enzymes in alfalfa. *Crop Science* 27, 627–631.

Joffre, R. (1990) Plant and soil nitrogen dynamics in Mediterranean grasslands: a comparison of annual and perennial species. *Oecologia* 85, 142–149.

John, U.P., Polotnianka, R.M., Sivakumaran, K.A., Chew, O., Mackin, L. *et al.* (2009) Ice recrystallization inhibition proteins (IRIPs) and freeze tolerance in the cryophilic Antarctic hair grass *Deschampsia antarctica* E. Desv. *Plant Cell and Environment* 32, 336–348.

Kaeppler, S.M., Parke, J.L., Mueller, S.M., Senior, L., Stuber, C. *et al.* (2000) Variation among maize inbred lines and detection of quantitative trait loci for growth at low phosphorus and responsiveness to arbuscular mycorrhizal fungi. *Crop Science* 40, 358–364.

Kanai, R. and Edwards, G.E. (1999) The biochemistry of C_4 photosynthesis. In: Sage, R.F. and Monson, R.K. (eds) *C_4 Plant Biology*. Academic Press, San Diego, California, pp. 49–87.

Kichey, T., Heumez, E., Pocholle, P., Pageau, K., Vanacker, H. *et al.* (2006) Combined agronomic and physiological aspects of nitrogen management in wheat (*Triticum aestivum* L.). Dynamic and integrated views highlighting the central role for the enzyme glutamine synthetase. *New Phytologist* 169, 265–278.

King, D.A. (1997) The functional significance of leaf angle in *Eucalyptus*. *Australian journal of Botany* 45, 619–639.

Kochian, L.V. and Jones, D.L. (1997) Aluminum toxicity and resistance in plants. In: Yokel, R. and Golub, M.S. (eds) *Research Issues in Aluminum toxicity*. Taylor and Francis, Washington, DC, pp. 69–90.

Kochian, L.V., Hoekenga, O.A. and Piñeros, A. (2004) How do crop plants tolerate acid soils? Mechanisms of aluminum tolerance and phosphorus efficiency. *Annual Review of Plant Biology* 55, 459–493.

Koyama, H., Kawamura, A., Kihara, T., Hara, T., Takita, E. and Shibata, D. (2000) Overexpression of mitochondrial citrate synthase in *Arabidopsis thaliana* improved growth on a phosphorus-limited soil. *Plant Cell Physiology* 41, 1030–1037.

Le Gouis, J., Bèghin, D., Heumez, E. and Pluchard, P. (2000) Genetic differences for nitrogen uptake and nitrogen utilization efficiencies in winter wheat. *European Journal of Agronomy* 12, 163–173.

Lemaire, G., Charrier, X. and Hèbert, Y. (1996) Nitrogen uptake and capacity of maize and sorghum crops in different nitrogen and water supply conditions. *Agronomie* 16, 231–246.

Lemaire, G., van Oosterom, E., Sheehy, J., Jeuffroy, M.H., Massignam, A. *et al.* (2007) Is crop N demand more closely related to dry matter accumulation or leaf area expansion during vegetative growth? *Field Crops Research* 100, 91–106.

Lewandovski, I., Clifton-Brown, J.C., Scurlock, J.M.O. and Huisman, W. (2000) *Miscanthus*; European experience with a novel energy crop. *Biomass and Bioenergy* 19, 209–277.

Li, X.F., Ma, J.F. and Matsumoto, H. (2000) Pattern of aluminum-induced secretion of organic acids differ between rye and wheat. *Plant Physiology* 123, 1537–1544.

Li, Y., Mi, G.H., Chen, F., Lao, X. and Zhang, F. (2001) Genotypic difference of nitrogen recycling between root and shoot of maize seedlings. *Acta Phytophysiologica Sinica* 27, 226–230.

Lian, X., Xing, Y., Xu, H.Y.C., Li, X. and Zhang, Q. (2005) QTLs for low nitrogen tolerance at seedling stage identified using a recombinant inbred line population derived from an elite rice hybrid. *Theoretical and Applied Genetics* 112, 85–96.

Liao, H., Yan, X., Rubio, G., Beebe, S.E., Blair, M.W. *et al.* (2004) Genetic mapping of basal root gravitropism and phosphorus acquisition efficiency in common bean. *Functional Plant Biology* 31, 959–970.

Lillo, C. (2004) Light regulation of nitrate uptake, assimilation and metabolism. In: Amancio, S. and Stule, I. (eds) *Plant Ecophysiology*, Vol. 3: *Nitrogen Acquisition and Assimilation in Higher Plants*. Kluwer, Dordrecht, the Netherlands, pp. 149–184.

Lin, H.X., Zhu, M.Z., Yano, M., Gao, J.P., Liang, Z.W. *et al.* (2004) Mapping of quantitative trait loci controlling low-temperature germinability in rice (*Oryza sative* L.). *Theoretical and Applied Genetics* 108, 794–799.

Lin, Y.R., Schertz, K.F. and Paterson, A.H. (1995) Comparative analysis of QTLs affecting plant height and maturity across the Poaceae, in reference to an interspecific sorghum population. *Genetics* 141, 391–411.
Liu, J., Samac, D.A., Bucciarelli, B., Allan, D.L. and Vance, C.P. (2005) Signaling of phosphorus deficiency-induced gene expression in white lupin requires sugar and phloem transport. *Plant Journal* 41, 257–268.
Liu, J., Allan, D.L. and Vance, C.P. (2010) Systemic signaling and local sensing of phosphate in common bean: cross-talk between photosynthate and micrRNA 399. *Moleular Plant* 3, 428–437.
Liu, L., Liao, H., Wang, X. and Yan, X. (2008) Adaptive changes of soybean genotypes with different root architectures to low phosphorus availability as related to phosphorus efficiency. *Scientia Agricultura Sinica* 41, 1089–1099.
Loneragan, J.F. (1978) The physiology of plant tolerance to low phosphorus availability. In: Jung, G.A. (ed.) *Crop Tolerance to Suboptimal Land Conditions.* ASA, CSSA and SSSA, Madison, Wisconsin, pp. 329–343.
Long, S.P. (1999) Environmental responses. In: Sage, R.F. and Monson, R.K. (eds) C_4 *Plant Biology.* Academic Press, San Diego, California, pp. 215–249.
Long, S.P., Postal, W.F. and Bohlår-Nordenkampf, H.R. (1993) Quantum yields for uptake of carbon dioxide in C-3 vascular plants of contrasting habitats and taxonomic groupings. *Planta* 189, 226–234.
Long, S.P., Humphries, S. and Falkowski, P.G. (1994) Photoinhibition and photosynthesis in nature. *Annual Review of Plant Physiology and Plant Molecular Biology* 45, 633–662.
Long, S.P., Zhu, X.G., Naidu, S.L. and Ort, D.R. (2006) Can improvement in photosynthesis increase crop yields? *Plant, Cell and Environment* 29, 315–330.
Loss, S.P. and Siddique, K.H.M. (1994) Morphological and physiological traits associated with wheat yield increases in Mediterranean environments. *Advances in Agronomy* 52, 229–276.
Loudet, O., Chaillou, S., Merigout, P., Talbotec, J. and Daniel-Vedele, F. (2003) Quantitative trait loci analysis of nitrogen use efficiency in *Arabidopsis. Plant Physiology* 131, 345–358.
Lücker, J., Schwab, W., van Hautum, B., Blaas, J., van der Plas, L. *et al.* (2004) Increased and altered fragrance of tobacco plants after metabolic engineering using three monoterpene synthase from lemon. *Plant Physiology* 134, 510–519.
Lynch, J.P. and Brown, K.M. (2001) Top soil foraging: an architectural adaptation of plants to low phosphorus availability. *Plant and Soil* 237, 225–237.
Ma, B.L. and Dwyer, M.L. (1998) Nitrogen uptake and use in two contrasting maize hybrids differing in leaf senescence. *Plant and Soil* 199, 283–291.
Ma, J.F. and Furukawa, J. (2003) Recent progress in the research of external Al detoxification in higher plants: a minireview. *Journal of Inorganic Biochemistry* 97, 46–51.
Ma, J.F. and Hiradate, S. (2000) Form of aluminum for uptake and translocation in buckwheat (*Fagopyrum esculentum* Moench). *Planta* 211, 355–360.
Ma, J.F., Hiradate, S. and Matsumoto, H. (1998) High aluminum resistance in buckwheat: II. Oxalic acid detoxifies aluminum internally. *Plant Physiology* 117, 753–759.
Ma, J.F., Ryan, P.R. and Delhaize, E. (2001) Aluminum tolerance in plants and the complex role of organic acids. *Trends in Plant Science* 6, 273–278.
Ma, J.F., Shen, R., Zhao, Z., Wissuwa, M., Takeuchi, Y. *et al.* (2002) Response of rice to Al stress and identification of quantitative trait loci for Al tolerance. *Plant Cell Physiology* 652–659.
Mae, T. (1997) Physiological nitrogen efficiency in rice; nitrogen utilization, photosynthesis, and yield potential. In: Ando, T. (ed.) *Plant Nutrition for Sustainable Food Production and Environment.* Kluwer Academic Publishers, Dordrecht, the Netherlands, pp. 51–60.
Mae, T., Makino, A. and Ohira, K. (1983) Changes in the amounts of ribulose biphosphate carboxylase synthesized and degraded during the life span of rice leaf (*Oryza sativa* L.). *Plant and Cell Physiology* 24, 1079–1086.
Magalhaes, J.V. (2002) Molecular genetic and physiological investigation of aluminum tolerance of sorghum (*Sorghum bicolor* L. Moench). PhD thesis, Cornell University, Ithaca, New York, 192 pp.
Martinez, J.M.D.F. and Estrella, L.H. (1999) Advance in understanding of aluminum toxicity and development of aluminum tolerant transgenic plants. In: Donald, L.P. (ed.) *Advances in Agronomy,* Vol.66. Academic Press, San Diego, California, pp. 103–120.
Martre, P., Porter, J.R., Jamieson, P.D. and Triboi, E. (2003) Modeling grain nitrogen accumulation and protein composition to understand the sink/source regulations of nitrogen utilization in wheat. *Plant Physiology* 133, 1959–1967.

Matsumoto, H., Yamamoto, Y. and Rama Devi, S. (2001) Aluminum toxicity in acid soils. Plant responses to aluminum. In: Prasad, D. (ed.) *Metals in the Environment.* Marcel Dekker, New York, pp. 329–343.

Matsuoka, M., Furbank, R.T., Fukayama, H. and Miyao, M. (2001) Molecular engineering of C4 photosynthesis. *Annual Review of Plant Physiology and Plant Molecular Biology* 52, 297–314.

Mauriat, M. and Moritz, T. (2009) Analyses of GA20ox- and GID1-overexpressing aspen suggest that gibberellin play two distinct roles in wood formation. *Plant Journal* 58, 989–1003.

McCarty, D.R., Settles, A.M., Suzuki, M., Tan, B.C., Latshaw, S. *et al.* (2005) Steady-state transposon mutagenesis in inbred maize. *Plant Journal* 44, 52–61.

McCollum, T., McCuistion, K. and Bean, B. (2005) *Brown midrib and photoperiod sensitive forage sorghums.* Publication number AREC 05-20. Texas Agricultural and Mechanical University, Agricultural Research and Extension Center, Amarillo, Texas, pp. 36–46.

McLean, E.O. (1976) Chemistry of soil aluminum. *Communications in Soil Science and Plant Analysis* 7, 619–636.

Melzer, S., Lens, F., Gennen, J., Vanneste, S., Rhode, A. *et al.* (2008) Flowering-time genes modulate meristem determinacy and growth from *Arabidopsis thaliana. Nature Genetics* 40, 1489–1492.

Mengel, K. (1985) Dynamics and availability of major nutrients in soils. *Advances in Soil Sciences* 2, 65–131.

Meiresonne, L., Schrijver, A.D. and De Vos, B. (2007) Nutrient cycling in a poplar plantation (*Populus trichocarpa* x *Populus deltoides* 'Beaupré') on former agricultural land in northern Belgium. *Canadian Journal of Forest Research* 37, 141–155.

Miflin, R.D. and Lea, P.J. (1976) The pathway of nitrogen assimilation in plants. *Phytochemistry* 15, 873–885.

Mitra, J. (2001) Genetics and genetic improvement of drought resistance in crop plants. *Current Science* 80, 758–763.

Moll, R.H., Kamprath, E.J. and Jackson, W.A. (1982) Analysis and interpretation of factors which contribute to efficiency of nitrogen utilization. *Agronomy Journal* 74, 562–564.

Monteith, J.L. (1977) Climate and efficiency of crop production in Britain. *Philosophical Transactions of the Royal Society of London* 281, 277–294.

Morgan, J.M. (1995) Growth and yield of wheat lines with differing osmoregulative capacity at high soil water deficit in seasons of varying evaporative demand. *Field Crops Research* 40, 143–152.

Morinaka, Y., Sakamoto, T., Inukai, Y., Agetsuma, M., Kitano, H. *et al.* (2006) Morphological alteration caused by brassinosteriod insensitivity increases the biomass and grain production of rice. *Plant Physiology* 141, 924–931.

Multani, D.S., Briggs, S.P., Chamberlin, M.A., Blakeslee, J.J., Murphy, A.S. *et al.* (2003) Loss of an MDR transporter in compact stalks of maize *br2* and sorghum *dw3* mutants. *Science* 302, 81–84.

Munns, R., James, R.A. and Läuchli, A. (2006) Approaches to increasing the salt tolerance of wheat and other cereals. *Journal of Experimental Botany* 57, 1025–1043.

Murata, Y. and Matsushima, S. (1975) Rice. In: Evans, L.T. (ed.) *Crop Physiology: Some Case Histories.* Cambridge University Press, Cambridge, UK, pp. 73–99.

Nagy, R., Vasconcelos, M.J.V., Zhao, S., McElver, J., Bruce, W. *et al.* (2006) Differential regulation of five *Pht1* phosphate transporters from maize (*Zea mays* L.). *Plant Biology (Stuttgart)* 8, 186–197.

Naidu, S.L., Moose, S.P., Al-Shoaibi, A.K., Raines, C.A. and Long, S.P. (2003) Cold tolerance of C_4 photosynthesis in *Miscanthus × gigantus*: adaptation in amounts and sequence of C_4 photosynthetic enzymes. *Plant Physiology* 132, 1688–1697.

Neumann, A.M., Ash, F., Wu, Y., Jensen, C.R., Naested, H. *et al.* (2002) Soluble invertase expression in an early target of drought stress during the critical, abortion sensitive phase of young ovary development in maize. *Plant Physiology* 130, 591–604.

Niu, J., Chen, F.J., Mi, G.H., Li, C.J. and Zhang, F.S. (2007) Transpiration, and nitrogen uptake and flow in two maize (*Zea mays* L.) inbred lines as affected by nitrogen supply. *Annals of Botany* 99, 153–160.

Nogueira, F.T.S., De Rosa, V.E. Jr, Menossi, M., Ulian, E.C. and Arruda, P. (2003) RNA expression profiles and data mining of sugarcane response to low temperature. *Plant Physiology* 132, 1–14.

Nye, P.H., Brewster, L. and Bhat, K.K.S. (1965) The possibility of predicting solute uptake and plant growth response from independently measured soil and plant characteristics. 1. The theoretical basis of the experiments. *Plant and Soil* 42, 161–170.

Orsel, M., Filleur, S., Frasier, V. and Daniel-Vedele, F. (2002) Nitrate transport in plants: which gene and which control? *Journal of Experimental Botany* 52, 825–833.

Ort, D.R. and Long, S.P. (2003) Converting solar energy into crop production. In: Chrispeels, M.J. and Sadava, D.E. (eds) *Converting Solar Energy into Crop Production.* American Society of Plant Biologist/Jones and Bartlett, Boston, Massachusetts, pp. 240–269.

Osawa, H. and Matsumoto, H. (2001) Possible involvement of protein phosphorylation in aluminum-responsive malate efflux from root apex. *Plant Physiology* 126, 411–420.
Otani, T., Ae, N. and Tanaka, H. (1996) Phosphorus uptake mechanisms of crops grown in soils with low P status. II. Significance of organic acids in root exudates of pigeon pea. *Soil Science and Plant Nutrition* 42, 553–560.
Pant, B.D., Buhtz, A., Kehr, J. and Scheible, W.R. (2008) MicroRNA 399 is a long-distance signal for the regulation of plant phosphate homeostasis. *Plant Journal* 53, 731–738.
Paszkowski, U. and Boller, T. (2002) The growth defect of lrt1, a maize mutant lacking lateral roots, can be complemented by symbiotic fungi or high phosphate nutrition. *Planta* 214, 584–590.
Pathan, M.S., Subudhi, P.K., Courtois, B. and Nguyen, H.T. (2004) Molecular dissection of abiotic stress tolerance in sorghum and rice. In: Nguyen, H.T. and Blum, A. (eds) *Physiology and Biotechnology Integration for Plant Breeding*. Marcel Dekker, New York, pp. 525–569.
Paterson, A.H., Lin, Y.R., Li, Z., Schertz, K.F., Doebley, J.F. *et al.* (1995) Convergent domestication of cereal crops by independent mutations at corresponding genetic loci. *Science* 269, 1714–1718.
Pereira, M.G. and Lee, M. (1995) Identification of genomic regions affecting plant height in sorghum and maize. *Theoretical and Applied Genetics* 90, 380–388.
Persson, S., Wei, H., Milne, J., Page, G.P. and Somerville, C.R. (2005) Identification of genes required for cellulose synthesis by regression analysis of public microarray data. *Proceedings of the National Academy of Sciences USA* 102, 8633–8638.
Peterson, A.H. (2008) Genomics of sorghum (a review). *International Journal of Plant Genomics*, 2008. Article ID 362451; doi:101155/2008/362451, 6 pp.
Peterson, G.A. and Frye, W.W. (1989) Fertilizer nitrogen management. In: Follet, R.F. (ed.) *Nitrogen Management and Ground Water Protection*. Elsevier, Amsterdam, pp. 183–219.
Piper, J.K. and Kulakow, P.A. (1994) Seed yield and biomass allocation in *Sorghum bicolor* and F_1 and backcross generations of *S. bicolor* x *S. halepense* hybrids. *Canadian Journal of Botany* 72, 468–474.
Pitterman, J. and Sage, R.F. (2000) Photosynthetic performance at low temperature of *Bouteloua gracilis* Lag., a high-altitude C4 grass from the Rocky Mountains, USA. *Plant Cell and Environment* 23, 811–823.
Presterl, T., Seitz, G., Landbeck, M., Thiemt, W., Schmidt, W. *et al.* (2003) Improving nitrogen use efficiency in European maize: estimation of quantitative parameters. *Crop Science* 43, 1259–1265.
Price, A.H., Townend, J., Jones, M.P., Audebert, A. and Courtois, B. (2002) Mapping QTLs associated with drought avoidance in upland rice grown in the Philippines and West Africa. *Plant Molecular Biology* 48, 683–695.
Raghothama, K.G. (1999) Phosphate acquisition. *Annual Review of Plant Physiology and Plant Molecular Biology* 50, 665–693.
Rami, J.F., Dufour, P., Trouche, G., Fliedel, G., Mestres, C. *et al.* (1998) Quantitative traits loci for grain quality, productivity, morphological and agronomical traits in sorghum (*Sorghum bicolor* L. Moench). *Theoretical and Applied Genetics* 97, 605–616.
Rauh, B.L., Basten, C. and Buckler, E.S. (2002) Quantitative trait locus analysis of growth response to varying nitrogen sources in *Arbidopsis thaliana*. *Theoretical and Applied Genetics* 104, 743–750.
Reich, P.B., Oleksyn, J., Modryzynski, J., Mrozinski, P., Hobbie, S.E. *et al.* (2005) Linking litter calcium, earthworms and soil properties: a common garden test with 14 tree species. *Ecology Letters* 8, 811–818.
Reymond, M., Muller, B., Leonardi, A., Charcosset, A. and Tardieu, F. (2003) Combining quantitative trait loci analysis and an ecophysiological model to analyze the genetic variability of the responses of maize to temperature and water deficit. *Plant Physiology* 131, 664–675.
Reymond, M., Svistoonoff, S., Loudet, O., Nussaume, L. and Desnos, T. (2006) Identification of QTL controlling root growth response to phosphate starvation in *Arabidopsis thaliana*. *Plant Cell Environment* 29, 115–125.
Ritter, E., Vesterdal, L. and Gundersen, P. (2003) Changes in soil properties after afforestation of former intensively managed soils with oak and Norway spruce. *Plant and Soil* 249, 319–330.
Rönnberg-Wästljung, A.C., Glynn, C. and Weih, M. (2005) QTL analyses of drought tolerance and growth for a *Salix dasyclados x Salix viminalis* hybrid in contrasting water regimes. *Theoretical and Applied Genetics* 110, 537–549.
Rooney, W.L., Blumenthal, J., Bean, B. and Mullet, J.E. (2000) Designing sorghum as a dedicated bioenergy feedstock. *Biofuels, Bioproducts and Biorefining* 1, 147–157.
Ryan, P.R., Delhaize, E. and Randall, P.J. (1995) Characterization of Al-stimulated efflux of malate from apices of Al-tolerant wheat roots. *Planta* 196, 103–110.
Sage, R.F. (2002) Variation in the k_{cat} of Rubisco in C_3 and C_4 plants and some implications for photosynthetic performance at high and low temperature. *Journal of Experimental Botany* 53, 609–620.

Sage, R.F. and Pearcy, R.W. (1987) The nitrogen use efficiency of C_3 and C_4 plants. II. Leaf nitrogen effects on the gas exchange characteristics of *Chenopodium album* L. and *Amaranthus retroflexus*. *Plant Physiology* 84, 959–963.
Sage, R.F. and Pearcy, R.W. (2000) The physiological ecology of C_4 photosynthesis. In: Leegood, R.C., Sharkey, T.D. and von Caemmerer, S. (eds) *Photosynthesis: Physiology and Metabolism*. Kluwer Academic, Dordrecht, the Netherlands, pp. 497–532.
Sage, R.F. and Zhu, X.-G. (2011) Exploiting the engine of C_4 photosynthesis. *Journal of Experimental Botany* 62, 2989–3000.
Sage, R.F., Pearcy, R.W. and Seeman Jr, R.W. (1987) The nitrogen use efficiency in C_3 and C_4 plants. *Plant Physiology* 85, 355–359.
Sagon, M., Ney, B. and Due, G. (1993) Plant symbiotic mutants as a tool to analyze nitrogen nutrition and yield relationship in field grown peas (*Pisum sativum* L.). *Plant and Soil* 153, 33–45.
Sakamoto, T. and Matsuoka, M. (2004) Generating high-yielding varieties by genetic manipulation of plant architecture. *Current Opinion in Biotechnology* 15, 144–147.
Sakamoto, T., Morinaka, Y., Ohnishi, T., Sunohara, H., Fujioka, S. *et al.* (2006) Erect leaves caused by brassinosteriod deficiency increase biomass production and grain yield in rice. *Nature Biotechnology* 24, 105–109.
Salehi, H., Ransom, C.B., Oraby, H.F., Seddighi, Z. and Sticklen, M.B. (2005) Delay in flowering and increase in biomass of transgenic tobacco expressing the Arabidopsis floral repressor gene FLOWERING LOCUS C. *Journal of Plant Physiology* 162, 711–717.
Salsac, L., Chaillou, S., Morot-Gaudry, J.F., Lesaint, C. and Jolivoe, E. (1987) Nitrate and ammonium nutrition in plants. *Plant Physiology and Biochemistry* 25, 553–558.
Samson, F.B. and Knof, F.L. (1994) Prairie conservation in North America. *Bioscience* 44, 418–421.
Sankar, B., Jaleel, C.A., Manivannan, P., Kishorekumar, A., Somasundaram, R. and Panneerselvam, R. (2008) Relative efficacy of water use in five varieties of *Abelmoschus esculentus* (L.) Moench under water limited conditions. *Colloids and Surfaces B: Biointerfaces* 62, 125–129.
Santamaria, J.M., Ludlow, M.M. and Fukai, S. (1990) Contribution of osmotic adjustment to grain yield in *Sorghum bicolor* (L.) Moench under water-limited conditions. 1. Water stress before anthesis. *Australian Journal of Agricultural Research* 41, 51–65.
Saranga, Y., Jiang, C.X., Wright, R.J., Yakir, D. and Paterson, A.H. (2004) Genetic dissection of cotton physiological response to arid conditions and their inter-relationship with productivity. *Plant Cell and Environment* 27, 263–277.
Sasaki, T., Yamamoto, Y., Ezaki, E., Katsuhara, M., Ahn, S.J. *et al.* (2004) A wheat gene encoding an aluminum-activated malate transporter. *Plant Journal* 37, 645–653.
Sawers, R.J. (2009) Breeding for reaction. In: Huttunen, N. and Sinisalo, T. (eds) *Plant Breeding*. Nova Publishers, New York, pp. 127–142.
Schachtman, D.P., Reid, R.J. and Ayling, S.M. (1998) Phosphorus uptake by plants; from soil to cell. *Plant Physiology* 116, 447–453.
Schrader, J., Nilsson, J., Mellerowicz, E., Berglund, A., Nilsson, P. *et al.* (2004) A high-resolution transcript profile across the wood-forming meristem of poplar identifies potential regulators of cambial stem cell identity. *Plant Cell* 16, 2278–2292.
Seki, M., Kamei, A., Yamaguchi-Shinozaki, K. and Shinozaki, K. (2003) Molecular responses to drought, salinity and frost: common and different paths for plant protection. *Current Opinion in Biotechnology* 14, 194–199.
Seemann, J.R., Badger, M.R. and Berry, J.A. (1984) Variations in the specific activity of ribulose-1,5,-biphosphate carboxylase between species utilizing differing photosynthetic pathways. *Plant Physiology* 74, 791–794.
Semiz, G., Heijari, J., Isik, K. and Holopainen, J.K. (2007) Variation in needle terpenoids among *Pinus sylvestris* L. (Pinaceae) provenances from Turkey. *Biochemical Systematics and Ecology* 35, 652–661.
Shi, H., Lee, B.H., Wu, S.J. and Zhu, J.K. (2003) Overexpression of a plasma membrane Na^+/H^+ antiporter improves salt tolerance in *Arabidopsis*. *Nature Biotechnology* 21, 81–85.
Sinclair, T.R. and Vadez, V. (2002) Physiological traits for crop yield improvement in low N and P environments. *Plant and Soil* 245, 1–15.
Singletary, G. and Below, F. (1990) Nitrogen induced changes in the growth and metabolism of developing maize kernels grown *in vitro*. *Plant Physiology* 92, 160–167.
Smidansky, E.D., Martin, J.M., Hannah, C.L., Fischer, A.M. and Giroux, M.J. (2003) Seed yield and plant biomass increases in rice are conferred by deregulation of endosperm ADP-glucose pyrophosphorylase. *Planta* 216, 656–664.

Smil, V. (2006) Nitrogen in crop production; an account of global flows. *Global Biogeochemical Cycles* 13, 647–662.
Smith, C.W. and Frederiksen, R.A. (2000) *Sorghum: Origin, History, Technology, and Production.* John Wiley and Sons, New York, 824 pp.
Smith, S.N. (1934) Response of inbred lines and crosses in maize to variations of nitrogen and phosphorus supplied as nutrients. *Journal of American Society of Agronomy* 26, 785–804.
Snyder, F.W. and Bunce, J.A. (1983) Use of plastochron index to evaluate effects of light, temperature and nitrogen on growth of soya bean (*Glycine max* L. Merr.). *Annals of Botany* 52, 895–903.
Springfield, G. and Salter, R. (1934) Differential response of corn varieties to fertility levels and seasons. *Journal of Agricultural Research* 49, 991–1000.
Stark, D.M., Timmerman, K.P., Barry, G.F., Preiss, J. and Kishore, G.M. (1992) Regulation of the amount of starch in plant tissues by ADP glucose pyrophosphorylase. *Science* 258, 287–292.
Steele, K.A., Price, A.H., Shashidhar, H.E. and Witcombe, J.R. (2009) Marker-assisted selection to introgress QTLs controlling root traits into an Indian upland rice variety. *Theoretical and Applied Genetics* 112, 208–221.
Steinkellner, S., Lendzemo, V., Langer, I., Schweiger, P., Khaosadd, T. *et al.* (2007) Flavonoids and strigolactones in root exudates as signals in symbiosis and pathogenic plant-fungus interactions. *Molecules* 12, 1290–1306.
Stevenson, F.J. and Cole, M.A. (1999) *Cycles of Soil: Carbon, Nitrogen, Phosphorus, Sulfur, Micronutrients,* 2nd edn. John Wiley, New York, 448 pp.
Sticklen, M.B. (2008) Plant genetic engineering for biofuel production: towards affordable cellulosic ethanol. *Nature Reviews Genetics* 9, 433–443.
Subedi, K.D. and Ma, B.L. (2005) Nitrogen uptake and partitioning in stay-green and leafy maize hybrids. *Crop Science* 45, 740–747.
Svecnjak, Z. and Rengel, Z. (2005) Canola cultivars differ in nitrogen utilization efficiency at vegetative stage. *Field Crops Research* 97, 221–226.
Svistoonoff, S., Creff, A., Reymond, M., Sigoillot-Claude, C., Ricaud, L. *et al.* (2007) Root tip contact with low phosphate media reprograms plant root architecture. *Nature Genetics* 19, 792–796.
Swank, J.C., Below, F.E., Lambert, R.J. and Hageman, R.H. (1982) Interaction of carbon and nitrogen metabolism in the productivity of maize. *Plant Physiology* 70, 1185–1190.
Talamè, V., Sanguineti, M.C., Chiapparino, E., Bahri, H., Ben Salem, M. *et al.* (2004) Identification of *Hordeum spontaneum* QTL alleles improving field performance of barley grown under rainfed conditions. *Annals of Applied Biology* 144, 309–319.
Tammisola, J. (2010) Towards much more efficient biofuel crops – can sugarcane pave the way? *GM Crops* 1, 181–198.
Tao, Y.Z., Henzell, R.G., Jordan, D.R., Butler, D.G., Kelley, A.M. *et al.* (2000) Identification of genomic regions associated with stay green in sorghum by testing RILs in multiple environments. *Theoretical and Applied Genetics* 100, 1225–1232.
Tesfaye, M., Temple, S.J., Allan, D.L., Vance, C.P. and Samac, D.A. (2001) Overexpression of malate dehydrogenase in transgenic alfalfa enhances organic acid synthesis and confers tolerance to aluminum. *Plant Physiology* 127, 1836–1844.
Tester, M. and Davenport, R. (2003) Na^+ tolerance and Na^+ transport in higher plants. *Annals of Botany* 91, 503–527.
Thompson, A., Cooper, J. and Ingram Jr, L.L. (2006) Distribution of terpenes in heartwood and sapwood of loblolly pine. *Forest Products Journal* 56, 46–48.
Thornton, B. and Bausenwein, U. (2000) Seasonal protease activity in storage tissue of the deciduous grass *Molinia caerulea. New Phytologist* 146, 75–81.
Thornton, B. and Millard, P. (1993) The effects of nitrogen supply and defoliation on the seasonal internal cycling of nitrogen in *Molinia caerulea. Journal of Experimental Botany* 44, 531–536.
Tian, Q., Chen, F.J., Zhang, F.S. and Mi, G.H. (2005) Possible involvement of cytokinin in nitrate mediated root growth in maize. *Plant and Soil* 277, 185–196.
Tian, Q.Y., Chen, F.J., Zhang, F.S. and Mi, G.H. (2006) Genotypic difference in nitrogen acquisition ability in maize plants is related to the coordination of leaf and root growth. *Journal of Plant Nutrition* 29, 317–330.
Ticconi, C.A., Lucero, R.D., Sakhonwasee, S., Adamson, A.W., Creff, A. *et al.* (2009) ER-resident proteins PDR2 and LPR1mediate the developmental response of root meristems to phosphate deficiency. *Proceedings of the National Academy of Sciences, USA* 106, 14174–14179.

Tiessen, H. (2008) Phosphorus in the global environment. In: White, P.J. and Hammonds, J.P. (eds) *Ecophysiology of Plant-Phosphorus Interactions*. Springer Science and Business Media, New York, pp. 1–8.
Tilman, D., Hill, J. and Lehman, C. (2006) Carbon-negative biofuels from low-input high diversity grassland biomass. *Science* 34, 1598–1600.
Tsarouhas, V., Gullberg, U. and Lagercrantz, U. (2004) Mapping of quantitative trait loci (QTLs) affecting autumn freezing resistance and phenology in *Salix. Theoretical and Applied Genetics* 108, 1335–1342.
Tuberosa, R. and Salvi, S. (2006) Genomic-based approaches to improve drought tolerance of crops. *Trends in Plant Science* 11, 405–412.
Tuberosa, R., Salvi, S., Sanguineti, M.C., Landi, P., Maccaferri, M. *et al.* (2002) Mapping qtls regulating morpho-physiological traits and yield: case studies, shortcomings and perspectives in drought-stressed maize. *Annals of Botany (London)* 89, 941–963.
Umezawa, T., Fujita, M., Fujita, Y., Yamaguchi-Shinozaki, K. and Shinozaki, K. (2006) Engineering drought tolerance in plants: discovering and tailoring genes to unlock the future. *Current Opinion in Biotechnology* 17, 113–122.
Urquiaga, S., Cruz, K.H.S. and Boddey, R.M. (1992) Contribution of nitrogen fixation to sugarcane: Nitrogen-15 and nitrogen balance estimates. *Soil Science Society of America Journal* 56, 105–114.
Vance, C.P. (2010) Quantitative trait loci, epigenetics, sugars, and microRNAS: quaternaries in phosphate acquisition and use. *Plant Physiology* 154, 582–588.
Vance, C.P., Uhde-Stone, C. and Allan, D.L. (2003) Phosphorus acquisition and use: critical adaptation by plants for securing a nonrenewable resource. *New Phytologist* 157, 423–447.
von Wirén, N., Gojon, A., Chaillou, S. and Rapier, D. (2001) Mechanism and regulation of ammonium transport in higher plants. In: Lea, P.J. and Morot-Gaudry, J.F. (eds) *Plant Nitrogen*. Springer-Verlag, Berlin, pp. 62–77.
Vos, J. and Biemond, H. (1992) Effects of nitrogen on the development and growth of the potato plant. 1. Leaf appearance, expansion growth, life span of leaves and stem branching. *Annals of Botany* 70, 27–35.
Vos, J. and van de Putten, P.E.L. (1998) Effect of nitrogen supply on leaf appearance, leaf growth, leaf nitrogen economy and photosynthetic capacity in potato. *Field Crops Research* 59, 63–72.
Vos, J., van de Putten, P.E.L. and Birch, C.J. (2005) Effect of nitrogen supply on leaf appearance, leaf growth, leaf nitrogen economy and photosynthetic capacity in maize (*Zea mays* L.). *Field Crops Research* 93, 64–73.
Wahid, A., Rao, A.R. and Rasul, E. (1997) Identification of salt tolerance traits in sugarcane lines. *Field Crops Research* 54, 9–17.
Wang, L., Liao, H., Yan, X., Zhuang, B. and Dong, Y. (2004) Genetic variability for root hair traits as related to phosphorus efficiency in soybean. *Plant and Soil* 261, 77–84.
Wang, Q., Zhang, Q.D., Zhu, X.G., Lu, C.M., Kunag, T.Y. and Li, C.Q. (2002) PSII photochemistry and xanthophylls cycle in two super-high-yield rice hybrids, Liangyoupeijiu and Hua-an 3 during photoinhibition and subsequent restoration. *Acta Botanica Sinica* 197–132.
Wang, W.X., Vincour, B. and Altman, A. (2003) Plant responses to drought, salinity and extreme temperature: towards genetic engineering for stress tolerance. *Planta* 218, 1–14.
Wang, X., Yan, X. and Liao, H. (2010) Genetic improvement for phosphorus efficiency in soybean: a radical approach. *Annals of Botany* 106, 215–222.
Wang, Y., Mi, G.H., Chen, F. and Zhang, F. (2003) Genotypic differences in nitrogen uptake by maize inbred lines its relation to root morphology *Acta Ecologica Sinica* 23, 297–302.
Weih, M., Rönnberg-Wästljung, A.C. and Glynn, C. (2006) Genetic basis of phenotypic correlations among growth traits in hybrid willow (*Salix dasyclados* x *Salix viminalis*) grown under two water regimes. *New Phytologist* 170, 467–477.
Weil, C.F. (2007) Tilling in grass species. *Plant Physiology* 145, 158–164.
Wilson, J.R., Ludlow, M.M., Fischer, M.J. and Schulze, E.D. (1980) Adaptation to water stress of the leaf water relations of four tropical forage species. *Australian Journal of Plant Physiology* 7, 207–220.
Wiren, N.V., Gazzarrini, S. and Frommer, W.B. (1997) Regulation of mineral nitrogen uptake in plants. In: Ando, T., Fujita, K., Mae, T., Tsumoto, H., Mori, S. and Sekiya, J. (eds) *Plant Nutrition for Sustainable Food Production and Environment.* Kluwer, Dordrecht, the Netherlands, pp. 41–49.
Wissuwa, M., Wegner, J., Ae, N. and Yano, M. (2002) Substitution mapping of *Pup1*: a major QTL increasing phosphorus uptake in rice. *Theoretical and Applied Genetics* 105, 890–897.
Wright, G.C., Smith, R.G.C. and McWilliam, J.R. (1983) Differences between two grain sorghum genotypes in adaptation to drought stress. I. Crop growth and yield responses. *Australian Journal of Agricultural Research* 34, 637–651.

Wu, L. and Birch, R.G. (2007) Doubled sugar content in sugarcane plants modified to produce a sucrose isomer. *Plant Biotechnology Journal* 5, 109–117.

Wu, Z.Y., Liang, F., Hong, B.M., Young, J.C., Sussman, M.R. *et al.* (2002) An endoplasmic reticulum-bound Ca^{2+}/Mn^{2+} pump, ECA1, supports plant growth and confers tolerance to Mn^{2+} stress. *Plant Physiology* 130, 128–137.

Xin, Z., Aiken, R. and Burke, J.J. (2009) Genetic diversity in sorghum transpiration efficiency. *Field Crops Research* 11, 74–80.

Yanagisawa, S., Akiyama, A., Kisaka, H., Uchimiya, H.T. and Miwa, T. (2004) Metabolic engineering with Dof1 transcription factor in plants: improved nitrogen assimilation and growth under low-nitrogen conditions. *Proceedings of the National Academy of Sciences USA* 101, 7833–7838.

Yeo, A.R., Yeo, M.E. and Flowers, T.J. (1988) Selection of lines with high and low sodium transport from within varieties of an inbreeding species, rice (*Oryza sativa*, L.). *New Phytology* 110, 13–19.

Yu, J. and Buckler, E.S. (2006) Genetic association mapping and genome organization in maize. *Current Opinion in Biotechnology* 17, 155–160.

Zhang, H.X. and Blumwald, E. (2001) Transgenic salt tolerant tomato plants accumulate salt in foliage but not in fruit. *Nature Biotechnology* 19, 765–768.

Zhang, H.X., Hodson, J.N., Williams, J.P. and Blumwald, E. (2001) Engineering salt-tolerant Brassica plants: characterization of yield and seed quality in transgenic plants with increased vacuolar sodium accumulation. *Proceedings of the National Academy of Sciences USA* 98, 12832–12836.

Zhang, J.Z., Creelman, R.A. and Zhu, J.K. (2004) From laboratory to field. Using information from *Arabidopsis* to engineer salt, cold, and drought tolerance in crops. *Plant Physiology* 135, 615–621.

Zhang, S.Z., Yang, B.P., Feng, C.L., Chen, R.K., Luo, J.P. *et al.* (2006) Expression of the *Grifola frondosa* trehalose synthase gene and improvement of drought tolerance in sugarcane (*Saccharum officinarum* L.). *Journal of Integrative Plant Biology* 48, 453–459.

Zheng, S.J., Ma, J.F. and Matsumoto, H. (1998) Continuous secretion of organic acids is related to aluminum resistance during relatively long-term exposure to aluminum stress. *Plant Physiology* 117, 745–751.

7 Deconstructing Plant Biomass: Cell Wall Structure and Novel Manipulation Strategies

**Caitlin S. Byrt,[1] Natalie S. Betts,[1]
Naser Farrokhi[2] and Rachel A. Burton[1]**
[1]*School of Agriculture, Food and Wine, University of Adelaide, Australia;*
[2]*Faculty of Agriculture, Shahrood University of Technology, Iran*

7.1 Introduction

Lignocellulosic biomass is primarily cell wall, composed of cellulose, non-cellulosic polysaccharides (hemicelluloses and pectin), lignin and other minor components such as proteins (Fig. 7.1). Cell wall polysaccharides can be converted to simple sugars and processed to generate a source of transportable energy known as biofuel, suitable for a wide range of end uses. For example, cellulose may be hydrolysed to glucose that may then be converted to bioethanol by fermentation using microorganisms such as yeast. In this chapter, we will outline the ways in which the structure and composition of the plant material, specifically the cell wall, impacts conversion to biofuel and will summarize many of the novel ways in which plant manipulation can enhance this conversion.

Lignocellulosic biofuel feedstock crops, such as woody species and grasses, may be compared on the basis of biomass production, where high dry matter yields per annum or per unit land area are favourable. However, biomass composition is also important, as this influences the yield of fermentable sugars that may be released during conversion.

Cellulose, present in the cell walls of all higher plants, is the most abundant terrestrial natural polymer, accounting for 15–50% of the cell wall of current biofuel feedstock crops (Pauly and Keegstra, 2008). Plants synthesize cellulose from glucose produced by photosynthesis. Chains of cellulose in the primary cell wall, the wall made during rapid growth, are less ordered than those in the secondary cell wall, which thickens once growth has ceased. In the secondary cell wall, the chains of cellulose are in a densely packed, highly crystalline arrangement that is more difficult to degrade.

Non-cellulosic polysaccharides are a heterogeneous mixture of molecules whose profile in the cell wall differs by species, tissue, age and even environmental conditions under which the plant was grown. They form a flexible gel-like matrix that links the rigid, crystalline cellulose with the protective lignin sheath while allowing water-soluble molecules to permeate the cell wall (Burton *et al.*, 2010a). Lignin is a non-linear molecule consisting predominantly of phenolic monomers. It provides structural support to the plant and its hydrophobicity plays an important role in water movement (Vanholme *et al.*, 2010).

7.2 Desirable Characteristics for Biomass Conversion

The main obstacle to using cell walls as a source of fermentable sugar is their recalcitrance: cell

©CAB International 2013. *Biofuel Crops: Production, Physiology and Genetics* (ed. B.P. Singh)

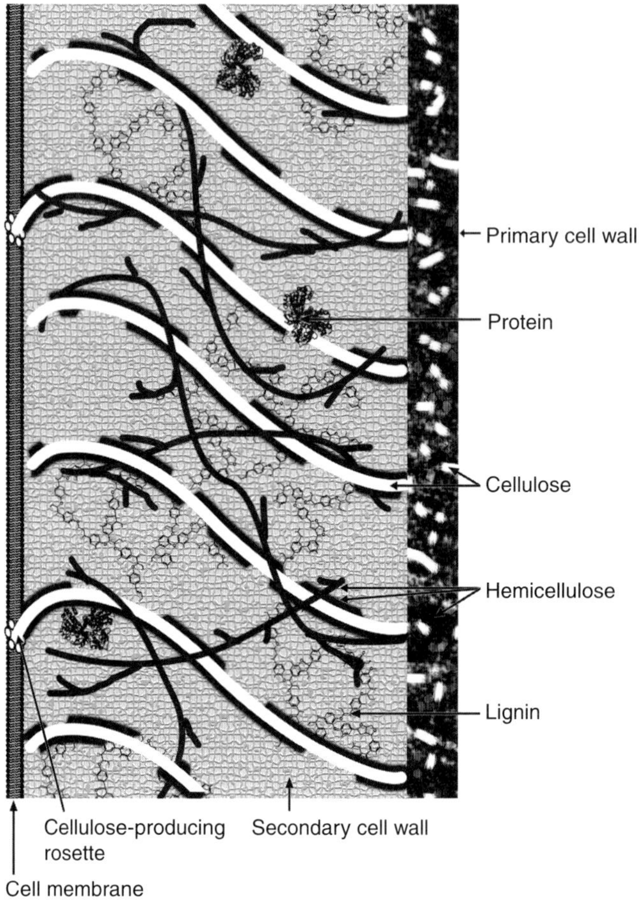

Fig. 7.1. The structure of the plant cell wall.

walls are inherently resistant to degradation, making it difficult to access the polysaccharides of interest. To improve accessibility, the biomass is usually subjected to a chemical, mechanical or biological pretreatment, which disrupts the bonds between lignin and cell wall polysaccharides, loosens chains of cellulose and hydrolyses polysaccharides to monosaccharides. However, this step is often expensive in time, energy, resources or money; inefficient or ineffective for hemicelluloses and lignin; or creates by-products that cannot be separated from the biomass or may be toxic to the ethanologenic organisms (reviewed by Carroll and Somerville, 2009). Adapting the composition of biomass to minimize pretreatment will therefore translate to significant process efficiencies and savings.

7.2.1 Cellulose

Structural features of cellulose

Cellulose is a glucan polymer, specifically, a chain of variable length of (1,4)-β-D-anhydroglucopyranose units (Fig. 7.2). The pyranose rings are in the conformation $^4C_1$ with hydroxyl groups in an equatorial position (O'Sullivan, 1997). Native cellulose has several hundred to over 10,000 glucopyranose units in one chain and the number of units is referred to as the degree of polymerization (Klemm *et al.*, 2005). The smallest repeating unit within cellulose is cellobiose. Cellobiose can be converted by enzymatic or acid hydrolysis into two glucose molecules.

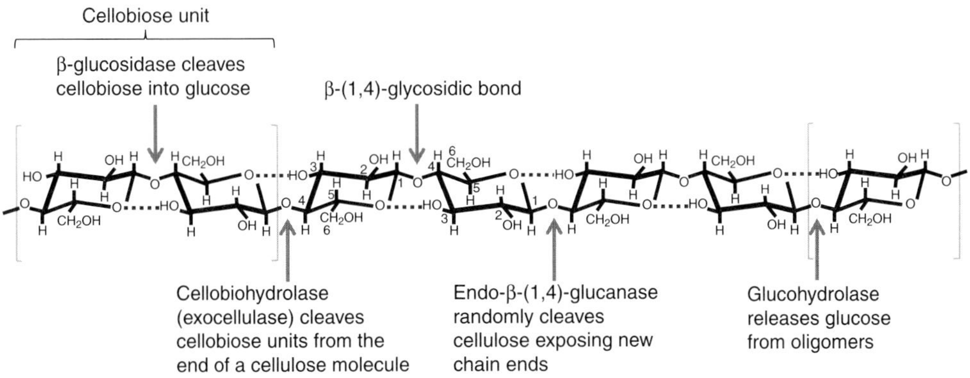

Fig. 7.2. A single chain of β-(1,4)-linked glucopyranose units. Glucose units are linked by β-(1,4)-glycosidic bonds. Endoglucanases insert water molecules into the β-(1,4) bond to hydrolyse the linkage and produce a new reducing and non-reducing end. Exoglucanases hydrolyse the β-(1,4) linkage at reducing or non-reducing ends only. β-glucosidases cleave cellobiose units into two glucose molecules.

Cellulose is synthesized by large membrane-bound complexes (Brown *et al.*, 1996) consisting of six subunits in a rosette shape (Kimura *et al.*, 1999) where each subunit contains six cellulose synthesis (CESA) proteins (Cosgrove, 2005). The rosette subunits synthesize six individual glucan chains which combine into an elementary microfibril and six elementary microfibrils form a cellulose microfibril 3–5 nm in width (Doblin *et al.*, 2002; Cosgrove, 2005). Cellulose microfibrils are assembled into a compact lattice and hydrogen bonding between the lattice sheets holds them in a crystal structure (see Nishiyama, 2009; Mazeau, 2011). This highly crystalline cellulose is the main component of the cell walls of higher plants but there are regions of amorphous, or less ordered cellulose. The amorphous form is easily hydrolysed to glucose whereas the crystalline cellulose is difficult to hydrolyse because it is water-insoluble and there is limited access to the β-(1,4)-glycosidic bonds. Thus, the relative crystallinity of cellulose influences the rate of enzymatic hydrolysis of lignocellulosic biomass (Laureano-Pérez *et al.*, 2005; Zhu *et al.*, 2008). For a summary of the ultrastructural characteristics of cellulose fibres that limit enzymatic hydrolysis see Mansfield *et al.* (1999).

Desirable cellulose characteristics

Ideally a biofuel feedstock crop would generate a high yield of glucan polymers in a non-crystalline form that can be efficiently hydrolysed by available pretreatments. Thus, increasing the non-crystalline and reducing the crystalline cellulose contents in the cell walls may improve the yields of fermentable sugar released during the pretreatment of biomass. A low degree of polymerization of cellulose is also ideal since there are more chain ends available to hydrolytic enzymes so the rate of hydrolysis is greater (Hallac and Ragauskas, 2011).

A comparison of cellulose content, degree of polymerization and crystallinity index in various biofuel feedstock crops is presented in Table 7.1. The percentage of cellulose in woody biomass species is 40–56% while in the grass species it has a wider range of 15–49% due to the twofold, or greater, variation in the cellulose content described for three of the grass species: sorghum, sugarcane and rice. This range may be due to varietal or methodological differences: if the differences are due to variety, there is potential to use plant breeding techniques to exploit inherent genetic variation to alter cellulose content.

7.2.2 Non-cellulosic polysaccharides

Hemicelluloses account for 12–35% of the cell wall of most plant feedstocks currently being considered for bioethanol production (reviewed by Pauly and Keegstra, 2008; Girío *et al.*, 2010).

Table 7.1. Comparison of the cell wall composition of various biomass feedstocks.

Potential biofuel feedstocks		Cellulose (% DW)	Cellulose (dp)	Cellulose crystallinity index	Hemicellulose (% DW)	Lignin (% DW)
Woody species	Beech (*Fagus* spp.)	40–55[a,b]	3063[e]	34[e]	32[b]	24–29[b,c]
	Eucalyptus spp.	48[c,d]			17[d]	27[d]
	Pine (*Pinus* spp.)	42–50[c,d]	1450[f]		23[d]	21–29[c,d]
	Poplar (*Populus* spp.)	39–40[d,g]	3500[h]	51[i]	21[d]	20–25[d,g]
	Spruce (*Picea* spp.)	40–50[a,b]			31[b]	28[b]
	Willow (*Salix* spp.)	56[g]			14[j]	19[g]
Grasses	Maize stover (*Zea mays*)	35–39[b,c,d,g]	2520–7000[h,f]	50[k]	28–33[b,d]	10–19[b,c,d,g]
	Miscanthus spp.	42[b]			27[b]	11–13[b,g]
	Sorghum spp. (whole pith and bark)	15–35[b,c,d,g]			12–18[b,d]	6–17[b,c,d]
	Sugarcane (*Saccharum* spp.)	22–49[b,c,d,g]	925[e,f]	37[e]	27–31[b,d]	7–23[b,c,d,g]
	Switchgrass (*Panicum* spp.)	33–46[b,c,d]			25–32[b,d]	6–18[b,c,d,g]
	Rice straw (*Oryza* sp.)	17–42[b,l]	1820[f]		20–32[b]	1–12[b,l]
	Wheat straw (*Triticum* sp.)	33–35[b,d]	1045–2660[e,f]	29[e]	23–25[b,d]	17–21[b,d,g]

[a]Sun and Cheng, 2002; [b]Pauly and Keegstra, 2008; [c]Huber *et al.*, 2006; [d]Carroll and Somerville, 2009; [e]Puri, 1984; [f]Hallac and Ragauskas, 2011; [g]Karp and Shield, 2008; [h]Kumar *et al.*, 2009; [i]Coleman *et al.*, 2009; [j]Szczukowski *et al.*, 2002; [k]Laureano-Pérez *et al.*, 2005; [l]Jahn *et al.*, 2011 and references therein.
% DW, % dry weight; dp, degree of polymerization.

The current model for synthesis suggests that molecules are synthesized in the Golgi, then exported, partial or complete, to the cell wall (Burton *et al.*, 2010a). Xyloglucans and pectic polysaccharides predominate in the cell walls of dicotyledonous and non-grass (*Poaceae*) monocotyledonous species. Xyloglucans consist of a β-(1,4)-glucan backbone, singly substituted at around 70% of residues with xylose. Xyloglucan may account for up to 25% of the dry weight of the primary wall (Scheller and Ulvskov, 2010). Pectins are a complex group of molecules with a backbone of galacturonic acid, and occasionally rhamnose, residues. Sugars from pectin are not directly converted to bioethanol, but pectins interact with cellulose and affect its saccharification (Srichuwong *et al.*, 2010).

Conversely, the major non-cellulosic polysaccharides in grass cell walls are arabinoxylans and (1,3;1,4)-β-D-glucans. Arabinoxylans (AXs) consist of a backbone of β-(1,4)-xylose residues that are randomly singly or doubly substituted with arabinose. The arabinose residues in turn may be substituted with: galactose; hydroxycinnamic acid moieties such as ferulic acid that may cross-link with lignin or other wall molecules; or acetyl groups that affect the solubility of the polysaccharide. Both ferulic acid cross-linking and acetylation adversely affect access of hydrolytic enzymes to xylan (reviewed by Gomez *et al.*, 2008). The (1,3;1,4)-β-D-glucans are much simpler linear molecules, consisting exclusively of glucose residues joined by either β-(1,4) or β-(1,3) linkages in a non-random but non-repeating manner. Grasses also contain mannan, xyloglucan and pectic polysaccharides in their cell walls, but to a much lesser extent (Burton *et al.*, 2010a).

Disrupting this hemicellulose matrix is a key element for efficient conversion of lignocellulosic biomass into bioethanol. Not only do these molecules represent a significant proportion of sugars available for fermentation, deconstructing them enables access of saccharification and fermentation enzymes to the cellulose lattice. Difficulties with pretreatment methodologies have already been described, but fermenting pentose sugars to ethanol is also proving challenging. The monosaccharides in the cell wall are predominantly glucose and xylose, with xylose contributing up to 22% in some biofuel feedstocks (Carroll and Somerville, 2009). Until the pentose sugar xylose – and to a lesser extent, arabinose – can be converted to ethanol (reviewed by Saha, 2003), the fermentation process must be considered relatively inefficient.

Thus the ideal biofuel feedstock should have non-cellulosic polysaccharides that are easily converted to monosaccharides, through either looser bonds to the cell wall matrix or bonds that are more easily disrupted; or alternatively, have an altogether reduced hemicellulose, or pentose, component.

7.2.3 Lignin

Lignin is the non-sugar fraction of the secondary cell wall that is mainly found in specialized tissues such as xylem and schlerenchyma. Lignin, a type of phenylpropanoid, is made up of hydrophobic phenolic monomers of guaiacyl propanol (G), *p*-hydroxyphenyl propanol (H) and syringyl propanol (S) (Boerjan *et al.*, 2003). Within the cell wall, the amorphous structure of lignin is linked to polysaccharides via oxidative coupling through phenol groups (Vogel, 2008). For example, ferulate and *p*-coumarate are ester- or ether-linked to arabinose residues of glucuronoarabinoxylans (GAXs). Lignin provides mechanical support and is a barrier to water movement, which is important for preventing lodging and conducting water throughout the plant, respectively (Himmel *et al.*, 2007). The biosynthesis of lignin involves the action of at least ten enzymes (Fig. 7.3; Vanholme *et al.*, 2010). Monolignol biosynthesis occurs in the cytosol and subsequent polymerization takes place within the cell wall (Simmons *et al.*, 2010).

The aromatic rings of the lignin monomers may contain no, one, or two methoxyl groups, defining three types of lignin: structures I, II and III, respectively. Type I lignin is mainly found in grasses; conifer woods are rich in type II lignin; and type III lignin is characteristic of deciduous woods (Gupta and Demirbas, 2010). Lignin in dicots is mainly composed of G and S lignins, whereas in monocots it is primarily made up of G, S and H units. G lignins tend to be particularly recalcitrant to chemical degradation and so the relative ratio of G to S units may significantly influence the digestibility of biomass for biofuel production (Vanholme *et al.*, 2008; Ambavaram *et al.*, 2011). The content and composition of lignin tends to be highly heritable as it is under relatively strong genetic control (Morreel *et al.*, 2006; Poke *et al.*, 2006).

During pretreatment of biomass, the bonds within lignin are broken and the structure depolymerized so that the physico-chemical interactions between lignin and plant cell wall polysaccharides are disrupted (Chang and Holtzapple, 2000; Hendriks and Zeeman, 2009; Gupta and Demirbas, 2010). However, lignin hinders biomass pretreatments because it: reduces the activity of hydrolysing enzymes such as cellulases; blocks the access of enzymes to sites where they may hydrolyse polysaccharides; and traps enzymes within the lignin structure (Chang and Holtzapple, 2000; Draude *et al.*, 2001; Charles *et al.*, 2003; Palonen *et al.*, 2004; Yang and Wyman, 2004). Modification of the lignin component of cell walls could be directed either at minimizing the lignin content as much as possible without compromising the biomass yield, altering the lignin residues, specifically lowering the G/S ratio (Casler and Jung, 1999; Boerjan *et al.*, 2003; Huntley *et al.*, 2003; Akin, 2007; Chen and Dixon, 2007), or modification of the lignin–polysaccharide linkages (Grabber *et al.*, 2002).

7.3 Novel Approaches to Improve Biomass Conversion

The challenge of the resistance of cell walls to degradation may be tackled in two ways: adapt the pretreatment to be more effective at

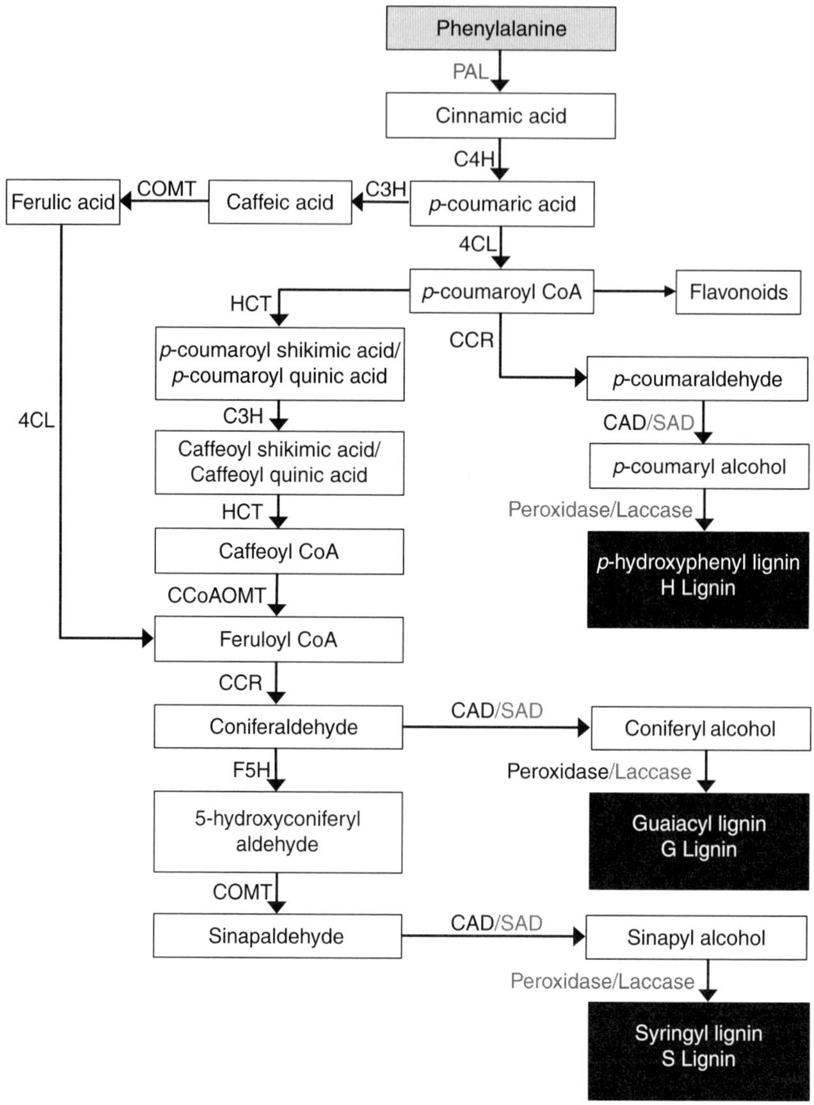

Fig. 7.3. Monolignol biosynthesis enzymes targeted for modification in biomass feedstock crops. Genes encoding enzymes in black text have been modified in biomass feedstock crops. PAL, phenylalanine ammonia-lyase; C4H, cinnamate 4-hydroxylase (Bjurhager *et al.*, 2010); 4CL, 4-coumarate-CoA ligase (Wagner *et al.*, 2009); CCR, cinnamoyl-CoA reductase (Leplé *et al.*, 2007; Wadenbäck *et al.*, 2008; Tu *et al.*, 2010); CAD, cinnamyl-alcohol-dehydrogenase (Baucher *et al.*, 1996; MacKay *et al.*, 1997; Halpin *et al.*, 1998); HCT, *p*-hydroxycinnamoyl-CoA: quinate shikimate *p*-hydroxycinnamoyl transferase (Wagner *et al.*, 2007); C3H, *p*-coumarate 3-hyroxylase (Coleman *et al.*, 2008); COMT, caffeic acid *O*-methyltransferase (Vignols *et al.*, 1995; Tsai *et al.*, 1998; Piquemal *et al.*, 2002; Tu *et al.*, 2010; Fu *et al.*, 2011); CCoAOMT, caffeoyl-CoA *O*-methyltransferase (van Doorsselaere *et al.*, 1993); F5H, ferulate-5-hydroxylase (Stewart *et al.*, 2009); SAD, sinapyl alcohol dehydrogenase; peroxidase (Li *et al.*, 2003b).

degrading the cell wall or modify the cell wall using conventional plant breeding or genetic engineering. Plant breeding programmes involve the identification of natural variation within a crop population, followed by crossing programmes where the trait of interest is selected for in each generation (Hatfield *et al.*, 1998; Sarath *et al.*, 2011; Struder *et al.*, 2011).

An alternative to using natural variation in breeding programmes is generating random genetic variation by chemical or radiation mutagenesis. Genetic engineering approaches involve adding, modifying or disrupting genes involved in cell wall biology such as the generation of loss-of-function mutants or lines over-expressing a particular gene. However, engineering approaches are limited by our understanding of cell wall biosynthesis. Changing one component of the cell wall can be ineffective if the plant employs compensatory mechanisms to counter the change (Hu *et al.*, 1999; Burton *et al.*, 2000; Hatfield and Vermerris, 2001; Grabber, 2005).

Strategies to modify plant biomass are directed at cellulose, non-cellulosic polysaccharides, lignin, or the linkages between these components. An obvious goal is to increase the amount of fermentable sugar present in the plant, or to make sugars more available for hydrolysis, or simply to make bigger plants. A less direct approach is to introduce genes encoding the enzymes necessary for self-deconstruction (Himmel *et al.*, 2007). The enzymes may be expressed in a specific organelle so as to be contained where they will not negatively impact plant growth. Alternatively, the genes encoding these enzymes could be designed to be activated by an external trigger or at a particular stage of plant development (Ragauskas *et al.*, 2006).

Modifications tend to be directed at increasing content or accessibility of hexose sugars, such as those found in cellulose or β-D-glucans. Most ethanologenic organisms preferentially or exclusively utilize hexose sugars; however, pentose-fermenting organisms are being developed (reviewed by Hahn-Hagerdal *et al.*, 2007; Bokinsky *et al.*, 2011). When these organisms are used in biomass processing, pentose sugars, such as xylans, become valuable. However, most research aims to increase hexose polymers and decrease pentose polymers such as xylans.

One of the challenges in modifying cell wall composition and structure is in maintaining plant fitness, such as resistance to lodging and disease (McCann and Carpita, 2008; Pauly and Keegstra, 2008). In a field environment, reduced fitness can lead to loss in total biomass, which is undesirable when the aim is to increase total yields.

7.3.1 Changing the structure and content of cellulose

Pretreatment of lignocellulosic materials is needed to partially disrupt the structure of cellulose and improve the reactivity of cellulose towards hydrolysis. The hydrolysis of the β-(1,4)-D-glycosidic linkages within cellulose may be catalysed by a class of enzymes called cellulases. There are five types of cellulases (see Kumar *et al.*, 2008):

1. Endoglucanases cleave internal β-(1,4)-glucosidic bonds by inserting a water molecule, thus disrupting the crystalline structure of cellulose and exposing individual cellulose chains.
2. Exoglucanases progressively act on the reducing and non-reducing ends of cellulose chains to release short-chain cello-oligosaccharides.
3. β-glucosidases hydrolyse the cello-oligosaccharides, such as cellobiose, to glucose.
4. Oxidative cellulases depolymerize cellulose by radical reactions.
5. Cellulose phosphorylases depolymerize cellulose using phosphates instead of water.

Many cellulases have a catalytic domain and a cellulose binding domain (CBD) that recognizes and binds the cellulose surface (Linder and Teeri, 1996). It may be possible to alter cellulose crystallinity by genetically modifying a feedstock crop to introduce a nucleotide fragment encoding a substrate-disrupting factor, such as a CBD or a novel expansion protein (Abramsom *et al.*, 2010). Incorporation of CBDs within cellulose during biosynthesis may disrupt the regular structure leading to reduced cellulose crystallinity (Levy *et al.*, 2002). The *in vitro* application of CBDs increased the rate of cellulose synthesis (Shpigel *et al.*, 1998); however, whether a similar result would be achieved by *in planta* expression of CBDs is yet to be demonstrated. Expansins are extracellular proteins involved in regulating wall extension and loosening (McQueen-Mason and Cosgrove, 1994; Cosgrove, 2000; Li *et al.*, 2003a). When the transcript level of an expansin gene was decreased using an antisense approach in *Petunia* plants, the amount of crystalline cellulose was reduced relative to control plants (Zenoni *et al.*, 2004).

It may be possible to change the content and the structure of the cellulose, in terms of the degree of crystallization or polymerization, by manipulating the *CesA* genes that encode the proteins in the cellulose synthase complex (Himmel *et al.*, 2007). Changing the number, the shape or the synthesizing activity of the rosettes could impact cellulose content (Turner *et al.*, 2001). For example, reduced levels of crystalline cellulose were observed in barley brittle stem mutants due to the presence of a retroelement in the *CesA4* gene (Burton *et al.*, 2010b).

Other examples of genes involved in cell wall formation for which genetic modification may influence cellulose content and structure in the wall include membrane-bound cellulases (Takahashi *et al.*, 2009) and genes encoding glycosylphosphatidylinositol (GPI)-anchored proteins involved in cell wall construction (Ching *et al.*, 2006). Two examples of the impact of altered endoglucanase activity on cellulose crystallinity include transgenic *Arabidopsis thaliana* plants over-expressing the poplar (*Populus tremula* L. × *tremuloides* Michx.) *PttCe9A1* cellulase, which had lower crystallinity indices (Takahashi *et al.*, 2009); and *Arabidopsis irx2-2* mutants in which KORRIGAN1 was mutated that had higher cellulose crystallinity indices (Szyjanowicz *et al.*, 2004). In maize (*Zea mays* L.) mutants where a putative GPI protein encoding gene was disrupted by a transposon, the cellulose content in the wall was significantly less than in wild-type siblings (Ching *et al.*, 2006).

Alternatively, regulatory networks coordinating cell wall biosynthesis may be targeted by genetically manipulating transcription factors. For example, expression of the *Arabidopsis* SHINE transcription factor in rice caused a 34% increase in cellulose and a 45% reduction in lignin (Ambavaram *et al.*, 2011). The authors suggest the SHINE transcription factor repressed a number of NAC transcription factors known to interact with downstream MYB transcription factors that influence the expression of cell wall biosynthetic genes. Recently a cellulase synthase interactive protein 1 (CSI1) has been described that mediates the interaction between microtubules and cellulose synthase (Li *et al.*, 2012), influencing the orientation in which microfibrils are deposited; modification of *CSI1*-like genes may therefore influence the crystalline structure of cellulose.

Supply of the precursor for cellulose biosynthesis, UDP-glucose, is another point of regulation influencing cellulose content and structure (Coleman *et al.*, 2009; Song *et al.*, 2011; Zhang, B. *et al.*, 2011). Transgenic poplar lines over-expressing sucrose synthase (SuSy) had 2–6% higher cellulose content per dry weight (Coleman *et al.*, 2009). SuSy catalyses the formation of UDP-glucose and fructose from sucrose, so higher SuSy activity may have led to an increase in the available UDP-glucose for the synthesis of cell wall polysaccharides. The transgenic poplar lines had increased cellulose, increased crystallinity and higher wood density (Coleman *et al.*, 2009). The activity of Golgi nucleotide sugar transporters (NSTs) can also influence the supply of UDP-glucose and thus influence the cellulose content. For example, loss of activity of an NST in a rice brittle culm mutant was associated with decreased cellulose content and altered wall structure (Zhang, B. *et al.*, 2011).

7.3.2 Modifying non-cellulosic polysaccharides and lignin

A major area of cell wall research is focused on modifying non-cellulosic polysaccharides and lignin to reduce either their quantity in the cell wall or the extent to which they bind cellulose. Loosening the cell wall matrix renders the cellulosic polysaccharides more accessible to hydrolytic enzymes and improves digestibility. One approach to altering xylan involved the down-regulation of an UDP-glucuronate decarboxylase, which makes a xylan precursor, in tobacco (*Nicotiana tabacum*; Bindschedler *et al.*, 2007). Xylose content was significantly reduced but the authors found that the cellulose seemed more tightly bound to the lignin. The RNAi silencing of a GT47 glycosyltransferase involved in glucuronoxylan synthesis in poplar also lowered xylan levels, and resulted in thinner secondary cell walls and improved cellulose digestibility (Lee *et al.*, 2009). Another attempt at loosening the cell wall matrix reduced the amount of a homogalacturonan pectin in *Arabidopsis*, tobacco and wheat cell walls, which also improved cellulose hydrolysis (Lionetti *et al.*, 2010).

Changing the structure and hydrophobicity of lignin can enable better penetrance of hydrolytic enzymes and improve the delignification of biomass (Grabber *et al.*, 2010). The structure of lignin could be changed by incorporating alternative phenolics or by changing the linkages between lignin and cell wall polysaccharides. For example, incorporation of coniferyl ferulate into the structure of lignin improved delignification efficiency (Leplé *et al.*, 2007; Grabber *et al.*, 2008; Ralph *et al.*, 2008) and reducing the ferulate cross-links between lignin and arabinoxylan reduced cell wall strength and increased digestibility (Buanafina *et al.*, 2010). Cell wall hydrolysis was also improved by substitution of common monolignols with feruloylquinic acid, methyl caffeate, or caffeoyl quinic acid (Grabber *et al.*, 2010).

Altered regulation of the genes involved in the biosynthesis of monolignols is associated with changes in lignin content, monomer ratios and biomass digestibility (Fig. 7.3). Natural mutations have occurred in the genes encoding cinnamyl-alcohol-dehydrogenase (CAD) and caffeic acid *O*-methyltransferase (COMT) and this reduced lignin content, increased digestibility of the cell wall and the mutation was associated with a brown-midrib phenotype (Cherney *et al.*, 1991; Barrière and Argillier 1993; Halpin *et al.*, 1998; Barrière *et al.*, 2004; Pedersen *et al.*, 2005). A CAD-deficient *brown-midrib* mutant of maize had 20% less lignin and was more digestible (Halpin *et al.*, 1998). Studies of these mutants have contributed to our understanding of the genes involved in lignin biosynthesis (Guillaumie *et al.*, 2008; Vanholme *et al.*, 2008) and many of these genes have been down-regulated in biofuel feedstock crops. In transgenic poplar, down-regulation of *CCR* decreased lignin and xylan contents (Leplé *et al.*, 2007) and down-regulation of *C3H* and *HCT* reduced G and S lignin and increased H lignin (Coleman *et al.*, 2008). Down-regulation of the gene encoding CAD reduced G and H lignin in rice (Li *et al.*, 2009) and altered the number of aldehyde units in poplar (Baucher *et al.*, 1996). Decreased COMT activity reduced both the total lignin content and particularly the S lignin in maize and poplar and improved the digestibility of maize biomass (Jouanin *et al.*, 2000; Piquemal *et al.*, 2002). Down-regulation of the G lignin anionic peroxidase reduced the G lignin content without affecting the S lignin in hybrid aspen (*Populus sieboldii* × *P. gradidentata;* Li *et al.*, 2003b). Lignin can also be altered by changing the activity of enzymes involved in acetylation (Del Rio *et al.*, 2007; Weng *et al.*, 2008), modifying transcription factors such as *NAC, SHINE, LIM* and *MYB* (Fornalé *et al.*, 2010; Ambavaram *et al.*, 2011; Zhong *et al.*, 2011; Wang and Dixon, 2012), or introducing microRNA molecules (*Corngrass1*; Chuck *et al.*, 2011).

7.3.3 Expressing hydrolytic enzymes *in planta*

To reduce the recalcitrance of the cell wall, plants may be engineered to express cellulolytic enzymes *in planta*. Making enzymes within the plant may significantly cut the cost of bioethanol production by reducing the requirement for expensive pretreatment steps and/or the need for additional hydrolytic enzymes (Sticklen, 2006). The targeted expression of the Cel5A endocellulase from *Acidothermus cellulolyticus* in the cell walls of tobacco and maize increased the digestibility of the transformed plants relative to controls (Brunecky *et al.*, 2011). A similar result was also observed in rice, where transgenic plants yielded 43% more sugar than wild-type (Chou *et al.*, 2011) and for heterologous expression of a similar enzyme, Cel5A from *Thermotoga maritima*, in tobacco that produced 33% more free sugar (Mahadevan *et al.*, 2011). Ethanol yields increased by 50% in transgenic maize lines expressing cellulases and xylanases (Zhang, D. *et al.*, 2011). Interestingly, poplar xylem over-expressing xyloglucanase also had significantly reduced cellulose crystallinity, suggesting that loosening the xyloglucans could enhance cellulose hydrolysis in wood (Kaida *et al.*, 2009).

However, for success in an industrial setting, enzymes should be chosen and targeted such that they can survive both the pretreatment process and long-term storage in potentially adverse environmental conditions. It is also important that the enzymes accumulate in sufficient amounts to obviate the need for additional enzyme without harming plant growth, development or pathogen resistance (reviewed by Sticklen, 2006 and Taylor *et al.*, 2008).

Untargeted or unregulated transgenic expression of hydrolytic enzymes can have severe, detrimental effects on plant health. The constitutive expression of bacterial xylanases in maize plants resulted in very stunted and deformed plants, while grain-specific expression produced shrivelled, sterile grains (Gray *et al.*, 2011a). To bypass this problem, recombinant enzymes may be targeted to plastids or the apoplast, which can also improve folding and accumulation. A fungal ferulic acid esterase gene expressed in *Lolium multiflorum* vacuoles improved the digestibility of the transgenic plants (Buanafina *et al.*, 2006). Expressing this same gene in different compartments of *Festuca arundinacea* cells had a similar effect on cell wall digestibility, with the best result derived from Golgi- and apoplast-targeted expression (Buanafina *et al.*, 2010).

Each cellular site confers its own advantage: the apoplast provides the most space for enzyme accumulation; the chloroplast has a transformable genome with its own transcriptional and translation machinery; and vacuoles allow enzymes to accumulate, but not be released, until cell lysis. Different cellular compartments also have their own local environments, such as pH, that may guide correct folding, post-translational modifications or enzyme activity (Sticklen, 2006). Furthermore, the targeting of multiple rather than single compartments has been shown to be more effective in protein accumulation (Hyunjong *et al.*, 2006). The *Thermotoga maritima* Cel5A was targeted to different tobacco subcellular compartments including the cytoplasm, apoplast and chloroplast, with the last showing the highest level of accumulation at 4.5% total soluble protein (Mahadevan *et al.*, 2011). A bacterial β-glucosidase was also successfully expressed in tobacco chloroplasts at levels up to 10% of leaf soluble protein (Gray *et al.*, 2011b). Two fungal cellobiohydrolase enzymes and one bacterial endoglucanase enzyme were expressed in transgenic sugarcane and targeted to the vacuole and chloroplast (Harrison *et al.*, 2011). The former accumulated to higher levels in the vacuole whereas the latter peaked when targeted to the chloroplast.

An alternative to the physical segregation of hydrolytic enzymes is temporal separation (Ragauskas *et al.*, 2006). The use of senescence-induced promoters, wound-inducible promoters or a promoter sequence that is activated after spraying with a chemical inducer could enable the controlled expression of cellulolytic enzymes at the end of plant developmental growth stages, thus avoiding undesirable agronomic effects (Torney *et al.*, 2007).

Biomass pretreatments may decrease the activity of hydrolytic enzymes expressed *in planta*, so the steps involved in processing transgenic lines may need to be adapted to preserve enzyme activity. When the catalytic domain of an endoglucanase from *Acidothermus cellulolyticus* was expressed in tobacco, over 65% of enzyme activity was lost after a pretreatment using the mild ammonia-fibre explosion (AFEX) method (Teymouri *et al.*, 2004). However, in subsequent experiments, the same transgenic protein was isolated from rice, frozen for 3 months, and then added to AFEX-treated rice and maize cellulose. Retention of the activity indicated that the enzyme can be crudely prepared from one biomass, stored, and then used to saccharify another pretreated biomass (Oraby *et al.*, 2007). Recently, two thermophilic bacterial endo-xylanases were expressed in the apoplast in *Arabidopsis* (Borkhardt *et al.*, 2010). After a heat treatment step compatible with current pretreatment technologies, there was a significant decrease in the molecular weight of xylans in the transgenic plants and xylanase activity remained stable after dry storage of plant material for 6 months.

To deconstruct lignin it should be possible to express *in planta* the lignase genes from white rot fungi such as *Phanerochaete chrysosporium* (Martinez *et al.*, 2005; Vanden Wymelenberg *et al.*, 2006), or those isolated from the microorganisms present in the hindgut of termites (Weng *et al.*, 2008).

7.4 Conclusion

Many of the crop plants that we currently use for the production of biofuels have been bred to provide the greatest amount of food rather than biomass. Notwithstanding the moral debate surrounding the use of foods as feedstocks, much of the biomass used to date for biofuel production has been generated as a secondary product in the pursuit of maximal crop yield. There is a now a shift becoming evident where specific feedstock plants are being sought with a focus on biomass quantity and composition. Desirable species include those plants able to grow in marginal areas with minimal inputs of fertilizer and water, with the potential to grow from long-term rootstocks and the ability to withstand higher levels of stress than conventional food crops. A previously underexploited resource for feedstock plants lies in natural genetic variation and many of the more obscure relatives of domesticated crops are now coming into their own in this arena. In addition to the wealth of possible growth habits provided by unexploited wild relatives is the ability to alter biomass structure and composition by design. The generation of transgenic feedstocks that maximize biomass, saccharify easily and ferment efficiently into biofuel is anticipated. Even if the uptake is slow, the sheer scale of financial investment and scientific endeavour that is being invested in the biofuels field must surely see a rapid and fundamental shift in the way that we grow and use plants for food, feed and fuel.

References

Abramsom, M., Shoseyov, O. and Shani, Z. (2010) Plant cell wall reconstruction toward improved lignocellulosic production and processability. *Plant Science* 178, 61–72.

Akin, D.E. (2007) Grass lignocellulose – strategies to overcome recalcitrance. *Applied Biochemistry and Biotechnology* 137, 3–15.

Ambavaram, M.M.R., Krishnan, A., Trijatmiko, K.R. and Pereira, A. (2011) Coordinated activation of cellulose and repression of lignin biosynthesis pathways in rice. *Plant Physiology* 155, 916–931.

Barrière, Y.A. and Argillier, O. (1993) *Brown mid-rib* mutants of maize: a review. *Agronomie* 13, 865–876.

Barrière, Y., Ralph, J., Mèchin, V., Guillaumie, S., Grabber, J.H. *et al.* (2004) Genetic and molecular basis of grass cell wall biosynthesis and degradability. II. Lessons from brown-midrib mutants. *Plant Biology and Pathology* 327, 847–860.

Baucher, M., Chabbert, B., Pilate, G., van Doorsselaere, J., Tollier, M.T. *et al.* (1996) Red xylem and higher lignin extractability by down-regulating a cinnamyl alcohol dehydrogenase in poplar. *American Society of Plant Biologists* 112, 1479–1490.

Bindschedler, L.V., Tuerck, J., Maunders, M., Ruel, K., Petit-Conil, M. *et al.* (2007) Modification of hemicellulose content by antisense down-regulation of UDP-glucuronate decarboxylase in tobacco and its consequences for cellulose extractability. *Phytochemistry* 68, 2635–2648.

Bjurhager, I., Olsson, A., Zhang, B., Gerber, L., Kumar, M. *et al.* (2010) Ultrastructure and mechanical properties of *Populus* wood with reduced lignin content caused by transgenic down-regulation of cinnamate 4-hydroxylase. *Biomacromolecules* 11, 2359–2365.

Boerjan, W., Ralph, J. and Baucher, M. (2003) Lignin biosynthesis. *Annual Review of Plant Biology* 54, 519–546.

Bokinsky, G., Peralta-Yahya, P.P., George, A., Holmes, B.M., Steen, E. *et al.* (2011) Synthesis of three advanced biofuels from ionic liquid-pretreated switchgrass using engineered *Escherichia coli*. *Proceedings National Academy Sciences* 13, 19949–19954.

Borkhardt, B., Harholt, J., Ulvskov, P., Ahring, B.K., Jørgensen, B. *et al.* (2010) Autohydrolysis of plant xylans by apoplastic expression of thermophilic bacterial endo-xylanases. *Plant Biotechnology Journal* 8, 363–374.

Brown, R.M., Saxena, I.M. and Kudlicka, K. (1996) Cellulose biosynthesis in higher plants. *Trends in Plant Science* 1, 149–156.

Brunecky, R., Selig, M.J., Vinzant, T.B., Himmel, M.E., Lee, D. *et al.* (2011) *In planta* expression of *A. cellulolyticus* Cel5A endocellulase reduces cell wall recalcitrance in tobacco and maize. *Biotechnology for Biofuels* 4, 1–10.

Buanafina, M.M. de O., Langdon, T., Hauck, B., Dalton, S.J. and Morris, P. (2006) Manipulating the phenolic acid content and digestibility of Italian ryegrass (*Lolium multiflorum*) by vacuolar-targeted expression of a fungal ferulic acid esterase. *Applied Biochemistry and Biotechnology* 129–132, 416–426.

Buanafina, M.M. de O., Langdon, T., Hauck, B., Dalton, S.J., Timms-Taravella, E. *et al.* (2010) Targeting expression of a fungal ferulic acid esterase to the apoplast, endoplasmic reticulum or Golgi can disrupt feruloylation of the growing cell wall and increase the biodegradability of tall fescue (*Festuca arundinacea*). *Plant Biotechnology Journal* 8, 316–331.

Burton, R.A., Gibeaut, D.M., Bacic, A., Findlay, K., Roberts, K. *et al.* (2000) Virus-induced silencing of a plant cellulose synthase gene. *The Plant Cell* 12, 691–706.

Burton, R.A., Gidley, M.J. and Fincher, G.B. (2010a) Heterogeneity in the chemistry, structure and function of plant cell walls. *Nature Chemical Biology* 6, 724–732.

Burton, R.A., Ma, G., Baumann, U., Harvey, A.J., Shirley, N.J. *et al.* (2010b) A customized gene expression microarray reveals that the brittle stem phenotype *fs2* of barley is attributable to a retroelement in the *HvCesA4* cellulose synthase gene. *Plant Physiology* 153, 1716–1728.

Carroll, A. and Somerville, C. (2009) Cellulosic biofuels. *Annual Review of Plant Biology* 60, 165–182.

Casler, M.D. and Jung, H.G. (1999) Selection and evolution of smooth bromegrass clones with divergent lignin or etherified ferulic acid concentration. *Crop Science* 39, 1866–1873.

Chang, V.S. and Holtzapple, M.T. (2000) Fundamental factors affecting biomass enzymatic reactivity. *Applied Biochemistry and Biotechnology* 84–86, 5–37.

Charles, N., Mansfield, S.D., Mirochnik, O. and Duff, S.J.B. (2003) Effect of oxygen delignification operating parameters on downstream enzymatic hydrolysis of softwood substrates. *Biotechnology Progress* 19, 1606–1611.

Chen, F. and Dixon, F.A. (2007) Lignin modification improves fermentable sugar yields for biofuel production. *Nature Biotechnology* 25, 759–761.

Cherney, J.H., Cherney, D.J.R., Akin, D.E. and Axtell, J.D. (1991) Potential of brown-midrib, low lignin mutants for improving forage quality. *Advances in Agronomy* 46, 157–198.

Ching, A., Dhugga, K.S., Appenzeller, L., Meeley, R., Bourett, T.M. *et al.* (2006) *Brittle stalk 2* encodes a putative glycosylphosphatidylinositol-anchored protein that affects mechanical strength of maize tissues by altering the composition and structure of secondary cell walls. *Planta* 224, 1174–1184.

Chou, H., Dai, Z., Hsieh, C. and Ku, M.S.B. (2011) High level expression of *Acidothermus cellulyticus* beta-1,4-endoglucanase in transgenic rice enhances the hydrolysis of its straw by cultured cow gastric fluid. *Biotechnology for Biofuels* 4, 58 (online).

Chuck, G.S., Tobias, C., Sun, L., Kraemer, F., Li, C. *et al.* (2011) Overexpression of the maize *Congrass1* microRNA prevents flowering, improves digestibility, and increases starch content of switchgrass. *Proceedings of the National Academy of Sciences* 108, 17550–17555.

Coleman, H.D., Park, J.Y., Nair, R., Chapple, C. and Mansfield, S.D. (2008) RNAi-mediated suppression of p-coumaroyl-CoA 3′-hydroxylase in hybrid poplar impacts lignin deposition and soluble secondary metabolism. *Proceedings of the National Academy of Sciences* 105, 4501–4506.

Coleman, H.D., Yan, J. and Mansfield, S.D. (2009) Sucrose synthase affects carbon partitioning to increase cellulose production and altered cell wall ultrastructure. *Proceedings of the National Academy of Sciences* 106, 13118–13123.

Cosgrove, D.J. (2000) New genes and new biological roles for expansins. *Current Opinion in Plant Biology* 3, 73–78.

Cosgrove, D.J. (2005) Growth of the plant cell wall. *Nature Reviews Molecular Cell Biology* 6, 850–861.

Del Rio, J.C., Marques, G., Rencoret, J., Martinez, A.T. and Gutierrez, A. (2007) Occurrence of naturally acetylated lignin units. *Journal of Agricultural Food Chemistry* 55, 5461–5468.

Doblin, M.S., Kurek, I., Jacob-Wilk, D. and Delmer, D.P. (2002) Cellulose biosynthesis in plants: from genes to rosettes. *Plant Cell Physiology* 43, 1407–1420.

Draude, K.M., Kurniawan, C.B. and Duff, S.J.B. (2001) Effect of oxygen delignification on the rate and extent of enzymatic hydrolysis of lignocellulosic material. *Bioresource Technology* 79, 113–120.

Fornalé, S., Shi, X., Chai, C., Encina, A., Irar, S. *et al.* (2010) ZmMYB31 directly represses maize lignin genes and redirects the phenylpropanoid metabolic flux. *The Plant Journal* 64, 633–644.

Fu, C., Mielenz, J., Xiao, X., Ge, Y., Hamilton, C.Y. *et al.* (2011) Genetic manipulation of lignin reduces recalcitrance and improves ethanol production from switchgrass. *Proceedings of the National Academy of Sciences* 108, 3803–3808.

Girío, F.M., Fonseca, C., Carvalheiro, F., Duarte, L.C., Marques, S. *et al.* (2010) Hemicelluloses for fuel ethanol: a review. *Bioresource Technology* 101, 4775–4800.

Gomez, L.D., Steele-King, C.G. and McQueen-Mason, S.J. (2008) Sustainable liquid biofuels from biomass: the writing's on the walls. *New Phytologist* 178, 473–485.

Grabber, J.H. (2005) How do lignin composition, structure, and cross-linking affect degradability? A review of cell wall model studies. *Crop Science* 45, 820–831.
Grabber, J.H., Ralph, J. and Hatfield, R.D. (2002) Model studies of ferulate-coniferyl alcohol cross-product formation in primary maize walls: implications for lignification in grasses. *Journal of Agricultural and Food Chemistry* 50, 6008–6016.
Grabber, J.H., Hatfield, R.D., Lu, F. and Ralph, J. (2008) Coniferyl ferulate incorporation into lignin enhances the alkaline delignification and enzymatic degradation of cell walls. *Biomacromolecules* 9, 2510–2516.
Grabber, J.H., Schatz, P.F., Kim, H., Lu, F. and Ralph, J. (2010) Identifying new lignin bioengineering targets: 1. Monolignol-substitute impacts on lignin formation and cell wall fermentability. *BMC Plant Biology* 10, 114–127.
Gray, B.N., Bougri, O., Carlson, A.R., Meissner, J., Pan, S. *et al.* (2011a) Global and grain-specific accumulation of glycoside hydrolase family 10 xylanases in transgenic maize (*Zea mays*). *Plant Biotechnology Journal* 9, 1100–1108.
Gray, B.N., Yang, H., Ahner, B.A. and Hanson, M.R. (2011b) An efficient downstream box fusion allows high-level accumulation of active bacterial beta-glucosidase in tobacco chloroplasts. *Plant Molecular Biology* 76, 345–355.
Guillaumie, S., Goffner, D., Barbier, O., Martinant, J., Pichon, M. *et al.* (2008) Expression of cell wall related genes in basal and ear internodes of silking brown-midrib-3, caffeic acid *O*-methyltransferase (COMT) down-regulated, and normal maize plants. *BMC Plant Biology* 8, 71 (online).
Gupta, R.B. and Demirbas, A. (2010) *Gasoline, Diesel, and Ethanol Biofuels from Grasses and Plants.* Cambridge University Press, New York.
Hahn-Hagerdal, B., Karhumaa, K., Fonseca, C., Spencer-Martins, I. and Gorwa-Grauslund, M.F. (2007) Towards industrial pentose-fermenting yeast strains. *Applied Microbiology and Biotechnology* 74, 937–953.
Hallac, B.B. and Ragauskas, A.J. (2011) Analyzing cellulose degree of polymerization and its relevancy to cellulosic ethanol. *Biofuels, Bioproducts and Biorefining* 5, 215–225.
Halpin, C., Holt, K., Chojecki, J., Oliver, D., Chabbert, B. *et al.* (1998) Brown-midrib maize (bm1) – a mutation affecting the cinnamyl alcohol dehydrogenase gene. *The Plant Journal* 14, 545–553.
Harrison, M.D., Geijskes, J., Coleman, H.D., Shand, K., Kinkema, M. *et al.* (2011) Accumulation of recombinant cellobiohydrolase and endoglucanase in the leaves of mature transgenic sugar cane. *Plant Biotechnology Journal* 9, 884–896.
Hatfield, R. and Vermerris, W. (2001) Lignin formation in plants: the dilemma of linkage specificity. *Plant Physiology* 126, 1351–1357.
Hatfield, R.D., Ralph, J. and Grabber, J.H. (1998) Cell wall cross-linking by ferulates and diferulates in grasses. *Journal of the Science of Food and Agriculture* 79, 403–407.
Hendriks, A.T.W.M. and Zeeman, G. (2009) Pretreatments to enhance the digestibility of lignocellulosic biomass. *Bioresource Technology* 100, 10–18.
Himmel, M.E., Ding, S.Y., Johnson, D.K., Adney, W.S., Nimlos, M.R. *et al.* (2007) Biomass recalcitrance: engineering plants and enzymes for biofuels production. *Science* 315, 804–807.
Hu, W.J., Harding, S.A., Lung, J., Popko, J.L., Ralph, J. *et al.* (1999) Repression of lignin biosynthesis promotes cellulose accumulation and growth in transgenic trees. *Nature Biotechnology* 17, 808–812.
Huber, G.W., Iborra, S. and Corma, A. (2006) Synthesis of transportation fuels from biomass: chemistry, catalysts, and engineering. *Chemical Reviews* 106, 4044–4098.
Huntley, S.K., Ellis, D., Gilbert, M., Chapple, C. and Mansfield, S.D. (2003) Significant increases in pulping efficiency in C4H–F4H transformed poplars: improved chemical savings and reduced environmental toxins. *Journal of Agricultural and Food Chemistry* 51, 6178–6183.
Hyunjong, B., Lee, D.S. and Hwang, I. (2006) Dual targeting of xylanase to chloroplasts and peroxisomes as a means to increase protein accumulation in plant cells. *Journal of Experimental Botany* 57, 161–169.
Jahn, C.E., Mckay, J.K., Mauleon, R., Stephens, J., McNally, K.L. *et al.* (2011) Genetic variation in biomass traits among 20 diverse rice varieties. *Plant Physiology* 155, 157–168.
Jouanin, L., Goujon, T., de Nadaï, V., Martin, M.T., Mila, I. *et al.* (2000) Lignification in transgenic poplars with extremely reduced caffeic acid *O*-methyltransferase activity. *Plant Physiology* 123, 1363–1373.
Kaida, R., Kaku, T., Baba, K., Oyadomari, M., Watanabe, T. *et al.* (2009) Loosening xyloglucan accelerates the enzymatic degradation of cellulose in wood. *Molecular Plant* 2, 904–909.
Karp, A. and Shield, I. (2008) Bioenergy from plants and the sustainable yield challenge. *New Phytologist* 17, 15–32.

Kimura, S., Laosinchai, W., Itoh, T., Cui, X., Linder, C.R. *et al.* (1999) Immunogold labelling of rosette terminal cellulose-synthesizing complexes in the vascular plant *Vigna angularis*. *The Plant Cell* 11, 2075–2086.
Klemm, D., Heublein, B., Fink, H. and Bohn, A. (2005) Cellulose: fascinating biopolymer and sustainable raw material. *Angewandte Chemie International Edition* 44, 3358–3393.
Kumar, R., Singh, S. and Singh, O.V. (2008) Bioconversion of lignocellulosic biomass: biochemical and molecular perspectives. *Journal of Industrial Microbiology and Biotechnology* 35, 377–391.
Kumar, R., Mago, G., Balan, V. and Wyman, C.E. (2009) Physical and chemical characterizations of corn stover and poplar solids resulting from leading pretreatment technologies. *Bioresource Technology* 100, 3948–3962.
Laureano-Pérez, L., Teymouri, F., Alizadeh, H. and Dale, B.E. (2005) Understanding factors that limit enzymatic hydrolysis of biomass. *Applied Biochemistry and Biotechnology* 121–124, 1081–1100.
Lee, C., Teng, Q., Huang, W., Zhong, R. and Ye, Z. (2009) Down-regulation of PoGT47C expression in poplar results in a reduced glucuronoxylan content and an increased wood digestibility by cellulose. *Plant & Cell Physiology* 50, 1075–1089.
Leplé, J.C., Dauwe, R., Morreel, K., Storme, V., Lapierre, C. *et al.* (2007) Downregulation of cinnamoyl-coenzyme A reductase in poplar: multiple-level phenotyping reveals effects on cell wall polymer metabolism and structure. *The Plant Cell* 19, 3669–3691.
Levy, I., Shani, Z. and Shoseyov, O. (2002) Modification of polysaccharide and plant cell wall by endo-1,4-β-glucanase and cellulose-binding domains. *Biomolecular Engineering* 19, 17–30.
Li, S., Lei, L., Somerville, C.R. and Gu, Y. (2012) Cellulose synthase interactive protein 1 (CSI1) links microtubules and cellulose synthase complexes. *Proceedings of the National Academy of Sciences* 109, 185–190.
Li, X., Yang, Y., Yao, J., Chen, G., Li, X., Zhang, Q. and Wu, C. (2009) *FLEXIBLE CULM1* encoding a cinnamyl-alcohol dehydrogenase controls culm mechanical strength in rice. *Plant Molecular Biology* 69, 685–697.
Li, Y., Jones, L. and McQueen-Mason, S. (2003a) Expansins and cell growth. *Current Opinion in Plant Biology* 6, 603–610.
Li, Y., Kajita, S., Kawai, S., Katayama, Y. and Morohoshi, N. (2003b) Down-regulation of an anionic peroxidase in transgenic aspen and its effect on lignin characteristics. *Journal of Plant Research* 116, 175–182.
Linder, M. and Teeri, T.T. (1996) The cellulose-binding domain of the major cellobiohydrolase of *Trichoderma reesei* exhibits true reversibility and a high exchange rate on crystalline cellulose. *Proceedings of the National Academy of Sciences* 93, 12251–12255.
Lionetti, V., Francocci, F., Ferrari, S., Volpi, C., Bellincampi, D. *et al.* (2010) Engineering the cell wall by reducing de-methyl-esterified homogalacturonan improves saccharification of plant tissues for bioconversion. *Proceedings of the National Academy of Sciences* 107, 616–621.
MacKay, J.J., O'Malley, D.M., Presnell, T., Booker, F.L., Campbell, M.M. *et al.* (1997) Inheritance, gene expression, and lignin characterization in a mutant pine deficient in cinnamyl alcohol dehydrogenase. *Proceedings of the National Academy of Sciences* 94, 8255–60.
Mahadevan, S.A., Wi, S.G., Kim, Y.O., Lee, K.H. and Bae, H. (2011) *In planta* differential targeting analysis of *Thermotoga maritima* Cel5A and CBM6-engineered Cel5A for autohydrolysis. *Transgenic Research* 20, 877–886.
Mansfield, S.D., Mooney, C. and Saddler, J.N. (1999) Substrate and enzyme characteristics that limit cellulose hydrolysis. *Biotechnology Progress* 15, 804–816.
Martinez, M.J., Gutierrez, A. and del Rio, J.C. (2005) Biodegradation of lignocellulosics: microbial, chemical, and enzymatic aspects of the fungal attack of lignin. *International Microbiology* 8, 195–204.
Mazeau, K. (2011) On the external morphology of native cellulose microfibrils. *Carbohydrate Polymers* 84, 524–532.
McCann, M.C. and Carpita, N.C. (2008) Designing the deconstruction of plant cell walls. *Current Opinion in Plant Biology* 11, 314–320.
McQueen-Mason, S. and Cosgrove, D.J. (1994) Disruption of hydrogen bonding between plant cell wall polymers by proteins that induce wall extension. *Proceedings of the National Academy of Sciences* 91, 6574–6578.
Morreel, K., Geominne, G., Storme, V., Sterck, L., Ralph, J. *et al.* (2006) Genetical metabolomics of flavonoid biosynthesis in *Populus*: a case study. *The Plant Journal* 47, 224–237.
Nishiyama, Y. (2009) Structure and properties of the cellulose microfibril. *Journal of Wood Science* 55, 241–249.
O'Sullivan, A.C. (1997) Cellulose: the structure slowly unravels. *Cellulose* 4, 173–207.

Oraby, H., Ventakesh, B., Dale, B., Ahmad, R., Ransom, C. *et al.* (2007) Enhanced conversion of plant biomass into glucose using transgenic rice-produced endoglucanase for cellulosic ethanol. *Transgenic Research* 16, 739–749.
Palonen, H., Tjerneld, F., Zacchi, G. and Tenkanen, M. (2004) Adsorption of purified *Trichoderma reesei* cellulases and their catalytic domains to steam pretreated softwood and isolated lignin. *Journal of Biotechnology* 107, 65–72.
Pauly, M. and Keegstra, K. (2008) Cell-wall carbohydrates and their modification as a resource for biofuels. *The Plant Journal* 54, 559–568.
Pedersen, J.F., Vogel, K.P. and Funnell, D.L. (2005) Impact of reduced lignin on plant fitness. *Crop Science* 45, 812–819.
Piquemal, J., Chamayou, S., Nadaud, I., Beckert, M., Barrière, Y. *et al.* (2002) Down-regulation of caffeic acid o-methyltransferase in maize revisited using a transgenic approach. *Plant Physiology* 130, 1675–1685.
Poke, F.S., Potts, B.M., Vaillancourt, R.E. and Raymond, C.A. (2006) Genetic parameters for lignin, extractives and decay in *Eucalyptus globulus*. *Annals of Forest Science* 63, 813–821.
Puri, V.P. (1984) Effect of crystallinity and degree of polymerization of cellulose on enzymatic saccharification. *Biotechnology and Bioengineering* XXVI, 1219–1222.
Ragauskas, A.J., Williams, C.K., Davison, B.H., Britovsek, G., Cairney, J. *et al.* (2006) The path forward for biofuels and biomaterials. *Science* 311, 484–489.
Ralph, J., Kim, H., Lu, F., Grabber, J.H., Leplé, J.C. *et al.* (2008) Identification of the structure and origin of a thioacidolysis marker compound for ferulic acid incorporation into angiosperm lignins (and an indicator for cinnamoyl CoA reductase deficiency). *The Plant Journal* 53, 368–379.
Saha, B.C. (2003) Hemicellulose bioconversion. *Journal of Industrial Microbiology and Biotechnology* 30, 279–291.
Sarath, G., Dien, B., Saathoff, A.J., Vogel, K.P., Mitchell, R.B. *et al.* (2011) Ethanol yields and cell wall properties in divergently bred switchgrass genotypes. *Bioresource Technology* 102, 9579–9585.
Scheller, H.V. and Ulvskov, P. (2010) Hemicelluloses. *Annual Review of Plant Biology* 61, 236–289.
Shpigel, E., Roiz, L., Goren, R. and Shoseyov, O. (1998) Bacterial cellulose-binding domain modulates *in vitro* elongation of different plant cells. *Plant Physiology* 117, 1185–1194.
Simmons, B.A., Loque, D. and Ralph, J. (2010) Advances in modifying lignin for enhanced biofuel production. *Current Opinion in Plant Biology* 13, 313–320.
Song, X., Zhang, B. and Zhou, Y. (2011) Golgi-localized UDP-glucose transporter is required for cell wall integrity in rice. *Plant Signalling Behaviour* 6, 1097–1100.
Srichuwong, S., Arakane, M., Fujiwara, M., Zhang, Z., Takahashi, H. *et al.* (2010) Alkali-aided enzymatic viscosity reduction of sugar beet mash for novel bioethanol production process. *Biomass and Bioenergy* 34, 1336–1341.
Stewart, J.J., Akiyama, T., Chapple, C., Ralph, J. and Mansfield, S.D. (2009) The effects on lignin structure of overexpression of ferulate 5-hydroxylase in hybrid poplar. *Plant Physiology* 150, 621–635.
Sticklen, M. (2006) Plant genetic engineering to improve biomass characteristics for biofuels. *Current Opinion in Biotechnology* 17, 315–319.
Struder, M.H., DeMartini, J.D., Davis, M.F., Sykes, R.W., Davison, B. *et al.* (2011) Lignin content in natural *Populus* variants affects sugar release. *Proceedings of the National Academy of Sciences* 108, 6300–6305.
Sun, Y. and Cheng, J. (2002) Hydrolysis of lignocellulosic materials for ethanol production: a review. *Bioresource Technology* 83, 1–11.
Szczukowski, S., Tworkowski, J., Klasa, A. and Stolarski, M. (2002) Productivity and chemical compostition of wood tissues of short rotation willow coppice cultivated on arable land. *Rostlinná Výroba* 48, 413–417.
Szyjanowicz, P.M.J., McKinnon, I., Taylor, N.G., Gardiner, J., Jarvis, M.C. *et al.* (2004) The *irregular xylem 2* mutant is an allele of *korrigan* that affects the secondary cell wall of *Arabidopsis thaliana*. *The Plant Journal* 37, 730–740.
Takahashi, J., Rudsander, U.J., Hedenström, M., Banasiak, A., Harholt, J. *et al.* (2009) *KORRIGAN1* and its aspen homolog *PttCel9A1* decrease cellulose crystallinity in *Arabidopsis* stems. *Plant Cell Physiology* 50, 1099–1115.
Taylor, L.E., Dai, Z., Decker, S.R., Brunecky, R., Adney, W.S. *et al.* (2008) Heterologous expression of glycosyl hydrolases *in planta*: a new departure for biofuels. *Trends in Biotechnology* 26, 413–424.
Teymouri, F., Alizaden, H., Laureano-Pérez, L., Dale, B. and Sticklen, M. (2004) Effects of ammonia fiber explosion treatment on activity of endoglucanase from *Acidothermus cellulolyticus* in transgenic plant. *Applied Biochemistry and Biotechnology* 116, 1183–1191.

Torney, F., Moeller, L., Scarpa, A. and Wang, K. (2007) Genetic engineering approaches to improve bioethanol production from maize. *Current Opinion in Biotechnology* 18, 193–199.

Tsai, C., Popko, J.L., Mielke, M.R., Hu, W., Podila, G.K. *et al.* (1998) Suppression of *O*-methyltransferase gene by homologous sense transgene in quaking aspen causes red-brown wood phenotypes. *Plant Physiology* 117, 101–112.

Tu, Y., Rochfort, S., Liu, Z., Ran, Y., Griffith, M. *et al.* (2010) Functional analyses of caffeic acid O-methyltransferase and cinnamoyl-CoA-reductase genes from perennial ryegrass (*Lolium perenne*). *The Plant Cell* 22, 3357–3373.

Turner, S.R., Taylor, N. and Jones, L. (2001) Mutations in the secondary cell wall. *Plant Molecular Biology* 47, 209–219.

van Doorsselaere, J., Dumas, B., Baucher, M., Fritig, B., Legrand, M. *et al.* (1993) One-step purification and characterization of a lignin-specific *O*-methyltransferase from poplar. *Gene* 133, 213–217.

Vanden Wymelenberg, A., Sabat, G., Mozuch, M., Kersten, P.J., Cullen, D. *et al.* (2006) Structure, organization, and transcriptional regulation of a family of copper radical oxidase genes in the lignin-degrading basidiomycete *Phanerochaete chrysosporium*. *Applied Environmental Microbiology* 72, 4871–4877.

Vanholme, R., Morreel, K., Ralph, J. and Boerjan, W. (2008) Lignin engineering. *Current Opinion in Plant Biology* 11, 278–285.

Vanholme, R., Demedts, B., Morreel, K., Ralph, J. and Boerjan, W. (2010) Lignin biosynthesis and structure. *Plant Physiology* 153, 895–905.

Vignols, F., Rigau, J., Torres, M.A., Capellades, M. and Puigdomènech, P. (1995) The brown midrib3 (bm3) mutation in maize occurs in the gene encoding caffeic acid *O*-methyltransferase. *The Plant Cell* 7, 407–416.

Vogel, J. (2008) Unique aspects of the grass cell wall. *Current Opinion in Plant Biology* 11, 301–307.

Wadenbäck, J., von Arnold, S., Egertsdotter, U., Walter, M.H., Grima-Pettenati, J. *et al.* (2008) Lignin biosynthesis in transgenic Norway spruce plants harboring an antisense construct for cinnamoyl CoA reductase (CCR). *Transgenic Research* 17, 379–392.

Wagner, A., Ralph, J., Akiyama, T., Flint, H., Phillips, L. *et al.* (2007) Exploring lignification in conifers by silencing hydroxycinnamoyl-CoA: shikimate hydroxycinnamoyltransferase in *Pinus radiata*. *Proceedings of the National Academy of Sciences* 104, 11856–11861.

Wagner, A., Donaldson, L., Kim, H., Phillips, L., Flint, H. *et al.* (2009) Suppression of 4-coumarate-CoA ligase in the coniferous gymnosperm *Pinus radiata*. *Plant Physiology* 149, 370–383.

Wang, H.Z. and Dixon, R.A. (2012) On-off switches for secondary cell wall biosynthesis. *Molecular Plant* 5, 297–303.

Weng, J., Li, X., Bonawitz, N.D. and Chapple, C. (2008) Emerging strategies of lignin engineering and degradation for cellulosic biofuel production. *Current Opinion in Biotechnology* 19, 166–172.

Yang, B. and Wyman, C.E. (2004) Effect of xylan and lignin removal by batch and flow through pretreatment on the enzymatic digestibility of corn stover cellulose. *Biotechnology and Bioengineering* 86, 88–98.

Zenoni, S., Reale, L., Tornielli, G.B., Lanfaloni, L., Porceddu, A. *et al.* (2004) Downregulation of the *Petunia hybrida* α-expansin gene *PhEXP1* reduces the amount of crystalline cellulose in cell walls and leads to phenotypic changes in petal limbs. *The Plant Cell* 16, 295–308.

Zhang, B., Liu, X., Qian, Q., Liu, L., Dong, G. *et al.* (2011) Golgi nucleotide sugar transporter modulates cell wall biosynthesis and plant growth in rice. *Proceedings of the National Academy of Sciences* 108, 5110–5115.

Zhang, D., VanFossen, A.L., Pagano, R.M., Johnson, J.S., Parker, M.H. *et al.* (2011) Consolidated pretreatment and hydrolysis of plant biomass expressing cell wall degrading enzymes. *Bioenergy Research* 4, 276–286.

Zhong, R., Lee, C., McCarthy, R.L., Reeves, C.K., Jones, E.G. *et al.* (2011) Transcriptional activation of secondary wall biosynthesis by rice and maize NAC and MYB transcription factors. *Plant Cell Physiology* 52, 1856–1871.

Zhu, L., O'Dwyer, J.P., Chang, V.S., Granda, C.B. and Holtzapple, M.T. (2008) Structural features affecting biomass enzymatic digestibility. *Bioresource Technology* 99, 3817–3838.

8 Sugarcane and Energycane

**N.K. Fageria,[1] A. Moreira,[2] L.A.C. Moraes,[2]
Anna L. Hale[3] and Ryan P. Viator[3]**
[1]*National Rice and Bean Research Center of Embrapa, Brazil;*
[2]*National Soybean Research Center of Embrapa, Brazil;*
[3]*USDA-ARS Sugarcane Research Unit, Louisiana, USA*

8.1 Sugarcane

N.K. Fageria, A. Moreira and L.A.C. Moraes

8.1.1 Introduction

Sugarcane is an important economic crop in the tropics and subtropics due to its high sucrose content and increasing interest in its bioenergy potential (Gilbert *et al.*, 2007; Fageria *et al.*, 2011; Glaz *et al.*, 2011). Sugarcane is the common name given to the sucrose-storing members of the genus *Saccharum* (Moore and Maretzki, 1996). Five species of sugarcane are known. *Saccharum officinarum* L. is the large-stalked, soft, juicy, sweet cane and varieties that come from it are called noble canes. The other four species are *Saccharum sinense* Roxb and *Saccharum barberi* Jeswiet, which are thin canes, and *Saccharum spontaneum* L. and *Sacchaharum robustum* Brandes, which are wild types (Nelson, 1980). Most modern sugarcane cultivars are interspecific hybrids of three species of the genus *Saccharaum* (*S. officinarum, S. spontaneum* and *S. robustum*) (Gascho *et al.*, 1993). Although *S. officinarum* was dispersed throughout Malaya, China, India, Micronesia and Polynesia during prehistoric times, it was not known by Europeans until it was collected by them on their exploration of the Pacific. *Saccharum sinense* Roxb., the sugarcane of China and *S. barberi* Jeswiet, the sugarcane of India, were known through folklore and mythology from 1000 to 500 BCE. These original sugarcanes (probably derived from natural hybrids between *S. officinaraum* and the wild canes of India and China) were spread by humans from the Orient through the Middle East, northern Africa and the Mediterranean to be delivered to the Americas by Columbus in 1493 (Moore and Maretzki, 1996).

Sugarcane is traditionally grown in monoculture for the production of crystalline sucrose. However, molasses, ethyl alcohol and fibre (bagasse) are important by-products of sugarcane (Fageria *et al.*, 2011). Sugarcane fibre is also the source of energy that is used to generate the electricity used by most sugarcane mills. In addition, fibre content of sugarcane is economically important when sugarcane is processed for sugar or ethanol, and this importance may increase as sugarcane is used for the production of cellulosic ethanol (Glaz *et al.*, 2011). Brazil, India and China have the three largest sugarcane industries. Other major sugarcane-producing countries are Thailand, Pakistan, Mexico, Colombia, Australia, the Philippines and the USA (FAO, 2009). Brazil has a long tradition of growing sugarcane. In the 16th century it was the world's major source of sugar (Hartemink, 2008). At present, sugarcane is also

©CAB International 2013. *Biofuel Crops: Production, Physiology and Genetics* (ed. B.P. Singh)

a major source of ethanol, and 1 ha of sugarcane land with a yield of 82 t ha^{-1} produces about 7000 l ethanol (Hartemink, 2008). Brazil currently produces about 31% of global production, and it is the largest producer, consumer and exporter of ethanol for fuel (Andrietta *et al.*, 2007; Hartemink, 2008). In 2006, the value of the sugar and ethanol industry reached about 17% of Brazil's agricultural output (Hartemink, 2008). Cultivation of sugarcane for bioethanol is increasing and the area under sugarcane is expanding in Brazil (Table 8.1).

8.1.2 Climate and soil requirements

Sugarcane is adapted to a range of tropical and subtropical climates. It is grown from 37°N in southern Spain to 31°S in the Republic of South Africa. It cannot tolerate freezing temperatures, and growth essentially ceases at mean minimum temperatures below about 12°C (Fageria *et al.*, 2011). Maximum photosynthetic rates occur at air temperatures of about 34°C (Alexander, 1973), and intact plants can survive temperatures in excess of 52°C (Irvine, 1983). The ideal climate for a 1-year crop would include at least 4 to 5 months with mean daytime temperatures of 30°C to 35°C to stimulate growth and 1.5 to 2 months of cooler temperatures prior to harvest to enhance sucrose accumulation (Gascho and Shih, 1982).

Sugarcane is grown successfully under a wide range of temperature, solar radiation, rainfall and soil conditions. If soil, water and

Table 8.1. Sugarcane, sugar and ethanol production in different states of Brazil. Data are for cropping year 2008/2009. (Source: A. Moreira, Brazil, 2011, personal communication.)

State	Sugarcane (t)	Sugar production (t)	Ethanol (1000 l)
Acre	0	0	0
Rondônia	106,292	0	7,224
Amazonas	303,350	14,320	7,963
Pará	626,865	13,726	44,908
Tocantins	55,456	0	2,801
Maranhão	2,280,160	15,335	181,559
Piauí	900,181	38,796	44,553
Ceará	122,355	0	9,241
Rio Grande do Norte	3,186,768	197,914	114,909
Paraíba	5,885,978	133,883	390,695
Pernambuco	18,949,518	1,521,275	530,467
Alagoas	27,309,285	2,200,862	845,363
Sergipe	1,831,714	82,099	89,832
Bahia	2,541,816	81,177	141,484
Minas Gerais	42,480,968	2,207,621	2,167,616
Espírito Santo	4,373,248	85,324	274,592
Rio de Janeiro	4,018,840	241,005	127,795
São Paulo	346,292,969	19,662,436	16,722,478
Paraná	44,829,652	2,459,512	2,048,752
Santa Catarina	0	0	0
Rio Grande do Sul	107,184	0	6,318
Amapá	0	0	0
Distrio Federal	0	0	0
Roraima	0	0	0
Mato Grosso	15,283,134	478,424	952,171
Mato Grosso do Sul	18,090,388	657,078	1,076,161
Goiás	29,486,508	958,419	1,726,080
Central and southern regions	504,962,891	26,749,819	25,101,963
North and north-eastern regions	64,099,738	4,299,387	2,410,999
Total Brazil	569,062,629	31,049,206	27,512,962

plant nutrition are adequate, temperature and/or solar radiation can be used to predict cane growth rates (Allen *et al.*, 1978). When water is limiting, rainfall and/or irrigation may be correlated with yields (Jones, 1980). Limited water resources restrict the amount of sugarcane grown in many regions throughout the world because sugarcane requires substantial amounts of water (Fageria *et al.*, 2011). Several studies have been conducted to determine the effect of different levels of water application on sugarcane (Wiedenfeld and Enciso, 2008). In a 3-year flood-irrigated study, when water inputs were reduced by 25 and 43%, yield reductions averaged 30 and 53%, respectively (Wiedenfeld, 1995). Smaller variation in water inputs of <20%, however, have not always produced significant differences in sugarcane yields (Wiedenfeld, 2004). Evaluations of water stress during different stages of plant growth have indicated that while the effect of water stress may be similar during all growth stages (Wiedenfeld, 2000), sugarcane has the capacity to compensate for brief periods of water stress given enough time for growth, therefore the impact of stress later in the growth cycle, but prior to ripening, may be more severe (Inman-Bamber, 2004).

Sugarcane can be grown on a wide variety of soils (Glaz and Gilbert, 2006; Suman *et al.*, 2006). It can also tolerate soil pH values ranging from about 4 to 9, but the optimum pH range is from 5.8 to 7.2 (Fageria *et al.*, 2011). Although some cultivars tolerate moderate salinity and seasonal flooding, good drainage and salinity management are normally required for high yields. Sugarcane is considered moderately susceptible to soil salinity; the salinity threshold (initial yield decline) has been reported to be 1.7 dS m^{-1}, and the yield decrease per unit increase in salinity beyond threshold is 5.9% (Maas and Hoffman, 1977). Serious yield reductions occur at a conductivity of 4 to 8 dS m^{-1}, and death or very little cane growth occurs above 10 dS m^{-1} (Bresler *et al.*, 1982).

8.1.3 Growth and development

Growth of a crop is basically a function of intercepted light (fraction of the sunlight intercepted by the crop), the dry matter produced from this light (which is controlled by radiation use efficiency) and the percentage of this dry matter transferred to the economic part of the plant (Loomis and Connor, 1992). Growth of the sugarcane plant, sucrose formation and storage is controlled by genetic, nutrition and water relationships and climatic factors. It is very rapid during the first few months after planting. Sugarcane is a tall, robust, clump-forming grass with culms, usually called stalks or stems, devoid of leaves below (Moore and Maretzki, 1996). Sugarcane is propagated vegetatively by planting stem cuttings (setts), from which auxiliary buds grow to produce erect primary stalks (main stems). Secondary and tertiary stalks (tillers) are produced at the base of the primary stalk. Sugarcane produces multiple tillers, each having numerous nodes separated by internodes. The internodes consist of sucrose-storing parenchyma cells and vascular tissue, with the stem being the major sink for photosynthates (sucrose) (Tejera *et al.*, 2007). Sugarcane leaf laminae are from 700 to 1200 mm long and up to 100 mm wide. Internodes are up to 250 mm long and from 20 to 60 mm in diameter. Only about ten fully expanded leaves are usually present on a stalk because both lamina and sheath are shed when they senesce. The inflorescence, a panicle (also known as the arrow or tassel), is produced under certain environmental conditions. Genotype, photoperiod, temperature, nutrition and water stress all affect panicle initiation and growth. Since the stalk ceases to grow and sucrose content declines after flowering, genotypes that flower readily under field conditions are avoided (Clements, 1980).

Photosynthesis in sugarcane is via the C_4 pathway, a factor contributing to extraordinary efficiency of this plant in converting solar energy into biomass. During the early months of sugarcane crop growth and development, the majority of the daily produced photo-assimilates are used by the plant to increase the mass of the plant body, the number and size of leaves, and the stalks and roots are all rapidly multiplying (Moore and Maretzki, 1996). On dry weight basis, sugarcane yields may be the highest among all crops. Factors that contribute to high yields of sugarcane are its perennial growth habit and continuous accumulation of sucrose in the vegetative plant structure.

Crop cycles vary from less than 10 months in temperate areas such as Pakistan and Louisiana, where killing frosts set rigid seasons, to 24 months in Peru and South Africa; even longer cycles are sometimes used in Hawaii (Moore and Maretzki, 1996). Most of the word's remaining sugarcane is grown in 14–18 month plant crops (first cycle) and 12-month ratoons (subsequent crops). Crop cycle averages just over 15 months for the highest producing countries. Sugar yields are also affected by the length of the harvest and milling seasons. In most countries, milling is limited to the 5 or 6 months producing the ripest cane. Hawaii does not have a ripening season and harvests over a 10–12 month period (Moore and Maretzki, 1996).

Leaf area index (LAI) is an important physiological parameter in determining growth of a sugarcane crop. Vigorous plants will carry 10–14 fully formed leaves for a two sides leaf area of more than 1.0 m^2 per stalk, providing an LAI of 5 to 7 (Van Dillewijin, 1952). The initial discovery of the C_4 photosynthetic pathway was in sugarcane and it is acclaimed as a leading performer in photosynthetic rates. Sugarcane photosynthetic capacity increases with increasing solar radiation up to full sunlight and is reported to have carbon fixation rates as high as 2.8 mg CO_2 m^{-2} s^{-1}. Under exceptional circumstances, total dry matter produced by sugarcane may average 40 g m^{-2} day^{-1} and exceed 150 t ha^{-1} $year^{-1}$. More than half of the biomass produced is portioned into the harvested stalk, which contains 30% dry matter composed of approximately 60% sucrose and 40% fibre (Moore and Maretzki, 1996).

Sugarcane production is generally measured by calculating its sucrose production per hectare, which is a function of the cane tonnage measured as a fresh weight and the sucrose content of the cane (Glaz *et al.*, 2011). Sucrose is translocated downward through the leaf blade and sheath to the stalk, then to the centre of the stalk, where it may remain or continue movement throughout the plant. The bulk of sucrose moves toward the plant base and roots while smaller amounts move upward toward the apical meristem and immature sink leaves (Hartt *et al.*, 1963; Hatch and Glasziou, 1964). In sugarcane, as in most plants, sucrose is the sugar translocated in the phloem (Hatch and Glasziou, 1964) to sinks, where it is used for cell growth, metabolism, respiration, or storage (Hawker, 1985).

8.1.4 Production practices

Production practices or cultural practices associated with sugarcane vary widely as a result of economic, social, climatic and soil conditions. In areas with high labour costs, essentially all cultural practices are mechanized. In other areas a great deal of hand labour continues to be used. Land preparation varies with soil type. Deep ploughing, deep ripping, or subsoiling is often used to disrupt compacted layers. Soil may be formed into beds approximately 1.5 m apart to facilitate furrow irrigation, to improve surface drainage, or both. However, such beds increase land preparation costs and, where possible, flat culture is used.

Liming and fertilization

Acidity is a major problem in sugarcane-producing regions worldwide. Tables 8.2 and 8.3 show soil pH and mineral content of the

Table 8.2. Principal soil chemical properties of Cerrado soils (0–20 cm depth) of Brazil. (Adapted from Fageria and Baligar, 2005.)

Soil property	Value[a]
pH in H_2O	5.2
Ca ($cmol_c$ kg^{-1})	0.6
Mg ($cmol_c$ kg^{-1})	0.6
Al ($cmol_c$ kg^{-1})	0.6
P (mg kg^{-1})	1.2
K (mg kg^{-1})	47.2
Cu (mg kg^{-1})	1.3
Zn (mg kg^{-1})	1.0
Fe (mg kg^{-1})	116.0
Mn (mg kg^{-1})	14.0
OM (g kg^{-1})	15.0
Base saturation (%)	17.0

[a]Data are average values of 200 soil samples collected from six states covering the Cerrado region. OM, organic matter.

Table 8.3. Principal soil chemical properties (0–20 cm depth) of sugarcane-producing areas in different municipalities in the state of São Paulo, Brazil. Most of these soils were classified as Dark Red Latosols according to the Brazilian system of soil classification and Oxisols according to USA Taxonomy classification. (Adapted from Rossetto *et al.*, 2004.)

	Municipality					
Soil property	Monte Mor	Santo Antônio da Posse	Santa Rosa do Viterbo	Elias Fausto	Lençóis Paulista	Macatuba
pH	4.6	4.3	4.4	4.5	4.1	3.9
P (mg kg^{-1})	6	21	4	11	5	10
K (mg kg^{-1})	43	59	47	47	51	20
Ca ($cmol_c$ kg^{-1})	0.9	0.9	0.6	0.6	0.6	0.4
Mg ($cmol_c$ kg^{-1})	0.5	0.4	0.2	0.2	0.1	0.1
H+Al ($cmol_c$ kg^{-1})	4.2	5.6	4	2.7	5.8	6.4
CEC ($cmol_c$ kg^{-1})	5.8	7.1	3.6	3.6	6.7	6.9
Base saturation (%)	28	21	25	25	13	7
OM (g kg^{-1})	21	22	20	11	22	21

CEC, cation exchange capacity; OM, organic matter.

major sugarcane-producing regions (central-southern) of Brazil, which are highly acidic in reaction and poor in soil fertility. The oxisols and ultisols dominate in sugarcane-growing regions of Brazil. Although low fertility is characteristic of acid soils, these vast areas have a large proportion of favourable topography for agriculture, adequate temperatures for plant growth throughout the year, sufficient moisture availability year round in 70% of the region and for 6–9 months in the remaining 30% of the region (Fageria and Baligar, 2008). When the chemical constraints are eliminated by liming and using adequate amounts of fertilizers, the productivity of oxisols and ultisols is among the highest in the world (Sanchez and Salinas, 1981).

Theoretically, soil acidity is quantified on the basis of hydrogen (H^+) and aluminium (Al^{3+}) concentrations of soils. For crop production, however, soil acidity is a complex of numerous factors involving nutrient/element deficiencies and toxicities, low activities of beneficial microorganisms and reduced plant root growth, which limits absorption of nutrients and water (Fageria and Baligar, 2003). In addition, acid soils have low water-holding capacity and are subject to compaction and water erosion (Fageria and Baligar, 2003). The situation is further complicated by various interactions among these factors (Foy, 1992). The components of the soil acidity complex have been thoroughly discussed in various publications (Foy, 1984, 1992; Kamprath and Foy, 1985).

Liming is a dominant and effective practice to overcome soil acidity-related constraints and improve crop production on acid soils. Lime is often called the foundation of crop production or 'workhorse' in acid soils. Lime requirement for crops grown on acid soils is determined by quality of liming material, status of soil fertility, crop species and cultivar within species, crop management practices and economic considerations (Fageria and Baligar, 2008). Liming improves pH, base saturation, and the Ca and Mg content of acid soils (Table 8.4). These chemical changes favour yield improvement of crop plants in acid soils. Soil pH, base saturation and aluminium saturation are important acidity indices, which are used as a basis for determination of liming rates for reducing plant constraints on acid soils. In addition, crop responses to lime rate are vital tools for making liming recommendations for crops grown on acid soils. In Brazil, liming recommendations for crops are mostly made on the basis of Al, Ca and Mg content of the soil. Base saturation is another criterion for making lime recommendations for crops. In oxisols, the desired optimum base saturation for sugarcane is reported to be around 60% (Rossetto *et al.*, 2010).

Table 8.4. Influence of lime on soil pH, base saturation, calcium and magnesium content of Oxisols of central Brazil (Fageria, 2001a,b).

Lime rate ($t\ ha^{-1}$)	pH in H_2O	Base sat. (%)	Ca ($cmol_c\ kg^{-1}$)	Mg ($cmol_c\ kg^{-1}$)
0	5.6	40	1.9	1.0
4	6.0	44	2.3	1.1
8	6.2	51	3.0	1.2
12	6.4	53	3.1	1.3
16	6.5	56	3.3	1.3
20	6.8	66	3.8	1.4
R^2	0.83**	0.80**	0.72**	0.23*

*, **Significant at the 5% and 1% probability level, respectively.

Because of its long growing season and high maximum growth rate, sugarcane has large nutrient requirements, but since it is a high-value crop, most sugarcane industries or farmers apply large amounts of fertilizers in an attempt to avoid nutrient deficiencies. Both soil and plant analyses have long been used to assess crop nutrition and soil fertility. Nitrogen requirements of sugarcane in south Texas, USA, have been found to be very low in the plant crop and increase in each second and subsequent ratoon (Wiedenfeld and Enciso, 2008). Soil testing for NO_3-N has not been found to be a useful indicator of sugarcane responses to N fertilization in this subtropical environment (Wiedenfeld and Enciso, 2008). Fertilizer recommendations are generally made on the basis of fertilizer trials under field conditions and the results of the soil analysis. Plant analyses are also used to check the adequacy of the fertilizer programme. Sufficient values of plant tissue analysis used to assess the nutritional status of sugarcane plants are given by Gascho *et al.* (1993), Reuter *et al.* (1997) and Fageria *et al.* (2011).

Planting

Although sugarcane can be grown from true seed, commercial plantings are always made using stem cuttings or setts, often called seed (Cock, 2003). Setts with two or more buds are cut by hand or mechanically, often from special areas maintained to minimize disease infestation. Setts may receive hot water and/or fungicide treatments to further reduce disease problems. They are planted in furrows and are covered with soil. Time to sprouting differs among cultivars and is temperature dependent, occurring after approximately 35°C days using a base temperature of 9°C (Keating *et al.*, 1999). Fertilizer may be applied broadcast, in the furrow, or through the irrigation system. In areas where poor drainage is common, the setts are planted on top of beds. Where furrow irrigation is used, they are often planted in the furrows between beds. Flat culture is practised whenever possible to reduce land preparation. Ratoon tillage operations are designed to remove postharvest compaction between rows and to control weeds.

In much of the world, sugarcane row spacing is about 1.5 m to facilitate mechanization. However, work on row spacing in Louisiana and Florida (USA) (Gascho and Shih, 1981) suggests that narrow row spacing could result in more rapid canopy development, increased light interception and higher yields in areas with short growing seasons.

Water requirements, irrigation and drainage

The sugar industries in Hawaii, Florida, Australia, South Africa and Taiwan have historically relied on pan evaporation as an indicator of potential evapotranspiration. Several studies suggest that potential evapotranspiration from a well-developed sugarcane canopy during the grand growth stage is approximately equal to evaporation from a standard US Weather Bureau class A pan (Fageria *et al.*, 2011). When calculated on a monthly basis, the ratio of evapotranspiration to pan evaporation usually varies from 0.8 to 1.2, even though ratios as low as 0.63 and as high as 1.59 have been reported (Kingston and Ham, 1975).

Sugarcane is most susceptible to drought stress during the first 3 to 4 months after planting, when severe stress can reduce stands and make replanting necessary. During tillering and canopy development, young tillers are more susceptible to drought than the main stem or older tillers. Leaves on young tillers begin to roll earlier than those on the primary stem, probably due to the less-developed root systems of the tillers (Clements, 1980). After a tiller begins to produce elongated internodes, the rate of stem elongation and final internode lengths are convenient means of assessing the effects of drought stress. Internodes that elongate during stress are permanently shortened relative to those of well-watered plants, and they serve as a record of the timing, length and severity of drought stress.

Irrigation by flooding, furrow, sprinklers, drip (trickle) and sub-irrigation (by water table adjustment) are all used for sugarcane. Several reviews of sugarcane irrigation practices are available (Gosnell and Pearse, 1971; Gibson, 1974; Leverington and Ridge, 1975; Thompson, 1977; Finkel, 1983). Four methods have been widely used to schedule irrigation: resistance blocks, tensiometers, the water balance and tissue moisture content. All are based on their ability to predict when incipient drought stress will occur, usually as indicated by a decrease in stem or leaf elongation. Experiments that compare cane and sugar yields from plots irrigated with different frequencies and/or amounts are used to validate these methods. Regardless which method is used to select ideal irrigation schedules, actual irrigation practices are modified in response to availability of water, personnel and equipment.

Jones (1980) analysed several irrigation experiments conducted in Hawaii. After eliminating treatments in which excessive water had reduced yields, presumably due to poor soil aeration and/or leaching of nutrients, a linear relationship was found between relative water use (RWU, the ratio of effective water applied to class A pan evaporation) and relative cane yield (Y, the ratio of actual yield to maximum yield in the experiment) in three experiments.

$$Y = 1.01\mathrm{RWU} + 0.03 \quad (R^2 = 0.90, n = 14) \tag{8.1}$$

Thompson (1976) selected data from studies in South Africa (Thompson and Boyce, 1967, 1968, 1971), Australia (Kingston and Ham, 1975), Hawaii (Campbell *et al.*, 1959) and Mauritius (Hardy, 1966) for which evapotranspiration had been estimated. For total crop evapotranspiration (ET) ranging from 660 to 3840 mm and cane yields (Y) ranging from 57 to 342 t ha^{-1}, he found the following relationship:

$$Y = 0.969\mathrm{ET} - 2.4 \quad (R^2 = 0.90, n = 91) \tag{8.2}$$

The slope of the relationship is similar to that reported by Jones (1980) for furrow and sprinkler irrigated experiments in Hawaii (0.103 t cane mm^{-1} effective water).

In many areas sugarcane is grown on nearly level soils in which the water table reaches the root zone during at least part of the year. In Louisiana, USA, sugarcane is grown on low, nearly level land near large bodies of water. High water tables are common, especially during the winter months, when sugarcane leaf area and evapotranspiration are low. In such cases, response to supplemental irrigation is rare (Carter and Floyd, 1973), probably because the crop can extract much of its water requirement from the water table. A similar situation occurs in western Taiwan, where water tables frequently rise near the soil surface in the rainy season. In the dry season the crop obtains 25–50% of its water requirement from a water table at approximately 1.6 m. This dramatically reduces irrigation demands (Yang and Chang, 1976; Hunsigi and Srivastava, 1977; Chang and Wang, 1983).

The most important use of water tables for irrigation occurs on organic soils of the south Florida Everglades (USA). During the dry season, growers pump supplemental water from a primary canal system into a system of farm canals, lateral ditches and field ditches. Lateral movement of water from the field ditches into the field is often facilitated by mole drains 15 cm in diameter, 70–90 cm deep and 2–3 m apart. The organic soils in this region are quite permeable, lateral movement is rapid, and the water table can easily be raised to provide supplemental water.

If the water table remains near the soil surface, poor aeration may cause setts to decay, reduce ratoon stalk populations (Carter *et al.*, 1985), reduce root growth (Banath and Monteith, 1966; Gosnell, 1971) and cause the cane to produce aerotropic (Srinivasan and Batcha, 1962) or floating (Sartoris and Belcher, 1949) roots. Rudd and Chardon (1977) reported a yield decline of 0.46 t cane ha^{-1} for each day the water table rose above 0.5 m.

Louisiana sugarcane farmers plant cane on high (0.30–0.45 m) ridges 1.8 m apart to enhance surface drainage and provide a small volume of aerated soil at almost all times. When winter rainfall is relatively low, these practices provide adequate aeration, but in many years excess rainfall raises water tables and reduces yields (Carter and Floyd, 1975; Carter and Camp, 1983).

A 7-year lysimeter study on fine-textured Louisiana soils indicated that drainage systems should reduce the water table to 1.2 m within 4 days in order to maximize cane and sugar yields. This maintains high redox potentials in the upper 0.5 m of soil (Carter, 1980), allows good ratoon growth early in the spring and increases stalk populations.

Sugarcane cultivars differ in sensitivity to high water tables (Srinivasan and Batcha, 1962; Andreis, 1976). Cultivars tolerant to waterlogging respond with production of aerotropic or floating roots (Srinivasan and Batcha, 1962). Some Florida cultivars exhibit better growth with water tables at 0.32 m than at 0.61 or 0.84 m; however, the growth of other cultivars is greatly inhibited by the 0.32 m water table (Gascho and Shih, 1979).

Harvest and ratoon growth

Sugarcane is normally grown as a perennial crop and is harvested several times before replanting. The first cycle is referred to as the plant crop and subsequent crops as ratoons (Cock, 2003). Cane fields are often burned before harvest to reduce the amount of trash and green leaves hauled to the mill. Harvesting may be labour intensive or highly mechanized. Good harvesting systems minimize cane breakage, transport little trash and other extraneous material to the mill, and minimize the time between harvest and cane processing. Cane that is left in the field after burning or cutting soon begins to deteriorate in quality, reducing sugar recovery in the mill.

Some sugar industries throughout the world are adopting green-cane harvesting, mainly due to public pressure to reduce smoke from burning, but also because burning can reduce total sugar recovery at the mill, reduce organic matter in the soil and increase erosion and water pollution (Richard *et al.*, 2001; Sampietro and Vattuone, 2006; Viator *et al.*, 2006). For many years, manual harvest of unburned cane has been practised in South Africa, where cane tops are left on the soil surface to conserve moisture and control erosion. With green-cane harvesting, 6 to 24 t ha^{-1} of postharvest residue, regardless of whether the sugarcane is manually or mechanically harvested, is deposited on the field surface (Viator *et al.*, 2006). Mechanized harvest of unburned cane is practised in Australia and some other areas, where cane trash is left in the field to conserve soil moisture and prevent erosion.

At harvest, the dry weight of a sugarcane stem usually consists of about 50–60% millable cane, 30–40% tops (leaves and immature stem) and 10% roots and stubble. Crops grown for more than 1 year, as in Hawaii, have much longer stems and usually have higher percentages of millable cane. Of the millable cane's fresh weight, about 70% is water. Of its dry weight, about 50% is sucrose. The percentage of sucrose in sugarcane juice, usually referred to as the polarization value (Pol), varies from 8 to 15% (Tewari *et al.*, 2003). Efficient factories can recover about 85% of the sucrose in the cane. Depending on the sucrose content of the cane and its recovery, from 8 to 15% of the millable cane harvested is recovered as raw (unrefined) sucrose (Clements, 1980; Jones, 1985).

Time of harvest is an important consideration in commercial sugarcane production because sugarcane harvested prior to complete ripening will not have reached its peak sucrose content and sugar yields will be reduced (Gilbert *et al.*, 2006). Sugarcane grown in subtropical areas with cool winter seasons is normally harvested over a period of several months as day-lengths decrease and temperatures cool. These conditions, along with the imposition of moderate drought stress by reduction in irrigation and/or application of chemical ripeners, cause stalk elongation to decrease relatively more than the decrease in photosynthesis. The result is increasing sucrose concentrations in stalk tissues and increased sucrose recovery in the mill. For example, in the Florida sugarcane industry harvest occurs over a 5-month period (October–mid-March) (Gilbert *et al.*, 2006).

Most of the world's sugarcane is harvested at an age of approximately 1 year. In subtropical areas the plant crop is often planted in the spring or summer season and is harvested 15–18 months later. Several 1-year ratoon crops usually follow the plant crop. Tillage and planting operations (including sett production for vegetative propagation) make the plant crop more costly to produce than ratoon crops (Salassi and Giesler, 1995). However, ratoon crop yields are typically less than plant crop yields. For example, in Louisiana yields typically decrease with each subsequent ratoon (Shrivastava *et al.*, 1992; Johnson *et al.*, 1993). The reasons for this decline are complex, but primarily relate to diseases, insects, weed competition, management practices and winter kill (Shrivastava *et al.*, 1992). Additionally, genotypes can vary substantially in their ability to produce good ratoon crop yields (Chapman, 1988; Chapman *et al.*, 1992). Ratooning ability can be enhanced by indirect selection for disease and insect resistance, as well as by direct selection of genotypes with high ratoon crop yields. Traits such as high stalk number, bud viability, vigorous root formation, high biomass accumulation and high light-use efficiency have been suggested as being indicative of better ratooning cultivars (Sundara, 1989). The importance of maintaining stalk weight in older crops has also been noted (Chapman, 1988; Chapman *et al.*, 1992).

Ratooning ability can be defined in either absolute or relative terms. In absolute terms, a good ratooning cultivar is one that produces high ratoon crop yields or several profitable ratoon crops. In relative terms, a good ratooning cultivar is one whose ratoon crop yields are a high percentage of its plant cane yields. Cultivars with high plant cane yields commonly produce high and numerous ratoon crop yields, but exceptions exist (Chapman, 1988; Sundara, 1989; Milligan *et al.*, 1996).

One-year sugarcane is bred and managed to avoid lodging, greatly facilitating both mechanical and manual harvesting. However, in the Hawaiian industry, stems of 2-year crops can reach a length of 10 m or more by growing upward, lodging, and then turning upward again. At harvest the 2-year crop consists of a mat of tangled stems that have lodged several times. This requires very different mechanical harvesting systems and selection of varieties adapted to a 2-year growth cycle (Clements, 1980).

8.1.5 Breeding

Sugarcane breeding started in Java (Indonesia) and Barbados (Caribbean) more or less simultaneously in the 19th century and presently there are several breeding programmes all over the world in the sugarcane-growing regions. The first interspecific crosses were made as early as 1893 in Java but success with this approach did not arrive until the 1920s (Simmonds, 1984). The most remarkable achievement in the breeding programme at that time was development of cultivar POJ 2878, the first of the so-called noble canes, now generally outclassed, but present in nearly every modern pedigree (Simmonds, 1984).

The main objectives of a crop breeding programme, including sugarcane, are developing cultivars of higher yield, higher sucrose content, early maturity, suitable to mechanized harvesting, adaptation to agroclimatic conditions and resistance to biotic and abiotic stresses (Mariotti, 1969). These characteristics are inherited, mainly by addition, i.e. the crossing of two productive varieties should preferably result in an even more productive variety. Selection of varieties is a continuous process, because due to the cultivation cycles in different edaphoclimatic conditions, the plants soon start to show signs of degeneration and susceptibility to pests and diseases (Stevenson, 1965; Nunes Júnior, 1987).

Breeding programmes require germplasm of wide genetic variability to incorporate the desired trait into the cultivar released for commercial production. Selective breeding currently relies on two international germplasm banks, located in Cannanore in India and Miami in the USA (Table 8.5). The objective of these two banks is to preserve the genes of the *Saccharum* genus in their original state for use in cross-breeding programmes to find plants resistant to pests and diseases (Cenisk and Miocque, 2004), such as occurred in response to sugarcane rust (*Puccinia melanocephala*), which wiped out over 60% of the cane fields in Cuba in the 1980s, when that country was one of the world's leading sugar producers (Tangley, 1986).

Because sugarcane originates from low latitudes – tropical regions near the Equator – it only flowers and produces viable seeds at high temperatures and humidity levels, a fact that restricts breeding efforts. Therefore, the main genetic improvement programmes are conducted in tropical regions: Java, Barbados, Guiana, Brazil (Camamu and Murici – north-eastern region,

Sergipe State), Australia (Queensland), Cuba, Mauritius, Hawaii, India (Coimbatore and Cannanore), Puerto Rico and Florida (Stevenson, 1965; Tangley, 1986; Cenisk and Miocque, 2004). These places are favourable for the flowering of most sugarcane varieties and also for the production of pollen with good fertility. The seedlings are planted in experimental fields at the research stations. The selection steps occur in the following order: (i) crossing; (ii) production of plantlets; (iii) selection and obtaining of clones; (iv) evaluation of the clones under different edaphoclimatic conditions; and (v) launch on the market of the cultivar(s). The efficacy of the selection in the initial phase can be enhanced if emphasis is given to the components of yield that are best correlated with the production of genotypes in future clonal generations. By following these steps in processes to select more productive varieties, Brazilian researchers have managed to increase the yield of stalks per hectare by an average of 97% over the past 25 years (Nunes Júnior, 1987; Oliveira *et al.*, 2010) (Table 8.6). Similar results have been achieved in other sugarcane-producing regions (Stevenson, 1965).

8.1.6 Sugarcane as a biofuel crop

Brazil is the largest producer of sugarcane as well as ethanol from sugarcane. Brazil began promoting the production of sugarcane ethanol in the mid-1970s after the first global energy crisis (Stone *et al.*, 2010). Within 10 years,

Table 8.5. Germplasm bank of species and genera of sugarcane in Cannanore (India) and Miami (USA) (Cenisk and Miocque, 2004).

Cannanore – India		Miami – USA	
Species	Number of clones	Species	Number of clones
S. officinarum	764	*S. officinarum*	661
S. barberi	43	*S. barberi*	121
S. sinense	29	*S. edule*	19
S. robustum	145	*S. sinense*	79
S spontaneum	79	*S. robustum*	178
Foreign hybrids	585	*S. spontaneum*	247
Indian hybrids	1020	Foreign hybrids	291
Hindu-American clones	130	American hybrids	50
Indian collection	393	Related genus	112
Related genus	152		
Total	3340		1758

Table 8.6. Increased production of stalks of different varieties of sugarcane in the period 1983 to 2007 in Brazil (Nunes Júnior, 1987; Oliveira *et al.*, 2010).

1983		2008	
Varieties	Stalks (t ha^{-1})	Varieties	Stalks (t ha^{-1})
NA 5679	85	SP 79-1011	155
IAC 52-150	85	RBS 13804	177
SP 70-1143	109	RBS 863129	181
SP 71-799	87	RBS 725552	176
SP 71-1081	103	RBS 943365	193
SP 71-1406	98	SP 78-4764	179
SP 71-6163	89	SP 81-3250	232
Average	94	Average	185

more than three-quarters of the nation's cars made in Brazil were able to run on ethanol (Stone *et al.*, 2010). Sugarcane's high concentration of sugar, which is readily available to microorganisms, makes it uniquely suitable for ethanol production. In addition, Brazil has the highest sugarcane production capacity due to availability of land and water and climatic conditions favourable for sugarcane production throughout the year. Brazil has the largest annual renewable freshwater supply in the world (8233 km^3 $year^{-1}$) (Stone *et al.*, 2010).

As biofuel crops should not compete with grain crops to have adverse effects on food supply, it would be ideal if they can be planted on agriculturally marginal/surplus lands and degraded or drastically disturbed soils. Restoring degraded soils through establishing biofuel crops would be a very attractive strategy. In Brazil there are 36 million ha of degraded pasture land which can be brought under sugarcane cultivation with the addition of lime and fertilizers and developing suitable irrigation facilities. This area can produce a significant amount of sugarcane and consequently ethanol.

The processing and converting of sugarcane to ethanol requires large amounts of water and is the second major use of water in Brazil. Water use in the production of ethanol can be divided between crop production and ethanol production. The water requirement for sugarcane production is approximately 8–12 mm t^{-1} of cane production. The sugarcane growing season is year round, and the annual water requirements for sugarcane production are approximately 1500–2500 mm $year^{-1}$ (Goldemberg *et al.*, 2008). The majority of sugarcane production plantations in Brazil rely on rainfall completed by partial ferti-irrigation, carried out mainly to manage water wastes. Most plantations limit their production to regions where reasonable rainfall occurs (Moreira, 2007). Therefore, irrigation use in Brazil for agricultural production is generally small. However, due to the increasing demand for ethanol and the high prices paid for it, sugarcane production is expanding to regions where irrigation would be needed to complement rainfall (Goldemberg *et al.*, 2008). Table 8.7 shows water requirements for ethanol production from sugarcane and other potential biofuel crops. Maize and sugarcane are the current major crops used for ethanol production. The water requirements to produce maize grain for ethanol are much higher than to produce sugarcane.

8.1.7 Conclusions

Sugarcane is the world's most important sugar as well as ethanol crop. It probably was domesticated in New Guinea in prehistoric times, and it is widely grown in tropical and subtropical environments from 36°N to 31°S latitude. Sugarcane's greatest asset is its ability

Table 8.7. Water requirements for sugarcane and other potential biofuel crops. (Adapted from Stone *et al.*, 2010.)

Crop species	Water requirements (m^3 water t^{-1} crop)	Biofuel conversion (l fuel t^{-1} crop)	Crop water requirement for biofuel (m^3 t^{-1} fuel)	Crop water requirement per unit energy (m^3 GJ^{-1})
		Ethanol		
World maize (grain)	833	409	2580	97
World sugarcane	154	334	580	22
Maize stover	634	326	2465	92
Switchgrass	525	336	1980	74
Grain sorghum	2672	358	9460	354
		Biodiesel		
Soybean	1818	211	9791	259
Canola	1798	415	4923	130

to produce sugar efficiently and in great quantity per unit land area. Sugarcane cultivation can substantially contribute to the supply of renewable energy, but improved crop husbandry and precision farming principles are needed to sustain and improve the resource base on which production depends. As a result of its C_4 photosynthetic pathway, maximum crop growth rates are over 40 g m^{-2} day^{-1}, and total dry matter production in many areas exceeds 40 t ha^{-1} $year^{-1}$. Maximum fresh cane yields are over 200 t ha^{-1} $year^{-1}$ with sucrose concentrations of from 7 to 13%. Commercial yields are usually less than half the maximum yields.

Most sugarcane is grown as a 1-year crop, and several ratoon crops are normally produced. The ideal environment for a 1-year crop would include at least 4 to 5 months with high solar radiation and mean daytime temperatures of 30–35°C to stimulate growth. Prior to harvest, 1.5 to 2 months of cooler temperatures help increase sucrose concentrations. Little growth occurs when mean air temperatures are below 20°C.

Sugarcane can be grown in a wide variety of soils and tolerates soil pH from 4 to 9, but the optimum pH range is 5.8 to 7.2. Cane and sugar yields are sensitive to drought, poor soil aeration, salinity, and nutrient deficiencies and toxicities. Because of its high value, sugarcane producers usually attempt to minimize stresses by application of irrigation, fertilizers, soil amendments and pesticides. However, moderate drought and nitrogen stresses are often imposed for 1 to 2 months prior to harvest to increase cane sucrose concentrations.

Because of its long duration, high growth rates and harvest of all millable cane (and sometimes tops and leaves), large amounts of nutrients are removed from the field. A mass of 1 t of fresh millable cane contains about 1.6 kg N, 1 kg P and 3.4 kg K. Since yields of 50 to 100 t millable cane are common, fertilization is required to achieve adequate yields and maintain soil fertility. Both soil and plant analyses are used to detect and correct nutrient deficiencies and toxicities. The yield of sugarcane has increased over the last 50 years on a worldwide basis, with some countries obtaining much greater increases than others.

8.2. Energycane

Anna L. Hale and Ryan P. Viator

8.2.1 Introduction

'Energycane' is a term that is used to describe sugarcane grown solely for the production of renewable energy. There is increasing interest in the production of fuels from renewable sources due, in part, to the worldwide escalation of oil prices (Van den Wall Bake *et al.*, 2009). Because it is not certain which type of conversion technology will be used to convert biomass to energy, two types of potential energycane feedstock have been proposed. A Type I energycane has somewhat lower sugar content (10–14%) and higher fibre content (14–20%) than a commercial sugarcane cultivar bred for sugar production. In contrast, a Type II energycane is selected for high biomass (total cane yield) and expresses a low sugar content (<10%) and high fibre content (>20%) (Tew and Cobill, 2008). Leafy material in energycane can range from 2 to 20%.

It is recognized that C_4 grasses, such as sugarcane, have the potential to be converted into different types of energy including liquid fuels, heat, steam, electricity, or gas (McLaughlin *et al.*, 1999). While Brazil has demonstrated the feasibility of producing sugar- and molasses-based ethanol from Type I feedstocks, this is currently not an economically feasible option for growers in the USA. With the enactment of the Energy Independence and Security Act of 2007, and the government mandate that at least 36 billion gallons of biofuel be produced by 2022, with at least 16 billion of these gallons resulting from cellulosic sources, sugarcane breeding programmes are expanding to include both Type I and Type II energycane in addition to their existing breeding programmes for sucrose production. Type II energycane feedstock is expected to be an important source of biomass for cellulosic energy conversion (Viator *et al.*, 2010).

Sustainable production of cane in non-traditional cane-growing areas will likely be a key to the success of a cane-based biofuel industry, and the broad genetic base of sugarcane will make it an attractive option for feedstock producers. *Saccharum spontaneum*, as a species, is known for its wide geographical distribution and adaptation and extreme genetic variability. Clones of *S. spontaneum* include ecotypes that are perennial, annual and semi-perennial, and show wild morphological variation. They are able to survive in environments ranging from forests, flood basins, rocks, fields, deserts, sandy flats, marshes and stagnant waters (Panje, 1964). Studies have shown that the species has extreme variation in resistance to red rot (caused by *Colletotrichum falcatum*) and that when used in crossing, it can pass on horizontal resistance to its progeny (Babu *et al.*, 2009; Hale *et al.*, 2010). Resistance to other diseases such as smut (caused by *Ustilago scitaminea* Syndow) and mosaic (caused by *Sugarcane mosaic virus* and *Sorghum mosaic virus*) has also been identified in *S. spontaneum* (Mendoza and Lantin, 1977; Grisham *et al.*, 1992). Whether bred for high sucrose or high biomass, *S. spontaneum*, and other species and genera related to sugarcane, have been shown to contribute to increased biotic and abiotic stress resistances as well as ratooning ability. Deren and Snyder (1991) demonstrated that canes derived from elite sugarcane × *S. spontaneum* crosses, when grown in a wetland environment, were able to rapidly establish, and were flood-tolerant, high-biomass canes with excellent ratooning ability.

Saccharum spontaneum will continue to be a major contributing source of genes for adaptability of the crop. It is recognized that a low number of parents have been used to initiate sugarcane breeding programmes, yet the coefficient of parentage between commercial sugarcane genotypes and ancestral clones is low and the genetic base is not as narrow as previously assumed. Factors contributing to the diversity likely include the polyploid nature of the crop, its interspecific origin, the unequal contribution of the parents when crossed, and the use of wide hybridization in modern breeding programmes (Deren, 1995; Lima *et al.*, 2002). Molecular markers indicate that a majority of alleles contributing to variability in sugarcane germplasm come from *S. spontaneum* even though a majority of the genetic make-up of sugarcane is of *S. officinarum* origin (Janoo *et al.*, 1999).

A dedicated basic breeding programme was established in Houma, Louisiana at the USDA-ARS Sugarcane Research Unit (SRU) in 1965 with the objective of broadening the genetic diversity in the parents used to produce new cane cultivars. Breaux (1979) reported the beneficial effects of hybridizing the *S. spontaneum* clone US 56-15-8 with commercial clones. Through crossing with this mosaic-resistant clone, Louisiana cultivars were developed that retained resistance to the disease and sucrose was largely restored by the BC_3 (third backcross) generation. This programme has been highly successful, and the *S. spontaneum* clones used in crossing have proven to be extremely useful parents in the sugarcane breeding programme by imparting resistance to biotic and abiotic stresses and contributing to ratooning ability. Though this basic breeding programme was established to introgress important traits from wild germplasm into commercial sugarcane, it is also the source of cane bred for the production of biomass.

While the germplasm enhancement programme and the biomass breeding programme run in parallel, special selection strategies are used for developing cane with different end uses. Divergent selection strategies for high biomass cane and cane for sucrose production were described by Breaux and Legendre (1981). When selecting cane for total biomass production, the most important traits to consider are biomass per hectare and total grams of fibre and soluble solids per kilogram of cane, but when selecting cane for sucrose production, sucrose per tonne of cane and sucrose per hectare are stressed. Multiple studies have demonstrated that in sugarcane breeding, the highest yields of fresh-weight and dry-matter biomass per hectare are observed in the first generation after wide crossing (F_1 generation of sugarcane spp. × *S. spontaneum*). With repeated backcrossing into elite sugarcane (BC_1, BC_2, etc.), sugar content increases, but a marked reduction in biomass yield components (e.g. fibre concentration, solids concentration and stalk number) is observed (Legendre and Burner, 1995). In addition, high sucrose content has been shown to be negatively correlated with traits contributing to ratooning ability (Jackson, 1994). Separate selection strategies for biomass cane and sugarcane are necessary because high fibre content inhibits sucrose recovery; thus genotypes selected for high sucrose are simultaneously selected for low fibre (Giamvala *et al.*, 1984). Furthermore, sugarcane bred and managed as an energy crop is expected to sustain a greater number of ratoons than sugarcane managed to maximize sucrose production (Tew and Cobill, 2008). Multiple harvests from a single planting for energycane is an advantage compared to annual crops and justifies the high cost of planting because costs associated with planting are the greatest expense to a grower.

8.2.2 Breeding

Sugarcane (*Saccharum* spp.; $2n$ = 100–130) is a genetically complex crop with a genomic makeup resulting from highly successful interspecific hybridization efforts, primarily involving *S. officinarum* L. and *S. spontaneum* L. (Ming *et al.*, 2006). Sugarcane clones are highly polyploid and aneuploid because they are derived from interspecific hybridization of species with varying numbers of chromosomes. The basic chromosome numbers of *S. officinarum* (x = 10) and *S. spontaneum* (x = 8) are different, and the chromosome complement is $2n$ = 80 and $2n$ = 40–128, respectively (Sreenivasan *et al.*, 1987; D'Hont *et al.*, 1995, 1996). When hybridized, the chromosome content of the resulting progeny is $2n + n$. Proper exploitation of the variability within the *Saccharum* complex is a complicated process because of the high ploidy levels and degree of heterozygosity of the crop. Despite the genetic complexity of the crop, long-term breeding efforts have been successful.

Most energycanes are derived from wide crosses between commercial sugarcane (a complex hybrid of *S. officinarum* L., *S. spontaneum* L., *S. barberi* Jeswiet and *S. sinense* Roxb. Amend. Jeswiet) genotypes and the wild species, *S. spontaneum* L. Progeny from wide crosses (F_1) are subsequently backcrossed to commercial sugarcane genotypes to generate backcross generations (BC_1, BC_2, BC_3, etc.). A majority of sugarcane breeding programmes are located in close proximity to the commercial industries in tropical and subtropical environments. These environments are conducive to flower induction, and stalks flower under natural photoperiod. In programmes such as the basic breeding programme in Houma, Louisiana, the breeding cane is induced to flower under an artificial photoperiod. The breeding clones are maintained on railcarts that are pushed in and out of photoperiod houses daily, commencing with the summer solstice, according to a set schedule. Using photoperiod chambers, days are shortened to a constant photoperiod of 12.5 h for 31 days. Following this constant day-length period, daylight is shortened by 1 min day^{-1} until 10 September (Hale *et al.*, 2012).

While flowers appear perfect, in certain genotypes, anthers may shed no viable pollen. Breeders therefore assign genotypes as 'male' or 'female' depending on the abundance of pollen. Tassels remain receptive for about 4 days, and once fertilization occurs, it takes from 3 to 5 weeks for the seed to mature (Tew and Cobill, 2008). In Louisiana, crossing is conducted between September and December, seed is sown in the greenhouse in February and early March, and seedlings are transplanted to the field in April. All energycane seedlings are selected in the first-ratoon crop based on vigour, stalk population, disease resistance, pith and stalk diameter. A two-stalk sample is collected from selected seedling plots and used to plant an unreplicated first line trial with 1.8 m plots. These plots are selected the following year and those that are advanced are planted into 4.9 m unreplicated trials (second line trials). The second line trials are evaluated in plant cane and first ratoon 4 and 5 years following the initial crossing. Type II energycane selections are made based on estimates of total biomass per hectare, total soluble solids per hectare and total fibre per hectare. Type I selections are made based on these criteria in addition to sugar per hectare and theoretical recoverable sugar. Selections made at this point are returned to the greenhouse for further crossing. The selections with the highest yields are tested in multiple locations to assess yield in adverse conditions (e.g. cold), or in areas where sugarcane is not traditionally grown. Advanced selections are tested for their ability to resist stalk degradation and to regrow in ratoon crops following below-freezing winter temperatures, durable disease resistance and ratooning ability.

To date, four energycane cultivars have been released from the energycane programme in Louisiana. These cultivars are L 79-1002 (released from the Louisiana State University Agricultural Center) (Gravois *et al.*, 2010) and

Ho 00-961, HoCP 91-552 and Ho 02-113, released by the USDA-ARS Sugarcane Research Unit (White *et al.*, 2011; Tew *et al.*, 2011; Hale *et al.*, 2012). HoCP 91-552 was released as a Type I energycane. This variety was derived from the commercial breeding programme, and possessed commercial sucrose levels and excellent ratooning ability. The high fibre level in this variety caused it to be dropped from further consideration as a commercial sugarcane variety. When released, the variety was considered resistant to brown rust, moderately susceptible to smut and moderately resistant to SMV and SCMV. The other varieties were released as Type II energycane. In addition to these released cultivars, there are other varieties under testing in the energycane breeding programme.

The varieties L 79-1002, Ho 00-961 and HoCP 91-552 were released in response to the need for a feedstock supply for a potential biofuel industry. Because of the immediate demand, these varieties were not as extensively tested for their response to disease as commercial varieties grown for sucrose production. When L 79-1002 was released, it was reported that the variety was moderately susceptible to smut, however when hectarage of the variety was increased, it became apparent that it was severely susceptible. Consequently, the variety is no longer being increased for energycane production, due in part to the heavy inoculum it provided to the nearby commercial sugarcane industry. When HoCP 00-961 was released, it was stated that it had a level of susceptibility to smut similar to the susceptible controls. This variety remains susceptible to the disease. To date, neither smut nor brown rust (caused by *Puccinia melanocephala* H. and P. Syndow) has been observed in Ho 02-113.

Because the energycane cultivars were recently released, there are few published reports documenting the superior yields of clones generated from wild crosses compared to conventional sugarcane cultivars. Early studies reported average gross cane yield (wet weight basis) of L 79-1002 surpassed that of the traditional sugarcane cultivar CP 65-357, with yields of 170 t ha^{-1} and 50.4 t ha^{-1} wet weight, respectively (Breaux *et al.*, 1974). The energycane also produced more leafy material than traditional sugarcane with 41.7 t ha^{-1} and 7.5 t ha^{-1}, respectively (Giamvala *et al.*, 1984). When compared to LCP 85-384 (15.9 t ha^{-1}), one of the leading Louisiana cultivars, Ho 00-961 yielded significantly more total dry biomass (23.6 t ha^{-1}) in the third ratoon harvest, as did L 79-1002 (28.3 t ha^{-1}) (White *et al.*, 2011). The most recently released energycane cultivar, Ho 02-113, surpassed L 79-1002 on average in plant-cane through third-ratoon crops for gross cane yield, total soluble solids, total fibre and total dry biomass per hectare by 21%, 33%, 17% and 21%, respectively. The average yield of dry biomass in this variety across the crop cycle was 45.3 t ha^{-1} when grown in the temperate Louisiana environment (Hale *et al.*, 2012).

Cold tolerance is one of the most important traits to consider when breeding cane for biomass feedstock. Since energycane is not expected to displace land currently in sugarcane production, target areas are on marginal lands outside the traditional cane-growing regions. The USDA-ARS Sugarcane Research Unit in Houma, Louisiana, has incorporated cold tolerance screening into their breeding programme. Wide crosses are routinely screened in northerly locations, such as Booneville, Arkansas (35°05′N, 93°59′W, elev. 150 m) to assess parental performance in progeny families. Differential survival of progeny gives an indication of breeding values of parents for the cold tolerance trait. Progeny from preselected families which were part of the original cold tolerance seedling screen in Booneville, showed family survival rates ranging between 10 and 84% over winter, when air temperatures stayed below freezing for 19 days total during the winter and soil temperatures were below freezing at a depth of 15 cm for 4 days. During one 5-day period temperatures did not rise above 0°C (unpublished). This work is preliminary and must be investigated over several seasons to ensure durability of the freeze tolerance. Experimental energycane varieties have also been tested for their ability to resist juice degradation following exposure to temperatures below −2.8°C for more than 6 h. The mannitol to Brix ratio is used as an indicator of juice degradation in early generation sugarcane hybrids (Eggleston and Legendre, 2003), and preliminary data show that energycane varieties selected from the breeding programme in Houma, Louisiana, have significantly lower mannitol to Brix ratios than commercial sugarcane cultivars grown in the same test (unpublished). Preliminary results from early studies suggest that variability exists within *S. spontaneum* germplasm that can be exploited for the development of varieties tailored to colder environments. When breeding energycane varieties, cold tolerance continues to be emphasized.

Intergeneric hybrids generated by crossing *Saccharum* with *Erianthus* and *Miscanthus* have been explored as possible sources of genes for cold tolerance. *Erianthus* is of interest to sugarcane breeders in China, Australia and the USA because in addition to its vigour, drought tolerance and ratooning ability the genus also possesses cold tolerance (Aitken *et al.*, 2006). Ram *et al.* (2001) investigated the winter growth rates of 18 intergeneric clones of *Saccharum* × *Erianthus* and reported that these hybrids had significantly more growth in winter months than commercial sugarcane varieties in the same test. The increased winter growth led to increased stalk yield and commercial recoverable sugar (CCS), indicating that *Erianthus* clones and their hybrids could be used for increasing chilling tolerance in sugarcane. *Miscanthus* represents another related species to energycane that can be used for cold-tolerance introgression. In a study conducted in Booneville, Arkansas, it was demonstrated that *Saccharum* × *Miscanthus* hybrids exhibited superior ratoon cold tolerance to that of clones derived from *S. spontaneum* (Burner *et al.*, 2009). This could be because freezing temperatures during growth have little effect on *Miscanthus* photosynthesis (Naidu *et al.*, 2003).

Sugarcane is a proven biofeedstock that accounts for 40% of biofuel production worldwide. It is one of the most efficient crops in converting solar energy into chemical energy and is among the most favourable in terms of energy output/input ratios (Heichel, 1974). Through future breeding efforts, it is expected that energycane selected for cold environments will be a major feedstock contributor to biofuel production.

References

Aitken, K., Li, J., Wang, L., Qing, C., Fan, Y.H. *et al.* (2006) Characterization of intergeneric hybrids of *Erianthus rockii* and *Saccharum* using molecular markers. *Genetic Resources and Crop Evolution* 54, 1395–1405.

Alexander, A.G. (1973) *Sugarcane Physiology*. Elsevier, Amsterdam.

Allen, R.J., Kidder, G., Jr and Gascho, G.J. (1978) Predicting tons of sugarcane per acre using solar radiation, temperature and percent plant cane, 1971 through 1976. *Proceedings of American Society of Sugarcane Technology* 7, 18–22.

Andreis, H.J. (1976) A water table study on everglades peat soil. *Sugar Journal* 39, 8–12.

Andrietta, M.G.S., Andrietta, S.R., Steckelberg, C. and Stupiello, E.N.A. (2007) Bioethanol: Brazil, 30 years of proalcool. *International Sugar Journal* 109, 195–200.

Babu, C., Koodalingam, K., Natarajan, U.S., Shanthi, R.M. and Govindaraj, P. (2009) Genetic enhancement of sugarcane (*Saccharum* spp. Hybrids) for resistance to red rot disease and economic traits. *Journal of Agricultural Science* 4, 97–107.

Banath, C.L. and Monteith, N.H. (1966) Soil oxygen deficiency and sugarcane root growth. *Plant Soil* 25, 143–149.

Breaux, R.D. (1979) Yield components in the F1 and backcross generations of *Saccharum spontaneum* L. (US 56-15-8) with commercial hybrids. *Sugar Y Azucar* 74, 69.

Breaux, R.D. and Legendre, B.L. (1981) Biomass and sucrose production in sugarcane: divergent selection strategies. *Proceedings of American Society of Sugar Cane Technologist* 8, 156 (abstract).

Breaux, R.D., Fanguy, H.P., Matherne, R.J. and Dunckelman, P.H. (1974) Registration of CP 65-357 Sugarcane (Reg No. 35). *Crop Science* 14, 605 (abstract).

Bresler, E., McNeal, B.L. and Carter, D.L. (1982) *Saline and Sodic Soils*. SpringerVerlag, Berlin.

Burner, D.M., Tew, T.L., Harvey, J.J. and Belesky, D.P. (2009) Dry matter partitioning and quality of *Miscanthus, Panicum*, and *Saccharum* genotypes in Arkansas, USA. *Biomass Bioenergy* 33, 610–619.

Campbell, R.B., Chang, J.H. and Cox, D.C. (1959) Evapotranspiration of sugarcane in Hawaii as measured by in field lysimeters in relation to climate. *Proceedings of International Society of Sugarcane Technology* 10, 637–645.

Carter, C.E. (1980) Redox potential and sugarcane yield relationship. *Transactions of the ASAE* 23, 924–927.
Carter, C.E. and Camp, C.R. (1983) Subsurface drainage of an alluvial soil increased sugarcane yields. *Transactions of the ASAE* 26, 426–429.
Carter, C.E. and Floyd, J.M. (1973) Subsurface drainage and irrigation for sugarcane. *Transactions of the ASAE* 16, 279–281.
Carter, C.E. and Floyd, J.M. (1975) Inhibition of sugarcane yields by high water table during dormant season. *Proceedings American Society of Sugarcane Technology* 4, 14–18.
Carter, C.E., Irvine, J.E., McDaniel, V. and Dunckelman, J. (1985) Yield response of sugarcane to stalk density and subsurface drainage treatments. *Transactions of the ASAE* 28, 172–178.
Cenisk, R. and Miocque, J. (2004) *Breeding of Sugarcane*. Embrapa, Brasília, Brazil.
Chang, Y.T. and Wang, P.L. (1983) Water consumption and irrigation requirements of sugarcane. *Taiwan Sugar Research Institute Annual Report*, pp. 15–23.
Chapman, L.S. (1988) Constraints to production in ratoon crops. *Proceedings Australian Society of Sugarcane Technology* 10, 189–192.
Chapman, L.S., Ferrias, R. and Ludlow, M.M. (1992) Ratooning ability of cane varieties: variation in yield and yield components. *Proceedings Australian Society of Sugarcane Technology* 14, 130–138.
Clements, H.F. (1980) *Sugarcane Crop Logging and Crop Control: Principles and Practice*. University of Hawaii Press, Honolulu.
Cock, J.H. (2003) Sugarcane growth and development. *International Sugarcane Journal* 105, 540–552.
Deren, C.W. (1995) Genetic base of US mainland sugarcane. *Crop Science* 35, 1195–1199.
Deren, C.W. and Snyder, G.H. (1991) Biomass production and biochemical methane potential of seasonally flooded inter-generic and inter-specific *Saccharum* hybrids. *Bioresource Technology* 36, 179–184.
D'Hont, A., Rao, P.S., Feldmann, P., Grivet, L., Islam-Faridi, N. *et al.* (1995) Identification and characterization of sugarcane intergeneric hybrids, *Saccharum officinarum* x *Erianthus arundinaceus*, with molecular markers and DNA *in situ* hybridization. *Theoretical and Applied Genetics* 91, 320–326.
D'Hont, A., Grivet, L., Feldmann, P., Rao, P.S., Berding, N. and Glaszmann, J.C. (1996) Characterisation of the double genomic structure of modern sugarcane cultivars, *Saccharum* ssp., by molecular cytogenetics. *Molecular and General Genetics* 250, 405–413.
Eggleston, G. and Legendre, B. (2003) Mannitol and oligosaccharides as a new criteria for determining cold tolerance in sugarcane varieties. *Food Chemistry* 80, 541–461.
Fageria, N.K. (2001a) Effect of liming on upland rice, common bean, corn, and soybean production in cerrado soil. *Pesquisa Agropecuária Brasileira* 36, 1419–1424.
Fageria, N.K. (2001b) Response of upland rice, dry bean, corn and soybean to base saturation in cerrado soil. *Revista Brasileira de Engenharia Agrícola e Ambiental* 5, 416–424.
Fageria, N.K. and Baligar, V.C. (2003) Fertility management of tropical acid soils for sustainable crop production. In: Rengel, Z. (ed.) *Handbook of Soil Acidity*. Marcel Dekker, New York, pp. 359–385.
Fageria, N.K. and Baligar, V.C. (2005) Enhancing nitrogen use efficiency in crop plants. *Advances in Agronomy* 88, 97–185.
Fageria, N.K. and Baligar, V.C. (2008) Ameliorating soil acidity of tropical Oxisols by liming for sustainable crop production. *Advances in Agronomy* 99, 345–399.
Fageria, N.K., Baligar. V.C. and Jones, C.A. (2011) Sugarcane. In: Fageria, N.K., Baligar. V.C. and Jones, C.A. (eds) *Growth and Mineral Nutrition of Field Crops*. CRC Press, Boca Raton, Florida, pp. 437–456.
FAO (2009) *Major Food and Agricultural Commodities and Producers: Sugarcane*. The Statistics Division, Rome, Italy.
Farrell, A.E., Pleven, R.J., Turner, B.T., Jones, A.D., O'Hare, M. and Kammen, D.M. (2006) Ethanol can contribute to energy and environmental goals. *Science* 311, 506–508.
Finkel, H.J. (1983) Irrigation of sugar crops. In: Finkel, H.J. (ed.) *Handbook of Irrigation Technology*. CRC Press, Boca Raton, Florida, pp. 119–135.
Foy, C.D. (1984) Physiological effects of hydrogen, aluminum and manganese toxicity in acid soils. In: Adams, F. (ed.) *Soil Acidity and Liming*, 2nd edn. American Society of Agronomy, Madison, Wisconsin, pp. 57–97.
Foy, C.D. (1992) Soil chemical factors limiting plant root growth. *Advances in Soil Science* 19, 97–149.
Gascho, G.J. and Shih, S.F. (1979) Varietal response of sugarcane to watertable. I. Lysimeter performance and plant response. *Soil Crop Science Society of Florida Proceedings* 38, 23–27.
Gascho, G.J. and Shih, S.F. (1981) Row spacing effects on biomass and composition of sugarcane in Florida. *Proceedings of American Society of Sugarcane Technology* 8, 72–76.

Gascho, G.J. and Shih, S.F. (1982) Sugarcane. In: Teare, I.D. and Peet, M.M. (eds) *Crop Water Relations*. Wiley, New York, pp. 445–479.

Gascho, G.J., Anderson, D.L. and Bowen, J.E. (1993) Sugarcane. In: Bennett, W.F. (ed.) *Nutrient Deficiencies and Toxicities in Crop Plants*. The American Phytopathological Society, St Paul, Minnesota, pp. 37–42.

Giamvala, M.J., Clarke, S.J. and Stein, J.M. (1984) Sugarcane hybrids of biomass. *Biomass* 6, 61–68.

Gibson, W. (1974) Hydraulics, mechanics, and economics of subsurface and drip irrigation of Hawaiian sugarcane. *Proceedings of International Society of Sugarcane Technology* 15, 639–648.

Gilbert, R.A., Shine, J.M., Miller, J.D., Rice, R.W. and Eainbolt, C.R. (2006) The effect of genotype, environment and time of harvest on sugarcane yields in Florida, USA. *Field Crops Research* 95, 156–170.

Gilbert, R.A., Rainbolt, C.R., Morris, D.R. and Bennett, A.C. (2007) Morphological responses of sugarcane to long term flooding. *Agronomy Journal* 99, 1622–1628.

Glaz, B. and Gilbert, R.A. (2006) Sugarcane response to water table, periodic flood, and foliar nitrogen on organic soil. *Agronomy Journal* 98, 616–621.

Glaz, B., Shine, J.M. Jr, Irey, M.S., Perdomo, R., Powell, G. *et al.* (2011) Seasonal fiber content of three sugarcane cultivars in three crop cycles on sand and muck soils. *Agronomy Journal* 103, 211–220.

Goldemberg, J., Coelho, S.T. and Guardabassi, P. (2008) The sustainability of ethanol production from sugarcane. *Energy Policy* 36, 2086–2097.

Gosnell, J.M. (1971) Some effects of watertable on the growth of sugarcane. *Proceedings International Society of Sugarcane Technology* 14, 841–849.

Gosnell, J.M. and Pearse, T.L. (1971) Methods of surface irrigation in sugarcane. *Proceedings of International Society of Sugarcane Technology* 14, 875–885.

Gravois, K.A., Bischoff, K.P., LaBorde, C.M., Hoy, J.W., Reagan, T.E., Pontif, M.J., Kimbeng, C.A., Hawkins, G.L., Sexton, D.R. and Fontenot, D.P. (2010) Registration of 'L 79-1002' Sugarcane. *Journal of Plant Regulation* 4, 183–188.

Grisham, M.P., Burner, D.M. and Legendre, B.L. (1992) Resistance to the H strain of sugarcane mosaic virus among wild forms of sugarcane and relatives. *Plant Disease* 76, 360–362.

Hale, A.L., Hoy, J.W. and Veremis, J.C. (2010) Identification of sources of resistance to sugarcane red rot. *Proceedings of American Society of Sugar Cane Technologist* 27, 1–8.

Hale, A.L., Dufrene, E.O., Tew, T.L., Pan, Y.-B., Viator, R.P. *et al.* (2012) Registration of 'Ho 02-113' Sugarcane. *Journal of Plant Regulation* 7, 51–57.

Hardy, M. (1966) Water consumption of the cane plant. *Annual Reporter Mauritius Sugar Industry Research Institute, pp.* 95–101.

Hartemink, A.E. (2008) Sugarcane for bioethanol: soil and environmental issues. *Advances in Agronomy* 99, 125–182.

Hartt, C.E., Kortschak, H.P., Forbes, A.J. and Burr, G.O. (1963) Translocation of C^{14} in sugarcane. *Plant Physiology* 38, 305–318.

Hatch, M.D. and Glasziou, K.T. (1964) Direct evidence for translocation of sucrose in sugarcane leaves and stems. *Plant Physiology* 39, 180–184.

Hawker, J.S. (1985) Sucrose. In: Dey, P.M. and Dixon, R.A. (eds) *Biochemistry of Storage Carbohydrates in Green Plants*. Academic Press, New York, pp. 1–51.

Heichel, G.H. (1974) Comparative efficiency of energy use in crop production. *Bulletin-Connecticut Agricultural Experiment Station New Haven* 739, 1–26.

Hunsigi, G. and Srivastava, S.C. (1977) Modulation of ET values of sugarcane because of high water table. *Proceedings of International Society of Sugarcane Technology* 16, 1557–1564.

Inman-Bamber, N.G. (2004) Sugarcane water stress criteria for irrigation and drying off. *Field Crops Research* 89, 107–122.

Irvine, J.E. (1983) Sugarcane. In: International Rice Research Institute. *Biological Basis, Physical Environment, and Crop Productivity*. IRRI, Los Baños, the Philippines, pp. 361–381.

Jackson, P. (1994) Genetic relationships between attributes in sugarcane clones closely related to *Saccharum spontaneum*. *Euphytica* 79,101–108.

Janoo, N., Grivet, L., Seguin, M., Paulet, F., Domaingue, R. *et al.* (1999) Molecular investigation of the genetic base of sugarcane cultivars. *Theoretical and Applied Genetics* 99, 171–184.

Johnson, J.L., Heagler, A.M., Zapata, H.O. and Ricaud, R. (1993) The impact of succession planting and a third ratoon crop on economic efficiency in sugarcane production in Louisiana. *American Society of Sugarcane Technology* 13, 28–32.

Jones, C.A. (1980) A review of evapotranspiration studies in irrigated sugarcane in Hawaii. *Hawaiian Planters' Record* 59, 195–214.

Jones, C.A. (1985) *C_4 Grasses and Cereals: Growth, Development, and Stress Response*. Wiley, New York.

Kamprath, E.J. and Foy, C.D. (1985) Lime-fertilize-plant interactions in acid soils. In: Engelstad, O.P. (ed.) *Fertilizer Technology and Use*, 3rd edn. Soil Science Society of America, Madison, Wisconsin, pp. 91–151.

Keating, B.A., Robertson, M.J., Muchow, R.W. and Huth, N.I. (1999) Modeling sugarcane production systems: I. Development and performance of the sugarcane module. *Field Crops Research* 61, 253–271.

Kingston, G. and Ham, G.J. (1975) Water requirements and irrigation scheduling of sugar cane in Queensland. *Proceedings of Queensland Society of Sugarcane Technology* 42, 57–65.

Legendre, B.L. and Burner, D.M. (1995) Biomass production of sugarcane cultivars and early-generation hybrids. *Biomass and Bioenergy* 8, 55–61.

Leverington, K.C. and Ridge, D.R. (1975) A review of sugar cane irrigation. *Proceedings of the Queensland Society of Sugarcane Technology* 42, 37–43.

Lima, M.L.A., Garcia, A.A.F., Oliveira, K.M., Matsuoka, S., Arizono, H. *et al.* (2002) Analysis of genetic similarity detected by AFLP and coefficient of parentage among genotypes of sugar cane (*Saccharum* spp.). *Theoretical and Applied Genetics* 104, 30–38.

Loomis, R.S. and Connor, D.J. (1992) Community concepts. In: Loomis, R.S. and Connor, D.J. (eds) *Crop Ecology: Productivity and Management in Agricultural Systems*. Cambridge University Press, Cambridge, UK, pp. 32–39.

Maas, E.V. and Hoffman, G.V. (1977) Crop salt tolerance: current assessment. *Journal of Irrigation Drainage Division American Society of Civil Engineering* 103, 115–134.

Mariotti, J.A. (1969) *El programa de mejoramiento varietal em caña de azucar em la Republica Argentina*. Universidade Nacional de Tucumãn, Tucumãn, Argentina, pp. 29–31.

McLaughlin, S., Bouton, J., Bransby, D., Conger, B., Coumpaugh, W. *et al.* (1999) Developing switchgrass as a bioenergy crop. In: Janick, J. (ed.) *Perspectives on New Crops and New Uses*. ASHS Press, Alexandria, Virginia, pp. 282–289.

Mendoza, T.C. and Lantin, M.M. (1977) Identification of *Saccharum spontaneum* clones resistant to sugarcane smut. *Philippine Journal of Crop Science* 2, 177–179.

Milligan, S.B., Gravois, K.A. and Martin, F.A. (1996) Inheritance of sugarcane ratooning ability and the relationship of younger crop traits to older crop traits. *Crop Science* 36, 45–50.

Ming, R., Moore, P.H., Wu, K.K., D'Hont, A., Glaszmann, J.C. *et al.* (2006) Sugarcane improvement through breeding and biotechnology. *Plant Breeding Reviews* 27, 17–118.

Moore, P.H. and Maretzki, A. (1996) Sugarcane. In: Zamski, E. and Schafeer, A.A. (eds) *Photoassimilate Distribution in Plants and Crops: Source-Sink Relationships*. Marcel Dekker, New York, pp. 643–669.

Moreira, J.R. (2007) Water use and impacts due ethanol production in Brazil. Available at: http://www.iwmi.cgiar.org/EWMA/files/papers/Jose_Moreira.pdf (accessed 6 May 2011).

Naidu, S.L., Moose, S.P., Al-Shoaibi, A.K., Raines, C.A. and Long, S.P. (2003) Cold tolerance of C_4 photosynthesis in *Miscanthus × giganteus*: adaptation in amounts and sequence of C_4 photosynthetic enzymes. *Plant Physiology* 132, 1688–1697.

Nelson, L.E. (1980) Phosphorus nutrition of cotton, peanut, rice, sugarcane, and tobacco. In: Dinauer, R. (ed.) *The Role of Phosphorus in Agriculture*. American Society of Agronomy, Madison, Wisconsin, pp. 693–736.

Nunes Júnior, M.S.D. (1987) Varieties of sugarcane. In: Paranhos, S.B. (ed.) *Sugarcane; Production and Use*. Fundação Cargill, Campinas, Brazil, pp. 187–259.

Oliveira, E.C.A., Freire, F.J., Oliveira, R.I., Freire, M.B.G.S., Simões Neto, D.E. *et al.* (2010) Nutrient extraction and export by fully irrigated sugarcane varieties. *Brazilian Journal of Soil Science* 34, 1343–1352.

Panje, R.R. (1964) Thoughts on *Saccharum spontaneum* sugarcane. *Proceedings of 5th All India Conference on Sugarcane Research Development Workers*, Coimbatore, India, pp. 379–387.

Ram, B., Sreenivasan, T.V., Sahi, B.K. and Singh, N. (2001) Introgression of low temperature tolerance and red rot resistance from *Erianthus* in sugarcane. *Euphytica* 122, 145–153.

Reuter, D.J., Edwards, D.G. and Wilhelm, N.S. (1997) Temperate and tropical crops. In: Reuter, D.J., Robinson, J.B. and Dutkiewicz, C. (eds) *Plant Analysis: An Interpretation Manual*, 2nd edn. CSIRO, Collingwood, Victoria, Australia, pp. 83–278.

Richard, C., Jackson, W. and Waguespack, H. Jr (2001) Harvester trials and extraneous matter in the Louisiana sugar industry. *Proceedings International Society of Sugarcane Technology* 24, 263–268.

Rossetto, R., Spironello, A., Cantarella, H. and Quaggio, J.A. (2004) Sugarcane response to liming and potassium fertilization. *Bragantia* 63, 105–119.

Rossetto, R., Cantarella, H., Dias, F.L.F., Vitti, A.C. and Tavares, S. (2010) Sugarcane. In: Prochnow, L.I., Casarin, V. and Stipp, S.R. (eds) *Good Practices for Fertilizer Use*. International Plant Nutrition, Piracicaba, Brazil, pp. 161–230.

Rudd, A.V. and Chardon, C.W.A. (1977) The effects of drainage on cane yields as measured by watertable heights in the Macknade mill area. *Proceedings of Queensland Society of Sugarcane Technology* 44, 111–117.
Salassi, M.E. and Giesler, G.G. (1995) *Projected Costs and Returns - Sugarcane, Louisiana, 1995.* AEA Information Series 132, Louisiana State University, Baton Rouge, Louisiana.
Sampietro, D.A. and Vattuone, M.A. (2006) Nature of the interference mechanism of sugarcane (*Saccharum officinarum* L.) straw. *Plant Soil* 280, 157–169.
Sanchez, P.A. and Salinas, J.G. (1981) Low-input technology for managing Oxisols and Ultisols in tropical America. *Advances in Agronomy* 34, 280–340.
Sartoris, G.B. and Belcher, B.A. (1949) The effect of flooding on flowering and survival of sugarcane. *Sugar* 44, 36–39.
Shrivastava, A.K., Ghosh, A.K. and Agnihotri, V.P. (1992) *Sugar Cane Ratoons.* Oxford Publishing, New Delhi.
Simmonds, N.W. (1984) Sugarcane. In: Simmonds, N.W. (ed.) *Evolution of Crop Plants.* Longman, New York, pp. 32–33.
Sreenivasan, T.V., Whloowalia, B.S. and Heinz, D.J. (1987) Cytogenetics. In: Heinz, D.J. (ed.) *Sugarcane Improvement Through Breeding.* Elsevier, Amsterdam, pp. 211–253.
Srinivasan, K. and Batcha, M.B.G.R. (1962) Performance of clones of *Saccharum* species and allied genera under conditions of waterlogging. *Proceedings of International Society of Sugarcane Technology* 11, 571–577.
Stevenson, G.C. (1965) *Genetics and Breeding of Sugar Cane.* Longmans, London.
Stone, K.C., Hunt, P.G., Cantrell, K.B. and Ro, K.S. (2010) The potential impacts of biomass feedback production on water resource availability. *Bioresource Technology* 101, 2114–2025.
Suman, A., Lal, M., Singh, A.K. and Gaur, A. (2006) Microbial biomass turnover in Indian subtropical soils under different sugarcane intercropping systems. *Agronomy Journal* 98, 698–704.
Sundara, B. (1989) Improving sugarcane productivity under moisture constraints and through cropping systems. In: Naidu, K.M. (ed.) *Proceedings of the International Symposium Sugarcane Varietal Improvement - Present and Future Strategies.* Sugarcane Breeding Institute, Coimbatore, India, pp. 221–256.
Tangley, L. (1986) Sugarcane: Cuba's 'noble crop'. *Bioscience* 36, 414–420.
Tejera, N.A., Rodes, R., Ortega, E., Campos, R. and Lluch, C. (2007) Comparative analysis of physiological characteristics and yield components in sugarcane cultivars. *Field Crops Research* 101, 64–72.
Tew, T.L. and Cobill, R.M. (2008) Genetic improvement of sugarcane (*Saccharum* spp.) as an energy crop. In: Vermerris, W. (ed.) *Genetic Improvement of Bioenergy Crops.* Springer, New York, pp. 249–272.
Tew, T.L., Dufrene, E.O., Cobill, R.M., Garrison, D.D., White, W.H., Grisham, M.P., Pan, Y.-B., Legendre, B.L., Richard, E.P. and Miller, J.D. (2011) Registration of 'HoCP 91-552' Sugarcane. *Journal of Crop Registrations* 5(2), 181–190.
Tewari, J., Mehrotra, R. and Irudayaraj, J. (2003) Direct near infrared analysis of sugarcane clear juice using a fibre-optic transmittance probe. *Journal of Near Infrared Spectroscopy* 11, 351–356.
Thompson, G.D. (1976) Water use by sugarcane. *South African Sugar Journal* 60, 593–600.
Thompson, G.D. (1977) Irrigation of sugarcane. *South African Sugar Journal* 61, 126–131.
Thompson, G.D. and Boyce, J.P. (1967) Daily measurements of potential evapotranspiration from fully canopied sugarcane. *Agriculture Meteorology* 4, 267–279.
Thompson, G.D. and Boyce, J.P. (1968) The plant crop results of two irrigation experiments at Pongola. *Proceedings of South African Sugar Technology Association* 42, 143–157.
Thompson, G.D. and Boyce, J.P. (1971) Comparisons of measured evapotranspiration of sugarcane from large and small lysimeters. *Proceedings of South African Sugar Technology Association* 45, 169–176.
Van den Wall Bake, J.D., Junginger, M., Faaij, A., Poot, T. and Walter, A. (2009) Explaining the experience curve: cost reductions of Brazilian ethanol from sugarcane. *Biomass and Bioenergy* 33, 644–658.
Van Dillewijin, C. (1952) *Botany of Sugarcane.* Cronica Botania, Waltham, Massachusetts.
Viator, R.P., Johnson, R.M., Grimm, C.C. and Richard, E.P. Jr (2006) Allelopathic, autotoxic, and hormetic effects of postharvest sugarcane residue. *Agronomy Journal* 98, 1526–1531.
Viator, R.P., White, P.M. and Richard, E.P., Jr (2010) Sustainable production of energycane for bioenergy in the Southeastern U.S. In: Eggleston, G. (ed.) *Sustainability of the Sugar and Sugar-Ethanol Industries.* ACS Symposium Series 1058, American Chemical Society, Washington, DC, pp. 147–161.
White, W.H., Tew, T.L., Cobill, R.M., Burner, D.M., Grisham, M.P. *et al.* (2011) Registration of 'Ho 00-961' Sugarcane. *Journal of Plant Regulation* 5, 332–338.

Wiedenfeld, R.P. (1995) Effects of irrigation and N fertilizer application on sugarcane yield and quality. *Field Crops Research* 43, 101–108.
Wiedenfeld, R.P. (2000) Water stress during different sugarcane growth periods on yield and response to N fertilization. *Agriculture Water Management* 43, 173–182.
Wiedenfeld, B. (2004) Scheduling water application on drip irrigated sugarcane. *Agriculture Water Management* 64, 169–181.
Wiedenfeld, B. and Enciso, J. (2008) Sugarcane response to irrigation and nitrogen in semiarid South Texas. *Agronomy Journal* 100, 665–671.
Yang, S.J. and Chang, Y.T. (1976) Studies on irrigation scheduling by using crop soil climatic data. *Taiwan Sugar Research Institute Annual Report*, pp. 13–14.

9 Sweet Sorghum: Genetics, Breeding and Commercialization

P. Srinivasa Rao,[1] A.V. Umakanth,[2] Belum V.S. Reddy,[1] Ismail Dweikat,[3] Sujata Bhargava,[4] C. Ganesh Kumar,[5] Serge Braconnier[6] and J.V. Patil[2]
[1]International Crops Research Institute for the Semi-Arid Tropics (ICRISAT), India; [2]Directorate of Sorghum Research (DSR), India; [3]University of Nebraska, USA; [4]University of Pune, India; [5]CSIR-Indian Institute of Chemical Technology (CSIR-IICT), India; [6]Centre de Coopération Internationale en Recherche Agronomique pour le Développement (CIRAD), France

9.1 Introduction

Sweet sorghum, similar to grain sorghum except for its juice-rich sweet stalk, has been grown in the USA (for syrup) and Africa (for fodder) for many centuries and is considered to be a potential bioethanol feedstock, and is expected to meet food, feed, fodder, fuel and fibre demands. Some sweet sorghum lines attain juice yields of about 78% of total plant biomass, containing from 15 to 23% soluble fermentable sugar (by comparison, sugarcane has 14–16%) (Reddy *et al.*, 2008). The sugar is composed mainly of sucrose (70–80%), fructose and glucose. Most of the sugars are distributed in the stalk, with about 2% in the leaves and inflorescences (Vietor and Miller, 1990), making the crop particularly amenable to direct fermentable sugar extraction.

9.2 Sweet Sorghum Characteristics and Utilization

The term sweet sorghum is used to distinguish varieties of sorghum with high concentration of soluble sugars in the plant stalk sap or juice compared to grain sorghum, which has relatively less sugar and juice in the stalks. Sweet sorghum is a C_4 species with wide, flat leaves and a round or elliptical head full of grain at the stage of maturity. It has, like grain sorghum, traditionally been under cultivation for nearly 3000 years. It can be grown successfully in the semi-arid tropics, where other crops fail to thrive, and is highly suitable for cultivation in tougher dryland growing areas. It can produce very high yields with irrigation. During very dry periods, sweet sorghum can go into dormancy, with growth resuming when sufficient moisture levels return (Gnansounou *et al.*, 2005). It can be grown easily on all continents, in tropical, subtropical, temperate, semi-arid regions as well as in poor quality soils. It is known as the sugarcane of the desert and also 'the camel among crops' for its drought-hardy characteristics (Sanderson *et al.*, 1992). It has higher drought tolerance and water use efficiency (WUE) compared to maize, and yields, like those of *Miscanthus*, range from 18 to 36 dry t ha^{-1} biomass $year^{-1}$ on low-quality soils with minimal inputs of fertilizer and water. In Indiana (USA), studies showed that sweet sorghum cultivars produce 25 to 40 t dry mass ha^{-1} with an input of from 0 to 60 kg ha^{-1} of nitrogen fertilizers. The high WUE and low N requirements of sorghum also

©CAB International 2013. *Biofuel Crops: Production, Physiology and Genetics* (ed. B.P. Singh)

provide significant advantages to the growers, because sorghum fits into a normal rotation scheme with maize and soybean, yet has lower production costs and employs similar production equipment. Its ratooning ability enables multiple harvests per season, a feature that could expand the geographical range of sorghum cultivation. For example, in Nebraska (USA), cold-tolerant sweet sorghum planted in April yielded 22 t dry biomass ha^{-1}, and a ratoon crop harvested from the same material in mid-October gave an additional 12 t ha^{-1} (Ali *et al.*, 2008). The grain stalk juice and bagasse (the fibrous residue that remains after juice extraction) can be used to produce food, fodder, ethanol and power. Owing to these favourable attributes, it is often referred to as a 'smart' crop (William D. Dar, Director General, ICRISAT) (Fig. 9.1).

Sweet sorghum candidate traits in relation to utilizable options are listed in Table 9.1. These important characteristics, along with its suitability for seed propagation, mechanized crop production and comparable ethanol production capacity compared to sugarcane and sugarbeet makes sweet sorghum a viable alternative source for ethanol production (Table 9.2).

Fig. 9.1. Improved sweet sorghum cultivar 'ICSV 25274'.

9.2.1 Food–fuel trade-off

It is often stated that sweet sorghum cultivars do not produce grain yield or the grain yield is very less compared to that of grain sorghum. Studies at ICRISAT showed that sweet sorghum hybrids had higher stem sugar yield (11%) and higher grain yield (5%) compared to grain sorghum types, while sweet sorghum varieties had 54% higher sugar yield and 9% lower grain yield compared to non-sweet stalk varieties in the rainy season. On the other hand, both sweet sorghum hybrids and varieties had higher stalk sugar yields (50% and 89%) and lower grain yields (25% and 2%) in the post-rainy season. Thus, there is little trade-off between grain and stalk sugar yields in the sweet sorghum hybrids in the rainy season while the trade-off is less in varieties in the post-rainy season (Srinivasa Rao *et al.*, 2009, 2010; Kumar *et al.*, 2010).

This is further supported by other published work (Zhao *et al.*, 2009) showing that there is significant soluble sugars content in the stems (79–94%) during the post-anthesis period, with the hybrids exhibiting significantly high soluble sugars content over varieties with the same maturity period, and effects of year, harvest time and genotype on calculated ethanol yield (CEY) are highly significant. The experimental data on the relationship between stalk sugar traits and grain yield showed that the regression coefficient of stalk sugar yield on grain yield is not significant, thereby indicating that the grain yield is not affected when selection is done for stalk sugar yield. Hence a selection programme can aim to improve both the traits simultaneously. When sweet sorghum complements sugarcane production, this does not require additional land area, but simply the insertion of a cycle of sweet sorghum between two cycles of sugarcane. The impact on food security is less, even if sweet sorghum replaces a cycle of legumes (groundnut, cowpea).

Table 9.1. Candidate traits of sweet sorghum as biofuel feedstock (Reddy *et al.*, 2005, 2008; Srinivasa Rao *et al.*, 2009, 2010).

As crop	As ethanol source	As bagasse	As raw material for industrial products
Short duration (3–4 months) C_4 dryland crop Good tolerance of biotic and abiotic constraints Meets fodder and food needs Non-invasive species Low soil N_2O and CO_2 emission Seed propagated	Amenable to ecofriendly processing Less sulfur in ethanol High octane rating Automobile friendly (up to 25% of ethanol–petrol mixture without engine modification)	High biological value Rich in micronutrients Use as feed, for power co-generation or biocompost Good for silage making	Cost-effective source of pulp for paper making Dry ice, acetic acid, fuel oil and methane can be produced from the co-products of fermentation Butanol, lactic acid, acetic acid and beverages can be manufactured

Table 9.2. Comparison of sweet sorghum with other bioethanol feedstocks (Reddy *et al.*, 2008; Almodares and Hadi, 2009; Srinivasa Rao *et al.*, 2009; Girase, 2010; Wortmann *et al.*, 2010).

Characteristics	Sugarcane	Sugarbeet	Maize	Sweet sorghum
Crop duration	12–13 months	5–6 months	3–4 months	4 months
Growing season	One season	One season	All seasons	All seasons
Propagation	Setts (40,000 ha^{-1})	Seed (3.6 kg ha^{-1}; pellet)	Seed (25 kg ha^{-1})	Seed (8 kg ha^{-1})
Soil requirement	Grows well in drained soil	Grows well in sandy loam; also tolerates alkalinity		All types of drained soil
Water management	Requires water throughout the year (36,000 m^3 ha^{-1})	Requires water, 40–60% compared to sugarcane (18,500 m^3 ha^{-1})	Requires water (12,000 m^3 ha^{-1})	Less water requirement; can be grown as rain-fed crop (8000 m^3 ha^{-1})
Crop management	Requires good management 250 to 400N: 125P:125K	Requires moderate management 120N:60P:60K	Requires good management 130N:60P:60K	Easy management; low fertilizer 90N:40P
[a]Stalk/beet/ grain yield (t ha^{-1})	60–85	85–100	5–10	45–65
Sugar content on weight basis	10–12%	15–18%		7–12%
Sugar yield (t ha^{-1})	5–12	11.25–18		3–7
Ethanol yield from juice (l ha^{-1})	4,350–7,000	7,100–10,500	2,150–4,300	2,475–3,500
Harvesting	Harvested mechanically	Harvested mechanically	Harvested mechanically	Very simple; predominantly manual and mechanical harvesting at pilot scale

[a]Stalk yield for sugarcane and sweet sorghum; beet yield for sugarbeet; and grain yield for maize.

9.3 Climate and Distribution

Sorghum (*Sorghum bicolor* (L.) Moench) is the fourth major cereal crop of the world in production and fifth in hectarage after wheat, rice, maize and barley. It is mostly grown in the semi-arid tropics (SAT) of the world wherein the production system is constrained by poor soils, low and erratic rainfall and low inputs resulting in low productivity. India is the largest sorghum grower in the world (7.7 Mha) followed by Sudan (5.6 Mha) and Nigeria (4.7 Mha). The USA is the largest producer followed by India and Nigeria (FAO, 2012). Sorghum is well adapted to the SAT and is one of the most efficient dryland crops to convert atmospheric CO_2 into sugar (Schaffert and Gourley, 1982). The crop can be grown in a wide range of climatic conditions as given below.

- Latitude: sweet sorghum can be grown between 40°N and 40°S latitude on either side of the Equator.
- Altitude: sorghum can be found at elevations between sea level and 1500 m. Most East African sorghum is grown between altitudes of 900 and 1500 m, and cold-tolerant varieties are grown between 1600 and 2500 m in Mexico.
- Temperature: sweet sorghum can be grown in the temperature range from 12 to 37°C and optimum temperature for growth and photosynthesis is from 32 to 34°C, day length from 10 to 14 h, optimum rainfall 550–800 mm and relative humidity 15–50%.
- Soils: alfisols (red) or vertisols (black clay loamy) with pH 6.5–7.5, organic matter >0.6%, depth >80 cm, bulk density <1.4 gcc, water holding capacity >50% field capacity, N >260 kg ha^{-1} (available), P >12 kg ha^{-1} (available), K >120 kg ha^{-1} (available) are optimal soil conditions for sorghum growth.
- Water: whilst sorghum will survive with a supply of less than 300 mm over the season of 100 days, it responds favourably with additional rainfall or irrigation water. Typically, sweet sorghum needs between 500 and 1000 mm of water (rain and/or irrigation) to achieve good yields, i.e. 50–100 t ha^{-1} total above ground biomass (fresh weight). Though sorghum is a dryland crop, sufficient moisture availability for plant growth is critically important for high yields. The great advantage of sorghum is that it can become dormant, especially in the vegetative phase, under adverse conditions and can resume growth after relatively severe drought. Early drought stops growth before panicle initiation and the plant remains vegetative; it will resume leaf production and flowering when conditions become favourable for growth again. Mid-season drought stops leaf development. Sorghum is susceptible to sustained flooding, but will survive temporary waterlogging much better than maize.
- Radiation: being a C_4 plant, sweet sorghum has high radiation use efficiency (RUE: about 1.3–1.7 g MJ^{-1}). It has been shown that taller sorghum types possess higher RUE because of better light penetration in the leaf canopy.
- Photoperiodism: most hybrids of sweet sorghum are relatively less photoperiod-sensitive. Traditional farmers, particularly in West Africa, use photoperiod-sensitive varieties. With photoperiod-sensitive types, flowering and grain maturity occurs almost during the same calendar days regardless of planting date, so that even with delayed sowing, plants mature before soil moisture is depleted at the end of the rainy season.

Based on the available literature, an attempt was made to depict diagrammatically the critical traits that have significant bearing on bioethanol productivity of the crop with regard to phenology (Fig. 9.2).

9.4 Taxonomy, Botanical Description and Reproductive Biology

9.4.1 Taxonomy and distribution

The genus *Sorghum* is a variable genus with 24 species divided into five subgeneric sections based on taxonomic differences: Eu-sorghum, Chaetosorghum, Heterosorghum, Para-sorghum and Stiposorghum (de Wet, 1978; Price *et al.*, 2006). *S. bicolor, S. arundinaceum, S. drummondii,*

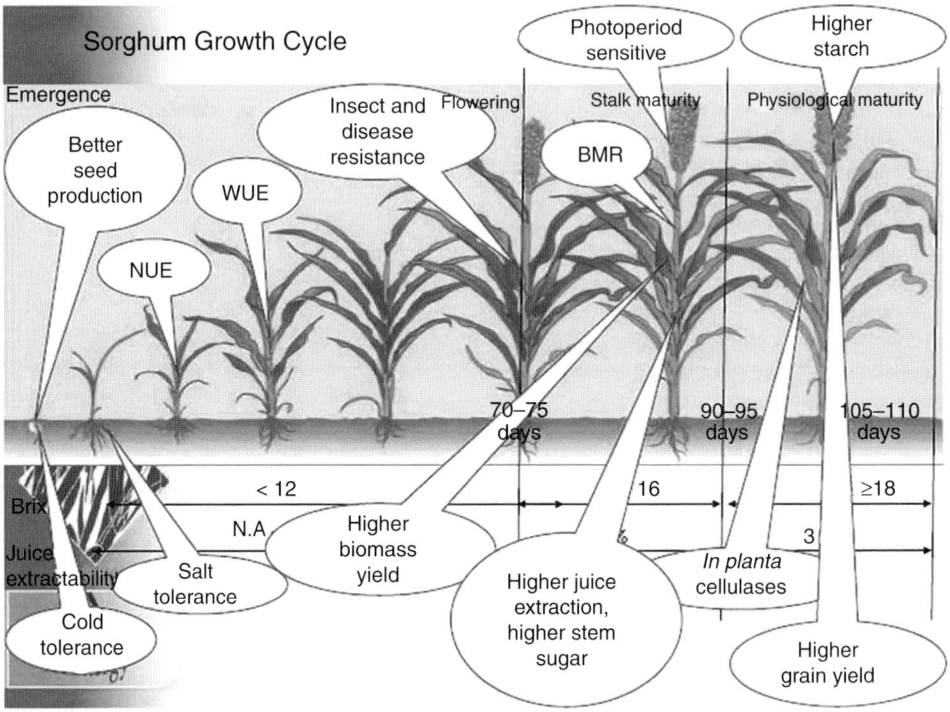

Fig. 9.2. Critical traits to be focused on for genetic enhancement in order to maximize bioethanol productivity in sweet sorghum (Gibson, 2010).

S. halepense, *S. propinquum* and *S. almum* form the section Eu-sorghum. The first three species form the primary gene pool while the latter three species form the secondary gene pool of sorghum. All species other than the Eu-sorghum section form the tertiary gene pool (Dahlberg, 2000). The species *S. bicolor* (2n = 20) is further divided into three subspecies *bicolor*, *arundinaceum* and *drummondii*. Subspecies *bicolor* includes all cultivated races and they are further divided into basic and intermediate races. The five basic races include bicolor, guinea, caudatum, kafir and durra and the ten intermediate races are those between any two of those types, classified primarily based on grain shape, glumes and panicle.

9.4.2 Reproductive biology

Breeding procedures that are used with a particular crop species are determined by its mode of reproduction. Understanding the details of phenology, i.e. floral biology, pollination, fertilization and seed development (Fig. 9.2) in a crop make it possible to develop orderly and efficient breeding procedures.

Panicle initiation

Sorghum is a short-day plant, and blooming is hastened by short days and long nights. However, varieties differ in their photoperiod sensitivity (Quinby and Karper, 1947). Tropical sweet sorghum varieties initiate the reproductive stage when day lengths return to 12 h. Floral initiation takes place 30 to 40 days after germination. Usually, the floral initial is 15–30 cm above the ground when the plants are about 50–75 cm tall (House, 1980). Floral initiation marks the end of the vegetative growth due to meristematic activity. The time required for transformation from the vegetative apex to reproductive apex is largely influenced by genetic characteristics and the environment (photoperiod and temperature). The grand period of growth in sorghum follows the formation of a floral bud and consists largely of cell enlargement. Hybrids take less time to reach panicle initiation and are relatively less influenced by photoperiod and temperature (Srinivasa Rao *et al.*, 2009).

Panicle emergence

During the period of rapid cell elongation, floral initials develop into an inflorescence. About 6–10 days before flowering, the boot will form as a bulge in the sheath of the flag leaf. This will occur, in a variety that flowers in 60–65 days, about 55 days from germination. Sorghum usually flowers in 55 to more than 70 days in warm climates, but flowering may range from 30 to more than 100 days. These observations are valid for tropical sweet sorghums, while temperate sorghums that mature in 5 months take 20–30 days longer for panicle emergence.

Panicle structure

The inflorescence is a raceme, which consists of one or several spikelets. It may be short, compact, loose or open, composed of a central axis that bears whorls of primary branches on every node. The racemes vary in length according to the number of nodes and the length of the internodes. Each primary branch bears secondary branches, which in turn bear spikelets. The spikelets usually occur in pairs, one being sessile and the second borne on a short pedicel, except the terminal sessile spikelet, which is accompanied by two pediceled spikelets. On the pediceled spikelet, the pedicels vary in length from 0.5 to 3.0 mm, and usually are very similar to the internodes. The first and second glumes of every spikelet enclose two florets; the lower one is sterile and is represented by a lemma, and the upper fertile floret has a lemma and palea. Two lodicules are placed on either side of the ovary at its base. The androecium consists of one whorl of three stamens. The anthers are attached at the base of the ovule by a very fine filament and are versatile and yellowish. The gynoecium is centrally placed and consists of two pistils with one ovule from which two feathery stigmas protrude. Many of these floral characters, such as anther colour, stigma colour, stigma length, length of pedicel, etc. are important traits for cultivar identification and classification.

Sessile spikelets

The sessile spikelet contains a perfect flower. It varies in shape from lanceolate to almost rotund and ovate and is sometimes depressed in the middle. The colour is green at flowering, which changes to different colours such as straw, cream, yellow, red, brown, purple, or almost black at grain maturity. The intensity and extent of colouring on the glumes is variable. Glumes vary from quite hairy to almost hairless. The seed may be enclosed by the glume or may protrude from it, being just visible to almost completely exposed.

Pedicelled spikelets

These are much narrower than the sessile spikelets, usually lanceolate in shape. They can be smaller, the same size, or longer than the sessile spikelets. They possess only anthers but occasionally have a rudimentary ovary and empty glumes.

Anthesis and pollination

Anthesis starts after panicle emergence from the boot leaf. Flowers begin to open 2 days after full emergence of the panicle. Floret opening or anthesis is achieved by swelling of the lodicules, and is followed by the exertion of anthers on long filaments and of stigmas from between the lemma and palea. The sorghum head begins to flower at its tip and flowers successively downward over a 4 or 5 day period. Flowering takes place first in the sessile spikelets from top to bottom of the inflorescence. It takes about 6 days for completion of anthesis in the panicle with maximum flowering at 3 or 4 days after anthesis begins. Flowering proceeds downwards to the base in a horizontal plane on the panicle. When flowering of the sessile spikelets is halfway down the panicle, pedicellate spikelets start to open at the top of the panicle and proceed downwards. The flowering phase of pedicellate spikelets overtakes the flowering phase of sessile spikelets before they reach the base of the inflorescence (Maiti, 1996). Anthesis takes place during the morning hours, and frequently occurs just before or just after sunrise, but may be delayed on cloudy damp mornings. It normally starts around midnight and proceeds until 10am depending on the cultivar, location and weather. Maximum flowering is observed between 6 and 8:30am. Because all heads in a field do not flower at the same time, pollen is usually available for a period of 10–15 days. At the time of flowering, the glumes open and all the three anthers fall free, while the two stigmas protrude, each on a stiff style. The anthers dehisce when they are dry and pollen is blown into air. Pollen remains viable several hours after shedding. Flowers remain open for 30–90 min. Dehiscence of the anthers for pollen diffusion takes place through the apical pore. The pollen drifts to the stigma, where it germinates; the pollen tube, with two nuclei, grows down the style, to fertilize the egg and form a 2n nucleus. Glumes close shortly after pollination, though the empty anthers and stigmas still protrude (except in the long glumed types). The florets of some of the very long-glumed types do not open for fertilization, a phenomenon known as cleistogamy.

Cytoplasmic male sterility has been found in sorghum (A1–A4 systems) and has made possible the development of a hybrid seed industry. A good male-sterile plant will not develop anthers, but in some instances dark-coloured shrivelled anthers with no viable pollen will appear. Partially fertile heads are also observed, and although the anthers frequently have viable pollen, the quantity is less than in normal plants.

Seed

The seed is a fruit or caryopsis. The ovule begins to develop as a light green, almost cream-coloured sphere; after about 10 days it begins to take size and becomes darker green. Maturity of grain follows a similar pattern to flowering. The development of grains follows a sequence of stages comprising milky, soft dough, hard dough to the final physiological maturity, when a black layer is formed at the hilar region due to the formation of callus tissue. It takes about 30 days for the seeds to reach maximum dry weight (physiological maturity). The seeds contain about 30% moisture at physiological maturity; they dry to about 10–15% moisture during the following 20–25 days (House, 1980). The crop can be preferably harvested at physiological maturity to take advantage of stalk sugars and grain. The seeds harvested and dried at physiological maturity have good quality and fetch a higher market price (Audilakshmi *et al.*, 2005). Seed size varies from very small (less than 1 g/100 seeds) to large (5 to 6 g/100 seeds).

9.5 Genetics

9.5.1 Genomics

Sorghum is an important target of genome analysis and genomic tool building among the C_4 grasses because the sorghum genome is relatively small (730 Mbp) (Paterson *et al.*, 2009), the cultivated species is diploid ($2n = 20$) and the sorghum germplasm is diverse (Dje *et al.*, 2000; Menz *et al.*, 2004; Casa *et al.*, 2005). The sorghum genome is appreciably smaller and less complex than the maize genome, and as a member of the *Saccharinae* subtribe, it is the ideal model for genetic studies as its fellow members sugarcane and *Miscanthus* are both polyploids that do not succumb easily due to sterility issues. As a consequence, numerous sorghum genetic, physical and comparative maps are available (Tao *et al.*, 1998; Boivin *et al.*, 1999; Peng *et al.*, 1999; Klein *et al.*, 2000; Haussmann *et al.*, 2002; Menz *et al.*, 2002; Bowers *et al.*, 2003). High-density reference maps of one intraspecific *S. bicolor* (Xu *et al.*, 1994; Bhattramakki *et al.*, 2000; Klein *et al.*, 2000; Menz *et al.*, 2002) and one interspecific *S. bicolor* × *S. propinquum* (Chittenden *et al.*, 1994; Bowers *et al.*, 2003) cross provide about 2600 sequence-tagged sites (based on low-copy probes that have been sequenced), 2454 AFLP and about 1375 sequence-scanned (based on sequences of genetically anchored BAC clones) loci. The two maps share one common parent (*S. bicolor* 'BT × 623') and are essentially co-linear (Feltus *et al.*, 2006). More than 800 markers mapped in sorghum are derived from other grasses (serve as comparative anchors), and additional sorghum markers have been mapped directly in other grass species or can be plotted based on sequence similarity. Anchoring of the sorghum maps to those of rice (Paterson *et al.*, 1995, 2004), maize (Whitkus *et al.*, 1992; Bowers *et al.*, 2003), sugarcane (Dufour *et al.*, 1997; Ming *et al.*, 1998), millet (Jessup *et al.*, 2003), switchgrass (Missaoui *et al.*, 2005), bermuda grass (Bethel *et al.*, 2006) and others provides for the cross-utilization of results to simultaneously advance knowledge of many important crops. To have a genetic, physical and cytological perspective of the *Sorghum bicolor* genome, Kim *et al.* (2005) selected 40 BAC clones that had different linkage group markers, 21 from linkage group 2 and 19 from linkage group 8. Multi-BAC probe cocktails were constructed for each chromosome from the landed BACs, which were also pre-evaluated for fluorescence *in situ* hybridization (FISH) signal quality, relative position and collective chromosome coverage. Comparison to the corresponding linkage map revealed full concordance of locus order between cytological and genetic analyses. A sorghum EST project (Pratt *et al.*, 2005) and associated microarray analyses of sorghum gene expression have been carried out (Buchanan *et al.*, 2005; Salzman *et al.*, 2005), a comprehensive analysis of sorghum chromosome architecture has been completed (Kim *et al.*, 2005), and an 8x draft of the sorghum genome sequences was released by the US Department of Energy's Joint Genome Institute and Center for Integrative Genomics in 2007 (http://www.phytozome.net; Paterson *et al.*, 2009). Globally, a total of 167,890 sorghum accessions are reported held in different germplasm collection centres representing about 86% of the total 194,250 accessions that have been documented in the Bioversity Germplasm Database (January 2006). Of the 167,890 accessions, the USDA germplasm collection maintains 42,614 sorghum accessions, of which more than 800 exotic landraces have been converted to day-length-insensitive lines to facilitate their use in breeding programmes. ICRISAT, for which sorghum is one of its mandated crops, is one of the major repositories, holding a total of 36,774 accessions (21.9%) from 91 countries (Reddy *et al.*, 2006). A preliminary survey indicated that the largest number of accessions (47,963; 28.6%) was held by gene banks in Asia, including China and India. An approximate total of 31,200 sorghum accessions (16.1%) were held in African gene banks, with East Africa holding larger collections of sorghum landraces than South and West Africa.

A set of mutation stocks, developed by the USDA Plant Stress and Germplasm Development Unit in Lubbock, Texas, USA (Xin *et al.*, 2008), is sufficiently extensive to allow identification of mutations in virtually every sorghum gene. Such genomic tools, already in place, will greatly facilitate the introduction of traits required to optimize sweet sorghum for bioenergy production schemes. The best known such mutations are the brown midrib (BMR). The brown midrib (*bmr*) mutations were first discovered in maize in 1926. Early studies revealed the trait resulted in lower fibre and lignin within the plant and could increase the conversion efficiency of sorghum biomass for lignocellulosic bioenergy. In sorghum, more than 19 *bmr* mutants were discovered by Porter *et al.* (1978). The *bmr* mutants are characterized by the reddish-brown coloration of the vascular tissue of the leaf blade, leaf sheath and stem, which is associated with alteration of secondary cell wall composition, especially lignin. Owing to the development of biocatalysts (e.g. genetically engineered enzymes, yeasts and bacteria), it is possible to produce ethanol from any plant or plant part containing lignocellulose biomass, including cereal crop residues (stovers). Sorghum stover also serves as an excellent feedstock for ethanol production. Stover contains lignin, hemicellulose and cellulose. Since the hemicelluloses and cellulose are enclosed by lignin (which is a phenolic polymer), it is difficult to convert them into ethanol, thereby increasing the energy requirement for processing. The *bmr* mutant sorghum, pearl millet (*Pennisetum glaucum*) and maize lines have significantly lower levels of lignin content (51% less in their stems and 25% less in their leaves). Purdue University research showed 50% higher yield of fermentable sugars from the stover of certain sorghum *bmr* lines after enzymatic hydrolysis. Therefore, the use of *bmr* cultivars would reduce the cost of biomass-based ethanol production. The *bmr* crop residues have higher rumen digestibility and palatability, making them good for fodder, too.

The availability of the genomic sequence for sorghum has made it possible to carry out genome-wide analyses. Whereas earlier studies on simple sequence repeats (SSR) marker development primarily utilized anonymous DNA fragments containing SSRs isolated from genomic libraries, more recent studies have used computational methods to detect SSRs in sequence data generated from genomic sequences projects (Manli *et al.*, 2009). In the sorghum genome, a total of 109,039 tandem repeats were detected, of which 15,194 were microsatellite (SSR) (Paterson *et al.*, 2009). Mining the frequency and density of the SSRs showed that the density was 154.98 counts Mbp^{-1} in sorghum genomes. Trinucleotide repeat (27.35%) motifs appear to be the most abundant type in sorghum, while the dinucleotide, tetranucleotide, hexanucleotide, pentanucleotide and mononucleotide repeats are 20.69%, 17.04%, 15.21%, 14.90% and 4.76%, respectively. The A-rich repeats are predominant in the most frequent SSRs. The SSR sizes are not evenly distributed, and hexa-motif gives the longest SSRs. This has provided a valuable resource to develop large numbers of SSR markers in sorghum. The genome sequence has also been used to investigate the diversity of *S. bicolor* resistance (R) genes and assess their importance in the mechanisms of disease resistance. The R genes were characterized based on their structural diversity, physical chromosomal location and phylogenetic relationships. Based on their N-terminal motifs and leucine-rich repeats (LRR), 50 non-regular nucleotide binding site (NBS) genes and 224 regular NBS genes were identified in 274 candidate NBS genes. The vast majority (97%) of NBS genes occurred in gene clusters, indicating extensive gene duplication in the evolution of *S. bicolor* NBS genes (Cheng *et al.*, 2010). Based on the location of individual NBS genes, 268 *Sorghum* NBS-encoding genes were mapped on the ten chromosomes.

Sorghum is extremely well-suited to association mapping methods (Hamblin *et al.*, 2005). Its largely self-pollinating mating system tends to preserve linkage relationships for longer periods than in largely outcrossing crops such as maize. Self-pollination also obviates the need to develop inbred lines. Hamblin *et al.* (2004) reported that linkage disequilibrium (LD) over very short distances in sorghum was more extensive than in maize, suggesting that sorghum may be suitable for LD mapping of genes underlying complex, agronomically important traits common to both species. More than 750 SSR alleles and 1402 single nucleotide polymorphic (SNP) alleles discovered in 3.3Mb of sequence (Gingle *et al.*, 2007) are freely available from the Comparative Grass Genomics Center relational database. In a grain sorghum panel of eight accessions, Nelson *et al.* (2011) reported 283,000 SNPs, which will greatly help in designing genotyping strategies. The sorghum SNP and indel data have been deposited in dbSNP under handle JCNLAB_KSU, with accession number ranges starting at 410962044 (SNPs) and 411578970 (indels).

Genetic synteny among C_4 grasses means that gene discovery in any one can readily be translated to genetic improvement of all bioenergy grass crops (Carpita and McCann, 2008; Vermerris, 2011). The diploid genomes of representative genotypes of sorghum (Paterson *et al.*, 2009) and maize (Schnable *et al.*, 2009) have been sequenced. Plant geneticists and breeders can now use genome-wide association studies (Gore *et al.*, 2009) to identify genes that are responsible for desirable phenotypes, and use genomic selection (Jannick *et al.*, 2010) to readily incorporate favourable alleles into breeding lines. Large panels of maize and sorghum accessions representing much of the worldwide genetic diversity for these species have been densely genotyped and offer powerful experimental systems for association genetics approaches to discover and validate key alleles relevant to biomass production (Buckler *et al.*, 2009; McMullen *et al.*, 2009).

Application of recent advances in phenotyping to the available populations for association genetics (Springer *et al.*, 2009; Springer and Jackson, 2010) can accelerate gene discovery (Tuberosa and Salvi, 2006; Famoso *et al.*, 2010), as demonstrated with the maize Nested Association Mapping resource. Such approaches are already in progress for traits important to bioenergy grasses. Spectroscopic and spectrophotometric screens offer a high-throughput method to characterize variation in biomass composition, even in the absence of visible phenotypes (McCann *et al.*, 2007;

Penning *et al.*, 2009). Grass species produce three primary forms of harvestable carbon: starchy grains, soluble sugar in the stem and lignocellulosic biomass. Molecular phenotyping has provided insights into the genetic basis for developmental and metabolic regulation of carbon partitioning and storage, which is essential to rapidly optimize feedstocks, either alone or in combination. Finally, detailed studies of root architecture help characterize novel traits such as root-system carbon mass, and provide greater insight into nutrient-use efficiency, drought tolerance and toxic-metal stress (Zhao *et al.*, 2004).

In a recent study, Murray *et al.* (2008b) have identified QTLs for grain and stem sugar composition and yield, and their results indicated that overall energy yields could be increased by concurrent improvement for both sorghum grain and sugar traits. Elucidating the genetic basis of stem sugar and stem juice accumulation, modifying cell wall composition so that sorghum biomass can be processed more efficiently, maximizing biomass yield for a given geographic area and production system, and understanding the different mechanisms underlying drought tolerance are the main focus areas among sorghum researchers who target bioenergy traits.

9.5.2 Genetic transformation

With the availability of the complete genome sequence in sorghum (Paterson *et al.*, 2009), it is also possible to explore functions of the reported genes using transgenic approaches. Transformation also offers a route to broaden the input and output traits for sorghum breeding programmes. The candidate traits for sweet sorghum improvement using genetic transformation are discussed briefly below.

Stalk sugar accumulation

Transgenic approaches to improve stem sugar accumulation have not been attempted in sweet sorghum. However, differential expression of some genes related to sucrose metabolism has been observed between sweet sorghum and grain sorghum (Qazi *et al.*, 2012). Further, mature internodes of sweet sorghum showed a lower expression of sucrose transporters suggesting that sucrose accumulation may result from lower transport of sucrose from sink tissues. These genes could serve as important candidate genes for transforming sorghum to achieve better stem sugar yields. However, genetic manipulation of some key enzymes involved in sucrose metabolism did not bring about greater sucrose accumulation in the mature internodes of sugarcane, suggesting their inadequacy in overcoming the osmotic limits of the sugar-storing vacuoles (Wu and Birch, 2010). Sugarcane transgenics expressing a bacterial sucrose isomerase gene or a sucrose–sucrose fructosyl transferase gene led to synthesis of isomaltulose and β-2,6-linked fructans, which are not normally found in sugarcane (Arruda, 2011). Synthesis of these less osmotically active carbohydrates in the vacuoles of storage parenchyma cells removed osmotic constraints and led to greater sucrose accumulation. A microRNA miR169 was recently shown to be involved in regulating sugar levels in sweet sorghum stems (Calvino *et al.*, 2011), suggesting epigenetic regulation of sucrose accumulation, which could be tackled using transgenic techniques.

Abiotic and biotic stress tolerance

The productivity of sweet sorghum is limited by water availability and the crop requires frequent irrigation during its growth period to achieve economically viable yields of grain and sugar. Stress-induced signalling intermediates and transcription factors are known to regulate expression of a large number of diverse downstream genes and have emerged as potential candidate genes for plant transformation. Transformation of sorghum with signalling intermediates such as calcium-dependent protein kinases did not help in improving abiotic stress tolerance, probably due to pleiotropic effects of this general signalling intermediate (Mall *et al.*, 2011). Transgenics expressing transcription factor coding genes *DREB* (Dubouzet *et al.*, 2003), *MYC, MYB* (Abe *et al.*, 2003) and *WRKY* (Wang *et al.*, 2007) were reported to show improved stress tolerance in crop plants, but no sorghum transgenics for these transcription factors has been reported so far. Cross-talk between drought and carbohydrate metabolism has been indicated by the discovery that promoters of drought-induced transcription factors (DREB proteins) show an over-representation of motifs related to sugar signalling (Srivastava *et al.*, 2010). Alteration in the expression levels of these transcription factors through genetic transformation may provide a useful means of improving drought tolerance and manipulating sugar metabolism in sorghum, which is yet to be tested.

Sweet sorghum plants are subjected to attack by a large number of pests, pathogens and parasitic plants such as striga (Aly, 2007; Maqbool *et al.*, 2001; Guo *et al.*, 2011). The success of the *Bt* technology in maize (Armstrong *et al.*, 1995; Barry *et al.*, 2000) and cotton (Cattaneo *et al.*, 2006) is a strong rationale for the evaluation of this technology in this feedstock as a means to combat specific target insect pests. Tolerance against insect pests (spotted stem borer) was reported in sorghum transgenics carrying the *Cry1Ac* gene (Girijashankar *et al.*, 2005). Methods to control fungal infestations such as anthracnose, including expression of chitinase genes from *Trichoderma*, have led to improved tolerance to this fungal disease of sorghum (Kosambo-Ayoo *et al.*, 2011). A number of viral agents have been shown to be capable of replication in sorghum (Jensen and Giorda, 2008), which includes sugarcane mosaic virus, maize dwarf mosaic virus and sorghum mosaic virus. Introduction of viral coat protein or replicase genes in transgenic plants has been shown to lead to virus resistance (Abel *et al.*, 1986; Nelson *et al.*, 1987; Stark and Beachy, 1989) and offers great potential for the introduction of durable virus resistance for sorghum. Striga is a parasitic plant species known to infect sorghum and other cereals (Aly, 2007). Silencing a critical gene in the parasitic plant's life cycle using RNAi technology has been successfully used in the *Orobanche aegyptiaca*/tomato host parasite interaction (Aly *et al.*, 2009), but such transgenic approaches have not been used in sorghum so far.

Delayed flowering

Delayed flowering enables sufficient biomass to be built before transition to the reproductive phase and therefore correlates to improved yields. MicroRNAs (miR172) that regulate flowering time have been identified in sorghum and over-expression of these miRNAs could be a useful strategy to delay flowering in sorghum (Calvino *et al.*, 2011). A PSEUDO RESPONSE REGULATOR 37 (SbPRR37) gene has been identified in sorghum, which inhibits floral induction through the suppression of expression of various floral activators. The expression of SbPRR37 was found to be light-dependent and under control of the circadian clock and provides the possibility of regulating flowering time through manipulating its expression (Calvino *et al.*, 2011). Genes associated with reproductive transition/inflorescence branching (*Sb-lfy* and *Sb-tfl*) and spikelet determinacy (*Sb-bd1* and *Sb-ids*) have been identified in sorghum, which also may serve as important candidate genes in manipulating flowering (Dwiwedi *et al.*, 2008).

Digestibility of sorghum grains

The grain-mould infected sorghum grain can be used for conversion to ethanol. The prolamin seed storage proteins of sorghum, called kafirins, are assembled into protein bodies, with a very defined pattern, which influence digestibility of sorghum protein. Oria *et al.* (2000) described a high digestible, enhanced lysine sorghum mutant in which the protein bodies were highly folded, with a redistribution of the different kafirins, which translated to a phenotype with increased digestibility (Duodu *et al.*, 2003). Like the maize *floury-2* and *opaque-2* mutants, the high digestible, enhanced lysine mutant of sorghum has value in both food and feed applications. However, there is a tendency of these altered prolamin grains to have reduced agronomic properties, and postharvest issues (Huang *et al.*, 2004). A transgenic approach to modulate the seed storage proteins, without negatively altering the endosperm characteristics, has emerged as an important target for improving digestibility of sorghum.

9.5.3 Genetic transformation methods

Genetic transformation of sorghum has been attempted using *Agrobacterium*-mediated

(Zhao *et al.*, 2000; Howe *et al.*, 2006; Arulselvi *et al.*, 2010) and particle bombardment methods (Girijashankar *et al.*, 2005; Raghuwanshi and Birch, 2010; Liu and Godwin, 2012). The explants used for transformation include immature embryo tissues (Howe *et al.*, 2006; Gurel *et al.*, 2009; Liu and Godwin, 2012) and shoot meristems (Devi *et al.*, 2004; Pandey *et al.*, 2010). Use of explant tissues having young cells, whose cell walls have a higher number of sites for *Agrobacterium* attachment (Verma *et al.*, 2008), heat shock treatments to the explants (Gurel *et al.*, 2009) and addition of polyvinyl pyrrolidone to the regeneration media on which the explants were cultured (Lu *et al.*, 2009) were shown to bring about more efficient transformation. Safer selectable markers for transformation have been used such as the bacterial phosphomannose isomerase (*Pmi*) gene, which enables the transformed but not the non-transformed plants to metabolize mannose (Gurel *et al.*, 2009). Methods to obtain marker-free transgenics of sorghum have also been developed (Lu *et al.*, 2009). Native sorghum promoters that would be more effective in driving the expression of transgenes have been identified, which include the sucrose synthase gene promoter (Sivasudha and Kumar, 2008), a meristem-specific promoter (Verma and Kumar, 2005) and a wound-inducible promoter (Girijashankar *et al.*, 2005). Transformation efficiency reported in sorghum ranges from 0.01 to 1.3% using biolistic transfer and from 0.8 to 7.6% using *Agrobacterium*-mediated transfer (Raghuwanshi and Birch, 2010), which is however still far lower than that achieved in other crops such as maize and rice.

With efficient sorghum transformation systems now being available, transgenic approaches to improve agronomic traits in sweet sorghum are not too distant a goal.

9.6 Breeding Sweet Sorghum

9.6.1 Breeding behaviour

Sorghum is basically a self-pollinating crop but natural cross-pollination varies from 0.6 to 6% depending on the cultivar. Sorghum has the advantage of possessing complete self-pollination due to its floral biology, cleistogamy and genetic and cytoplasmic genetic male sterility. Breeding methods relevant to self- as well as cross-pollinated crops are therefore applied to breed pure line varieties, hybrids and populations in sorghum. Stigmas exposed before the anthers dehisce are subjected to cross-pollination. Hand pollination should begin around 9:30 or 10am and can be extended up to 11:30am to 12:30pm on a foggy morning (House, 1980).

9.6.2 Candidate traits and variability

The major characteristics that a sweet sorghum cultivar should possess are:

1. High biomass productivity (45–80 t ha^{-1}).
2. High Brix% (18–20%).
3. Thick stems and juicy internodes with maintenance of stem juiciness until maturity.
4. Photo- and thermo-insensitivity so that it can be grown throughout the year and fit into diversified cropping systems.
5. Tolerance to shoot pests and diseases.
6. Good digestibility of residues when used as forage or for lignocellulosic ethanol production.
7. Tolerance to mid-season and terminal drought.
8. High water and nitrogen-use efficiencies.
9. Suitability for specific conversion technologies (*bmr*).
10. Grain yield (3.0–5.0 t ha^{-1}).

Ayyangar *et al.* (1936) suggested a single dominant gene conferring the non-sweet character. Guiying *et al.* (2000) reported that stalk sugar is under the control of recessive genes with additive and dominance effects. On the contrary, later studies provided support for the existence of multiple genes with additive effects. Continuous variation in the amount of extractable juice was observed in juicy genotypes and inbred progeny of juicy × dry lines, suggesting multiple genes may be involved in controlling the trait (Saballos, 2008).

Recent studies suggest the involvement of several genes affecting the biofuel traits in the sweet sorghum background. The evaluation of four promising sweet sorghum lines ('Keller', 'BJ 248', 'Wray' and 'NSSH 104' ('CSH 22SS') along with the check 'SSV 84' indicated substantial genotypic differences for extractable juice, total sugar content, fermentation efficiency and alcohol production (Ratnavathi *et al.*, 2003). An analysis of 53 ICRISAT-bred elite hybrids in both the rainy and post-rainy seasons showed that the correlation and regression coefficients are significantly high for all the component traits of sugar yield (Brix%, stalk yield, juice weight and juice volume) (Srinivasa Rao *et al.*, 2009). Knowing general (GCA) and specific (SCA) combining ability effects of genetic materials is of practical value in breeding programmes. GCA effects represent the fixable component of genetic variance, and are important to develop superior genotypes. SCA represents the non-fixable component of genetic variation and it is important to provide information on hybrid performance. The line × tester analysis of 171 hybrids along with their parents in both rainy and post-rainy seasons showed that the magnitude of SCA variance was higher, suggesting the importance of non-additive gene action in inheritance of sugar yield-related traits though both additive and dominant genes controlled overall sugar yield during both rainy and post-rainy seasons in tropical sweet sorghums. Hence, selection in early generations would be ineffective and recurrent selection with periodic intercrossing is advocated. However, breeding for good combining restorer parents can result in high sugar yield in the post-rainy season. There is an indication of existence of transgressive segregation for sugar yield that can be exploited (Reddy *et al.*, 2011). The heritability for traits such as stem juiciness, sugar concentration in stems, total sugar, juice glucose, juice fructose and juice sucrose was low (Murray *et al.*, 2008a).

The predominant role of non-additive gene action for plant height, stem girth, total soluble solids, millable stalk yield and extractable juice yield, substantial magnitude of standard heterosis for all the traits related to ethanol production (stem girth: up to 5.3%, total soluble solids: up to 7.4%, millable stalk yield: up to 1.5% and extractable juice yield: up to 122.6%) indicates the importance of heterosis breeding for improving these traits (Sankarapandian *et al.*, 1994). The significant positive correlation of GCA effects with per se performance of parents in sweet sorghum facilitates quicker identification and development of sugar rich, high biomass yielding hybrid parents (Selvi and Palanisamy, 1990). Makanda *et al.* (2009) and Srinivasa Rao *et al.* (2009) reported significant GCA effect for stem Brix and associated traits, implying the importance of additive gene action.

The generation mean analysis of two crosses has shown predominant additive gene action for traits such as sucrose percentage and Brix% of juice. However, for cane and juice yield, dominance gene action and dominance × dominance gene interaction were of higher magnitude in both the crosses. Since the traits important for high sugar content have dominance and over-dominance inheritance, utilization of hybrid vigour by developing sweet sorghum hybrids is an attractive option. Also one of the parents with high sucrose content will suffice in obtaining good hybrids with high sugar and juice yield (AICSIP, 2007).

From these studies, it is quite evident that significant diversity exists in traits important for biofuel production and this opens up excellent opportunities for sweet sorghum improvement. Biofuel traits are governed by multiple genes and both additive and dominance components of gene action have to be exploited while breeding for high stalk sugar and juice-yielding genotypes.

9.6.3 Breeding objectives

In general, the sweet sorghum breeding programmes would aim at development of hybrid parents and hybrids which can address both first and second generation (lignocellulosic feed stock development) biofuel production issues. The breeding objectives would be:

1. To develop sweet sorghum female parents with high stalk sugar and grain yield apart from brown midrib trait.
2. To develop restorer lines/varieties with high sugar content, brown midrib trait and resistance to stem borer and shootfly.
3. To develop and identify sorghum hybrids (amenable for mechanical harvesting) with high biomass suitable for use in bioethanol and bioenergy production.

9.6.4 Breeding methods

The most commonly used programmes in sweet sorghum improvement are short-term programmes (pedigree method and backcross) and long term programmes (population improvement methods).

Short-term approach

The most common short-term approach in sweet sorghum breeding has been elite × elite crosses followed by pedigree selection. Breeding new female lines, B and R-lines has increasingly become dependent on crossing elite by elite lines, B × B and in some cases such as improving for resistance B × R lines. In case of male lines (R-lines) improvement, it is R × R crosses. This process progressively narrows the genetic base of breeding programmes and requires new traits, especially resistances, to be brought in by pre-breeding and often backcrossing. The success of a backcrossing programme depends on the precision with which the desired trait can be identified and thus introgressed into the recurrent parent through backcrossing.

PEDIGREE METHOD. The pedigree breeding method is the most commonly used method of breeding in sorghum, where the selection begins in the F_2 generation targeting superior plants that are expected to produce the best progenies. Hybrids between diverse parents segregate for a large number of genes and every F_2 individual is genetically different from each other individual. The population size becomes crucial for the success of recovering desirable genotypes, when several genes are involved. In this method (Fig. 9.3), superior individual plants are selected in successive segregating generations from the selected families and a complete record of parent progeny relationship is maintained. Identifying a potentially good cross is essential since best F1 parents give better yielding F_4 progenies. The selection in segregating generations should be based on: (i) performance of the families of the selected cross on the whole; and (ii) the individual plant's performance within the selected family.

Selection for many of the specific selection criteria encompassing various traits such as tallness, stem thickness, juice yield, etc. can be rapidly applied in the first two or three segregating generations since crosses between elite lines produce a high proportion of progeny with desirable specific values. Once the promising lines have been identified, they can

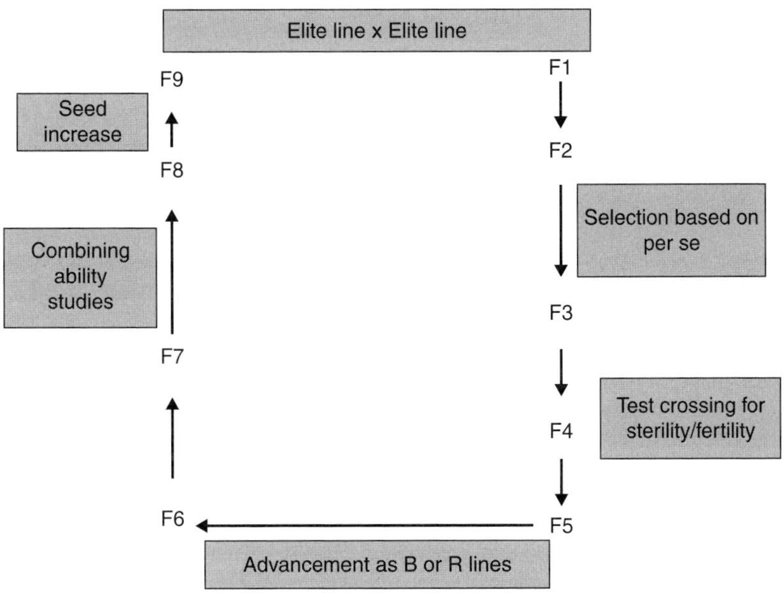

Fig. 9.3. Development of new sweet sorghum genotypes by the pedigree method.

be test crossed on to male-sterile lines for checking fertility restoration and may be classified as B or R lines. Lines with high biomass yield and other desirable agronomic characters can be released as varieties. The pedigree method has been utilized to create new recombinants, transfer of few to many genes governing resistances to various insects, diseases, cold tolerance etc. in sorghum. In India, the important sweet sorghum genotypes released through the pedigree method of selection are 'SSV 84' and 'CSV 19SS'.

BACKCROSS METHOD. This method does not offer an opportunity to provide new recombinants and thus they cannot be fixed. However, it can be utilized to incorporate BMR or specific defence (e.g. stem borer resistance) (Fig. 9.4) or improve other traits such as seed size, seed shape and cold tolerance through repeated backcrosses. The backcross method has also been successfully employed in the Indian and ICRISAT breeding programmes for transfer of BMR genes and genes that confer high digestibility into elite dual-purpose varieties. Several BMR lines in sweet sorghum backgound, stacked BMR mutants, stem borer-tolerant lines etc. have been developed through this method. Several staygreen QTLs (*stgB, stg2 and stg3*) are being introgressed into elite sweet sorghum cultivars by deploying this method.

Long-term approach

When the objective is to introgress new desirable genes distributed in many source lines, population improvement methods are used. Population improvement methods, besides offering greater opportunities for recombination to break linkages between desired and undesired traits, provide scope for increased utilization of biotic and abiotic stress resistant, but agronomically non-elite source germplasm lines. The population improvement provides a long-term breeding strategy to derive diverse and broad genetic-based superior varieties/ hybrid parents. Therefore, a comprehensive crop improvement strategy has to combine both short- and long-term progress for continuous improvement of economic traits (Reddy

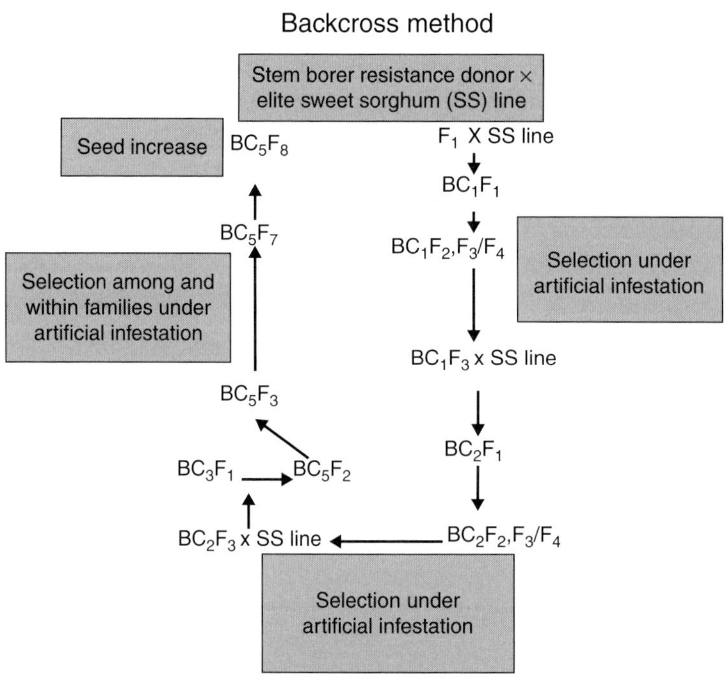

Fig. 9.4. Transfer of resistance genes by the backcross method into elite sweet sorghum lines.

et al., 2006). The population improvement procedure involves selection of component parents with high GCA, incorporation of genetic male sterility, intercrossing and random mating among parents and applying appropriate recurrent selection schemes. At ICRISAT-Patancheru, 24 sorghum populations encompassing characters such as grain mould, good grain, photo-insensitive, early dual-purpose, etc. were developed and maintained. Recently ICRISAT has started developing a sweet sorghum population with ms_3 gene for applying recurrent selection. While population improvement programmes are not the most common in sweet sorghum breeding, they are an important source of genetic variation and improved traits (Rooney and Smith, 2000).

9.7 Crop Agronomy and Value Chain Integration

9.7.1 Crop agronomy

The already standardized agronomic practices for grain sorghum are not entirely applicable to sweet sorghum because sweet sorghums produce more biomass along with sugars. Developing improved ecoregion-specific agrotechnology and pre- and post-harvesting stalk juice quality studies are the urgent priority. Moreover, the commercial viability of industry hinges upon raw material (sweet sorghum) availability for most part of the year. The adaptation (general and specific) of improved cultivars to different regions and seasons needs to be identified owing to high genotypic main effect plus genotype-by-environment (GGE) interaction of sugar yield (Srinivasa Rao *et al.*, 2011a) and its competent traits as described earlier. Standardization of optimized spacing (45 × 15 cm/60 × 15 cm/75 × 15 cm), fertilizer application (80–100 kg N ha^{-1}, 30–50 kg P_2O_5 ha^{-1}), intercultural operations (thinning, weeding, soil mulch), irrigation schedule (both alfisols and vertisols are rain fed in the rainy season and require up to five irrigations in the post-rainy season), harvesting timing and methodology will greatly enhance the productivity of sweet sorghum. In some areas response to micronutrients (B, Zn and S) in juice yield and quality was observed (Srinivasa Rao *et al.*, 2011b). Even though cultivars differ in biomass and grain production and hence differ in N uptake, the crop seems to be insensitive to the mineral nitrogen supply and seems to have a great potentiality in semi-arid environments in terms of yield production (Cosentino *et al.*, 2012). The grain and sugar yields are highest in the rainy and summer seasons, whereas in the post-rainy season the grain yield is high, but with less stalk and sugar yield. However, the results from tropical and temperate crosses have helped in deriving few post-rainy season cultivars at ICRISAT. In Brazil efforts are being made to grow sweet sorghum in a period where stalks are harvested before and after the sugarcane season so as to extend the period of operation of the distillery. The present-day multi-feedstock distilleries can successfully run on a variety of feedstocks. Therefore, studies on intercropping or relay cropping with cassava, sugarcane, sugarbeet, soybean, jatropha, pongamia, etc. are required to enhance the period of raw material availability. Agronomic and physiological measures aiding in increasing the period of industrial utilization (PIU) of sweet sorghum (e.g. customized fertilizer application, irrigation at physiological maturity, spraying gibberellic acid (GA), ethrel, solubar, etc., or soil application of micronutrients or other amendments to delay maturity, etc.) will further strengthen sweet sorghum as a biofuel/industrial crop. Rapid sugar accumulation immediately after flowering and its retention for a longer period for staggered feedstock supply is another area of research that deserves immediate attention.

As the global climate is gradually changing to higher temperatures and sweet sorghum is bound to grow in new areas, thermo-photo-insensitive non-lodging cultivars that are resistant to multiple pests and diseases will need to be developed. Breeding of short, mid–late and late maturing genotypes is necessary in order to have a broad harvest window in sweet sorghum, and thus providing raw material to the distillery over a long period. Proper planning of sowing a mix of these cultivars in the catchment area of a distillery would help to achieve more commercial stalk sugar/ethanol. When cultivars with different maturity groups are grown in an area, pests such as shootfly and midge may be likely to infest late-maturing cultivars. Therefore, breeding for tolerant cultivars for these insects is needed. The sorghum crop is traditionally challenged by marginal lands with poor fertility status and poor moisture-holding capacity and sweet sorghum also encounters similar problems. Sporadic water inundation due to excessive rains/ floods also becomes an unforeseen constraint. The self-fermentation of juice inside the stalk prior to juice extraction is a major concern, particularly when juice extraction is delayed after harvest due to the long distance between factory and the field. Preliminary results indicated that there will be reduction of sugar yield by 16.8% if the juice extraction is delayed by 48 h (Srinivasa Rao *et al.*, 2012). Research should address the postharvest losses in terms of juice quality and quantity.

9.7.2 Value chain integration

The sweet sorghum feedstock supply chains have two primary models of operation.

1. The centralized model: the sweet stalk is directly supplied to the plant from the farmers' fields, and the juice is extracted and fermented to ethanol and allied co-products. Its operational area is generally limited to a 40–50 km radius around the plant owing to high transportation costs involved in bulky raw material supply. Examples of such centralized plants include Tata Chemicals Ltd, Nanded, Maharashtra, India and ZTE Ltd, Inner Mongolia, China. A few of the sugar mills in Brazil started utilizing the sweet sorghum for bioethanol production in 2011/12.

2. The decentralized model: the decentralized crushing units (DCU) are key to centralized distillery as supply of syrup during the off season will augment the longer period of operation and this is viable in case of crop production >50 km away from the central distillery. In simple terms a DCU comprises the crusher and boiling unit, and essentially crushes the stalks to extract juice. The extracted juice is either concentrated to syrup or fermented *in situ* to alcohol. Sweet sorghum is a seasonal crop that in India can be cultivated in three seasons a year (rainy, post-rainy and summer) to supply raw material for 3–4 months annually for ethanol production (Kumar *et al.*, 2010). The grain and sugar yields are best in the rainy and summer seasons, whereas in the post-rainy season the grain yield is high, but with less stalk and sugar yield. A commercial ethanol distillery requires feedstock year round – for at least 10 months annually – for economical operation. However, in regions with short harvest windows, smaller hectarages or with low plantation densities, a typical centralized model with a 30 kilolitres (kl) day^{-1} processing plant dedicated to sweet sorghum ethanol production could operate only seasonally, requiring a high capital investment that might not be cost effective. In areas with low plantation densities, the transportation costs associated with supplying the plant with sweet sorghum feedstock become prohibitive. Transportation costs are a significant cost factor in all sweet sorghum models studied, with costs ranging from US$34 to US$107 t^{-1} fermentable carbohydrates (Bennett and Anex, 2009). In view of the need for a regular supply of feedstock to the distillery, it is widely believed that DCUs help in sustainability of the supply chain by feeding syrup (60% Brix) to the distillery in the off season. The following gaps were identified in DCU sustainability.

- At present, there is a very limited period of operation of the crushing unit (less than 20–25 days) as the cultivar maturity window is not large. Research should aim at developing sweet sorghum genotypes with adaptability across seasons and months of the year.
- DCUs are being operated only for the rainy season crop (June–September). The post-rainy and summer season crops require an assured irrigation source, thereby increasing the cost of cultivation. Hence, the need for developing post-rainy season-adapted cultivars with drought tolerance.
- Juice extraction efficiency and syrup conversion efficiency are low. A scenario analysis conducted at ICRISAT showed that improving these even by 5% has significant bearing on the economics of the whole value chain.
- As syrup is the main product of a DCU, its quality parameters need to be improved to

meet the requirements of diverse end users (such as suitability for use in food, beverage and pharmaceutical industries). Research also needs to focus on improving organoleptic characteristics.

- Commercial dairies are increasingly using the fresh bagasse, after chopping, to feed cattle. Education and training is needed for farmers to raise awareness of the multiple uses of bagasse, such as for feed block making, ensiling or biocomposting.
- Little or no information is available on the utilization of co-products such as vinasse, steam, foam and froth. Therefore, efforts are needed in using steam for heating or boiling the juice, and in exploring the use of nutrient-rich vinasse, foam and froth as livestock feed and biofertilizers.
- Capacity building of staff at every step – not only syrup production, but also co-product utilization – would go a long way toward improving the operational efficiency and economic viability of DCUs.
- The varied products and co-products of the DCU need to be positioned to exploit locally existing market opportunities, i.e. an inclusive market-oriented development (IMOD) approach, as this brings the DCU closer to the rural farming communities.

9.8 Commercialization – Status and Bottlenecks

Experiences with sweet sorghum in India, the USA and China have shown that the crop has high potential as a bioenergy feedstock, with several opportunities for immediate use as a complementary feedstock and seasonal low-cost feedstock (mould-affected grain). Regions with a warm climate, large tracts of land and a system similar to sugarcane processing should work well for sweet sorghum. Further, the crop can be grown in regions of the world where sugarcane cannot be cultivated. Government policy support is necessary for utilization of this novel feedstock for commercial bioethanol production. Since 2007, a handful of distilleries across the world have started using sweet sorghum for ethanol production on a commercial scale. M/s Rusni distillery, the first sweet sorghum distillery, was established in 2007 at Sangareddy, Medak district of Andhra Pradesh, India, which is amenable to multiple feedstock. It has a capacity of 40 kilolitres (kl) day^{-1} and produces fuel ethanol (99.6% alcohol), Extra Neutral Alcohol (ENA) (96%) and pharma alcohol (99.8%) from agro-based raw materials such as sweet sorghum stalks (juice), moulded grains, broken rice, cassava and rotten fruits. This is not under operation currently for different reasons. Another, 30 kl day^{-1} Tata Chemicals Limited distillery in Nanded, Maharashtra, India, started operations in 2009 solely based on sweet sorghum and produced 90 kl of transport grade ethanol in 2010. Several companies in China (including Liaoning Guofu Bioenergy Development Company Limited, Binzhou Guanghua Biology Energy Company Ltd, Jiangxi Qishengyuan Agri-Biology Science and Technology Company Ltd, Xinjiang Santai Distillery, Jilin Fuel Alcohol Company Limited, Heilongjiang Huachuan Siyi Bio-fuel Ethanol Company Ltd, ZTE Agribusiness Company Limited and Fuxin Green BioEnergy Corporation) have conducted large-scale sweet sorghum trials. In 2010, ZTE Agribusiness Company Limited, Wuyuan County, Inner Mongolia and Fuxin Green BioEnergy Corporation, Heishan County, Shenyang province used sweet sorghum as feedstock to produce ethanol. The Chinese government is encouraging sweet sorghum processing industries by offering a subsidy to both growers and processors. In the Philippines San Carlos Bio-Energy Incorporated, Negros produced ethanol from sweet sorghum. In 2010, grain sorghum production in the USA was 11.6 Mt (http://faostat.fao.org), equivalent to approximately 457 million bushels, and 20% of those were used for ethanol production (http://www.sorghumgrowers.com). EnviroFuels LLC in the USA is currently in the process of developing a 20 million gallon (75,708,235 l) per year sugar-based ethanol plant in Highlands County, Florida using sweet sorghum as the primary feedstock. Sugarcane will be used to supplement the feedstock base in the winter months when sweet sorghum is not available. The facility will be designed for a future expansion to 40 million gallons (151,416,471 l) per year. In Tennessee (USA), BioDimensions Industrial Sugar Platform Development Company planted and harvested about 75 ha of sweet sorghum. A portion of the sugars was fermented to ethanol, while the bagasse was used for both fuel pellets and animal feed. The group expects to distill about 5000 gallons (18,927 l) of hydrous ethanol, much of which will be used in industrial ethanol engines. In Brazil, seed company Ceres Inc. has established a subsidiary focusing on developing sweet sorghum as a feedstock for the ethanol industry. The company's goal is to be the first supplier of new hybrids with high levels of sugar. The company is currently working with multiple ethanol mills, technology providers and equipment companies to facilitate the introduction of sweet sorghum hybrids into existing ethanol mills.

To make the value chain sustainable, some of the major challenges to be addressed include the following:

1. Convincing and educating farmers on the cultivation of sweet sorghum for higher returns.
2. Mechanization in sweet sorghum cultivation and crushing activities.
3. Supply chain innovations to reduce the time lag between the harvesting and crushing of stalks both in centralized and decentralized areas and continuous supply of sweet sorghum stalk and syrup to the distillery to run at optimum capacity.
4. Increasing the juice storability and fermentation efficiency, increasing process efficiency and reduction in investment and operational costs of syrup and ethanol production.
5. Exploring alternative markets for syrup such as food additives, pharmaceuticals, beverages, bakery and confectionery units.

9.9 Sustainability and the Way Forward

9.9.1 Sustainability

The major components of the biofuel life cycle are feedstock production, transportation to refinery, processing and conversion and distribution to retailers and customers (Fig. 9.5).

The amount of greenhouse gas (GHG) emission reduction and how this is calculated is also a matter of much debate. Authors such as Searchinger *et al.* (2008) and Fargione *et al.* (2008) argue that life cycle studies have failed to factor in land use change effect, such as carbon storage and sequestration sacrificed by diverting land from its existing uses for biofuels harms the environment rather than helping it. It is obvious that the above parameters of life cycle assessment (LCA) vary widely with respect to the location,

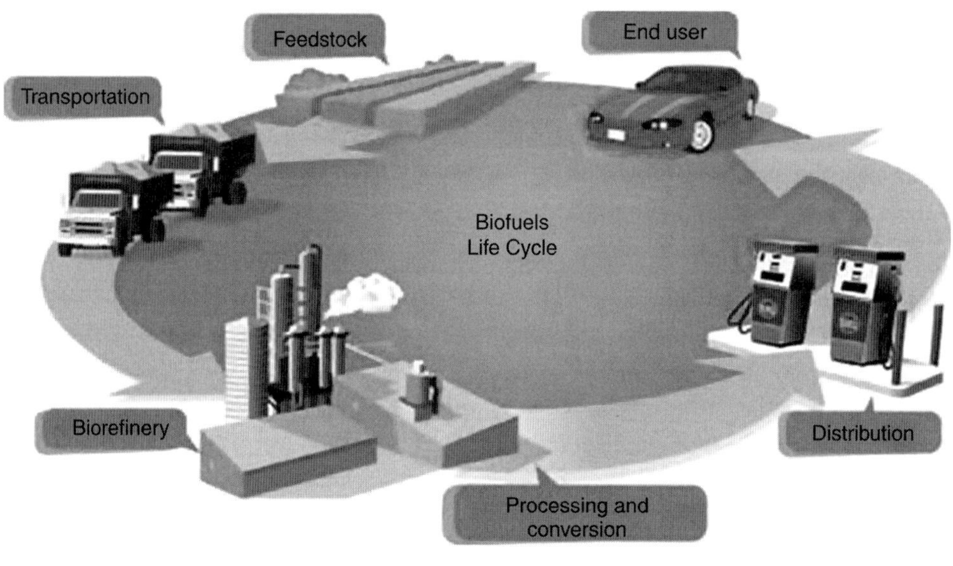

Fig. 9.5. Biofuel life cycle (US Department of Energy).

technology, methodology and several other factors either directly or indirectly affecting the value chain. Net energy ratio and GHG balance primarily decides the benefits of an energy value chain. However, it was reported that sweet sorghum has a higher net energy balance (3.63) compared to grain sorghum (1.50) and maize (1.53) (Wortmann *et al.*, 2008). Another report estimated an energy balance of 8 and carbon emission reduction by 86% (CII-DBT study, 2010). First- and second-generation bioethanol from sweet sorghum can contribute significantly to the conservation of fossil resources and to the mitigation of GHGs. If the crop is used for the production of ethanol (from grains and sugar) and green electricity (from surplus bagasse), 3500 l crude oil equivalents can be saved per hectare cultivation area. If both food from grains and ethanol from the juice are produced, 2300 l crude oil equivalents can be saved per hectare cultivated area. Regarding GHGs, between 1.4 and 22 kg CO_2 equivalents can be saved depending on yield, production methods and the land cover prior to sweet sorghum cultivation (Köppen *et al.*, 2009). For both categories, the exact values vary greatly with specific scenarios and local conditions. In general, the following parameters determine the results: type and efficiency of conversion technology, the use of byproducts (e.g. bagasse), the crop yield per cultivation area, land-use changes, as well as the type of fossil energy carriers that are replaced. Even if the seeds were used as food, bioethanol from the stem's sugar juice still shows clear advantages over fossil fuels. If both sugar and seeds were used as food, the respective conversion related energy and GHG expenditures could be compensated by producing second-generation ethanol from the bagasse. Even though the ethanol yield per unit weight of feedstock is lower for sweet sorghum compared to sugarcane, the much lower production costs and water requirement for this crop more than compensates for the difference, and hence it still returns a competitive cost advantage in the production of ethanol in India (Farrell *et al.*, 2006).

9.9.2 Way forward

Sweet sorghum has a low water demand and is especially advantageous in areas with water shortage (Srinivasa Rao *et al.*, 2012). It has a lower nitrogen fertilizer demand possibly due to traits such as BNI-capacity, reduces the risk of nutrient leaching and thus soil and water pollution, as well as making it well suited for small-scale farming. Its relatively short vegetation cycle allows sweet sorghum to be grown in double-cropping systems based on water availability, which in turn can lead to greater agrobiodiversity and a reduced demand for fertilizers and pesticides. Under intensive production practices, sweet sorghum production risks similar disadvantages as other intensive monocultures, such as soil degradation and loss or soil and water pollution due to more fertilizer and pesticide use. Establishing new sweet sorghum cultivation sites instead of integrating the crop into existing agricultural systems may lead to a loss of biodiversity, which is more detrimental for species-rich ecosystems. Like many other biofuel feedstocks, the sweet sorghum-based bioethanol value chain has some limitations with regards to location-specific LCA, certain emissions compared to its fossil equivalent due to poor productivity owing to poor management of the crop or submarginal ecologies where the crop is cultivated.

A limiting factor for its widespread cultivation is the limited availability of varieties/hybrids adapted to different agroclimatic conditions resisting both biotic and abiotic stresses, including colder climate. Consequently, research should address the optimization of sweet sorghum as an energy crop through breeding for enhanced productivity under limited available resources. Genetic improvement should focus on stalk sugar, biomass quantity and quality and general agronomic traits (such as water and nutrient use efficiency) and, in particular, adaptation of sweet sorghum to colder, arid saline, and alkaline conditions. Further improvement in Brix%, juice volume and stalk yield (≥45 t ha^{-1} with hybrids) should be targeted in sweet sorghum to help improve the benefits to the industry and farmers without any detrimental effect on grain yield. The juice volume should not be compromised while increasing the Brix%. The best way of selecting genotypes will be based on sugar yield per ha (a function of juice yield and Brix%) with 18–20% Brix as the base level in the restorers and 14–17% in the female parents in the rainy season, and 15–17% in R-lines/varieties and 12–16% in the female parents in the post-rainy season. Going by the high variability present in sorghum germplasm for Brix% (up to 23%) with a low Brix% observed in female (seed) parents (12–15%), there is an urgent need to improve the sugar content (Brix%) in seed parents through genetic enhancement. There is also a need to develop and evaluate cultivars producing high stalk yield per unit time, inputs, energy and land area in different agroclimatic areas of the country. Other research areas on quality and processing that need immediate attention include high ethanol yield, fermentation efficiency, diffusion, and diversified products from bagasse (power, pulp, biomanure, cattle feed, etc.).

Bioethanol from sweet sorghum (sorganol) is potentially a win–win solution. Sorganol will not be the unique solution, but will compliment other renewable sources of energy and contribute to addressing some of the problems, e.g. reduction of GHG emissions, improving air quality in large cities, reducing dependency on imported oil, creating jobs in rural areas and improving quality of life in developing countries. From an industry viewpoint, it is imperative to improve field and distillery integration as well as significant enhancement in processing and fermentation efficiency. From the experience gained over the last 3–4 years, a viable agro-enterprise of this massive scale will be commercially viable if it is a multi-feedstock distillery backed up by government policy support.

As the demand for biofuels rapidly expands, its associated production systems and supply chains are consolidating. Forward-thinking management systems could significantly enhance ecological sustainability and livelihood development, particularly for poor farmers in the developing world. International trade will be crucial to enlarge the share of bioethanol in future transport energy demand. All nations, irrespective of the development index, should join hands in formulation of policies that target the entire innovation chain to ensure that the development and use of biofuels in general and sorganol in particular follow an integrated pathway, which simultaneously targets climate change mitigation and adaptation, energy security and all-round sustainable economic development. The potential for significant genetic improvement of sorghum as a biofuel crop is excellent. Full exploration of the available genetic resources through plant breeding with the aid of molecular tools could dramatically increase biomass yield of sorghum and thus meet the demand of feedstocks for biofuel production without a significant impact on our food supply and natural environment.

References

Abe, H., Urao, T., Ito, T., Seki, M., Shinozaki, K. *et al.* (2003) *Arabidopsis* AtMYC2 (bHLH) and AtMYB2 (MYB) function as transcriptional activators in abscisic acid signaling. *Plant Cell* 15, 3–78.

Abel, P.P., Nelson, R.S., De, B., Hoffmann, N., Rogers, S.G., Fraley, R.T. *et al.* (1986) Delay of disease development in transgenic plants that express the tobacco mosaic virus coat protein gene. *Science* 232, 738–743.

AICSIP (All India Coordinated Sorghum Improvement Project) (2007) Sweet sorghum and physiology. *All India Coordinated Sorghum Improvement Project Annual Progress Report for 2006–2007.* AICSIP Tech. Publication No. 3, Sweet Sorghum and Physiology 2007 (Book 3 of 3-agm07 pre-meet), National Research Centre for Sorghum, Rajendranagar, Hyderabad, A.P., India, pp. 102.

Ali, M.L., Rajewski, J.F., Baenziger, P.S., Gill, K.S., Eskridge, K.M. *et al.* (2008) Assessment of genetic diversity and relationship among a collection of US sweet sorghum germplasm by SSR markers. *Molecular Breeding* 21, 497–509.

Almodares, A. and Hadi, M.R. (2009) Production of bioethanol from sweet sorghum: A review. *African Journal of Agricultural Research* 4, 772–780.

Aly, R. (2007) Conventional and biotechnological approaches for control of parasitic weeds. *Vitro Cell Developmental Biology* 43, 304–317.

Aly, R., Cholakh, H., Joel, D.M., Leibman, D., Steinitz, B. *et al.* (2009) Gene silencing of mannose 6-phosphate reductase in the parasitic weed *Orobanche aegyptiaca* through the production of homologous dsRNA sequences in the host plant. *Plant Biotechnology Journal* 7, 487–498.

Armstrong, C.L., Parker, G.B., Pershing, J.C., Brown, S.M., Sanders, P.R. *et al.* (1995) Field evaluation of European corn borer control in progeny of 173 transgenic corn events expressing an insecticidal protein from *Bacillus thuringiensis*. *Crop Science* 35, 550–557.

Arruda, P. (2011) Genetically modified sugarcane for bioenergy generation. *Current Opinion in Biotechnology* 23, 1–8.

Arulselvi, P.I., Michael, P., Umamaheswari, S. and Krishnaveni, S. (2010) *Agrobacterium* mediated transformation of *Sorghum bicolor* for disease resistance. *International Journal of Pharmacology and Biological Sciences* 1, 272–281.

Audilakshmi, S., Aruna, C., Garud, T.B., Nayakar, N.Y., Atale, S.B. *et al.* (2005) A technique to enhance the quality and market value of rainy season sorghum grain. *Crop Protection* 24, 251–258.

Ayyangar, G., Ayyar, M., Rao, V. and Nambiar, A. (1936) Mendelian segregation for juiciness and sweetness in sorghum stalk. *Madras Agricultural Journal* 24, 247–248.

Barry, B.D., Darrah, L.L., Huckla, D.L., Antonio, A.Q., Smith, G.S. *et al.* (2000) Performance of transgenic corn hybrids in Missouri for insect control and yield. *Journal of Economic Entomology* 93, 993–999.

Bennett, A.S. and Anex, R.P. (2009) Production, transportation and milling costs of sweet sorghum as a feedstock for centralized bio-ethanol production in the upper Midwest. *Bioresource Technology* 100, 1595–1607.

Bethel, C.M., Sciara, E.B., Estill, J.C., Bowers, J.E. and Hanna, W. (2006) A framework linkage map of Bermuda grass (*Cynodon dactylon* × *transvaalensis*) based on single-dose restriction fragments. *Theoretical and Applied Genetics* 112, 727–737.

Bhattramakki, D., Dong, J.M., Chhabra, A.K. and Hart, G.E. (2000) An integrated SSR and RFLP linkage map of *Sorghum bicolor* (L.) Moench. *Genome* 43, 988–1002.

Boivin, K., Deu, M., Rami, J.F., Trouche, G. and Hamon, P. (1999) Towards a saturated sorghum map using RFLP and AFLP markers. *Theoretical and Applied Genetics* 98, 320–328.

Bowers, J.E., Abbey, C., Anderson, S., Chang, C., Draye, X. *et al.* (2003) A high-density genetic recombination map of sequence-tagged sites for Sorghum, as a framework for comparative structural and evolutionary genomics of tropical grains and grasses. *Genetics* 165, 367–386.

Buchanan, C., Lim, S., Salzman, R.A., Kagiampakis, I., Morishige, D.T. *et al.* (2005) *Sorghum bicolor*'s transcriptome response to dehydration, high salinity and ABA. *Plant Molecular Biology* 58, 699–720.

Buckler, E.S., Holland, J.B., Bradbury, P.J., Acharya, C.B., Brown, P.J. *et al.* (2009) The genetic architecture of maize flowering time. *Science* 325, 714–718.

Calvino, M., Bruggmann, R. and Messing, J. (2011) Characterization of the small RNA component of the transcriptome from grain and sweet sorghum stems. *BMC Genomics* 12, 356–67.

Carpita, N.C. and McCann, M.C. (2008) Maize and sorghum: genetic resources for the bioenergy grasses. *Trends in Plant Science* 13, 415–420.

Casa, A.M., Mitchell, S.E., Hamblin, M.T., Sun, H., Bowers, J.E. *et al.* (2005) Diversity and selection in sorghum: concurrent studies with simple sequence repeats (SSRs). *Theoretical and Applied Genetics* 111, 23–30.

Cattaneo, M.G., Yafuso, C., Schmidt, C.W., Huang, C.-Y., Rahman, M. *et al.* (2006) Farm-scale evaluation of the impacts of transgenic cotton on biodiversity, pesticide use, and yield. *Proceedings of the National Academy of Sciences, United States of America* 103, 7571–7576.

Cheng, X., Jiang, H., Zhao, Y., Qian, Y., Zhu, S. *et al.* (2010) A genomic analysis of disease-resistance genes encoding nucleotide binding sites in *Sorghum bicolor. Genetics and Molecular Biology* 33, 292–297.

Chittenden, L.M., Schertz, K.F., Lin, Y.R., Wing, R.A. and Paterson, A.H. (1994) A detailed RFLP map of *Sorghum bicolor* × *S. propinquum*, suitable for high-density mapping, suggests ancestral duplication of sorghum chromosomes or chromosomal segments. *Theoretical and Applied Genetics* 87, 925–933.

CII-DBT Report (2010) *Estimation of energy and carbon balance for Bio-fuels in India*. 54 pp.

Cosentino, S.L., Mantineo, M. and Testa, G. (2012) Water and nitrogen balance of sweet sorghum (*Sorghum bicolor* Moench (L.)) cv. Keller under semi-arid conditions *Industrial Crops and Products* 36, 329–342.

Dahlberg, J.A. (2000) Classification and characterization of sorghum. In: Smith, C.W. and Frederiksen, R.A. (eds) *Sorghum: Origin, History, Technology and Production*. John Wiley and Sons, New York, USA, pp. 99–129.

de Wet, J.M.J. (1978) Systematics and evolution of *Sorghum* sect. *Sorghum* (*Gramineae*). *American Journal of Botany* 65, 477–484.

Devi, P., Zhong, H. and Sticklen, M. (2004) Production of transgenic sorghum plants with related *HVA1* gene. In: Seetharama, N. and Godwin, I.D. (eds), *Sorghum Tissue Culture and Transformation*, Oxford Publishers, New Delhi, India, pp. 75–79.

Dje, Y., Heuertz, M., Lefebvre, C. and Vekemans, X. (2000) Assessment of genetic diversity within and among germplasm accessions in cultivated sorghum using microsatellite markers. *Theoretical and Applied Genetics* 100, 918–925.

Dubouzet, J.G., Sakuma, Y., Ito, Y., Kasuga, M., Dubouzet, E.G. *et al.* (2003) OsDREB genes in rice, *Oryza sativa* L., encode transcriptional activator that function in drought, high salt and cold responsive gene expression. *Plant Journal* 33, 751–763.

Dufour, P., Deu, M., Grivet, L., Dhont, A., Paulet, F. *et al.* (1997) Construction of a composite sorghum genome map and comparison with sugarcane, a related complex polyploid. *Theoretical and Applied Genetics* 94, 409–418.

Duodu, K.G., Taylor, J.R.N., Belton, P.S. and Hamaker, B.R. (2003) Factors affecting sorghum protein digestibility. *Journal of Cereal Science* 38, 117–131.

Dwivedi, S., Perotti, E. and Ortiz, R. (2008) Towards molecular breeding of reproductive traits in cereal crops. *Plant Biotechnology Journal* 6, 529–559.

FAO (2012) Crops – FAOSTAT. Available at: http://faostat.fao.org/site/567/DesktopDefault.aspx?PageID=567#ancor (accessed 25 May 2012).

Famoso, A., Clark, R., Shaff, J., Craft, E., McCouch, S.R. *et al.* (2010) Development of a novel aluminum tolerance phenotyping platform used for comparisons of cereal Al tolerance and investigations into rice Al tolerance mechanisms. *Plant Physiology* 153, 1678–1691.

Fargione, J., Hill, J., Tilman, D., Polasky, S. and Hawthorne, P. (2008) Land clearing and the biofuel carbon debt. *Science* 319, 1235–1238.

Farrell, A.E., Plevin, R.J., Turner, B.T., Jones, A.D., O`Hare, M. *et al.* (2006) Ethanol can contribute to energy and environmental goals. *Science* 311, 506–508.

Feltus, F., Hart, G., Schertz, F., Casa, A., Brown, P. *et al.* (2006) Genetic map alignment and QTL correspondence between inter- and intra-specific sorghum populations. *Theoretical and Applied Genetics* 112, 1295–1305.

Gibson, S. (2010) Developing and improving sorghum as an energy crop. Presented at the *International workshop on Role of Lignin on Sorghum Biomass Quality for Energy and Animal Utilization*, 19–20 April 2010, EMBRAPA, Sate Lagoas, MG, Brazil.

Gingle, A.R., Yang, H., Chee, P., May, O.L., Rong, J. *et al.* (2007) An integrated web resource for cotton. *Crop Science* 46, 1998–2007.

Girase, J.R. (2010) Evaluation of the economic feasibility of grain sorghum, sweet sorghum, and switch grass as alternative feedstocks for ethanol production in the Texas panhandle. MS thesis, West Texas A&M University, Canyon, Texas, USA. pp. 101.

Girijashankar, V., Sharma, H.C., Sharma, K.K., Swathisree, V., Prasad, L.S. *et al.* (2005) Development of transgenic sorghum for insect resistance against the spotted stem borer (*Chilo partellus*). *Plant Cell Reports* 24, 513–522.

Gnansounou, E., Dauriat, A. and Wyman, C.E. (2005) Refining sweet sorghum to ethanol and sugar: economic trade-offs in the context of North China. *Bioresource Technology* 96, 985–1002.

Gore, M.A., Wright, M.H., Ersoz, E.S., Bouffard, P., Szekeres, E.S. *et al.* (2009) Large-scale discovery of gene-enriched SNPs. *Plant Genome* 2, 121–133.

Guiying, L., Weibin, G., Hicks, A. and Chapman, K.R. (2000) *A Training Manual for Sweet Sorghum*. FAO/CAAS/CAS, Bangkok, Thailand.

Guo, C., Cui, W., Feng, X., Zhao, J. and Lu, G. (2011) Sorghum insect problems and management. *Journal of Integrative Plant Biology* 53, 178–192.

Gurel, S., Gurel, E., Kaur, R., Wong, J., Meng, L. *et al.* (2009) Efficient, reproducible *Agrobacterium*-mediated transformation of sorghum using heat treatment of immature embryos. *Plant Cell Reports* 28, 429–444.

Hamblin, M.T., Mitchell, S.E., White, G.M., Gallego, J., Kukatla, R. *et al.* (2004) Comparative population genetics of the panicoid grasses: Sequence polymorphism, linkage disequilibrium and selection in a diverse sample of *Sorghum bicolor. Genetics* 167, 471–483.

Hamblin, M.T., Fernandez, M.G.S., Casa, A.M., Mitchell, S.E., Paterson, A.H. *et al.* (2005) Equilibrium processes cannot explain high levels of short- and medium-range linkage disequilibrium in the domesticated grass sorghum bicolor. *Genetics* 171, 1247–1256.

Haussmann, B.I.G., Hess, D.E., Seetharama, N., Welz, H.G. and Geiger, H.H. (2002) Construction of a combined sorghum linkage map from two recombinant inbred populations using AFLP, SSR, RFLP, and RAPD markers, and comparison with other sorghum maps. *Theoretical and Applied Genetics* 105, 629–637.

House, L.R. (1980) *A Guide to Sorghum Breeding*. ICRISAT, Patancheru, Andhra Pradesh, India, pp. 238.

Hoskinson, R.L. (2007) Screening for sugar and ethanol processing characteristics from anatomical fractions of wheat stover. *Biomass and Bioenergy* 31, 585–592.

Howe, A., Sato, S., Dweikat, I., Fromm, M. and Clemente, T. (2006) Rapid and reproducible *Agrobacterium*-mediated transformation of sorghum. *Plant Cell Reports* 25, 784–791.

Huang, S., Adams, W.R., Zhou, Q., Malloy, K.P., Voyles, D.A. *et al.* (2004) Improving nutritional quality of maize proteins by expressing sense and antisense zein genes. *Journal of Agricultural and Food Chemistry* 52, 1958–1964.

Jannick, J.-L., Lorenz, A.J. and Iwata, H. (2010) Genomic selection in plant breeding: from theory to practice. *Brief Functional Genomics* 9, 166–177.

Jensen, S.G. and Giorda, L.M. (2008) Virus diseases of sorghum and millet in the Americas and Australia. In: Leslie, J.F. (ed.) *Sorghum and Millets Diseases*. Iowa State Press, Ames, Iowa. DOI 10.1002/9780470384923.

Jessup, R.W., Burson, B.L., Burow, G., Wang, Y.W., Chang, C. *et al.* (2003) Segmental allotetraploidy and allelic interactions in buffel grass (*Pennisetum ciliare* (L.) Link syn. *Cenchrus ciliaris* L.) as revealed by genome mapping. *Genome* 46, 304–313.

Kim, J.-S., Klein, P.E., Klein, R.R., Price, H.J., Mullet, J.E. *et al.* (2005) Chromosome identification and nomenclature of *Sorghum bicolor. Genetics* 169, 1169–1173.

Klein, P.E., Klein, R.R., Cartinhour, S.W., Ulanch, P.E., Dong, J. *et al.* (2000) A high-throughput AFLP-based method for constructing integrated genetic and physical maps: progress toward a sorghum genome map. *Genome Research* 10, 789–807.

Köppen, S., Reinhardt, G. and Gärtner, S. (2009) *Assessment of Energy and Greenhouse Gas Inventories of Sweet Sorghum for First and Second Generation Bioethanol.* Food and Agriculture Organization of the United Nations, Rome, 77 pp.

Kosambo-Ayoo, L.M., Bader, M., Loerz, H. and Becker, D. (2011) Transgenic sorghum (*Sorghum bicolor* L. Moench) developed by transformation with chitinase and chitosanase genes from *Trichoderma harzianum* expresses tolerance to anthracnose. *African Journal of Biotechnology* 10, 3659–3670.

Kumar, C.G., Fatima, A., Srinivasa Rao, P., Reddy, B.V.S., Rathore, A. *et al.* (2010) Characterization of improved sweet sorghum genotypes for biochemical parameters, sugar yield and its attributes at different phenological stages. *Sugar Tech* 12, 322–328.

Liu, G. and Godwin, I.D. (2012) Highly efficient sorghum transformation. *Plant Cell Reports* DOI 10.1007/s00299-011-1218-4.

Lu, L., Wu, X., Yin, X., Morrand, J., Chen, X. *et al.* (2009) Development of marker-free transgenic sorghum [*Sorghum bicolor* (L.) Moench] using standard binary vectors with *bar* as a selectable marker. *Plant Cell Tissue and Organ Culture* 99, 97–108.

Maiti, R. (1996) *Sorghum Science*. Oxford and IBH Publishing, New Delhi, 352 pp.

Makanda, I., Tongoona, P. and Derera, J. (2009) Combining ability and heterosis of sorghum germplasm for stem sugar traits under off-season conditions in tropical lowland environments. *Field Crops Research* 114, 272–279.

Mall, T.K., Dweikat, I., Sato, S.J., Neresian, N., Xu, K. *et al.* (2011) Expression of the rice CDPK-7 in sorghum: molecular and phenotypic analyses. *Plant Molecular Biology* 75, 467–479.

Manli, L., Yuyama, N., Luo, L., Hirata, M. and Cai, H. (2009) *In silico* mapping of 1758 new SSR markers developed from public genomic sequences for sorghum. *Molecular Breeding* 24, 41–47.

Maqbool, S.B., Devi. P. and Sticklen, M.B. (2001) Biotechnology: advances for the genetic improvement of sorghum (*Sorghum bicolor* (L.) Moench). *In vitro Cellular and Developmental Biology – Plant* 37, 504–515.

McCann, M.C., Defernez, M., Urbanowicz, B.R., Tewari, J.C., Langwisch, T. *et al.* (2007) Neural network analyses of infrared spectra for classifying cell wall architectures. *Plant Physiology* 143, 1314–1326.

McMullen, M.D., Kresovich, S., Villeda, H.S., Bradbury, P., Li, H.H. *et al.* (2009) Genetic properties of the maize nested association mapping population. *Science* 325, 737–740.

Menz, M.A., Klein, R.R., Mullet, J.E., Obert, J.A., Unruh, N.C. *et al.* (2002) A high-density genetic map of *Sorghum bicolor* (L.) Moench based on 2926 AFLP(R), RFLP and SSR markers. *Plant Molecular Biology* 48, 483–499.

Menz, M.A., Klein, R.R., Unruh, N.C., Rooney, W.L., Klein, P.E. *et al.* (2004) Genetic diversity of public inbreds of sorghum determined by mapped AFLP and SSR markers. *Crop Science* 44, 1236–1244.

Ming, R., Liu, S.C., Lin, Y.R., da Silva, J., Wilson, W. *et al.* (1998) Detailed alignment of *Saccharum* and *Sorghum* chromosomes: comparative organization of closely related diploid and polyploid genomes. *Genetics* 150, 1663–1682.

Missaoui, A., Paterson, A.H. and Bouton, J.H. (2005) Investigation of genome organization in switchgrass (*Panicum virgatum* L.) using DNA markers. *Theoretical and Applied Genetics* 110, 1372–1383.

Murray, S.C., Rooney, W.L., Mitchell, S.E., Sharma, A., Klein, P.E. *et al.* (2008a) Genetic improvement of sorghum as a biofuel feedstock: II. QTL for stem and leaf structural carbohydrates. *Crop Science* 48, 2180–2193.

Murray, S.C., Sharma, A., Rooney, W.L., Klein, P.E., Mullet, J.E. *et al.* (2008b) Genetic improvement of sorghum as a biofuel feedstock I: QTL for stem sugar and grain nonstructural carbohydrates. *Crop Science* 48, 2165–2179.

Nelson, J., Wang, S., Wu, Y., Li, X., Antony, G. *et al.* (2011) Single-nucleotide polymorphism discovery by high-throughput sequencing in sorghum. *BMC Genomics* 12, 352.

Nelson, R.S., Abel, P.P. and Beachy, R.N. (1987) Lesions and virus accumulation in inoculated transgenic tobaco plants expressing the coat protein gene of tobacco mosaic virus. *Virology* 158, 126–132.

Oria, M.P., Hamaker, B.P., Axtell, J.D. and Huang, C.-P. (2000) A highly digestible sorghum mutant cultivar exhibits a unique folded structure of endosperm protein bodies. *Proceedings of the National Academy of Sciences, United States of America* 97, 5065–5070.

Pandey, A.K., Bhat, B.V., Balakrishna, D. and Seetharama, N. (2010) Genetic transformation of sorghum (*Sorghum bicolor* (L.) Moench.). *International Journal of Biotechnology and Biochemistry* 6, 45–53.

Paterson, A., Lin, Y., Li, Z., Schertz, K., Doebley, J. *et al.* (1995) Convergent domestication of cereal crops by independent mutations at corresponding genetic loci. *Science* 269, 1714–1718.

Paterson, A.H., Bowers, J.E. and Chapman, B.A. (2004) Ancient polyploidization predating divergence of the cereals, and its consequences for comparative genomics. *Proceedings of the National Academy of Sciences, United States of America* 101, 9903–9908.

Paterson, A.H., Bowers, J.E., Bruggmann, R., Dubchak, I., Grimwood, J. *et al.* (2009) The *Sorghum bicolor* genome and the diversification of grasses. *Nature* 457, 551–556.

Peng, Y., Schertz, K.F., Cartinhour, S. and Hart, G.E. (1999) Comparative genome mapping of *Sorghum bicolor* (L.) Moench using an RFLP map constructed in a population of recombinant inbred lines. *Plant Breeding* 118, 225–235.

Penning, B.W., Hunter, C.T., Tayengwa, R., Eveland, A.L., Dugard, C.K. *et al.* (2009) Genetic resources for maize cell wall biology. *Plant Physiology* 151, 1703–1728.

Porter, K.S., Axtell, J.D., Lechtenberg, V.L. and Colenbrandu, V.F. (1978) Phenotype fiber composition and *in vitro* dry matter disappearance of chemically induced brown–midrib (*bmr*) mutants of sorghum. *Crop Science* 18, 205–208.

Pratt, L.H., Liang, C., Shah, M., Sun, F., Wang, H.M. *et al.* (2005) *Sorghum* expressed sequence tags identify signature genes for drought, pathogenesis, and skotomorphogenesis from a milestone set of 16,801 unique transcripts. *Plant Physiology* 139, 869–884.

Price, H.J., Hodnett, G.L., Burson, B.L., Dillon, S.L., Stelly, D.M. *et al.* (2006) Genotype dependent interspecific hybridization of *Sorghum bicolor. Crop Science* 46, 2617–2622.

Qazi, H.A., Paranjpe, S. and Bhargava, S. (2012) Stem sugar accumulation in sweet sorghum – activity and expression of sucrose metabolizing enzymes and sucrose transporters. *Journal of Plant Physiology* 169, 605–613.

Quinby, J.R. and Karper, R.E. (1947) The effect of short photoperiod on sorghum varieties and first generation hybrids. *Journal of Agricultural Research* 75, 295–300.

Raghuwanshi, A. and Birch, R.G. (2010) Genetic transformation of sweet sorghum. *Plant Cell Reports* 29, 997–1005.

Ratnavathi, C.V., Dayakar Rao, B. and Seetharama, N. (2003) Sweet sorghum stalk: a suitable raw material for fuel alcohol production. DSR/NRCS Report Number 12/2003. NATP (DSR) Series No. 1, National Research Center for Sorghum (NRCS), Rajendranagar, Hyderabad, Andhra Pradesh, India, 8 pp.

Reddy, B.V.S., Ramesh, S., Sanjana Reddy, P., Ramaiah, B., Salimath, P.M. *et al.* (2005) Sweet sorghum – a potential alternative raw material for bioethanol and bio-energy. *International Sorghum and Millets Newsletter* 46, 79–86.

Reddy, B.V.S., Ramesh, S., Ashok Kumar, A., Wani, S.P., Ortiz, R. *et al.* (2008) Biofuel crops research for energy security and rural development in developing countries. *Bioenergy Research* 1, 248–258.

Reddy, P.S., Reddy, B.V.S. and Srinivasa Rao, P. (2011) Genetic analysis of traits contributing to stalk sugar yield in *Sorghum. Cereal Research Communications* 39, 453–464.

Reddy, V.G., Upadhyaya, H.D. and Gowda, C.L.L. (2006) Current status of sorghum genetic resources at ICRISAT: Their sharing and impacts. *SAT eJournal* 2(1), Available at http://www.icrisat.org/journal/cropimprovement/v2i1/v2i1currentstatus.pdf (accessed 23 March 2012).

Rooney, W.L. and Smith, C.W. (2000) Techniques for developing new cultivars. In: Smith, C.W. and Frederiksen, R.A. (eds) *Sorghum: Origin, History, Technology and Production*. John Wiley and Sons, New York, pp. 329–347.

Saballos, A. (2008) Development and utilization of sorghum as a bioenergy crop. In: Vermerris W. (ed.) *Genetic Improvement of Bioenergy Crops*. Springer-Verlag, New York, pp. 211–248.

Salzman, R.A., Brady, J.A., Finlayson, S.A., Buchanan, C.D., Sun, F. *et al.* (2005) Transcriptional profiling of sorghum induced by methyl jasmonate, salicylic acid, and aminocyclopropane carboxylic acid reveals cooperative regulation and novel gene responses. *Plant Physiology* 138, 352–368.

Sanderson, M.A., Jones, R.M., Ward, J. and Wolfe, R. (1992) Silage sorghum performance trial at Stephenville. Forage Research in Texas. *Report PR-5018*. Texas Agricultural Experimental Station, Stephenville, Texas.

Sankarapandian, R., Ramalingam, J., Pillai, M.A. and Vanniarajan, C. (1994) Heterosis and combining ability studies for juice yield related characteristics in sweet sorghum. *Annals of Agricultural Research* 15, 199–204.

Schaffert, R.E. and Gourley, L.M. (1982) Sorghum as an energy source. In: *Sorghum in the Eighties: Proceedings of the International Symposium on Sorghum*, 2–7 November 1981, ICRISAT Center, India. International Crops Research Institute for the Semi-Arid Tropics, Patancheru 502 324, Andhra Pradesh, India, pp. 605–623.

Schnable, P.S., Ware, D., Fulton, R.S., Stein, J.C., Wei, R.S. *et al.* (2009) The B73 maize genome: complexity, diversity, and dynamics. *Science* 326, 1112–1115.

Searchinger, T., Heimlich, R., Houghton, R.A., Dong, F., Elobeid, A. *et al.* (2008) Use of US croplands for biofuels increases greenhouse gases through emissions from land-use change. *Science* 319, 1238–1240.

Selvi, B. and Palanisamy, S. (1990) Heterosis and combining ability for grain yield in sweet sorghum. *Madras Agricultural Journal* 77, 493–496.

Sivasudha, T. and Kumar, P.A. (2008) Isolation, sequencing and in silico analysis of sorghum (*Sorghum bicolor*) sucrose synthase promoter. *Journal of Plant Sciences* 3, 203–215.

Springer, N.M. and Jackson, S.A. (2010) Realizing the potential of genomics for crop improvement. *Brief Functional Genomics* 9, 93–94.

Springer, N.M., Ying, K., Fu, Y., Ji, T., Yeh, C.-T. *et al.* (2009) Maize inbreds exhibit high levels of copy number variation (CNV) and presence/absence differences (PAV) in genome content. *PLoS Genetics* 5(11): e1000734. doi:10.1371/journal.pgen.1000734

Srinivasa Rao, P., Rao, S.S., Seetharama, N., Umakanth, A.V., Sanjana Reddy, P. *et al.* (2009) *Sweet Sorghum for Biofuel and Strategies for its Improvement*. Information Bulletin No 77. International Crops research Institute for the Semi-Arid Tropics, Patancheru 502 324, Andhra Pradesh, India, 80 pp, ISBN: 978-92-9066-518-2, Order Code: IBE 077.

Srinivasa Rao, P., Reddy, B.V.S., Blümmel, M., Subbarao, G.V., Chandraraj, K. *et al.* (2010) Sweet sorghum as a biofuel feedstock: can there be food-feed-fuel trade-offs? Available at: http://www.wgcrop.icidonline.org/sweet%20sorghum dec09.pdf (accessed 23 March 2012).

Srinivasa Rao, P., Sanjana Reddy, P., Rathore, A., Reddy, B.V.S. and Panwar, S. (2011a) Application of GGE biplot and AMMI model to evaluate sweet sorghum hybrids for Genotype × Environment interaction and seasonal adaptation. *The Indian Journal of Agricultural Sciences* 81, 438–444.

Srinivasa Rao, P., Jayalakshmi, M., Kumar, C.G., Kamal, A. and Reddy, B.V.S. (2011b) Response of fertilizer treatments on agronomic and biochemical traits in main and ratoon crops of sweet sorghum (*Sorghum bicolor* (L.) Moench) cultivar ICSV 93046. In: Li, Y.-R., Srivastava, M.K., Rao, G.P., Singh, P. and Solomon, S. (eds) *Balancing Sugar and Energy Production in Developing Countries: Sustainable Technologies and Marketing Strategies, Proceedings of the 4th IAPSIT International Sugar Conference*, New Delhi, India, 21–25 November 2011, pp. 160–164.

Srinivasa Rao, P., Kumar, C.G., Malapaka, J., Kamal, A. and Reddy, B.V.S. (2012) Feasibility of sustaining sugars in sweet sorghum stalks during post-harvest stage by exploring cultivars and chemicals: a desk study. *Sugar Tech* 14, 21–25.

Srivastava, A., Mehta, S., Lindlof, A. and Bhargava, S. (2010) Overrepresented promoter motifs in abiotic stress-induced DREB genes of rice and sorghum and their probable role in regulation of gene expression. *Plant Signaling and Behavior* 5, 775–784.

Stark, D.M. and Beachy, R.N. (1989) Protection against potyvirus infection in transgenic plants: evidence for broad spectrum resistance. *Nature Biotechnology* 7, 1257–1262.

Tao, Y.Z., Jordan, D.R., Henzell, R.G. and Mcintyre, C.L. (1998) Construction of a genetic map in a sorghum RIL population using probes from different sources and its comparison with other sorghum maps. *Australian Journal of Agricultural Research* 49, 729–736.

Tuberosa, R. and Salvi, S. (2006) Genomics-based approaches to improve drought tolerance of crops. *Trends in Plant Science* 11, 405–412.

Verma, A. and Kumar, P.A. (2005) Isolation and deletion analysis of meristem specific promoter from *Sorghum bicolor* (L.) Moench. *Indian Journal of Biotechnology* 4, 516–521.

Verma, A., Nain, V., Kumari, C., Singh, S.K., Lakshmi Narasu, M. *et al.* (2008) Tissue specific response of *Agrobacterium tumefaciens* attachment to *Sorghum bicolor* (L) Moench. *Physiology and Molecular Biology of Plants* 14, 307–313.

Vermerris, W. (2011) Survey of genomics approaches to improve bioenergy traits in maize, sorghum, and sugarcane. *Journal of Integrative Plant Biology* 53, 105–119.

Vietor, D. and Miller, F. (1990) Assimilation, partitioning, and nonstructural carbohydrates in sweet compared with grain sorghum. *Crop Science* 30, 1109–1115.

Wang, W., Wang, J., Yang, C., Li, Y., Liu, L. *et al.* (2007) Pollen-mediated transformation of *Sorghum bicolor* plants. *Biotechnology and Applied Biochemistry* 48, 79–83.

Whitkus, R., Doebley, J. and Lee, M. (1992) Comparative genetic mapping of sorghum and maize. *Genetics* 132, 1119–1130.

Wortmann, C., Ferguson, R. and Lyon, D. (2008) Sweet sorghum as a biofuel crop in Nebraska. Paper presented at the 2008 Joint Annual Meeting, Celebrating the international Year of Planet Earth, 5–9 October 2008, Houston, Texas. Available at: http://crops.confex.com/crops/2008am/techprogram/P44581.htm (accessed 23 March 2012).

Wortmann, C.S., Liska, A.J., Ferguson, R.B., Lyon, D.J., Klein, R.M. and Dweikat, I. (2010) Dryland performance of sweet sorghum and grain crops for biofuel. *Agronomy Journal* 102, 319–326.

Wu, L. and Birch, R.G. (2010) Physiological basis for enhanced sucrose accumulation in an engineered sugarcane cell line. *Functional Plant Biology* 37, 1161–1174.

Xin, Z., Wang, M.L., Barkley, N.L., Burow, G.B., Franks, C.D. *et al.* (2008) Applying genotyping (TILLING) and phenotyping analyses to elucidate gene function in a chemically induced sorghum mutant population. *BMC Plant Biology* 8, 103. doi:10.1186/1471-2229-8-103.

Xu, G.-W., Magill, C.W., Schertz, K.F. and Hart, G.E. (1994) A RFLP linkage map of *Sorghum bicolor* (L.) Moench. *Theoretical and Applied Genetics* 89, 139–145.

Zhao, J., Fu, J.B., Liao, H., He, Y., Nian, H. *et al.* (2004) Characterization of root architecture in an applied core collection for phosphorus efficiency of soybean germplasm. *Chinese Science Bulletin* 49, 1249–1257.

Zhao, Y.L., Dolat, A., Steinberger, Y., Wang, X., Osman, A. *et al.* (2009) Biomass yield and changes in chemical composition of sweet sorghum cultivars grown for biofuel. *Field Crops Research* 111, 55–64.

Zhao, Z., Cai, T., Taglini, L., Miller, M., Wang, N. *et al.* (2000) *Agrobacterium*-mediated sorghum transformation. *Plant Molecular Biology* 44, 789–798.

10 Switchgrass

John H. Fike and David J. Parrish
Crop and Soil Environmental Sciences, Virginia Polytechnic Institute and State University, Virginia, USA

10.1 Introduction

The first members of the grass family (*Poaceae* or *Gramineae*) appeared 55 to 70 million years ago (mya) (Kellogg, 2001) and, since the much more recent appearance of human beings, our connections to the grasses have become extensive. Grasses produce the great majority of the calories consumed in our diets. Forage grasses feed livestock that provide additional dietary components. Other grasses serve as sources of fibre, turf and ornamentation (Casler, 2006). Switchgrass (*Panicum virgatum* L.), which is a very recent addition to the list of useful grasses, has been valued primarily as a forage species but is now being widely considered as a bioenergy crop.

10.2 Origin, Botany and History

10.2.1 Biological origins and family affiliations of switchgrass

The approximately 10,000 grass species have been grouped into a few hundred genera (Watson and Dallwitz, 1994). The *Panicum* genus, to which switchgrass belongs, is large and quite diverse, having C_3, C_4, and C_3/C_4 intermediate species and base chromosome numbers ranging from 8 to 15 (Aliscioni *et al.*, 2003). For those and other reasons, many taxonomists feel the *Panicum* genus is polyphyletic, i.e. that its species did not all evolve from a single *Panicum* ancestor. This limits their ability to determine the full lineage of the *Panicum* species of greatest interest to us, *P. virgatum*.

Using molecular marker studies, the Grass Phylogeny Working Group (GPWG, 2001) looked at 62 grasses representing the breadth of the family. The 62 included switchgrass and several other crops. Base-pair variations for six nuclear genes as well as chloroplast restriction site data were used to produce a 'family tree' showing the most likely relationships of the 62 species. The phylogeny placed switchgrass as a close relative (adjacent branch, or clade) of pearl millet (*Pennesetum alopecuroides* (L.) Spreng.), and that pair was next most closely related to maize (*Zea mays* L.) and miscanthus (*Miscanthus japonicus* Andersson).

The timeline for evolutionary branching points, or nodes, in the lineage of switchgrass and other grasses has been investigated with molecular clock methods (Huang *et al.*, 2003; Paterson *et al.*, 2010; Zhang *et al.*, 2011a). Such work shows, for example, that the progenitors of switchgrass and maize diverged from a shared ancestor about 22 to 23 mya (million years ago). Subsequently, the first *Panicum* species evolved and were believed to have been diploid; but a polyploid line developed – putatively from the crossing of

©CAB International 2013. *Biofuel Crops: Production, Physiology and Genetics* (ed. B.P. Singh)

two diploid species – about 5 to 10 mya (Zhang *et al.*, 2011a). That polyploid *Panicum* line continued to evolve, and the gene pool associated with the species now recognized as *P. virgatum* appears to have been assembled by about 2 mya (Huang *et al.*, 2003), i.e. the species is thought to be about 2 million years old.

After switchgrass arose some 2 mya, it presumably radiated and adapted from an unknown point of origin. Based on its current range of occurrence/adaptation, it likely evolved lineages that colonized significant portions of North America. However, repeated periods of glaciation in the last 2 million years would undoubtedly have driven those lineages into extinction or into more southern, ice-free climates. The survivors would presumably have followed the ice northward during interglacial periods, only to repeat the retreat/recolonize cycle again and again (McMillan, 1959; Zhang *et al.*, 2011a,b).

When Europeans arrived in the New World, switchgrass's range included Central America and eastern North America (Hitchcock, 1935); but we would have to suppose that it attained that distribution only in the last 11,000 years – after the last period of continental glaciation. McMillan (1959) suggested switchgrass and other prairie grasses retreated to refugia during the Ice Ages. He posited three regional refugia during the most recent Pleistocene glaciation. Recent molecular marker studies examining simple sequence repeats (SSRs) of switchgrass cultivars and accessions (Zalapa *et al.*, 2011) provide support for this three-refugia theory. Most – perhaps all – of today's cultivars can be sorted into three groups based on SSRs, with each group harking back to one of the three refugia. To follow this line of reasoning, we must look more closely at the notion of switchgrass 'ecotypes'.

10.2.2 Switchgrass ecotypes and their origins

Essentially all cultivars, lines, or accessions of switchgrass have been placed into one of two categories: upland or lowland. A few 'intermediate' or 'ambiguous' types, which are not readily assigned to one of these two categories, may represent hybrids of the two (Zhang *et al.*, 2011b). The upland and lowland groups are usually described as 'ecotypes'. Ecotypes typically differ in ways that make them better suited for different environments, but they are able to interbreed and produce fertile offspring. The ecotype groupings of switchgrass have also been termed upland and lowland 'cytotypes', referring to diagnostic DNA sequences carried in their plastids (Vogel, 2004). The two ecotypes, which were initially distinguished only by phenotype, can now be grouped into U and L cytotypes using various molecular markers (Hultquist *et al.*, 1996; Missaoui *et al.*, 2006; Vogel and Mitchell, 2008; Zalapa *et al.*, 2011).

Switchgrass plants belonging to the upland ecotype are typically finer stemmed and shorter than those identified as lowlands. As the upland designation implies, these lines are generally better adapted to drier and colder habitats, while plants belonging to the lowland ecotype tend to thrive in warmer, wetter habitats. Most of the lowland lines are accessions (or derived from accessions) from the southern USA, and the upland genotypes are more generally associated with the northern Great Plains. Cultivars and lines from the lowland ecotype are essentially all tetraploid ($2n = 4x = 36$), whereas the upland ecotype consists of genotypes that are both tetraploid and octoploid ($2n = 8x = 72$) (Zalapa *et al.*, 2011). Recently, possible octoploid lowland plants have been discovered (Zhang *et al.*, 2011a,b).

Using molecular clock calculations based on cpDNA sequences, estimates of the upland–lowland divergence range from 0.5 to 1.3 mya (Young *et al.*, 2011; Zhang *et al.*, 2011a). Clear separation of tetraploid and octoploid lineages within the upland ecotype suggests that some octoploid lineages must have arisen early after the appearance of the original *P. virgatum* (Zhang *et al.*, 2011a).

To survive centuries-long periods of glaciation during the Ice Ages, switchgrass lineages of both ecotypes would have to have been endemic to or immigrated to ice-free areas. As suggested by McMillan (1959), in the most recent period of glaciation, three areas in North America may have provided sufficiently warm growing seasons to serve as refugia for many grassland species. Casler and colleagues have looked carefully at the distribution of North American populations of the two switchgrass ecotypes and the morphological and genetic similarities and differences between and within those populations (Casler and Boe, 2003; Casler *et al.*, 2007a,b; Zalapa *et al.*, 2011; Zhang *et al.*, 2011a,b). Other researchers (e.g. Cortese *et al.*, 2010) have provided similar or additional evidence that the current populations of North American switchgrasses can be placed into a few groups associated with particular geographies, or provenances. These analyses reveal clusters of allelic similarities that reflect provenance. Lowland cultivars can be grouped by allelic similarities into two clusters; cultivars in one come from the Eastern Gulf Coast region, and those in the other are from the Southern Great Plains. The nine octoploid upland cultivars fall into three allelic clusters, or genetic pools, each with a unique provenance: those associated with the Central Great Plains, the Northern Great Plains, and the Eastern Savannah (Zalapa *et al.*, 2011).

The findings of Zalapa *et al.* (2011) provide support for the three Ice Age refugia posited by McMillan (1959). Zalapa *et al.* (2011) hypothesize that the two lowland allelic (and geographic) genetic pools noted above each descended from McMillan's similarly named refugia, i.e. Lowland/Southern Great Plains and Eastern Gulf Coast. They suggest also that at least two of the upland genetic pools may be the descendants of plants that survived in an Upland Plains refugium. Zalapa *et al.* (2011) also offer a reasonable model for arriving at the current situation where octoploids are the more frequent ploidy level for upland cultivars. It builds on the notion that the duplicated genome offers more grist for the evolutionary mill, a notion reflected in the writings of others (e.g. Paterson *et al.*, 2009, 2010).

In sum, we can suggest that our 'modern' switchgrasses, i.e. those that emerged from and radiated after the last Ice Age, may have come from a relatively small number of survivors. Those survivors included a few – maybe only two – groups representing the lowland genetic pool and perhaps a few more groups carrying the upland gene set. What we see today reflects the rather remarkable ability of those few survivors/pioneers to radiate, adapt and recolonize two-thirds of a continent in a scant 11,000 years.

10.2.3 The agronomic history of switchgrass

Panicum virgatum pre-existed *Homo sapiens*, of course, but only very recently has it been adopted and adapted to human purposes. It has been a 'crop' in the usual sense of that word only since the middle of the 20th century. Switchgrass is now used in several roles. In addition to use as a forage, it has been adopted or is under consideration for a broad range of other purposes (Parrish and Fike, 2005). The list of well-established functions includes: forage (grazing, hay, or haylage), erosion control (waterways, levees, stream margins, etc.), vegetative filter strips (to reduce runoff of soil and nutrients), reclamation and/or stabilization (sand dunes, disturbed areas, etc.) and wildlife habitat. Other roles under study or in early adoption include: energy feedstock (combustion, co-firing, conversion to liquid or gaseous forms, etc.); value-added products from biorefineries; fibre or pulp; phytoremediation (smelter and mining sites, etc.); pharmaceuticals, biomaterials, plastics, etc.; and substrate for mushroom culture.

We have surveyed the history of switchgrass studies using CAB Direct, the bibliographic database of CABI. We searched CAB Direct for the occurrences of *Panicum virgatum*, switchgrass, or switch grass in titles, abstracts, key words and CABI's coding descriptors and identifiers. Our survey does not provide an exhaustive or comprehensive list of publications dealing with switchgrass, but it offers a good indication of the trend or trajectory for such publications. Figure 10.1 plots the annual total of CABI-indexed reports referring to switchgrass (as 'switchgrass', 'switch grass' or '*Panicum virgatum*'), the number of those looking only at switchgrass (or comparing it with only one other species) and the number mentioning switchgrass as a potential energy crop. It shows switchgrass first appears as a subject of scientific investigation about 100 years ago, and then interest in it increases slowly at first and then sharply, especially in the last three decades.

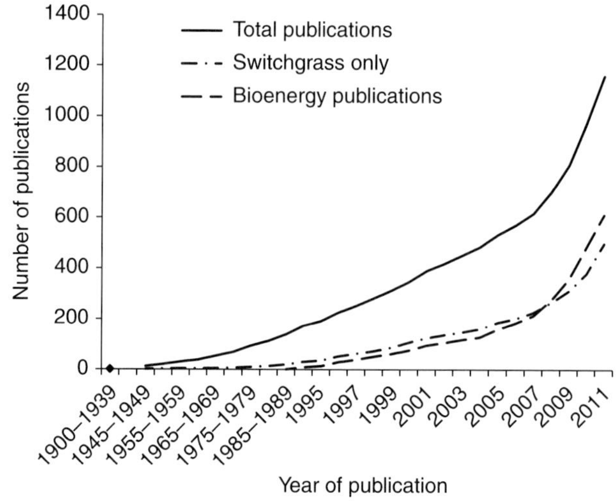

Fig. 10.1. Number of publications in the CAB Abstracts database containing a reference to switchgrass (search terms included 'switchgrass', 'switch grass' or '*Panicum virgatum*') as a function of year of publication.

The first indexed occurrence of switchgrass in the CABI database came in 1914, where the species was mentioned as not being a host for an aphid. The next appearance was in a 1931 article discussing 'prairie grasses', where switchgrass was discussed as occupying 'less desirable lowland soils' (Weaver, 1931). That publication and most of the few that followed over the next 20 years alluded to switchgrass as one of the species in mixtures of 'prairie grasses', 'prairie hay', 'native grasses', 'range grasses', 'mixed grasses', 'warm-season grasses', etc. Switchgrass began to emerge from the anonymity of being 'just' a prairie grass in the 1940s. A 1947 agricultural experiment station report referred to studies of switchgrass and other prairie grasses in pure stands established in 1937 (Newell and Keim, 1947). During the late 1940s and 1950s, reports on selection and breeding studies with switchgrass appear in a few agricultural experiment station annual reports. Overall, though, the species received scant scientific attention.

Studies with switchgrass as the major or sole focus of work began to appear in the middle of the 20th century. Reports dealt mostly with the species as forage, either from an agronomic or an animal nutrition perspective. These included cultivar releases discussing forage value. However, as early as the 1940s, switchgrass began to be mentioned for erosion control; and the species' value in reclamation was noted in several articles from the 1960s.

The 1980s brought an increase in the study of switchgrass. In that decade, the number of reports dealing primarily or solely with switchgrass more than doubled the previous 50 years' total. At the close of the decade came the first peer-reviewed article written on switchgrass as an energy crop (Lowenberg-DeBoer and Cherney, 1989), and that publication presaged an explosion in work on the species.

In sum, the history of interest in and culture of switchgrass is very short. During the second half of the 20th century the species gradually moved from being a 'prairie grass' to being a crop that might be grown in monoculture. For the first 40 years of its very short agronomic history, the volume of work on switchgrass averaged less than one report per year dedicated primarily or solely to it. Since 1930, more than 1600 reports that index to switchgrass have been published; but two-thirds of that total has appeared only in the 21st century, and two-thirds of the work done in the 21st century has focused on the species as an energy crop. Were it not for its selection as a model energy crop, it might well still rank with the other two of the once 'big three' prairie grasses, big bluestem (*Andropogon gerardii* Vitman) and indiangrass (*Sorghastrum nutans* L.). In 2010, those latter two species combined for a total of five CAB Direct citations where they were a primary focus of the work.

10.2.4 The origins of switchgrass as an energy crop

Biomass use for energy is prehistoric, and wood remains a primary energy source in some countries. Developed nations have become increasingly dependent on fossil fuels, but even they occasionally refocus on biomass energy sources. For example, the 'energy crisis' that began in 1973 spurred interest and work on energy cropping and bioenergy in the USA, in Europe (Hall, 1979) and in the United Nations (El-Hinnawi, 1981). Out of those efforts came many reports, including one in the USA that mentioned switchgrass as a possible biomass source (Wedin and Helsel, 1980). However, there was a loss of continuity and momentum in the biomass-for-energy narrative in the late 1970s when oil prices returned to 'pre-crisis' levels.

In 1984, the Oak Ridge National Laboratory (ORNL) of the US Department of Energy (DOE) issued a request for proposals (RFP) to screen herbaceous, lignocellulosic species as energy crops. The RFP stipulated that the work was to be done on 'marginal croplands' (Wright and Turhollow, 2010). In 1985, the first five subcontracts were awarded for the Herbaceous Energy Crops Program (HECP), and two more were added in 1988 (Wright, 2007). Seven states from eastern and central USA were represented. Each of the subcontractors proposed a region-appropriate list of species to be screened, but there was no benchmark species to allow cross-region comparisons of biomass productivity of over 30 disparate species that would be grown at over 30 disparate locations. (Data in this and the next few paragraphs were compiled from Wright, 2007.)

The eight species proposed by the Virginia subcontractor (Virginia Tech) included switchgrass. Their proposal noted that switchgrass is a native that will 'produce better growth and cover on droughty, infertile, eroded soils than most introduced grasses'. At the April 1985 meeting of subcontractors, Dale Wolf, the forage scientist who chose switchgrass for Virginia Tech's proposal, suggested to ORNL administrators that switchgrass should serve well as the desired benchmark species. So, switchgrass appeared in all seven subcontractors' screening studies – but only after it was added to most.

When the final reports of the initial 5-year screening studies were compiled by ORNL, switchgrass had proved to be one of the best biomass producers across all seven states. It, in fact, did well soon enough in the 5-year cycle that some of the subcontractors carried out switchgrass-specific management studies (Virginia Tech) and cultivar screening (Auburn University) (Wright, 2007).

Switchgrass, which had been somewhat serendipitously chosen as the benchmark species for HECP's initial 5-year, seven-state, 36-species screening studies, did so well in those studies that ORNL subsequently invited proposals to study only switchgrass. (By that point, the HECP had become part of ORNL's Bioenergy Feedstock Development Program (BFDP).) The RFP described switchgrass as a 'model species' (McLaughlin and Kszos, 2005; Wright, 2007; Wright and Turhollow, 2010). Lessons learned from studies of a model species could be applied or adapted to other promising biomass species when they might appear. The decision to focus on a single herbaceous species was made at a time when ORNL's budgets for biofuels work were shrinking, suggesting that the decision to focus on a single herbaceous species was perhaps partially a pragmatic one (Wright, 2007; Wright and Turhollow, 2010). The upshot of the switchgrass-as-a-model-species decision was that DOE essentially stopped looking for new herbaceous energy crops after the 5-year screening study.

The second and then third 5-year rounds of DOE-funded subcontract work – focused solely on switchgrass – began in 1992 and 1997. The work included long-term varietal and management studies, some of which ultimately received 10 years of DOE support (e.g. Fike *et al.*, 2006a,b). Programmes with field studies focused on cultivar selection, establishment and management (McLaughlin and Kszos, 2005). Switchgrass breeding efforts received DOE support, as did tissue culture and transgenic work (McLaughlin and Kszos, 2005). Also included in these rounds of DOE/ORNL funding were collaborations with USDA personnel based at various public universities and USDA (US Department of Agriculture) facilities (McLaughlin and Kszos, 2005).

The DOE/ORNL/BFDP funding for extramural bioenergy research was ended in 2002 based on 'decisions made within DOE' (McLaughlin and Kszos, 2005). Following loss of DOE funding, USDA first slowly and then more vigorously assumed the leadership for research on switchgrass. For example, the USDA Agricultural Research Service (ARS) developed a national intramural programme on Bioenergy and Energy Alternatives that includes major studies with switchgrass at several USDA facilities (http://www.ars.usda.gov/research/programs/programs.htm?np_code=307). Some of the post-DOE funding for efforts on switchgrass has come from the Sun Grant Program, which is overseen by the US Department of Transportation (DOT) with inputs from USDA and DOE. Various regional studies on switchgrass and other bioenergy species have been developed and funded by Sun Grant (http://www.sungrant.org).

Also stepping into the biofuels arena increasingly in the first decade of the 21st century has been the private sector. Some major petroleum companies have invested in biofuels research, in some cases via centres established at public universities. A number of new companies that hope to capitalize on switchgrass's and other species' bioenergy potential have also appeared. Another major participant in switchgrass-for-energy studies has been the private, not-for-profit Noble Foundation in Oklahoma, which has expanded its long-standing efforts on forages into studies aimed specifically at the energy-crop potential of switchgrass (http://www.noble.org/Research/Biofuels/index.html).

Switchgrass is native to southern portions of Canada, and Canadian workers became involved in switchgrass-for-energy studies early on, but it was largely a one-institution project based at McGill University (e.g. Madakadze *et al.*, 1998a,b). After these studies, Canadian work on switchgrass-for-energy was taken up and expanded to an international scale by scientists at Resource Efficient Agricultural Production (REAP) Canada, which is particularly interested in growing switchgrass for conversion into densified units (e.g. pellets) that can be used for heating (Samson *et al.*, 2005). The REAP site (http://www.reap-canada.com) has a comprehensive list of their work and recommendations, including growing guides and information on pelletizing and burning switchgrass.

The number of countries in which switchgrass has been studied is growing. In a tally done in 2005, the species had been investigated in 11 countries (Parrish and Fike, 2005). Our more recent CABI-indexed survey produced a list of 18: Argentina, Australia, Canada, China, Colombia, England, Germany, Greece, Italy, Japan, Korea, Mexico, Pakistan, Poland, Spain, Sudan, the USA and Venezuela.

In sum, switchgrass has moved in a very short time from being an obscure prairie grass to being one of the most widely cited energy crop species. It has gained such attention rather serendipitously. After an admittedly very limited screening effort, it was chosen as a model species and received significant, well-orchestrated funding from DOE for 10 years, in a time when little other work was being done on herbaceous biomass species. In some sense then, it perhaps stole a march on possibly better-suited species. Some other species are now beginning to have significant amounts of effort devoted to their study, but switchgrass's lead on the herbaceous, lignocellulosic field remains rather substantial. Logic would suggest that there may well be other species that might serve us equally well or better as a lignocellulosic feedstock; but switchgrass certainly deserves full consideration. We feel it is as strong a candidate – especially on 'its own territory' – as any that has been so far put forward.

10.3 Crop Characteristics: Adaptation, Physiology, Morphology and Phenology

Clearly the evolutionary history of switchgrass has provided the species the genetic capacity needed to contend for a role as a lead player on the stage of herbaceous energy cropping systems. But what are those character traits, and how do they lend themselves to a coming bioeconomy? Typically, the 'first pass' factors for choosing an energy crop are yield and return to inputs (production efficiency). Here we discuss switchgrass's robustness in terms of broad adaptability (or lower sensitivity) to a range of conditions.

10.3.1 Adaptation to varied edaphic conditions

Few highly productive species are as widely adapted to the conditions of North America's (and other continents') humid, temperate zones as switchgrass. No doubt the plant's extensive range reflects the large genetic variation within the species. However, describing the species with such broad brush strokes obscures the fact that ecotypes display greater suitability to a particular set of conditions. As one example, switchgrass displays longitudinal adaptations in North America; cultivars originating in more humid eastern regions often have lower productivity when moved west to drier regions of similar latitude, and vice versa. For 'eastbound' cultivars in North America, this lower productivity has in part been pegged to lower pathogen resistance (Vogel, 2000), as their origins in drier climates has limited the need for such adaptations. Latitude of origin is also a key component of adaptation – one that is of great utility when energy cropping – and of survival, and we will give this further consideration in the section below on 'cultivar selection'.

Along with its broad geographic adaptation, switchgrass is also quite adapted to a wide range of edaphic and fertility conditions within a given region. This suitability follows broadly along the lines of lowland and upland ecotypes, with lowland cultivars being better suited to hydric bottom lands. Upland cultivars, which typically occur on more mesic, upslope positions, are (not surprisingly) less sensitive to moisture deficits than their lowland cousins.

Soil type itself appears to have limited effects on switchgrass performance (Sanderson *et al.*, 1999). Soil texture, inasmuch as it relates to water holding capacity, can greatly affect establishment and subsequent yield. Both excessive drainage (in sandy soils) or poor internal drainage (as in depositional sites on the landscape; e.g. Thelemann *et al.* (2010)) may limit switchgrass productivity. Slope also may limit productivity: we have reported a case in which two sites with the same soil type and separated by only 100 m had about 40% greater biomass yield on the site with less slope (Fike *et al.*, 2006a). Presumably both water infiltration and subsequent availability were affected, but slope aspect may have had some effect as well; the more productive site was slightly more south facing.

Although many switchgrass production guides suggest amending soil pH to 6.0 or more, rarely does soil acidity limit switchgrass production. Tolerance to pH extremes has been reported in several studies, and roots have been observed growing at pH 3.7 (Stucky *et al.*, 1980), although the pH range for optimum germination is narrower (Hanson and Johnson, 2005). Switchgrass also has high tolerance for, and in many cases, the ability to remediate or extract soil contaminants such as herbicides (e.g. atrazine (Murphy and Coats, 2011)) and heavy metals (e.g. cadmium (Juang and Lee, 2010)). Such characteristics, coupled with the plant's productivity on drought-prone and poorly structured soils have established switchgrass's reputation as a useful species for phytoremediation schemes.

Part of the species' productivity in difficult soils and growing conditions comes from its C_4 physiology, which we will discuss shortly. Switchgrass also receives benefit from fungal mutualisms that may mediate responses to stresses such as drought, contaminants and limited water and nutrient availability. Switchgrass often has greater yields with mycorrhizal associations, although the benefits of mycorrhizae vary by strain and source (Koslowsky and Boerner, 1989; Clark, 2002). While mycotrophy's benefits are often greater at lower pH (Clark *et al.*, 1999a,b; Clark, 2002) and under conditions of environmental stress, increased plant production has been reported even in non-stressed environments (Ghimire and Craven, 2011). Such results suggest opportunities to exploit plant–microbe interactions for greater nutrient use efficiency, or even to tap systems of biological nitrogen (N) fixation, which have been hinted at in some preliminary research (Ker *et al.*, 2010).

10.3.2 The C_4 pathway and its contributions to switchgrass adaptations

The C_4 photosynthetic pathways – at least four biochemical variations occur – arose multiple times in North America, and recent evidence suggests that a drying climate was the primary driver for the shift from C_3 to C_4 dominance on the Great Plains (Strömberg and McInerney, 2011). More energetically expensive than the archetypic C_3 pathway, the C_4 pathways can convey competitive advantages over the C_3 photomechanism when species are put in head-to-head competition under conditions of constraint. Although a range of adaptive capacity exists within the two syndromes, C_4 species generally are more heat and drought tolerant and have higher N- and water-use efficiencies; they also do not undergo photorespiration. These factors, which conveyed competitive advantage to C_4 species on the court of natural selection, have also been important selection criteria in humans' search for bioenergy crops.

Grass species with C_4 photosynthesis are more loosely referred to as 'warm-season' grasses, and productivity at higher temperatures is an important manifestation of the syndrome. For switchgrass, minimal growth occurs below about 12°C (Kiniry *et al.*, 2005; Grassini *et al.*, 2009) but differences in this cardinal temperature occur between northern- and southern-adapted ecotypes and cultivars. Interestingly, and despite higher base growth temperatures, switchgrass of southern lineage is often observed beginning growth earlier in the season than switchgrass of northern derivation. The delayed onset of growth observed for northern varieties likely is a survival mechanism that reduces the probability of frost damage in the event of a late-spring freeze. This response has great implications for cultivar selection.

Among candidate C_4 bioenergy species, switchgrass does not have the greatest photosynthetic rates (Dohleman *et al.*, 2009), and this may place the species at some disadvantage for selection as an energy crop – at least within specific geographic regions. High leaf area and radiation use efficiency serve to compensate for observed lower CO_2 exchange rates (Kiniry *et al.*, 1999). Within switchgrasses, some studies suggest photosynthetic rates are greater with higher ploidy, but photosynthetic rate can change over the course of the growing season (Wullschleger *et al.*, 1996). Ecotypes and cultivars also vary in functions such as CO_2 exchange rate, maintenance of leaf water potential during drought and more rapid recovery following drought (Nickell, 1973). Greater drought tolerance in upland switchgrasses seems a likely adaptive measure to support growth in more mesic environments.

Switchgrass development closely correlates with growing degree days, which heretofore has been the primary basis for modelling switchgrass production (Brown *et al.*, 2000; Kiniry *et al.*, 2005). However, the nature of this relationship varies by cultivar and location, and this has limited the ability to develop predictive equations that could be broadly applied (Parrish and Fike, 2005). More recent modelling efforts have used empirical relationships between cultivar-specific maximum development rates and temperature optima (among other model inputs) to successfully predict switchgrass development and yield (Grassini *et al.*, 2009).

Modelling efforts can have practical applications for growers, processors and policy makers who might want to predict switchgrass yields and availability (Grassini *et al.*, 2009). Such predictions would be especially important in the siting of biorefineries. However, models also provide means to explore potential outcomes across a range of changing climatic conditions; having such predictive ability can help to ensure that expanded plantings of energy crops are 'climate proofed' (Oliver *et al.*, 2009).

We began this section thinking about the conditions leading to the rise of C_4 grasslands. As a key setting on the evolutionary stage, prehistoric climate changes likely served as a trigger for the advent and rise of warm-season grasses. There is perhaps some irony in climate change's reprise role as the spur that may prod humans to explore the value of warm-season grasses as sources of alternative energy – and to mitigate that very climate change.

10.3.3 Switchgrass morphology and phenology and implications for management

Switchgrass's diverse morphological presentation compares with that of its diversity of adaptation. Canopy heights typically range from 0.5 to 3.0 m, and roots may extend to 3 m depth (Porter, 1966; Moser and Vogel, 1995). Upland types commonly are distinguishable from lowland types by their shorter stature and the finer nature of their leaves and stems (Fig. 10.2). The taller, coarser lowland ecotypes generally have larger leaves and panicles, and the leaves frequently have a bluer coloration (Casler, 2005). Leaves generally are erectophile, and for both types, lamina lengths increase with successive appearance to about the seventh leaf. Subsequent leaf laminae are shorter, coincident with the initiation of floral development. This leaf pattern gives switchgrass the distinctive form that is the basis for its name.

Upland plants generally have thicker roots than in lowlands, and upland plants also generally have longer rhizomes, root internodes and overall root lengths. The bunch-grass form, more typical of plants with short rhizomes, is common in lowland types (Vogel, 2000) while upland types are more likely to form sod (Beaty *et al.*, 1978). New shoots originate from rhizome buds on lowland plants but also from basal nodes of culms in the upland ecotypes (Porter, 1966).

Seed size may vary by cultivar; this normally has little agronomic implication except perhaps in cases of excessive planting depth (Zhang and Maun, 1991b; Smart and Moser, 1999). Seedling morphology, rather than seed size, may be more important for the development and survival of new switchgrass plants, as plants with elevated crowns could in theory have less chance for adventitious root formation; but this line of reasoning has not borne out in field studies to date (Elbersen *et al.*, 1999). Similarly, selecting for increased tillering in seedlings may be effective for developing divergent genotypes but has not proven a suitable morphological selection trait for improving plant establishment or yield (Smart *et al.*, 2003, 2004).

Switchgrass acts as a determinate plant, producing a flush of tillers that become reproductive after exposure to the appropriate environmental cues, and biomass accumulation ceases with floral development. Although perhaps not under complete photoperiod control (Esbroeck *et al.*, 2003), day-length appears to be the primary driver for inflorescence development. Factors reported to delay or inhibit reproductive development include flooding, excessive K fertilization and low temperatures (Porter, 1966; Balasko and Smith, 1971; Friedrich *et al.*, 1977).

Fig. 10.2. Even with senesced switchgrass that has overwintered, distinct differences are visible among ecotypes and varieties. This photo, taken in late March in Blacksburg, Virginia, USA, shows lowland ('Alamo' – far left; 'Kanlow' – far right) along with several upland ecotypes in between. Note differences in height, leafiness and lodging among varieties. The child in the photo is about 0.90 m tall.

The short-day photoperiodic flowering response presents management with an opportunity to increase switchgrass biomass production. A cultivar's required photoperiod (some minimum critical night length actually) is genetically programmed and keyed to its latitude of origin. To turn an old phrase to a new purpose, 'you can take the cultivar out of the photoperiod, but you can't take the photoperiod out of the cultivar'. If a cultivar of southern origin is moved to more northern latitudes (where shorter nights occur during the summer), it will remain in a vegetative (biomass-producing) stage for longer than it would in its original locale. It will continue producing stems and leaves until it experiences nights longer than the minimum critical dark period that triggers flowering. Conversely, moving a northern line south will hasten its exposure to longer night lengths, thus reducing its growing season – and biomass accumulation. (Note, our point of reference for explaining this phenomenon is the northern hemisphere.)

10.4 Plant Breeding

Compared to most crops, very little breeding has been done with switchgrass. The majority of named switchgrass cultivars are essentially wild types or 'natural track': accessions collected from the wild, screened for a few years in a 'common garden' (and sometimes in field trials), increased for seed production, and then named and released (Casler *et al.*, 2012). A few cultivars have been developed following limited crosses, but it would seem that only the hem of the genotypic garment has been touched, given the great phenotypic diversity seen within the species. In short, the species is essentially undomesticated, and much remains to be learned about how to advance it genetically; but genetic improvement efforts, which began in the 1950s, have greatly increased in the last few years.

10.4.1 Reproductive biology

Switchgrass is an open-pollinating, self-incompatible species. Its self-incompatibility system is quite effective; bagged panicles set seed in <1% of fertile florets (Talbert *et al.*, 1983). The species' genome size varies widely, with somatic cells, e.g. root tips, containing from 18 to 108 chromosomes (Nielsen, 1944; Barnett and Carver, 1967). The basic chromosome number is $x = 9$, with ploidy levels of individuals ranging from $2n = 2x = 18$ (diploid, very rare), to $2n = 4x = 36$ (tetraploid), to $2n = 6x = 54$ (hexaploid, very rare), to $2n = 8x = 72$ (octoploid) and rarely higher. The higher ploidy levels obvious in chromosome counts that total 90 or 108 may occur only in somatic cells – not in those producing gametes. For those individuals that are tetraploid, the chromosomes act as 18 divalent pairs during meiosis, i.e. 18 linkage groups; and this disomic pairing pattern has been observed in octoploid individuals as well, i.e. 36 linkage groups (Riley and Vogel, 1982; Martínez-Reyna *et al.*, 2001; Okada *et al.*, 2010). Aneuploid and higher polyploid individuals are more likely to show polyvalent or polysomic behaviours (more than two chromosomes 'paired' during the coupling phase of meiosis) (Casler *et al.*, 2012).

As noted previously, essentially all lowland switchgrass lines are tetraploid, and most upland lines are octoploid. Crosses between octoploid and tetraploid genotypes are highly – if not totally – incompatible. By analogy with some racquet sports, it would appear switchgrass can 'play' with equally matched single or double (and even quadruple) sets of chromosomes in its gametes, but cutthroat is not allowed. Crosses between the two ecotypes, or cytotypes, are 'allowed', however. In a greenhouse setting, tetraploid cultivars from each of the two ecotypes have been crossed and produced fertile offspring (Martínez-Reyna *et al.*, 2001; Vogel and Mitchell, 2008). However, though tetraploid – and perhaps octoploid (Casler *et al.*, 2012) – lines of each ecotype may occur sympatrically, i.e. side by side in natural settings, upland–lowland crosses are unlikely without human intervention. The barrier is not genetic but physiological; the uplands tend to flower well before the lowlands do, eliminating the possibility for cross-pollination (Casler *et al.*, 2012).

These genetic and physiological factors must be taken into account in breeding programmes.

10.4.2 Breeding methodology

Switchgrass breeding programmes generally begin with acquiring accessions adapted to the locality for which the improved lines would be developed. Beginning with regionally adapted lines or 'gene pools' is crucial, because synchronous flowering (due to photoperiodic mechanisms, which are not well understood nor readily manipulated genetically) and suitable hardiness (whose causes are even less well understood) are essential to ultimate success (Vogel *et al.*, 2005; Casler *et al.*, 2007a,b). Crosses must be between those with like ploidy but with synchronized or overlapping anthesis as well.

The grist for the mills of switchgrass breeders is typically accessions, individual plants or a population, taken from the wild; but a named cultivar may sometimes be used as a starting point, since the open-pollinating, self-incompatible, polyploid nature of the species produces very heterozygous genotypes and heterogeneous phenotypes (Casler *et al.*, 2012). Many switchgrass accessions have been conserved at various Plant Materials Centers of the USDA's Natural Resources Conservation Service. The primary public collection of switchgrass germplasm is at Griffin, Georgia, USA, and contains about 500 accessions, of which about one-third are available in small quantities to requestors (http://www.ars-grin.gov). Many times that number of switchgrass accessions may be found in various institutional and private collections, but such materials are not generally, or readily, available and most have not even been characterized phenotypically as to whether they might be upland or lowland (Casler *et al.*, 2012).

Breeding populations for launching improvement efforts can come from accessions in appropriate regional collections or from plants collected by the breeder at a local site, given the high degree of heterogeneity and heterozygosity inherent in the species. At other times, crosses and subsequent selections are made by compositing populations from diverse sites to broaden the germplasm pool, and the cultivars resulting from a few cycles of selection are sometimes called 'synthetics'.

Before any crosses are made, the breeding population(s) will often be screened in spaced-plant nurseries, and more promising individuals may be selected and propagated either by seed or vegetatively. Subsequently, desired crosses can be made by allowing intercrossing *in situ* in the nursery or by juxtaposing the parent plants in various patterns. The F_1 and subsequent generations of crosses are typically then planted in a nursery for phenotypic observation, further selection and additional crosses. Often the offspring are planted in spaced, 'family' (or, more often, 'half-sib') rows, where all of the offspring from a particular cross (or from the seed-producing parent of a particular cross) can be looked at both collectively and individually. More promising families, and especially more promising individuals from those families, then become the candidates for further crosses or release. Spaced plantings, to include more intricate arrangements such as honeycombs, are used in these selection steps, and such space-planted selections have shown themselves to perform well when planted in swards (Casler, 2010).

10.4.3 Breeding objectives: traits of importance for a bioenergy feedstock

Early breeding efforts with switchgrass focused heavily on forage quality, with good progress made in improving the digestibility of switchgrass. The initial screenings and selections were done not on animals but with an *in vitro* dry matter digestibility (IVDMD) laboratory assay. The higher IVDMD lines provided improved animal gains and profitability. Interestingly, higher IVDMD might also translate to improved feedstock value when switchgrass would be used for bioenergy. The increases in IVDMD achieved by early breeders were largely due to reductions in levels of lignin, which interferes with ruminant digestion. As it turns out, lignin can also limit fermentation in biorefinery operations. Further reduction of lignin could be of interest to breeders who wish to improve switchgrass for bioenergy; but that objective will be tempered by the conversion technology to be employed. In thermochemical biorefineries – those relying more on 'straight' chemistry than on biochemistry – lignin may not be an anti-quality factor (Fike *et al.*, 2007). Since biorefinery technologies are still a moving target, breeders may want to await further developments to know whether lignin reduction or genetic modification of the level of some other biomass constituents will be desirable traits for a feedstock.

Certainly, increasing biomass yield is a primary interest for breeders, and switchgrass offers great promise in this regard. While few 'biomass' cultivars bred and selected specifically for increased biomass yield have been released, more are in the 'pipeline'. The breeding, selection, field trials, seed increase, etc. inherent in developing a new switchgrass cultivar take essentially a decade.

Biomass yield is a complex product of many complex agronomic, morphological, physiological and biochemical traits. Among the agronomic traits of interest is success in establishment. Switchgrass has a reputation for being slow or difficult to establish (Parrish and Fike, 2005). Sometimes the difficulty stems from improper seed placement or poor weed control, but the biology of switchgrass seeds and seedlings could perhaps be 'tweaked' genetically to increase the odds of success. For example, seed dormancy can be a major problem in switchgrass plantings (Parrish *et al.*, 2008; Parrish and Fike, 2009). In a simple demonstration, overall germinability of seed from 'Cave-in-Rock' switchgrass was increased by about 5% (from an original 5%) in each of three selection cycles (Dale Wolf, unpublished data). Similar reductions in dormancy of selections from 'Alamo' switchgrass have been achieved as well (Burson *et al.*, 2009). Other approaches to improving establishment have included selecting for larger seeds, for more vigorous seedlings, and for seedlings with lower crown node placement (via reduced elongation of the mesocotyl). The first two approaches have shown some promise for improving establishment or stand density – parameters that potentially can translate to increased biomass yields.

The biomass hauled to a switchgrass biorefinery is essentially chopped up tillers; so increases in biomass might logically be achieved by producing more or larger tillers. Efforts to select for tiller number in seedlings have shown that the trait is heritable and that tiller numbers in mature plants parallel alterations in the seedling (Smart *et al.*, 2004). Seedlings with fewer tillers developed into adult plants with fewer, but larger, tillers. The net result was greater biomass per plant. Some breeders have had success in increasing tiller biomass by looking at tillers as repeating morphological units, or phytomers, and increasing their number and/or size (Boe and Casler, 2005).

We have ignored until now the yield implications of the obvious morphological differences between upland and lowland ecotypes. Where they are equally adapted, i.e. where each can tolerate the climate and diseases present, a lowland cultivar will essentially always out-yield an upland cultivar. With their thicker and taller tillers (and often with their longer periods of vegetative growth), lowland cultivars can produce >20% more biomass than uplands when each is cut once per season (e.g. Fike *et al.*, 2006a,b). This does not mean, however, that biomass improvement efforts have been or should be focused strictly on the larger ecotype. The lowland germplasm lacks hardiness to survive winters in much of switchgrass's range. Lines suitable as starting points for breeding efforts must be regionally adapted; and, for much of the northern portion of the USA, that means only uplands. In switchgrass's southern range, lowland cultivars or accessions would be the overwhelming choice for breeding lines.

Besides the physiological limitations imposed by cold hardiness, the physiology of flowering, which was alluded to in the previous paragraph, must be taken into account in breeding for biomass yield. The biology of switchgrass is such that it essentially ceases biomass production after it flowers and produces seed (Parrish and Fike, 2005), and the plateau of biomass yield may be reached with many weeks left in the growing season. During the course of the growing season, plants produce both vegetative and reproductive (panicle-bearing) tillers, but growth of both types essentially ceases as panicles are produced (unless the sward is mowed). That can mean significant portions of the growing season are not producing biomass – especially if northern-derived lines are planted at lower latitudes. Growing farther south means exposure to shorter summer days (actually, longer summer nights), which trigger earlier flowering, stopping the production of biomass-building stems and leaves even sooner. Conversely, plants from a lower-latitude germplasm pool will flower later (and potentially produce more biomass) at a slightly higher latitude; but winter hardiness limitations with more southern germplasm pools restricts the use of this approach to perhaps one USDA plant hardiness zone (Casler *et al.*, 2012). Finding and breaking the linkage between photoperiodic control of flowering and the physiological or biochemical mechanisms for cold hardiness is an obviously desirable, if complex, objective.

Additional breeding objectives in programmes seeking to improve switchgrass for bioenergy purposes include pest resistance. Although the species has relatively few serious disease and insect pests, it is likely they will increase if/when the crop is grown on much expanded hectarages (Parrish and Fike, 2005). Tolerance of abiotic stress, e.g. heat and drought, will also likely be increasingly important as the climate changes.

10.4.4 Hybrid and molecular approaches to switchgrass improvement

Development of hybrid maize contributed tremendously to the improvement of that open-pollinating, heterozygous species. Switchgrass could potentially benefit similarly and for the same genetic reasons. Hybrid vigour increases, or heterosis, of 30–38% (when compared to best parental lines) have been observed in crossings between (Vogel and Mitchell, 2008) but not within (Martínez-Reyna and Vogel, 2008) upland and lowland cultivars. The crosses that provide the dramatic heterotic effects presumably work because evolutionary divergence between the two ecotypes has resulted in good allelic complementation.

Commercialization of hybrid switchgrass to the point that hybrid maize now enjoys will be a protracted process at best for genetic and logistical reasons. The open-pollinated, highly heterozygous nature of the species again will frustrate breeders, even while it holds so much potential. Parental lines for hybrid crosses could be heterozygous or they could be inbred and more nearly homozygous. The latter approach will take many years of inbreeding (which would be complicated by the species' great self-incompatibility) and the trial-and-error determination of heterotic patterns. Crossing defined heterozygous parental lines with good heterotic complementarity will likely be a quicker route to commercialization of hybrid switchgrass; but producing clones of parental lines in sufficient numbers for large-scale plantings will pose challenges, as will getting anthesis of the two lines to coincide. 'Alamo' has been regenerated from tissue cultures, but most other cultivars have not lent themselves so readily to *in vitro* propagation or cloning. When these logistic and genetic hurdles can be overcome, hybrid seed technology could do for switchgrass what it did for maize.

Molecular approaches to improving switchgrass are in their infancy. Marker-assisted selection holds promise (Casler *et al.*, 2012), but the complexity and semi- or quasi-redundancy of the genome, e.g. multiple ploidy levels to include many octoploids, make the approach more difficult. Transgenic switchgrass plants have been produced, but logistical and likely legal/environmental issues make wide-scale or commercial use of a transgenic approach to improvement rather problematic. The problems include the difficulty of regenerating plants from non-'Alamo' tissue cultures after transgenes have been inserted and, given the highly outcrossing nature and the frequency of native/wild switchgrass, it is highly unlikely that cultivars producing transgenic pollen could be released. However, cytoplasmic male sterility such as that employed in hybrid maize production might be inserted into a transgenic line to overcome this environmental risk (Martínez-Reyna and Vogel, 2008). If so, transgenic approaches to switchgrass improvement could become commercially feasible (Casler *et al.*, 2012).

10.5 Cultivar Selection for Biofuel Production

Considering its short history as a crop, and especially as a potential bioenergy crop, it should not be surprising that no switchgrass cultivar bred specifically for biomass production has been registered or is commercially available as of 2012. Historically, almost all switchgrass cultivars have been selected for their potential as forage crops, and most recent registrations describe cultivars that can serve dual purposes as feedstock for livestock and biorefineries alike – although management to these ends will likely be quite dissimilar. Most breeding efforts have striven to couple greater yield with greater digestibility for livestock; thus, their use in either enterprise would seem appropriate, although we will discuss a caveat in the section on conversion.

Regardless of end use, a switchgrass cultivar should be chosen for a given site by aligning its characteristics with the conditions 'on the ground'. Factors for consideration would include the need for drought or flooding tolerance or heat or cold stress resistance. In this context, we revisit the idea of moving southern (i.e. lower-latitude lines) northward (to higher latitudes). This can produce yield gains due to a delayed photoperiod signal for inflorescence development, but there is some risk involved. In order to develop sufficient freeze resistance, switchgrass must have adequate time to senesce and 'harden off' (Hope and McElroy, 1990; Casler *et al.*, 2004). Developing cold tolerance depends on the plant receiving the appropriate signal in time; thus, the northern march strategy can be lethally overplayed when southern-adapted lines are planted too far north or at too-high elevation (e.g. Casler *et al.*, 2004). Keeping cultivars within 500 km of their locale of origin serves as a useful, though general, rule-of-thumb (Moser and Vogel, 1995). Rules-of-thumb being rough practical guides, we note that exceptions to the '500-km rule' have occurred at high latitudes with moderate winters (e.g. Christian *et al.*, 2001) or at sites with low elevation.

10.6 Cultural Practices for Switchgrass in Biofuel Production Systems

Of all the things that need doing to promote success in agronomic production systems, perhaps the most critical and overarching factor is getting the work done in a timely fashion. Switchgrass has a reputation for difficult establishment, but adequate, timely preparation can greatly allay this concern. In the following sections we will discuss methods for successful establishment and management, with the caveat that no practice is likely to bear much fruit if applied in an untimely fashion.

10.6.1 Site selection and preparation

Given the wide adaptability of switchgrass – taking for granted appropriate cultivar selection – one might assume that switchgrass can be planted almost anywhere in agricultural landscapes. Indeed, the presumption about energy cropping has been that this new enterprise would be introduced to agriculture's 'marginal' lands, thus minimizing conflicts with existing production systems and quieting 'food versus fuel' concerns. It perhaps seems contradictory, then, that the USDA NRCS technical field note on planting switchgrass for bioenergy production recommends that fields be chosen using 'criteria typically used for row crop agriculture' (Douglas *et al.*, 2009, p. 1) in order to avoid steep, irregular terrain and hydric sites among others. Thus, we note that while switchgrass has broad suitability to sites, not all sites may be well-suited to switchgrass (or other crops) for bioenergy production – whether for low productivity, environmental fragility, logistic constraints, or a combination of such factors.

Sites selected for switchgrass production ideally should be free of excess herbage residue at the time of planting. This may be removed by killing and tilling, by allowing sufficient time for decay after chemical burn down, or by direct combustion (fire). In addition to removing residue, burning may kill small weeds and help reduce weed seed loads in the soil seed bank, thus reducing competition for the new seedlings (Wolf and Fiske, 1995). This may be particularly true in old pastures and abandoned 'grow back' fields. Converting existing cropland to switchgrass production can present real advantages in terms of a clean seedbed and minimal tillage needs, although caution must be taken to avoid residual herbicides from the previous production system that could inhibit seedling germination (Douglas *et al.*, 2009).

10.6.2 Seeding methods and planting depth

Stand establishment can be successful (or not) with both conventional and no-till seeding methods (Parrish and Fike, 2005). Tillage can be used to good effect for site preparation, but one should first consider the 'fragility' of the site; tillage should be avoided on those sites susceptible to erosion. In our experience, switchgrass emerges best when soils are quite firm. Thus, tilled fields should be firmly packed either before seed are drilled or after broadcast, which is distinctly less desirable than placing seeds at depth in conventional systems. No-till establishment also can be quite effective. No-till systems allow for soil moisture conservation, prevent erosion losses, save fuel and allow farmers into the field earlier after rainfall (Parrish and Fike, 2005; Douglas *et al.*, 2009) These systems also can leave some residue on the field surface, which further helps retain soil moisture; but, if residues are present in excess, the drill coulter can push residues into the drill furrow, preventing good soil–seed contact and reducing establishment success (Wolf and Fiske, 1995).

Planting depth is critical for successful stand establishment, and poor seed placement can produce stand failure with both conventional (especially broadcast) and no-till seeding methods. General recommendations are to plant to a depth of about 1±0.5 cm, although emergence from greater depth is possible in sandy soils (Zhang and Maun, 1990). Soil texture is the chief consideration for this variable recommendation. Greater depth improves establishment in coarser soils, where moisture limitations can slow imbibition, germination and emergence (Newman and Moser, 1988; Evers and Butler, 2000; Evers and Parsons, 2003). Conversely, heavier, fine-textured soils can impede emergence of low-vigour seeds, especially when they are planted too deep.

10.6.3 Seeding rates and seed dormancy

Seeding rate recommendations typically fall somewhere between 4 and 10 kg ha^{-1} (Vassey *et al.*, 1985; Moser and Vogel, 1995; Wolf and Fiske, 1995; Vogel, 2000; Teel *et al.*, 2003). Such recommendations ultimately could be pushed quite a bit lower provided sufficient agronomic technique and pest control. Recent research comparing multiple seeding rates suggests rates as low as 2.8 kg ha^{-1} can be successful, although biomass yields in year 1 may be reduced (Mooney *et al.*, 2009). Similarly, limited effects of drill row spacing have been observed (Sanderson and Reed, 2000; Muir *et al.*, 2001) and, in some cases, yields actually may be better with wider spacing (Ma *et al.*, 2001). Although high biomass yields in the planting year may be desirable, they are not the final arbiter in a stand's success (i.e. its productivity) over a multi-year timeframe. In a multi-site study, Schmer *et al.* (2006) investigated thresholds for successful establishment using frequency grids. Stand frequencies of ≥40% were considered successful plantings for biofuel purposes.

Switchgrass seeds often have high levels of dormancy, and this has been a common cause of stand failures (Parrish and Fike, 2005). With recently harvested seed, dormancy rates may approach 95%. The seed is sold on a Pure Live Seed (PLS) basis – determined as % purity × % viability × 100. Unfortunately, this number does not predict 'practical' germinability, i.e. the percentage of seeds likely to germinate when placed in the soil, which is a critical factor for establishing plantings during the growing season. We make this 'growing season' distinction because dormancy may not be evident when seeds are planted during the winter or early spring and given opportunity to stratify in the soil. Stratification breaks dormancy and greatly increases seed germinability by the time soils warm to germination temperatures in mid- to late spring (e.g. see Sanderson *et al.*, 1996). 'Off-season' plantings can be successful both in late autumn or early spring, but success rates depend on many factors including ground cover conditions and moisture (Vassey *et al.*, 1985; Zarnstorff, 1990; Teel *et al.*, 2003). Weed control following 'dormant plantings' also can be a significant challenge to good establishment, because many weed species may germinate and grow before switchgrass does (Hsu *et al.*, 1985b).

Planting after soils have warmed to about 15 or 20°C – around the optimum for switchgrass germination (Hsu *et al.*, 1985a) – can improve germination conditions and avoid some of the weed competition associated with dormant season plantings. However, this will require non-dormant seed to have fullest effect. Many researchers have explored ways to artificially break seed dormancy (Parrish and Fike, 2005) in order to remove the requirement for stratification from Mother Nature – and thus gain greater control of establishment in the process. Although several techniques can work, the simplest interventions involve soaking dormant seeds in water, draining off the water, and holding at 5–10°C for at least 1 month (Wolf and Fiske, 1995). This method, though quite feasible on the small scale, has several practical limitations and may prove impossible for planting very large hectarages. Anthropogenic stratification also can fail, e.g. mishandled seed can heat or ensile, and seed can revert to dormancy if they have insufficient exposure time. After-ripening, a method that relies on time to break dormancy, involves holding seed at moderate to slightly elevated temperatures for a period of up to several months (Shen *et al.*, 1999). While seed sellers may be reluctant to maintain inventory for multiple years to break dormancy, a grower could purchase seed in the harvest year, hold at moderate temperatures for a year, and thereby break most of the dormancy.

Selection, breeding and inoculation techniques may also be deployed to reduce or eliminate dormancy and to increase the speed of germination. While smaller-sized switchgrass seeds can be slower to develop, this generally has not been seen as a limitation to establishment (Zhang and Maun, 1991a; Aiken and Springer, 1995; Smart and Moser, 1999). However, selecting seedlings from neoteric (freshly harvested) seeds that readily germinate can effectively reduce dormancy in just a few selection cycles (Sanderson *et al.*, 1996). Adding a wrinkle to that approach, Burson *et al.* (2009) first selected plants from rapidly germinating seeds and then used these plants to establish a crossing nursery to breed for rapid germination. Other research suggests that inoculating seeds with specific mycorrhizal fungi can improve both seed germination and seedling growth, although this work was with scarified seed (Ghimire *et al.*, 2009).

10.6.4 Weed control

Beyond the issues of seedbed preparation and seed dormancy, controlling weeds has been one of the biggest challenges associated with switchgrass establishment (Moser and Vogel, 1995a; Evers and Butler, 2000; Mitchell and Britton, 2000). Although weed species vary by region, annual warm-season grasses such as the foxtails (*Setaria*) and crabgrasses (*Digitaria*) are some of the most broadly adapted and competitive species for switchgrass establishment. For example, crabgrass (*D. sanguinalis*) grows more rapidly than switchgrass at temperatures on the low side of each species' growth range (Hsu *et al.*, 1985b; Hsu and Nelson, 1986a,b), presenting a strong challenge to establishment.

Herbicides such as glyphosate or grammoxone can be used to good effect in preparing plots for planting and may give temporary control over existing weeds at the time of planting, but they have no effect on annual grass weed seeds waiting to germinate. Of possible herbicidal solutions, atrazine (6-chloro-*N*-ethyl-*N'*-(1-methylethyl)-1,2,5-triazine-2,4-diamin) and other triazine herbicides have been some of the most effective for weed control (Martin *et al.*, 1982; McKenna *et al.*, 1991; Mitchell *et al.*, 2010). Label restrictions have limited atrazine use, although this is not banned universally. Even where it is currently restricted, some states have authorized use for switchgrass-to-bioenergy plantings (Teel *et al.*, 2003), and regulatory changes might be expected as a large-scale industry develops. Quinclorac (3,7-dichloro-8-quinolinecarboxylic acid) has also proven a useful tool for grassy weed and broadleaf control, especially in combination with atrazine when applied at planting (Mitchell *et al.*, 2010), although application to new seedlings can cause injury and reduce biomass production (Boydston *et al.*, 2010).

Imidazol herbicides ((±)-2-(-isopropyl-4methyl-5-oxo-2-imidazolin-2-yl)-5-methyl-nicotinc acids) can be useful alone or in combination with sulfometuron (2-(((((4,6-dimethyl-2-pyrimidinyl)amino)-carbonyl) amino) sulfonyl) benzoate) when applied in autumn prior to spring switchgrass seedings (Masters *et al.*, 1996; Masters and Nissen, 1998). We know of some planters who use the imidazol herbicides during the spring of switchgrass establishment, but they are known to cause injury to seedlings (Peters *et al.*, 1989; Wilson, 1995; Masters *et al.*, 1996). This problem may be countered in part by increasing seeding rates. Damage and reduced production in the establishment year may be tolerable, as long as there is good establishment and productivity in succeeding seasons. However, reduced growth in the first season can delay achievement of mature stand yields, thus imidazols are not recommended (Mitchell *et al.*, 2010). Sulfometuron applications have been used successfully for post-emergence weed control (Peters *et al.*, 1989).

In addition to herbicides, mechanical control can be used to some effect in new stands. Mowing weeds just above the height of newly establishing switchgrass seedlings prevents the over-topping of seedlings. This may require several clippings in a season; but, if the seedlings are successfully established in the planting year, stands are generally less susceptible to weed competition in the second growing season.

10.6.5 Fertility considerations for establishment

Fertility management has important implications for switchgrass production costs, and it also interacts with harvest management practices, which we will discuss shortly. On the 'front end' of establishment, appropriate fertility practices – particularly with respect to N – can help minimize weed pressures and improve stand success. For example, added N has little impact on switchgrass yields at establishment (Sanderson and Reed, 2000), but the added fertility does support the aggressive growth typical of many weeds, greatly increasing competition with the switchgrass seedlings. Thus, the standard recommendation for N fertility calls for no establishment-year N applications (Jung *et al.*, 1988; Wolf and Fiske, 1995; Brejda, 2000; Sanderson and Reed, 2000).

Data for soil pH and other nutrients at establishment are not unequivocal. Potassium (K) has not been implicated in boosting seedling growth at establishment. Phosphorus (P) can improve plant size and first-season yield, but the effects may be transitory (McKenna and Wolf, 1990). Switchgrass is quite tolerant of low pH, and liming effects on establishment in acidic soils or mine tailings have not been observed in several studies (Harper and Spooner, 1983; Bona and Belesky, 1992; Hopkins and Taliaferro, 1997), although yield gains were reported for the combination of lime and P (McKenna and Wolf, 1990).

10.6.6 Fertility for mature plantings: N management

Among nutrients to manage for mature, productive switchgrass-to-bioenergy systems, N gets top billing because of its potential impacts on crop growth, feedstock quality, economic returns and environmental outcomes. Many studies testing switchgrass's response to N have been conducted, and the results and recommendations that have followed have been nearly as numerous (Parrish and Fike, 2005). The broad recommendations follow a diverse set of responses to N. For example, in a summary of yield responses to added N (versus a 0-N control), Brejda (2000) found biomass yield increased from 0 to 6.2 Mg ha^{-1}. Drivers of this N-response variability include harvest management, low nutrient demand/internal nutrient cycling, soil types (and inherent N fertility), precipitation and atmospheric N deposition.

As noted above, part of the variation in response to N comes from end-use considerations. Nitrogen requirements vary greatly depending on whether the material is harvested for forage or for bioenergy feedstock or both, and a strategy of replacement (i.e. N in for N out) is perhaps most appropriate for long-term stand (and environmental) sustainability (Lemus *et al.*, 2008). Harvest and removal of green biomass during the growing season under management typical for hay production will require higher N inputs to maintain stand productivity. This is due to the greater amounts of N removed with forage production than is typical of bioenergy systems (Parrish and Fike, 2005; Guretzky *et al.*, 2011).

Switchgrass has low nutrient demand (relative to many other potential energy crops) and the plant is able to 'recycle' nutrients (Beaty *et al.*, 1978; Lemus *et al.*, 2008; Garten *et al.*, 2011) after the growing season. Such capacity to 'sink' N into below-ground biomass at the end of the growing season appears to be a common feature in plants from North America's tall-grass prairie (Hargrave and Seastedt, 1994) and is one of many advantages for perennial, herbaceous biomass production systems. While this ability to recycle does not reduce the crop's N requirements *per se*, it does effectively reduce the amounts of N that will be harvested (and thus required for replacement) under post-growing-season harvest management. Biomass N levels decline through the growing season (Waramit *et al.*, 2011) and generally are quite low in senescent feedstock, particularly with a harvest a few weeks after a killing frost. Nitrogen concentrations in the realm of 5–8 g kg^{-1} are typical for such single, end-of-season harvests (Madakadze *et al.*, 1999; Fike *et al.*, 2006a; Guretzky *et al.*, 2011).

Along with effects on production and N concentrations, fertility also affects plant morphology. With high N rates, switchgrass partitions greater amounts of nutrients to above-ground growth. This reduces root:shoot ratios (Heggenstaller *et al.*, 2009; Garten *et al.*, 2011) and coincides with the observation that switchgrass managed under high-N, once-per-season cutting regimes has fewer tillers (Muir *et al.*, 2001; Fike *et al.*, 2006a). Although above-ground biomass production may not be affected (Muir *et al.*, 2001) by changes in canopy architecture, the decrease in below-ground biomass has implications for carbon sequestration and greenhouse gas emissions (Garten *et al.*, 2010). This issue will bear further research as increasing N applications may have offsetting benefits in terms of increased soil organic carbon stocks (Jung and Lal, 2011).

Requirement of switchgrass for N may vary substantially with soil type and biology. Responses to N typically are greater in sandy soils (e.g. Ma *et al.*, 2001; Muir *et al.*, 2001), and minimal responses may occur in soils that have high levels of N in the soil organic pool (Stout and Jung, 1995). Some research is suggestive of biological N fixation (Tjepkema, 1975; Riggs *et al.*, 2002; Ker *et al.*, 2010) and, if correct, would help further explain the limited response to N inputs commonly observed in many environments (Parrish and Fike, 2005). Even without such associations, the plant's ability to extract N from soils and to internally sequester and recycle that N are attractive features relative to minimizing N inputs.

Alternative nutrient sources may play a role in further minimizing the need for fertilizer N inputs. Manure applications can support greater switchgrass production (Sanderson *et al.*, 2001; Lee *et al.*, 2007, 2009) and also may improve stand composition (Lee *et al.*, 2007). Legumes and alternative N sources also may provide the N needed to fuel switchgrass growth; but compatibility and long-term productivity, particularly when used for bioenergy production, remain in question (Springer *et al.*, 2001; Bow *et al.*, 2008). For example, despite apparently supplying switchgrass with N, the early-season growth of legumes such as lucerne (*Medicago sativa* L.) and red clover (*Trifolium pratense* L.) can compete with switchgrass, shifting its growing season and reducing yield (El Hadj, 2000).

10.6.7 Fertility for mature plantings: P, K and lime

Similar to N, the data are varied regarding response to P applications. Most reports on switchgrass's response to P suggest that little is needed when managed in a biomass-for-bioenergy setting (Hall *et al.*, 1982; Muir *et al.*, 2001).

In some establishment studies, P has supported first-year growth but was ineffective in subsequent growing seasons (McKenna and Wolf, 1990). In contrast, positive responses to up to 45 kg ha^{-1} have been reported in a forage production setting (Rehm, 1990).

The variable responses to P may in part reflect the variable relationship between switchgrass and mycorrhizae. Switchgrass's ability to extract P in acidic, low-P and high-aluminium soil environments is greatly enhanced by mycorrhizae (Koslowsky and Boerner, 1989; Boerner, 1992a,b; Clark *et al.*, 1999a,b; Clark, 2002). Research from 'opposite directions' has shown the value of mycorrhizae for plant P fertility relations: adding mycorrhizae to sterilized soil eliminated a response to P (Brejda, 2000), while eliminating mycorrhizae reduced switchgrass growth unless P was added (Bentivenga and Hetrick, 1991).

Not unexpectedly, repeated annual harvests with no nutrient returns will draw down available soil P in switchgrass production systems (Schmer *et al.*, 2011). Schmer *et al.* (2011) reported that annual losses in the top 30 cm averaged 1.5 kg P ha^{-1} $year^{-1}$ during on-farm trials where production averaged about 5.8 Mg ha^{-1}. However, many of the fields in this study were harvested near anthesis, likely resulting in greater P removal, as P would be expected to decline with maturity and following senescence (Parrish and Fike, 2005; Lemus *et al.*, 2009).

Switchgrass has displayed little response to K in greenhouse and field studies (Friedrich *et al.*, 1977; Smith and Greenfield, 1979; Hall *et al.*, 1982). Thus, recommended K applications, similar to P, call for levels in the medium range for soil tests (Teel *et al.*, 2003; George *et al.*, 2008; Douglas *et al.*, 2009).

Lime may have benefit for switchgrass plantings on waste sites with very low pH (Jung *et al.*, 1988) and may benefit seedling growth in the establishment year (McKenna and Wolf, 1990). As with other mineral nutrients, response limitations suggest lime will not be needed except in very acidic soils. A potential caveat is worth mentioning; some switchgrass lines have displayed obvious productive advantage in terms of adaptability to soil acidity (Bona and Belesky, 1992). Thus, screening for adaptability to acid soils may be a useful practice to find advanced lines that could be productive on truly marginal sites with limited inputs.

10.6.8 Harvest strategies, feedstock yield, quality interactions and logistics considerations

The early reports regarding switchgrass harvest research considered the crop's management in forage production systems – a familiar motif, indeed – and much of that work indicated that multiple harvests reduce productivity and stand survival (McMurphy *et al.*, 1975; Beaty and Powell, 1976; Balasko *et al.*, 1984; Haferkamp and Copeland, 1984; George and Obermann, 1989; Belesky and Fedders, 1995). Some bioenergy-based research comparing upland and lowland varieties under single and multiple (two) harvests per season suggests that added harvests with lowland ecotypes achieved only modest, if any, yield gains, but the added harvest produced substantial (30–40%) yield gains for upland ecotypes (Fike *et al.*, 2006a,b). In contrast, Madakadze *et al.* (1999) reported reduced yields of uplands managed with three cuts per season in Canada. Whether the difference in upland ecotype response between these studies is a function of more frequent cuttings, a shorter growing season, or both, cannot be determined with these limited data.

Part of the response to cutting frequency involves the timing of harvest. In two-cut systems, harvesting switchgrass near the time of anthesis provides the best opportunity for regrowth and appears to be a function of reserve carbohydrate status (George *et al.*, 1989). Early-season growth depends on the mobilization of storage carbohydrates and proteins in roots and stem bases; in-season harvests that occur before switchgrass replenishes these reserves may affect subsequent regrowth, ultimately weakening stands and negatively affecting future yields (Smith, 1975). Cutting frequency or timing effects may be further compounded by high defoliation intensity. Intense defoliation during the growing season can reduce the number of growing points (axillary buds) available for regrowth. Although lower cutting heights during the growing season may increase biomass yields in the short term (George and Obermann, 1989), sparing the axillary buds (by cutting at a greater height) during the growing season supports greater regrowth and will likely have long-term positive benefits in terms of overall stand health and productivity.

Although cutting twice per season may increase yield, harvesting during the growing season removes more nutrients than if switchgrass is harvested only after senescence (Vogel *et al.*, 2002; Fike *et al.*, 2006a,b; Guretzky *et al.*, 2011). Nitrogen and minerals harvested in 'green' biomass are anti-quality components for thermochemical process systems, and their removal increases production system costs given the need to replace greater quantities of nutrients. Because delaying harvest past the point of senescence allows for drying and nutrient translocation to storage organs and improves overall feedstock quality, a more typical harvest strategy, then, will be to take a single annual cutting at the end of the growing season. While this may not render the maximum recoverable biomass, it serves to optimize yields relative to input costs. For example, a recent analysis suggests the economic optimum for switchgrass management would be annual inputs of about 69 kg N ha^{-1} with a single annual harvest (Aravindhakshan *et al.*, 2011).

While we suggest that switchgrass is best harvested after completion of the growing season, there may be compelling reasons for harvesting through a much broader window of time. For example, the sheer volume of biomass needed to supply a significant portion of the existing liquid fuel market: gathering this in a narrow harvest window becomes most problematic in the face of variable weather and field conditions; in some regions, the 'post-senescence' window for harvest may be small enough to compel a farmer to harvest during or before the end of the growing season. While it may negatively affect fuel yield or quality per unit biomass input, a biorefinery system could reduce costs by using both 'in-season' and 'post-season' harvests to reduce the storage needed to maintain a year-round supply of feedstock. Spreading harvests through time would also reduce the costs in manpower and equipment needed to collect and distribute biomass to the process site, particularly for contract harvesters and haulers. As should be clear, there are a number of interacting factors all along a bioenergy supply chain (Fike *et al.*, 2007), and the balance of such benefits and disadvantages will certainly need determining as bioenergy systems come online.

10.7 Sustainability and Related Issues

Landowners' justification for growing a particular crop is usually economic: to earn a return on their land by providing society a saleable commodity. Important as this end is, an overemphasis on economic outcomes can result in broader ethical and ecological (and even economic) dysfunctions. We contend that the cultivation of crops – especially those that might be grown on a large scale – should satisfy ethical, ecological and economic criteria; and, for an energy crop, there is also an energetic criterion. In this context, many feel that producing fuel from maize grain fails to meet three of these four criteria. It has been an economic boon for producers, but energetic, ecological and ethical outcomes may be less sanguine. Before switchgrass is grown on millions of hectares, we should try to determine if that can be done in ways that satisfy all four criteria. We shall try to make such an analysis primarily as 'thought experiments', rather than trying to sort through often conflicting information from the literature. This is somewhat in the manner of Patzek (2010) who, by using a probabilistic analysis, reckoned that growing switchgrass for ethanol might not be as desirable a policy strategy as improving photovoltaic and battery technology. Perhaps not coincidentally, that author is in a department of petroleum and geosystems engineering. Perhaps not coincidentally, we, who are in a department of crop and soil environmental sciences, will come to a more charitable conclusion with regards to switchgrass's potential for helping to alleviate our fossil fuel dependency. That being said, we do not suppose or propose that biofuels can be sustainably produced in sufficient quantities to satisfy even 20% of the world's current fuel needs, let alone future needs. We concur at least partially with the assessment of others that over-zealous attempts to obtain our energy needs from dedicated energy crops (or biomass in general) could have negative consequences (Chaves, 2009).

10.7.1 Economic dimensions of sustainability

With regards to the economics of switchgrass-for-energy production, many studies have now examined what price farmers might receive for biomass produced, and it is highly variable depending on region and other factors (e.g. Chang *et al.*, 2005; Bransby *et al.*, 2008; Perrin *et al.*, 2008; Jain *et al.*, 2010). Some studies have modelled vertically to determine what margins might be achievable at various levels within switchgrass-based bioenergy production systems (Laser *et al.*, 2009; Sendich and Dale, 2009; Sokhansanj *et al.*, 2009). Whether such systems will provide sufficient incentives (profits) at each level to spur development depends on two volatile variables: the price of energy and governmental policies. As energy prices (US$ kW^{-1}) go up, all energy sources (to include non-conventional methods for extracting fossil fuels) will rise on the same tide, eventually making some economically viable. (In our view, switchgrass-based systems are perhaps currently in a not-yet viable energy price range.) With regard to policy environments, western Europe has taken steps to make bioenergy more viable, and that has spurred interest in switchgrass (Lychnaras and Schneider, 2011); and US energy policies have helped, albeit erratically, to drive interest in ethanol-from-maize (McGee and Hilton, 2011). Carbon offset policies could have dramatic effects on biomass (and other crop) markets wherever they might be instituted (Popp and Nalley, 2011). If externalities of fuel choice such as climate change and human health effects are monetized, the advantages of cellulosic ethanol or other 'drop-in' fuels can far outdistance both gasoline and ethanol from maize grain (Hill *et al.*, 2009). In summary, we acknowledge that economic sustainability is a *sine qua non*. Switchgrass will be grown for energy purposes only if it is profitable, and it will become profitable if/when energy prices and public policy converge to make it so.

10.7.2 Energy dimensions of sustainability

With regards to energy balance, switchgrass-based fuels would appear to be clear net energy producers (Schmer *et al.*, 2008; Vadas *et al.*, 2008). This stands in rather jarring contrast to widely varying analyses done on the energy input:output ratio for ethanol produced from maize grain (e.g. Pimentel, 1991; Mamani-Pati *et al.*, 2011). However, while hard data are available for doing energy analyses of grain- or cane-sugar-based ethanol-producing systems, some assumptions or estimates must be made for biomass-based systems. The ones that are perhaps most volatile are related to biorefinery efficiency (e.g. litres liquid fuel produced Mg^{-1} biomass) and energy consumption (e.g. kcal (or kW) consumed kg^{-1} biomass). Indeed, we do not yet know what fundamental chemical processes will be employed in successful biorefineries: biochemical, thermochemical, or pyrolytic/gasification. Until commercially successful biorefineries are online and willing to share information about their efficiencies, we must rely on educated guesses about conversion yields (e.g. the often-cited 600–700 l Mg^{-1} biomass) and energy input:output ratios, but we expect the numbers to suggest switchgrass-for-energy enterprises are net producers of energy.

(We note parenthetically that work coming from Resource Efficient Agricultural Production (REAP) Canada is championing direct combustion of pelletized switchgrass biomass for space heating (e.g. Samson *et al.*, 2005). When considered as a strictly energy-producing enterprise (with no value-added 'by-product' stream such as might come from a biorefinery), such systems are clear 'winners'. They have energetic and economic costs for densifying biomass, but they avoid the economic costs and energy losses associated with a biorefinery. Unfortunately however, they do not produce liquid fuels needed for transportation.)

10.7.3 Environmental dimensions of sustainability

Our 'thought experiments' on sustainability continue with a consideration of the ecological/

environmental implications of potentially employing switchgrass-for-energy on a large scale. Would such an enterprise be sustainable in the sense that it could be carried out *ad infinitum* without doing damage to or exhausting key environmental services? We concede immediately that all agroecosystems sacrifice or diminish some ecosystem services in the name of magnifying the food provisioning service. The well-known ecological observation/principle of old-field succession suggests agroecosystems are inherently unstable and not sustainable without regular inputs, i.e. human management. If abandoned, farm fields will – using only natural, renewable inputs – produce a metastable community, i.e. one that can continue in perpetuity. Such climax communities meet the definition of sustainable in a deep ecological sense. Any other state, to include intermediate successional series, is unstable and therefore unsustainable, i.e. cannot be maintained *ad infinitum* without conscious inputs or management.

Of course, this *über* ecological definition of sustainability would put sustainability out of the reach of all agriculture, which, in its very name, has an implicit human component. However, some argue that humans are merely another biological component of agroecosystems, making their inputs 'natural' also. So the question can perhaps more suitably be framed somewhat Hippocratically: can humans manage switchgrass-based agroecosystems in ways that minimize their potential to do harm to the rest of the environment? The environment can be described simply as living and non-living matter plus energy. The life forms and energy are rather dynamic – in flux spatially and temporally. By contrast, non-living matter makes up the stage (soil, air and water) on which life and energy act. Accordingly, we can examine the sustainability of energy cropping systems by looking at their potential to do harm to other living things, to soil, to air and to water. When switchgrass is held up to this metric, we feel that it fares well in at least three of the four instances.

Seeking biodiversity within a monoculture might seem oxymoronic, but several studies have examined avian fauna in switchgrass plantings (e.g. Roth *et al.*, 2005; Robertson *et al.*, 2011). A few studies have looked at other vertebrates, insects and soil organisms (e.g. Bliss *et al.*, 2010; Werling *et al.*, 2011). In all such cases, one sees a less diverse fauna than might be present in a more diverse plant community. Species that prefer tall grasses may be more abundant, but many other species may be essentially excluded. A priori reasoning suggests that planting switchgrass (or any other species in monoculture) on large tracts of land would have an adverse effect on biodiversity within that area. Likewise, the harvest management of switchgrass could potentially adversely affect even those species that might favour its habitat. Conversely, properly managed fields (or agroforestal interplantings) interspersed with other habitats might encourage greater biodiversity – especially compared to maize or soybean (*Glyine max* L. Merr.) grown from horizon to horizon. In sum, growing switchgrass monocultures on very large scales – especially if done by putting non-cropland or marginal cropland into production – would likely reduce biodiversity over that same scale. As a minimum, however, we can suggest that it should be no more harmful than other large-scale monocultures.

With regards to the abiotic trinity of the environment (soil, water and air), switchgrass culture could help conserve or even improve all three – again when compared to other agroecosystems, but not necessarily when compared to 'natural' ecosystems. Because switchgrass is perennial and a native prairie grass, well established and managed stands should minimize erosion (Nelson *et al.*, 2006) and even contribute to soil-building processes (Schmer *et al.*, 2011). Water quality could certainly be improved also – again relative to annual, conventionally tilled crops. Indeed, switchgrass is often cited as a desirable vegetative filter strip, able to reduce both soil and chemical runoff from agricultural land (Mersie *et al.*, 2003; Blanco-Canqui *et al.*, 2004). As a C_4 species, it has inherently higher water use efficiency than C_3 species – making more units of biomass per unit of water transpired.

One recent report has suggested that switchgrass culture may be unsustainable – or at least perturb hydrologic cycles – because it draws more and deeper water due to its longer (than maize) growing season (Le *et al.*, 2011).

We question the concern: water must be extracted from the soil profile if plants are to make biomass. Tall-grass prairie ecosystems, dominated by deep-rooted grasses such as switchgrass (Porter, 1966), are prolific perennial biomass producers in regions that receive >50 cm of rainfall annually. Such natural ecosystems would seem to meet all the criteria for sustainability. It is indeed ironic that in a bioenergy setting, such perennial energy crops – grown in part to alleviate climate change – would be considered a problem rather than a partial solution to this issue.

The converse argument is that atmospheric (and climatic) consequences of large-scale switchgrass production could be salubrious (Hill *et al.*, 2009). Ideally, biofuels would be at least CO_2 neutral, releasing into the atmosphere only the CO_2 recently sequestered in biomass, but some fossil fuels will inevitably be used in biofuel production. Life cycle analysis suggests using E85 fuel would release up to 65% less CO_2 compared to gasoline (Yu *et al.*, 2010). By some analyses, deploying switchgrass-for-energy on a large scale could result in a short-term reduction in emissions of this greenhouse gas, as CO_2 sequestered in root biomass would turn over and build organic matter in the soil. This might be a limited reversal at best, however, because organic matter can accumulate only to some upper limit, which varies with location/soil and is determined by a variety of factors (Sartori *et al.*, 2006). However, such biotic limits to carbon sequestration may be surpassed if renewable liquid fuel systems develop around pyrolitic conversion. Such systems may ultimately supply large quantities of recalcitrant carbon as a biochar co-product (Laird, 2008). This also has implications for other greenhouse gases, e.g. N_2O, that may be released as a result of agronomic and/or biorefinery operations. Clearly, systems – both agroecosystems and biorefinery systems – should be examined with an eye to minimizing emissions (Spatari and MacLean, 2010).

In summary, we posit that switchgrass-for-energy might be employed on the millions-of-hectares scale being proposed by some without doing harm to the environment. At a very minimum, we would argue that it would do less harm than the ethanol-from-maize system currently employed within the USA (Fike *et al.*, 2007) and may convey environmental gains.

10.7.4 Ethical dimensions of sustainability

With regards to the ethics of producing energy from biomass, the most serious and most frequent concern raised can be summed up as the 'food versus fuel' dilemma. By their very nature, biomass-to-energy enterprises do not divert actual food; but a case can be made that they might indirectly reduce food availability. If the biomass used to make ethanol or heat a home could have fed livestock instead, it would have produced much-needed protein for humans; or, if the biomass was grown on 'good cropland', it might have displaced food-producing crops. These scenarios or eventualities pose an ethical dilemma for many, and we include ourselves in that group. We have simply defined ethics as considerations of how the actions of some affect the wellbeing of others. Taking food from some (in a virtual, if not actual, sense) so that others might drive cars or fly jets (or heat fine homes) clearly deserves consideration as 'unethical'.

We do not wish to set ourselves up as arbiters of this ethical dilemma. We would point out, however, that switchgrass can grow on land that is 'marginal'. Indeed, it was selected as DOE's model species after being screened only on marginal sites at DOE's request (McLaughlin and Kszos, 2005), an implicit, early effort to avoid such ethical (and economic) concerns. In addition, switchgrass has been regularly planted on Conservation Reserve Program land – hectarages that are erodible or otherwise unsuitable for row crops (Mulkey *et al.*, 2006; Hartman *et al.*, 2011; Robertson *et al.*, 2011). By some estimates, there are many millions of hectares worldwide that are not currently used for row-crop agriculture (or that should not be) and that could potentially grow a species such as switchgrass (Wolf *et al.*, 2003; Gopalakrishnan *et al.*, 2011). Growing switchgrass on non-cropland avoids direct competition with food-producing efforts, but some bioenergy scenarios project using rangeland or pastureland for growing bioenergy crops; and such efforts could reduce the availability of livestock-based food production.

In sum, we concede that growing switchgrass for energy purposes is not sustainable in a 'purist' or 'deep ecology' sense. Except perhaps for a rotational slash-and-burn, or swidden system, agricultural systems are not closed. They must have external inputs of more than sunlight and precipitation, and some of those inputs are drawn from exhaustible sources, e.g. nutrient-rich deposits and fossil fuels. That is the nature of modern agriculture. When hunter-gatherers became agriculturalists, ecosystems and natural resources were inevitably put at risk. That being said, growing switchgrass for biofuels on non-cropland can minimize the impacts on natural resources – especially when compared with impacts of some other crops. A recent, extensive review of the potential ecological effects of switchgrass cultivation for bioenergy concurs (Hartman *et al.*, 2011). Furthermore, this can be done while simultaneously meeting economic, energetic and ethical criteria that further define a sustainable system.

References

Aiken, G.E. and Springer, T.L. (1995) Seed size distribution, germination, and emergence of 6 switchgrass cultivars. *Journal of Range Management* 48, 455–458.

Aliscioni, S.S., Giussani, L.M., Zuloaga, F.O. and Kellogg, E.A. (2003) A molecular phylogeny of *Panicum* (Poaceae: Paniceae): tests of monophyly and phylogenetic placement within the Panicoideae. *American Journal of Botany* 90, 796–821.

Aravindhakshan, S.C., Epplin, F.M. and Taliaferro, C.M. (2011) Switchgrass, Bermudagrass, Flaccidgrass, and Lovegrass biomass yield response to nitrogen for single and double harvest. *Biomass and Bioenergy* 35, 308–319.

Balasko, J.A. and Smith, D. (1971) Influence of temperature and nitrogen fertilization on the growth and composition of switchgrass (*Panicum virgatum* L.) and timothy (*Phleum pratense* L.). *Agronomy Journal* 63 853–856.

Balasko, J.A., Burner, D.M. and Thayne, W.V. (1984) Yield and quality of switchgrass grown without soil amendments. *Agronomy Journal* 76, 204–204.

Barnett, F.L. and Carver, R.F. (1967) Meiosis and pollen stainability in switchgrass, *Panicum virgatum* L. *Crop Science* 7, 301–304.

Beaty, E.R. and Powell, J.D. (1976) Response of switchgrass (*Panicum virgatum* L.) to clipping frequency. *Journal of Range Management* 29, 132–135.

Beaty, E.R., Engel, J.L. and Powell, J.D. (1978) Tiller development and growth in switchgrass. *Journal of Range Management* 31, 361–365.

Belesky, D.P. and Fedders, J.M. (1995) Warm-season grass productivity and growth rate as influenced by canopy management. *Agronomy Journal* 87, 42–48.

Bentivenga, S.P. and Hetrick, B.A.D. (1991) Relationship between mycorrhizal activity, burning, and plant productivity in tallgrass prairie. *Canadian Journal of Botany* 69, 2597–2602.

Blanco-Canqui, H., Gantzer, C.J., Anderson, S.H., Alberts, E.E. and Thompson, A.L. (2004) Grass barrier and vegetative filter strip effectiveness in reducing runoff, sediment, nitrogen, and phosphorus loss. *Soil Science Society of America Journal* 68, 1670–1678.

Bliss, T., Powers, T.O. and Brassil, C.E. (2010) The spatial influence of aboveground diversity on belowground communities. *Ecosphere* 1, 1–12.

Boe, A. and Casler, M.D. (2005) Hierarchical analysis of switchgrass morphology. *Crop Science* 45, 2465–2472.

Boerner, R.E.J. (1992a) Plant life span and response to inoculation with vesicular- arbuscular mycorrhizal fungi. I. Annual versus perennial grasses. *Mycorrhiza* 1, 153–161.

Boerner, R.E.J. (1992b) Plant life span and response to inoculation with vesicular- arbuscular mycorrhizal fungi. III. Responsiveness and residual soil P levels. *Mycorrhiza* 1, 169–174.

Bona, L. and Belesky, D.P. (1992) Evaluation of switchgrass entries for acid soil tolerance. *Communications in Soil Science and Plant Analysis* 23, 1827–1841.

Bow, J.R., Muir, J.P., Weindorf, D.C., Rosiere, R.E. and Butler, T.J. (2008) Integration of cool-season annual legumes and dairy manure compost with switchgrass. *Crop Science* 48, 1621–1628.

Boydston, R.A., Collins, H.P. and Fransen, S.C. (2010) Response of three switchgrass (*Panicum virgatum*) cultivars to mesotrione, quinclorac, and pendimethalin. *Weed Technology* 24, 336–341.

Bransby, D.I., Samson, R., Parrish, D.J. and Fike, J.H. (2008) Harvest and conversion systems for producing energy from switchgrass: logistic and economic considerations. *Forage and Grazinglands* FG-2008-0722-01-RV.
Brejda, J.J. (2000) Fertilization of native warm-season grasses. Native warm-season grasses: research trends and issues. *Proceedings of the Native Warm-Season Grass Conference and Expo,* Des Moines, Iowa, 12–13 September 1996, pp.177–200.
Brown, R.A., Rosenberg, N.J., Hays, C.J., Easterling, W.E. and Mearns, L.O. (2000) Potential production and environmental effects of switchgrass and traditional crops under current and greenhouse-altered climate in the central United States: a simulation study. *Agriculture, Ecosystems and Environment* 78, 31–47.
Burson, B.L., Tischler, C.R. and Ocumpaugh, W.R. (2009) Breeding for reduced post-harvest seed dormancy in switchgrass: registration of TEM-LoDorm switchgrass germplasm. *Journal of Plant Registrations* 3, 99–103.
Casler, M.D. (2005) Ecotypic variation among switchgrass populations from the Northern USA. *Crop Science* 45, 388–398.
Casler, M.D. (2006) Perennial grasses for turf, sport and amenity uses: evolution of form, function and fitness for human benefit. *Journal of Agricultural Science* 144, 189–203.
Casler, M.D. (2010) Changes in mean and genetic variance during two cycles of within-family selection in switchgrass. *BioEnergy Research* 3, 47–54.
Casler, M.D. and Boe, A.R. (2003) Cultivar x environment interactions in switchgrass. *Crop Science,* 43 2226–2233.
Casler, M.D., Vogel, K.P., Taliaferro, C.M. and Wynia, R.L. (2004) Latitudinal adaptation of switchgrass populations. *Crop Science* 44, 293–303.
Casler, M.D., Stendal, C.A., Kapich, L. and Vogel, K.P. (2007a) Genetic diversity, plant adaptation regions, and gene pools for switchgrass. *Crop Science* 47, 2261–2273.
Casler, M.D., Vogel, K.P., Taliaferro, C.M., Ehlke, N.J., Berdahl, J.D. *et al.* (2007b) Latitudinal and longitudinal adaptation of switchgrass populations. *Crop Science* 47, 2249–2260.
Casler, M.D., Mitchell, R.B. and Vogel, K.P. (2012) Switchgrass. In: Kole, C., Joshi, C.P. and Shonnard, D.R. (eds) *Handbook of Bioenergy Crop Plants.* Taylor and Francis, London, pp. 563–590.
Chang, H., Liao, H.-F., Lee, C.-C. and Shieh, C.-J. (2005) Optimized synthesis of lipase-catalyzed biodiesel by Novozym 435. *Journal of Chemical Technology and Biotechnology* 80, 307–312.
Chaves, A.R. (2009) Biofuels and the lessons of Easter Island. *Sustainability* 1, 335–345.
Christian, D.G., Riche, A.B. and Yates, N.E. (2001) The yield, composition and production costs for seven varieties of switchgrass and one panic grass grown as a biofuel. *Aspects of Applied Biology* 65, 199–204.
Clark, R.B. (2002) Differences among mycorrhizal fungi for mineral uptake per root length of switchgrass grown in acidic soil. *Journal of Plant Nutrition* 25, 1753–1772.
Clark, R.B., Zeto, S.K. and Zobel, R.W. (1999a) Arbuscular mycorrhizal fungal isolate effectiveness on growth and root colonization of *Panicum virgatum* in acidic soil. *Soil Biology and Biochemistry* 31, 1757–1763.
Clark, R.B., Zobel, R.W. and Zeto, S.K. (1999b) Effects of mycorrhizal fungus isolates on mineral acquisition by *Panicum virgatum* in acidic soil. *Mycorrhiza* 9, 167–176.
Cortese, L.M., Honig, J., Miller, C. and Bonos, S.A. (2010) Genetic diversity of twelve switchgrass populations using molecular and morphological markers. *BioEnergy Research* 3, 262–271.
Dohleman, F.G., Heaton, E.A., Leakey, A.D.B. and Long, S.P. (2009) Does greater leaf-level photosynthesis explain the larger solar energy conversion efficiency of *Miscanthus* relative to switchgrass? *Plant, Cell and Environment* 32, 1525–1537.
Douglas, J., Lemunyon, J., Wynia, R. and Salon, P. (2009) *Planting and Managing Switchgrass as a Biomass Energy Crop.* USDA-NRCS Technical Note No. 3. US Department of Agriculture Natural Resources Conservation Service.
El Hadj, M. (2000) *Compatibility, yield, and quality of warm-season grass-legume mixtures* MSc thesis, Virginia Tech, Blacksburg, Virginia.
El-Hinnawi, E.E. (1981) *The Environmental Impacts of Production and Use of Energy: an Assessment.* Tycooly Press, Dublin, Ireland.
Elbersen, H.W., Ocumpaugh, W.R., Hussey, M.A., Sanderson, M.A. and Tischler, C.R. (1999) Field evaluation of switchgrass seedlings divergently selected for crown node placement. *Crop Science* 39, 475–479.
Esbroeck, G.A.V., Hussey, M.A. and Sanderson, M.A. (2003) Variation between Alamo and Cave-in-Rock switchgrass in response to photoperiod extension. *Crop Science* 43, 639–643.

Evers, G.W. and Butler, T.W. (2000) Switchgrass establishment on coastal plain soil. *Proceedings/Reports of the American Forage and Grassland Council, 37th North American Alfalfa Improvement Conference,* Madison, Wisconsin, 16–19 July 2000, pp. 150–154.

Evers, G.W. and Parsons, M.J. (2003) Soil type and moisture level influence on Alamo switchgrass emergence and seedling growth. *Crop Science* 43, 288–294.

Fike, J.H., Parrish, D.J., Wolf, D.D., Balasko, J.A., Green, J.T. Jr *et al.* (2006a) Long-term yield potential of switchgrass-for-biofuel systems. *Biomass and Bioenergy* 30, 198–206.

Fike, J.H., Parrish, D.J., Wolf, D.D., Balasko, J.A., Green, J.T. Jr *et al.* (2006b) Switchgrass production for the upper southeastern USA: influence of cultivar and cutting frequency on biomass yields. *Biomass and Bioenergy* 30, 207–213.

Fike, J.H., Parrish, D.J., Alwang, J. and Cundiff, J.S. (2007) Challenges for deploying dedicated, large-scale, bioenergy systems in the USA. *CAB Reviews: Perspectives in Agriculture, Veterinary Science, Nutrition and Natural Resources* 2, 28.

Friedrich, J.W., Smith, D. and Schrader, L.E. (1977) Herbage yield and chemical composition of switchgrass as affected by N, S and K fertilization. *Agronomy Journal* 69, 30–33.

Garten, C.T. Jr, Smith, J.L., Tyler, D.D., Amonette, J.E., Bailey, V.L. *et al.* (2010) Intra-annual changes in biomass, carbon, and nitrogen dynamics at 4-year old switchgrass field trials in west Tennessee, USA. *Agriculture, Ecosystems and Environment* 136, 177–184.

Garten, C.T. Jr, Brice, D.J., Castro, H.F., Graham, R.L., Mayes, M.A. *et al.* (2011) Response of 'Alamo' switchgrass tissue chemistry and biomass to nitrogen fertilization in West Tennessee, USA. *Agriculture, Ecosystems and Environment* 140, 289–297.

George, J.R. and Obermann, D. (1989) Spring defoliation to improve summer supply and quality of switchgrass. *Agronomy Journal* 81, 47–52.

George, J.R., Obermann, D.J. and Wolf, D.D. (1989) Seasonal trends for nonstructural carbohydrates in stem bases of defoliated switchgrass. *Crop Science* 29, 1282–1287.

George, N., Tungate, K., Hobbs, A., Fike, J.H. and Atkinson, A. (2008) *A Guide for Growing Switchgrass as a Biofuel Crop in North Carolina*. NC Solar Center, North Carolina State University, Raleigh, North Carolina, 21.

Ghimire, S.R. and Craven, K.D. (2011) Enhancement of switchgrass (*Panicum virgatum* L.) biomass production under drought conditions by the ectomycorrhizal fungus *Sebacina vermifera. Applied and Environmental Microbiology* 77, 7063–7067.

Ghimire, S.R., Charlton, N.D. and Craven, K.D. (2009) The mycorrhizal fungus, *Sebacina vermifera*, enhances seed germination and biomass production in switchgrass (*Panicum virgatum* L). *BioEnergy Research* 2, 51–58.

Gopalakrishnan, G., Negri, M.C. and Snyder, S.W. (2011) A novel framework to classify marginal land for sustainable biomass feedstock production. *Journal of Environmental Quality* 40, 1593–1600.

GPWG (2001) Phylogeny and subfamilial classification of the grasses (Poaceae). *Annals Missouri Botanical Garden* 88, 373–457.

Grassini, P., Hunt, E., Mitchell, R.B. and Weiss, A. (2009) Simulating switchgrass growth and development under potential and water limiting conditions. *Agronomy Journal* 101, 564–571.

Guretzky, J.A., Biermacher, J.T., Cook, B.J., Kering, M.K. and Mosali, J. (2011) Switchgrass for forage and bioenergy: harvest and nitrogen rate effects on biomass yields and nutrient composition. *Plant and Soil* 339, 69–81.

Haferkamp, M.R. and Copeland, T.D. (1984) Shoot growth and development of Alamo switchgrass as influenced by mowing and fertilization. *Journal of Range Management* 37, 406–412.

Hall, D.O. (1979) Solar energy use through biology-past, present and future. *Solar Energy* 22, 307–328.

Hall, K.E., George, J.R. and Riedl, R.R. (1982) Herbage dry matter yields of switchgrass, big bluestem and indiangrass with N fertilization. *Agronomy Journal* 74, 47–51.

Hanson, J.D. and Johnson, H.A. (2005) Germination of switchgrass under various temperature and pH regimes. *Seed Technology* 27, 203–210.

Hargrave, B.S. and Seastedt, T.R. (1994) Nitrogen concentrations of senescent foliage in a relict tall-grass prairie. *Prairie Naturalist* 26, 61–65.

Harper, J. and Spooner, A.E. (1983) Establishment of selected herbaceous species on acid bauxite minesoils. *Proceedings. Symposium on Surface Mining, Hydrology, Sedimentology and Reclamation*, pp. 413–422.

Hartman, J.C., Nippert, J.B., Orozco, R.A. and Springer, C.J. (2011) Potential ecological impacts of switchgrass (*Panicum virgatum* L.) biofuel cultivation in the Central Great Plains, USA. *Biomass and Bioenergy* 35, 3415–3421.

Heggenstaller, A.H., Moore, K.J., Liebman, M. and Anex, R.P. (2009) Nitrogen influences biomass and nutrient partitioning by perennial, warm-season grasses. *Agronomy Journal* 101, 1363–1371.

Hill, J., Polasky, S., Nelson, E., Tilman, D., Huo, H. *et al.* (2009) Climate change and health costs of air emissions from biofuels and gasoline. *Proceedings of the National Academy of Sciences of the United States of America* 106, 2077–2082.

Hitchcock, A.S. (1935) *Manual of the Grasses of the United States*. United States Department of Agriculture, Washington, DC.

Hope, H.J. and McElroy, A. (1990) Low-temperature tolerance of switchgrass (*Panicum virgatum* L.). *Canadian Journal of Plant Science* 70, 1091–1096.

Hopkins, A.A. and Taliaferro, C.M. (1997) Genetic variation within switchgrass populations for acid soil tolerance. *Crop Science* 37, 1719–1722.

Hsu, F.H. and Nelson, C.J. (1986a) Planting date effects on seedling development of perennial warm-season forage grasses. I. Field emergence. *Agronomy Journal* 78, 33–38.

Hsu, F.H. and Nelson, C.J. (1986b) Planting date effects on seedling development of perennial warm-season forage grasses. II. Seedling growth. *Agronomy Journal* 78, 38–42.

Hsu, F.H., Nelson, C.J. and Matches, A.G. (1985a) Temperature effects on germination of perennial warm-season forage grasses. *Crop Science* 25, 215–220.

Hsu, F.H., Nelson, C.J. and Matches, A.G. (1985b) Temperature effects on seedling development of perennial warm-season forage grasses. *Crop Science* 25, 249–255.

Huang, S.X., Su, X.J., Haselkorn, R. and Gornicki, P. (2003) Evolution of switchgrass (*Panicum virgatum* L.) based on sequences of the nuclear gene encoding plastid acetyl-CoA carboxylase. *Plant Science* 164, 43–49.

Hultquist, S.J., Vogel, K.P., Lee, D.J., Arumuganathan, K. and Kaeppler, S. (1996) Chloroplast DNA and nuclear DNA content variations among cultivars of switchgrass, *Panicum virgatum* L. *Crop Science* 36, 1049–1052.

Jain, A.K., Khanna, M., Erickson, M. and Huang, H.X. (2010) An integrated biogeochemical and economic analysis of bioenergy crops in the Midwestern United States. *GCB Bioenergy* 2, 217–234.

Juang, K. and Lee, D. (2010) Prediction of cadmium removal from highly contaminated soils by phytoextraction with different switchgrass cultivars. *19th World Congress of Soil Science Symposium 3.5.1*, Brisbane, Australia, 1–6 August 2010, pp. 197–200.

Jung, G.A., Shaffer, J.A. and Stout, W.L. (1988) Switchgrass and big bluestem responses to amendments on strongly acid soil. *Agronomy Journal* 80, 669–676.

Jung, J.Y. and Lal, R. (2011) Impacts of nitrogen fertilization on biomass production of switchgrass (*Panicum virgatum* L.) and changes in soil organic carbon in Ohio. *Geoderma* 166, 145–152.

Kellogg, E.A. (2001) Evolutionary history of the grasses. *Plant Physiology* 125, 1198–1205.

Ker, K., Fyles, J.W., Driscoll, B.T., Seguin, P. and Smith, D. (2010) Development of a low-input switchgrass-microbe system to improve energy ratio and production of biofuels. *ASA-CSSA-SSSA 2010 International Annual Meetings,* Long Beach, California.

Kiniry, J.R., Tischler, C.R. and Van Esbroeck, G.A. (1999) Radiation use efficiency and leaf CO2 exchange for diverse C4 grasses. *Biomass and Bioenergy* 17, 95–112.

Kiniry, J.R., Cassida, K.A., Hussey, M.A., Muir, J.P., Ocumpaugh, W.R. *et al.* (2005) Switchgrass simulation by the ALMANAC model at diverse sites in the southern US. *Biomass and Bioenergy* 29, 419–425.

Koslowsky, S.D. and Boerner, R.E.J. (1989) Interactive effects of aluminium, phosphorus and mycorrhizae on growth and nutrient uptake of *Panicum virgatum* L. (Poaceae). *Environmental Pollution* 61, 107–125.

Laird, D.A. (2008) The charcoal vision: a win-win-win scenario for simultaneously producing bioenergy, permanently sequestering carbon, while improving soil and water quality. *Agronomy Journal* 100, 178–181.

Laser, M., Jin, H.M., Kemantha, J., Dale, B.E. and Lynd, L.R. (2009) Projected mature technology scenarios for conversion of cellulosic biomass to ethanol with coproduction thermochemical fuels, power, and/or animal feed protein. *Biofuels, Bioproducts and Biorefining* 3, 231–246.

Le, P.V.V., Kumar, P. and Drewry, D.T. (2011) Implications for the hydrologic cycle under climate change due to the expansion of bioenergy crops in the Midwestern United States. *Proceedings of the National Academy of Sciences of the United States of America* 108, 15085–15090.

Lee, D.K., Owens, V.N. and Doolittle, J.J. (2007) Switchgrass and soil carbon sequestration response to ammonium nitrate, manure, and harvest frequency on conservation reserve program land. *Agronomy Journal* 99, 462–468.

Lee, D.Y., Owens, V.N., Boe, A. and Koo, B. (2009) Biomass and seed yields of big bluestem, switchgrass, and intermediate wheatgrass in response to manure and harvest timing at two topographic positions. *GCB Bioenergy* 1, 171–179.

Lemus, R., Parrish, D.J. and Abaye, O. (2008) Nitrogen-use dynamics in switchgrass grown for biomass. *BioEnergy Research* 1, 153–162.

Lemus, R., Parrish, D.J. and Wolf, D.D. (2009) Nutrient uptake by 'Alamo' switchgrass used as an energy crop. *BioEnergy Research* 2, 37–50.

Lowenberg-DeBoer, J. and Cherney, J.H. (1989) Biophysical simulation for evaluating new crops: the case of switchgrass for biomass energy feedstock. *Agricultural Systems* 29, 233–246.

Lychnaras, V. and Schneider, U.A. (2011) Multi-farm economic analysis of perennial energy crops in Central Greece, taking into account the CAP reform. *Biomass and Bioenergy* 35, 700–715.

Ma, Z., Wood, C.W. and Bransby, D.I. (2001) Impact of row spacing, nitrogen rate, and time on carbon partitioning of switchgrass. *Biomass and Bioenergy* 20, 413–419.

Madakadze, I., Coulman, B.E., Stewart, K., Peterson, P., Samson, R. *et al.* (1998a) Phenology and tiller characteristics of big bluestem and switchgrass cultivars in a short growing season area. *Agronomy Journal* 90, 489–495.

Madakadze, I.C., Coulman, B.E., Mcelroy, A.R., Stewart, K.A. and Smith, D.L. (1998b) Evaluation of selected warm-season grasses for biomass production in areas with a short growing season. *Bioresource Technology* 65, 1–12.

Madakadze, I.C., Stewart, K.A., Peterson, P.R., Coulman, B.E. and Smith, D.L. (1999) Cutting frequency and nitrogen fertilization effects on yield and nitrogen concentration of switchgrass in a short season area. *Crop Science* 39, 552–557.

Mamani-Pati, F., Clay, D.E., Carlson, G. and Clay, S.A. (2011) Production, economic, and energy life cycle analysis can produce contrary results for corn used in ethanol production. *Journal of Plant Nutrition* 34, 1278–1289.

Martin, A.R., Moomaw, R.S. and Vogel, K.P. (1982) Warm-season grass establishment with atrazine. *Agronomy Journal* 74, 916–920.

Martínez-Reyna, J.M. and Vogel, K.P. (2008) Heterosis in switchgrass: spaced plants. *Crop Science* 48, 1312–1320.

Martínez-Reyna, J.M., Vogel, K.P., Caha, C. and Lee, D.J. (2001) Meiotic stability, chloroplast DNA polymorphisms, and morphological traits of upland × lowland switchgrass reciprocal hybrids. *Crop Science* 41, 1579–1583.

Masters, R.A. and Nissen, S.J. (1998) Revegetating leafy spurge (*Euphorbia esula*)-infested rangeland with native tallgrasses. *Weed Technology* 12, 381–390.

Masters, R.A., Nissen, S.J., Gaussoin, R.E., Beran, D.D. and Stougaard, R.N. (1996) Imidazolinone herbicides improve restoration of Great Plains grasslands. *Weed Technology* 10, 392–403.

McGee, C. and Hilton, A.B.C. (2011) Analysis of federal and state policies and environmental issues for bioethanol production facilities. *Environmental Science and Technology* 45, 1780–1791.

McKenna, J.R. and Wolf, D.D. (1990) No-till switchgrass establishment as affected by limestone, phosphorus, and carbofuran. *Journal of Production Agriculture* 3, 475–479.

McKenna, J.R., Wolf, D.D. and Lentner, M. (1991) No-till warm-season grass establishment as affected by atrazine and carbofuran. *Agronomy Journal* 83, 311–316.

McLaughlin, S.B. and Kszos, L.A. (2005) Development of switchgrass (*Panicum virgatum*) as a bioenergy feedstock in the United States. *Biomass and Bioenergy* 28, 515–535.

McMillan, C. (1959) The role of ecotypic variation in the distribution of the central grassland of North America. *Ecological Monographs* 29, 285–308.

McMurphy, W.E., Denman, C.E. and Tucker, B.B. (1975) Fertilization of native grass and weeping lovegrass. *Agronomy Journal* 67, 233–236.

Mersie, W., Seybold, C.A., Mcnamee, C. and Lawson, M.A. (2003) Abating endosulfan from runoff using vegetative filter strips: the importance of plant species and flow rate. *Agriculture Ecosystems and Environment* 97, 215–223.

Missaoui, A.M., Paterson, A.H. and Bouton, J.H. (2006) Molecular markers for the classification of switchgrass (*Panicum virgatum L.*) germplasm and to assess genetic diversity in three synthetic switchgrass populations. *Genetic Resources and Crop Evolution* 53, 1291–1302.

Mitchell, R.B. and Britton, C.M. (2000) Managing weeds to establish and maintain warm-season grasses. Native warm-season grasses: research trends and issues. *Proceedings of the Native Warm-Season Grass Conference and Expo*, Des Moines, Iowa, 12–13 September 1996, pp. 159–176.

Mitchell, R.B., Vogel, K.P., Berdahl, J. and Masters, R.A. (2010) Herbicides for establishing switchgrass in the Central and Northern Great Plains. *BioEnergy Research* 3, 321–327.

Mooney, D.F., Roberts, R.K., English, B.C., Tyler, D.D. and Larson, J.A. (2009) Yield and breakeven price of 'Alamo' switchgrass for biofuels in Tennessee. *Agronomy Journal* 101, 1234–1242.
Moser, L.E. and Vogel, K.P. (1995) Switchgrass, big bluestem and indiangrass. In: Barnes, R.F., Miller, D.A. and Nelson, C.J. (eds) *Forages,* Vol 1. *An Introduction to Grassland Agriculture.* Iowa State University Press, Ames, Iowa.
Muir, J.P., Sanderson, M.A., Ocumpaugh, W.R., Jones, R.M. and Reed, R.L. (2001) Biomass production of 'Alamo' switchgrass in response to nitrogen, phosphorus, and row spacing. *Agronomy Journal* 93, 896–901.
Mulkey, V.R., Owens, V.N. and Lee, D.K. (2006) Management of switchgrass-dominated conservation reserve program lands for biomass production in South Dakota. *Crop Science* 46, 712–720.
Murphy, I.J. and Coats, J.R. (2011) The capacity of switchgrass (*Panicum virgatum*) to degrade atrazine in a phytoremediation setting. *Environmental Toxicology And Chemistry* 30, 715–722.
Nelson, R.G., Ascough, J.C., II and Langemeier, M.R. (2006) Environmental and economic analysis of switchgrass production for water quality improvement in northeast Kansas. *Journal of Environmental Management* 79, 336–347.
Newell, L.O. and Keim, F.D. (1947) Effects of mowing frequency on the yield and protein content of several grasses grown in pure stands. *University of Nebraska Agricultural Experiment Station Research Bulletin* No. 150, 36 pp.
Newman, P.R. and Moser, L.E. (1988) Grass seedling emergence, morphology, and establishment as affected by planting depth. *Agronomy Journal* 80, 383–387.
Nickell, G.L. (1973) The physiological ecology of upland and lowland *Panicum virgatum. Dissertation Abstracts International B* 33, 3512.
Nielsen, E.L. (1944) Analysis of variation in *Panicum virgatum. Journal of Agricultural Research* 69, 327–353.
Okada, M., Lanzatella, C., Saha, M.C., Bouton, J., Wu, R.L. and Tobias, C.M. (2010) Complete switchgrass genetic maps reveal subgenome collinearity, preferential pairing and multilocus interactions. *Genetics* 185, 745–760.
Oliver, R.J., Finch, J.W. and Taylor, G. (2009) Second generation bioenergy crops and climate change: a review of the effects of elevated atmospheric CO2 and drought on water use and the implications for yield. *GCB Bioenergy* 1, 97–114.
Parrish, D.J. and Fike, J.H. (2005) The biology and agronomy of switchgrass for biofuels. *Critical Reviews in Plant Sciences* 24, 423–459.
Parrish, D.J. and Fike, J.H. (2009) Selecting, establishing, and managing switchgrass (*Panicum virgatum*) for biofuels. In: Mielenz, J.R. (ed.) *Methods in Molecular Biology: Biofuels, Methods and Protocols.* Springer/Humana Press, New York.
Parrish, D.J., Fike, J.H., Bransby, D.I. and Samson, R. (2008) Establishing and managing switchgrass as an energy crop. *Forage and Grazinglands,* FG-2008-0220-01-RV.
Paterson, A.H., Bowers, J.E., Feltus, F.A., Tang, H.B., Lin, L.F. and Wang, X. (2009) Comparative genomics of grasses promises a bountiful harvest. *Plant Physiology* 149, 125–131.
Paterson, A.H., Freeling, M., Tang, H.B. and Wang, X.Y. (2010) Insights from the comparison of plant genome sequences. *Annual Review of Plant Biology* 61, 349–372.
Patzek, T.W. (2010) A probabilistic analysis of the switchgrass ethanol cycle. *Sustainability* 2, 3158–3194.
Perrin, R., Vogel, K., Schmer, M. and Mitchell, R. (2008) Farm-scale production cost of switchgrass for biomass. *BioEnergy Research* 1, 91–97.
Peters, T.J., Moomaw, R.S. and Martin, A.R. (1989) Herbicides for postemergence control of annual grass weeds in seedling forage grasses. *Weed Science* 37, 375–379.
Pimentel, D. (1991) Ethanol fuels: energy security, economics and the environment. *Journal of Agricultural and Environmental Ethics* 4, 1–13.
Popp, M. and Nalley, L.L. (2011) Modeling interactions of a carbon offset policy and biomass markets on crop allocations. *Journal of Agricultural and Applied Economics* 43, 399–411.
Porter, C.L. Jr (1966) An analysis of variation between upland and lowland switchgrass *Panicum virgatum* L. in central Oklahoma. *Ecology* 47, 980–992.
Rehm, G.W. (1990) Importance of nitrogen and phosphorus for production of grasses established with no-till and conventional planting systems. *Journal of Production Agriculture* 3, 333–336.
Riggs, P.J., Moritz, R.L., Chelius, M.K., Dong, Y., Iniguez, A.L. *et al.* (2002) Isolation and characterization of diazotrophic endophytes from grasses and their effects on plant growth. *Nitrogen fixation: global perspectives. Proceedings of the 13th International Congress on Nitrogen Fixation,* Hamilton, Ontario, 2–7 July 2001, pp. 263–267.

Riley, R.D. and Vogel, K.P. (1982) Chromosome-numbers of released cultivars of switchgrass, Indiangrass, big bluestem, and sand bluestem. *Crop Science* 22, 1082–1083.
Robertson, B.A., Doran, P.J., Loomis, L.R., Robertson, J.R. and Schemske, D.W. (2011) Perennial biomass feedstocks enhance avian diversity. *GCB Bioenergy* 3, 235–246.
Roth, A.M., Sample, D.W., Ribic, C.A., Paine, L., Undersander, D.J. and Bartelt, G.A. (2005) Grassland bird response to harvesting switchgrass as a biomass energy crop. *Biomass and Bioenergy* 28, 490–498.
Samson, R., Mani, S., Boddey, R., Sokhansanj, S., Quesada, D. *et al.* (2005) The potential of C4 perennial grasses for developing a global BIOHEAT industry. *Critical Reviews in Plant Sciences* 24, 461–495.
Sanderson, M.A. and Reed, R.L. (2000) Switchgrass growth and development: water, nitrogen, and plant density effects. *Journal of Range Management* 53, 221–227.
Sanderson, M.A., Reed, R.L., Mclaughlin, S.B., Wullschleger, S.D., Conger, B.V. *et al.* (1996) Switchgrass as a sustainable bioenergy crop. *Bioresource Technology* 56, 83–93.
Sanderson, M.A., Reed, R.L., Ocumpaugh, W.R., Hussey, M.A., Van Esbroeck, G.A. *et al.* (1999) Switchgrass cultivars and germplasm for biomass feedstock production in Texas. *Bioresource Technology* 67, 209–219.
Sanderson, M.A., Jones, R.M., Mcfarland, M.J., Stroup, J., Reed, R.L. and Muir, J.P. (2001) Nutrient movement and removal in a switchgrass biomass-filter strip system treated with dairy manure. *Journal of Environmental Quality* 30, 210–216.
Sartori, F., Lal, R., Ebinger, M.H. and Parrish, D.J. (2006) Potential soil carbon sequestration and CO2 offset by dedicated energy crops in the USA. *Critical Reviews in Plant Sciences* 25, 441–472.
Schmer, M.R., Vogel, K.P., Mitchell, R.B., Moser, L.E., Eskridge, K.M. and Perrin, R.K. (2006) Establishment stand thresholds for switchgrass grown as a bioenergy crop. *Crop Science* 46, 157–161.
Schmer, M.R., Vogel, K.P., Mitchell, R.B. and Perrin, R.K. (2008) Net energy of cellulosic ethanol from switchgrass. *Proceedings of the National Academy of Sciences of the United States of America* 105, 464–469.
Schmer, M.R., Liebig, M.A., Vogel, K.P. and Mitchell, R.B. (2011) Field-scale soil property changes under switchgrass managed for bioenergy. *GCB Bioenergy* 3, 439–448.
Sendich, E.D. and Dale, B.E. (2009) Environmental and economic analysis of the fully integrated biorefinery. *GCB Bioenergy* 1, 331.
Shen, Z.X., Welbaum, G.E., Parrish, D.J. and Wolf, D.D. (1999) After-ripening and aging as influenced by anoxia in switchgrass (*Panicum virgatum* L.) seeds stored at 60 deg C. *Acta Horticulturae*, 191–197.
Smart, A.J. and Moser, L.E. (1999) Switchgrass seedling development as affected by seed size. *Agronomy Journal* 91, 335–338.
Smart, A.J., Vogel, K.P., Moser, L.E. and Stroup, W.W. (2003) Divergent selection for seedling tiller number in big bluestem and switchgrass. *Crop Science* 43, 1427–1433.
Smart, A.J., Moser, L.E. and Vogel, K.P. (2004) Morphological characteristics of big bluestem and switchgrass plants divergently selected for seedling tiller number. *Crop Science* 44, 607–613.
Smith, D. (1975) Trends of nonstructural carbohydrates in stem bases of switchgrass. *Journal of Range Management* 28, 389–391.
Smith, D. and Greenfield, S.B. (1979) Distribution of chemical-constituents among shoot parts of timothy and switchgrass at anthesis. *Journal of Plant Nutrition* 1, 81–99.
Sokhansanj, S., Mani, S., Turhollow, A., Amit, K., Bransby, D. *et al.* (2009) Large-scale production, harvest and logistics of switchgrass (*Panicum virgatum* L.) – current technology and envisioning a mature technology. *Biofuels, Bioproducts and Biorefining* 3, 124–141.
Spatari, S. and MacLean, H.L. (2010) Characterizing model uncertainties in the life cycle of lignocellulose-based ethanol fuels. *Environmental Science and Technology* 44, 8773–8780.
Springer, T.L., Aiken, G.E. and Mcnew, R.W. (2001) Combining ability of binary mixtures of native, warm-season grasses and legumes. *Crop Science* 41, 818–823.
Stout, W.L. and Jung, G.A. (1995) Biomass and nitrogen accumulation in switchgrass: effects of soil and environment. *Agronomy Journal* 87, 663–669.
Strömberg, C.A.E. and McInerney, F.A. (2011) The Neogene transition from C3 to C4 grasslands in North America: assemblage analysis of fossil phytoliths. *Paleobiology* 37, 50–71.
Stucky, D.J., Bauer, J.H. and Lindsey, T.C. (1980) Restoration of acidic mine spoils with sewage sludge: I. Revegetation. *Reclamation Review* 3, 129–139.
Talbert, L.E., Timothy, D.H., Burns, J.C., Rawlings, J.O. and Moll, R.H. (1983) Estimates of genetic parameters in switchgrass. *Crop Science* 23, 725–728.

Teel, A., Barnhart, S. and Miller, G. (2003) Management guide for the production of switchgrass for biomass fuel in southern Iowa. Iowa State University Extension, PM 1710. Available at: http://www.extension.iastate.edu/publications/PM1710.pdf (accessed 3 March 2012).

Thelemann, R., Johnson, G., Sheaffer, C., Banerjee, S., Cai, H.W. and Wyse, D. (2010) The effect of landscape position on biomass crop yield. *Agronomy Journal* 102, 513–522.

Tjepkema, J. (1975) Nitrogenase activity in the rhizosphere of *Panicum virgatum*. *Soil Biology and Biochemistry*, 7, 179–180.

Vadas, P.A., Barnett, K.H. and Undersander, D.J. (2008) Economics and energy of ethanol production from alfalfa, corn, and switchgrass in the Upper Midwest, USA. *BioEnergy Research* 1, 44–55.

Vassey, T.L., George, J.R. and Mullen, R.E. (1985) Early-, mid-, and late-spring establishment of switchgrass at several seeding rates. *Agronomy Journal* 77, 253–257.

Vogel, K.P. (2000) Improving warm-season forage grasses using selection, breeding, and biotechnology. Native warm-season grasses: research trends and issues. *Proceedings of the Native Warm-Season Grass Conference and Expo,* Des Moines, Iowa, 12–13 September 1996, pp. 83–106.

Vogel, K.P. (2004) Switchgrass. In*:* Moser, L.E., Burson, B.L. and Sollenberger, L.E. (eds) *Warm-Season (C4) Grasses*. ASA-CSSA-SSSA, Madison, Wisconsin.

Vogel, K.P. and Mitchell, R.B. (2008) Heterosis in switchgrass: biomass yield in swards. *Crop Science* 48, 2159–2164.

Vogel, K.P., Brejda, J.J., Walters, D.T. and Buxton, D.R. (2002) Switchgrass biomass production in the Midwest USA: harvest and nitrogen management. *Agronomy Journal* 94, 413–420.

Vogel, K.P., Schmer, M.R. and Mitchell, R.B. (2005) Plant adaptation regions: ecological and climatic classification of plant materials. *Rangeland Ecology and Management* 58, 315–319.

Waramit, N., Moore, K.J. and Heggenstaller, A.H. (2011) Composition of native warm-season grasses for bioenergy production in response to nitrogen fertilization rate and harvest date. *Agronomy Journal* 103, 655–662.

Watson, L. and Dallwitz, M.J. (1994) *The Grass Genera of the World*. CABI, Wallingford, UK.

Weaver, J.E. (1931) Who's who among the prairie grasses. *Ecology* 12, 623–632.

Wedin, W.F. and Helsel, Z. (1980) Plant species for biomass production on marginal sites. In*:* Robinson, J.S. (ed.) *Fuels From Biomass: Technology and Feasibility*. Noyes Data Corporation, Park Ridge, New Jersey.

Werling, B.P., Meehan, T.D., Gratton, C. and Landis, D.A. (2011) Influence of habitat and landscape perenniality on insect natural enemies in three candidate biofuel crops. *Biological Control* 59, 304–312.

Wilson, R.G. (1995) Effect of imazethapyr on perennial grasses. *Weed Technology* 9, 187–191.

Wolf, D.D. and Fiske, D.A. (1995) *Planting and Managing Switchgrass for Forage, Wildlife, and Conservation*. Virginia Cooperative Extension, 418–013.

Wolf, J., Bindraban, P.S., Luijten, J.C. and Vleeshouwers, L.M. (2003) Exploratory study on the land area required for global food supply and the potential global production of bioenergy. *Agricultural Systems* 76, 841–861.

Wright, L. (2007) Historical perspective on how and why switchgrass was selected as a 'model' high-potential energy crop. ORNL/TM-2007/109. Oak Ridge National Laboratory. Available at: http://www1.eere.energy.gov/biomass/pdfs/ornl_switchgrass.pdf (accessed 25 May 2012).

Wright, L. and Turhollow, A. (2010) Switchgrass selection as a 'model' bioenergy crop: a history of the process. *Biomass and Bioenergy* 34, 851–868.

Wullschleger, S.D., Sanderson, M.A., Mclaughlin, S.B., Biradar, D.P. and Rayburn, A.L. (1996) Photosynthetic rates and ploidy levels among populations of switchgrass. *Crop Science* 36, 306–312.

Young, H.A., Lanzatella, C.L., Sarath, G. and Tobias, C.M. (2011) Chloroplast genome variation in upland and lowland switchgrass. *PLoS ONE*, e23980.

Yu, B., Lin, L. and van der Voit, E. (2010) Life cycle assessment of switchgrass-derived ethanol as transport fuel. *International Journal of Life Cycle Assessment* 15, 468–477.

Zalapa, J.E., Price, D.L., Kaeppler, S.M., Tobias, C.M., Okada, M. *et al.* (2011) Hierarchical classification of switchgrass genotypes using SSR and chloroplast sequences: ecotypes, ploidies, gene pools, and cultivars. *Theoretical and Applied Genetics* 122, 805–817.

Zarnstorff, M.E. (1990) Physioecological effects of cultural and pesticide management on no-till establishment of switchgrass and seed dormancy. PhD thesis, North Carolina State University, Raleigh, North Carolina.

Zhang, J. and Maun, M.A. (1990) Sand burial effect on seed germination, seedling emergence and establishment of *Panicum virgatum*. *Holarctic Ecology* 13, 56–61.

Zhang, J.H. and Maun, M.A. (1991a) Effects of partial removal of seed reserves on some aspects of seedling ecology of 7 dune species. *Canadian Journal Of Botany – Revue Canadienne De Botanique* 69, 1457–1462.

Zhang, J.H. and Maun, M.A. (1991b) Establishment and growth of *Panicum virgatum* L. seedlings on a Lake Erie sand dune. *Bulletin of the Torrey Botanical Club* 118, 141–153.

Zhang, Y., Zalapa, J.E., Jakubowski, A.R., Price, D.L., Acharya, A. *et al.* (2011a) Post-glacial evolution of *Panicum virgatum*: centers of diversity and gene pools revealed by SSR markers and cpDNA sequences. *Genetica* 139, 933–948.

Zhang, Y., Zalapa, J.E., Jakubowski, A.R., Price, D.L., Acharya, A. *et al.* (2011b) Natural hybrids and gene flow between upland and lowland switchgrass. *Crop Science* 51, 2626–2641.

11 *Miscanthus* Species

Kossonou Guillaume Anzoua and Toshihiko Yamada
Field Science Center for Northern Biosphere, Hokkaido University, Japan

11.1 Introduction

The genus *Miscanthus* consists of giant grasses belonging to the C_4 group, wild and perennial, which contains diverse species that are mostly unknown. Globally, the genus has attracted attention as a likely lignocellulosic biomass crop for biofuel because of its high dry matter production capability. It is also receiving considerable attention for its amenability to sustainable production methods requiring less inputs and ability to grow on marginal soils. The grass has been globally known through a single species, *Miscanthus* × *giganteus*, a natural sterile hybrid, from a cross between *Miscanthus sinensis* and *Miscanthus sacchariflorus*. Although *Miscanthus* is more distributed in warm tropical lands, the genus includes some species that could distribute naturally or with human intervention in different climates (Koyama, 1987; Chiang *et al.*, 2003; Chen and Renvoize, 2006). For example, *M. sinensis* is able to grow in tropical as well as temperate environments as native or adapted habitats regardless of land altitude (Schwarz, 1983; Jones and Walsh, 2001; Scally *et al.*, 2001; Lewandowski and Schimdt, 2006). *Miscanthus* has been grown successfully under European conditions for more than a century (Andersson, 1855) and has recently been introduced to the USA (Heaton *et al.*, 2008; Khanna *et al.*, 2008a; Villamil *et al.*, 2008; Pyter *et al.*, 2009). Agronomically, *Miscanthus* exhibits multiple desirable features that place it advantageously over other renewable energy crops. Being a perennial grass, it bypasses the 'food versus fuel' debate. *Miscanthus* is also superior in terms of productivity (volume and quality) (Lewandowski *et al.*, 2003; Adler *et al.*, 2006) due to its efficient physiological functions under normal conditions and is adaptable also to stressful conditions including low temperatures of cool climates under which many other biofuel crops are unable to function (Boehmel *et al.*, 2008; Dohleman and Long, 2009). As a result, it has come to be seen as one of the best dedicated feedstock energy crops. Recently cultivation of *Miscanthus* as an energy crop has gained momentum and hectarage has been grown mostly in Europe and in some parts of the USA. However, further expansion to other geographical areas of suitable climate has not yet occurred. The adoption of *Miscanthus* as a biofuel feedstock on an industrial scale is still awaited.

The full success of *Miscanthus* as an energy crop depends on: (i) understanding its agronomic aspects relating to cultivation for feedstock production; (ii) comprehending its interaction with the environment, impact on landscape, land utilization and biodiversity; (iii) analysing economic viability of feedstock production; and (iv) improving the genus genetically.

©CAB International 2013. *Biofuel Crops: Production, Physiology and Genetics* (ed. B.P. Singh)

11.2 Botanical Characteristics

Although a native of eastern Asia, *Miscanthus* was first described by Andersson (1855) and thereafter variously documented (Kobayashi, 1981; Mutoh *et al.*, 1985; Numata, 1988; Tang *et al.*, 1988; Clifton-Brown *et al.*, 2001, 2008; Chen and Renvoize, 2006; Christian *et al.*, 2008). Taxonomically, the *Miscanthus* genus comes in the family *Gramineae* (*Poaceae*), tribe *Andropogoneae* and subtribe *Saccharinae*. It contains many species, including *M. sinensis* and *M. sacchariflorus*. *Miscanthus* × *giganteus* is a natural sterile hybrid between *M. sinensis* and *M. sacchariflorus*. It is also known as giant grass and commonly used as an ornamental plant, but has recently drawn attention for other economic purposes.

Morphologically, *Miscanthus* is a vertically tall plant (up to 4~5 m) with long (50–120 cm) and broad (8–30 mm) leaves (Watson and Dallwitz, 1992) (Fig. 11.1). *Miscanthus sinensis* is the shorter member of the genus, ranging between 2 and 3.5 m in height but its tillering capacity (50–150 stems per plant) and compact crown diameter (40–60 cm) is larger than *M.* × *giganteus*. *Miscanthus sacchariflorus* is taller (3.5–5 m), but inferior in terms of tillering capacity (1–2 stem) to *M.* × *giganteus*. The rhizomes of *M. sacchariflorus* are potentially invasive. *M.* × *giganteus* is ranged in height (3.5–4 m) between *M. sinensis* and *M. sacchariflorus* in terms of morphological structure. Botanical characteristics of *Miscanthus* have been reported in many documents and recently by Stewart *et al.* (2009).

Field observations during 2009–2012 in Northern Japan (cold region) encompassing native cold and native warm environment species showed genotypic variability in morphological growth. Genotypes native to warm regions grew more vigorously than genotypes belonging to cold regions. Flowering is relatively earlier in native species from cold areas than those from warmer regions. In general, flowering started from midsummer (mid-July), and when various ecotypes of *M. sinensis* differing in their origins were grown together side-by-side, flowering occurred through the whole autumn until killing frost in November. The inflorescence is 20–30 cm long and produces bisexual spikelets, usually all alike (see additional detailed description in Ohwi, 1964; Watson and Dallwitz, 1992).

In grasses, morphological variability among species represents a source of difference in productivity, while variations in flowering influences maturity, harvesting time and other physiological process relating to biomass quality (Adler *et al.*, 2006). Implicitly, it means these

Fig. 11.1. *Miscanthus* × *giganteus* in field, summer 2010, Experimental Farm, Field Science Center for Northern Biosphere, Hokkaido University, Sapporo, Japan.

parameters must be considered or combined carefully to set up the necessary beneficial cultivation system in order to realize the full biomass feedstock potential, quality and all other agronomic factors (harvesting, conservation, conversion quality and efficiency). Owing to the fact that research in bioenergy and particularly that based on *Miscanthus* is still in its infancy, there is a necessary opportunity to investigate the members of the genus, evaluate their adaptability and then determine growth parameters for high and efficient biomass productivity. In *Miscanthus*, as in most members of the grass family, vigorous and long stems, profuse tillering and thick stems enable plants to accumulate high amounts of dry matter in vegetative parts.

11.3 Origin and Distribution

It has been largely reported that *Miscanthus* originated in East Asia, throughout tropical and temperate areas as well as at both high and low altitudes, with some species observed in the Philippines (Watson and Dallwitz, 1992), India, Malaysia and Polynesia (Ohwi, 1964). Several plants were also observed in some parts of South-east Asia, Himalaya, New Guinea and somewhere in southern Africa (Wikberg, 1990), but their identity was not always discovered due to similarity in appearance to other species. Geographically, the distribution of *Miscanthus* has been updated in many recent studies as in Clifton-Brown *et al.* (2008) and Stewart *et al.* (2009). *Miscanthus* can grow in diverse conditions including disturbed, abandoned and/or marginal lands (unfertilized poor soils) and stressful dry or cool climates (Lewandowski *et al.*, 2003). However, it prefers open areas without shade. Wetlands are also suitable if soils have good drainage characteristics. In all habitats, *Miscanthus* remains a perennial species. However, the perenniality is managed differently depending on the region. In semi-natural grasslands, the crop is maintained by mowing or annual burning (Moore, 1964).

Ability to adapt to various climates, soils and topography gives *Miscanthus* flexibility to expand to new geographical regions and might have also facilitated establishment throughout Europe (Schwarz, 1993; Lewandowski *et al.*, 2000; Lewandowski and Schmidt, 2006), into America (Heaton *et al.*, 2008) and elsewhere. However, potential productivity in most regions or habitats remains unknown. Its wide adaptability will be an advantage in extending its production to new areas for biofuel and will open opportunities to identify new high potential genotypes for selection for breeding purposes.

11.4 Diversity in Genus *Miscanthus*

Wide climatic differences in distribution areas, variations in growing habitat and differences in morphological growth all point to significant diversity in the species belonging to the genus *Miscanthus*. Different polyploidy is observed among species, *M.* × *giganteus* is allotriploid ($3n = 57$), *M. sinensis* a diploid ($2n = 38$) and *M. sacchariflorus* a tetraploid ($4n = 76$) (Nishiwaki *et al.*, 2011). Many species of *Miscanthus* remain unknown because few studies have been conducted on their genetic background. Often, similarity between species has led to incorrect characterization. For example, *M.* × *giganteus* was wrongly identified as *M. sinensis*. At best, the genus presents a confused and complex identity that has been only partially elucidated (Chou *et al.*, 1987; Von Wuhlisch *et al.*, 1994).

Miscanthus remains unimproved in its wild genetic makeup (Clifton-Brown *et al.*, 2008). However, it is speculated that natural hybridization between *M. sacchariflorus* and *M. sinensis* formed the basis of the origin of several complex species within the genus (Adati and Shiotani, 1962; Hodkinson *et al.*, 2002). That might be the case of *M.* × *giganteus*, a non-fertile hybrid, as well as for several other species existing in the biodiversity.

Hodkinson *et al.* (1997) observed that about 20 *Miscanthus* species could be listed. Scally *et al.* (2001) noted that at least 14 species are clearly differentiated. Collaborative taxonomical studies by the Royal Botanic Garden, Kew, England and Missouri Botanic Garden, St Louis, Missouri, USA, and relying on opinion have generated the list of already found *Miscanthus* species shown in Table 11.1.

Table 11.1. *Miscanthus* species (Source: http://www.theplantlist.org/; contributors@theplantlist.org, 2010).

Miscanthus changii
Miscanthus depauperatus
Miscanthus ecklonii
Miscanthus floridulus
Miscanthus fuscus
Miscanthus × giganteus
Miscanthus junceus
Miscanthus lutarioriparius
Miscanthus nepalensis
Miscanthus nudipes
Miscanthus oligostachyus
Miscanthus paniculatus
Miscanthus sacchariflorus
Miscanthus sinensis[a]
Miscanthus tinctorius
Miscanthus violaceus

[a]*M. sinensis* spp. *condensatus* is the latest species identified and added to this list by Koyama (1987).

The reason that so few species have been identified might be due to the methods used. DNA analysis and genetic polymorphism based on identification methods, of course, are highly accurate but costly, time consuming and complex. Identification based on field observation is another alternative, although lacking the sophistication and authenticity of DNA analysis, and provides a direct and rapidly applicable tool to establish a database. Field observations can partially establish species identity and assess the characteristics of high potential species under field conditions (Greef *et al.*, 1997; Iwata *et al.*, 2005). Therefore, field evaluation provides a beneficial platform to grow plants from different origins side-by-side, select high yield potential species, and then to understand genotypic behaviours, predict future improvement direction and also identify high productive environments. Knowledge of these key characteristics is vital when selecting *Miscanthus* as an energy crop (Kossonou *et al.*, 2011).

11.4.1 Biofuel *Miscanthus* species

Biofuel feedstock candidate species are selected based on specific criteria required to attain the status of an energy crop. The desired species should show high agronomic potentialities (cultivation, low input, efficiency, yield and viability) and could be cultivated without possible interaction with food-crop production (Henry, 2010). To date, *M.* × *giganteus* is the only *Miscanthus* species that has drawn attention for wide scale production. It has been broadly tested in different regions, under different climates and even compared with other crops (maize, switchgrass) in terms of productivity (Dohleman and Long, 2009) in Europe and in the USA (Heaton *et al.*, 2008). However, because they were not included among the species evaluated, the potential of two other promising *Miscanthus* species, *M. sinensis* and *M. sacchariflorus*, remain unexplored. Recently, field trials, not only in cold regions in northern Japan (2009–2011) but also in Europe and the USA, showed that *M. sinensis* particularly and also *M. sacchariflorus* were comparable to *M.* × *giganteus* in terms of productivity and *M. sinensis* could out-yield *M.* × *giganteus*. The morphological growth performance of *M. sinensis* in the wild has been found to be close to that of *M.* × *giganteus* (Clifton-Brown *et al.*, 2001; Christian *et al.*, 2008; Stewart *et al.*, 2009). Considering that morphological growth determines sink source and sink capacity, then it is reasonable to speculate that *M. sinensis* would also be a species of choice with an added advantage that it is a fertile species lending itself easier to genetic improvement. It is risky to consider one species, *M.* × *giganteus* alone, in a diverse genus of high biomass potential and to pretend to have found a viable energy crop alternative. New selections are possible from the existing list or from wild habitats or collection sites, which should then be tested for final selection. It is very likely that many species suitable for biofuel feedstocks exist in *Miscanthus* but unfortunately have not been identified yet. Species diversity in itself is a beneficial character, as they provide the required gene pool to improve developed cúltivars.

11.5 Agronomical Aspects

Understanding agronomy of *Miscanthus* is essential to deciding whether its field scale planting presents a viable feedstock option for biofuel production in a given situation. It is also important to develop better production methods in an efficient agricultural system to ensure optimum longevity of this perennial grass.

11.5.1 Growth habit and propagation

Growth habit

In the wild, *Miscanthus* grows in a wide range of soil types, from fertile to unfertile, tolerates drought well, and is distributed from warm to cold climates (Stewart *et al.*, 2009). It can also grow in heavy rainfall regions if the soil is well drained (Clifton-Brown and Lewandowski, 2002). Although possessing C_4 photosynthesis, *Miscanthus* grows well in temperate cold regions and cool summer seasons, and where heavy snowfall dominates the winter season (e.g. Sapporo, Japan). It is commonly known that under such cool climate environments many annual plants species would not even survive. *Miscanthus* species and particularly *M. sinensis* have a great tolerance to low temperature. Recently, in northern Japan (border area of natural habitat of *M. sinensis*), a field study (still running 2009–2012) demonstrated that *Miscanthus* could survive at low temperatures if soil is covered by snow. In cold areas, however, *Miscanthus* density diminishes as the temperatures decrease and become severe (Numata and Mitsudera, 1969), indicating an environmental effect. Pittermann and Sage (2001), Naidu and Long (2004) and Sage and Kubien (2007) have reported genotypic differences in response to overwintering generally in the first year of field establishment. *Miscanthus* × *giganteus* was found to be most sensitive, but impact on potential productivity was not reported. In a field established in northern Japan using 43 accessions of *M. sinensis* differing in their origins (including native from warm areas), only 5% were missing after winter. But at the end of the growing season and in subsequent years, genotypic differences among accessions were mostly due to the difference in morphological growth, which resulted in a higher biomass potential in native genotypes from warm areas. Interestingly, across accessions, it seemed that *Miscanthus* could adapt and grow well regardless of original habitat. Also, this would mean that *Miscanthus* might have a specific physiological ability to adapt to different environmental conditions and produce vigorous growth and high productivity independent of exterior factors. Accordingly, we assume that globally *Miscanthus* will be a suitable energy crop for growing in various habitats including cool areas and marginal land unsuited to food production (Tilman *et al.*, 2006; Cadoux *et al.*, 2011). Data are required to confirm environmental index in many habitats to geographically diversify its production for biofuel feedstock.

Propagation

Understanding the propagation method is necessary to know how to derive planting materials, and the cheapest method of producing it in sufficient quantities to meet large hectarage needs. Propagation of *Miscanthus* can be achieved via seeds or rhizomes. To date, most evaluation field trials were established based on vegetative propagation of *M.* × *giganteus* owing to its sterility. Propagation by rhizomes poses problems of scaling up for industrial planting because large numbers of plants are needed to produce the needed number of rhizomes and excavating, splitting the rhizomes to generate separate plants and replanting is a complex and costly affair and presents a bottleneck for *Miscanthus* commercialization.

Propagation of *Miscanthus* by seeds is mostly done naturally in the wild and for research purposes. In general, the direct seed-method for field establishment is unreliable because seeds are too small to support sufficient carbohydrate to ensure good germination and healthy seedlings (Clifton-Brown *et al.*, 2008). As a result, seedling mortality is high. Consequently, *Miscanthus* establishment in the field using direct-seeding is mostly unknown. The propagation via rhizomes, on the other hand, provides healthy seedlings and establishes easily. There are many propagation centres in Europe. Vegetative propagation provides propagules at a high cost, while cost is less via seeds. Finding ways to improve germination of seeds and tillage methods to support small seed sprouting and early growth or seed germination in a nursery environment and subsequent field transplanting may reduce start-up costs and ensure availability of adequate quantities of planting material. Tissue-cultured plants, if economically produced and transplanted, may provide another option.

11.5.2 Field establishment

Cultivation of *Miscanthus* is possible via transplanting or the combined sowing-transplanting method. DEFRA (2001) have reported protocols to cultivate *Miscanthus* with particular attention on aspects critical for establishment and growth.

Seeding-transplanting

The practice of seedling-transplanting (STP) is less frequently used to establish *Miscanthus* in the field. Seeds are sown, raised in a controlled house then thinned after emerging and watered for vigorous growth of young plants. Healthy, vigorous and mature seedlings are selected for field transplanting. Although STP is more beneficial in cold regions, the practice is independent of the environment. When using STP in cold regions, earlier seeding is better for sufficient vegetative growth and seedling vigour before transplanting.

Transplanting

Common practice consists of planting mature healthy seedlings from seeds or rhizomes. It is customary to transplant directly a portion of rhizome (5 cm containing 2–3 live buds) after excavation. Direct transplanting of rhizomes involves less work, is fast and easy to carry out before summer. In cold regions, the ideal transplanting period is when soil temperatures (5–10°C) have sufficiently risen for reshooting. It is not recommended to extend the period of transplanting, in order to avoid the late season effect. Planting is possible manually or mechanically on well drained soils and in non-shaded areas (sufficient distribution of sunlight). Commonly applied planting density is 1.5 m × 1.5 m (4444 plants ha^{-1}) or 2 m × 2 m (2500 plants ha^{-1}), as this factor has an impact on productivity. Plant structure (Fig. 11.2) and harvesting technique also depend on planting density. A spacing of 0.5 m × 0.5 m (40,000 plants ha^{-1}) or 1.0 m × 1.0 m (10,000 plants ha^{-1}) is too dense and creates crowding.

11.5.3 Field management, morphological growth and harvest

Field management

Owing to the target 'produce high, efficiently at low cost for high return profit' in bioenergy feedstock production, field management is an important factor in the viability of the whole cultivation and production chain of energy crop. Being a perennial crop, the task of field management remarkably reduces from year 1, when field establishment is carried out, to subsequent years. However, as *Miscanthus* is

(a) (b)

Fig. 11.2. Young field of *Miscanthus sinensis* at Field Science Center for Northern Biosphere in Sapporo, Hokkaido, Japan, established in 2009. (a) General view showing planting density and (b) young plants growing well with control of invasive weeds.

planted in wide row spacings, the open ground between rows opens an opportunity for weed competition at an early growth stage. As a result, weed control is common practice during first 2 years; after this time the crop will have developed enough vigour to outcompete any weed. Reports about fertilization to support plant growth are ambiguous and inconsistent. Although in wild conditions *Miscanthus* grows without fertilization, in semi-natural or semi-domestic conditions some studies report no response to additional fertilizer input while others show a positive effect of low level of fertilization (Boehmel *et al.*, 2008). Dependence of *Miscanthus* on added water has been reported, but its drought tolerance level is still unclear and under consideration (Clifton-Brown *et al.*, 2002). There is no specific recommendation about irrigation.

Morphological growth

In general, even under good field management *Miscanthus* is less vigorous in year 1 of establishment and is slow to establish. Owing to its perennial character, in subsequent growing seasons, however, morphological growth begins vigorously in both above-ground and below-ground portions and plant height, tillering capacity, crown diameter, stems and leaf size increase. On an individual basis, *M.* × *giganteus, M. sacchariflorus* and *M. sinensis* differ in growth. But in general, *Miscanthus* spp. can produce about 14–18 leaves during one growing season. Although some *Miscanthus* species produce fertile flowers, the threat of unwanted range expansion is not sufficient to confer on them a high risk potential invasive status. *Miscanthus sinensis*, a fertile species, might be invasive but at a very low risk, and does not preclude its cultivation in environments where it is most suited. There is no one *Miscanthus* species suited for all environments and the species most suited for each environment must be identified and adopted.

Flowering time provides important genetic information on which production period, soil moisture management, biomass quality, maturity and harvesting time are dependent (Jensen *et al.*, 2011). It is also a source of genetic manipulation. Flowering appears at the end of all vegetative processes, which is earlier in species native to cold regions (earlier types) and later in those native to warm regions (later types). By growing genotypes from different origins together, flowering will occur through midsummer (July) and the entire autumn season until a killing frost, and is useful in staggering harvest. In cold regions, *M. sinensis* generally flowers earlier than *M.* × *giganteus* (relatively late) and finally *M. sacchariflorus* (very late autumn). Owing to its longer vegetative growth duration, *M. sacchariflorus* seldom flowers in cold regions but does so in warm regions. Thus, later flowering types generally have longer growth duration for biomass accumulation.

11.5.4 Harvesting and biomass yield

Harvesting

Currently *Miscanthus* is grown essentially to provide biomass as feedstock for biofuel extraction. Potential harvest amount depends on biomass accumulated by the plant throughout the growth period leading to maturity. Actual harvest amount, however, is contingent upon total biomass collected after maturity due to limitations of environmental condition (rain etc.), timing (number of times per season), technique (machinery adaptation and efficiency), materials and species (lodging and physical properties). The aim of the *Miscanthus* harvest is to collect from the field the maximum amount of dry matter possible (raw material of bioenergy) (Fig. 11.3).

Miscanthus is usually harvested once a year at the end of the current season growth. According to evidence from the field in several regions, late harvesting (in winter) seems to be the best scenario (Adler *et al.*, 2006). However, current harvesting practice was designed based on a single species' (*M.* × *giganteus*) behaviour in environments with less snow (mostly in Europe), and was to be harvested in winter to give time for the plant to fully senesce and translocate nutrients from above- to below-ground organs (Jørgensen, 1997). That practice seems unsuited to conditions offered in temperate cold regions having heavy snow (Fig. 11.4). Accordingly, it is difficult to establish a standard harvesting time for all climates. In general, harvest in autumn could be early for high yielding species, while spring harvest may cause yield loss to weather and could harm reshooting

Fig. 11.3. *Miscanthus sinensis* in field, summer 2010, Experimental Farm, Field Science Center for Northern Biosphere, Hokkaido University, Sapporo, Japan. (a) Early harvest of *M. sinensis* in temperate cold region due to heavy snowfall in winter, Hokkaido, northern Japan; (b and c) transportation of *M. sinensis* to conservation centre after harvest; (d) conservation technique of *M. sinensis* after harvest and before utilization.

(Clifton-Brown *et al.*, 2007). The conventional machine used for maize is also suitable to mow *Miscanthus* (DEFRA, 2001). Late harvesting is not feasible in heavy snow areas because there are no adequate machines for harvest under snow cover. For these climates, harvest remains a problem to be solved.

Biomass yield and quality

The species influences biomass potential. Our research revealed that *Miscanthus* production can vary because of environmental conditions, but potential difference between genotypes does not necessarily depend on response to environment or agronomic practices (Kossonou *et al.*, 2011). As the main end use for *Miscanthus* is the conversion of its biomass into biofuel, annual harvest starting in year 2 has been found to be the most acceptable procedure. *Miscanthus* achieves maturity by then and produces about twice as many tillers as it did in year 1. Crop yield keeps increasing annually as the underground rhizomes are establishing until it reaches a stable growing process. Across studies from many locations under current agronomic

Fig. 11.4. Field conditions of *Miscanthus sinensis*, winter 2010 (a) in November and (b) December, at the Experimental Farm, Field Science Center for Northern Biosphere, Hokkaido University, Sapporo, Japan.

practices, *Miscanthus* biomass dry yield ranged between 10 and 35 t ha^{-1} mostly based on the performance of one species *M.* × *giganteus* compared to switchgrass ranging between 5 and 13 t ha^{-1} (McLaughlin *et al.*, 1997; Brummer *et al.*, 2000; Vogel *et al.*, 2002; Casler and Boe, 2003). Recently, some studies reported that *M. sinensis* harvested later (in winter time), could yield 1.1 times the production of *M.* × *giganteus* (Lewandowski *et al.*, 2003). In a cold climate without fertilization, *M. sinensis* (across 41 accessions) yield of 20–50 dry t ha^{-1} was achievable, but the field was harvested manually in late autumn (Kossonou *et al.*, 2011). Species native to warm areas yield about 50% higher than species native to cold areas. Although biomass loss due to delaying harvest was not quantitatively measured, it was estimated to average in the realm of 33%. If this loss can be avoided by devising an efficient winter harvesting method, *Miscanthus* should be able to provide at least 7–17 t ha^{-1} more biomass in heavy snow areas. In Europe and in many other places, *Miscanthus* is mowed, large-baled then transported to a facilities centre for conservation or immediate usage.

Crop suitability for biomass production is dependent not only on the quantity of biomass but also on its quality, which refers to its chemical composition (Baumbach *et al.*, 1997; Jørgensen, 1997; Lewandowski and Kicherer, 1997; Hartmann *et al.*, 1999; Siegle and Spliethoff, 1999; Lewandowski *et al.*, 2003; Boateng *et al.*, 2006). For biofuel purposes, high cellulose content and low lignin content, minimum moisture, ash and other mineral elements (N, P, K, Cl) are desired biomass chemical characteristics to improve conversion efficiency (Obernberger *et al.*, 1997; Sander *et al.*, 1997). So, in deciding harvest time and methods, quality considerations must also be kept in mind.

11.6 Why *Miscanthus*?

Bioenergy contributes weakly to the current global energy system because feedstock source is limited or is not sufficiently explored. Food crops were the first source of feedstock, but their benefits in net energy gain and greenhouse energy effect were widely questioned and the 'food versus fuel' conflict arose from dual-purpose crop use. The second generation crops are intended to be produced solely for energy production purpose. Candidate plants must be highly productive in terms of biomass and produce predictable and stable amounts of feedstock in the short as well as the long term (Norby *et al.*, 2005). Growth of crops can be tempered by interactions with environmental factors (Oren *et al.*, 2001), e.g. energy crops such as napiergrass and energycane are affected by cold temperatures and photoperiod could alter juice yield and Brix content of sweet sorghum. *Miscanthus* is able to perform in various stressful environments, including cool climate conditions (Somerville *et al.*, 2010), unlike most C_4 species whose growth function drastically diminishes as temperature cools. Statistically, *Miscanthus* yield averaged 30–40 t ha^{-1} dry matter corresponding to 12,000 l biofuel. Beale and Long (1995) reported that *Miscanthus* can generate annually biomass to produce 2.5 times more biofuel than food crops per unit land area. This means *Miscanthus* will need only 37.5% of land area compared to food crops to produce the same amount of fuel. Moreover, *Miscanthus* can be produced on marginal soils or land rested from food crop cultivation. Unlike sugarcane, whose use for biofuel has not raised much alarm because of the frequently prevailing excess supply, but nevertheless subjugated to demands for other purposes, *Miscanthus* is an entirely non-edible lignocellulosic plant, thus a dedicated biofuel crop. If *Miscanthus* high biomass yield could be combined with an efficient conversion technology and production chain able to convert the entire plant biomass into bioenergy, it would implicitly raise biofuel output. It has been stated that biomass feedstock is profitable only if an energy plant can be grown for at least 12 years (University of Minnesota, 2011). As it has a perennial habit, *Miscanthus* species should be able to provide biomass feedstock continuously and sufficiently for 20–25 years without need for replanting, which is a substantial gain for energy industries and other related structures. *Miscanthus* adjusts to different environments and functions efficiently and is currently the only C_4 biofuel crop adapted to cool climate production. Furthermore, as *Miscanthus* is still essentially in its wild form, the possibilities of genetic improvement for biofuel suitable traits in this crop could be immense.

11.6.1 Environmental Impact

In the context of global climate change, to cultivate *Miscanthus* as an energy crop will be beneficial for global mitigation in terms of atmospheric CO_2, greenhouse gas emissions and environmental pollution. *Miscanthus*' ability to grow with low or no external input will reduce or eliminate harmful nitrous oxide emission from fertilizer application for feedstock production and its high biomass will sequester considerable volumes of CO_2 (Schneider and McCarl, 2003; Kim and Dale, 2005; Farrell *et al.*, 2006; Herrera, 2006; Abraham *et al.*, 2007; Balat and Balat, 2009). Part of the sequestered CO_2 will be deposited in roots, which upon decomposition may reside in soil for a long period. With minimal carbon expenditure for cultural needs, the above-ground biomass should also provide benefit in terms of carbon input–output balance. *Miscanthus* cultivation as an energy crop is expected to increase because of its high yield potential (Holm-Nielsen *et al.*, 2006; EEA, 2007). This obviously will provide new farming options to farmers with the obvious advantage of diversification (Faaji, 2006). In terms of landscape, cultivation of an energy crop will change the composition and the spatial configuration of crops, relieving pressure on needed farmland biodiversity (Tilman *et al.*, 2001). Based on its agrophysiological-ecological performance, *Miscanthus* can be used to restore many marginal lands.

Ecological concerns regarding renewable energy crops and a need for an environmental impact assessment before large scale production have been put forward. *Miscanthus* is distributed in different environments with different life histories (Mutoh *et al.*, 1985; Naito and Nakagoshi, 1995; Chiang *et al.*, 2003; Chen and Renvoize, 2006; Yi, 2006). Recently invasiveness of *Miscanthus* has been reported in the USA (Quinn *et al.*, 2010), but that has not been substantiated by common agreement. *Miscanthus*, in any case, is easily removable from land if it becomes unnecessary, without the need to spray herbicide. Overall, more detailed information is necessary to truly evaluate its impact level on the community composition (Jepsen *et al.*, 2005; Raghu *et al.*, 2006; Schröder *et al.*, 2008). Somerville *et al.* (2010) stated that biodiversity of *Miscanthus* is beneficial to support ecosystem health, maximize tolerance to both biotic and abiotic stresses and beneficial to identify best management practice, species and the implication of allocation of land for bioenergy production.

Agronomically, perennial energy crops present an ecological advantage in term of mitigating land pollution via nutrient leaching. Generally, high economic value crops perform best at high fertilization and management levels because plant species and cultivars developed were selected to respond accordingly. However, these practices present an important source of pollution from N, P and pesticide runoff to water systems and disaster by emission of NO_3. Inversely, *Miscanthus* is one of the newest crops among economic value-containing species, highly productive and showing less dependence on additional fertilizer or other chemical input. Such characters contribute to reduced environmental pollution (Christian and Haase, 2001) while maintaining a reasonable return from farm enterprise. As *Miscanthus* is a wild species growing naturally under varied conditions, it should possess a trait for optimal growth at poor soil fertility conditions. Additionally, if agricultural management is well placed (harvest after senescence), it should be possible to greatly lower or eliminate the need for fertilization altogether, reducing investment and environmental pollution (Baumbach *et al.*, 1997; Siegle and Spliethoff, 1999). Thus, we summarize that to cultivate a biofuel crop such as *Miscanthus* presents a nutrient recycling and carbon sequestration mechanism with the advantage of changing soil organic carbon and mitigating atmospheric CO_2 by emitting neutral carbon.

11.6.3 Economic impact

Economic viability of bioenergy crops depends on production costs, which includes costs of land, inputs, equipment, storage and transportation, and the subsequent return. *Miscanthus* once established in the field does not require replanting in subsequent years except for some replacements due to missing stands from the winter of the first year, thus the propagule is a one-time fixed cost (Heaton *et al.*, 2004; Long *et al.*, 2007). *Miscanthus* does not respond significantly to fertilization after year 1 (Christian and Haase, 2001) and herbicide application is also unnecessary for weed control after establishment. Existing farm machinery can be adapted for planting. Equipment (machinery) replacements are necessary mostly for harvesting (mowing), baling and conservation. The cost of management after year 1 did not exceed 25% of that in year 1 and varies depending on item (Brummer *et al.*, 2000; Duffy and Nanhou, 2001; Schnitkey *et al.*, 2003). Harvesting cost is stable, while baling, transportation and conservation costs change with dry matter weight, distance, bale size and moisture content. The storage cost is low because biomass products are stored outside crushed, a method found to be most cost-effective (Turhollow, 2000). Land cost also needs to be considered in the cost of production. Land cost is considered as the forsaken profits from the most profitable alternative use of the land that is converted from that use to a perennial grass. Land cost is also variable depending on land availability and agricultural systems (Khanna *et al.*, 2008b). For example, cultivation on marginal land costs less than using arable land. Considering that investments decrease as activities in field management decrease in subsequent years after establishment, implicitly production cost also is reduced. This fact makes cultivation of *Miscanthus* as a perennial energy crop beneficial. The price structure of *Miscanthus* will depend on the supply and demand equation and will be driven by market forces. A reasonable estimate of return expectation can only be arrived at once the cellulosic biomass industry becomes viable and flourishing.

11.6.4 Beneficial traits

Miscanthus is a highly productive plant species in term of biomass. Obviously, that performance comes from beneficial traits that confer upon it the status of energy crop. Some of these traits are summarized in Table 11.2.

11.7 Crop Improvement Research Needs

Screening and selection under field conditions has been the common method used to evaluate genetic resources of *Miscanthus* species. However, detailed information of production

Table 11.2. *Miscanthus* traits important for status as an energy crop.

Traits	Comments
Non-edible	Less impact on food production and security (Hill *et al.*, 2006; Borland *et al.*, 2009; Pimentel *et al.*, 2009) More productive because reshoot earlier; longer growth duration and higher biomass than annuals because they senesce (Dohleman and Long, 2009)
Persistence	Perennial (productive for about 20–25 years); regrows each season; no requirement for replanting Problem found: possibility of invasiveness (Vitousek *et al.*, 1995; Rejmanek, 2000; Richardson *et al.*, 2000; Colautti and MacIsaac, 2004; Raghu *et al.*, 2006; Radosevich *et al.*, 2007; Barney and DiTomaso, 2008; Howard and Ziller, 2008; Valéry *et al.*, 2008; Buddenhagen *et al.*, 2009; Quinn *et al.*, 2010; http://www.invasivespeciesinfo.gov)
Efficiency	Less dependence on chemical and mechanical inputs after planting year; fertilization and herbicide are not required in subsequent years; continuous yearly harvest; efficient use of solar radiation; productive at low temperature; drought tolerance unclear; low production cost owing to low cost investment after planting year; high biofuel yield potential (Bryan *et al.*, 2010)
Carbon mitigation, sequestration and nutrient recycle tool	Carbon sequestration, a beneficial process for global climate warming and environment pollution (Long, 2009); neutral carbon emission meaning recycles atmospheric CO_2 re-utilizing CO_2 emitted in biomass production; recycles soil organic carbon (Ceulemans and Mousseau, 1994; Drake *et al.*, 1997; Saxe *et al.*, 1998; Ainsworth and Rogers, 2007; Toma *et al.*, 2011, 2012)
Adaptability	Tolerates and grows on poor, stressful, marginal land not used for crop production (McKnight and Hess, 2000; Long, 2009)
Effect on stability and value for life	Provides cover for wildlife (Bomb *et al.*, 2007; Bindraban *et al.*, 2009); additional economic resource for farmers
Propagation	Via seeds and/or vegetatively via rhizomes; via tissue culture (Wang *et al.*, 2011)

processes is needed as a guide for 'future' breeding purposes, crop modelling to explain sites, and annual variation in terms of potential owing to environmental conditions. *Miscanthus* contains a great genetic diversity within species, distributed widely in different environmental conditions and many species in the genus are still unknown (Hodkinson *et al.*, 2002; Chou, 2009). This biodiversity represents a potential resource within which selections are possible. While assessing genetic difference, it is also possible to identify new species in the wild environment with better agronomic characteristics. Thus, all distribution areas represent a potential genetic resource. Continuous genetic improvement is essential to assure stable feedstock production and availability.

Bioenergy targets for yield depend on crop development and maturity, harvesting time, transportation and conservation system. Any improvement made in these variables will impact positively on energy crop productivity and bioenergy production. These call for manipulation of some specific traits governing growth and productivity. *Miscanthus* is a completely unimproved genus. Successful improvement is possible through combining existing agronomical and physiological data to come up with the criteria for designing new cultivars which meet the qualifications of an ideal energy crop (Kiniry *et al.*, 2005). In the short term, selection of *Miscanthus* empirically based on specific morphological traits suitable for various stressful conditions will indirectly improve plant performance (Ceulemans and Mousseau, 1994; Gunderson and Wullschleger, 1994; Lindroth *et al.*, 1994; Lindroth and Cienciala, 1996; Drake *et al.*, 1997; Curtis and Wang,

1998; Hall *et al.*, 1998; Saxe *et al.*, 1998; Norby *et al.*, 1999; Condon *et al.*, 2004; Finch *et al.*, 2004; Pennisi, 2008). For efficient productivity, selection and test of eligible species must integrate both poor and stressful conditions; this imples that final selection must be done in environments with stressful characteristics.

Genetic modification used in many other crops can also be applied as a way to improve *Miscanthus* genetically. Recently a transgenic approach has been developed using *M. sinensis* (Wang *et al.*, 2011) and provides opportunity to modify some desired traits which are otherwise not possible (Tuskan *et al.*, 2006). Caution must be observed while implementing genetic modification, especially in grass plants, because of associated negative consequences such as the appearance of new diseases (Schrotenboer *et al.*, 2011) or increased potential invasiveness of grass plants by inadvertence (Rejmanek, 2000; Richardson *et al.*, 2000; Colautti and MacIsaac, 2004; Radosevich *et al.*, 2007; Valéry *et al.*, 2008; Quinn *et al.*, 2010).

11.8 Economic Viability

If it is possible to grow *Miscanthus* on a large scale utilizing currently available non-food-crop agricultural land, then it would be realistic to think that it could be an effective alternative bioenergy feedstock. However, development of energy crops such as *Miscanthus* depends strongly on its production chain cost, which comes from the production system. It means that the simpler the production chain, the lower the cost and the greater its viability as a bioenergy crop will be. In general, it is beneficial to convert biomass from lignocellulosic plants into biofuel because it costs less and requires fewer investments than using food crops. It is also possible to reduce production cost via conversion efficiencies and cost savings generated by improvements in harvesting, collecting and transporting biomass. Owing to perenniality, the cost of *Miscanthus* production decreases as some of the work involved in field management disappears in subsequent years after establishment. Renewability and sustainability are an essential part in successful transition to renewable energy. It means that, even when land is available, the price to pay for the production is an important aspect to clarify from the farmer's angle. Khanna *et al.* (2008b) reported that the break-even delivered cost for electricity generation (US\$44–80 t^{-1} DM) was higher than coal-energy equivalent (US\$20–22 t^{-1} DM) in Illinois, USA. Comparing two energy crops, *Miscanthus* and switchgrass, by integrating cost of harvesting, land and transportation to the existing power centre, production cost was largely cheaper (2/3) for *Miscanthus* than switchgrass. To incentivize biofuel production, there is a need for policies that will provide incentives to farmers for producing crops such as *Miscanthus* based on their environmental benefit. Cellulosic biofuel refineries are on the way and will become more prevalent with time. Environmental concerns, pressures for oil independence, and energy policy are creating a strong market for renewable energy and bode well for energy plants such as *Miscanthus*.

References

Abraham, M.C., Lu, Y. and Shaham, S. (2007) A morphologically conserved non-apoptotic program promotes linker cell death in *C. elegans*. *Developmental Cell* 12, 73–86.

Adati, S. and Shiotani, I. (1962) The cytotaxonomy of the genus *Miscanthus* and its phylogenic status. *Bulletin of the Faculty of Agriculture, Mie University* 25, 1–24.

Adler, R.P., Sanderson, A.M., Akwasi, B.A., Weimer, J.P. and Jung, H.-J.G. (2006) Biomass yield and biofuel quality of switchgrass harvested in fall or spring. *Agronomy Journal* 98, 1518–1525.

Ainsworth, E.A. and Long, S.P. (2005) What have we learned from 15 years of free-air CO_2 enrichment (FACE)? A meta-analytic review of the responses of photosynthesis, canopy properties and plant production to rising CO_2. *New Phytologist* 165, 351–372.

Ainsworth, E.A. and Rogers, A. (2007) The response of photosynthesis and stomatal conductance to rising (CO_2): mechanisms and environmental interactions. *Plant, Cell and Environment* 30, 258–270.

Andersson, N.J. (1855) Om de med *Saccharum* beslägtade genera. *Öfvers Kungl. Vet. Adad. Förh. Stockholm* 12, 151–168.

Balat, M. and Balat, H. (2009) Recent trends in global production and utilization of bio-ethanol fuel. *Applied Energy* 86, 2273–2282.

Barney, J.N. and DiTomaso, J.M. (2008) Global climate niche estimates for bioenergy crops and invasive species of agronomic origin: potential problems and opportunities. *PLoS ONE* 6, e17222. doi:10.1371/journal.pone.0017222.

Baumbach, G., Zuberhübler, U. and Siegle, V. (1997) *Emissions from Industrial and Commercial Combustion of Biomass and Wood.* Ecomed, Landsberg (Lech), Germany (in German).

Beale, C.V. and Long, S.P. (1995) Can perennial C_4 grasses attain high efficiencies of radiant energy conversion in cool climates? *Plant, Cell and Environment* 18, 641–650.

Bindraban, P.S., Bulte, E.H., Conijn, J.G., Eickhout, B., Hoogwijk, M. *et al.* (2009). Can biofuels be sustainable by 2020? An assessment for an obligatory blending target of 10% in The Netherlands. Report to the Scientific Assessment and Policy Analysis on Climate Change, The Netherlands. Report No. 500102024, pp. 94. Available at: http://www.pri.wur.nl/UK/ publications/Plant+Research+International publications / New+publications (accessed 14 March 2012).

Boateng, A.A., Hicks, K.B. and Vogel, K.P. (2006) Pyrolysis of switchgrass (*Panicum virgatum*) harvested at several stages of maturity. *Journal of Analytical and Applied Pyrolysis* 75, 55–64.

Boehmel, C., Lewandowski, I. and Claupein, W. (2008) Comparing annual and perennial energy cropping systems with different management intensities. *Agricultural Systems* 96, 224–236.

Bomb, C., McCormick, K., Deurwaarder, E. and Kalberger, T. (2007) Biofuels for transport in Europe: lessons from Germany and the UK. *Energy Policy* 35, 2256–2267.

Borland, A.M., Griffiths, H., Hartwell, J. and Smith, J.A.C. (2009) Exploiting the potential of plants with crassulacean acid metabolism for bioenergy production on marginal lands. *Journal of Experimental Botany* 60, 2879–2896.

Brummer, E.C., Burras, C.L., Duffy, M.D. and Moore, K.J. (2000) *Switchgrass Production in Iowa: economic analysis, soil suitability, and varietal performance*. Iowa State University, Ames, Iowa. Prepared for Bioenergy Feedstock Development Program, Oak Ridge National Laboratory, Oak Ridge, Tennessee.

Bryan, B.A., King, D. and Wang, E. (2010) Biofuels agriculture: landscape-scale trade-offs between fuel, economics, carbon, energy, food, and fiber. *Global Change Biology Bioenergy* 2, 330–345.

Buddenhagen, C.E., Chimera, C. and Clifford, P. (2009) Assessing biofuel crop invasiveness: a case study. *PLoS One* 4, 1–6.

Cadoux, S., Riche, A.B., Yates, N.E. and Marchet, J. (2011) Nutrient requirements of *Miscanthus* × *giganteus*: conclusions from a review of published studies. *Biomass and Bioenergy* 38, 14–22.

Casler, M.D. and Boe, A.R. (2003) Cultivar × environment interactions in switchgrass. *Crop Science* 43, 2226–2233.

Ceulemans, R. and Mousseau, M. (1994) Tansley Review no. 71: Effects of elevated atmospheric CO_2 on woody plants. *New Phytologist* 127, 425–446.

Chen, S.L. and Renvoize, S.A. (2006) *Miscanthus*. *Flora of China* 22, 581–583.

Chiang, Y.-C., Schaal, A.B., Chou, C.H., Huang, S. and Chiang, T.-Y. (2003) Contrasting selection modes at the *Adh1* locus in outcrossing *Miscanthus sinensis* vs. inbreeding *Miscanthus condensatus* (Poaceae). *American Journal of Botany* 90, 561–570.

Chou, C.-H., Hwang, S.Y. and Chang, F.C. (1987) Population study of *Miscanthus floridulus* (Labill.) Warb. *Botanic Bulletin Academia Sinica* 28, 247–281.

Chou, C.-H. (2009) *Miscanthus* plants used as an alternative biofuel material: the basic studies on ecology and molecular evolution. *Renewable Energy* 34, 1908–1912.

Christian, D.G. and Haase, E. (2001) Agronomy of *Miscanthus*. In: Jones, M.B. and Walsh, M. (eds) *Miscanthus for Energy and Fibre*. James and James (Science Publishers), London, pp. 1–45.

Christian, D.G., Riche, A.B. and Yates, N.E. (2008) Growth, yield and mineral content of *Miscanthus* × *giganteus* grown as a biofuel for 14 successive harvests. *Industrial Crops and Products* 28, 320–327.

Clifton-Brown, J.C. and Lewandowski, I. (2002) Screening *Miscanthus* genotypes in field trials to optimise biomass yield and quality in Southern Germany. *European Journal of Agronomy* 16, 97–110.

Clifton-Brown, J.C., Lewandowski, I., Andersson, B., Basch, D.G.C., Kjeldsen, J. *et al.* (2001) Performance of 15 *Miscanthus* genotypes at five sites in Europe. *Agronomy Journal* 93, 1013–1019.

Clifton-Brown, J.C., Lewandowski, I., Bangerth, F. and Jones, M.B. (2002) Comparative responses to water stress in stay-green, rapid and slow senescing genotypes of the biomass crop, *Miscanthus*. *New Phytologist* 154, 335–345.

Clifton-Brown, J.C., Breuer, J. and Jones, M.B. (2007) Carbon mitigation by the energy crop, *Miscanthus. Global Change Biology* 13, 2296–2307.

Clifton-Brown, J., Chiang, Y.C. and Hodkinson, T.R. (2008) *Miscanthus*: Genetic resources and breeding potential to enhance bioenergy production. In: Vermerris, W. (ed.) *Genetic Improvement of Bioenergy Crops*. Springer Science, New York, pp. 273–294.

Condon, A.G., Richards, R.A., Rebetzke, G.J. and Farquhar, G.D. (2004) Breeding for high water-use efficiency. *Journal of Experimental Bot*any 55, 2447–2460.

Colautti, R.I. and MacIsaac, H.J. (2004) A neutral terminology to define 'invasive' species. *Diversity and Distributions* 10, 135–141.

Curtis, P.S. and Wang, X. (1998) A meta-analysis of elevated CO_2 effects on woody plant mass, form, and physiology. *Oecologia* 113, 299–313.

DEFRA (2001) Planting and growing *Miscanthus* – best practice guidelines. Available at: http://www.naturalengland.org.uk/Images/miscanthus-guide_tcm6-4263.pdf (accessed 15 February 2012).

Dohleman, F.G. and Long, S.P. (2009) More productive than maize in the midwest: How does *Miscanthus* do it? *Plant Physiology* 150, 2104–2115.

Drake, B.G., Gonzàlez-Meler, M.A. and Long, S.P. (1997) More efficient plants: a consequence of rising atmospheric CO_2? *Annual Review of Plant Physiology and Plant Molecular Biology* 48, 609–639.

Duffy, M.D. and Nanhou, V.Y. (2001) *Costs of producing switchgrass for biomass in Iowa*. Available at: http://www.extension.iastate.edu/Publications/PM1866.pdf (accessed 15 February 2012).

EEA (European Environment Agency) (2007) *Report 27 November 2007*. OPOCE (Office for Official Publications of the European Communities).

Faaji, A.P.C. (2006) Bio-energy in Europe: changing technology choices. *Energy Policy* 34, 322–342.

Farrell, A.D., Clifton-Brown, J.C., Lewandowski, I. and Jones, M.B. (2006) Genotypic variation in cold tolerance influences the yield of *Miscanthus. Annals of Applied Biology* 149, 337–345.

Finch, J.W., Hall, R.L., Rosier, P.T.W., Clark, D.B., Stratford, C. *et al.* (2004) *The Hydrological Impacts of Energy Crop Production in the UK*. B/CR/000783/00/00. Department of Trade and Industry, London.

Greef, J.M., Schondelmaier, J. and Deuter, M. (1997) Genetic diversity of European *Miscanthus* species revealed by AFLP fingerprinting. *Aspects of Applied Biology* 49, 231–235.

Gunderson, C.A. and Wullschleger, S.D. (1994) Photosynthetic acclimation in trees to rising atmospheric CO_2: a broader perspective. *Photosynthesis Research* 39, 369–388.

Hall, R.L., Allen, S.J., Rosier, P.T.W. and Hopkins, R. (1998) Transpiration from coppiced poplar and willow measured using sap-flow methods. *Agricultural and Forest Meteorology* 90, 275–290.

Hartmann, H., Maier, L. and Böhm, T. (1999) Quality of solid biofuels – database and field trials. In: Overend, R.P. and Chornet, E. (eds) *Proceedings of Biomass Conference of the Americas*, Oakland, California, 29 August–2 September 1999, pp. 273–279.

Heaton, E., Voigt, T. and Long, S.P. (2004) A quantitative review comparing the yields of two candidate C_4 perennial biomass crops in relation to nitrogen, temperature and water. *Biomass and Bioenergy* 27, 21–30.

Heaton, E.A., Dohleman, F.G. and Long, S.P. (2008) Meeting US biofuel goals with less land: the potential of *Miscanthus. Global Change Biology* 14, 2000–2014.

Henry, R.J. (2010) Evaluation of plant biomass resources available for replacement of fossil oil. *Plant Biotechnology Journal* 8, 288–293.

Herrera, S. (2006) Bonkers about biofuels. *Science Biotechnology* 24, 755–760.

Hill, J., Nelson, E., Tilman, D., Polasky, S. and Tiffany, D. (2006) Environmental, economic, and energetic costs and benefits of biodiesel and ethanol biofuels. *Proceedings of the National Academy of Sciences of the United States of America* 103, 11206–11210.

Hodkinson, T.R., Renvioze, S.A. and Chase, M.W. (1997) Systematics of *Miscanthus*. In: Bullard M.J. *et al.* (eds) *Biomass and Bioenergy Crops*. Aspects of Biology 49. Association of Applied Biologists, Warwick, UK, pp. 189–197.

Hodkinson, T.R., Chase, M.W., Takahashi, C., Leitch, I.J., Bennett, M.D. *et al.* (2002) The use of DNA sequencing (ITS and trnL–F), AFLP and fluorescent *in situ* hybridisation to study allopolyploid *Miscanthus* (Poaceae). *American Journal of Botany* 89, 279–286.

Holm-Nielsen, J.B., Madsen, M. and Popiel, P.O. (2006) Predicted energy crop potentials for bioenergy, worldwide and for EU-25. In: *World Bioenergy Conference and Exhibition on Biomass for Energy*, Jonkoping, Sweden, 30 May–1 June 2006.

Howard, G. and Ziller, S. (2008) Alien alert – plants for biofuel may be invasive. *Bioenergy Business* 14, 6.

Iwata, H., Kamijo, T. and Tsumura, Y. (2005) Genetic structure of *Miscanthus sinensis* ssp. *condensatus* (Poaceae) on Miyake Island: implications for revegetation of volcanically devastated sites. *Ecological Research* 20, 233–238.

Jensen, E., Farrar, K., Thomas-Jones, S., Hastings, A., Donnison, I. *et al.* (2011) Characterization of flowering time diversity in *Miscanthus* species. *Global Change Biology Bioenergy* 3, 387–400.

Jepsen, J.U., Topping, C.J., Odderskær, P. and Andersen, P.N. (2005) Evaluating consequences of land-use strategies on wildlife populations using multiple-species predictive scenarios. *Agriculture, Ecosystems and Environment* 105, 581–594.

Jones, M.B. and Walsh, M. (eds) (2001) *Miscanthus for Energy and Fibre*. James and James, London.

Jørgensen, U. (1997) Genotypic variation in dry matter accumulation and content of N, K and Cl in *Miscanthus* in Denmark. *Biomass and Bioenergy* 12, 155–169.

Khanna, M., Ando, A. and Taheripour, F. (2008a) Welfare effects and unintended consequences of ethanol subsidies. *Review of Agricultural Economics* 30, 411–421.

Khanna, M., Dhungana, B. and Clifton-Brown, J. (2008b) Costs of producing miscanthus and switchgrass for bioenergy in Illinois. *Biomass and Bioenergy* 32, 482–493.

Kim, S. and Dale, B. (2005) Life-cycle assessment of various cropping systems utilized for producing biofuels: bio-ethanol and bio-diesel. *Biomass and Bioenergy* 29, 426–439.

Kiniry, J.R., Cassida, K.A., Hussey, M.A., Muir, J.P., Ocumpaugh, W.R. *et al.* (2005) Switchgrass simulation by the ALMANAC model at diverse sites in the southern US. *Biomass and Bioenergy* 29, 419–425.

Kobayashi, K. (1981) Studies on the growth and maintenance of patch of *Miscanthus sinensis* Andress. Formation process of the patch and its ecological characteristics in a warm region. *Journal of the Japanese Society of Grassland Science* 27, 71–78 (in Japanese with English summary).

Kossonou, G.A., Kajihara, Y., Toma, Y., Iizuka, N. and Yamada, T. (2011) Potentiality of four cool season grasses and *Miscanthus sinensis* for feedstock in the cool regions of Japan. *Journal of the Japan Institute of Energy* 90, 59–65.

Koyama, T. (1987) *Grasses of Japan and its Neighboring Regions: an identification manual*. Kodansha Ltd, Tokyo.

Lewandowski, I. and Kicherer, A. (1997) Combustion quality of biomass: Practical relevance and experiments to modify the biomass quality of *Miscanthus × giganteus*. *European Journal of Agronomy* 6, 163–177.

Lewandowski, I. and Schmidt, U. (2006) Nitrogen, energy and land use efficiencies of *Miscanthus*, reed canary grass and triticale as determined by the boundary line approach. *Agriculture, Ecosystems and Environment* 112, 335–346.

Lewandowski, I., Clifton-Brown, J.C., Scurlock, J.M.O. and Huisman, W. (2000) *Miscanthus*: European experience with a novel energy crop. *Biomass and Bioenergy* 19, 209–227.

Lewandowski, I., Scurlock, J.M.O., Lindvall, E. and Christou, M. (2003) The development and current status of perennial rhizomatous grasses as energy crops in the US and Europe. *Biomass and Bioenergy* 25, 335–361.

Lindroth, A. and Cienciala, E. (1996) Water-use efficiency of short rotation *Salix viminalis* at leaf, tree and stand scales. *Tree Physiology* 16, 257–262.

Lindroth, A., Verwijst, T. and Halldin, S. (1994) Water-use efficiency of willow: variation with season, humidity and biomass allocation. *Journal of Hydrology* 156, 1–19.

Long, S.P. (2009) First to know the nature of bioenergy. *Global Change Biology Bioenergy* 1, 1.

Long, S.P., Dohleman, F., Jones, M.B., Clifton-Brown, J. and Jørgensen, U. (2007) *Miscanthus* – panacea for energy, security and the Midwest economy or another kudzu? *The Illinois Steward Magazine* 16, 1. Available at: http://ilsteward.nres.uiuc.edu/Issues/2007/Spring/energy.htm (accessed 15 March 2012).

Mcknight, T.L. and Hess, D. (2000) *Climate Zones and Types: The Köppen System*. Prentice Hall, Upper Saddle River, New Jersey.

McLaughlin, S.B. (1997) Forage crops as bioenergy fuels: evaluating the status and potential. In: *XVIII International Grassland Congress*, 8–19 June 1997, Winnipeg, Manitoba, Canada.

Moore, C.W.E. (1964) Distribution of grasslands. In: Bernard, C. (ed.) *Grasses and Grasslands*. Macmillan, New York, pp. 182–205.

Mutoh, N., Kimura, M., Oshima, Y. and Iwaki, H. (1985) Species diversity and primary productivity in *Miscanthus sinensis* grasslands. 1. Diversity in relation to stand structure and dominance. *Botanical Magazine Tokyo* 98, 159–170.

Naidu, S.L. and Long, S.P. (2004) Potential mechanisms of low-temperature tolerance of C-4 photosynthesis in *Miscanthus × giganteus*: an *in vivo* analysis. *Planta* 220, 145–155.

Naito, K. and Nakagoshi, N. (1995) The conservation ecology of *Iris rossii* Baker (Iridaceae), a threatened plant in rural Japan. *Journal of Plant Research* 108, 477–482.

Nishiwaki, A., Mizugutl, A., Kuwabara, S., Toma Y., Ishigaki G. *et al.* (2011) Discovery of natural *Miscanthus* (Poaceae) triploid plants in sympatric populations of *Miscanthus sacchariflorus* and *Miscanthus sinensis* in southern Japan. *American Journal of Botany* 98, 154–159.

Norby, R.J., Wullschleger, S.D., Gunderson, C.A., Johnson, D.W. and Ceulemans, R. (1999) Tree responses to rising CO_2 in field experiments: implications for the future forest. *Plant, Cell and Environment* 22, 683–714.

Norby, R.J., DeLucia, E.H. and Gielen, B. (2005) Forest response to elevated CO_2 is conserved across a broad range of productivity. *Proceedings of the National Academy of Sciences of the United States of America* 102, 18052–18056.

Numata, M. (1988) Stage and phase dynamics in early dynamics of secondary succession. *Shukutoku University Bulletin* 22, 17–24.

Numata, M. and Mitsudera, M. (1969) Efficient environmental factors in relation to the growth and production of the *Miscanthus sinensis* meadows in Japan. *Japanese Journal of Botany* 20, 135–151.

Obernberger, I., Bidermann, F., Widmann, W. and Riedl, R. (1997) Concentrations of inorganic elements in biomass fuels and recovery in the different ash fractions. *Biomass and Bioenergy* 12, 211–224.

Ohwi, J. (1964) *Flora of Japan*. Smithsonian Institution, Washington, DC.

Oren, R., Ellsworth, D.S., Johnsen, K.H., Phillips, N., Ewers, B.E. *et al.* (2001). Soil fertility limits carbon sequestration by forest ecosystems in a CO2-enriched atmosphere. *Nature* 411, 469–472.

Pennisi, E. (2008) Plant genetics: the blue revolution, drop by drop, gene by gene. *Science* 320, 171–173.

Pimentel, D., Marklein, A., Toth, M.A., Karpoff, M.N., Paul, G.S. *et al.* (2009) Food versus biofuels: environmental and economic costs. *Human Ecology* 37, 1–12.

Pittermann, J. and Sage, R.F. (2001) The response of the high altitude C4 grass *Muhlenbergia Montana* (Nutt.) A.S. Hitchc. to long- and short-term chilling. *Journal of Experimental Botany* 52, 829–838.

Pyter, R.J., Dohleman, F.G. and Voigt, T.B. (2009) Effects of rhizome size, depth of planting and cold storage on *Miscanthus giganteus* establishment in the Midwestern USA. *Biomass and Bioenergy* 34, 1466–1470.

Quinn, L.D., Allen, D.J. and Stewart, J.R. (2010) Invasiveness potential of *Miscanthus sinensis*: implications for bioenergy production in the United States. *Global Change Biology Bioenergy* 2, 310–320.

Radosevich, S.R., Holt, J.S. and Ghersa, C.M. (2007) *Ecology of Weeds and Invasive Plants: Relationship to Agriculture and Natural Resource Management*, 3rd edn. John Wiley and Sons, New York, 454 pp.

Raghu, S., Anderson R.C., Daehler, C.C., Davis, A.S., Wiedenmann, R.N. *et al.* (2006) Adding biofuels crops to the invasive species fire? *Science* 313, 1742.

Rejmanek, M. (2000) Invasive plants: approaches and predictions. *Austral Ecology* 25, 497–506.

Richardson, D.M., Pysek, P., Rejmánek, M., Barbour, M.G., Panetta, F.D. *et al.* (2000) Naturalization and invasion of alien plants: concepts and definitions. *Diversity and Distributions* 6, 93–107.

Sage, R.F. and Kubien, D.S. (2007) The temperature response of C_3 and C_4 photosynthesis. *Plant, Cell and Environment* 30, 1086–1106.

Saxe, H., Ellsworth, D.S. and Heath, J. (1998) Tree and forest functioning in an enriched CO_2 atmosphere. *New Phytologist* 139, 395–436.

Scally, L., Hodkinson, T. and Jones, M.B. (2001) Origins and taxonomy of *Miscanthus*. In: Jones, M.B. and Walsh, M. (eds) Miscanthus *for Energy and Fibre*. James and James, UK, pp. 1–9.

Schneider, U.A. and McCarl, B.A. (2003) *Implications of a carbon based energy tax for U.S. Agriculture*. FNU-17, Centre for Marine and Climate Research, Hamburg University, Hamburg, Germany.

Schnitkey, G., Lattz, D. and Siemens, J. (2003) Machinery cost estimates: forage field operation. In: Farm Business Management Handbook. Available at: http://www.farmdoc.uiuc.edu/manage/pdfs/Mach_forages_2003.pdf (accessed 14 March 2012).

Schröder, B., Rudner, M., Biedermann, R., Kögl, H. and Kleyer, M. (2008) A landscape model for quantifying the trade-off between conservation needs and economic constraints in the management of a semi-natural grassland community. *Biological Conservation* 141, 719–732.

Schrotenboer, A.C., Allen, M.S. and Malmstrom, C.M. (2011) Modification of native grasses for biofuel production may increase virus susceptibility. *Global Change Biology Bioenergy* 3, 360–374.

Schwarz, H. (1993) *Miscanthus sinensis* 'Giganteus' production on several sites in Austria. *Biomass and Bioenergy* 5, 413–419.

Siegle, V. and Spliethoff, H. (1999) Physical and chemical basis for combustion. In: Fachagentur Nachwachsende Rohstoffe e.V. (ed.) *Leitfaden Biomasse*. Gülzow, Germany (In German.), pp. 88–94.

Somerville, C., Youngs, H., Taylor, C., Davis, C.S. and Long, S.P. (2010) Feedstocks for lignocellulosic biofuels. *Science* 329, 790–792.

Stewart, J.R., Toma, Y., Fernandez, F.G., Nishiwaki, A., Yamada, T. *et al.* (2009) The ecology and agronomy of *Miscanthus sinensis*, a species important to bioenergy crop development, in its native range in Japan: a review. *Global Change Biology Bioenergy* 1, 126–153.
Tang, Y.H., Washitani, I., Tsuchiya, T. and Iwaki, H. (1988) Fluctuation of photosynthetic photon flux density within a *Miscanthus sinensis* canopy. *Ecological Research* 3, 253–266.
Tilman, D., Peter, B.R., Knops, J., Wedin, D., Mielke, T. and Lehman, C. (2001) Diversity and productivity in a long-term grassland experiment. *Science* 294, 843–845.
Tilman, D., Hill, J. and Lehman, C. (2006) Carbon-negative biofuels from low-input high-diversity grassland biomass. *Science* 314, 1598–1600.
Toma, Y., Fernández, F.G., Sato, S., Izumi, M., Hatano, R. *et al.* (2011) Carbon budget and methane and nitrous oxide emissions over the growing season in a *Miscanthus sinensis* grassland in Tomakomai, Hokkaido, Japan. *Global Change Biology Bioenergy* 3, 116–134.
Toma, Y., Armstrong, K., Stewart, J.R., Yamada, T., Nishiwaki, A. *et al.* (2012) Carbon sequestration in soil in a semi-natural *Miscanthus sinensis* grassland and *Cryptomeria japonica* forest plantation in Aso, Kumamoto, Japan. *Global Change Biology Bioenergy*, DOI: 10.1111/j.1757-1707.2012.01160.x.
Turhollow, A. (2000) Costs of producing biomass from riparian buffer strips. Available at: http://www.ornl.gov/~webworks/cpr/v823/rpt/108548.pdf (accessed 15 February 2012).
Tuskan, G.A., DiFazio, S., Jansson, S. *et al.* (2006) The genome of black cottonwood, *Populus trichocarpa* (Torr. & Gray). *Science* 313, 1596–1604.
University of Minnesota (2011) Miscanthus as a Biomass Fuel Source. Available at: http://www.University of Minnesota/Research/Bioenergy grass/Miscanthus (accessed 12 November 2012).
Valéry, L., Fritz, H., Lefeuvre, J.C. and Simberloff, D. (2008) In search of a real definition of the biological invasion phenomenon itself. *Biological Invasions* 10, 1345–1351.
Villamil, M.B., Silvis, A.H., Bollero, G.A. (2008) Potential *Miscanthus* adoption in Illinois: information needs and preferred information channels. *Biomass & Bioenergy* 32, 1338–1348.
Vitousek, P.M., Aplet, G.H., Raich, J.R. and Lockwood, J.P. (1995) Mauna Loa as a model system for ecological research. *Geophysical Monograph* 92, 117–126.
Vogel, K.P., Brejda, J.J., Walters, D.T. and Buxton, D.R. (2002) Switchgrass biomass production in the Midwest USA: harvest and nitrogen management. *Agronomy Journal* 94, 413–420.
Von Wuhlisch, G., Deuter, M. and Muhs, H.J. (1994) Identifizierung verschiedener Miscanthus Sorten mittels Isoenzymen. *Journal of Agronomy and Crop Science* 172, 247.
Wang, X., Yamada, T., Kong, F.-J., Abe, Y., Hoshino, Y., Sato, H., Takamizo, T., Kanazawa, A. and Yamada, T. (2011) Establishment of an efficient *in vitro* culture and particle bombardment-mediated transformation systems in *Miscanthus sinensis* Anderss., a potential bioenergy crop. *GCB Bioenergy* 3, 322–332.
Watson, L. and Dallwitz, M.J. (1992) The grass genera of the world. Available at: http://delta-intkey.com/grass (accessed 14 March 2012).
Wikberg, S. (1990) *The Genus* Miscanthus*: a summary of available literature*. The University of Lund, Sweden, Projekt Agro-Fiber, 42 pp.
Yi, Y.N. (2006) *New Flora of Korea*. Kyo-Hak Publishing Company, Seoul, Korea (in Korean).

12 *Arundo donax*

C.M.J. Williams,[1] T.K. Biswas,[2] L. Márton[3] and M. Czakó[3]
[1]*Formerly South Australian Research and Development Institute, Australia (Retired);* [2]*Murray-Darling Basin Authority, Australia;* [3]*Department of Biological Sciences, University of Southern Carolina, South Carolina, USA*

12.1 Introduction

Arundo donax L. (giant reed), *Poaceae*, is a robust, perennial, rhizomatous grass, from 3 to 9 m tall, growing in multi-stemmed cane-like clumps. Its native range extends from eastern Asia to southern Europe including tropical, warm temperate and Mediterranean environments (Lewandowski *et al.*, 2003). *A. donax* in different countries has been given several common names including bamboo reed, Carrizo cane, Italian reed, false bamboo, nalgrass, Spanish reed and wild cane. The science community has adopted the common name of giant reed.

Over many centuries, *A. donax* has been introduced around different parts of the world because of its many uses, as a windbreak, an ornamental, for wind musical instruments, for pulp and paper (Lewis and Jackson, 2002; Paul and Williams, 2006; Shatalov and Pereira, 2006), garden stakes, building material (Perdue, 1958) and for erosion control as a plant to stabilize channel banks (often becoming invasive, Virtue *et al.*, 2010a). *A. donax* was first introduced to Australia in 1788, by the first fleet bringing the British to Australia for colonization (Lee, 2009). They collected Spanish reed at Cape Town in transit to Sydney Cove (Lee, 2009). In Australia, it has been used as above, especially for windbreaks around horticulture crops and to stabilize sand drifts, and the immature canes cut when shorter than 1.8 m for use as a drought reserve grass for ruminants (Spafford, 1941; Williams and Biswas, 2010).

Although *A. donax* has been widely cultivated for centuries, limited scientific information has been published on its biology, genetics, yield, agronomy, weed risk, etc. for its use as a biofuel crop. In recent years, *A. donax* has been evaluated for its potential as a high yield cellulose and biomass crop for biofuels or pulp/paper (in the USA by Mack, 2008; Kering *et al.*, 2012; in Australia by Williams *et al.*, 2007, 2008a,b, 2009; in Europe by Lewandowski *et al.*, 2003; in Italy by Angelini *et al.*, 2009). In Europe, the USA and Australia, in general, shoot dry matter (DM) yields have been greater than other grass or woody biomass crops when *A. donax* was grown with similar or lesser water and nutrient inputs (Hidalgo and Fernandez, 2001; Williams *et al.*, 2007, 2008a; Angelini *et al.*, 2009; Williams and Biswas, 2010; Kering *et al.*, 2012). In South Australia, shoot DM yields of >45 oven dry t ha^{-1} $year^{-1}$ were recorded under sewage or winery wastewater irrigation and 12 t ha^{-1} $year^{-1}$ under an average of 455 mm of natural rainfall between annual clear fell (complete removal of top growth) harvests (Williams *et al.*, 2008a, 2009; Williams and Biswas, 2010).

Expanded use of *A. donax* is controversial and investment in research has been limited as it has an international reputation of being a

©CAB International 2013. *Biofuel Crops: Production, Physiology and Genetics* (ed. B.P. Singh)

weed of riparian environments (Bell, 1997; Dudley, 2000; Lowe *et al.*, 2000), but it has shown good potential as a biomass crop for biofuels and pulp/paper (Angelini *et al.*, 2005, 2009; Williams *et al.*, 2007, 2008a; Biswas and Williams, 2009; Williams and Biswas, 2010; Kering *et al.*, 2012). Detailed analysis of the weed risk of *A. donax* in Australia by Virtue *et al.* (2010a) indicated that while *A. donax* is clearly a serious weed threat to riparian ecosystems, it can be grown as a low weed-risk biofuel crop in non-riparian zones under a set of strict management guidelines.

This chapter reviews the scientific literature and presents innovative findings on the production and utilization of *A. donax* biomass feedstock for biofuels and other uses. A subsidiary aim is to provide an overview of the requirements and characteristics of *A. donax* for profitable and sustainable production systems (physiology, breeding, agronomy, salt tolerance, weed risk management and economics) and biofuel conversion yields.

12.2 Growth, Physiology and Genetics

Arundo donax is a tall, C_3 grass species of the *Poaceae* family. It is a species of tropical and warm climates and is hardy to Zone 6 (Miller and Forney, 2011). The mature, hollow stems, 2 to 5 cm thick, look like bamboo canes. Leaves are lance shaped and 30–50 cm long when mature, and are evenly spaced and alternate in two rows on tan-coloured stems. *A. donax* leaves form a ribbed sheath at the stem junction, whereas bamboo leaves have a short stem (petiole like) structure at the stem junction. It produces an extensive network of rhizomes (usually varying from 3 to 8 cm wide and 10 to 25 cm in length), from which grow tough, fibrous, long taproots. The rhizomes usually form close to the soil surface, from 5 to 30 cm deep (unless deposited in eroded mud where they can exist metres deep in certain moist soils), while most roots are more than 100 cm long (Bell, 1997; Sharma *et al.*, 1998). In the Mediterranean region, where the warm, temperate climate is characterized by mild, wet winters and hot, dry summers, *A. donax* new shoots emerge from buds on rhizomes in late winter/early spring, achieving maximum growth rates in midsummer. Under optimal conditions stems can grow 10 cm day^{-1}, placing it among the fastest growing plants (Perdue, 1958; Bell, 1997). Thus, clear fell harvest during midsummer is likely to be detrimental to maximum biomass yield per year, since this would reduce the duration of optimal leaf area index for high growth during the growing season, as new shoots would need to grow from rhizomes (Sharma *et al.*, 1998; Williams and Biswas, 2010). *A. donax* grows on a wide variety of soil types, from clays to sands, but is most productive in well-drained soils where moisture is abundant (Perdue, 1958; Dudley, 2000; Williams and Biswas, 2010). It is a hydrophyte, growing along the banks of watercourses (Dudley, 2000), it can tolerate saline conditions and access water in shallow water tables and survive up to 6 months with the water level at the top of the soil profile (Rezk and Edany, 1979). However, this plant does not thrive in soils that are waterlogged or submerged year round. Well established plants can also survive extended periods of water stress due to their extensive rhizome and fibrous root systems (Spafford, 1941; Williams and Biswas, 2010).

Shoot moisture content decreases during autumn when anthesis occurs, the rate varying depending on seasonal temperature conditions. Significant proportions of nutrients and carbohydrates are translocated from shoots to the rhizomes and roots, especially during autumn for storage during winter and are remobilized to new shoots the next spring (Williams *et al.*, 2010a,b; Nassi o Di Nasso *et al.*, 2011). During winter, *A. donax* often becomes dormant because of low temperatures and regrowth occurs the following spring (Williams and Biswas, 2010). Minimum temperature range reported for the production of new ramets (new stems bases) from rhizomes ranged from 7 to 12.7°C, thereafter rate of sprouting increased linearly to the optimum temperature with the greatest rate at 28–30°C (Spencer and Ksander, 2006; Graziani and Steinmaus, 2009). The relationship between carbon dioxide exchange and leaf temperature showed that *A. donax* has a broad photosynthetic temperature optimum between 24 and 30°C (Rossa *et al.*, 1998). *A. donax* has a high photosynthetic capacity with high carbon dioxide exchange rates (19.8–36.7 μmol m^{-2} s^{-1} under natural conditions) compared to other C_3 and C_4 species

(Rossa *et al.*, 1998). Despite being a C_3 species, the biomass yield of *A. donax* was higher than the other C_4 species tested by Angelini *et al.* (2009) and Kering *et al.* (2012). *A. donax* is grown commercially in Mediterranean Europe, South America, Caribbean islands, Australia and China (Williams and Biswas, 2010; Miller and Forney, 2011).

12.2.1 Genetics and plant breeding

No viable seeds were produced by *A. donax* in European conditions (Lewandowski *et al.*, 2003; Mariani *et al.*, 2010), in Asia (Bhanwra *et al.*, 1982; Popov and Belyaeva, 1987) or in the USA (Balogh *et al.*, 2010). In Australia, Williams *et al.* (2008b) found no viable embryos for 400 seeds sampled from each of five wild populations in South Australia. Since it is an asexually reproductive species due to seed sterility, its genetic variability and the chances for finding new genotypes or varieties are low (Lewandowski *et al.*, 2003). The European Giant Reed Network studied the genetics of a large number of wild populations grown across Europe. They found indications of low genetic polymorphism among the studied populations and a rather uniform genetic pool. However, there was clustering of populations around their geographical origin, reflecting restricted migration of germplasm for local adaptation (Lewandowski *et al.*, 2003). The genetic fingerprinting of *A. donax* samples from 167 wild populations in South Australia and five from other states identified two clades present in Australia (Virtue *et al.*, 2010a,b). Most of the SA samples, a sample from Brisbane (Queensland) and a sample from Kununurra (Western Australia) were grouped into a single clade with low genetic divergence (less than 1% genetic diversity), indicating clonality. A second clade consisted of two samples from Sydney and four samples from SA. This indicates at least two separate introductions of *A. donax* into Australia. Similar clonality has been measured in genetic studies in the USA (e.g. Ahmad *et al.*, 2008). Mariani *et al.* (2010) found more clades when sampling on a wider geographical scale; some far Asian samples showed the greatest genetic divergence.

Since genetics, breeding and bioenergy scientific studies on *A. donax* are comparatively recent and limited, there is potential for improvements in feedstock biomass qualities for different bioenergy end uses (e.g. biochemical, pyrolysis and complete combustion), increased yield and adaption to certain climate zones and low-input production systems.

Since it is a sterile plant without viable seeds, *A. donax* requires *in vitro* cell and tissue culture for both genetic improvement and large-scale propagation for plantations. In the USA, Márton and Czakó (2004, 2007) developed a universal cell culture procedure that established embryonic cell cultures, micropropagation and genetic engineering protocols for successful use in many species of monocots. The sustained totipotent embryonic culture system allowed efficient somaclonal breeding and selection of elite lines for trichlorophenol dehaloperoxidation and for salt tolerance (Czakó and Márton, 2011). Somaclonal breeding could be utilized for improving many other qualities of *A. donax*. Transgenic approaches to mercury remediation and biomass quality are in progress (Czakó *et al.*, 2006; Dhir *et al.*, 2010). An *A. donax* micropropagation laboratory with a 20,000 m^2 nursery unit can produce 60 million plantlets (for 6000 ha of land) with one laboratory technician, then one nursery technician is needed to grow-on bare-rooted propagules to produce 2 million field ready plants per year (Márton and Czakó, 2004, 2007; Fári *et al.*, 2009; Williams *et al.*, 2009). See Section 12.7.1 for further details on propagation.

12.3 Yield, Carbon Sequestration and Nutrient Removal

The maximum productivity of *A. donax* recorded to date, with an annual clear-fell harvest, is approximately 60–63 t oven DM ha^{-1} $year^{-1}$ in warm climates with sufficient irrigation and nutrients, but marginal soils (Hidalgo and Fernandez, 2001; Williams and Biswas, 2010). The potential biomass yield of *A. donax* on well drained fertile soils under optimum biotic and abiotic management conditions in warm climates is estimated to be over 100 dry t ha^{-1} $year^{-1}$.

The flood-irrigated 'Loveday' rootstock of *A. donax*, that received winery wastewater, produced the highest biomass yields of 45.2 t ha^{-1} of dry tops (Fig. 12.1), including 35.9 t ha^{-1} of dry stems (leaves removed) in the first year (Table 12.1). In comparison, the 'Henley Beach' rootstock produced 29 t ha^{-1} of dry tops including 17.7 t ha^{-1} of dry bare stems. These exceptionally high biomass yields in the first year of growth by the 'Loveday' rootstock (in South Australia) were similar to those reported in Spain (Lewandowski *et al.*, 2003), in Italy (Angelini *et al.*, 2009) and at Roseworthy, SA (Williams *et al.*, 2010a). Williams *et al.* (2010a) conducted their Roseworthy studies on arable land with good quality irrigation water.

The biomass yields of the irrigated 'Loveday' rootstock (South Australia) far exceed (by up to 300%) those traditional biomass crops grown on arable land with *ad libitum* treated sewage irrigation (Myers *et al.*, 1999; Biswas *et al.*, 2002). For example, Biswas *et al.* (2002) reported 15 t ha^{-1} of dry tops removed per season for forage sorghum grown on arable land with subsoil drainage system and *ad libitum* irrigation with secondary treated sewage (~1 dS m^{-1}) near Griffith, New South Wales, Australia.

Fig. 12.1. A biomass yield of 45.2 dry t ha^{-1} was produced in the first year by the 'Loveday' ecotype of *A. donax* irrigated on a saline soil at Barmera, South Australia.

Photosynthesis by *A. donax* during the first year was likely the main mechanism for the large sequestration of carbon, with 20.6 and 22.2 t ha^{-1} in the dry tops for the irrigated 'Loveday' and 'Roseworthy' rootstocks, respectively (Table 12.1). The high carbon yields per year of *A. donax* in Table 12.1 are far greater than the rates of accumulation of carbon in tree biomass, of 3.7 to 4.9 t ha^{-1} $year^{-1}$ for *Eucalyptus cladocalyx* (sugar gum) plantations in southern Australia (Paul *et al.*, 2008). Furthermore, the irrigated 'Loveday' rootstock had 26.6 dry t ha^{-1} of underground rhizomes carryover at the end of year 1, which contained 12 dry t organic carbon ha^{-1} when all shoot growth was removed, and this came into a dynamic equilibrium of 20.4 to 19.8 dry t organic carbon ha^{-1} at years 2 and 3 harvests' of shoots, respectively (Williams *et al.*, 2010a). Similar results were recorded at Roseworthy, South Australia, where the carryover irrigated *A. donax* rhizome yields reached a dynamic equilibrium of 20.5–25.9 dry t organic carbon ha^{-1} in years 2–4, under an annual shoot harvest regime (Williams *et al.*, 2010a). Riffaldi *et al.* (2010) compared the impacts after 14 years of three long-term agricultural systems on organic matter characteristics and microbial biomass in the topsoil (0–15 cm) in Tuscany, Italy: continuous and untilled *A. donax* harvested each year (AD), untilled natural grassland (NG) – the usual baseline for soil quality – and a cereal/legume crop rotation (CR). AD showed the greatest value of soil total organic carbon (TOC), of 18.8 g kg^{-1} dry soil, which was significantly greater ($P < 0.05$) than the NG value of 13.8 g kg^{-1} and the CR had significantly less ($P < 0.05$) TOC, 9.8 g kg^{-1} dry soil than either

Table 12.1. Biomass yield and nutrient removal by *A. donax* in South Australia (Williams *et al.*, 2009).

	Biomass yield (t ha^{-1} $year^{-1}$)			*A. donax* tops removed (kg ha^{-1} $year^{-1}$)			
Rootstock/land type	Leaf	Stem	Tops	Org C[a]	N	P	K
Barmera (saline soil) irrigated							
Loveday	9.3	35.9	45.2	20,577	528	22	664
Henley Beach	11.3	17.7	29.0	13,698	448	19	472
Roseworthy (arable land)							
Irrigated	10.3	40.7	51.0	22,200	773	40	832
Dryland	4.0	11.4	15.4	6,500	282	17	331

[a]Org C, organic carbon.

untilled treatment. Also, the absolute amounts of microbial biomass in the topsoil were significantly ($P < 0.05$) more in AD and NG than in the tilled CR site. This pivotal work by Riffaldi *et al.* (2010) showed that the AD cropping system, with large amounts of plant residues from the large, underground rhizome systems (Williams *et al.*, 2010a) and low tillage, was the pinnacle system tested to improve the amount and quality of soil organic matter (carbon compounds), at least in the Mediterranean climatic area. Therefore, *A. donax* crops, especially the rhizomes and roots, could qualify for carbon (C) credits in such programmes introduced in Australia in 2012 (AUS$23 t^{-1} carbon) and if introduced in future in the USA and elsewhere.

Nutrient uptake of nitrogen (N), phosphorus (P) and potassium (K) in the above-ground biomass of *A. donax* was 528, 22 and 664 kg ha^{-1} $year^{-1}$, respectively, during the first year of growth (Table 12.1). This achieved the aim to 'nutrient strip' the treated sewage used for irrigation (as per Mavrogianopoulos *et al.*, 2002), the high nutrient uptake was due to high nutrient supply (Williams and Biswas, 2010). The wastewater could supply most of the N requirements for *A. donax* growth in year 1 (Williams *et al.*, 2008a,b). In contrast, large rates of K applied in the wastewater were likely to be high to excessive for *A. donax* growth. Zinc (Zn) and copper (Cu) were the main micronutrients removed in the *A. donax* tops with 182 and 112 g ha^{-1}, respectively (Williams *et al.*, 2008a,b, 2010a). In a South Carolina (USA) study, nutrient removal was dependent on plant age at harvest and nutrient supply (Table 12.2) (Williams *et al.*, 2009). For example, from the high input nutrient site (irrigated with diluted sludge, at Cayce), the pre-flowering stems contained higher concentrations of N, P, S and K than those at the low nutrient site (Table 12.2; Williams *et al.*, 2009).

For a rain-fed experiment, on a sandy loam soil in southern Oklahoma (USA), where it was desirable to conserve nutrients, *A. donax* had the highest biomass yield of six species tested (33.2 dry t ha^{-1} at 168 kg N ha^{-1}) and the highest nutrient removal rates: 162–228 kg N ha^{-1}, 23–25 kg P ha^{-1} and 121–149 kg K ha^{-1} (Kering *et al.*, 2012). This led them to suggest that although giant reed had 'tremendous' biomass production, its higher nutrient removal rates indicate a potential for increased fertilizer needs over time. However, if the biomass is used as feedstock to make bioethanol (which contains only C, O and H), then virtually all the plant nutrients in the biomass feedstock can be returned to the *A. donax* plantations through the waste mulch after the 14–18 h fermentation/batch process time (EnEnergy Ltd, Norway, personal communication, 2009). This should greatly reduce the use of post-planting fertilizers, making *A. donax*

Table 12.2. Combustion and chemical analyses[a] of *A. donax* stems from US sites.

	Sample[b]		
	#1	#2	#3
Combustion analysis			
Volatile matter (%)	73.82	74.88	77.01
Fixed carbon (%)	21.53	20.54	19.65
Ash (%)	4.65	4.58	3.34
Sulfur (%)	0.034	0.33	0.12
Calories g^{-1}	4507	4478	4563
MJ kg^{-1}	18.87	18.75	19.10
Chemical analysis (%)			
Carbon	46.94	46.61	47.67
Hydrogen	5.85	5.82	5.81
Nitrogen	1.17	0.63	0.13
Phosphorus	0.07	0.05	0.03
Potassium	1.27	1.01	0.31
Chlorine	0.45	0.22	0.05

[a]% dry basis, Hampton Roads Testing Labs, Inc. (Hampton, Virginia, USA).
[b]#1, 1-year-old live post-flowering stem, sludge-irrigated land, August 2007, Cayce, South Carolina; #2, 1-year-old live post-flowering stem (Blossom stand), August 2007, Columbia, South Carolina; #3, dead stem, same location as #2.

an even more competitive, low input, pinnacle biofuel crop for non-riparian zones.

12.4 Managing the Weed Risk of Cultivated *Arundo donax*

The risk of invasion of biofuel crops into natural ecosystems, with subsequent effects on biodiversity, is a growing international concern. The International Union for the Conservation of Nature has produced guidelines for species selection and management (IUCN, 2009). *A. donax* has an international reputation as a significant invader of riparian ecosystems outside its native range, with adverse impacts on biodiversity, access, fire regime, water use and infrastructure. However, Virtue *et al.* (2010a,b) found the potential weed risk of *A. donax* in Australia to be considerably less in non-riparian areas.

A comparative weed risk analysis tool used for prioritizing species for weed management programmes (Virtue, 2010) was applied to *A. donax* (Virtue *et al.*, 2010a,b), comparing the potential for invasion into riparian and non-riparian natural ecosystems. The analysis found that *A. donax* posed a very high weed risk to riparian areas but a negligible risk to terrestrial natural ecosystems (Table 12.3). The key reason for the difference was limited capacity for dispersal in the absence of floodwaters and substantially slower growth due to lower moisture and nutrient levels in non-riparian areas (Boland, 2006). In riparian areas, *A. donax* poses a high risk of invasion due to the capacity for flood events to disperse vegetative root and shoot fragments to new areas of suitable moist habitat, in which it can form tall, dense monocultures with lateral spread via layering of stems on wet substrates (Boland, 2006). In the absence of layering and root disturbance *A. donax* tends to form discrete clumps, with only slow radial expansion from short, horizontal rhizomes.

Given its substantial weed risk to riparian areas, Virtue *et al.* (2010a,b) recommended that *A. donax* plantations should be located well away from such areas (beyond a 1 in 50 year flood zone) and managed to minimize the inadvertent dispersal of vegetative material during its establishment, growth, harvest and transport. Its lack of seed production (Balogh *et al.*, 2010) makes its containment in intensively managed production systems much more feasible than many other invasive biofuel crops.

The initial development of an *A. donax* industry is an opportune time to establish formal containment procedures, for which Virtue *et al.* (2010a,b) advised needed to be incorporated within an ongoing, government regulatory framework. Invasion of cultivated *A. donax* could be adequately prevented through location of plantings outside areas subject to flooding, buffer zones around plantings, annual surveys to detect and remove any escapees, and hygiene protocols for harvest, transport and processing. In an Australian context, Virtue *et al.* (2010a,b) recommended that such protocols should be made mandatory via declaration of *A. donax* under noxious weed laws and then use of government permits with strict management conditions allowing cultivation at low risk sites in non-riparian landscapes.

In the USA, an Environmental Assessment National Environmental Policy Act by the US Department of Agriculture, Rural Business-Cooperative Service of a biofuel plantation in a non-riparian zone involving *A. donax* found no significant impact on the natural or human environment under the mandatory conditions including buffer zones, monitoring of escapees, and best management practices (USDA, 2011).

The weed risk analysis and management recommendations by Virtue *et al.* (2010a,b) for *A. donax* would have broad applicability to other countries. However, the extent of government

Table 12.3. Weed risk assessment of *A. donax* for Australia using the SAWRMS for riparian and terrestrial natural ecosystems (Virtue *et al.*, 2010a,b).

Assessment criteria	Riparian	Terrestrial
Invasiveness	4.7	3.3
Impacts	6.8	2.1
Potential distribution	6.0	0.5
Comparative weed risk	192 (very high)	4 (negligible)
Control costs	4.7	2.7
Current distribution	0.9	0.9
Persistence	3.6	1.8
Feasibility of containment	16 (high)	4 (very high)
Management action	Destroy infestations	Monitor

regulation and the capacity for its effective enforcement needs consideration in establishing a sustainable industry (IUCN, 2009). *A. donax* is not the only biofuel crop with weed risk concerns and the issue should be considered for any proposed species.

12.5 Salt-tolerant and Bioremediation Crop

Halophytes (salt-loving plants) as defined by Flowers *et al.* (1986) are those plants that can complete their normal annual life cycle under conditions of over 150 mM root-zone salinity (equivalent to 15 dS m^{-1}). Williams *et al.* (2009) showed that *A. donax* can be classified as a halophyte due to its tolerance to salinity of up to 25 dS m^{-1}, which is equivalent to 250 mM in the soil water for prolonged periods. To further assess the salt tolerance of *A. donax*, the ratios of potassium (K):sodium (Na) in *A. donax* organs were calculated. The Loveday rootstock ecotype had K:Na ratios for leaf, stem and rhizomes of 62:1, 15:1 and 8:1, respectively, at the final harvest in year 1 (June 2007). These results indicated that the Loveday ecotype leaves and stems were excluding Na with a preference for K more than the rhizomes. This is one possible mechanism of salt tolerance of halophytes, as described by Flowers *et al.* (1977, 1986). An ecotype found in a saline environment in the USA, where the K:Na ratio of the soil was very low, 0.063:1, produced 49.2 dry t ha^{-1} biomass yield (Balogh *et al.*, 2011). This ecotype excluded Na from the above-ground parts, especially leaves, more than the ecotype from a low salinity environment where it maintained a much higher K:Na ratio in the leaves than in the roots (Balogh *et al.*, 2011). Williams and Biswas (2009) were the first to define the upper limits of salt tolerance of *A. donax* grown under saline irrigated conditions (in a series of greenhouse pot trials using sodium chloride-enriched solutions). They conducted replicated pot trials using topsoil from Loxton, South Australia, a loamy sand and showed that 75% of maximum yield of oven dry top growth of *A. donax* was recorded at soil saturated paste extract electrical conductivity (ECe) of 8.3 dS m^{-1} and a 50% of maximum yield at 12 dS m^{-1} at both clear fells (cut 1 results are shown in Fig. 12.2).

12.5.1 Guidelines for using saline wastewaters for *Arundo donax* production and for bioremediation

For closed systems, where salt inputs equal salt outputs, such as the pot trials conducted by Williams and Biswas (2009), results suggest: (i) *A. donax* can be grown with soil ECe up to 12 dS m^{-1} (approximates the upper limits of soil salinity for *A. donax* in systems, as at 12 dS m^{-1} a 50% yield loss is predicted) with over a 20% leaching fraction under the conditions of the pot trials; (ii) exceeding soil salinities above 25 dS m^{-1} will kill *A. donax* plants in a few weeks when there is no precipitation for leaching; and (iii) for sustainable salt management systems, water and soil salinity must be monitored and managed so that: 'salt into the root zone is equal or less than salt out of the root zone'. To achieve salt balance in the root system, drainage management is critical via the use of subsurface drainage and disposal of the concentrated leachate. Estimated costs of such a drainage system are AUS$5000 ha^{-1} (for laser levelling, 65 mm diameter porous pipes spaced from 10 up to 50 m apart (clay versus sandy soils), approximately 1 m deep and connected to a sump outlet (drainage for reuse or evaporation or fish farms)). Further details can be found in Biswas and Williams (2009).

The Serial Biological Concentration (SBC) biosystem, as shown in Fig. 12.3, replicates three stages of a land filter system (cell) to produce drainage waters having different salinities at the end of each filtration event, due to the use of 33% leaching fraction/cell (Blackwell *et al.*, 2005). Therefore, by having a sequence of filter cells, as described by Biswas *et al.* (1999, 2002) and Jayawardane *et al.* (2001), the SBC biosystem can be used to grow salt-tolerant crops, such as *A. donax*, and at the same time achieving both a reduction of drainage volumes of wastewaters and maximization of financial returns from the crops produced (Biswas and Williams, 2009). The filter cell uses a system of flood irrigation and subsurface agricultural drains to process sewage effluent or other wastewaters by stripping out nutrients, and mitigating pathogens including algae, suspended solids and high oxygen demand water using a combination of volatilization, oxidation, reduction (i.e. denitrification), soil adsorption and plant uptake processes. The sequential process is terminated when the drainage water is too salty to sustain economic crop production.

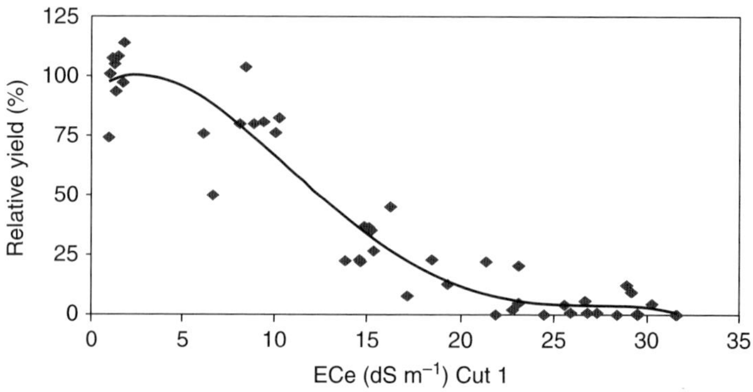

Fig. 12.2. Relative yield (RY) of dry tops of *A. donax* in response to the salinity of the saturated soil extract (ECe). Irrigated with sodium chloride solutions in a pot trial (Williams and Biswas, 2010).

Then the highly salty drainage water can be used in the final three components of the SBC system as shown in Fig. 12.2. These final three cells can consist of production of a range of aquatic species in ponds of varying salinities, and/or salt gradient solar ponds to produce energy and/or evaporation basins to produce pure salts for industrial use and/or stockpiling for disposal. It is estimated that a well-planned SBC system can generate enough electricity through the solar ponds to run all the pumps for irrigation and drainage required for running that SBC biosystem (Biswas and Williams, 2009).

Preliminary results indicated that *A. donax* was as or more effective compared to the recommended plant species, such as *Phragmites australis* (common reed), for the treatment of wastewaters in wetlands (Rezk and Edany, 1979; Cronk, 1996) and in non-riparian zones (Biswas and Williams, 2009; Williams and Biswas, 2009), where *A. donax* can be managed as a low-weed-risk crop (Virtue *et al.*, 2010a,b).

12.6 *Arundo donax*, a Biofuel Crop for Saline Lands and Waters

Lignocellulosic biomass from *A. donax* and other grasses and trees can produce heat, electricity or transport fuels using several technologies (Nassi o Di Nasso *et al.*, 2011; for details, see Chapter 2, this volume). *A. donax* biomass from the 1-year-old regrowth (stems and leaves) from the Loveday rootstock, in South Australia, contained 42.5% cellulose and 31.2% hemicellulose on a dry weight basis (Williams and Biswas, 2010). Similar high and stable values were reported for *A. donax* grown in Italy of 38% cellulose and 25% hemicellulose from midsummer onwards (Nassi o Di Nasso *et al.*, 2011). Use of *A. donax* as a pinnacle biomass yield and lignocellulosic biofuel crop, in non-riparian zones (where it can be managed as a low-weed-risk crop, Virtue *et al.*, 2010a,b) to utilize degraded, saline soils and wastewaters would be a highly attractive strategy, provided this was done under strict weed-risk management guidelines.

In field trials in Italy over 12 years, on a site with a shallow water table (<120 cm deep) Angelini *et al.* (2009) calculated the net energy yield (energy output minus energy input per hectare) and energy production efficiency (the ratio between energy output and input per hectare) of *A. donax* compared to *Miscanthus* × *giganteus*. Their results showed that 1 ha of *A. donax* produced 37.7 dry t^{-1} $year^{-1}$ of biomass, net energy of 637 GJ ha^{-1} and could substitute for 14 t ha^{-1} and 20 t ha^{-1} of petroleum and coal, respectively. Such yields surpassed *M. giganteus* by over 25% per annum, grown under the same conditions. They also calculated annual crops require approximately 50% of the total energy produced for production (e.g. ploughing, sowing, harvest), whereas *A. donax* requires only 1.9% and *M. giganteus* 2.6% per year.

There is a pressing need to reduce atmospheric greenhouse gases and associated temperatures back below potentially dangerous thresholds (Lowe *et al.*, 2009). An important driver for second generation biofuel crops is the potential to reduce life cycle carbon dioxide (CO_2) emissions by replacing fossil fuels. They could reduce oil well-to-vehicle

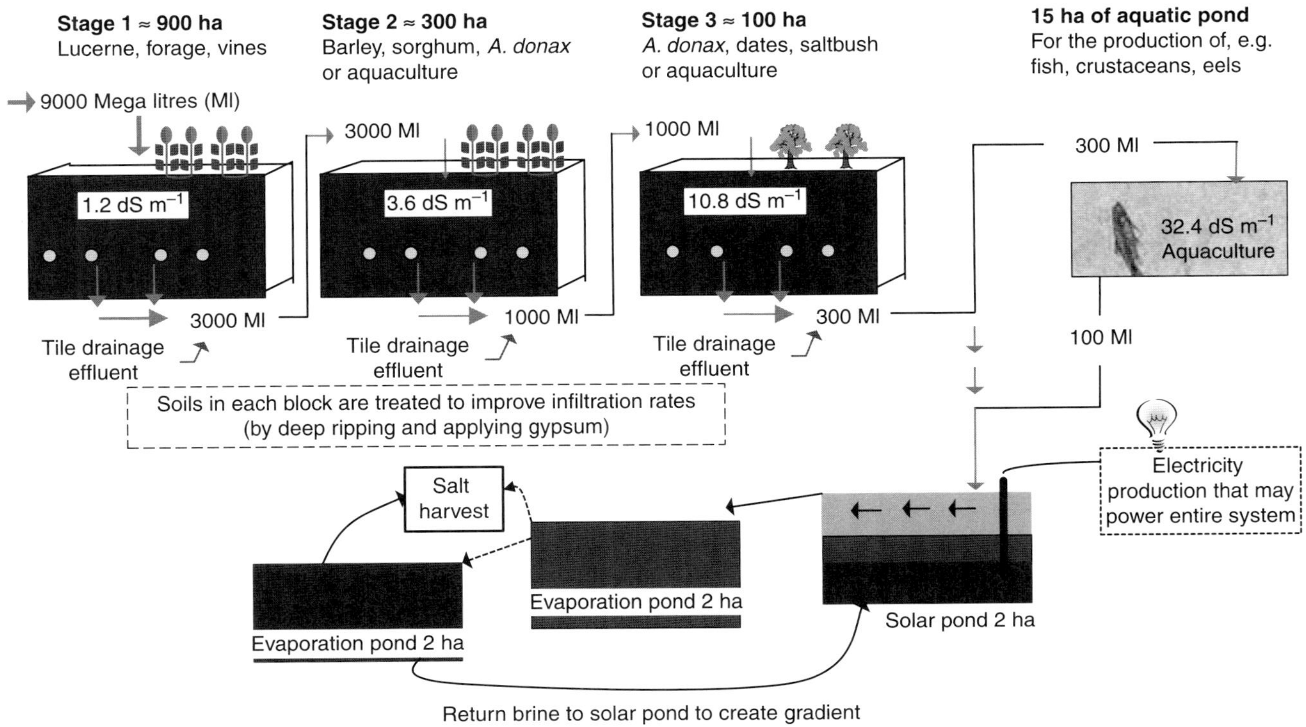

Fig. 12.3. Schematic representation of flows, concentration and potential use of saline drainage from *A. donax* growing with saline water (modified from Blackwell *et al*., 2005 and Biswas and Williams, 2009).

wheels CO_2 production by 60 to 100% compared to fossil fuels (Eisentraut, 2010). Such biofuel may be one of the most cost-effective methods of renewable, low-carbon energy for vehicle transport (Wright and Brown, 2007; Eisentraut, 2010). Although costs of production are comparable for grain ethanol and cellulosic biofuel, the current estimated higher capital costs of factories processing cellulose may be a short term barrier to their commercialization (Wright and Brown, 2007). However, new technologies for biomass-to-liquid (BTL) fuel are advancing rapidly for second generation biofuels. For example, a biofuel company recently processed (in laboratory scale tests in Norway) samples of Australian grown *A. donax* into BTL ethanol. *A. donax* produced 299 l ethanol t^{-1} of shoot dry matter in a process time of less than 24 h (Biswas and Williams, 2009). This equates to 11,000 l ha^{-1} of bioethanol (Fig. 12.4) from *A. donax* (from 36.8 t ha^{-1} of dry tops $year^{-1}$). In comparison, maize kernels (*Zea mays*) produce 4400 l ha^{-1}, switchgrass (*Panicum virgatum*) 4600 l ha^{-1} and sugarcane (*Saccharum officinarum*) 8800 l ha^{-1} (Bourne, 2007) (Fig. 12.4).

Williams *et al.* (2008b) analysed the harvested biomass of *A. donax* for fuel properties. A calorific value of 19 MJ kg^{-1} of oven dry *A. donax* biomass burnt is equivalent to 75% of that produced from 1 kg of combusted coal. One tonne of dry *A. donax* biomass combustion generating 19 GJ is equivalent to 5320 kWh, which is capable of providing electricity for 266 homes for 1 day.

12.7 Production Systems for *Arundo donax* Biofuel Crops

12.7.1 Establishment

Plantation establishment is the most critical part of *A. donax* cultivation as it has major effects on biomass yield and economic viability (Lewandowski *et al.*, 2003). Three main factors affecting crop establishment success and costs are propagation material, plant density and available soil moisture post-planting. It has to be propagated vegetatively due to lack of viable seed. Rhizome planting compared to whole stems and stem cuttings has been the most successful establishment method (Christou *et al.*, 2001a). Planting large rhizome pieces at 12,500 plants ha^{-1} (i.e. 1.25 rhizomes m^{-2}) with well-developed buds directly into the field, in early spring in Greece, produced establishment rates of nearly 100% and maximum yields per hectare per year at clear fell harvest in the second winter of growth (Christou *et al.*, 2001a). They recorded no yield advantage of using a higher density of 25,000 plants ha^{-1}. Similarly in central Italy, Angelini *et al.* (2005) found higher dry matter yields with

Fig. 12.4. Comparative potential bioethanol production (l ha^{-1} $year^{-1}$) from food and non-food crops. (First three crops modified from Bourne, 2007; *A. donax* from Biswas and Williams, 2009.)

planting rhizomes equivalent to 20,000 plants ha^{-1} compared to 40,000 plants ha^{-1}. For Southern Europe, lower populations of from 12,500 to 20,000 plants ha^{-1} were suggested due to less cost of establishment (Christou *et al.*, 2001a). However, Williams *et al.* (2010a) planted large rhizome pieces at 3 m^{-2} (i.e. 30,000 plants ha^{-1}) in May (early winter in the southern hemisphere, at Barmera, South Australia) and with a high irrigation regime obtained a high shoot yield of 45.2 dry t ha^{-1} at the first clear-fell harvest, 12 months after planting (Fig.12.1; Table 12.1).

The main limitations of using rhizomes compared with stem cuttings and whole stem planting for propagation are the higher costs to source and plant rhizomes, and lack of refined mechanical planting machinery (i.e. that can plant large areas efficiently). One method to rapidly establish initial demonstration biofuel plantations (e.g. to provide commercial scale start-up biofuel feedstocks for batch testing) would be to source planting material, first from wild populations (with samples of such material disease tested well before planting). When larger areas are to be sown, disease-tested plantlets from tissue culture should be used (as described in Section 12.2.1).

In order to rapidly establish initial new biofuel plantations, we suggest that rhizomes and stems could be harvested from wild populations nearby and/or those posing significant weed risks in accessible riparian zones, where careful rhizome removal will not disturb the ecosystem and release fragments of *A. donax* nor release other significant weeds. Such opportunities may arise with permission of landowners and local weed regulatory authorities. Giessow *et al.* (2011) have estimated costs of control of *A. donax* average US$25,000 ha^{-1} and total US$196m over a decade for California's southern river systems. Removal of *A. donax* to use as new planting stock, from accessible, negotiated areas as described above has potential to reduce costs of control. However, in all cases care should be taken to use only wild material that has been tested and found free of viruses of agricultural importance.

Few studies of density and/or yield of micropropagated plantlets of *A. donax* in the field have been reported. Micropropagated plantlets from an embryonic cell line (by the procedure of Márton and Czakó, 2004, 2007, 2011, 2012a,b,c and Fari *et al.*, 2009) performed well in multiple locations and times, e.g. South Carolina, Italy and Hungary. Micropropagated plantlets from the sustained (over 12 years) embryogenic cell culture of the 'Blossom' line were planted in South Carolina, where *A. donax* grew 6–8 m tall by the end of year 4 and produced 181.6 wet t ha^{-1} at 50% moisture content, equal to 90.8 dry t ha^{-1} (Márton and Czakó, unpublished data). The first year yield of an *A. donax* plantation in Debrecen, Hungary from micropropagated plants was 30 dry t ha^{-1} $year^{-1}$ in year 1 (Fig. 12.5) and 48 dry t ha^{-1} in year 2, and in another location (Tiszasas) produced 39.6 dry t ha^{-1} in the second year (I. Gyói and M. Fari, University of Debrecen, Hungary, 2010, personal communication). For an approximate cost of US$0.26 per plantlet (http://www.cellulosefarming.com), the total cost of micropropagated disease-tested plantlets for 10,000 plantlets ha^{-1} is US$2600 for plantations of 1 ha with an expected perennial longevity of at least 35 years (Merinotti, 1941; Focherini, 2000; Williams and Biswas, 2010). For example, *A. donax*, in a dryland planting by Italian prisoners of World War II completed by 1945, at Loveday, South Australia was still highly productive in 2006, after 61 years from planting, as reported by Williams and Biswas (2010). The *A. donax* plantation for rayon production in Torviscosa, Italy was established in 1938 and abandoned in 1965 (Merinotti, 1941; Focherini, 2000). The unmanaged Blossom stand of *A. donax* in South Carolina, USA, has been in place for at least 35 years (J. Morris, personal communication).

Besides the multiplication potential, another advantage of clonal mass propagation *in vitro* from meristems via somatic embryogenesis and/or shoot multiplication is that it can eliminate viral, bacterial and fungal pathogens (Agindotan *et al.*, 2010; Chomič *et al.*, 2010, Ramgareeb *et al.*, 2010; Snyman *et al.*, 2011; Subba Reddy and Sreenivasulu, 2011; Tóth *et al.*, 2011), and can supply safe *A. donax* planting material for large scale plantations, which has been tested, serologically and by RNA/DNA-based detection assays (Tóth *et al.*, 2011). Such plantations, established with pathogen-free tested initial plantlets, can be expected to persist as perennial biofuel crops for 35 years or more, without replanting (Merinotti, 1941; Focherini, 2000; Williams and Biswas, 2010).

Fig. 12.5. First year growth of the *A. donax* plantation established from micropropagated 'Blossom' line in Debrecen, Hungary (with Prof. L. Márton).

12.7.2 Weeds, pests and diseases

Herbicide application and cultivation between the rows of *A. donax* may be required in the first year to control weeds. The high growth rates and large leaf area of *A. donax* usually out-competes most weeds from the second year on (Williams and Biswas, 2010). *A. donax* mature stems and leaves contain a wide range of chemicals that likely help protect it from most insects and diseases. These chemicals include silica, triterpines, sterols, cardiac glycosides, curare-like indoles, hydroxamic acid and many other alkaloids (Perdue, 1958; Ghosal *et al.*, 1972; Miles *et al.*, 1993a, b; Bell, 1997; Khuzhaev, 2004; Jia *et al.*, 2008; Hong *et al.*, 2010). A common finding from long term field trials, in different parts of the world, was of no need for pesticide applications to *A. donax* (Angelini *et al.*, 2005, 2009; Williams *et al.*, 2009; Williams and Biswas, 2010; Kering *et al.*, 2012; US EPA, 2012).

The arundo wasp (*Tetramesa romana*), a native insect to southern Europe, was released in southern Texas in 2009, the first natural insect pest to be used against *A. donax* (Goolsby and Moran, 2009; Racelis *et al.*, 2010). It was followed by another native insect of Europe, armoured scale (*Rhizaspidiotus donacis*), which was released in southern Texas in 2011 (Goolsby *et al.*, 2011). The armoured scale insect has low dispersal ability (Cortés *et al.*, 2011). Simultaneous infestation by the wasp and the scale insects, as above, are additive and cause thinning and reduction of the shoot and rhizome mass of *A. donax* (Moore *et al.*, 2010).

Extreme care and knowledge of pest migration and host range is applied when new biological control agents of *A. donax* are released in targeted, new ecosystems in the USA (e.g. release of the arundo wasp along riparian corridors in the bi-national Rio Grande Basin (Goolsby and Moran, 2009; Racelis *et al.*, 2010).

Commercial plantations of *Arundo donax* grown in non-riparian zones are gaining acceptance as potential biofuel and/or fibre crops in the USA (USDA, 2011; Kering *et al.*, 2012), in southern Europe (Angelini *et al.*, 2009; Mantineo *et al.*, 2009) and in South Australia (Williams *et al.*, 2009a; Williams and Biswas, 2010), when grown under weed risk management guidelines (such as in Section 12.4). Any newly introduced biological control agents of *A. donax* may attack such commercial plantations. However, it is possible to control the newly released arundo wasp, and armoured scale by foliar or soil drenching application of pesticides that are efficacious on both these insects (Elbert et al., 2008; Xu *et al.*, 2009);

however, that will add costs to the *A. donax* biofuel production system. Several diseases of *A. donax* have been reported in the USA, including root rot, lesions, crown rust and stem speckle, but none has seriously affected the plant yield (Bell, 1997; Zheng *et al.*, 2005). If it is desirable to terminate an *A. donax* stand, application of the herbicide glyphosate to freshly cut stem bases in autumn is highly effective (Bell, 1997; Spencer *et al.*, 2008).

12.7.3 Plant nutrition, fertilizers and irrigation

Fertilizer recommendations for precision agriculture are usually based on fertilizer experiments conducted on major soil types in production areas, which calibrate soil and plant tissue nutrient analyses as diagnostic tools to define deficient, adequate and excessive concentrations for each nutrient per plant species, from the relationship between each nutrient concentration and percentage of maximum yield (as per Reuter and Robinson, 1997 and Williams *et al.*, 1999, 2004 for other species). After fertilizer applications, soil and plant tissue nutrient analyses are again used to assess the adequacy of the fertilizer programmes (Reuter and Robinson, 1997; Williams *et al.*, 1999, 2010b). We found no reports published in which such soil or plant tissue nutrient tests have been calibrated as diagnostic tools for *A. donax* target biomass yields (i.e. no defined concentrations of each nutrient in soil or plant tissue tests for adequate or deficient growth and yield, as per the methods of Reuter and Robinson, 1997).

In the 'Giant Reed Network' project in Europe, Christou *et al.* (2001b) found in trials in Greece that annual applications of N at 40 and 120 kg ha^{-1} did not influence biomass yields of *A. donax*. However, in experiments in central Italy, Angelini *et al.* (2009) found N:P:K fertilizer at 200:80:200 kg ha^{-1} compared to the zero rate increased biomass yield from 23 to 27 dry t ha^{-1} (years 1–6 mean value). In Oklahoma, USA, Kering *et al.* (2012) reported biomass yields of *A. donax* increased in a linear fashion from 19.4 dry t ha^{-1} at 0 kg N ha^{-1} to 33.2 dry t ha^{-1} at 168 kg N ha^{-1} (with supplementary irrigation). Such information suggests that moderate N fertilization of *A. donax* for increased biomass yields may be needed, in certain regions, for economic and environmental reasons. Soil and plant nutrient tissue tests, to our knowledge, have not been calibrated for *A. donax* yield response nor published. We suggest prospective growers, to obtain an estimate of the nutrient status of their *A. donax* crops, compare the published concentrations of nutrients in the leaf, stem or rhizome fractions of high yielding giant reed crops (e.g. Mavrogianopoulos *et al.*, 2002; Williams *et al.*, 2010b) with their crops (using the same tissue, sampled at the same stage of growth, for each nutrient).

A. donax usually responds to increased soil water supply by producing significantly more biomass yield per hectare (Christou *et al.*, 2001b; Mantineo *et al.*, 2009). The highest irrigation applied by Christou *et al.* (2001b) was 8.6–14 Ml ha^{-1} $year^{-1}$, and produced the highest yields each year. For example, in year 2, the no irrigation treatment yielded 32 dry t ha^{-1} compared to 45 dry t ha^{-1} for the 859 mm irrigated plots (Christou *et al.*, 2001b). Among the three energy crops Mantineo *et al.* (2009) studied in Sicily, *A. donax* had the highest values of dry biomass and net energy yield per hectare and water use efficiency, calculated both on dry biomass and on the output energy bases, compared to *Miscanthus* × *gigantus* and *Cynara cardunculus*.

12.7.4 Ruminant fodder, fire risk management and harvest methods

In Australia, *A. donax* stands occur in every mainland state (Virtue *et al.*, 2010b). Numerous stands exist as it has been used for many purposes including summer green fodder for ruminants – especially in times of drought (Spafford, 1941). When feeding to ruminants, rather than grazing, it is better to cut *A. donax* canes before they reach 1.8 m in length, when they start lignifying. Chemical analyses of *A. donax* showed the composition of fresh, green, young shoots up to 1.8 m in length to be closely comparable to that of green oats, barley and wheat plants at flowering time (Spafford, 1941). For example, green canes (1.8 m long) were composed of: 10.6% carbohydrates, 1.9% crude protein, 10.8% fibre, 1.5% ash and 75% water (Spafford, 1941). Mature stands of giant reed are not very palatable to cattle, but they will eat it during dry seasons (Wynd *et al.*, 1948; Talapatra, 1950; Williams and Biswas, 2010). Goats will also eat it (Daar, 1983).

Stems of *A. donax* usually live between 2 and 4 years, then die and often remain upright and dry (usually less than 15% moisture content) for many years, accumulating in wild populations and in neglected farm plantations that have not been harvested annually (Bell, 1997; McWilliams, 2004; Williams and Biswas, 2010). Therefore, it is not surprising that Bell (1997) and others stated that giant reed, in the context of wild and most non-harvested populations, is 'extremely flammable' throughout most of the year in California, USA. However, in managed, irrigated commercial plantations of *A. donax*, which are clear felled at least once each year, there is likely to be minimal fire risk, since at least 95% of stems remain green and alive (Williams and Biswas, 2010) and usually contain a minimum 40–55% moisture content up to each annual harvest (Mantineo *et al.*, 2009; Williams and Biswas, 2010). For example, a 30-year-old *A. donax* windbreak stand at Roseworthy, South Australia was first clear felled on 7 June 2005 to initiate an annual, clear fell regime experiment by Williams *et al.* (2010a). This first clear fell of this stand yielded 82.1 dry t ha^{-1} of green, live stems (>40% moisture content) plus 60.3 dry t ha^{-1} of dry, dead stems (<15% moisture content) (Williams *et al.*, 2010b). For the next 4 clear-fell years (i.e. removal of all top growth each winter, each year) to 2009, green live stems always made up over 95% of the biomass harvested and were over 40% moisture content at all harvests, for both irrigated and non-irrigated treatments (Williams *et al.*, 2010a). It is essential to state that if prolonged droughts occur in areas where non-irrigated plantations of *A. donax* are grown for biofuels, increased stem mortality rates are highly likely to occur and such crops should be clear-fell harvested regularly, more than twice a year to remove the entire dead dry biomass fuel load from the field (Williams *et al.*, 2010a). Experience in Texas confirms that significant amounts of dead, dry stems of *A. donax* pose a high fire risk in dry seasons (Dr R. Pollock, USA, personal communication). We suggest such drought-stricken crops or wild stands, with a high percentage of dead stems, could be harvested first for bioenergy or animal bedding straw, before the high fire risk season.

A. donax can be harvested each year or every second year depending on its uses (Lewandowski *et al.*, 2003) and irrigation supply. Two harvests per year are feasible if non-irrigated crops are drought stricken, to greatly reduce fire risks as described above. Repeated cutting per year did not sustain high growth rates and total yield declined (Sharma *et al.*, 1998), presumably because of reduced optimum leaf area duration as new canes must grow from rhizomes (Spencer and Ksander, 2006). In southern Europe, a late winter annual clear-fell harvest is recommended to attain a modest reduction in the moisture content of stems (Lewandowski *et al.*, 2003), presumably for combustion end use. However, if the crop is to be used as feedstock for pretreatment and fermentation to bioethanol, high crop moisture content may not be an issue (Hendriks and Zeeman, 2009). An annual clear-fell harvest of *A. donax* in late winter is recommended as the optimum time in southern Australia for rapid spring regrowth, and for a modest decrease in stem moisture content (Williams and Biswas, 2010). However, it is feasible to extend the duration of *A. donax* harvest season by 5 months from mid-autumn to the end of winter (for the southern hemisphere, April to August), provided the field harvested in mid-autumn in year 1 is harvested at the end of winter in year 2 and vice versa. Additional applied nutrients may be required by *A. donax* in such a two-phase harvest regime.

12.8 Costs and Returns from Plantation *Arundo donax*

A. donax has not been grown in large areas as yet as a feedstock for biofuels mainly due to its global reputation as a serious riparian invasive weed (Dudley, 2000; Lowe *et al.*, 2000). The profitability of growing *A. donax* in non-riparian zones and converting it to usable products is of paramount importance in determining whether or not this possible new feedstock for industry has a future.

The cost of growing *A. donax* in the upper south-east of South Australia, in the Meningie Downs area, was assessed by Black and Williams (2010). The concept was of a large complex of plantations on rented land managed by a growing company in contract to a conversion factory. It was assumed the production company required a 15% internal rate of return (IRR) on funds invested to be viable. Three types of production systems were envisaged: non-irrigated (rain-fed), conventionally irrigated and 'naturally irrigated' through the plants' roots reaching the moderately saline, shallow water table (Rezk and Edany, 1979). With nutrients commercially supplied, the average factory gate price of the three systems was AUS$65 per oven dry t of *A. donax* (Table 12.4). If a substantial part of the plant's nutrient requirements were supplied by *A. donax* biofuel fermentation waste mulch from the processing factory at no cost to the company, the average price was calculated to be AUS$53 per oven dry t.

At AUS$60 per dry t at the factory gate and 500,000 oven-dry t year^{-1} supplied to a conversion factory, *A. donax* shows potential to offer South Australia a new industry to produce either bioethanol and lignin, or pulp/paper, provided 3 years of near-market agronomy research and development and upscaling is funded and conducted. Based on central price estimates, the preliminary estimates given in Table 12.5 indicate an IRR of 22% per annum for the ethanol and lignin enterprise and 18% per annum for the pulp/paper enterprise. Electricity generation plus biochar production using *A. donax* as the feedstock at a cost of AUS$60 per oven dry t and a slow pyrolysis process did not appear to represent a sound private enterprise investment (Black and Williams, 2010). However, the economics of electricity generation from *A. donax* feedstocks may improve if other combustion processes are used. For example, electricity generation using a fast pyrolysis process that does not produce biochar as a co-product may be a more efficient conversion process for electricity generation alone (R. Tonkin, personal communication, 2009).

It is expected that the commercial potential of non-food, cellulosic crop feedstocks grown on marginal lands for conversion to biofuels will increase in future if the price of fossil fuels rises significantly, as noted in The Economist (2009). Further income per hectare is likely to increase if the perennial crops can attract carbon credits (for soil and root carbon accumulation) and/or price subsidies for bioethanol (carbon neutral). Also, the yield of ethanol from each dry tonne of biomass is expected to increase in future and process time to decrease as improved technology becomes available (The Economist, 2008).

Table 12.4. Preliminary factory gate prices, from a case study, for oven dry tops (AUS$ t^{-1}) to achieve 15% internal rate of return (IRR) to a growing company and target, mature *A. donax* plantation yields (t ha^{-1} year^{-1} of oven dry tops) (Black and Williams, 2010).

System	Price	Mature yield
Rain-fed only,[a] fertilizer purchased	67	15
Rain-fed only,[a] free nutrient return from factory	52	15
Roots in water table, fertilizer purchased	49	30
Roots in water table, free nutrient return from factory	37	30
Irrigated, fertilizer purchased	80	40
Irrigated, free nutrient return from factory	70	40

[a]450 mm annual rainfall, Mediterranean environment.

12.9 Constraints, Future Research and Prospects

A number of research and development gaps have been identified.

- Areas that have moderately saline to saline groundwater and saline soils in non-riparian ecosystems in Mediterranean to subtropical regions have potential for commercial *A. donax* plantations. Further work is needed to validate experimental plot findings, in small-scale commercial plantations of *A. donax* of *c*.5 ha with proper drainage (e.g. using SBC biosystems) to demonstrate production systems and weed risk biosecurity guidelines. The three proposed production systems are: (i) rain-fed (non-irrigated); (ii) roots self-irrigated by the shallow water table; and (iii) a saline, flood irrigation/drainage biosystem (e.g. SBC as in Fig. 12.3).
- Modify and develop commercial-scale machines for efficient planting and harvest operations for *A. donax*. This could include improvements to transplanters for

Table 12.5. IRR results, from a case study, for conversion factories using *A. donax* feedstock sited in the south-east of South Australia (central price estimates shown first in each series) (Black and Williams, 2010).

Product	Product price (AUS$)	Capital cost (AUS$, millions)	IRR (%)
Electricity and biochar	100 KWh^{-1}, 120 t^{-1}	200	8
Electricity and biochar	120 KWh^{-1}, 120 t^{-1}	200	14
Electricity and biochar	80 KWh^{-1}, 120 t^{-1}	200	0
Bioethanol and lignin	0.6 l^{-1}, 50 t^{-1}	275	22
Bioethanol and lignin	0.8 l^{-1}, 50 t^{-1}	275	33
Bioethanol and lignin	0.4 l^{-1}, 50 t^{-1}	275	9
Bioethanol, lignin and	0.6 l^{-1}, 50 t^{-1}	275	21
electricity, biochar	100 KWh^{-1}, 120 t^{-1}	160	
Pulp/paper	1100 t^{-1}	550	18
Pulp/paper	1400 t^{-1}	550	28
Pulp/paper	800 t^{-1}	550	5

micropropagated plantlets. Further, for whole stem and/or rhizome plantings (Christou *et al.*, 2001a), modifications of sugarcane or potato planting equipment are needed. For harvest of *A. donax*, the equipment such as silage and forage harvesters and sugarcane harvesters may need to be modified (e.g. hardened steel cutting blades) for rapid harvest of large-scale plantations.

- The definition of the minimum nutrient (calibrate soil and plant tissue tests as diagnostic tools for each nutrient as per section 12.7.3) and irrigation requirements of *A. donax* for target biomass yields for a range of environments are required. This should include assessment of wastewaters of different qualities and quantities on the survival and productivity of *A. donax*. Further work is needed on carbon sequestration by *A. donax* to extend the initial pivotal findings of Riffaldi *et al.* (2010). They found when it is grown as a low till, long term perennial crop, it was the pinnacle system to improve the amount and quality of soil organic matter in the one region and soil type tested. Such work is required to improve the potential for carbon credit payments to growers.
- Plant species posing significant weed risks can be regulated through noxious weed laws of governments in tandem with industry containment protocols. These are policy decisions for each government and broad and clear guidance on the matter for biofuel crops is warranted.
- It is desirable to obtain funds and conduct an international forum on: 'Potential and barriers to develop *A. donax* and other lignocellulosic crops for biofuels or pulp/paper'. This would greatly facilitate the integration of best practices and technologies to help establish new, second-generation biofuels industries.

12.10 Conclusions

This chapter provides for the perennial, rhizomatous grass, *Arundo donax* (giant reed), baseline data and guidelines for sustainable production systems, salt tolerance, weed risk management, potential biomass and biofuel yields, and carbon sequestration when grown for lignocellulosic feedstocks for biofuels. *A. donax* has displayed remarkable potential for high biomass yield, longevity and carbon sequestration. It was usually the pinnacle species, often producing over 35 dry t h^{-1} $year^{-1}$, for many years, under high rainfall or irrigated systems in warm temperate to subtropical climates. Whilst *A. donax* is clearly a serious weed threat to riparian systems, a detailed weed risk analysis indicated it could be grown as a low weed-risk, biofuel crop in non-riparian zones, if it is done under strict management guidelines within a noxious weed legislative framework.

Establishment of plants is the most critical period of *A. donax* cultivation as it has major effects on biomass yield, crop longevity and economic viability. Rhizomes, the traditional planting material for *A. donax*, can be costly to locate, dig and replant. In addition, planting material needs to be disease tested before use in large-scale plantations. Recent developments using tissue culture and mass micropropagation techniques means that disease-tested, safe planting material of *A. donax* can now be produced for both plant breeding purposes and large-scale plantations to grow biofuel feedstock.

Some have suggested that although giant reed had 'tremendous' potential for biomass production, its higher nutrient removal rates means increased fertilizer needs over time. However, if the biomass is used as feedstock for bioethanol (which contains only carbon, oxygen and hydrogen), then virtually all the plant nutrients in the biomass feedstock can be returned to the *A. donax* plantations, in the form of waste (ferment) mulch left from the daily, batch time, factory process (biomass-to-bioethanol). This should greatly reduce the use of post-planting fertilizers, making *A. donax* an even more competitive, very low input pinnacle, feedstock crop for biofuels and for soil carbon sequestration.

A. donax is a moderately salt-tolerant plant tolerating up to 12 dS m^{-1} soil saturated paste extract electrical conductivity (a 50% yield loss is predicted at 12 dS m^{-1}). With increasing demand for arable lands and good quality water for food production, the underutilized resources of moderately saline wastewater and saline lands can be used to grow *A. donax* in a contained, integrated biosystem, such as 'Serial Biological Concentration'. *A. donax* has been found to be more effective than the recommended common reed (*Phragmites australis*) in treating wastewaters.

A. donax can produce 299 l ethanol t^{-1} of dry shoot matter in a process time of less than 24 h; equating to a potential production of 11,000 l ha^{-1} $year^{-1}$, a 2.5 times higher yield than maize (*Zea mays*) seed biofuel production, from an equivalent land area. The yield of ethanol from each dry tonne of biomass is expected to increase in future and process time and cost to decrease as improved technology becomes available.

A. donax, together with other lignocellulosic feedstocks, has good potential to form the basis of a new biofuel industry in many countries. For example, preliminary estimates from a case study indicated an internal rate of return to be 22% per annum (attractive to investors) for a potential *A. donax* to bioethanol factory in South Australia, when bioethanol was priced at AUS$0.6 l^{-1} at the factory gate. Included in this analysis is a 15% per annum rate of return to the *A. donax* growing company.

It is expected that the commercial potential of non-food, lignocellulosic crop feedstocks grown on marginal or surplus lands for conversion to biofuels will increase in future as fossil fuels become scarcer and higher in price. Additional income is likely if the perennial crops, such as *A. donax*, can attract carbon credits (for its proven ability in a low till long-term cropping system to increase the amount and quality of soil organic matter, at least in the Mediterranean climatic zones) and/or price subsidies for bioethanol (carbon neutral).

Acknowledgements

The authors acknowledge funding from the Rural Industries Research and Development Corporation of Australia (RIRDC), FibreCell Australia Pty Ltd, South Australian Murray-Darling Basin Natural Resources Management Board and the South Australian Research and Development Institute (SARDI). The authors thank Dr John Virtue, Biosecurity SA, South Australia for help on the weed risk assessment and management strategies section. We thank Dr Ian Black, former Principal Economist, SARDI, Adelaide, South Australia for economic analyses of growing *A. donax* in South Australia. Drs Czakó and Márton acknowledge funding from National Institute of Environmental Health Sciences (USA), SouthernSun Biosystems, Inc. (Hodges, South Carolina, USA), Biomass Gas & Electric, LLC (Atlanta, Georgia, USA), White Technology, LLC (Clinton, Indiana, USA), and the University of South Carolina Research Foundation.

References

Agindotan, B.O., Ahonsi, M.O., Domier, L.L., Gray, M.E. and Bradley, C.A. (2010) Application of sequence-independent amplification (SIA) for the RNA viruses in bioenergy crops. *Journal of Virological Methods* 169, 119–128.

Ahmad, R., Liow, P.S., Spencer, D.F. and Jasieniuk, M. (2008) Molecular evidence for a single genetic clone of invasive *Arundo donax* in the United States. *Aquatic Botany* 88, 113–120.

Angelini, L.G., Ceccarini, L. and Bonari, E. (2005) Biomass yield and energy balance of giant reed (*Arundo donax* L.) cropped in central Italy as related to different management practices. *European Journal of Agronomy* 22, 375–389.

Angelini, L.G., Ceccarini, L., Di Nasso, N.N.O. and Bonari, E. (2009) Comparison of *Arundo donax* L. and *Miscanthus* x *giganteus* in a long-term field experiment in Central Italy: Analysis of productive characteristics and energy balance. *Biomass and Bioenergy* 33, 635–643.

Balogh, E., Herr Jr, J.M., Czakó, M. and Márton, L. (2010) Defective development of male and female gametophytes in *Arundo donax* L. (Poaceae). *Southeastern Biology* 57, 292–292.

Balogh, E., Hann, T.C., Ghoshroy, S., Herr Jr, J.M., Czakó, M. and Márton, L. (2011) Differential distribution of sodium, potassium and chloride ions in tissues of high-salinity and low salinity ecotypes of *Arundo donax* using environmental scanning electron microscopy and energy-dispersive X-ray spectroscopy. *Association of Southeastern Biologists (ASB) Conference, April 13–16, 2011*. Huntsville, Alabama.

Bell, G. (1997) Ecology and management of *Arundo donax*, and approaches to riparian habitat restoration in Southern California. In: Brock, J.H., Wade, M., Pysek, P. and Green, D. (eds) *Plant Invasions: Studies from North America and Europe*. Blackhuys Publishers, Leiden, the Netherlands, pp. 103–113.

Bhanwra, R.K., Choda, S.P. and Kumar, S. (1982) Comparative embryology of some grasses. *Proceedings of the Indian National Science Academy* B48, 152–162.

Biswas, T.K. and Williams, C.M.J. (2009) Biosystems for growing a non-food biofuel crop (*Arundo donax*) with saline wastewater. In: Nair, J., Furedy, J., Hoysala, C. and Doelle, H. (eds) *Technologies and Management for Sustainable Biosystems*. Nova Science Publishers, New York, pp. 1–11.

Biswas, T.K., Jayawardane, N.S., Blackwell, J., Christen, E.W. and Cook, F.J. (1999) The 'FILTER' technique for year round treatment of wastewater. In: Patterson, R.A. (ed.) *Proceedings of On-site '99 Conference: Making On-site Wastewater Systems Work*, 13–15 July 1999 University of New England. Lanfax Laboratories, Armidale, New South Wales, Australia, pp. 59–65.

Biswas, T.K., Jayawardane, N.S., Blackwell, J. and Tull, D. (2002) A Land Filter System – Turning Griffith's Sewage Effluent into an Asset. In: *Proceedings of the ANCID Conference*, September 1–4, Griffith, Australia. ANCID, PO Box 58, Berrigan, New South Wales, Australia, pp. 1–4.

Black, I. and Williams, C.M.J. (2010) *Arundo donax* in the upper south east of South Australia. In: Williams, C.M.J. and Biswas, T.K. (eds) *Commercial Potential of Giant Reed (*Arundo donax*) for Pulp/Paper and Biofuel Production*. Final report No. 10/215 of Rural Industries Research and Development Corporation. RIRDC, Canberra, Australia, pp. 82–88. Available at: http://www.rirdc.gov.au (accessed 17 January 2012).

Blackwell, J., Jayawardane, N.S., Biswas, T.K. and Christen, E. (2005) Evaluation of a sequential biological concentration system in natural resource management of a saline irrigated area. *Australian Journal of Water Resources* 9, 169–175.

Boland, J.M. (2006) The importance of layering the rapid spread of *Arundo donax* (giant reed). *Madroño* 53, 303–312.

Bourne, J.K. (2007) Biofuels: boon or boondoggle? *National Geographic* 212, 38–59.

Chomič, A., Pearson, M.N., Clover, G.R.G., Farreyrol, K., Saul, D., Hampton, J.G. and Armstrong, K.F. (2010) A generic RT-PCR assay for the detection of Luteoviridae. *Plant Pathology* 59, 429–442.

Christou, M., Mardikis, M. and Alexopoulou, E. (2001a) Propagation material and plant density effects on the *Arundo donax* yields. In: Kyritsis, S., Beenackers, A.A.C.M., Helm, P., Grassi, A. and Chiaramonti, D. (eds) *1st World Conference on Biomass for Energy and Industry*, 5–9 June, 2000 Sevilla, Spain. James and James (Science Publishers) Ltd, London, pp. 1622–1625.

Christou, M., Mardikis, M. and Alexopoulou, E. (2001b) Research on the effect of irrigation and nitrogen upon growth and yields of *Arundo donax* L. in Greece. *Aspects of Applied Biology: Biomass and Energy Crops II* 65, 47–55.

Cortés, E., Kirk, A.A., Goolsby, J.A., Moran, P.J., Racelis, A.E. and Marcos-Garcia, M.A. (2011) Impact of the Arundo scale *Rhizaspidiotus donacis* (Hemiptera: Diaspididae) on the weight of *Arundo donax*

(Poaceae: Arundinoideae) rhizomes in Languedoc southern France and Mediterranean Spain. *Biocontrol Science and Technology* 21, 1369–1373.
Cronk, J.K. (1996) Constructed wetlands to treat wastewater from dairy and swine operations: a review. *Agriculture Ecosystems and Environment* 58, 97–114.
Czakó, M. and Márton, L. (2011) Subtropical and tropical reeds for biomass. In: Halford, N.G. and Karp, A. (eds) *Energy Crops*. Royal Society of Chemistry, London and Cambridge, UK, pp. 322–340.
Czakó, M., Feng, X., He, Y., Liang, D., Pollock, R. and Márton, L. (2006) Phytoremediation with transgenic plants. *Acta Horticulturae* 725, 753–770.
Daar, S. (1983) Using goats for brush control. *The IPM Practitioner* 5, 4–6.
Dhir, S., Knowles, K., Pagan, C.L. and Mann, J. (2010) Optimization and transformation of *Arundo donax* L. using particle bombardment. *African Journal of Biotechnology* 9, 6460–6469.
Dudley, T. (2000) *Arundo donax*. In: Bossard, C., Randall, J. and Hochovsky, M. (eds) *Invasive Plants of California's Wildlands*. US Press, Berkeley, California, pp. 53–57.
Eisentraut, A. (2010) Sustainable production of second- generation biofuels: potential and perspectives in major economies and developing countries. Information paper. Available at: http://www.iea.org/papers/2010/second_generation_biofuels.pdf (accessed 14 April 2012).
Elbert, A., Haas, M., Springer, B., Thielert, W. and Nauen, R. (2008) Applied aspects of neonicotinoid uses in crop protection. *Pest Management Science* 64, 1099–1105.
Fári, M.G., Kertész, T. and Paluska, F. (2009) *Berendezés és eljárás növények tömeges, programozható mikroszaporítására [Programmable instrument and method for mass micropropagation of plants]*. Hungary, patent application P0900018.
Flowers, T.J., Troke, P.F. and Yeo, A.R. (1977) The mechanism of salt tolerance in halophytes. *Annual Review of Plant Physiology* 28, 89–121.
Flowers, T.J., Hajibagheri, M.A. and Clipson, N.J.W. (1986) Halophytes. *The Quarterly Review of Biology* 61, 313–337.
Focherini, F. (2000) *La battaglia del Rayon: Torviscosa – la città della cellulosa*. Available at: http://digilander.libero.it/trombealvento/vari/rayon.htm (accessed 14 April 2012).
Ghosal, S., Chaudhur, R.K., Bhattach, S.K. and Dutta, S.K. (1972) Occurrence of curarimimetic indoles in flowers of *Arundo donax*. *Planta Medica* 21, 22–22.
Giessow, J.H., Casanova, J., Leclerc, R., MacArthur, R., Fleming, G. and Giessow, J. (2011) *Arundo donax* Distribution and Impact Report March 2011. Available at: http://www.cal-ipc.org/ip/research/arundo/Arundo%20Distribution%20and%20Impact%20Report_Cal-IPC_March%202011.pdf (accessed 3 April 2012).
Goolsby, J.A. and Moran, P. (2009) Host range of *Tetramesa romana* Walker (Hymenoptera: Eurytomidae), a potential biological control of giant reed, *Arundo donax* L. in North America. *Biological Control* 49, 160–168.
Goolsby, J.A., Kirk, A.A., Moran, P.J., Racelis, A.E., Adamczyk, J.J., Cortes, E., Garcia, M.A.M., Jimenez, M.M., Summy, K.R., Ciomperlik, M.A. and Sands, D.P.A. (2011) Establishment of the armored scale, *Rhizaspidiotus donacis*, a biological control agent of *Arundo donax*. *Southwestern Entomologist* 36, 373–374.
Graziani, A. and Steinmaus, S.J. (2009) Hydrothermal and thermal time models for the invasive grass, *Arundo donax*. *Aquatic Botany* 90, 78–84.
Hendriks, A.T.W. and Zeeman, G. (2009) Pretreatments to enhance the digestibility of lignocellulosic biomass. *Bioresource Technology* 100, 10–18.
Hidalgo, M. and Fernandez, J. (2001) Biomass production of ten populations of giant reed (*Arundo donax* L.) under the environmental conditions of Madrid (Spain). In: Kyritsis, S., Beenackers, A.A.C.M., Helm, P., Grassi, A. and Chiaramonti, D. (eds) *1st World Conference on Biomass for Energy and Industry*, 5–9 June 2000, Sevilla, Spain. James and James (Science Publishers) Ltd, London, pp. 1881–1884.
Hong, Y., Hu, H.Y., Sakoda, A. and Sagehashi, M. (2010) Isolation and characterization of antialgal allelochemicals from *Arundo donax* L. *Allelopathy Journal* 25, 357–367.
IUCN (2009) *Guidelines on Biofuels and Invasive Species*. International Union for Conservation of Nature and Natural Resources, Gland, Switzerland.
Jayawardane, N.S., Biswas, T.K., Blackwell, J. and Cook, F.J. (2001) Management of salinity and sodicity in a land FILTER system for treating saline wastewater on a saline-sodic soil. *Australian Journal of Soil Research* 39, 1247–1258.
Jia, A.L., Ding, X.Q., Chen, D.L., Chao, Z.Z., Liu, Z.Y. and Chao, R.B. (2008) A new indole alkaloid from *Arundo donax* L. *Journal of Asian Natural Products Research* 10, 105–109.

Kering, M.K., Butler, T.J., Biermacher, J.T. and Guretzky, J.A. (2012) Biomass yield and nutrient removal rates of perennial grasses under nitrogen fertilization. *Bioenergy Research* 5, 61–70.

Khuzhaev, V.U. (2004) Alkaloids of *Arundo donax*. XVIII. Nitrogenous bases in flowers of cultivars. *Chemistry of Natural Compounds* 40, 516–517.

Lee, I. (2009) *The Coming of the British to Australia 1788-1829*. Longmans, Green and Company, London, UK. Available at: http://gutenberg.net.au/ebooks09/0900091h.html#ch-1 (accessed 14 January 2012).

Lewandowski, I., Scurlock, J.M.O., Lindvall, E. and Christou, M. (2003) The development and current status of perennial rhizomatous grasses as energy crops in the US and Europe. *Biomass and Bioenergy* 25, 335–361.

Lewis, M. and Jackson, M. (2002) Nalgrass: a non wood fibre suitable for existing US pulp mills. In: Janick, J. and Whipkey, A. (eds) *Trends in New Crops and New Uses*. ASHS Press, Alexandria, Virginia, pp. 371–376.

Lowe, J.A., Huntingford, C., Raper, S.C.B., Jones, C.D., Liddicoat, S.K. and Gohar, L.K. (2009) How difficult is it to recover from dangerous levels of global warming? *Environmental Research Letters* 4, 1–9.

Lowe, S., Brown, M., Boudjelas, S. and De Poorter, M. (2000) 100 of the World's Worst Invasive Alien Species: A selection from the Global Invasive Species Database. The Invasive Species Specialist Group of the World Conservation Union. Available at: http://www.k-state.edu/withlab/consbiol/IUCN_invaders.pdf (accessed 16 January 2012).

Mack, R.N. (2008) Evaluating the credits and debits of a proposed biofuel species: Giant Reed (*Arundo donax*). *Weed Science* 56, 883–888.

Mantineo, M., D'Agosta, G.M., Copani, V., Patane, C. and Cosentino, S.L. (2009) Biomass yield and energy balance of three perennial crops for energy use in the semi-arid Mediterranean environment. *Field Crops Research* 114, 204–213.

Mariani, C., Cabrini, R., Danin, A., Piffanelli, P., Fricano, A., Gomarasca, S., Dicandilo, M., Grassi, F. and Soave, C. (2010) Origin, diffusion and reproduction of the giant reed (*Arundo donax* L.): a promising weedy energy crop. *Annals of Applied Biology* 157, 191–202.

Márton, L. and Czakó, M. (2004) *Sustained totipotent culture of selected monocot genera*. USA, patent 6,821,782.

Márton, L. and Czakó, M. (2007) *Sustained totipotent culture of selected monocot genera*. USA, patent 7,303,916.

Márton, L. and Czakó, M. (2011) *Method for micropropagation of monocots based on sustained totipotent cell cultures*. Europe (EPO), patent application EP07794611.9.

Márton, L. and Czakó, M. (2012a) *Method for micropropagation of monocots based on sustained totipotent cell cultures*. Australia, patent 2007352625.

Márton, L. and Czakó, M. (2012b) *Method for micropropagation of monocots based on sustained totipotent cell cultures*. USA, patent 8,105,835.

Márton, L. and Czakó, M. (2012c) *Method for micropropagation of monocots based on sustained totipotent cell cultures*. Malaysia, patent application PI 20094650.

Mavrogianopoulos, G., Vogli, V. and Kyritsis, S. (2002) Use of wastewater as a nutrient solution in a closed gravel hydroponic culture of giant reed (*Arundo donax*). *Bioresource Technology* 82, 103–107.

McWilliams, J.D. (2004) *Arundo donax*. In: Fire effects information system. US Department of Agriculture, Forest Service, Rocky Mountain Research Station, Fire Sciences Laboratory. Available at: http://www.fs.fed.us/database/feis/ (accessed 11 April 2012).

Merinotti, F. (1941) L'utilizzazione della canna gentile (*Arundo donax*) per la produzione autarchica di cellulosa nobile per raion. *La Chimica nell'industria, nell'agricolturia, nella biologia, nelle relazzioni autartiche e corporative* 8, 349–355.

Miles, D.H., Tunsuwan, K., Chittawong, V., Hedin, P.A., Kokpol, U., Ni, C.Z. and Clardy, J. (1993a) Agrochemical activity and isolation of n-(4'-bromophenyl)-2,2-diphenylacetanilide from the Thai plant *Arundo donax*. *Journal of Natural Products* 56, 1590–1593.

Miles, D.H., Tunsuwan, K., Chittawong, V., Kokpol, U., Choudhary, M.I. and Clardy, J. (1993b) Boll-weevil antifeedants from *Arundo donax*. *Phytochemistry* 34, 1277–1279.

Miller, G. and Forney, T. (2011) Oregon Department of Agriculture Plant Pest Risk Assessment for Giant Reed *Arundo donax* L. February 2011. Available at: http://www.westerninvasivesnetwork.org/pdfs/finalarundoriskasses2182011.pdf (accessed 14 April 2012).

Moore, G.W., Watts, D.A. and Goolsby, J.A. (2010) Ecophysiological responses of Giant Reed (*Arundo donax*) to herbivory. *Invasive Plant Science and Management* 3, 521–530.

Myers, B.J., Bond, W.J., Benyon, R.A., Falkiner, R.A., Polglase, P.J., Smith, C.J., Snow, V.O. and Theiveyanathan, S. (1999) *Sustainable effluent-irrigated plantations: An Australian guideline*. CSIRO Forestry and Forest Products, Canberra, Australia.

Nassi o Di Nasso, N., Roncucci, N., Triana, F., Tozzini, C. and Bonari, E. (2011) Seasonal nutrient dynamics and biomass quality of giant reed (*Arundo donax* L.) and miscanthus (*Miscanthus* x *giganteus* Greef et Deuter) as energy crops. *Italian Journal of Agronomy*, 6. Available at: http://www.agronomy.it/index.php/agro/rt/printerFriendly/ija.2011.e24/352 (accessed 16 January 2012).
Paul, D. and Williams, C.M.J. (2006) *Arundo donax* – Plantation establishment and pulp quality in Australia. In: *5th International Conference of Non-Wood Fibre Pulp and Paper Making Conference*, 8–10 November, Guangzhou, China. South China University of Technology Press, Guangzhou, China, pp. 36–39.
Paul, K.L., Jacobsen, K., Koul, V., Leppert, P. and Smith, J. (2008) Predicting growth and sequestration of carbon by plantations growing in regions of low-rainfall in southern Australia. *Forest Ecology and Management* 254, 205–216.
Perdue Jr, R.E. (1958) *Arundo donax* – source of musical reeds and industrial cellulose. *Economic Botany* 12, 368–404.
Popov, K.P. and Belyaeva, N.S. (1987) Developmental rhythm and the cause of infertility in the grass *Arundo donax* L. *Izvestii'a Akademii Nauk Turkmenskoi' SSR. Serii'a Biologicheskikh Nauk (Ashkhabad)* 1, 51–53.
Racelis, A.E., Goolsby, J.A., Penk, R., Jones, W.K. and Roland, T.J. (2010) Development of an inundative, aerial release technique for the arundo wasp, biological control agent of the invasive *Arundo donax* L. *Southwestern Entomologist* 35, 495–501.
Ramgareeb, S., Snyman, S.J., van Antwerpen, T. and Rutherford, R.S. (2010) Elimination of virus and rapid propagation of disease-free sugarcane (*Saccharum* spp. cultivar NCo376) using apical meristem culture. *Plant Cell Tissue and Organ Culture* 100, 175–181.
Reuter, D.J. and Robinson, J.B. (1997) *Plant Analysis: An Interpretation Manual*. CSIRO Publishing, Collingwood, Australia.
Rezk, M.R. and Edany, T.Y. (1979) Comparative responses of two reed sp. to water table levels. *Egyptian Journal of Botany* 22, 157–172.
Riffaldi, R., Saviozzi, A., Cardelli, R., Buller, F. and Angelini, L. (2010) Comparison of soil organic-matter characteristics under the energy crop giant reed, cropping sequence and natural grass. *Communications in Soil Science and Plant Analysis* 41, 185–192.
Rossa, B., Tuffers, A.V., Naidoo, G. and von Willert, D.J. (1998) *Arundo donax* L. (Poaceae) – a C_3 species with unusually high photosynthetic capacity. *Botanica Acta* 111, 216–221.
Sharma, K.P., Kushwaha, S.P.S. and Gopal, B. (1998) A comparative study of stand structure and standing crops of two wetland species, *Arundo donax* and *Phragmites karka*, and primary production in *Arundo donax* with observations on the effect of clipping. *Tropical Ecology* 39, 3–14.
Shatalov, A.A. and Pereira, H. (2006) Papermaking fibers from giant reed (*Arundo donax* L.) by advanced ecologically friendly pulping and bleaching technologies. *Bioresources* 1, 45–61.
Snyman, S.J., Meyer, G.M., Koch, A.C., Banasiak, M. and Watt, M.P. (2011) Applications of *in vitro* culture systems for commercial sugarcane production and improvement. *In Vitro Cellular and Developmental Biology – Plant* 47, 234–249.
Spafford, W.J. (1941) The bamboo reed (*Arundo donax*) in south Australian agriculture. *The Journal of the Department of Agriculture of South Australia* 45, 75–83.
Spencer, D.F. and Ksander, G.G. (2006) Estimating *Arundo donax* ramet recruitment using degree-day based equations. *Aquatic Botany* 85, 282–288.
Spencer, D.F., Wailun, T., Pui-Sze, L., Ksander, G.G., Whitehand, L.C., Weaver, S., Olson, J. and Newhouser, M. (2008) Evaluation of glyphosate for managing giant reed (*Arundo donax*). *Invasive Plant Science and Management* 1, 248–254.
Subba Reddy, C.V. and Sreenivasulu, P. (2011) Generation of sugarcane streak mosaic virus-free sugarcane (*Saccharum* spp. hybrid) from infected plants by *in vitro* meristem tip culture. *European Journal of Plant Pathology* 130, 597–604.
Talapatra, S.K. (1950) The nutritive value of the indigenous grasses of Assam. 3, The semiaquatic grasses as cattle feeds. *Indian Journal of Veterinary Science* 20, 229–240.
The Economist (2008) 'Grow your own'. In: The power and the glory, A special report on energy. *The Economist* 387, 13–16.
The Economist (2009) Briefing: the outlook for the oil price. *The Economist* 391, 65–67.
Tóth, E.K., Kriston, K., Nyerges, K., Nardin, F., Czakó, M., Márton, L. and Fári, M. (2011) Virological investigation of third-generation biomass plants – *Arundo donax* L. as an example. *Növényvédelem (Budapest)* 47, 451–454.
US EPA (2012) Regulation of fuels and fuel additives: identification of additional qualifying renewable fuel pathways under the renewable fuel standard program; direct rinal rule. *Federal Register* 77, 700–727.

USDA (2011) Finding of No Significant Impact (FONSI), Chemtex Project Alpha, Clinton, North Carolina by the U.S. Department of Agriculture, Rural Business-Cooperative Service. Available at: http://www.rurdev.usda.gov/rhs/pss/Assessments/ChemtexFONSIandEA.pdf (accessed 21 March 2012).

Virtue, J.G. (2010) South Australia's Weed Risk Management System. Proceedings of 2nd International Weed Risk Assessment Workshop. *Plant Protection Quarterly* 25, 75–79.

Virtue, J.G., Reynolds, T., Malone, J., Preston, C. and Williams, C.M.J. (2010a) Managing the weed risk of cultivated *Arundo donax* L. In: Zydenbos, S.M. (ed.) *17th Australasian Weeds Conference*, 26–30 September, Christchurch, New Zealand. New Zealand Plant Protection Society, Christchurch, New Zealand, pp. 176–179.

Virtue, J.G., Reynolds, T., Malone, J., Preston, C. and Williams, C.M.J. (2010b) Weed risk management guidelines for *A. donax* plantations in Australia. In: Williams, C.M.J. and Biswas, T.K. (eds) *Commercial potential of giant reed (Arundo donax) for pulp/paper and biofuel production*. Final report No. 10/215 of Rural Industries Research and Development Corporation. RIRDC, Canberra, Australia, pp. 41–68. Available at: http://www.rirdc.gov.au (accessed 3 February 2012).

Williams, C.M.J. and Biswas, T.K. (2009) *Salinity Tolerance, Nutrient Needs, Weed Risk Management and Carbon Sequestration of Arundo Donax for Remediation of Highly Saline Wastewaters for Biofuel or Pulp/Paper Feedstock Production*. Final Report Project No. 56/2008 for the South Australian Murray-Darling Basin Natural Resources Management Board, SARDI, Adelaide, Australia.

Williams, C.M.J. and Biswas, T.K. (2010) Commercial potential of giant reed (Arundo donax) for pulp/paper and biofuel production. Final report No. 10/215 of Rural Industries Research and Development Corporation, RIRDC, Canberra, Australia. Available at: http://www.rirdc.gov.au (accessed 15 January 2012).

Williams, C.M.J., Vitosh, M.L., Maier, N.A. and MacKerron, D.K.L. (1999) Nutrient management strategies for sustainable potato (*Solanum tuberosum*) in the Southern and Northern Hemispheres. In: Nee, M., Lester, R.N. and Jessop, J.P. (eds) *Solanaceae IV*. Royal Botanic Gardens, Kew, UK, pp. 443–458.

Williams, C.M.J., Maier, N.A. and Bartlett, L. (2004) Effect of molybdenum foliar sprays on yield, berry size, seed formation, and petiolar nutrient composition of 'Merlot' grapevines. *Journal of Plant Nutrition* 27, 1891–1916.

Williams, C.M.J., Biswas, T.K., Glatz, P. and Kumar, M. (2007) Use of recycled water from intensive primary industries to grow crops within integrated biosystems. *Agricultural Science* 21, 34–36. (September issue, Parkville, Australia.)

Williams, C.M.J., Biswas, T.K., Harris, P. and Heading, S. (2008a) Use of giant reed (*Arundo donax*) to treat wastewaters for resource recycling in South Australia. *Acta Horticulturae* 792, 701–707.

Williams, C.M.J., Biswas, T.K., Black, I. and Heading, S. (2008b) Pathways to Prosperity: Second generation biomass crops for biofuels using saline lands and wastewater. *Agricultural Science* 21, 28–31. (February issue, Parkville, Australia.)

Williams, C.M.J., Biswas, T.K., Harris, P., Heading, S., Márton, L., Czakó, M., Pollock, R. and Virtue, J.G. (2009) Use of poor quality water to produce high biomass yields of giant reed (*Arundo donax* L.) on marginal lands for biofuel or pulp/paper. *Acta Horticulturae* 806, 595–600.

Williams, C.M.J., Biswas, T.K., Chvyl, L., Harris, P. and Dyson, C. (2010a) Dry matter yield, carbon accumulation and biochar from *Arundo donax* grown in South Australia. In: Williams, C.M.J. and Biswas, T.K. (eds) Commercial potential of giant reed (*Arundo donax*) for pulp/paper and biofuel production. Final report No. 10/215 of Rural Industries Research and Development Corporation. RIRDC, Canberra, Australia, pp. 4–19. Available at: http://www.rirdc.gov.au (accessed 15 January 2012).

Williams, C.M.J., Biswas, T.K., Chvyl, L. and Dyson, C. (2010b) Salt tolerance and nutrient dynamics of *Arundo donax*. In: Williams, C.M.J. and Biswas, T.K. (eds) Commercial potential of giant reed (*Arundo donax*) for pulp/paper and biofuel production. Final report No. 10/215 of Rural Industries Research and Development Corporation. RIRDC, Canberra, Australia, pp. 20–41. Available at: http://www.rirdc.gov.au (accessed 15 January 2012).

Wright, M. and Brown, R. (2007) Comparative economics of biorefineries based on the biochemical and thermo-chemical platforms. *Biofuels, Bioproducts, Biorefineries* 1, 49–56.

Wynd, F.L., Steinbauer, G.P. and Giaz, N.R. (1948) *Arundo donax* as a forage grass in sandy soils. *Lloydia* 11, 181–184.

Xu, T., Jacobsen, C.M., Hara, A.H., Li, J. and Li, Q.X. (2009) Efficacy of systemic insecticides on the gall wasp *Quadrastichus erythrinae* in wiliwili trees (*Erythrina* spp.). *Pest Management Science* 65, 163–169.

Zheng, H., Wu, Y., Ding, J., Binion, D., Fu, W. and Reardon, R. (2005) *Invasive Plants of Asian Origin Established in the United States and Their Natural Enemies*, Vol. 2. Chinese Academy of Agricultural Sciences and USDA Forest Service Forest Health Technology Enterprise Team. Available at: http://www.fs.fed.us/foresthealth/technology/bcpubs.shtml (accessed 3 April 2012).

13 Elephantgrass

Bharat P. Singh, Hari P. Singh and Eric Obeng
Fort Valley State University, Georgia, USA

13.1 Introduction

Elephantgrass is also called merker grass, napiergrass, Uganda grass and by several other names in different parts of the world. It carries the name napiergrass to recognize the contribution of Colonel Napier of Rhodesia who first wrote to the Rhodesian Agricultural Department to make them aware of the value of this plant.

Elephantgrass (*Pennisetum purpureum* Schumach) is a monocot belonging to the family *Poaceae* (grass family) and genus *Pennisetum* (*Pennisetum* Rich. ex Pers.; fountain grass). The *Pennisetum* genus is very diverse consisting of a heterogeneous group of approximately 10,000 species, with basic chromosome number of 5, 7, 8 or 9 and ploidy ranging from diploid to octoploid with annual, biennial or perennial life cycle (Martel *et al.*, 1997). Surprisingly though, the species genetically remain closely related and gene content of different species does not vary greatly (Bennetzen and Freeling, 1997), variations in ploidy levels often accounting for differentiation. Elephantgrass is an allopolyploid with chromosome constitution of $2n = 4x = 48$ (Hanna, 1981). The allopolyploid carries two genomes A′A′ BB in which chromosomes of A′ genome are larger in size than the B genome (Jauhar, 1981). The B genome contributes perennial growth habit to elephantgrass.

Being non-edible, elephantgrass escapes the 'food versus fuel' debate. It has an advantage over tree species in that only after several years of growth can trees can be harvested for biomass, while elephantgrass is harvested in the first year after planting. Furthermore, the lignin content of woody biomass is 25% or more (Sjöström, 1993) compared to lignin in the range of 6 to 10% (Ramírez *et al.*, 2010) in the herbaceous biomass of grasses. Lignin is a hindrance to the fermentation of biomass to ethanol and has to be separated first from cellulose for fermentation to proceed.

Ra *et al.* (2012) compared the annual biomass yield of seven grass species, viz. *Erianthus*, johnsongrass, switchgrass, sugarcane, maize, sorghum and elephantgrass and found elephantgrass to out-yield others by a significant margin. According to Karlsson and Vasil (1985), elephantgrass is the fastest growing plant in the world. A study in Florida reported a dry biomass yield of 45 dry t ha^{-1} $year^{-1}$ (Woodard and Sollenberger, 2008). Elephantgrass has for long been an important forage crop in the tropics because of its high yields and nutrient value.

Being a grass, elephantgrass shares the common trait with other members of the grass family for the ability to grow with low input, and thus is suited for sustainable production. Grasses in general are known for their associative relationship with nitrogen-fixing bacteria, and because of this reason are able to grow without added nitrogen (James and Olivares,

©CAB International 2013. *Biofuel Crops: Production, Physiology and Genetics* (ed. B.P. Singh)

1997). Elephantgrass has drawn particular attention for N-sustainability because of the discovery of the presence in the plant of the same N-fixing endophytic bacteria, *Gluconacetobacter diazotrophicus,* that has been found to contribute significant amounts of N to sugarcane (Döbereiner, 1988). The suitability and drawbacks of elephantgrass for biofuel feedstock are presented in Table 13.1.

13.2 Origin and Distribution

Elephantgrass is a tropical grass with its origin in humid tropical Africa. Being a perennial grass possessing underground stems or rhizomes and thereby having its growth nodes protected by the soil during a ground fire, elephantgrass in its native range developed as a fire sub-climax to broadleaf tropical forests. It has been naturalized to Central and South America, tropical parts of Asia, Australia, the Middle East and Pacific islands for use as animal fodder. Elephantgrass was introduced into the USA from Africa in 1913 (Burton, 1990). The introduction of elephantgrass in some cases has been direct but in others through a naturalized source; for example, it was introduced into Brazil from Cuba around 1920 (Daher *et al.*, 2002). Once the fast-growing habit and potential for fodder especially for dairy cows became known outside Africa, spread of elephantgrass to other continents was expeditious and in some of the countries it became an important animal fodder.

13.3 Ecology

The most suited climate for elephantgrass growth is that similar to its native range typified by light showers alternating with bright sunshine. Day temperatures of 30–35°C are optimum for its growth (Ferrais, 1978) and no growth takes place below 10°C (Bogdan, 1977). Elephantgrass grows well in all tropical and subtropical regions. It is adaptable to a wide range of annual rainfall between 750 and 2500 mm and altitudes from sea level to 2100 m, but is susceptible to frost damage (Skerman and Riveros, 1990). In the tropical climate, elephantgrass grows throughout the year. The aboveground portion of the plant is killed wherever frost occurs, but the soil assures its continuity by protecting the underground rhizomes, which again sprout giving out new tillers as and when weather becomes favourable for growth. Burton (1989) observed that rhizomes of elephantgrass cultivar 'Merkeron' survived winter temperatures as low as −18°C and remerged for growth during the next spring. Ito and Inanaga (1988a,c) contend that although elephantgrass requires warm weather for optimum growth, because of the ability to recover rapidly it can also adapt successfully to mild temperate climates (~35°N). In temperate climates, early growth is severely restricted due to low spring temperatures, but the plant recoups rapidly with the warming temperatures of summer. The biomass yields of approximately 80 dry t ha^{-1} $year^{-1}$ (Vicente-Chandler *et al.*, 1959) in the tropics and

Table 13.1. Advantages and shortcomings of elephantgrass as biofuel feedstock.

Advantages	Shortcomings
High yield	Low cold-tolerance limits production range
Moderate cost of production	High mortality of first year overwinter plants in temperate climates
Suited for marginal soils	Recent outbreaks of both below-ground and above-ground diseases impacting yield
Existence of probable associative N-fixation traits	Can become noxious weed because of invasiveness
High soil carbon sequestration because of rhizomatous root system	High transport cost to refineries because of bulk
Better adapted than sugarcane to hilly areas	Logistics: handling more difficult than wood
Low erosion potential	Amount of ash and minerals in the feedstock
Once planted lasts for multiple years under once-a-year harvest	
Multiple uses especially for animal fodder and biofuel	
Suitable for production by farmers with smallholdings because of diversified uses	

exceeding 45 dry t ha^{-1} $year^{-1}$ (Prine *et al.*, 1988) at ~30°N latitude (Gainesville, Florida, USA) have been reported. The overwinter survival of elephantgrass has been reported as far as ~36°N at Oak Ridge, Tennessee (USA) (Turhollow, 1991). However, yield appeared to have suffered because of the shortened summer (~30 dry t ha^{-1} for 'N-51' and 22 dry t ha^{-1} for 'PI 300086').

Nagasuga (2005) and Nagasuga and Kubota (2006) demonstrated that elephantgrass was well adapted for temporary shade conditions similar to the rainy season cloudy weather common to tropical climates. Elephantgrass continued leaf area expansion during shade (30% full sunlight for 23 days starting 45 days after planting) at the expense of stem thickness and root weight. After restoration of full sunlight, the shaded plants reached similar total dry weight as those exposed to full sunlight through the entire period. While most plants when shifted from shade to sunlight show leaf wilting due to declined water conductivity (Shultz and Mathews, 1997) and require time to adjust to the new conditions, elephantgrass resumes growth quickly without showing stress symptoms (Nagasuga and Kubota, 2006). The flexibility of total hydraulic resistance (R) of the stem allowed leaves to maintain similar R in shade and sun, thus avoiding water-imposed limitation to growth while under shade. This may help explain, at least in part, the ability of elephantgrass to maintain higher productivity compared to other tropical plant species. However, elephantgrass cannot grow as understorey of closed tree canopy.

Elephantgrass can adapt to a wide range of soil types from sandy to clayey. It can also grow in soils in the pH range of highly acidic to alkaline (Center for New Crops and Plant Products, 2002). The maximum growth is attained on well-drained loamy soils with high organic matter content. It is susceptible to waterlogging.

13.4 Botanical Description

Elephantgrass has a perennial life cycle and is propagated vegetatively. It has a profuse root system, penetrating deep into the soil, and an abundance of fibrous roots spreading into the top soil horizons. The rhizomes (underground stem) are short and creeping and nodes develop fine roots and culms. There are two plant types, giant (tall) and dwarf. The plant forms clumps and grows upwards, profusely branched with thick cane-like stems (Fig. 13.1). The culms are erect and tall, varying in height from 2 to 6 m.

Fig. 13.1. Field plots at Fort Valley, Georgia (USA), showing vigorous upward growth habit of elephantgrass plants.

It has long, tapering and stiff leaves. The leaf edges have bristle-like ridges. The upper leaf surface is covered with stiff hairs and leaf sheaths are also very hairy (Fig. 13.2). A few morphological characteristics based on the average of 12 cultivars grown in Kenya as provided by Nyambati *et al.* (2010) are as follows: tillers number 60; tiller diameter 5.4 cm; tiller angle (stem to the ground angle) 79.2°; leaf length 61.2 cm; and leaf width 2.0 cm. The plant may have up to 20 nodes (Henderson and Preston, 1997). The above-ground dry weight is distributed ~54% in leaf and 46% in the stem (Ansah *et al.*, 2010). The plant grows vigorously starting from the very first year. In this respect elephantgrass differs from many other vigorous growing grasses such as *Miscanthus*, which needs the first year for establishment.

In tropical climates, elephantgrass flowers from autumn through winter (Holm *et al.*, 1977). According to Hanna *et al.* (2004), elephantgrass requires a day length of 11 h or less to flower. It produces no or very few flowers when grown in sub-temperate regions since, by the time days shorten for flowering, the temperature dips too low to support further growth. The reproductive part of the plant is the panicles, characterized by tawny or purplish colour, sessile fascicles, and sparsely plumose bristles. The stigma in elephantgrass flowers is receptive before anther exertion, facilitating cross-pollination. The exertion period of the stigma lasts for 3–4 days starting at the top of the inflorescence, followed by anthers shedding pollen for a similar period. Self-fertilization is also prevented because of the presence of a substantial level of self-incompatibility. However, certain genotypes such as 'Merkeron' have been found to produce seed by selfing or produce seed on maternal parents in crosses with other genotypes (Hanna *et al.*, 2004). The seed set is poor and seeds are prone to shattering (Cheng, 1991). The seeds are very small in size with about 3.8 million seeds kg^{-1} (Skerman and Riveros, 1990) and are wind dispersed. The seeds often have a low germination rate and seedlings develop very slowly.

Fig. 13.2. Illustration of profuse hair on leaf sheath and leaf surface, and bristle-like leaf edge of elephantgrass plant.

13.4.1 Growth habit

Elephantgrass possesses the ideal ideotype for large biomass accumulation (Kubota *et al.*, 1994). It tillers vigorously, providing the extensive stem structure for leaf formation needed for photosynthesis. The plant is tall with multiple nodes providing sites for bearing long leaves. Long internodal intervals permit efficient light distribution through the canopy. Leaves are long and narrow, thus providing the leaf area without hampering canopy-wide light distribution. Ito and Inanaga (1988b) reported that unfolded elephantgrass leaves had a higher rate of photosynthesis compared to other C_4 species. Masuda *et al.* (1991) has itemized the morphological traits behind the high productivity of elephantgrass: (i) rapid increase in stem number (~100 m^{-2}) after emergence and horizontal orientation at the early growth stage of stem and leaves to cover ground fully for high light interception at low leaf area index (LAI) (Fig. 13.3); (ii) at the middle stage of growth there is fast expansion of LAI (~13), self-thinning of stems (~25 m^{-2}), accelerated elongation of stem and change of elongation angle from horizontal to vertical allowing elevated net assimilation of CO_2 and high crop growth rate (Figs 13.4 and 13.5); and (iii) ability to hold relatively high LAI at

Fig. 13.3. Illustration of horizontal orientation of stems and leaves at the early growth stage of elephantgrass.

Fig. 13.4. Illustration of the long internodal length and acute leaf angle at the later growth stage of elephantgrass.

the late stage of growth by continued new leaf formation to compensate for lethality of old leaves, thus continuing the biomass accumulation at the late growth stage.

Saeki (1960) using a model showed that light utilization efficiency of a canopy of high LAI would be greater at low K (light extinction coefficient) while the reverse would be true for a low LAI canopy. As leaf area is essential for photosynthesis and dry matter accumulation, a plant with greater leaf area and canopy that permits light penetration to lower leaves should be able to assimilate more dry matter than one with similar leaf area but poorer light distribution or a lower leaf area plant. Light utilization efficiency of elephantgrass canopy is ideal (Fig. 13.6), K value is high at low LAI and as the canopy develops and LAI increases, there is an accompanying decline in K value. Masuda *et al.* (1991) observed large differences in K among the different strata of elephantgrass canopy and considered it advantageous to its dry matter production. Ishii *et al.* (2005) also found K in elephantgrass to be negatively regressed with LAI. Kubota *et al.* (1994) attribute these favourable canopy K value changes with LAI to changes in the angle of stem inclination and stem lengths with growth (Figs 13.7 and 13.8).

Matsuda *et al.* (1991) found growth rate of elephantgrass superior to that of maize (*Zea mays*). The final dry weight of elephantgrass was twice that of maize. Considering the

Fig. 13.5. Elephantgrass clump showing open architecture and narrow long leaves to permit efficient canopy light penetration.

enormous amount of research effort and resources spent towards making maize more productive to where it is now, chances of large initial biomass gain in elephantgrass with concentrated research effort are promising.

There are several avenues to further improve growth habit of elephantgrass. Ishii *et al.* (2005) reported similar K among different elephantgrass cultivars. It has been demonstrated that the variation in K is related to canopy structure, especially to leaf angle (Kasanaga and Monsi, 1954; Loomis *et al.*, 1968). Thus, breeding new cultivars for more-erect leaves should further improve light penetration in the elephantgrass canopy. Variation for leaf angle exists in the elephantgrass population as evidenced by Sunusi *et al.* (1999) of more erect leaves in cultivar 'Wruk wona' than 'Merkeron'. Elephantgrass plant canopy is composed of a large number of tillers with severe competition between them. Therefore synchronous growth would provide all tillers with a similar growth opportunity and would lead to a robust canopy (Kubota *et al.*, 1994). Ishii *et al.* (2005) observed varietal difference in tiller growth, 'Wruk wona' and 'Taiwan A146' having more synchronous growth than 'Merkeron'. Furthermore, 'Merkeron' was more sensitive to shading than the other two cultivars. Cunha *et al.* (2011) showed a close relationship between dry matter production, basal tiller density and plant height. They also observed existence of significant variation for these traits among the eight genotypes tested. Thus, tiller number and plant height among growth characters can also be targeted for improvement. The leaf/stem ratio is another area of research on which to focus, as a greater proportion of stem to leaf is desirable in biofuel feedstock. Anderson *et al.* (2008) observed a range of 0.12–0.87 for stem to leaf ratio among 100 elephantgrass accessions indicating the existence of sufficient variation for this character. Thick-stem cultivars assemble more of their dry matter in the stem in the form of cellulose and this is a desired trait for biofuel feedstock. Mannetje (1992) has reported existence of a considerable amount of variation among elephantgrass accessions for stem thickness as well as another desirable trait, plant height.

13.5 Cultivation

13.5.1 Cultivar selection

As per Kretschemer and Pitman (2001), somewhere around 25 cultivars of elephantgrass and 16 of hybrids with millet are under cultivation. Cultivars adapted to different parts of the world are available. Farmers have developed a preference for dwarf cultivars for forage because they adapt themselves better to hand-harvesting practised in developing countries and are also preferable than normal height cultivars for grazing (Williams and Hanna, 1995). Mukhtar *et al.* (2003) observed that dwarf cultivars possessed higher tiller number, LAI, dry blade percentage of leaf blade but lower plant height, mean tiller dry weight and total dry weight than normal cultivars, but the gap in total plant dry weight narrowed at higher plant densities. 'Mott' is a popular dwarf elephantgrass cultivar. In East Africa, 'Bana', 'French Cameroon', 'Clone 13' and 'Pakistan hybrid' are common cultivars, with 'Bana' being the most popular because of short stems and broad leaves and less stem herbage at maturity (Nyambati *et al.*, 2010).

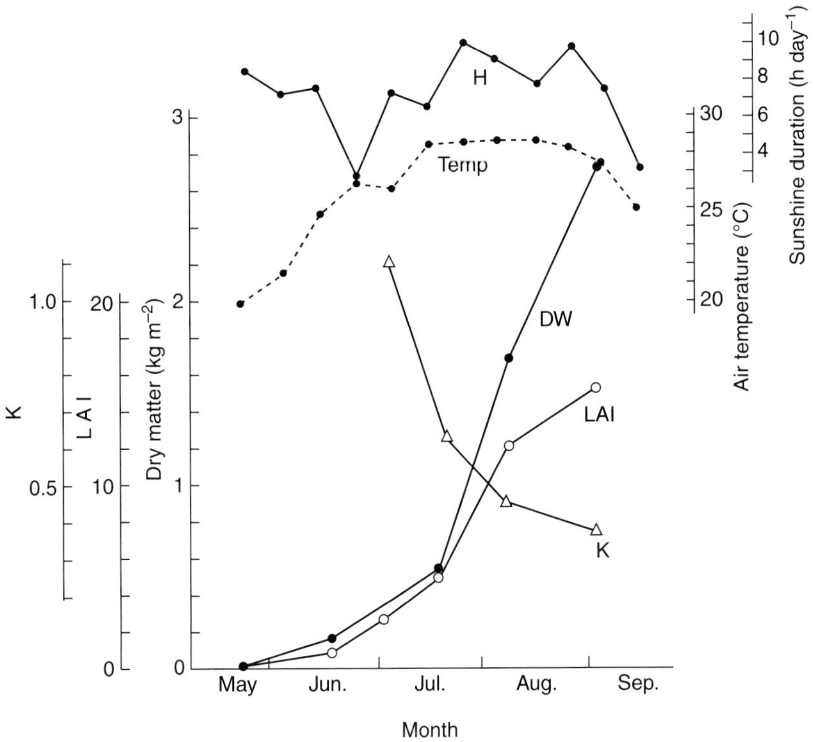

Fig. 13.6. Climatic factors and changes in plant weight, LAI and K with growth. (DW, dry matter weight; LAI, leaf area index; K, light extinction coefficient of the elephantgrass canopy.) (Kubota *et al.*, 1994; with permission.)

Cultivars with thick stems should be more suitable as biofuel feedstock because the stem is the sink for cellulose. 'Wrunk wona' is described as a thick-stem superior-yielding elephantgrass cultivar with high *in vitro* dry matter digestibility (Ishii *et al.*, 2005). 'Taishigrass No.2' is another cultivar with favourable characteristics for biofuel feedstock. It is fast growing, adjusts well to a short growing season and is thus widely adaptable and yields high amounts of biomass (Tsai, 2009).

13.5.2 Planting

Elephantgrass is not propagated by seeds for several reasons. First, due to cross-pollination the seeds are a mixed lot, thus the plants produced from them will not be uniform and their performance not predictable. The seeds often germinate poorly and seedlings are weak. Moreover, poor seed-setting and shattering makes seed availability problematic. Thus, vegetative propagation of elephantgrass is a routine practice. The plant lends itself to easy propagation and the method assures germplasm stability and quality preservation.

The propagation of elephantgrass is by stem cuttings (setts), rooted stems or splitting of rhizomes. The most common way of planting is using setts. The mature stems in the basal 2/3 zone with well-developed leaf buds are cut into sections of three nodes and planted upright by burying two basal nodes into the soil. An alternative and more commonly used method is to place cut sections horizontally into shallow trenches (5–10 cm) and cover the trenches. Knoll and Anderson (2012) observed that cuttings taken from the basal portion of the plant were more suited for planting than those from the younger upper portion of the plant. Furthermore, in subtemperate climates, five- and ten-node cuttings had a higher rate of

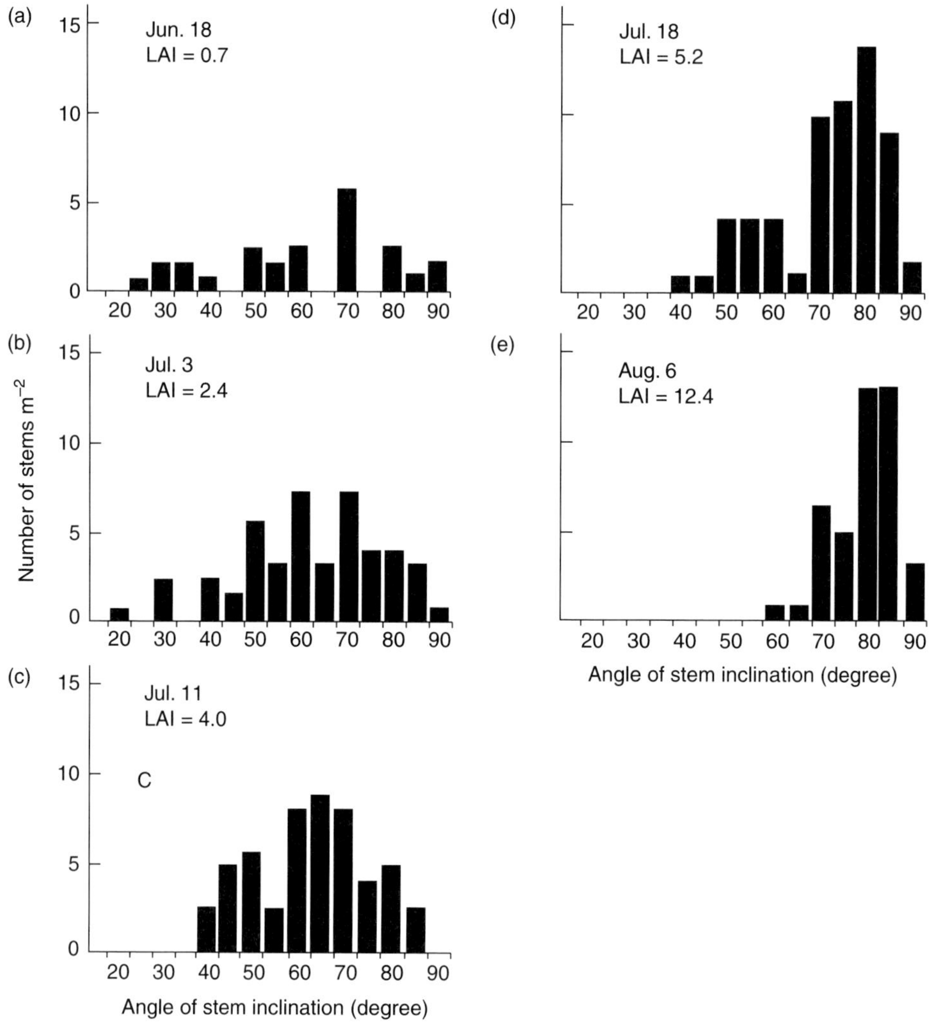

Fig. 13.7. Distributions of stem inclination angle at various growth stages of elephantgrass (Kubota *et al.*, 1994; with permission).

after-winter emergence than one- or two-node cuttings. Woodward *et al.* (1985) suggested cutting stems into short pieces as apical dominance kept the bud at the middle node dormant while buds on the nodes at the two ends started normal growth. However, orientation of stem cutting in the furrow had no effect on emergence or winter survival. A wide range of planting densities has been reported from different parts of the world depending on soil fertility, cultivar and harvesting purpose. For forage production, intra-row and inter-row spacing of 0.5 m × 1 m in Ghana (Ansah *et al.*, 2010), 1 m × 1 m in Kenya (Nyambati *et al.*, 2010) and 50 cm × 50 cm in Japan (Ishii *et al.*, 2005) have been suggested. Mukhtar *et al.* (2003) found that where plants have to go through overwintering, high population density impacted adversely the number of stubble tillers emerging from underground stems.

The guiding principle in determining spacing for all crops is similar; to attain rapid full groundcover without creating excessive interplant competition for nutrients and soil water at the minimum population density, because considerable expenditure is involved in raising

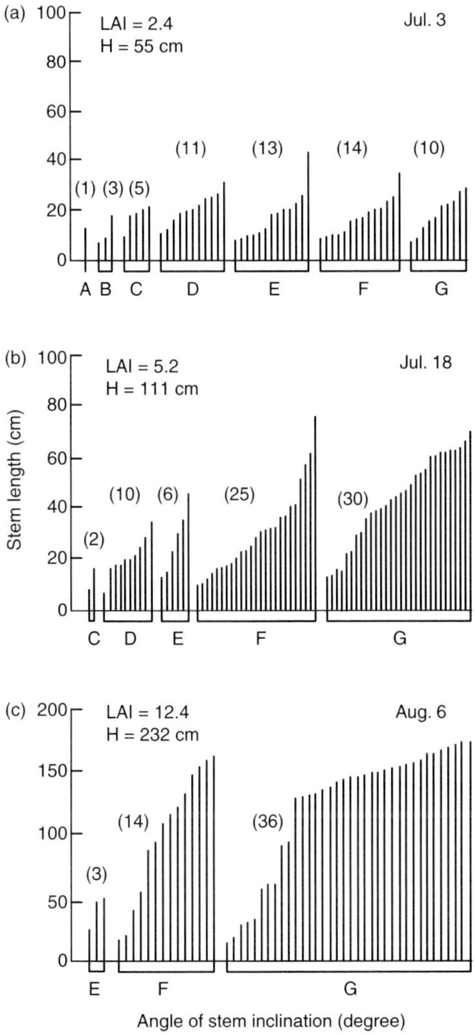

Fig. 13.8. Distributions of stem inclination angle and length at various growth stages of elephantgrass. (H = natural height of plant; between parentheses, the number of stems. Stem inclination angle: A = 20–30°; B = 30–40°; C = 40–50°; D = 50–60°; E = 60–70°; F = 70–80°; G = 80–90°.) (Kubota *et al.*, 1994; with permission.)

or purchase of planting material and there is added cost in labour and machinery with increased population density.

In general, vegetative propagation is more costly and labour intensive than planting by seed; propagation of elephantgrass is no exception. As of now, the primary purpose of growing this grass species has been for cut-and-carry fodder by small producers. These farmers use manual labour to plant the setts. As the production is scaled up for biofuel feedstock, mechanization of the farming operations will be needed. As elephantgrass is similar to sugarcane in stem structure, sugarcane planters should be amenable to modification for its planting.

Elephantgrass is usually replanted every 5–6 years to keep the stand vigorous under repeated cuttings for fodder. The interval for replanting when crop is harvested only once a year as in case of biofuel feedstock after the plant has reached maturity and had an opportunity to store sufficient carbohydrate in the rhizome and translocate mineral nutrients back to the root should be much greater but has not been determined yet.

13.5.3 Fertilization

The rate of fertilization varies with the soil and is dependent on the existing fertility level. The reported range of optimum fertilization rate especially of N is wide. While Mohammad *et al.* (1988) found N rate of 80 kg N ha^{-1} to be adequate, Miyagi (1983) recorded forage yield increases to rates up to 600 kg N ha^{-1}. Sunusi *et al.* (1999) even reported continuous increase in dry matter yield by increasing N supply up to 1200 kg N ha^{-1}. On the other hand, Walsmsley *et al.* (1978) did not find any advantage to yield above 170 kg N ha^{-1}.

There are many reports of planting of elephantgrass in a mixture with legumes to increase the yield and nutrient value of fodder (Mureithi *et al.*, 1995; Njoka-Njiru *et al.*, 2006). In the tropics, trailing or shrub legumes are recommended for companion cropping as low growing types may be shaded out by the elephantgrass canopy. Tiley (1989) found a combination of desmodium (*Desmodium intortum* (Mill.) Urb.) and 120 kg ha^{-1} fertilizer N produced maximum yield of elephantgrass. We are in the process of evaluating the soil fertilization as well as carbon sequestration value of clover and lupine grown between rows as winter cover (Fig. 13.9). In this trial, the cover crop is allowed to mature in the field, die and become mulch. It has proven effective in providing

Fig. 13.9. Crimson clover winter cover between elephantgrass rows at Fort Valley, Georgia (USA).

ground cover to prevent spring weed problems. The cover crop reseeds itself after first year planting. We have not yet established the extent of supplemental seeding that will be required to maintain full winter cover-crop stand.

Quesada (2005) noted while fertile soil is needed to produce high protein elephantgrass forage, this plant has the ability to assimilate similar amounts of dry matter but with lower N composition in limited N soils. As low N content is a desired characteristic for biofuel feedstock, this provides a unique opportunity for infertile land utilization. The soils in semi-arid climates have low organic matter content (Jones, 1973). With a soil containing 0.03% N and 4% annual mineralization rate (Wetselaar, 1967), the amount of N mineralization equals 50 kg ha^{-1}. But Quesada (2005) recorded accumulations in elephantgrass biomass of 270 kg ha^{-1} N in 15 months without N fertilization, which is much greater than that available from N mineralization in tropical soils. This brings into the equation the possibility of the presence of additional N supply source and importance to the suggestions of Baldani *et al.* (1998) that biological nitrogen fixation might be an important source of N for elephantgrass in poor soils. The report by Kirchhof *et al.* (2001) of the presence in this grass of N-fixing bacteria belonging to *Herbaspirillum* genus reinforces this possibility. De Morais *et al.* (2009) found that N uptake by elephantgrass varied with the genotype and the N accumulation that they attributed to bacterial associative N fixation varied from 240 to 343 kg ha^{-1}. Quesada (2005) also observed genotypic variation in the amount of N accumulation that can be assigned to biological N fixation. Observance of higher biological N under low soil N compared to higher soil N is similar to legume symbiotic N fixation. Lowering of the need for N results in both saving in cost of farming operation and to the mitigation of aerial CO_2. According to Robertson and Grace (2004), avoiding the use of each 100 kg of N fertilizer accounts towards mitigating 450 kg CO_2 generated during processing, transportation and fertilizer application.

Ra *et al.* (2012) report an interesting study regarding the effect of planting density on N use. According to this report, higher density planting from 20,000 and 40,000 plants ha^{-1} did not change yield, but helped cut down on N absorption from 775.1 kg ha^{-1} to 536.7 kg ha^{-1}. The reason for such a decrease was not very clear. Lower N absorption while maintaining the dry matter production is important to the farmer for cost management and also for saving fossil fuel required for N fertilizer manufacture. It is of especial importance in the case of producing elephantgrass for biofuel feedstock, as low plant N content is a desirable attribute. Further confirmation of these results should be helpful in N management decision making.

In addition to N, elephantgrass also has substantial need for other mineral nutrients. These needs may require the addition of these elements in the form of fertilizer. As elephantgrass is grown widely in soils in Africa, in South America and potentially in future in the southeastern USA in acidic soil, P availability may become a problem. Inoculation with mycorrhizae can assist in the plant's uptake of P and Hung *et al.* (1990) obtained greater growth in inoculated plants across P rates.

13.5.4 Diseases and pests

Elephantgrass commonly does not suffer from serious diseases and pests (Boonman, 1993). Lennè (1990) listed 66 fungal species associated with the plant but there was no reference to any serious damage. However, elephantgrass is reported to harbour pests of other cultivated crops (Farrell *et al.*, 2002).

Moisture can be trapped on the leaf because of its hairy nature and lead to white mould (*Spharaedea beniwoskia*), but this does not cause serious damage to the yield (Nyambati *et al.*, 2010). In East Africa, the existing cultivars have become vulnerable to head smut (*Ustilago kamerunensis*) (Farrell *et al.*, 2001) and stunting disease caused by phytoplasma (Jones *et al.*, 2004), which results in severe yield losses. Thus, the need for breeding cultivars resistant to these serious diseases is obvious. Cultivars 'Kakamega 1' and 'Clone 13' have shown resistance to head smut (Mwendia *et al.*, 2006).

13.5.5 Harvesting and yield

The yield estimates vary depending on the climate, soil, cultivar and cultural practices. As elephantgrass is susceptible to frost and low temperature, and in such climates the percentage of overwinter plant decreases if cut near the ground surface (Ishii *et al.*, 1995), Mukhtar *et al.* (2003) recommended a cutting height of at least 10 cm above the ground surface for the plant's overwinter survival. The harvesting of elephantgrass has been mostly carried out manually by smallholders. A silage chopper or some modification may be suited for mechanical harvesting.

Elephantgrass as biofuel feedstock will probably be harvested once a year to accumulate maximum amount of dry matter in the stem and ensure maximum return of minerals back to the soil. When grown for forage, elephantgrass is very palatable to animals at the early leafy stage but not much liked after stem development. The harvesting frequency differs by climatic conditions but should coincide with the timing of high forage quality and the duration between two cuttings depending on the time needed by the plant to regrow. The dry matter yield is higher if the plant is permitted to develop stems, but it decreases leaf blade to stem ratio, reducing the forage quality as a result of lowered protein content and digestibility (Sunusi *et al.*, 1997). In the subtropical southern USA (Florida), several reports suggest two cuttings per year for maximum biomass yield and significantly low yields when cut three times per season (Calhoun and Prine, 1985; Woodard *et al.*, 1991). The yield of dwarf elephantgrass has been reported significantly lower than the tall types (Cheng, 1991; Woodard *et al.*, 1991). Sollenberg *et al.* (1989) reported that dwarf elephantgrass was also suitable for continuous grazing and the live-weight gain for beef cattle of nearly 1 kg day^{-1} was greater than obtained from grazing on bahiagrass (*Paspalum notatum* Flügge).

Cheng (1991) has compared the yield and chemical composition of elephantgrasses (both tall and dwarf) with three other tropical grasses, viz., pangolagrass (*Digitaria decumbens*), pigenon grass (*Setaria sphacelata*) and guineagrass (*Panicum maximum*). Both dwarf and tall elephantgrass produced significantly higher fresh and dry weights and protein than the other three tropical grasses, somewhat lower in crude fibre, higher in P and K content. In comparison, dwarf elephantgrass hybrid strain '7301' produced lower fresh and dry weights (262.9 versus 319.5 t fresh weight and 39.2 versus 51.8 t dry weight), higher crude protein (17.4% versus 11.4%), similar crude fibre (23.9 versus 24.9) content and higher P and K percentages than tall cultivar 'A146'. Ansah *et al.* (2010) reported the following chemical composition for elephantgrass 120 days after planting (all metrics in g kg^{-1} dry matter): organic matter 928.6, ash 55.0, crude protein 79.9, NDF (neutral detergent fibre) 765.1, ADF (acid detergent fibre) 511.8, hemi cellulose 259.6, cellulose 354.6 and acid detergent lignin 151.0.

13.6 Uses

Elephantgrass is principally used for cut-and-carry fodder for animals. It is also sometimes cut for hay and to ferment into silage for dry-season feeding. Other than fodder and now potentially biofuel, elephantgrass has been used for multiple purposes (Table 13.2).

13.7 Genetics and Breeding

Exploitation of variation is the basis of all breeding. The most convenient way to import variation is from within species and therefore breeding begins with the acquisition of a genetic base population of diverse traits. In the case of elephantgrass, several research centres around the world, including Brazil, China, Puerto Rico, the USA, several African countries and India, maintain germplasm collections, but these collections are not well characterized. The breeder has to make sure that the genotypes in their germplasm bank are not just duplicates and the assessment of variability among the genotypes of the collection is thus essential.

Van de Wouw *et al.* (1999) studied eight agronomic and 20 morphological traits of 53 accessions of *Pennisetum purpureum* and *P. purpureum* × *P. glaucum* from International Livestock Research Institute (ILARI) field gene bank maintained at Zwai, Ethiopia in order to determine their variability for these traits. Several accessions were similar for all traits and appeared to be wrongly separated, other accessions varied significantly in both agronomic and morphological traits. They grouped the accessions into five, based on agronomic, and six based on morphological traits. Lowe *et al.* (2003) analysed 56 ILARI accessions of *P. purpureum, P. glaucum* and *P. purpureum* × *P. glaucum* using random amplified polymorphic DNA markers (RAPD). Of them, 49 showed none or very low intra-accession variability indicating the purity of clonally propagated genotypes. Several accessions had similar RAPD profiles and could be grouped. Logically, only one accession out of every group would be needed to preserve the diversity of the original collection. However, as Virk *et al.* (1995) cautioned, low RAPD diversity is not able to reflect the diversity at the quantitative trait loci (QTLs) controlling various important traits, and exercise of eliminating accessions from a particular collection should be corroborated by morphological/agronomic data (van de Wouw *et al.*, 1999). There was considerable amount of variability among groups. The neighbour-joining (NJ) dendogram clearly separated *P. purpureum, P. glaucum* and their hybrids. The main group of *P. purpureum* was separable into five significantly different (18.85 of total variance) subgroups. The germplasm collection had an overall Shannon's diversity index of 0.304 indicating a wide genetic base of the collection.

Researchers (Daher *et al.*, 2002; Xie and Liu, 2005) from different countries have reported significant differences among elephantgrass cultivars utilizing RAPD. Xie *et al.* (2009) analysed the variability among and within five existing existing elephantgrass cultivars in China including one elephantgrass × pearl millet hybrid using SRAP (sequence-related amplified polymorphism) assay. The cultivars included introductions from the USA and Indonesia and two as a result of local breeding. The results showed the presence of significant genetic variations within cultivars, but the degree of this type of variation differed with the cultivar. The variation among cultivars was greater than within cultivars. The percentage of polymorphic bands (PPB) at species level was 72.8%, indicating high level of variation among genotypes. They attributed it to the propagation of elephantgrass through clones and thus limited opportunity for sharing of genes as is the case among sexually propagated out-crossing species. Babu *et al.* (2009) analysed the genetic diversity among 30 elephantgrass accessions of wide geographical distribution. The percentage of polymorphic marker loci for RAPD and ISSR (inter-sequential simple repeat) markers were 86.6 and 90.3%, respectively. They attributed higher diversity values observed in their study than found by others to the selection of accessions of diverse origins. Summarizing, it can be stated that there is considerable variation in

Table 13.2. Miscellaneous uses of elephantgrass.

Use	Reference	Remark
Natural fibre	Reddy *et al.* (2012)	High tensile strength of the alkali-treated fibre suited for use in reinforcement materials
Phytoremediation	Ayotamuno *et al.* (2006)	Phytoremediation of petroleum-contaminated soils
	Zhang *et al.* (2010)	Phytoextraction of Cd and Zn from contaminated soils
Organic functional juice	Tsai *et al.* (2008)	Mostly in South-east Asia, favoured because of antioxidant properties, leaves high in polyphenols
Pulp for paper	Madakadze *et al.* (2010)	Has been found to be suitable raw material but still in experimental stage
Windbreak	Karschon and Heth (1958)	To protect field crops and bananas from high wind damage
Hedge	Magcale-Macandog *et al.* (1998)	Erosion control; prevent nutrient runoff
Alley crop in agroforestry	Magcale-Macandog *et al.* (1998)	Harvested for forage
Household fence and screen		Mostly in Africa; for privacy
Mixed with mud to build house walls		The grass reinforces the mud for strength
Soup and stew	Burkill (1994)	Young leaves and stem make the ingredient
Crop mulch	Nakagawa *et al.* (1992); Nishimoto (1994)	Mostly used in East Africa; thick mulch laid out for weed control in field crops; mulch also adds organic matter to the soil; mostly horticultural crops such as coffee, plantain, lettuce, etc. mulched
Herbal medicine	Burkill (1994); CHCRB (2003)	In Africa and China; nine cultivars registered with Chinese Herbage Cultivar Registration Board
Diuretic	Burkill (1994)	Mostly in Africa, plant extract highly diuretic and used as medicine for relieving stomach congestion
Feedstock for paper	Zhou *et al.* (2007)	Mostly in China
Culture substrate for mushroom production	Zhou *et al.* (2007)	Mostly in China
Fish-food ingredient	Chikafumbwa *et al.* (1993)	Africa and Asia
Trap plant in the management of African stem borer (*Busseola fusca*)	Khan *et al.* (2007)	Major cereal pest
Growth support for vine crops	Matias and Matias (1995)	Support for *Teramnus labialis* (L.) in Cuba and other countries

the germplasm collections around the world to facilitate future breeding for improved traits.

The first breeding of elephantgrass was undertaken in Africa during the 1930s and many of the selections from that period are still in production. Thereafter several cultivars such as 'Merkeron', 'Mott', 'Minna', 'Taiwan', 'Napier', 'Puerto Rico', 'Merker', 'Pusa Napier No.1' and 'Cameroon' have been released from different parts of the world. Several breeding options are available for the improvement of elephantgrass germplasm. Open-pollinated progeny selection, hybridization and selection from F_1 generation, hybridization and selection from F_2 generation have resulted in improved cultivars (Grof, 1969; Reynolds and Sini, 1976; Hanna and Monson, 1988). Selfing is another alternative for obtaining superior recombinants and transgressive segregants. A large population may be required for evaluation because common offspring should be expected to suffer from inbreeding depression because of the high level of parental heterozygosity. Burton (1990) developed a widely distributed tall elephantgrass cultivar 'Merkeron', from many crosses made in 1934 between a high performing tall selection No.1 and a very leafy dwarf No. 208 and selecting the best among many hybrids to obtain a vegetatively propagated F_1 hybrid with desirable forage traits. Cultivar 'A146' has been an important cultivar in Taiwan. Subsequently, strain 7262 from an open-pollinated population of cultivars 'A146', 'A148' and 'A149' was developed that produced 20% more dry matter than 'A146', has very few hairs on leaves, flowers late and grows erect (Cheng, 1991). Selfed progeny of 'Merkeron' has produced a number of leafy dwarf releases, 'Mott' being the best known. 'Mott' is a selection made by W.W. Hanna in 1977 (Hanna and Monson, 1988). It has high yield, is very nutritious and possesses good forage quality. It is also suitable for grazing with good animal weight gain.

The aim of elephantgrass breeding programmes has been to improve its value as forage. The objectives have been mainly twofold: (i) increase the leafiness of the plant by dwarfing and thus reducing the proportion of stem in relation to the leaf; and (ii) increase the palatability of the leaves by reducing the coarseness, hairiness and sharpness of the leaf edges. In breeding for biofuel feedstock, the goal will be to develop cultivars with higher proportion of stem at harvest than leaves, as stems have the two desired constituuents, higher cellulose percentage and lower ash content than the leaves (Ansah *et al.*, 2010). There has not been much breeding specifically for disease resistance. The selected offspring is propagated by clones as the plant is cross-pollinated producing a very heterozygous population of not true-to-type seedlings.

The chromosomes in the A'A' genome of elephantgrass is homologous to the AA genome of pearl millet (*Pennisetum glaucum*) (Jauhar, 1981). These two species hybridize naturally. The hybrids are triploids (AA' B genome) and as a result sterile. Many hybrids between the two species have been produced to assemble together desired traits and the hybrids are popularly called 'Bana'. Osgood *et al.* (1997) describes a scheme for producing a elephantgrass × pearl millet hybrid based on exploiting the cytoplasmic-nuclear male sterility (CMS) in pearl millet. Alternate rows of elephantgrass and CMS pearl millet are planted and seeds produced from this arrangement are an interspecific hybrid. Varied frequencies of hybrids are lethal. Elephantgrass parents that produce very low percentage or no lethal hybrid should be selected as parent. Although pearl millet inbreds have been used in crosses, the hybrids are highly variable because of the heterozygosity of elephantgrass clones. The elephantgrass × pearl millet hybrids are favoured for forages they combine the superior forage quality of pearl millet with the high yielding ability of elephantgrass (Gupta and Mhere, 1997). A number of pearl millet × elephantgrass hybrids were released from India between 1989 and 2000; 'CO-3', 'PBN-233' and '616' were among the releases during this period (Premaratne and Premalal, 2006). 'Gimu-1', a triple hybrid (pearl millet cv. 'Tift 23A' × 'N51') x Mott, was registered in China during 2000 (CHCRB, 2001). The hybrid resulted from a cross between pearl millet and 'N51' elephantgrass again crossed with 'Mott' to bring in more of the genetic material of elephantgrass for higher yield. How useful pearl millet × elephantgrass crosses will be for developing biofuel feedstocks is uncertain as pearl millet has been sought for forage quality and not high dry matter production.

Efforts have also been made to transfer the apomictic trait present in the genus *Pennisetum* to elephantgrass in order to produce true-breeding seed-propagated cultivars of elephantgrass or of elephantgrass × pearl millet hybrids. Hanna *et al.* (2004) reported success in crossing *Pennisetum squamulatum*, an apomictic species with low level of sexuality, to produce a partially seed-fertile apomictic interspecific hybrid. It should be possible to transfer the gene controlling the apomictic trait into elephantgrass through repeated backcrossing.

13.8 Invasiveness

What makes a certain plant species a weed? Essentially, it is the rapid rate of growth right from the start to out-compete the cultivated species and, if not controlled at the early stage, capacity to completely take over the crop. Weeds also are hardy, able to grow under adverse soil and climatic conditions, resistant to insects and diseases and possess the ability to respond aggressively to inputs. These are exactly the same characteristics desired in potential biomass feedstock crop species in order to produce high amounts of biomass. No wonder, then, that several of the cellulosic biomass feedstocks have been categorized as noxious weeds. Elephantgrass, one of the fastest growing grass species, was destined to occupy that distinction especially when produced in new environments where even the minimum levels of resistance which it encountered in the native environment were absent.

Elephantgrass has been recognized as a weed infesting 19 crops in 25 countries (Holm *et al.*, 1977). Infestation by elephantgrass of sugarcane in Hawaii, fruit orchards in Israel, maize and rice in Nigeria and coffee in Kenya has been reported (Farrell *et al.*, 2002). The threat posed to natural vegetation by elephantgrass in the Amazon basin has been narrated by D'Antonio and Vitousek (1992). The well-known report of its weedy character comes from south Florida (USA), where it poses a major concern to sugarcane cultivation, the main crop of the region. The Florida Exotic Pest Council lists it as an invasive species. Elephantgrass was introduced to the area by forage breeders as a new high-yielding forage species, but it did not find favour with the local cattle farmers over the existing species and was abandoned but not destroyed, and left uncared for. Due to its low seed potential the elephantgrass did not spread by seed, but the swampy terrain of south Florida provided an ideal way of expansion through plant segments carried to different locations by water. Once anchored in soil, elephantgrass can easily create its exclusive colony devoid of any other plant species. Elephantgrass can be seen growing on the banks of canals and waterways, and along roadsides throughout south Florida. The plant impedes water flow because of dense biomass, clogs waterways and is a hindrance to flood control. The weed infestation by elephantgrass of sugarcane fields comes from the similar growth habits of the two species, both being cane, and stems being the main mode of propagation. It was thus easy for elephantgrass growing in sugarcane field to get bundled with sugarcane cuttings intended for planting of new fields and thus expand the infestation.

When starting a large hectarage of elephantgrass cultivation for biofuel, it will be important to keep in mind that such plantations must not be situated near riparian zones, and farm sanitation procedures must be put in place to avoid spread of cuttings by farm machinery or biomass transportation vehicles. Farm machineries such as tractors, cultivators and mowers should be thoroughly cleaned before moving from elephantgrass sites to other crop areas. As elephantgrass propagates mostly from germination at the node, the harvesting procedure should be based on shredding or chipping so that nodes are disintegrated and unfit for growth. There should also be a crop escape response plan in place where large hectarages are to be planted to a biofuel species, probably in the vicinity of biorefineries. The response should be built across two components, rapid containment and then eradication spreading to several years. DiTomaso (2000) breaks down the eradication methods to mechanical and chemical. While mechanical methods may be effective in smallholdings, it may prove to be expensive and time consuming on larger areas. Usually, chemical control is a preferred option. In the case of elephantgrass, glyphosate, a non-selective herbicide or chemicals disrupting the synthesis of essential amino acids such as imidiazolinone and sulfonylurea should provide needed control.

13.9 Concluding Remarks

Elephantgrass is a fast-growing and high biomass-producing cellulosic source and thus a choice candidate for biofuel feedstock. However, it has a limitation placed on its edaphic distribution because of high susceptibility to frost. Due to its fast rate of growth, the plant manages to produce high biomass yield in sub-temperate climates where soil does not freeze in winter, killing the rhizomes. The fate of elephantgrass is similar to other high-yielding grass species dependent on the ability of biofuel to compete with the fossil fuels. Currently, their trade is highly dependent upon various factors outside of the market price support (e.g. subsidy, incentive, mandate, etc.). High yield of biomass feedstocks, establishment of a reliable supply chain, postharvest handling to minimize transportation cost and an efficient conversion process will all contribute towards improved competitiveness and eventual success of cellulosic biofuel. Elephantgrass biomass yield at present is higher than other feedstocks but scope for improvement is promising as the crop is basically unexploited for the high biomass trait. Cultural practices developed at present suit forage production. There is a need to separately establish cultural practices, including planting density, fertilization rate and time and frequency of harvest of elephantgrass intended for biofuel. Improved cold tolerance will enable the range expansion of this grass to wider geographical areas. Breeding for thickness of stem, plant ideotype for enhanced light distribution throughout the canopy, and close associative relationship with N-fixing bacteria to harness all available N sources should add to the value of elephantgrass as a biofuel feedstock. The next step to lower the cost of feedstock is through postharvest size reduction and densification at or near the centre of production to lower the cost of transport. However, the cost of densification at present is substantial due to the high energy need of the process; but efforts are being made to bring down this requirement. The pretreatment of lignocellulosic biomass is carried out in order to obtain fermentative sugars from the polymers (celluloses and hemicelluloses). (See Chapter 23, this volume, for recent innovations in densification and pretreatment.) Attempts are also being made to combine saccharification with fermentation, and genetically engineer *E. coli*, which ferments pentose as well as hexose sugars, and these applications have been tried on elephantgrass but optimal ethanol yield has not been achieved (Yasuda *et al.*, 2012). Lin *et al.* (2010) reported using *Klebsiella oxytoca* THLC0409 to generate ethanol directly from micro-sized elephantgrass powder and obtained an ethanol yield of 82 g kg^{-1}. These efforts will eventually yield a cost-effective conversion process for lignocellulosic biomass. Because of high yield, low input requirement, efficient N use and substantial CO_2 mitigation, elephantgrass undoubtedly will feature prominently among feedstocks of the flourishing biofuel industry.

Acknowledgement

The biofuel research at the Fort Valley State University is supported by a grant from USDA-NIFA-AFRI (GEOX-2010-03868) and authors are grateful for the support.

References

Anderson, W., Casler, M. and Baldwin, B. (2008) Improvement of perennial forage species as feedstock for bioenergy. Publications from USDA-ARS/UNL Faculty. Paper 248. Available at: http://digitalcommons.unl.edu/usdaarsfacpub/248 (accessed 22 June 2012).

Ansah, T., Osafo, E.L.K. and Hansen, H.H. (2010) Herbage yield and chemical composition of four varieties of Napier (*Pennisetum purpureum*) grass harvested at three different days after planting. *Agricultural and Biology Journal of North America* 1, 923–929.

Ayotamuno, J.M., Kogbara, R.B. and Egwuenum, P.N. (2006) Comparison of corn and elephant grass in the phytoremediation of a petroleum-hydrocarbon-contaminated agricultural soil in Port Harcourt, Nigeria. *Journal of Food, Agriculture and Environment* 3/4, 218–222.

Babu, C., Sundarmoorthi, J., Vijayakumar, G. and Ganesh Ram, S. (2009) Analysis of genetic diversity in Napier grass (*Pennisetum purpureum* Schum) as detected by RAPD and ISSR markers. *Journal of Plant Biochemistry and Biotechnology* 18, 181–187.

Baldani, J.I., Olivares, F.L., Hemerly, A., Reis, F.B. Júnior, Oliveira, A.L.M. *et al.* (1998) Nitrogen-fixing endophytes: recent advances in the association with graminaceous plants grown in the tropics. In: Elmerich, C., Kondorosi, A. and Newton, W.E. (eds) *Biological Nitrogen Fixation for the 21st Century*. Kluwer Academic Publishers, the Netherlands, pp. 203–206.

Bennetzen, J.L. and Freeling, M. (1997) The unified grass genome: synergy in synteny. *Genome Research* 7, 301–306.

Bogdan, A.V. (1977) *Tropical Pasture and Fodder Crops*. Tropical Agriculture Series. Longman, London.

Boonman, J.G. (1993) *East Africa Grasses and Fodders: Their Ecology and Husbandry*. Kluwer (Springer), Dordrecht, the Netherlands.

Burkill, H.M. (1994) *The Useful Plants of West Tropical Africa*. Royal Botanic Gardens, Kew, London.

Burton, G.W. (1989) Registration of 'Merkeron' napiergrass. [Reg. No. 119, PI 531087]. *Crop Science* 29, 1327.

Burton, G.W. (1990) Grasses: new and improved. In: Janick, J. and Simon, J.E. (eds) *Advances in New Crops*. Timber Press, Portland, Oregon, pp. 174–177.

Calhoun, D.S. and Prine, G.M. (1985) Response of elephantgrass to harvest interval and method of fertilization in the cooler subtropics. *Proceedings - Soil & Crop Science Society of Florida* 44, 111–115.

Center for New Crops and Plant Products (2002) *Pennisetum purpureum* K. Schumach. Perdue University, 4 pp. Available at: http://hort.purdue.edu/newcrop/duke_energy/Pennisetum_purpurium.html (accessed 22 June 2012).

Cheng, Y.K. (1991) Forage breeding in Taiwan. *Asian Journal of Arts and Sciences* 4, 203–209.

Chikafumbwa, F.J.K.T., Costa Pierce, B.A., Jamu, B.A., Kadongola, W.K. and Balarin, J.D. (1993) Investigations on the use of on-farm resources as pond inputs to culture *Tilapia rendall* and *Ochreochromis shiranus* on smallholder farms in rural Malawi. *Aquaculture* 117, 261–271.

Chinese Herbage Cultivar Registration Board (CHCRB) (2001) Registration of new cultivar of forage. *Acta Agrestia Sinica* 9, 311–318.

CHCRB (Chinese Herbage Cultivar Registration Board) (2003) Registration of new cultivar of forage. *Acta Agrestia Sinica* 11, 269–276.

Cunha, M.V., Andrade Lira, M., Santos, M.V.F., Freitas, E.V., Dubeux, J.C.B. Jr *et al.* (2011) Association between the morphological and productive characteristics in the selection of elephant grass clones. *Revista Brasileira de Zootecnia* 40, 482–488.

Daher, R.F., Pereira, M.G., Pereira, A.V. and Amraral, A.T., Jr (2002) Genetic divergence among elephantgrass cultivars assessed by RAPD markers in composite samples. *Scientia Agricola* 59, 623–727.

D'Antonio, C.M. and Vitousek, P.M. (1992) Biological invasions by exotic grasses, the grass/fire cycle, and global change. *Annual Review of Ecology and Systematics* 23, 63–87.

de Morais, R.F., de Souza, B.J., Leite, J.M., de Barros Soares, L.H., Alves, B.J.R., Boddey, R.M. and Urquiaga, S. (2009) Elephant grass genotypes for bioenergy production by direct biomass combustion. *Pesquisa Agropecuária Brasileira* 44,133–140.

DiTomaso, J.M. (2000) Invasive weeds in rangelands: species, impacts, and management. *Weed Science* 48, 255–265.

Döbereiner, J. (1988) Isolation and identification of root associated diazotrophs. *Plant and Soil* 110, 207–212.

Farrell, G., Simons, S.A. and Hillocks, R.J. (2001) Aspects of the biology of *Ustilago kamerunensis*, a Smut pathogen of Napier grass (*Pennisetum purpureum* K. Schum.). *Journal of Phytopathology* 140, 739–744.

Farrell, G., Simons, S.A. and Hillocks, R.J. (2002) Pests, diseases and weeds of Napier grass, *Pennisetum purpureum*: a review. *International Journal of Pest Management* 48, 39–48.

Ferrais, R. (1978) The effect of photoperiod and temperature on the first crop and ratoon growth of *Pennisetum purpureum* Schum. *Australian Journal of Agricultural Research* 29, 941–950.

Grof, B. (1969) Elephant grass for warmer and wetter lands. *Queensland Agricultural Journal* 94, 227–234.

Gupta, S.C. and Mhere, O. (1997) Identification of superior pearl millet by Napier hybrids and napiers in Zimbabwe. *African Crop Science Journal* 5, 229–237.

Hanna, W.W. (1981) Method of reproduction in napiergrass and the 3x and 6x alloploid hybrids with pearl millet. *Crop Science* 21, 123–126.

Hanna, W.W. and Monson, W.G. (1988) Registration of dwarf Tift N75 napiergrass germplasm. [GP-55, PI 517947]. *Crop Science* 28, 870–871.

Hanna, W.W., Chaparro, C.J., Mathews, B.W., Burns, J.C. and Sollenberger, L.E. (2004) Perennial *Pennisetums*. In: Moser, L.E., Burson, B.L. and Sollenberger, L.E. (eds) *Warm-season (C_4) Grasses*. Agronomy Monograms 45. ASA, CSSA, and SSSA, Madison, Wisconsin, pp. 503–545.

Holm, L.G., Plucknett, D.L., Pancho, J.V. and Herberger, J.P. (1977) *The World's Worst Weeds*. East-West Center, Honolulu, Hawaii, 609 pp.

Hung, L.L., Sylvia, D.M. and O'Keefe, D.M. (1990) Isolate selection and phosphorus interaction of vascular-arbuscular mycorrhizal fungi in biomass crops. *Soil Science Society of America Journal* 54, 762–768.

Ishii, Y., Ito, K. and Numaguchi, H. (1995) Effects of cutting date and cutting height before overwintering on the spring regrowth of summer planted napiergrass (*Pennisetum purpureum* Schumach). *Japanese Journal of Grassland Science* 40, 396–409.

Ishii, Y., Yamaguchi, N. and Idota, S. (2005) Dry matter production and *in vitro* dry matter digestibility of tillers among napiergrass (*Pennisetum purpureum* Schumach) varieties. *Japanese Journal of Grassland Science* 51, 153–163.

Ito, K. and Inanga, S. (1988a) Studies on dry matter production of napiergrass. I. Comparison of dry matter productivities and growth parameters between Tokyo and Miyazaki. *Japanese Journal of Crop Science* 57, 90–96.

Ito, K. and Inanaga, S. (1988b) Studies on dry matter production of napiergrass. III. Areas and light-photosynthesis relations in single leaves from different positions on a stem at Tokyo and Miyazaki. *Japanese Journal of Crop Science* 57, 431–437.

Ito, K. and Inanga, S. (1988c) Studies on dry matter production of napiergrass. IV. Direct-and after-effects of temperature on leaf growth and dry matter production. *Japanese Journal of Crop Science* 57, 699–707.

James, E.K. and Olivares, F.L. (1997) Infection and colonization of sugarcane and other graminaseous plants by endophytic diazotrophs. *Critical Reviews in Plant Science* 17, 77–119.

Jauhar, P.P. (1981) *Cytogenetics and Breeding of Pearl Millet and Related Species*. Alan R. Reiss, New York.

Jones, M.J. (1973) The organic matter content of the savanna soils of west Africa. *Journal of Soil Science* 24, 42–53.

Jones, P., Devonshire, B.J., Holman, T.J. and Ajanga, S. (2004) Napier grass stunt: a new disease associated with a 16Srx1 group phytoplasma in Kenya. *New Disease Reports* 9, 14.

Karlsson, S.B. and Vasil, I.K. (1985) Growth, cytology and flow cytometry of embryogenic cell suspension cultures of *Panicum maximum* Jacq. and *Pennisetum purpureum* Schumach plant. *Bulletin of the Institute of Tropical Agriculture, Kyushu University* 28, 15–20.

Karschon, R. and Heth, D. (1958) Wind speed, wind-borne salt and agricultural crops as affected by wind-breaks. *La Yaaran* 8, 8–13.

Kassanga, H. and Monsi, M. (1954) On the light-transmission of leaves, and its meaning for the production of matter in plant communities. *Japanese Journal of Botany* 14, 304–324.

Khan, Z.R., Midega, C.A.O., Wadhams, L.J., Pickett, J.A. and Mumuni, A. (2007) Evaluation of Napier grass (*Pennisetum purpureum*) varieties as trap plants for the management of African stemborer (*Busseola fusca*) in a push-pull strategy. *Entomologia Experimenatlis et Applicata* 124, 201–211.

Kirchhof, G., Eckert, B., Stoffels, M., Baldani, J.I., Reis, V.M. and Hartmann, A. (2001) *Herbaspirillum frisingense* sp. nov., a new nitrogen-fixing bacterial species that occurs in C4-fibre plants. *International Journal of Systematic and Evolutionary Microbiology* 51, 157–168.

Knoll, J.E. and Anderson, W.F. (2012) Vegetative propagation of napiergrass and energycane for biomass production in the Southeastern United States. *Agronomy Journal* 104, 518–522.

Kretschemer, E.A. and Pitman, W.D. (2001) Germplasm resources of tropical forage grasses. In: Sotomayor-Rios, A. and Pitman, W.D. (eds) *Tropical Forage Plant: Development and Use*. CRC Press, Gainesville, Florida, pp. 27–40.

Kubota, F., Matsuda, Y., Agata, W. and Nada, K. (1994) The relationship between canopy structure and high productivity in napier grass, *Pennisetum purpureum* Schumack. *Field Crop Research* 38, 105–110.

Lennè, J.M. (1990) *A World List of Fungal Diseases of Tropical Pasture Species*. Phytopathological Papers no. 31, CAB International, Wallingford, UK.

Lin, C.W., Tran, D.T., Lai, C.Y., Yet-Pole, I. and Wu, C.H. (2010) Response surface optimization for ethanol production from *Pennisetum alopecoider* by *Klebsiella oxytoca* THLC0409. *Biomass and Bioenergy* 34, 1922–1929.

Loomis, R.S., Williams, W.A., Duncan, W.G., Dovrat, A. and Nunez, A. (1968) Quantitative descriptions of foliage display and light absorption in field communities of corn plants. *Crop Science* 8, 352–356.

Lowe, A.J., Thorpe, W., Teale, A. and Hanson, J. (2003) Characterisation of germplasm accessions of Napier grass (*P. purpureum* and *P. purpureum* x *P. glaucum* hybrids) and comparison with farm clones using RAPD. *Genetic Resources and Crop Evolution* 50, 121–132.

Madakadze, I.C., Masamvu, T.M., Radiotis, T., Li, J. and Smith, D.L. (2010) Evaluation of pulp and paper making characteristics of elephant grass (*Pennisetum purpureum* Schumack) and switchgrass (*Panicum virgatum* L.). *African Journal of Environmental Science and Technology* 47, 465–470.

Magcale-Macandog, D.B., Predo, C.D., Menz, K.M. and Calub, A.D. (1998) Napier grass strips and livestock: a bioeconomical analysis. *Agroforestry Systems* 40, 41–58.

Mannetje, L. 't (1992) *Pennisetum purpureum* Schumack. In: Mannetje, L. 't and Jones, R.M. (eds) *Plant Resources of Southeast Asia, Forages*. Pudoc Science Publishing, Wageningen, the Netherlands.

Martel, E., De Nay, D., Siljak-Yakovlev, S., Brown, S. and Sarr, A. (1997) Genome size variation and basic chromosome number in pearl millet and fourteen related *Pennisetum* species. *The Journal of Heredity* 88, 139–143.

Masuda, Y., Kubota, F., Agata, W. and Ito, K. (1991) Analytical study on high productivity of napier grass (*Pennisetum purpureum* Schumach) 1. Comparison of the characteristics of dry matter production between napier grass and corn plants. *Journal of Japanese Grassland Science* 37, 150–156.

Matias, C. and Matias, Y. (1995) Effect of support on seed production in *Teramnus labialis* cv. Semilla Clara I. Choice of supports. *Pastos y Farrajes* 18, 51–57.

Miyagi, E. (1983) Studies on the productivity and feeding value of napier grass (*Pennisetum purpureum* Schumach) 2. The effect of nitrogen fertilization on the nutritive value napier grass. *Journal of Japanese Grassland Science* 29, 232–240.

Mohammad, N., Butt, N.M. and Qamar, I.A. (1988) Effect of nitrogen fertilization and harvesting intervals on the yield and nutritional value of Napier grass. *Pakistan Journal of Agricultural Research* 9, 478–482.

Mukhtar, M., Ishii, Y., Tudsri, S., Idota, S. and Sonoda, T. (2003) Dry matter productivity and overwintering ability of dwarf and normal napiergrass as affected by planting density and cutting frequency. *Plant Production Science* 6, 65–73.

Mureithi, J.G., Taylor, R.S. and Thorpe, W. (1995) Productivity of alley farming with leuceaena (*Leucaena leucocephala* Lam. De Wit) with Napier Grass (*Pennisetum purpureum* K. Schum.) in coastal lowland Kenya. *Agroforestry Systems* 31, 59–78.

Mwendia, S.W., Wanyoike, M., Nguguna, J.G.M., Wahome, R.G. and Mwangi, D.M. (2006) Evaluation of Napier Grass cultivars for resistance to Napier head smut. *Proceedings of the 10th KARI Biennial Scientific Conference*, KARI, Nairobi, Kenya, pp. 85–97.

Nagasuga, K. (2005) Acclimatization of biomass productivity to growth light intensity condition in napiergrass (*Pennisetum purpureum* Schumach) plant. *Bulletin of the Institute of Tropical Agriculture, Kyushu University* 28, 15–20.

Nagasuga, K. and Kubota, F. (2006) Change in hydraulic resistance and shoot morphology of napiergrass (*Pennisetum purpureum* Schumach) under shaded condition. *Plant Production Science* 9, 364–368.

Nakagawa, J., Prochnow, L.I., Bull, L.T. and Boas, R.L.V. (1992) Effects of organic compost on a lettuce (*Lactuca sativa* L.) crop. Part I. *Ceientifica Jaboticabal* 20, 173–180.

Nishimoto, R.K. (1994) Weed control in coffee plantations. In: Labrada, R., Caseley, J.C. and Parker, C. (eds) *Weed management for Developing Countries*. FAO Plant Production and Protection Paper 120. Food and Agriculture Organization of the United Nations, Rome, Italy.

Njoka-Njiru, E.N., Njarui, M.G., Abdulrazak, S.A. and Mureithi, J.G. (2006) Effect of intercropping herbaceous legumes with Napier grass on dry mater yield and nutritive value of the feedstuffs in semi-arid region of eastern Kenya. *Agricultura Tropica et Subtropica* 39, 255–267.

Nyambati, E.M., Muyekho, F.N., Onginjo, E. and Lusweti, C.M. (2010) Production, characterization and nutritional quality of napier grass [*Pennisetum purpureum* (Schum.)] cultivars in Western Kenya. *African Journal of Plant Science* 4, 496–502.

Osgood, R.V., Hanna, W.W. and Tew, T.L. (1997) Hybrid seed production of pearl millet x napiergrass triploid hybrids. *Crop Science* 37, 998–999.

Premaratne, S. and Premalal, G.G.C. (2006) Hybrid Napier (*Pennisetum purpureum* x *Pennisetum americanum*) Var. CO-3: A resourceful fodder grass for dairy development in Sri Lanka. *The Journal of Agricultural Science* 2, 22–33.

Prine, G.M., Dunavin, L.S., Brecke, B.J., Stanley, R.L., Mislevy, P. *et al.* (1988) Model crop systems: sorghum, napiergrass. In: Smit, W.H. and Frank, J.R. (eds) *Methane from Biomass: A System Approach*. Elsevier, New York, pp. 83–102.

Quesada, D.M. (2005) Parâmetros quantitativos e qualitativosda biomassa de genótipos de capim-elefante (*Pennisetum purpureum* Schum.) com potencial para uso energético,na forma de carvão vegetal. Tese (Doutorado) Universidade Federal Rural do Rio de Janeiro, Seropédica, 86 pp.

Ra, K., Shiotsu, F., Abe, J. and Morita, S. (2012) Biomass yield and nitrogen use efficiency of cellulosic energy crops for ethanol production. *Biomass and Bioenergy* 37, 330–334.

Ramírez, R.G., Haenlei, G.F.W., García-Castillo, C.G. and Núñez-González, M.A. (2010) Protein, lignin and mineral contents and *in situ* dry matter digestibility of native Mexican grasses consumed by range goats. *Advances Investigación en Ingeniería* 13, 98–101.

Reddy, K.O., Maheshwari, C.M., Shukla, M. and Rajulu, A.V. (2012) Chemical composition and structural characterization of napier grass fibers. *Materials Letters* 67, 35–38.

Reynolds, S.G. and Sini, O.V. (1976) A comparison of local and hybrid elephantgrass. *Alfan. Agricultural Bulletin* 1, 1–4.

Robertson, G.P. and Grace, P.R. (2004) Greenhouse gas fluxes in tropical and temperate agriculture: the need for a full-cost accounting of global warming potentials. *Environment, Development and Sustainability* 6, 51–63.

Saeki, T. (1960) Interrelationships between leaf amount, light distribution and total photosynthesis in a plant community. *Botanical Magazine, Tokyo* 73, 55–63.

Shultz, H.R. and Mathews, M.A. (1997) High vapor pressure deficit exacerbates xylem cavitation and photoinhibition in shade grown *Piper auritum* H.B. &K. during sunflecks. I Dynamics of plant water relations. *Oecologia* 110, 312–319.

Sjöström, E. (1993) *Wood Chemistry: Fundamentals and Applications*. Gulf Publishing, Houston, Texas, 293 pp.

Skerman, P.J. and Riveros, F. (1990) *Tropical Grasses*. FAO Plant Production and Protection Series 23. Food and Agriculture Organization of the United Nations, Rome, 832 pp.

Sollenberg, L.E., Prine, G.M., Ocumpaugh, W.R., Hanna, W.W., Jones, C.S. *et al.* (1989) Registration of 'Mott' dwarf elephantgrass. *Crop Science* 29, 827–828.

Sunusi, A.A., Ito, K., Tanaka, S., Ishii, Y., Uneno, M. and Miyagi, E. (1997) Yield and digestibility of napiergrass (*Pennisetum purpureum* Schumach) as affected by the level of manure input and cutting interval. *Journal of Japanese Grassland Science* 43, 209–217.

Sunusi, A.A., Ito, K., Ishii, Y., Uneno, M. and Miyagi, E. (1999) Effects of the level of fertilizer input on dry matter productivity of two varieties of napiergrass (*Pennisetum purpureum* Schumach). *Journal of Japanese Grassland Science* 45, 35–41.

Tiley, G.E.D. (1989) Factors affecting the productivity of elephantgrass. In: Desroches, R. (ed.) *Proceeding16th International Grassland Congress*, 4–11 October 1989, Nice, France. The Grassland Society, Versailles, France, pp. 599–600.

Tsai, W.T. (2009) Coupling of energy and agricultural policies on promoting production of biomass energy crops and grasses in Taiwan. *Renewable Sustainable Energy Review* 13, 1495–1503.

Tsai, P.-J., Wu, S.-C. and Cheng, Y.-K. (2008) Role of polyphenols in antioxidant capacity of napiergrass from different growing seasons. *Food Chemistry* 106, 27–32.

Turhollow, A.F. (1991) Screening herbaceous lignocellulosic energy crops in temperate regions of the United States. *Bioresource Technology* 36, 247–252.

van de Wouw, M., Hanson, J. and Luethi, S. (1999) Morphological and agronomic characterization of a collection of Napier grass (*Pennisetum purpureum*) and *P. purpureum* x *P. glaucum*. *Tropical Grasslands* 33, 150–158.

Vicente-Chandler, J., Silva, S. and Figarella, J. (1959) The effect of nitrogen fertilization and frequency of cutting on the yield and composition of three tropical grasses. *Agronomy Journal* 51, 202–206.

Virk, P.S., Ford-Lioyd, B.V., Jackson, M.T. and Newbury, H.J. (1995) Use of RAPD for the study of diversity within plant germplasm collections. *Heredity* 74, 170–179.

Walsmsley, D.V., Sergeant, A.L. and Doorkeran, M. (1978) Effect of fertilizer on growth and composition of elephant grass (*Penniestum purpureum*) in Tobago. *West Indies Tropical Agriculture (Trinidad)* 55, 329–334.

Wetselaar, R. (1967) Determination of the mineralization of coefficient of soil organic nitrogen on two soils at Katherine, N.T. *Australian Journal of Experimental Agriculture and Animal Husbandry* 7, 266–274.

Williams, M.J. and Hanna, W.W. (1995) Performance and nutritive quality of dwarf and semi-dwarf elephantgrass genotypes in the south-eastern USA. *Tropical Grassland* 29, 122–127.

Woodard, K.R. and Sollenberger, L.E. (2008) *Production of biofuel crops in Florida: elephantgrass*. SS-AGR-297. Institute of Food and Agricultural Sciences, Gainesville, Florida.

Woodard, K.R., Prine, G.M. and Ocumpaugh, W.R. (1985) Techniques in the establishment of elephantgrass (*Pennisetum purpureum* Schumach). *Soil Crop Science Society of Florida, Proceedings* 44, 216–221.

Woodard, K.R., Prine, G.M., Bates, D.B. and Chynoweth, D.P. (1991) Preserving elephantgrass and energycane biomass as silage for energy. *Bioresource Technology* 36, 253–259.
Xie, X.M. and Lu, X.L. (2005) Analysis of genetic relationships of cultivars in *Pennisetum* by RAPD markers. *Acta Prataculturae Sincia* 14, 52–56.
Xie, X.M., Zhou, F., Zhang, X.Q. and Zhang, J.M. (2009) Genetic variability and relationship between MT-1 elephant grass and closely related cultivars assessed by SRAP markers. *Journal of Genetics* 88, 281–289.
Yasuda, M., Miura, A., Shiragami, T., Matsumoto, J., Kamei, I., Ishii, Y. and Ohta, K. (2012) ethanol production from non-pretreated napiergrass through a simultaneous saccharification and fermentation process followed by a pentose fermentation with *EscheriColi* K011. *Journal of Bioscience and Bioengineering* 20, 1–5.
Zhang, X., Xia, H., Li, Z., Zhuang, P. and Gao, B. (2010) Potential of four forage grasses in remediation of Cd and Zn contaminated soils. *Bioresource Technology* 101, 2063–2066.
Zhou, F., Dong, Z.X. and Xie, X.M. (2007) Review on the research and utilization of *Pennisetum purpureum* cv. Mott in tropical and subtropical areas. *Grassland and Turf* 3, 76–82.

14 Bast and Leaf Fibre Crops: Kenaf, Hemp, Jute, *Agave*, etc.

Pratik Satya[1] and Ratikanta Maiti[2]
[1]*Central Research Institute for Jute and Allied Fibres, India;*
[2]*Vibha Seeds, India*

14.1 Introduction

Rapid depletion of fossil fuel sources, exponentially increasing consumption of fuels in the automobile sectors and negative impacts of fossil fuels on environment and health make the search for alternative sources of fuel an absolute necessity (Farrell *et al.*, 2006; Scragg, 2009). The green fuels, more commonly known as biofuels, offer the most potential substitutes for petrol and diesel, and release lesser amount of greenhouse gases (GHG) compared to fossil fuel. Production of biofuel from crop biomass started in the early 20th century in Brazil through mixing of ethanol derived from sugarcane into fuel. In the USA, a biofuel blend from corn alcohol was used in petrol (gasoline) in the 1920s. However, discoveries of large deposits of fossil fuel sources, particularly in central Asia, made petroleum-based fuels cheaper and readily available during the post-World War period. Since then, continuous depletion of fossil fuel, changes in geopolitical scenarios and environmental crises have, again, increased attention to biological resources for fuel production during the last decades of the 20th century.

The International Energy Agency has set targets and roadmaps to replace 27% of the transport fuel by biofuel before 2050, which would also reduce carbon dioxide (CO_2) emission by 2.1 gigatonnes (Gt) (Fairley, 2011). However, the current use of biofuel to drive automobiles is limited by a number of policies, driven from biofuel economics, ethical issues and precautionary measures adopted by different countries. The United Nations' report on Sustainable Bioenergy (UN-Energy, 2007) observes that the recent rise in food prices along with increased population pressure have increased the debates on driving food resources to biofuel production. Today, about one-third of the US maize harvest is channelled to biofuel production. Although conversion of food to biofuel is not the only factor responsible for escalation of food prices, it is no doubt a major concern limiting further production of biofuel from grain crops. In addition, extensive cultivation of these first generation biofuel crop species requires higher inputs, causes more GHGs and often tends to destabilize land use pattern (Graham-Rowe, 2011). To meet the rising demands of biofuel in the near future, large amounts of plant biomass feedstock are required as input. Considering the food versus fuel debate, supply of a large amount of renewable lignocellulosic plant biomass can only be obtained from non-food crop species. The fast-growing fibre crops synthesize additional lignocellulosic biomass in the form of vegetable fibres containing higher proportions of cellulose and hemicellulose compared to non-fibre

©CAB International 2013. *Biofuel Crops: Production, Physiology and Genetics* (ed. B.P. Singh)

crop species; hence they are extremely useful as biomass feedstock for producing biofuels such as biodiesel, ethanol or butanol.

The principal bast and leaf fibre crops of the world are jute (*Corchorus capsularis* L. and *Corchorus olitorius* L.), kenaf (*Hibiscus cannabinus* L.), roselle (*H. sabdariffa* L.), flax (*Linum usitatissimum* L.), sunnhemp (*Crotalaria juncea* L.), hemp (*Cannabis sativa* L.), ramie (*Boehmeria nivea* L. (Gaud.)) and various species of *Agave*, such as sisal (*Agave sisalana*), lechuguilla (*A. lechuguilla*), henequen (*A. fourcroydes*) and blue agave (*A. tequilana*). Among these, kenaf, roselle, hemp, jute and *Agave* have high potential as second generation lignocellulosic biofuel crops.

14.2 Origin, Ecology and Distribution

The bast and leaf fibre-producing species mostly belong to different plant families. The origin, distribution and ecology of each fibre crop is different. *Hibiscus cannabinus* ($2n = 2x = 36$), commonly known as kenaf, mesta, Deccan jute or Guinea hemp, originated in Africa, possibly from another wild progenitor *Hibiscus surattensis* before land migration and was domesticated there about 4000 BP (before present) (Maiti *et al.*, 2010). The origin of *H. sabdariffa* (roselle, mesta or Bimli jute), an allotetraploid species ($2n = 4x = 72$), is less defined; it probably evolved much later through separate polyploidization events in Africa and the Indian subcontinent. Kenaf and roselle are cultivated as fibre, paper pulp or as biofuel crops in China, India, Indonesia, Thailand, Vietnam, Malaysia, Brazil and Argentina.

There are two common types of *C. sativa* ($2n = 2x = 20$). The agricultural type is known as hemp or industrial hemp, while the narcotic/medicinal type is known as marijuana (marihuana). Hemp originated in central Asia and spread to northern Europe and Russia. In Europe, it was first cultivated in Finland, Sweden and Norway (Laitinen, 1996; Svennerstedt and Svensson, 2006). Fibres and clothes of hemp dating back to 3000–2700 BCE have also been recovered from Korea, suggesting it was also known for fibre in Asia (Clarke, 2006). Due to the narcotic properties of the plant, its cultivation is banned in many countries. Hemp production is primarily limited to temperate regions of the world, particularly in China, European countries (UK, France, Spain, the Netherlands, Denmark, Norway, Finland, Hungary, Romania and Poland), Canada, Turkey, Ukraine, North and South Korea and Chile. Russia, which was earlier one of the largest producers of hemp, banned its cultivation during the 1960s, but has relaxed restrictions in recent years. Similarly, any type of *C. sativa* cultivation is banned in the USA. The laws for banning narcotic hemp complicate popularization of industrial hemp cultivation. Recognizing the multiple use of industrial hemp, a number of countries such as Canada and Germany have amended drug control acts during recent decades to allow hemp cultivation. Presently about 30 countries allow cultivation of industrial hemp (Johnson, 2010).

The two jute species also exhibit wide variation in origin and distribution. *Corchorus olitorius* ($2n = 2x = 14$), the golden jute or tossa jute and most other wild *Corchorus* species, originated in Africa, while *C. capsularis* ($2n = 2x = 14$), the white jute, is proposed to have evolved in south China (Maiti *et al.*, 2010). The Indo-Burma region is considered as a secondary centre of origin of both *C. olitorius* and *C. capsularis* (Kundu, 1956). Both these crops are widely distributed in different parts of the world and have high adaptability to diverse climatic and soil conditions. However, large scale cultivation of jute is primarily limited to the Indian subcontinent including India, Bangladesh, Nepal and parts of China, Thailand, Myanmar and Vietnam.

The *Agave* species (*A. sisalana, A. lechuguilla, A. angustifolia, A. fourcroydes, A. salmiana, A. tequilana* and *Furcrea gigantia*) originated in Central America, particularly in Mexico. The word sisal finds its origin from the Yucatan port of Sisal, a place of *Agave* fibre export (Batra, 2007). *Agave* is native to tropical and subtropical America and Caribbean islands with the maximum concentration in Mexico (10°N to 25°N latitude). Sisal (*A. sisalana*) is more widely distributed in different countries of the world, particularly in the tropics. It is considered to be a natural hybrid between *A. angustifolia* and *A. kewensis* (Gentry, 1982). Lechuguilla and henequen cultivation are mainly limited to Mexico and Cuba while sisal is cultivated primarily in Brazil, Ethiopia, Kenya, Tanzania, Madagascar, India and Indonesia. Since the 1960s, the area under sisal has considerably fallen upon invention of synthetic polypropylene compounds. *Agave tequilana* is used to produce a very popular Mexican beverage known as tequila. Many other *Agave* species are also used to produce local beverages such as mescal and pulque.

14.3 Importance as Biofuel Crop

Due to restrictions in production of biofuels from food crops, alternative non-edible plant sources for biofuel production are a priority in current biofuel research. Instead of engulfing farm lands dedicated for providing food and nutrition, sustainable biofuel production emphasizes cultivation of bioenergy crops and maximization of waste biomass utilization for biofuel production. The miscellaneous fibre crops (kenaf, jute, hemp, *Agave*, etc.) fit adequately to these criteria, either as whole biomass or as waste by-products. In all these fibre crops, the harvestable main product (fibre) is well below 20% of the biomass. For example, the ratio of dry fibre weight to green biomass in kenaf is around 1:20, while the ratio of dry fibre weight to dry biomass is 1:6. The rest is either wasted or used for energy-inefficient direct combustion. Thus these crops have a potential biomass reserve, which can be captured for biofuel production, reducing requirement of food crops. Moreover, the convertible cellulosic materials of the whole or waste biomass from fibre crops are either higher or comparable to many dedicated biofuel and bioenergy crops. Recent progress in enzymatic degradation of lignocellulosic plant material is an additional boost to increase biofuel production efficiency from these crops. Considering these, a number of countries have already initiated commercial ventures in biofuel production from these species.

14.4 Botanical, Physiological and Genetic Features

14.4.1 Vegetative and reproductive morphology

Kenaf

Both *H. cannabinus* and *H. sabdariffa* are annual herbaceous crop plants. The taproot is pivotal and secondary and tertiary roots are derived to give support to the crop and absorb nutrients. The stem could be round, flat, bristled/bristleless with different degrees of pigmentation. The plants exhibit little branching or no branching at close spacing and may reach a height of 2–5 m, depending on the environmental conditions and genotype (Fig. 14.1). The leaves are polymorphic, palmate, extremely lobulate with three or five lobes, and alternate on the stem with very deep venation. The plants are generally ramified, but some wild types have high ramification. *Hibiscus sabdariffa* has a longer vegetative period (180–240 days) than *H. cannabinus* (160–200 days), is more tolerant to drought and can grow well under low precipitation (<500 mm annual rainfall). However, *H. cannabinus* is more tolerant to waterlogging. The flowers of both species are bisexual, big and of cream, yellow or pink colour, purple at the centre, reddish and yellow at the mouth. The capsules are five-locular with four to five seeds in locules. Each capsule contains 18–20 seeds. The nature of the apex of the bract, the calyx lobe, the form of calyx lobe, and the size and colour of the hilum are the distinctive characteristics of these species.

Kenaf is a photosensitive crop flowering under short-day conditions. Critical day-length for *H. cannabinus* is 12.5 h while for *H. sabdariffa* it is 14 h (Maiti *et al.*, 2010). To facilitate year-round production of kenaf for supplying to the paper pulp industries, photo-insensitive varieties of kenaf have been developed in China (Li *et al.*, 1996).

Hemp

Hemp is an annual plant, erect, that reaches a height of 1.5–4.5 m and exhibits sexual dimorphism. The plants are usually dioecious, but also produce monoecious male plants under

Fig. 14.1. (a) Kenaf (*Hibiscus cannabinus*) crop and (b) flower, (c) leaf and pod of roselle (*Hibiscus sabdariffa*).

adverse conditions, which are taller, slender and mature early (Moliterni *et al.*, 2004). The taproot of hemp can reach great depths under adequate conditions of humidity and soil, but if the prevalent conditions are not ideal, the root system is affected. The stem is thin, erect and hollow, and produces few branches in the apical part of the plant; the diameter varies from 4 to 20 mm in accordance with the climatic conditions and the variety. The leaves are compound and palmate, and have indentations in the margins. Each leaf has 7 to 10 leaflets. The flowers are dioecious, although sometimes monoecious plants appear; inflorescence occurs first in the stamens, but the pistillate flowers remain until the seeds mature (Maiti *et al.*, 2010). The fruit is classified as an achene; it does not present albumen and has one seed. At maturity, the pericarp breaks and separates from the testa of the seed.

Hemp is a photo- and thermosensitive crop. The vegetative growth phase is divided into a temperature-dependent basic vegetative phase and day-length-dependent photo-inductive phase (Lisson *et al.*, 2000). Hemp is a short-day plant, with a critical day length of 8–10 h. At a shorter photoperiod below the critical day-length, flowering time is influenced by temperature, but at longer day-length, flowering is considerably delayed (Lisson *et al.*, 2000). Monoecious genotypes flower earlier than dioecious genotypes, and among dioecious plants, female plants flower earlier than male flowers (Pahkala *et al.*, 2008).

Jute

Corchorus olitorius is an annual herbaceous plant. In its natural habitat the species is branched with cylindrical stems growing up to 90–120 cm. In cultivated conditions, the plant is unbranched with an erect, smooth, cylindrical stem, being slightly tapered at the top. The plant can grow up to 3.5–4.5 m within a short vegetative period of 4 months (Kundu, 1956) (Fig. 14.2). Leaves are elliptical-lanceolate or ovate with serration in the leaf margin. *Corchorus capsularis* is slightly shorter than *C. olitorius* under both natural habitat and cultivated conditions. The stem is erect and conical from the base and varies in plant height among the varieties, with a range between

Fig. 14.2. (a) Jute (*Corchorus olitorius*) crop, (b) flower and (c) pod of *C. olitorius* and (d) *Corchorus capsularis*.

3 and 4 m under cultivated conditions. Stem colour varies from green to dark red with various shades. Flowers are small, yellow and contain numerous anthers. The leaves have a bitter flavour due to the presence of compounds such as capsularin. Another compound imparting bitter taste is corchorin, which is present in seed of both *C. capsularis* and *C. olitorius*. Since the plant type is tall erect, support by a vigorously growing root system is essential for crop stand, therefore the primary and lateral roots of *C. olitorius* penetrate deeper into the soil than *C. capsularis* roots. *Corchorus capsularis* roots are less penetrating, but have numerous secondary roots, providing a bushy appearance; this helps withstand waterlogged conditions. Tossa jute yields more fibre per unit area compared to white jute. The fibre is softer, more lustrous and less rooty than white jute.

Agave

The *Agave* plants form a rosette of leaves arising initially from below the soil surface; the growing stem or the trunk pushes up the rosette of leaves. The roots are fibrous, fasciculated and extended. Stems are about 20–50 cm in diameter and can grow up to 2–3 m (Fig. 14.3). The leaves are green, sessile, flat, succulent, thick and fleshy, with small curved thorns. Mature sisal leaves grow 1–2 m with a width of 10–20 cm. The lechuguilla produces smaller leaves (30–50 cm long, 3–5 cm wide). *Agave salmiana* produces more fleshy and succulent leaves that bend down at the tip. Depending on species, climatic conditions and stress, *Agave* plants live from 6 to 35 years, at the end of which a terminal inflorescence axis with height varying from 4 to 8 m is produced. This process is termed as pole formation. Pole formation in sisal takes place between 6 and 35 years depending on climatic conditions. The flowers are yellow, with short pedicels. Upon completing the flowering phase, the *Agave* plants die and new plants are produced from seeds and bulbils.

14.4.2 Fibre development and maturation

The bast fibre in kenaf, jute and hemp is a cellulosic structure present in the stem, derived mainly from the cambial activity and the growth in length by the apical meristem. Fibre bundles are arranged radially in superimposed arcs along radii, tangential in the axis of the

Fig. 14.3. *Agave sisalana* (a) plant and (b) inflorescence.

stem, and the bundles are traversed by phloem rays (Maiti, 1979). In jute, 90% of the total fibre of the plant is developed in the secondary phloem by the activity of the cambium and the remaining 10% is formed in the protophloem region (Maiti *et al.*, 2010). The vegetative phase represents the period of greatest cambial activity. The fibre cells developed from cambium undergo rapid change in length and breadth. The secondary wall of the young fibre cells becomes thicker by the gradual deposition of secondary layers of cellulose materials. The outermost layers of the fibre bundles in the fibre wedge are protophloeic in origin (Maiti and Satya, 2009). The sieve tubes and companion cells are obliterated by the expansion and elongation of the surrounding cells. The procambium in the protophloeic region undergoes cell division and modification leading to the protophloeic fibres. This process continues towards the inner side, thereby crushing and obliterating all the sieve tubes and the companion cells and the whole of the protophloem is covered in solid fibre patches. The fibre strands make their way through the stem in a zigzag pattern and become connected here and there, forming a meshy structure of jute and other bast fibres. The degree of meshiness is associated with tangential expansion and radial growth of the stem. The meshes are looser towards the top and compact towards the basal region. The ultimate fibre cells, which form the building skeleton of the fibre filament, are pointed at the apex and have a lumen at the centre.

The hemp fibre has been classified by Dickison (2000) as primary cortical fibres and secondary phloem fibres. The primary fibre is thicker and longer than the secondary fibre. Unlike jute and kenaf, hemp fibres do not show a radial arrangement. The length of the fibres is influenced by internode length. Secondary fibre growth starts at a later phase after internode elongation has been completed.

In kenaf, the structure of the bundle surface is more irregular and has more intense reticulation than jute. The pattern of orientation of the fibre bundles is similar to that of jute, but kenaf varies in the structure and form of the fibre bundle surface. In *H. sabdariffa* the anatomical structure of the fibre cells has many interconnections among fibre bundles, resulting in a more abundant reticulation. Comparative anatomy of *H. cannabinus* fibre shows that the number of fibre cells per bundle and the number of fibre bundles per fibre wedge are related to the yield of fibre under similar environmental conditions (Maiti *et al.*, 2010). The morphology of ultimate fibre cells varies in several genotypes. In general, it is regular; the lumen varies from wide to narrow in accordance with the genotypes. Also, the tips of the ultimate fibre could be pointed or round, and the cell wall varies in thickness and undulations. The fibre bundle tenacity (force) increases with an increase in the length:breadth ratio (L/B) of the ultimate fibre.

Agave fibre grows longitudinally through the leaf, ranging from 1 to 2.5 m depending on climate, maturity period and leaf thickness. A mature leaf contains about 300–400 g of dry fibre (Batra, 2007). Two types of fibres are visible. The 'ribbon' fibres run through the central portion of the entire length of the leaf, split longitudinally during extraction, and are coarser. The mechanical fibres are in the peripheral regions, which also run continuously, but are nearly round and do not split easily during extraction. All the bast fibres contain higher amount of cellulose and hemicellulose and lower amount of lignin than the common biomass sources for biofuel such as wood, maize stover and wheat straw. Even the compositions of lignocellulosic materials in the core of these fibre crops are comparable to wood, maize or wheat straw. The bark or fibre regions of all the bast fibre crops have higher cellulose and hemicellulose content than the core, while the lignin content increases in the core region. Among these fibre crops, cellulose and hemicellulose content of hemp is highest, while these are lowest in leaf-fibre crops (Table 14.1).

14.4.3 Lignocellulosic biomass composition

The lignocellulosic biomass primarily comprises three major residues, cellulose, hemicellulose and lignin. The cellulose component, a polymer of D-glucose subunits, is linked by β-1,4 glycosidic bonds and is the simplest and largest polymeric component of the biomass. The cellulose strains are bundled together forming cellulose microfibrils, which are weakly linked together. Among these fibre crops, cellulosic components are high in hemp, which makes it a more efficient biofuel crop. The other fibre crops such as kenaf, jute and *Agave* also have high cellulose content. Hemicellulose, on the other hand, is a complex polymer of different types of pentose and hexose sugars and their acids. The dominant type of hemicellulose in fibre crops (both in bast and core) is xylan (Table 14.2). The xylan component of hemicellulose is easily extractable, which is a plus point in downstream processing of lignocellulosic biomass from these species. The hemicellulose component maintains linkage with cellulose and lignin with various secondary structure and bonding forces. The high xylan content compared to other complex hemicellulosic components (mannan, arabinan, etc.) of *Agave*, kenaf and jute is useful for conversion to biofuels such as ethanol and butanol.

Lignin is an amorphous polymer of polypropane (*p*-coumaryl, coniferyl and sinapyl alcohol) (Hendriks and Zeeman, 2009). Of the three components of the biomass, lignin has the lowest solubility and is most difficult for downstream processing. Thus high lignin content adds to the cost of biofuel production. The lignin content in bast and core of these fibre crops is much lower than that of wood

Table 14.1. Major cellulosic composition (% dry weight) of bast and leaf fibre species in comparison to wood and grasses (Han, 1998; Morrison *et al.*, 1999; Bakker and Elbersen, 2005; Sur and Amin, 2010).

	Chemical constituents			
Biomass type	Cellulose (%)	Hemicellulose (%)	Lignin (%)	Ash (%)
Kenaf fibre	58–63	14–24	10–14	2–5
Kenaf core	37–49	18–24	15–21	2–4
Roselle fibre	59–61	16–28	9–10	0.5
Hemp fibre	57–77	14–17	9–13	0.8
Hemp core	50–60	15–20	12–14	3–4
Jute fibre	58–63	18–21	12–14	0.5–1.2
Jute core	40–48	18–22	21–24	0.6–0.8
Agave	47–62	21–24	7–9	0.7
Maize stover	45	35	15–19	5–7
Wheat straw	33–40	20–25	16–20	6.5
Wood	40–49	7–26	23–34	0.2–0.5

Table 14.2. Principal carbohydrate content of kenaf, hemp, jute and lechuguilla in comparison to peat moss (values in mg g^{-1}) (Han, 1998; Morrison *et al.*, 1999).

Plant material	Glucose	Galactose	Xylan	Arabinan
Kenaf core	410–500	8.3	142.4	4.9
Kenaf fibre	460–520	4.6	108.0	20.4
Jute core	390.9	4.1	173.5	1.1
Jute fibre	568.7	4.9	121.7	1.1
Hemp, degum	838.1	13.4	11.8	2.6
Lechuguilla	418.4	10.3	173.3	4.5
Peat moss	191.6	25.4	27.7	2.5

(Table 14.1), therefore substitution of logwood components in biomass feedstock has the potential to reduce the cost of biofuel.

14.4.4 Genetics and biochemistry of important characters

Kenaf

Limited information is available on the genetics of economically important characters in kenaf and roselle. A linkage map of kenaf has recently been constructed and a few genes such as stem colour and petiole colour; leaf shape, petal dimension and petal shape have been mapped on linkage groups 5 and 6, respectively. Stem pigmentation in the roselle is controlled by a single dominant gene. Flower colour in kenaf is also controlled by a dominant gene, although digenic inheritance has been reported for pink flower (Falusi, 2008). Insensitivity to photoperiod in kenaf is controlled by a single dominant gene (Li *et al.*, 1996). The main determinants of fibre yield in kenaf are plant height and stem diameter. It has been reported that these two characters exhibit high general as well as specific combining ability and are controlled by both additive and dominance gene actions (Chen *et al.*, 2004). Additive gene action and high heritability is also observed for green weight per plant. However, dry stem weight and dry bark weight is reported to be controlled by dominant gene action (Chen *et al.*, 2004; Qi *et al.*, 2005). The chemical properties of the whole kenaf plant are indeed appealing for biofuel production (Table 14.1). The principal monomeric carbohydrate in kenaf is glucose. On maturity (180 days) kenaf fibre and core contain 460–520 mg g^{-1} and 410–500 mg g^{-1} glucose, respectively (Morrison *et al.*, 1999), higher than that of tree wood.

Hemp

The most important characters studied in hemp are sex types and THC (δ-9 tetrahydrocannabinol) content. Sex in dioecious hemp is determined by a pair of sex chromosomes, while in monoecious hemp it is controlled by autosomal genes. In dioecious hemp nine pairs of autosomes and one pair of sex chromosomes are present, the male being XY and female being XX type (Moliterni *et al.*, 2004). A 390 bp sequence characterized amplified region (SCAR) marker linked to the male phenotype has been developed for rapid identification of male plants (Mandolino *et al.*, 1999). The ratio of cannabinoids and THC was proposed to be controlled by a single gene (Becu *et al.*, 1998) or two codominant genes (Meijer *et al.*, 2003), although the environment exerts a lot of influence in the THC profile. The biofuel-related properties of hemp biomass, such as high energy efficiency, low ash content on burning and net calorific values, make it a highly suitable feedstock for biofuel production (Table 14.3).

Jute

The three major plant characters that have influenced evolution of present day elite jute cultivars are non-shattering property of pod, enhanced tolerance to premature flowering and non-branching plant type. Shattering habit

Table 14.3. Properties of hemp biomass (Burczyk *et al.*, 2008; Rice, 2008).

Property	Hemp whole biomass
Gross calorific value (MJ kg^{-1})	18.5
Net calorific value (MJ kg^{-1})	13.4
Ash content (% by weight)	3–4
Total carbon (% by weight)	43.1
Energy value (GJ t^{-1})	17–18
Energy efficiency (GJ ha^{-1})	254–259

is controlled by a single gene, non-shattering type being dominant over shattering type. *Corchorus capsularis* genotypes are comparatively more tolerant to premature flowering than *C. olitorius* genotypes. Resistance to premature flowering in *C. olitorius* was first identified in an African accession 'Sudan Green', which was then transferred to the new cultivars. Although the inheritance of this character is not well understood, insensitivity to photoperiod in jute is controlled by a single dominant gene. The branching habit in jute is controlled by a single gene, branched being dominant over non-branched type. Most of these reports suggest that fibre yield is controlled predominantly by dominance gene action, although additive gene actions are also important. Dry matter production in jute is controlled by both additive and dominance gene actions. Fibre strength and fibre fineness are controlled by both additive and dominant gene action but exhibit low heritability.

14.5 Genetic Improvement

14.5.1 Germplasm resources and traits of importance

Kenaf

The Institute of Bast Fiber Crops, China, maintains over 1800 accessions of kenaf, while the Central Research Institute for Jute and Allied Fibres, India, maintains over 600 working collections of *H. cannabinus* and 700 working collections of *H. sabdariffa*. Plant height and stem diameter are two major traits of importance, both of which exhibit high heritability and good response to selection. About 400 accessions are present in the US National Plant Germplasm System's regional Plant Introduction Station, Griffin, Georgia.

Hemp

Genetic variation in industrial hemp is limited. The largest hemp germplasm collection is in the Vavilov Institute (VIR) in Russia, holding about 500 accessions (Ranalli, 2004). Other notable collections are present in the USA, the Netherlands, Ukraine, Canada and Romania. About 45 varieties are registered as industrial hemp in Europe (Ranalli, 2004) and 38 varieties are registered in Canada (Health Canada, 2011). Systematic breeding efforts in the former USSR have led to improvement of hemp fibre yield by 35–37%. Dioecious hemp produces more fibre than monoecious hemp. Growth rate of hemp depends on photoperiod length. The plants flower under short-day conditions, thus a continuous long photoperiod will prolong the vegetative phase, which will increase biomass of the crop. For a fibre crop, indefinite growth may hamper growth and quality of fibre, but as a biofuel crop, plants continuously grown under long photoperiod are expected to accumulate more biomass. Plants with low branching are more suitable as a fibre crop, but greater branching and wider spacing are beneficial for better solar energy use and higher biomass production.

Jute

There are about 40–100 species in the genus *Corchorus*. However, only two species, *C. olitorius* and *C. capsularis*, are economically important as a fibre crop. A number of other *Corchorus* species such as *C. aestuans*, *C. fascicularis* and *C. tridens* are also used as green leafy vegetables in parts of Africa and Asia (Westfall-Stevels, 1986). Both *C. olitorius* and *C. capsularis* have high value as leafy vegetable crops, particularly in India, Bangladesh, South-east Asia and Africa. About 3000 accessions of jute are being maintained in the working germplasm collection of the Central Research Institute for Jute and Allied Fibres in India.

Agave

The *Agave* species exhibit high morphological similarity and are evolutionarily related. The primary genetic resources of *Agave* species are available in Mexico and parts of South America. Although these species have been transported and established elsewhere in the world, little genetic variability is observed in other parts of the world. *Agave tequilana* is a derivative of *A. angustifolia*.

14.5.2 Breeding methods and achievements

Kenaf

China has made considerable progress in developing high yielding and hybrid kenaf varieties. The popular kenaf varieties in China are 'Fuhong 952', '74-3', '71-4', 'T14', 'Taihong 763', 'J-1-113', 'Hongyin 135', 'BG52-135', 'Yue74-3', 'Xuan No.2', 'Minghong 298', 'Minghong 82/34', etc. A new variety of kenaf, 'China Kenaf 13', has recently been developed with high fibre yield (4.25 t ha^{-1}), disease and lodging resistances and good fibre quality. Hybrid breeding in kenaf has resulted in development of high yielding hybrids such as 'KB2' and 'KB11'.

Kenaf breeding efforts in the USA are rather recent, comprising a span of 25–30 years. During early breeding efforts in the USA, two popular kenaf varieties 'Everglades 41' and 'Everglades 71' were developed by the United States Department of Agriculture, Agricultural Research Service (USDA-ARS) researchers in Florida (Wilson *et al.*, 1965). These two varieties became popular not only in the USA but also in European countries. Both the varieties are resistant to anthracnose disease, produce high biomass (15–20 t ha^{-1} dry weight) and are suitable for biofuel production. A variety 'SF459' has been developed in the USA that exhibits high yield potential with tolerance to the southern root-knot nematode/soil-borne fungi complex. In India, 80% of the mesta (kenaf and roselle) is *H. sabdariffa* (roselle). Consequently it has received more attention from Indian plant breeders. A good number of varieties have been developed using pureline selection and pedigree breeding methods (Table 14.4). Using ethylmethane sulfonate (EMS)-induced chemical mutagenesis, a non-bristled variety 'Non Bris-4' has also been developed from 'HS 4288'. Among the Indian kenaf (*H. cannabinus*) varieties, 'HC 583' is most popular. A kenaf variety, 'MT 150' suitable for paper pulp production, has also been developed in India.

Hemp

The first successful breeding attempt in hemp was made by Bredemann in 1924, who targeted to improve performance of male plants in the population by early elimination of low-fibre-containing male plants. In dioecious species, identification of high-fibre-containing male parents is important for increasing population performance, since they will serve as pollen donors. This is popularly known as Bredemann's principle (Bredemann, 1924; Bocsa and Karus, 1998). Systematic crop improvement efforts around the world in hemp has led to substantial improvement of hemp fibre yield (>40%). The breeding methods in hemp are determined by the sexual status of the crop. Self-pollination (>20%) in monoecious genotypes leads to inbreeding following reduction in fibre yield. Intervarietal cross followed by population improvement following Bredemann's principle is the most successful breeding approach in hemp (Bocsa and Karus, 1998).

Cultivars 'Kompolti', 'Felina', 'Fedrina', 'Futura', 'Fedora' and 'Fibrimon' are widely grown in Europe (Table 14.4). Most of the French and Romanian cultivars are dual purpose, being suitable for both fibre and seed production. A number of varieties such as 'YUSO 14' and 'YUSO 16' were popular in the former USSR. 'YUSO 14' has been bred by family selection from 'YUSO1', while 'YUSO 16' has been developed from French genotype 'Fibrimon 56'. Although high heterosis is reported in hemp, hybrid breeding has only been successful in Hungary. A variety 'Uniko-B' has been developed by crossing 'Kompolti' and 'Fibrimon 21'.

Development of hemp varieties with low narcotic properties is a major objective in industrial hemp breeding. The industrial hemp

Table 14.4. Varieties of kenaf, hemp and jute.

Region	Kenaf	Hemp	White jute	Tossa jute
Europe	G-4, Cuba 108, SF 459, Everglades 41, Everglades 71, El Salvador, Endora	Fedora, Carmagnola, Epsylon, Felina 34, Fedrina, Futura 75, Futura 77, Fasamo, Chamaeleon, Silesia, Delta Llosa, Delta 405, Fibrimon 56, Beniko, Tygra, YUSO-14, Kompolti	Not cultivated	Not cultivated
Asia	Hongyin 135, KB2, KB11, Xuan No.2, 8310, Zhe Xiaoma No.1, Minghong 298, Minghong 82/34, Fuhong 952, Fuhong 4, HC-583, GR-27, AMC-108, MT-150	Generally not allowed for cultivation	JRC 321, JRC 212, JRC-7447, JRC-4444, UPC-94, Padma, KC-1, JRC-698, Bidhan Pat-3, Bidhan Pat-1, Bidhan Pat-2, JRC -80, C517, C 532, RRPS-27-C-3, NDC 2008, JBC-5, CVL-1, CVE-3, CC-45, BJC-7370, BJC-83	JRO 524, JBO-1, CO-58, JBO-2003-H, AAU-OJ-1, S-19, JRO-128, JRO-8432, JRO-66, KOM-62, JRO-3690, TJ-40, JRO-7835, JRO-878, JRO-632, JRO 204, O-4, O-9897, OM-1, O-72
North America	Everglades 41, Everglades 71, Tainung 1, Tainung 2, Cuba 2032, Guatemala 4, SF459, Gregg, Dowling	Anka, CanMa, Petera, Carmen, Fedora, Felina 34, USO-14, Kompolti, Fedrina, Futura 75	Not cultivated	Not cultivated

varieties of Europe contain low cannabinoids. Hemp breeders identify low cannabinoid breeding lines by thin layer chromatography of leaf extracts. 'Kuban', a dioecious hemp cultivar having very low cannabinoid content, has been developed in Russia by multiphase family/group selection.

The seed varieties are bred for high oil content (>35%), linoleic acid (>20%) and linolenic acid (>15%) content. Hemp oil is a rich source of polyunsaturated essential fatty acids (81%) (http://naihc.org). Hemp is primarily grown in North America (Canada) as seed crop. Health Canada (http://www.hc-sc.gc.ca) publishes every year the approved industrial hemp varieties for cultivation that contain low cannabinoids (less than 0.3% THC). Most of the Canadian varieties are European in origin, such as 'Fedrina', 'Felina', 'Ferimon', 'Futura', 'Fibrimon', 'Kompolti', 'USO 14', etc. However, new breeding efforts have led to development of a number of hemp varieties in Canada, such as 'Anka', 'CanMa', 'Petera' and 'Carmen'.

Jute

Although jute has been reported as a fibre crop in India since the medieval period, its cultivation was very limited and localized to small pockets in eastern India (Kundu, 1956).

Systematic jute improvement was started by R.S. Finlow in 1904 in the Burdwan district of West Bengal, India. From a local landrace, the first variety, 'Kakya Bombai', was developed in 1916. *Corchorus olitorius* was less preferred by farmers during these periods and the cultivation was limited to certain pockets. The first tossa jute variety, 'D-38', commonly known as 'Chinsurah Green' (CG), was developed in 1915 through pureline selection. It was the only tossa jute variety in general cultivation for a long period. Presently, *C. olitorius* is grown in 90% of the jute area of the world.

Major successes in jute breeding have been achieved by pureline selection, pedigree breeding and mutation breeding techniques releasing a good number of varieties (Table 14.4). In white jute, two very popular varieties, 'JRC-212' (fibre yield 2.0–2.5 t ha^{-1}) and 'JRC-321' (fibre yield 1.8–2.0 t ha^{-1}), were developed in India through selection and were released during 1954. The first mutant white jute variety 'JRC-7447' ('Shyamali') was developed through X-ray mutagenesis of 'JRC 212'. In tossa jute, several varieties have been developed through pedigree breeding using 'Sudan Green', a donor genotype for tolerance to premature flowering. Of these, 'JRO 524' ('Navin'), developed in 1977, is still the most popular jute variety in both India and Bangladesh. Even the recently developed varieties such as 'JBO-2003-H' and 'JBO-1' have been developed using 'Sudan Green' as a donor parent.

In Bangladesh, about 8–10 varieties of jute have been developed, although the principal jute variety grown is 'JRO 524', which is imported from India (Worldjute Report, 2011). White jute varieties 'CVL-1', 'CVE-3', 'CC-45', 'BJC-7370', 'BJC-83' and tossa jute varieties 'O-4', 'O-9897', 'OM-1' and 'O-72' have been developed, having the capability of producing 3–4.5 t ha^{-1} jute fibre.

Agave

Most of the *Agave* cultivars are of local origin. Some common local cultivars of henequen in Mexico are 'Sac Ki' (white henequen), 'Yaax Ki' (green henequen) and 'Chelem' (wild henequen). Long, unpredictable time gaps in flowering and difficulties in sexual hybridization are major problems in selective breeding of *Agave*. A sisal hybrid was developed in Tanzania in 1937 by crossing *A. amaniensis* with *A. angustifolia* and backcrossing the F_1 with *A. amanuensis*, which has higher productivity and matures earlier than sisal. This hybrid is widely cultivated all over the world. In India, an improved type of the hybrid ('Bamra Hybrid 1') is cultivated. Reduction in crop duration and increase in fibre yield are primary targets of sisal improvement.

14.6 Cultivation Practices

The objectives and methodologies adopted for cultivation of fibre crops and biofuel crops are different for each of the bast fibre crop species. In the case of fibre crops, the industrial requirement is long strands of fibre without disconnections, rootiness or hard regions. This is achieved by closer spacing so that the stems are non-branched, tall and erect, favouring harvest of long strands of fibre. In the case of a dedicated biofuel crop, whole crop biomass is more important, which will be increased by wider spacing. The same principle of wider spacing applies to crops produced for paper pulp industries, where whole biomass is the main harvest. While hemp and kenaf are grown for fibre, paper pulp or biofuel crops, the cultivation of jute is primarily limited as a fibre crop. However, given the capability of high cellulosic biomass production, jute also has a high potential as a biofuel crop. Under a favourable optimum cultivation environment jute produces more biomass per unit area than kenaf or hemp.

Kenaf

Kenaf grows better in a warm humid climate with a good loamy soil. Low temperature (<15°C) during germination adversely affects germination and seedling growth. For fibre production, a row spacing of 30–40 cm with plant density of 40–50 plants m^{-2} is recommended. For paper pulp production higher spacing (40–50 cm) and lower plant density (20–25 m^{-2}) gives better biomass yield. Kenaf gives a good response to nitrogen. Application of 40–100 kg N ha^{-1} gives good results depending on location and soil conditions

(Chew *et al.*, 1982). Due to its deep taproot system it can better remove residual nutrients from the previous crop. It requires a good amount of rainfall (500–700 mm) during the vegetative growth period. Prolonged water stress leads to rolling and drooping of leaves and reduces biomass yield. Like the other bast fibre crops jute and hemp, kenaf exhibits rapid vegetative growth and can outperform weed growth particularly after 25–40 days, but in warm humid climates it needs protection from weeds during early seedling growth. Kenaf biomass yield varies widely (6–26 dry t ha^{-1}) depending on climate, soil, growing season, plant density and crop management (Alexopoulou *et al.*, 2000). Kenaf is also suitable for multiple harvests by ratooning twice or thrice. Increase in duration of vegetative growth beyond 6–7 months and higher seed rate increases biomass yield, but negatively affects fibre yield and quality (Wood, 1978).

Hemp

The cultivation practices of fibre hemp differ in different countries according to environmental and soil conditions. Fibre hemp is sown either by broadcasting or by line sowing (row to row spacing 20–30 cm) using mechanical devices, with varying plant density of 100–300 plants m^{-2} (Dempsey, 1975; Martinov *et al.*, 1996). However, optimum plant density is reported at around 100 plants m^{-2}, as it provides maximum interception of solar radiation. Due to high interplant competition, total biomass at flowering stage does not change significantly with increase in plant density. For higher biomass production, plant density of 30–45 plants m^{-2} is more appropriate, which also reduces seed cost (Amaducci *et al.*, 2002). Among the mineral nutrients, nitrogen is most important for high biomass and fibre production. Higher applications of nitrogenous fertilizer increase plant height, biomass, high fibre content and ratio of male to female flowers, but reduce fibre quality. Phosphorus increases fibre yield and strength, and also increases the efficiency of nitrogen uptake. Pests and diseases are not a major problem in hemp cultivation. Harvesting of fibre hemp is done just before flowering or at initiation of flowering as the maximum amount of good quality fibre is obtained at this stage (Mediavilla *et al.*, 2001). Fibre productivity usually varies from 1 to 2.5 t ha^{-1} depending on the climatic conditions. Sowing during late March and early April with a seed rate of 25–30 kg ha^{-1} usually produces 10–12 t ha^{-1} dry matter under European conditions (Pahkala *et al.*, 2008; Rice, 2008). However, low temperature in northern European countries may delay sowing up to June.

Jute

Jute can grow on diverse soil types, but thrives best on well-drained alluvial loamy soils. It is sown by the line sowing method (row to row distance 25–30 cm) using a seed drill or is broadcast evenly. About 15–21 days after germination, extra seedlings are uprooted to maintain a plant to plant distance of 5–7 cm. Weeding is the most essential cultural operation for initial establishment and proper growth of the seedling. First weeding should be done within 15–21 days after germination either by hand weeding or application of herbicides. Application of N:P:K at the rate of 40–80: 40–60:40–60 kg ha^{-1} is optimum for crop growth and high fibre yield. The major pests of jute are jute semilooper (*Anomis sabulifera*), hairy caterpillar (*Spilosoma obliqua*), yellow mite (*Polyphagotarsonemus latus*) and stem weevil (*Apion corchori*), while stem rot caused by *Macrophomina phaseolina* is the major disease. The crop is harvested within 110–150 days after sowing depending on agroclimatic conditions. Late harvesting negatively affects fibre yield and quality. The fibre is extracted through retting in free-flowing water for 10–18 days. The retting process removes fibre from other parts of the stem through microbial digestion. It is highly influenced by the quality of retting water, materials used for retting and temperature of water. Fibre yield varies from 2.5 to 4.5 t ha^{-1}, while dry biomass yield ranges from 10 to 25 t ha^{-1}.

Agave

Agave plants can thrive under low rainfall (200–500 mm year^{-1}) conditions and grow well in fallows and hilly regions where food crops cannot be grown. In Mexico, the temperature in the *Agave* growing regions may go well below freezing point. The species grow well on sandy soil with pebbles; however, adverse growing conditions reduce fibre productivity (Maiti *et al.*, 1990). Sisal is asexually propagated by bulbils as seeds are rarely formed. Henequen reproduces sexually by seed and asexually by bulbils and suckers. The planting materials are put in furrows at a row distance of 2–4 m with a spacing of 1–1.5 m between plants. Sisal requires a good amount of fertilizer (N:P:K: 100–140:40–80: 150–200 kg ha^{-1}). Little fertilizer is applied to henequen by local farmers in Mexico during first years of establishment. Farmers grow other intercrops for a better return. Henequen grows more slowly than sisal. The leaves (15–20) are cut manually with a sickle at an angle of inclination of 40° or more (Villarreal and Maiti, 1991). The terminal and marginal thorns are removed in order to facilitate fibre extraction. Fibre yield of henequen and lechuguilla ranges from 1 to 1.5 t ha^{-1}, depending on the ecological conditions, age of the plant and methods of extraction. Yield of sisal fibre is around 800–2200 kg ha^{-1} (McGovern, 1990). About 10–12 cuttings (12–15 leaves in each cutting) are obtained in a life cycle (8–12 years) of a plantation having about 5000 plants ha^{-1}. Since 1 t of sisal fibre also yields 25 t of residue dry biomass, the sisal fibre waste is a big source for residue biomass utilization.

14.7 Primary Use of the Crops and Rationale for Use as Biofuel Feedstock

The bast fibre crops such as jute, kenaf, hemp and *Agave* are primarily grown for non-textile raw fibres for diversified uses, such as ropes, bags, twines, hessian, cordages, carpets, etc. In recent years, efforts have been initiated to increase alternative use of these fibres, including fibre-reinforced composites, geotextiles, textile blends and high aesthetic value products. Kenaf and hemp are also used to a limited extent in paper pulp industries. Oil from seeds of hemp and kenaf also has industrial applications. However, the demand for these fibre crops is industry driven, facing tough competition from synthetic fibres. The uncertainty in demand and small industrial requirement are the main limitations for increase in area and production of these crops. Many of these crops like kenaf and *Agave* are also cultivated only for limited local consumption as raw fibre used in daily household purposes. This demand-driven economy makes these crops very suitable for channelling to other industrial applications such as biofuel and paper pulp production. For example, after allowing hemp cultivation in Canada, the area expanded 15 times within a short period of 5 years (2001–2006), due to a requirement for hemp oil in the cosmetic industry. However, in Europe the hemp area remained stagnant over a long period as the primary industrial product is fibre, which has a limited demand due to high competition.

The bast and leaf fibre crops can be used as a dedicated biofuel crop, or the waste product can be channelled for biofuel production. Agricultural waste is a major source of eco-friendly biomass feedstock for the biofuel industry (Oliveira and Franca, 2009). In these fibre crops, non-fibre components comprise more than 75% of biomass, and are generally wasted or used primarily as a burning fuel or consumed for local household use. The residue coefficient (residue crop ratio) is considered as an important indicator of biomass waste, which is much higher in fibre crops than other cereal and non-cereal crops used as a food source. For example, the residue coefficients of hemp (2.5) and jute stick are much higher (1.85) compared to other lignocellulosic biofuel sources such as rice straw (0.62–1.53), sorghum stover (0.74) and sugarcane bagasse (0.25) (Cuiping *et al.*, 2004; Purohit *et al.*, 2006). Given the potential of biofuel production of these second generation biofuel crops, whole or waste biomass from fibre crops provide a good source of lignocellulosic feedstock. Moreover, the lignin components can additionally be channelled to production of valuable by-product, which can sustain the biofuel production.

A comparative estimate of biofuel and other bioenergy crops reveals that the fibre crops share most of the advantages of these crops, and can supply a good proportion of biomass feedstock in future biofuel production

(Table 14.5). The biomass yield of jute, kenaf and *Agave* is comparable to the major bioenergy crops such as poplar, switchgrass and *Miscanthus*, with better biomass quality (cellulosic material). The major advantages of using fibre crops over other plant biomass sources for biofuel production are as follows.

1. High adaptability of fibre crops such as kenaf and hemp to be cultivated in different environmental and soil conditions, thus allowing increase in cultivation area, and particularly the identification of new areas.
2. Non-grain source of biofuel, thus reducing the share of food-grain in biofuel production.
3. High percentage of cellulose and hemicellulose in biomass compared to non-fibre crop species and tree species.
4. Environmental sustainability, continuous year-round supply of biomass and long-term benefits (carbon credit, bioremediation, water conservation, etc.).
5. Fits well to regional and local biofuel production system and rotation with other annual bioenergy crops such as rapeseed-mustard or legumes, thus also benefiting small and marginal farmers.
6. Lower cost of biofuel extraction than tree species due to soft, herbaceous biomass.
7. Less time of establishment compared to perennial tree biofuel crops.

Agricultural biomass from fibre crops is a largely untapped sector for biofuel production. Although having high potential, the biomass potential of these crops has largely been neglected so far. The 'Asian Biomass Handbook', published by the Japan Institute of Energy in 2008, does not mention the potential of kenaf, hemp, jute or sisal in production of biofuel. Similarly, although a lot of private- and government-funded researchers are investigating the biofuel potential of bast fibre crops in the USA, the workshop summary on 'Expanding Biofuel Production and the Transition to Advanced Biofuels' published by The National Academies Press (Koshel and McAllister, 2010) bears no discussion on these fibre crops,

Table 14.5. A comparative study of fibre crops with major bioenergy crops (Pignatelli and Robertiello, 2007).

	Hemp	Kenaf	Jute	*Agave*	Poplar	Miscanthus	Switchgrass
Biomass yield ($t\ ha^{-1}\ year^{-1}$)	8–15	10–20	10–25	25–30	10–15	15–25	10–25
Establishment period (year)	<1	<1	<1	3+	3+	3+	2–3+
Fertilizer	Medium	Medium	Medium	Medium	Medium	Medium	Low
Water requirement	Moderate	Moderate	Moderate	Low	Moderate	Moderate	Low
Plant protection adopted by farmers	Minimum	Low	Minimum	Minimum	Low	Low	Low
Establishment cost	Low	Low	Low	Medium	High	Very high	High
Ability to grow in fallow lands	High	Medium	Low	High	High	High	High
Ability to fit in low input agriculture	Medium	High	High	High	Low	Low	Low
Energy output ($GJ\ ha^{-1}\ year^{-1}$)	128–270	155–326	na	na	105	260–530	174–435
Ability of farmers to shift to other crop	High	High	High	Medium	Low	Low	Low

na, data not available.

indicating less concern and low research focus on these crops. In comparison to established biofuel crops like *Jatropha* or maize grain, little research has been carried out.

The potential for hemp as a biofuel crop has been investigated only during the past few years. Earlier studies indicated that hemp as an energy crop may not be suitable due to high transport and raw material cost and only residue can be used as biomass for energy production (Brodersen *et al.*, 2002). However, later studies pointed out the suitability of conversion of high lignocellulosic biomass from hemp. In Europe, annual hemp dry biomass production has been reported to be 10–20 t ha^{-1}, with an average production of 16 t ha^{-1} (Finell *et al.*, 2006). This equates to an amount of 80 GJ ethanol calculated on the basis of fermentation of hexose (Sipos *et al.*, 2010). Anaerobic digestion of ground fresh hemp (16 t) is also estimated to produce 136 GJ of methane (Kreuger *et al.*, 2010). However, there is a need for improvement of the efficiency of conversion of the hemp biomass to biofuel. New conversion technologies are being developed to increase the efficiency and reduce the cost of chemical conversion. Using steam distillation processes, Kreuger *et al.* (2011) could recover 82% of the hexoses and 46% of the pentoses from hemp biomass. Co-production of ethanol and methane from steam pretreated hemp provided 2600–3000 l ethanol and 2800–2900 m^3 methane (171–180 GJ) from 1 ha. Moxley *et al.* (2008) also recovered 96% of glucose from hemp cellulosic herd (waste fibre). Hempseed oil, having an oil content of 26–38%, also holds a high possibility for biodiesel production. The oil can be converted to certified grade biodiesel with >99.5% conversion and 97% product yield (Li *et al.*, 2010). Compared to these crops, almost no study has been taken up in jute and *Agave* on their suitability as a biofuel source. Although jute is primarily used for combustion in the household kitchens of eastern India and Bangladesh, no reliable report is available on the suitability of jute as a biofuel crop. The burn rate of jute stick (1.1 kg h^{-1}) is comparatively lower than straw and the emission of fine particulates (1.7 g kg^{-1}) is estimated to be the lowest among several wood and non-wood fuel sources (Stone *et al.*, 2010). The lignocellulosic composition of jute is comparable to kenaf and hemp (Table 14.1), indicating that the energy output and biofuel conversion rate will be similar in these crops. About 3–4 million t of jute stick is produced in India alone, which can efficiently be channelled for biofuel production. Since these stalks are almost dry, the transport costs to processing plants are expected to be much lower than whole green biomass. The growth rate of jute under a favourable climate is higher than hemp or kenaf, reaching up to 4.5 m within 90–110 days, suggesting more biomass per unit area can be harvested from jute, particularly in subtropical humid climates. Under such ecological conditions, jute has good potential as a dedicated or waste biomass crop.

14.8 Current Status as Biofuel Crop

Kenaf is presently being used as a source of biofuel in several countries, including the USA and China. A number of US-based biofuel companies have been successfully producing biodiesel from whole kenaf biomass that can fuel not only cars and heavy duty vehicles, but also aeroplanes. For example, 'ReadiJet', a jet fuel commercially available in the USA, is produced from feedstock using kenaf and *Camelina* (http://www.ara.com/fuels). The south Florida district (USA) also has set up a large (>4000 acres) area for planting of kenaf variety 'Everglades' and biofuel crop *Camelina* to be used as biomass feedstock. Kenaf is also being promoted by the United States Department of Agriculture (USDA) for large scale biofuel production (Webber *et al.*, 2011). The kenaf crop in Florida is estimated to capture 25 t of carbon, which makes kenaf more environmentally friendly for biofuel production (USCJO Inc., 2012). Recently in the Heilongjiang province of China, Bio-Dynamic Group and Shanghai Zhongwei Biochemistry teamed up to produce cellulosic ethanol from kenaf. Their target is to produce between 28 and 56 million gallons (100,000 to 200,000 t) of ethanol per year by 2013 (BiofuelsDigest, 2011). In India, the Center for Jatropha Promotion and Biodiesel has identified kenaf as a potential source of biofuel.

Since hemp cultivation is prohibited in the USA, and Canada primarily cultivates hemp for seed purposes, the major focus of using hemp as a biofuel crop is in European countries. The ABE industrial process industries in the former USSR used hemp as a source of lignocellulosic biomass for butanol and acetone production (Zverlov *et al.*, 2006).

Both hemp and kenaf have been identified as potential biofuel crops in Europe (Tuck *et al.*, 2006). Monti and Zatta (2009) suggested that hemp may be a better biofuel crop in European countries than kenaf. They argued that hemp productivity is higher than kenaf and the crop is well known to European farmers. However, kenaf dry biomass productivity is higher under favourable climates, surpassing hemp by 3–5 t ha^{-1} (Smith-Heisters, 2008). Since hemp is more tolerant than kenaf to cold stress, the option of growing kenaf as a biofuel crop is limited primarily to the Mediterranean regions, while hemp can be grown further north, up to 65°N. Introducing kenaf to these regions would require development of cold-tolerant varieties. However, due to global warming and climate changes, a substantial area for kenaf farming can be identified in these non-traditional areas in future.

14.7 Future Prospects

Future biofuel production estimates consider dedicated bioenergy crops and biowaste as a major source for future biofuel production. Diverting cereal grains for biofuel production is inevitably pressurizing the food economy, resulting in the rise of food prices and debates all over the world (Oliveira and Franca, 2009). Forest tree species are a good source of cellulosic ethanol, but large scale deforestation for biofuel production poses enormous environmental risk. Additionally, the tree species contain less cellulose and high lignin and are less energy efficient in biofuel production than fibre crop species. The non-food fibre crops have potential to sustain a fine balance between the dedicated food crops and tree species, which would not disturb the food basket and would also be energy efficient for biofuel production. These fibre crops are widely adapted to different parts of the world, can be grown on limiting soil conditions and also in fallow lands, thereby not affecting agricultural area. The fibre crops produce more fuel per unit of agricultural land used, require less chemical and energy input per unit produce and are environmentally sustainable resulting in a higher yield in terms of net energy than from tree species (Table 14.5). However, several researchable issues have to be addressed for increasing efficiency of biofuel production from these crops, such as improving efficiency of extraction process, reducing cost of biofuel production, transport of large amounts of biomass feedstock to processing industries, environmental impacts of large scale cultivation of a single crop on natural habitat, socio-economic impacts of introduction of fibre crops to new areas, issues related to subsidization of biofuel cultivation as well as commercial and legal issues related to farm-based industries. Due to supply limitations, the use of a mixture of several complementary biomass types is a new attractive approach in advanced biofuel production. The fibre crop biomasses, with high cellulosic biomass coupled with low lignin, fit well in such biofuel production systems.

References

Alexopoulou, E., Christou, M., Mardikis, M. and Chatziathanassiou, A. (2000) Growth and yields of kenaf in central Greece. *Industrial Crops and Products* 11, 163–172.

Amaducci, S., Errani, M. and Venturi, G. (2002) Plant population effects on fibre hemp morphology and production. *Journal of Industrial Hemp* 7, 33–60.

Bakker, R.R. and Elbersen, H.W. (2005) Managing ash content and quality in herbaceous biomass: an analysis from plant to product. *14th European Biomass Conference*, 17–21 October 2005, Paris, France, pp. 210–213.

Batra, S.K. (2007) Other long vegetable fibers: abaca, banana, sisal, henequen, flax, ramie, hemp, sunn, and coir. In: Lewin, M. (ed.) *Handbook of Fibre Chemistry*. CRC Press, Taylor and Francis Group, Boca Raton, Florida.

Becu, D.M.S., Mastebroek, H.D. and Marvin, H.J.P. (1998) Breeding for root knot nematode resistance in hemp. *Proceedings of Bast Fibrous Plants Today and Tomorrow*, St Petersburg, Russia, 149 pp.

BioFuelsDigest (2011) Bio-Dynamic, Shanghai Zhongwei Biochemistry JV for China cellulosic ethanol in 2011. Available at: http://www.biofuelsdigest.com/bdigest/tag/shanghai-zhongwei-biochemistry (accessed 11 November 2011).

Bócsa, I. and Karus, M. (1998) *The Cultivation of Hemp. Botany, varieties, cultivation and harvesting.* Hemptech, Sebastopol, California.

Bredemann, G. (1924) Beiträge zur Hanfzüchtung II. Anslese faserreicher. Männchen zur Befruchtung durch Faser-bestimmung an der lebeden Pflanze vor deu Blüte. *Angewandte Botanik* 6, 348–360.

Brodersen, C., Drescher, K. and McNamara, K. (2002) Energy from Hemp? Analysis of the competitiveness of hemp using a geographical information system. In: van Ierland, E.C. and Lansink, A.O. (eds) *Economics of Sustainable Energy in Agriculture*. Kluwer Academic Publishers, Dordrecht, the Netherlands, pp. 121–134.

Burczyk, H., Grabowska, L., Kołodziej, J. and Strybe, M. (2008) Industrial hemp as a raw material for energy production. *Journal of Industrial Hemp* 13, 37–48.

Chen, Y.Y., Lin, L.H., Wu, J.M., Qi, J.M. and Zhou, R.Y. (2004) Genetic effect analysis of some field and quality traits of kenaf hybrid and parents. *Plant Fibers and Products* 26, 261–266.

Chew, W.Y., Abdul-Malek, M.A. and Ramli, K. (1982) Nitrogen and potassium fertilization of Congo Jute (*Urena lobata*) and kenaf (*Hibiscus cannabinus*) on Malaysian peat. *MARDI Research Bulletin* 10, 317–322.

Clarke, R.C. (2006) Hemp (*Cannabis*) cultivation and use in the Republic of Korea. *Journal of Industrial Hemp* 11, 51–86.

Cuiping, L., Yanyongjie, Chuangzhi, W. and Haitao, H. (2004) Study on the distribution and quantity of biomass residues resource in China. *Biomass and Bioenergy* 27, 111–117.

Dempsey, J.M. (1975) *Fiber Crops*. University Press of Florida, Gainesville, Florida.

Dickison, W.C. (2000) *Integrative Plant Anatomy*. Academic Press, San Diego, California, 533 pp.

Fairley, P. (2011) Next generation biofuels. *Nature* 474 (7352), S2–S5.

Falusi, O.A. (2008) Inheritance of characters in kenaf (*Hibiscus cannabinus*). *African Journal of Biotechnology* 7, 904–906.

Farrell, A.E., Pelvin, R.J., Turner, B.T., Jones, A.D., O'Hare, M. *et al.* (2006) Ethanol can contribute to energy and environmental goals. *Science* 311, 506–508.

Finell, M., Xiong, S. and Olsson, R. (2006) *Multifunktionell industrihampa för norra Sverige [Multifunctional industrial hemp for Northern Sweden]*. Unit for Biomass Technology and Chemistry, Swedish University of Agricultural Sciences, Umeå, Sweden. BTK-rapport 2006:13 (in Swedish).

Gentry, H.S. (1982) *Agaves of Continental North America*. University of Arizona Press, Tucson, Arizona.

Graham-Rowe, D. (2011) Beyond food versus fuel. *Nature* 474, S6–S8.

Han, J.S. (1998) Properties of nonwood fibers. In: *Proceedings of the Korean Society of Wood Science and Technology Annual Meeting*, Seoul, Korea, 24–25 April 1998, pp. 3–12.

Health Canada (2011) List of approved cultivars for the 2011 growing season *Cannabis sativa* L. Available at: http://www.hc-sc.gc.ca/hc-ps/alt_formats/hecs-sesc/pdf/pubs/precurs/list _cultivars- liste2011/list_cultivars-liste2011-eng.pdf (accessed 23 March 2011).

Hendriks, A.T.W.M. and Zeeman, G. (2009) Pretreatments to enhance the digestibility of lignocellulosic biomass. *Bioresource Technology* 100, 10–18.

Johnson, R. (2010) Hemp as an agricultural commodity. Congressional Research Service Report RL32725. Available at: http://assets.opencrs.com/rpts/RL32725_20101222.pdf (accessed 13 October 2011).

Koshel, P. and McAllister, K. (2010) *Expanding Biofuel Production and the Transition to Advanced Biofuels Lessons for Sustainability from The Upper Midwest*. The National Academies Press, Washington, DC.

Kreuger, E., Prade, T., Escobar, F., Svensson, S.E., Englund, J.E. *et al.* (2010) Anaerobic digestion of industrial hemp – effect of harvest time on methane energy yield per hectare. *Biomass and Bioenergy* 35, 893–900.

Kreuger, E., Sipos, B., Zacchi, G., Svensson, S.E. and Björnsson, L. (2011) Bioconversion of industrial hemp to ethanol and methane: the benefits of steam pretreatment and co-production. *Bioresource Technology* 102, 3457–3465.

Kundu, B.C. (1956) Jute – World's foremost bast fibre. 1. Botany, agronomy pests and diseases. *Economic Botany* 10, 103–133.
Laitinen, E. (1996) History of hemp in Finland. I. *Journal of the International Hemp Association* 3, 34–37.
Li, D.F., Liu, W.J. and Tan, S.L. (1996) Investigations and exploitation of material insensitive to a short photoperiod in kenaf. *Acta Agronomica Sinica* 22, 50–54.
Li, S.Y., Stuart, J.D., Li, Y. and Parnas, R.S. (2010) The feasibility of converting *Cannabis sativa* L. oil into biodiesel. *Bioresource Technology* 101, 8457–8460.
Lisson, S.N., Mendham, N.J. and Carberry, P.S. (2000) The development of a hemp (*Cannabis sativa* L.) simulation model: II The flowering response of two hemp cultivars to photoperiod. *Australian Journal of Experimental Agriculture* 40, 413–417.
Maiti, R.K. (1979) A study of the microscopic structure of the fiber strands of common Indian Bast fibers and its economic implications. *Economic Botany* 33, 78–87.
Maiti, R.K. and Satya, P. (2009) Fibre bundle anatomy determines the yield potentials, and fibre quality of bast fibre (long fibre): a hypothesis. *International Journal of Agriculture Environment and Biotechnology* 1, 41–46.
Maiti, R.K., Martínez, M.A. and Mercado, H.R. (1990) Quantitative description of morphoanatomical characters and productivity of lechuguilla (*Agave lecheguilla* torr.) (Agavaceae), Villa de García, Nuevo León, México. *Publicaciones biológicas. F.C.B.* 4, 29–34.
Maiti, R., Satya, P. and Rodríguez, H.G. (2010) *Horizon of World Plant Fiber: an insight*. Pushpa Publishing House, Kolkata, India.
Mandolino, G., Carboni, A., Forapani, S., Faeti, V. and Ranalli, P. (1999) Identification of DNA markers linked to the male sex in dioecious hemp (*Cannabis sativa* L.). *Theoretical and Applied Genetics* 98, 86–92.
Martinov, M., Markovic, D., Tesic, M. and Grozdanic, N. (1996) Hemp harvesting mechanisation. *Agricultural Engineering* 2, 23–38.
McGovern, J.N. (1990) *Vegetable Fibers, Polymers-fibers and Textiles – A Compendium*. University of Wisconsin Press, Madison, Wsconsin.
Mediavilla, V., Leupin, M. and Keller, A. (2001) Influence of the growth stage of industrial hemp on the yield formation in relation to certain fibre quality traits. *Industrial Crops and Products* 13, 49–56.
Meijer de, E.P.M., Bagatta, M., Carboni, A., Crucitti, P., Moliterni, V.M.C. *et al.* (2003) The inheritance of chemical phenotype in *Cannabis sativa* L. *Genetics* 163, 335–346.
Moliterni, V.M.C., Cattivelli, L., Ranalli, P. and Mandolino, G. (2004) The sexual differentiation of *Cannabis sativa* L.: a morphological and molecular study. *Euphytica* 140, 95–106.
Monti, A. and Zatta, A. (2009) Root distribution and soil moisture retrieval in perennial and annual energy crops. *Agriculture, Ecosystem and Environment* 132, 252–259.
Morrison, W.H. III, Akin, D.E., Archibald, D.D., Dodd, R.B. and Raymer, P.L. (1999) Chemical and instrumental characterization of maturing kenaf core and bast. *Industrial Crops and Products* 10, 21–34.
Moxley, G., Zhu, Z. and Zhang, Y.P. (2008) Efficient sugar release by the cellulose solvent-based lignocellulose fractionation technology and enzymatic cellulose hydrolysis. *Journal of Agricultural and Food Chemistry* 56, 7885–7890.
Oliveira, L.S. and Franca, A.S. (2009) From solid biowastes to liquid biofuels. In: Ashworth, G.S. and Azevedo, P. (eds) *Agricultural Wastes*. Nova Science Publishers, New York.
Pahkala, K., Pahkala, E. and Syrjälä, H. (2008) Northern limits to fiber hemp production in Europe. *Journal of Industrial Hemp* 13, 104–116.
Pignatelli, V. and Robertiello, A. (2007) Biomass for renewable energy. In: Amadei, C. (ed.) *Encyclopaedia of Hydrocarbons*. Vol. III. *New Developments: Energy, Transport, Sustainability*. Istituto Della Enciclopedia Italiana Fondata Da Giovanni Treccani, Italy, pp. 611–632.
Purohit, P., Tripathi, A.K. and Kandpal, T.C. (2006) Energetics of coal substitution by briquettes of agricultural residues. *Energy* 31, 1321–1331.
Qi, J., Chen, Y., Zhou, R., Lin, L., Liang, K., Wu, J. and Fang, P. (2005) Genetic effects and heterosis analysis for yield and quality traits in kenaf (*Hibiscus cannabinus* L.). *Zuo Wu Xue Bao* 31(4), 469–475. Available at: http://europepmc.org/abstract/CBA/505050/reload=0;jsessionid=It25c1zQwFHJA5j4VZC9.4 (accessed 15 May 2012).
Ranalli, P. (2004) Current status and future scenarios of hemp breeding. *Euphytica* 140, 121–131.
Rice, B. (2008) Hemp as a feedstock for biomass to energy conversion. *Journal of Industrial Hemp* 13, 145–156.
Scragg, A.H. (2009) *Biofuels Production Application and Development*. CAB International, Wallingford, UK.

Sipos, B., Kreuger, E., Svensson, S.E., Réczey, K., Björnsson, L. and Guido, Z. (2010) Steam pretreatment of dry and ensiled industrial hemp for ethanol production. *Biomass and Bioenergy* 34, 1721–1731.

Smith-Heisters, S. (2008) Illegally green: environmental costs of hemp prohibition. *Policy Study* 367. Reason Foundation, Los Angeles, California.

Stone, E.A., Schauer, J.J., Pradhan, B.B., Dango, P.M., Habib, G. *et al.* (2010) Characterization of emissions from South Asian biofuels and application to source apportionment of carbonaceous aerosol in the Himalayas. *Journal of Geophysical Research* 115, Doi:10.1029/2009JD011881.

Sur, D. and Amin, N. Md. (2010) Physics and Chemistry of Jute. In: Sur, D. (ed.) *Jute Basics*. International Jute Study Group, Dhaka, Bangladesh, pp. 35–56.

Svennerstedt, B. and Svensson, G. (2006) Hemp (*Cannabis sativa* L.) trials in southern Sweden 1999–2001. *Journal of Industrial Hemp* 11, 17–25.

Tuck, G., Glendining, M.J., Smith, P., House, J.I. and Wattenbach, M. (2006) The potential distribution of bioenergy crops in Europe under present and future climate. *Biomass and Bioenergy* 30, 183–197.

UN-Energy (2007) Sustainable bioenergy – a framework for decision makers. Available at: http://www.un-energy.org/sites/default/files/share/une/susdev.biofuels.fao_.pdf (accessed 6 November 2012).

USCJO Inc. (2012) Biomass and biofuel feedstock farming. Available at: http://www.uscjo.com (accessed 19 May 2012).

Villarreal, R.L. and Maiti, R.K. (1991) Características morfoanatómicas y productividad de fibra en *Agave lecheguilla* Torr., en Nuevo León, México. *Turrialba* 41, 423–429.

Webber III, C.L., Robert, V.K. and Bledsoe, R.E. (2011) United States kenaf (*Hibiscus cannabinus* L.) cultivar review. In: Webber, C.L. III and Liu, A. (eds) *Plant Fibers as Renewable Feedstocks for Biofuel and Bio-based Products*. CCG International, St Paul, Minnesota, pp. 117–126.

Westphal-Stevels, J.M.C. (1986) Local vegetables in Cameroon: *Corchorus* species used as a vegetable. In: First International Symposium on Taxonomy of Cultivated Plants. *Acta Horticulturae* 182, 423–425.

Wilson, F.D., Summers, T.E. and Joyner J.F. (1965) *'Everglades 41' and 'Everglades 71' – two new varieties of kenaf (*Hibiscus cannabinus *L.) for fiber and seed.* Florida Agricultural Experiment Station Circular S-168.

Wood, I.M. (1978) Preliminary experiments on the growth of kenaf (*Hibiscus cannabinus*) for paper pulp production in Ord irrigation area, Western Australia. *Australian Journal of Experimental Agriculture and Animal Husbandry* 18, 97–106.

Worldjute Report (2011) Available at: http://www.worldjute.com/jute_bangladesh/bangladesh_jute_news.html (accessed 15 October 2011).

Zverlov, V.V., Berezina, O., Velikodvorskaya, G.A. and Schwarz, W.H. (2006) Bacterial acetone and butanol production by industrial fermentation in the Soviet Union: use of hydrolyzed agricultural waste for biorefinery. *Applied Microbiology and Biotechnology* 71, 587–597.

15 Jatropha (*Jatropha curcas* L.)

Suhas P. Wani
International Crops Research Institute for the Semi-Arid Tropics (ICRISAT), India

15.1 Introduction

Biodiesel is an important renewable biofuel that offers great potential to ease and mitigate energy shortage, supplement/replace fossil fuel while at the same time minimize/reduce carbon emissions. Biodiesel production is expanding rapidly because of demand, necessary policy support and technological availability. However, shortage of raw material to produce biodiesel is a major constraint (Wani *et al.*, 2006b). Oils from plants are the main source for the production of biodiesel. Of the 100–300 oil-bearing plant species, 30 hold promise for biodiesel production (Hegde, 2003). Many developed countries are using edible oilseed crops such as soybean, rapeseed, groundnut and sunflower for production of biodiesel. Biodiesel production from vegetable oils during 2004/2005 was estimated to be 2.36 Mt globally. Of this, EU countries (1.93 Mt with expectation of 30% annual increase) and the USA (0.14 Mt) together accounted for 88% and the rest of the world (0.29 Mt) for the remaining 12% (Parikh, 2005). However, for developing countries like India, having a huge shortage of edible oil (6.31 Mt $year^{-1}$) for consumption, there is no question of using edible oil as a source of biofuel.

The non-edible oilseeds such as pongamia (*Pongamia pinnata*) and jatropha (*Jatropha curcas*), which along with providing biodiesel can also meet the additional criteria of greening the wastelands without compromising the food, fodder security and improved livelihoods, are more suited as biofuel oil crops for developing countries (Wani *et al.*, 2006a; Reddy *et al.*, 2008). Jatropha, which meets the American and European biofuel standards, has gained importance in tropical and subtropical countries and has been promoted heavily (Tiwari *et al.*, 2007). The committee on biofuels constituted by the Planning Commission, Government of India (2003), out of a large number of oilseed-bearing tree species identified jatropha and pongamia as potential candidates best suited to Indian conditions.

Jatropha (*Jatropha curcas* L.), commonly known as 'purging nut' or 'physic nut', is also referred to by various other names in different countries. The plant is naturally distributed in Mexico, Central America and parts of South America (Trabucco *et al.*, 2010). Carried to different parts of the world by the Portuguese colonial power, jatropha now grows in many African and South Asian countries. Jatropha is a tropical, perennial, C_3 plant belonging to the family *Euphorbiaceae* (Tatikonda *et al.*, 2009; Divakara *et al.*, 2010). The plant is a deciduous tree with a succulent stem and leaves and a soft wood (Borchert, 1994; Foidl *et al.*, 1996). It is adapted to semi-arid tropical conditions. As a plant of the dry regions of Meso-America, jatropha is suitable to grow on degraded drylands.

©CAB International 2013. *Biofuel Crops: Production, Physiology and Genetics* (ed. B.P. Singh)

It has a liking for well-drained soils and is susceptible to collar rot disease in high rainfall humid areas or excessive irrigation or to waterlogging. It produces fruits with a high content of non-edible oil that can be used for several purposes. As with many members of the family *Euphorbiaceae*, it contains compounds that are highly toxic and is thus not browsed by livestock (Wani and Sreedevi, 2007).

15.2 Jatropha, a Wonder Plant – Myth or Reality?

The claim of jatropha being a wonder plant is based on its perceived ability to grow successfully in semi-arid environmental conditions and on degraded soils. Therefore a special deliberation on these features of the plant is presented below.

15.2.1 Drought tolerance

There is limited knowledge about the actual water requirement of jatropha in different agro-ecological regions. However, minimum and optimum rainfall to produce harvestable jatropha fruits has been assessed at 500–600 and 1000–1500 mm year^{-1} in arid and semi-arid tropics, respectively (Achten *et al.*, 2008; da Schio, 2010; Trabucco *et al.*, 2010).

Water use assessment of jatropha plantations in the semi-arid tropical location at ICRISAT (Indian Crop Research Institute for the Semi-Arid Tropics), Patancheru, India indicated that evapotranspiration (ET) under no moisture stress conditions varied from 1410 to 1538 mm year^{-1} during 2006–2009 (ICRISAT, 2010). Under the actual field conditions, the ET requirements varied from 614 to 930 mm depending on the atmospheric demand, rainfall and crop phenological stage. Patterns of soil water depletion indicated that the 2-year-old jatropha plantation was able to extract water up to a soil depth of about 100 cm and later at the age of 5 years it was able to extract up to 150 cm. Monthly water use of jatropha varied from 10 to 20 mm (leaf shedding period) to 140 mm depending on water availability and environmental demand (Kesava Rao *et al.*, 2012). Thus, contrary to the belief that jatropha needs less water, under favourable soil moisture conditions the tree could use large amounts of water for luxurious growth and high yield.

The low transpiration coefficients (0.51–0.60) of jatropha compared to that of a reference plant (0.51–0.70) and limited stem sap flow of 12-year-old adult trees in a semi-arid region of South Africa supports its conservative water use (Holl *et al.*, 2007; Achten *et al.*, 2010a). Jatropha can survive under 200 mm water year^{-1}. The extent and timing of its leaf drop is acutely influenced by drought (Maes *et al.*, 2009; Achten *et al.*, 2010a). The remaining plant leaves are known to exhibit parahelionasty under drought stress as a drought response. Jatropha is also known to change the anatomy of the freshly produced leaves to suit the low water availability situations under drought. It is capable of rapid adjustment to drought or to escape drought well, but not continuously, nor is it equipped for high productivity under drought. Maes *et al.* (2009) are of the opinion that jatropha has evolved and adapted to plant survival under water-deficit environments rather than to maintain high levels of plant productivity under stress.

Jatropha root traits and their significance for drought tolerance

Root distribution of jatropha, both in terms of weight and length density, was recorded by Krishnamurthy *et al.* (2012) in the top 0–30 cm soil horizon. The highest root length density distribution was noted in the top 15 cm soil profile measuring close to 0.5 cm cm^{-3} of soil with decreasing root length density distribution to <0.1 cm cm^{-3} in deeper layers below 60 cm depth (Fig. 15.1). An exposed trench wall showed the rooting to extend up to 1.4 m (Fig. 15. 2). However, this is in contradiction to the descriptions quoted by Foidl *et al.* (1996) who observed four to five branches on a vertical taproot that penetrates up to 5 m deep under Nicaraguan conditions. The difference in findings of Foidl *et al.* (1996) from Krishnamurthy *et al.* (2012) could be due to enormous resistance to the root system growth under high soil bulk density of the red soil with heavy, compact and shallow murram layer below 60–75 cm soil depth where the

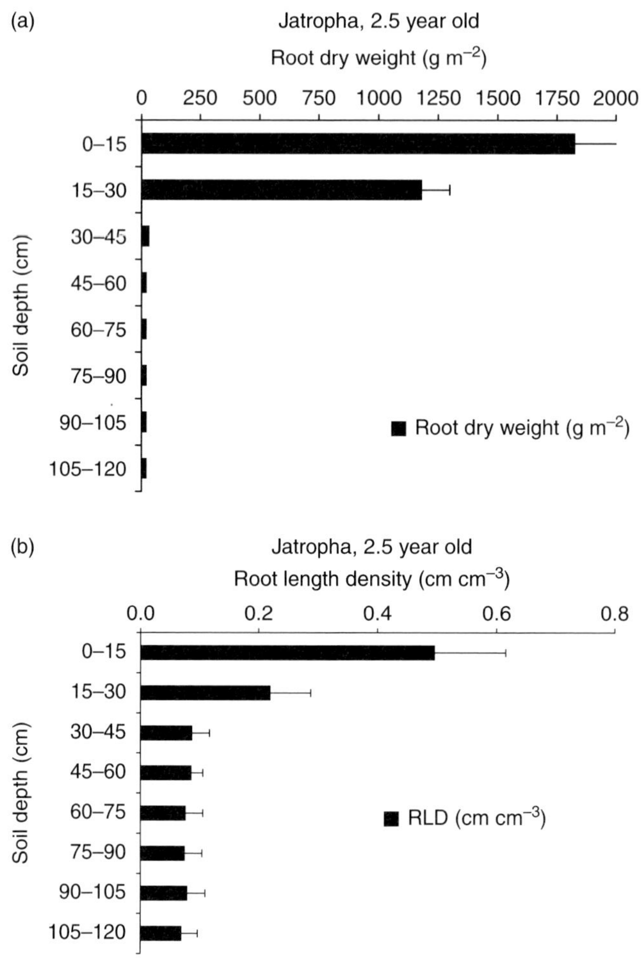

Fig. 15.1. The root distribution in jatropha as measured through a monolith method. Changes in (a) root dry weight across various soil depths and (b) in root length density across various soil depths in a 2.5-year-old plant. Such a large difference in distribution weight or length is primarily caused by the small section of taproot at the 0–30 cm soil horizon (Krishnamurthy *et al.*, 2012).

later experiment's root extraction was carried out. In annual crop plants such as wheat, a root length density greater than 0.5 cm cm^{-3} was estimated to be adequate for complete extraction of available soil water (Passioura, 1982). Though measurements of rooting depth and root length density need not necessarily give an estimated ability of the plant to extract soil water, still a root length density of <0.1 cm cm^{-3} in jatropha was found to be grossly suboptimal. Root dry weight distribution up to the depth 0–30 cm accounted for almost all the roots of the jatropha plant. In deeper soils below 30 cm the root dry weights ranged from 21 to 2 g m^{-2}. Root length distribution was dense only in the surface soil regions.

The allocation of dry matter for root to shoot was 0.37, but this ratio was reduced to a mere 0.03 when the root weight of the taproot at the 0–30 cm soil depth was not included for this estimation, indicating that the major root biomass is distributed for the purpose of anchoring rather than for absorption of soil moisture or nutrients from the soil (Krishnamurthy *et al.*, 2012). Closely comparable values of root/shoot ratio, ranging from 0.41 to 0.27 depending on the soil water deficit regimes, had been reported for 104-day-old

LC7	LC6	LC5	LC4	LC3	LC2	LC1	C	RC1	RC2	RC3	RC4	RC5	RC6	RC7	
20	18	33	16	18	20	7	15	20	18	18	18	14	23	11	270
11	8	18	11	14	16	12	8	21	17	18	16	15	5	11	202
7	6	3	8	3	8	13	10	15	21	9	5	3	7	1	123
8	4	3	3	5	6	5	7	6	10	3	13	5	8	8	94
7	2	3	4	7	3	8	1	3	12	5	4	4	3	5	71
1	3	5	8	7	6	9	4	6	6	1	5	1	4	8	75
4	3	5	1	5	6		1	2	3	3	2	3	3	6	47
2				7	3	6			1		2	1	6	3	31
1	1	2		2				1			2	1			10
	1						3								4
1			1				3		1					1	7
		2					5	1						2	13
								1							1
					1	1									2

Fig. 15.2. The root distribution in jatropha as measured through a trench wall method. Number of root intercepts that were measured in a 10 cm soil horizon and across 150 cm (70 cm on either side of the plant's base). After exposing a trench wall close to the plant base, the wall was washed with a fine spray of water and the protruding roots were trimmed before recording the observations. Only a single plant root growth was measured for this purpose. C in the X scale means centre grid, LC_n means to the left of centre and RC_n to the right of centre. Each grid denotes the number of root intercepts within each 100 cm^2 grid and numbers on the right *y* axis are the total number of intercepts in each 10 cm soil horizon (Krishnamurthy *et al.*, 2012). Black, >10 intercepts per grid; dark grey, 5–10 intercepts per grid; medium grey, 3–4 intercepts per grid; light grey, 1–2 intercepts per grid.

jatropha seedlings by Achten *et al.* (2010b). These values were found to be the lowest compared with ten other tropical deciduous woody and shrub species (*Celaenodendron mexicanum* Standl, *Trichilia trifolia* L., *Caesalpinia eriostachys* Bench, *C. platyloba* S. Wats, *Plumeria rubra* L., *Apoplanesia paniculata* Presl., *Cordia alliodora* (Ruiz & Pav.) Oken, *Cochlospermum vitifolium* (Willd) Spreng, *Ipomoea wolcottiana* Rose, *Heliocarpus pallidus* Rose (Huante and Rincon, 1998)). Under extreme drought, jatropha started shedding its leaves and stopped growing (Achten *et al.*, 2010b). For a perennial growth habit and for a shrub, the root system distribution in jatropha is shallow, root proliferation poor and root biomass allocation to the absorbing roots meagre. But the roots were found to be well equipped for rapid branching whenever growth opportunity arose after rain and adapted for absorbing minerals from highly secluded pockets of soil. Krishnamurthy *et al.* (2012) found that under conditions of water replenishment the lateral root growth quickly resumed because of the large number of branch primordia on the thicker roots.

Krishnamurthy *et al.* (2012) observed the presence of a limited number of narrow xylem vessels and a well-defined cortex in jatropha indicative of potentially slow absorption of water needed to conserve soil water under drought. Therefore it can be concluded that jatropha has a relatively small root system for a tree to be a good producer under limited soil conditions but a xylem vessel system that enables it to survive under drought.

15.2.2 Jatropha plantations' impact on hydrological processes

Evaluation of how downstream hydrological processes and sediment transport at the mesoscale (10–10,000 km^2) will be affected by large-scale jatropha planting is so far lacking. Even so, from the perspective of water, jatropha cultivation is considered an option for making productive use of wastelands while at least partly avoiding conflicts with downstream environmental flow requirements. Beneficial effects such as less erosive storm floods and lower sediment loads in riverine ecosystems, and larger groundwater formation as a result of improved infiltration is envisioned (Garg *et al.*, 2011).

Water balance

The water balance differs substantially depending on land use and amount of annual average rainfall (Fig. 15.3a). In general, a larger share of the total rainfall forms runoff during wetter years compared with drier years. A comparison of the different land management scenarios showed that more than 50% of the non-productive soil evaporation in the wasteland was shifted into productive transpiration in the case of both current and long-term jatropha plantations. The total amount of ET was relatively similar in all three cases, except during dry seasons when ET was higher in current jatropha plantation than wasteland, and increased further under improved soil conditions in the long-term plantation. Groundwater recharges doubled in the current and quadrupled in the long-term plantation compared to the wasteland. As a result of higher ET and groundwater formation, runoff decreased in the jatropha plantings, in particular during dry years. In the case of wasteland, runoff constituted around 40% of the total rainfall during dry years while the corresponding figure for the current jatropha plantation was around 30%, and even lower (down to 20%) for the long-term plantation. Such a large reduction in outflows from the watershed at a time when the average rainfall amount is low might have negative impacts on downstream ecosystems and water users.

The distribution of the water balance components within the year also varied with land use (Fig. 15.3b). While the total ET was lower for the two jatropha plantations during the dry season (December–March), it became higher during the wetter parts of the year. This means that the annual fluctuations in runoff and groundwater generation are smaller in the jatropha plantations compared with the wasteland.

Runoff generated from the watershed consists of two components: (i) surface runoff; and (ii) base flow generation. It was found that even though the total runoff was significantly lower with jatropha plantations compared with the wasteland condition, base flow was in fact higher with jatropha plantations (Fig. 15.3c). On average, the total amount of base flow generation in the wasteland was only 70% of the base flow in the jatropha plantations. However, total runoff was 40% larger for the wasteland compared with the established jatropha.

Land management also affects runoff intensity. In general, higher runoff intensities were predicted for the wasteland state, compared with jatropha plantations (Fig. 15.3d). The results show that the average daily runoff intensity decreased by 12% for the current jatropha plantation, compared with the wasteland condition, and is likely to decrease even further with continued jatropha cropping (the established jatropha plantation had 39% lower runoff intensity than the wasteland).

A comparison of water balance components between the well-managed ICRISAT BL3 watershed and another site at the rural community of Velchal (current jatropha plantation) shows that a larger part of the rainfall formed green water flows (i.e. ET) at the well-managed site (80–90% compared with 40–60%, respectively) (Table 15.1). This means that only a small fraction (10–20%) of the total rainfall generated blue water flows (runoff and groundwater recharge) at the ICRISAT BL3 location. During dry years, blue water generation was lower than green water generation at both sites.

Sediment transport and soil loss

Currently, the estimated average soil loss in the Velchal watershed is between 10 and 15 t ha^{-1} $year^{-1}$. Because the soil depth is low and the available water holding capacity is poor in the watershed, large runoff is commonly generated during heavy rain with the capacity to carry large amounts of sediments. Soil loss was found to increase exponentially with rainfall intensity, and varied with land use (Fig. 15.4a). The highest soil loss occurred at high rainfall intensities under wasteland conditions. Cumulative soil loss generated at the watershed outlet over a 10 year period showed that jatropha cultivation resulted in a reduction of the total soil loss amount of nearly 50% compared to the wasteland state (Fig. 15.4b).

5.2.3 Jatropha plantations' impact on soil and environmental quality

Carbon and nutrient sequestration

Jatropha plants have a mechanism of drought avoidance by shedding their leaves during the stress period to minimize the water requirement,

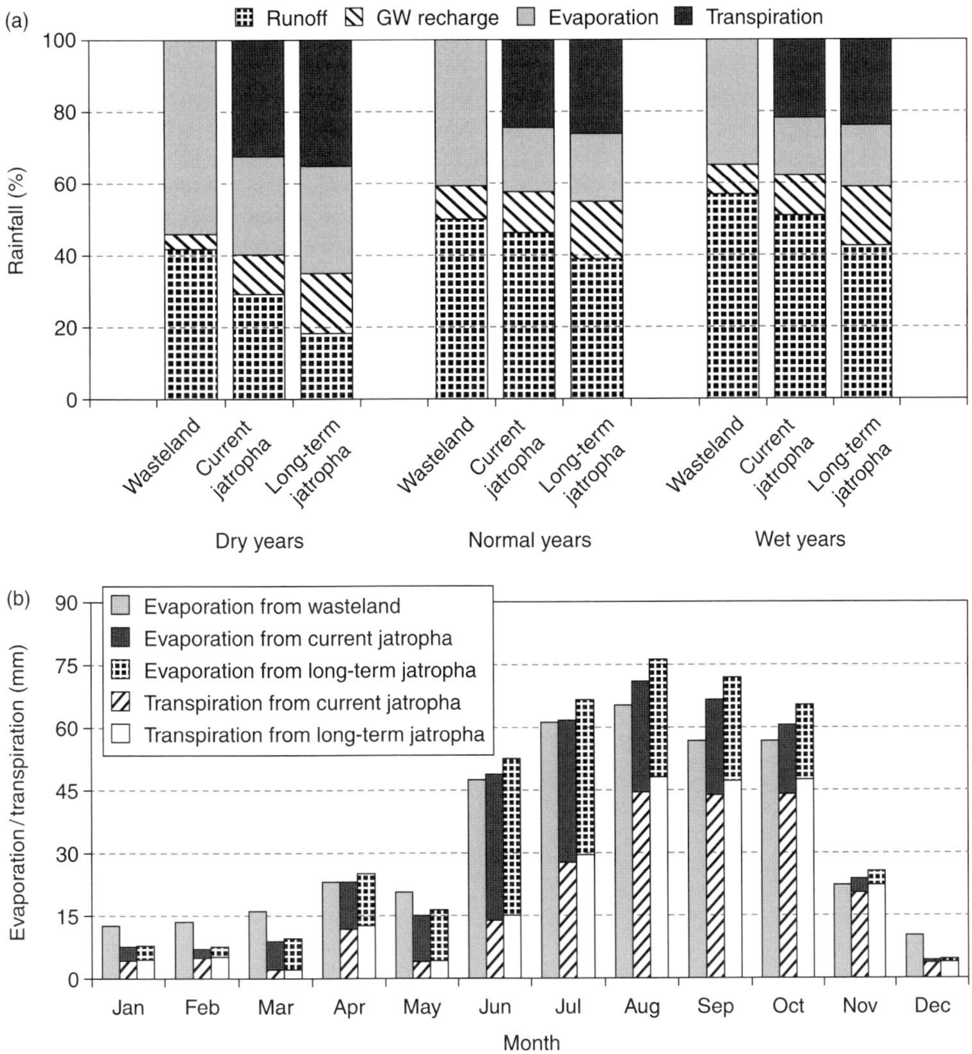

Fig. 15.3. (a) Water balance components of different land management scenarios during dry, normal and wet years (data from 2001 to 2010). (b) Monthly soil evaporation and transpiration for three different land management scenarios in Velchal watershed. (c) Total runoff generation from the watershed, divided into base flow and surface runoff, for three different land management scenarios during dry, normal and wet years (data from 2001 to 2010). (d) Frequency of daily runoff intensity for three different land management scenarios (data from 2001 to 2010).

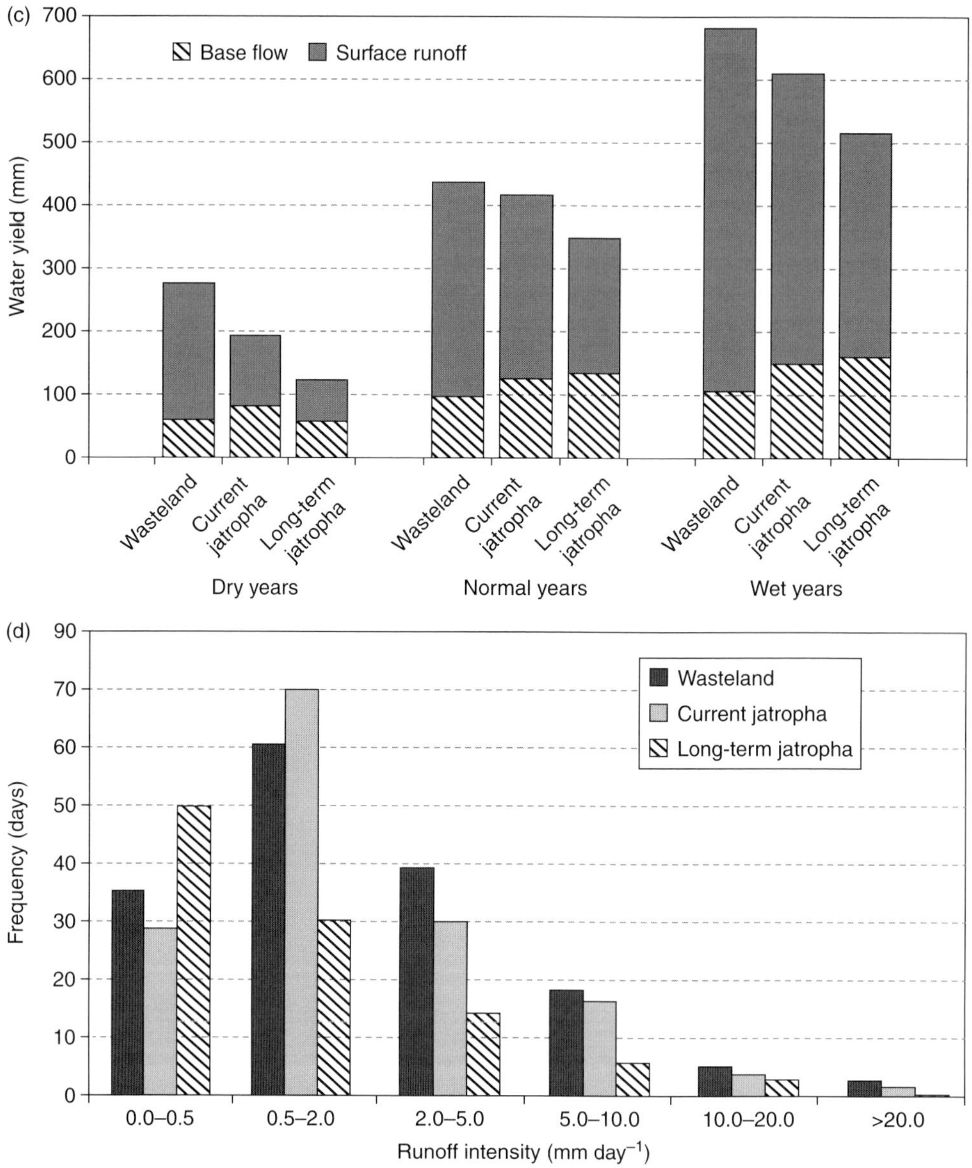

Fig. 15.3. Continued.

which is desirable in the rehabilitation of degraded lands through C addition to the soil. Data from the on-station study at two different ages of the plantation revealed an annual leaf fall of 552 g per plant in a 1-year-old plantation and 1451 g per plant in a 3-year-old plantation (Table 15.2), which corresponds to 305 kg C sequestrated ha^{-1} $year^{-1}$ through leaves in a 1-year-old plantation and 800 kg C sequestrated ha^{-1} $year^{-1}$ in a 3-year-old plantation comprising 1667 plants ha^{-1} at 2×3 m spacing. Jatropha pruned loppings generated around 410 kg ha^{-1} dry biomass and it was another source for adding C to the soil at the rate of 150 kg C ha^{-1}. The 3-year-old plants dropped around 2.6 times more leaves as compared to 1-year-old plants. The falling leaves in a 3-year-old plantation returned to the soil around 20 kg each of N and K and around 2 kg of P. Available P in the surface soil layer under jatropha was 5.2-fold

Table 15.1. Comparison of different hydrological components and crop yields between the ICRISAT BL3 watershed and the Velchal watershed ('current jatropha' scenario) (Garg *et al.*, 2011).

	Dry year (2007)		Wet year (2008)	
Variable (unit)	ICRISAT watershed, BL3	Velchal watershed	ICRISAT watershed, BL3	Velchal watershed
Inputs				
Available water (cm^3 cm^{-3}) (soil moisture at FC-PWP)	0.13	0.07	0.13	0.07
Soil depth (cm)	300	35	300	35
Annual average rainfall (mm)	707	707	1105	1105
Outputs				
Evaporation (mm)	251 (36%)	188 (27%)	265 (24%)	180 (16%)
Transpiration (mm)	400 (57%)	263 (37%)	606 (55%)	262 (24%)
Outflow (mm)	ND	162 (23%)	ND	550 (50%)
Groundwater recharge/deep percolation (mm)	ND	95 (13%)	ND	111 (10%)
Jatropha seed yield (t ha^{-1})	0.9	0.5	1.1	0.5

FC, field capacity; PWP, permanent wilting point; ND, not determined.

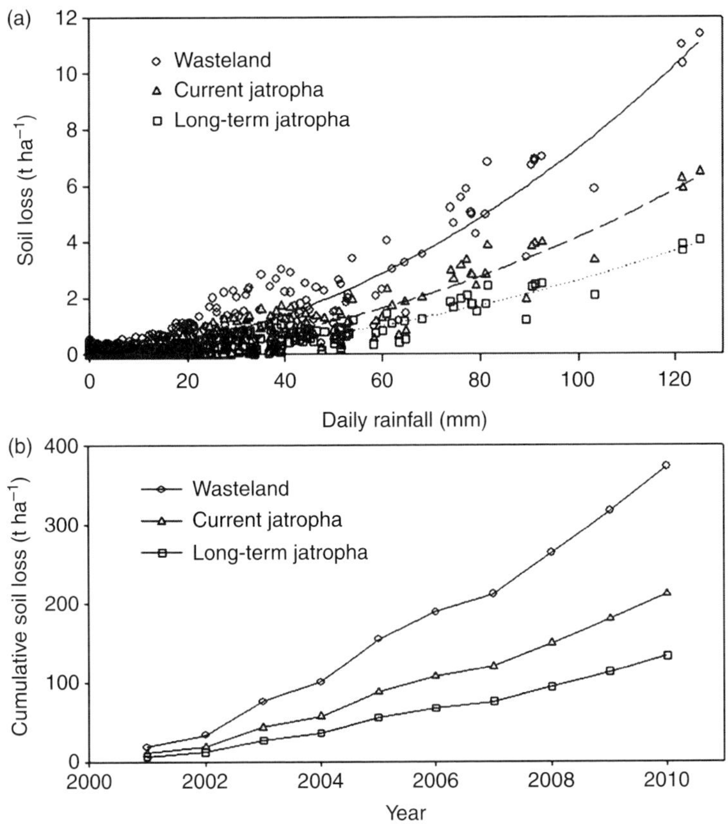

Fig. 15.4. (a) Impact of land management practices on sediment transport under different land management conditions (data from year 2001 to 2010). (b) Cumulative soil loss (t ha^{-1}) under different land management conditions (data from year 2001 to 2010).

as compared to the adjacent uncultivated control grasslands. A positive relationship was found between soil organic C and available P, implying the role of increased organic matter in enhancing soil P (Wani *et al.*, 2003).

In the on-station study, the 4-year-old jatropha plantation recorded 1290 kg ha^{-1} seed yield in unfertilized control plots. The seed productivity varied from 1290 to 1610 kg ha^{-1} without and with fertilization and sequestered 580–725 kg C ha^{-1} year^{-1}. Out of the C sequestered in the seed, 185–230 kg C ha^{-1} year^{-1} is attributed to biodiesel/oil C replacement in the fossil fuel based on the amount of C in fossil fuel-based biodiesel equivalent to jatropha oil-produced biodiesel. After oil extraction, the by-product de-oiled seed cake is a good source of organic matter for improving soil health through organic carbon additions. The seed cake biomass returned 395–495 kg C ha^{-1} year^{-1} into the soil. This by-product is a good source of plant nutrients as well and has been evaluated for soybean and maize crops (Wani *et al.*, 2006a; Osman *et al.*, 2009).

The live plant (shoot and root) biomass standing in the fields is a good sink for C. The findings of the on-station study at Patancheru revealed that standing jatropha plants (4 years old) sequestrated 3.07 kg C per plant in above- and below-ground biomass, which corresponds to 5120 kg C ha^{-1} for a plant population of 1667 plants ha^{-1}. Similarly, the on-farm study at Kothlapur recorded 2.44 kg C-sequestration per plant, which translates to 6100 kg C ha^{-1} for a plant population of 2500 plants ha^{-1} at 2 × 2 m spacing.

Based on multiple study results, we have been able to conclude that a 3–5-year-old jatropha plantation in a tropical Indian location with 800 mm annual rainfall recycled around 4000 kg plant biomass or 1450 kg ha^{-1} equivalent organic C into the soil (Table 15.3). Further, biodiesel/oil extracted replaced 230 kg C ha^{-1} in the fossil fuel. In addition to the C in the recycled plant material and C replacement in fossil fuel, the live plantation also contained more than 5100 kg C equivalent ha^{-1} in biomass. In the tropics, identified agricultural C sequestering systems with the inclusion of pigeonpea and other legume and horticulture based cropping systems with improved management increased carbon sequestration by 330 kg C ha^{-1} year^{-1} (Wani *et al.*, 2003; Bhattacharyya *et al.*, 2007, 2009; Wani and

Table 15.2. Content and amounts of nutrients returned through fallen leaves during 2007 at ICRISAT, Patancheru (Wani *et al.*, 2009).

Age of the plant (years)	Dry leaf (g/plant)	% N	% P	% K	Nutrient recycling by leaf litter (kg ha^{-1})		
					N	P	K
1	552.5	1.14	0.06	1.1	15.7	0.8	15.2
3	1451.1	0.86	0.08	0.95	20.8	2	23

Table 15.3. Total C sequestered through jatropha plantation as C returned to soil, biodiesel C replacement per year and live plant C.

C through jatropha plantation	Plant part involved	Organic C (kg ha^{-1})
C returned back to soil	Leaf fall	800[a]
	Pruned loppings	150[a]
	De-oiled cake	495[b]
C replacement in fossil fuel	Jatropha oil	230[b]
C in live plant	Shoots and roots	5120

[a]Leaf and stem prunings added C every year.
[b]Jatropha oil C (fuel replacement) and de-oiled cake added C from 4th year onwards every year.

Chaliganti, 2007). Addition of substantial amounts of C to degraded soil through jatropha would kick-start biological activity and nutrient release. Continuous addition of C to the soil would also bring an improvement in the water-holding capacity of the soil and would benefit the growing plants (Zhu *et al.*, 2010). Such large quantities of C sequestrated in plant biomass is a great ecosystem service we can have and, as per the Kyoto protocol, these C credits can be traded with the countries or regions who are not able to manage their C credits (D'Silva *et al.*, 2004).

Soil biology in the rhizosphere of Jatropha

Sustainability of plant–soil systems is largely governed by the floral and faunal biodiversity in the rhizosphere soils of the plant. Microbial biomass is an undifferentiated parameter of different microorganisms in soil, which play an important role in controlling nutrient cycling and availability to plants.

A study to evaluate the effect of jatropha plants on the soil biological activity at different locations and characterizing the difference between rhizosphere and non-rhizosphere soils of the plants was undertaken (Table 15.4). Soil samples (rhizosphere and non-rhizosphere) from four different locations (ICRISAT campus, Velchal, Siddhapur, Kothalapur) all in Andhra Pradesh, India, were studied for microbial counts, microbial biomass and dehydrogenase enzyme activity in the soil samples. The rhizosphere soil in locations planted with jatropha recorded high populations of bacteria, fungi and actinomycetes as well as dehydrogenase activity.

High microbial populations in rhizosphere soil indicates increased soil biological activity, which was also observed with increased microbial biomass C and N. The number of bacteria and actinomycetes in the rhizosphere soils of jatropha were more by 40 and 50%, respectively, than from the non-rhizosphere soil samples. Numbers of fungi in the rhizosphere soil samples were more by twofold than the non-rhizosphere soil samples. Although jatropha plants are non-edible for human beings as well as animals, microbes were not adversely affected but were stimulated due to rhizosphere effect of jatropha.

15.3 Cultivation

15.3.1 Field planting

Seedling multiplication and transplanting

Quality of the seedlings affects survival, growth and yield of the crop. Therefore raising healthy seedlings assumes importance. Jatropha seedlings can be grown by two methods, by the bare root and container method (polythene bag). In the bare-root method, a nursery bed is prepared by mixing farm yard manure (FYM), soil and sand in equal parts. Soaked seeds are sown at a row spacing of 25 cm and plant to plant spacing of 5 cm. The plants are ready for transplanting 6 weeks after germination; they are then carefully uprooted from nursery beds, wrapped in wet gunny bags and transplanted within 24 h. Before transplanting it should be ensured that enough moisture is available in the pit receiving bare-root seedlings.

Seedlings of jatropha can also be raised in poly-bags (10 cm × 18 cm, 150 gauge for 3–4-month-old seedlings) filled with 2 kg medium comprising equal parts of soil, sand and FYM. Diammonium phosphate (DAP) may be added at the rate of 1.0 g per poly-bag.

Table 15.4. Microbial population as influenced by jatropha plantation at Velchal, Andhra Pradesh, India (Susanna, 2009).

Microbial parameters	Non-jatropha plantation soil	Jatropha plantation soil	Coefficient of variation
Bacteria (cfu g^{-1} soil)	8×10^4	1×10^{5}*	54.6
Fungi (cfu g^{-1} soil)	1×10^3	2×10^{3}*	35.6
Actinomycetes (cfu g^{-1} soil)	8×10^2	8×10^{2}*	80.0

*Significant at 1%.

Good quality seeds having 80% germination should be sown at a rate of one seed per bag at 2–3 cm depth for getting higher percentage germination; 1 g of mycorrhizae (mixed culture) may be placed below the seed at the time of sowing to enhance growth of seedlings. In a study, jatropha seedlings treated with mycorrhizae had higher plant height, girth and number of leaves compared to control, when sampled 85 days after sowing. Grading and root pruning is suggested to promote uniform growth of seedlings.

The study on evaluation of the effects of propagation techniques in jatropha showed more than 80% survival rate in plots planted with poly-bag seedlings, followed by bare-root cuttings as observed 1 year after the planting. Planting of stem cuttings and direct sowing of seeds in the main field proved less effective and the survival rate was less than 20%.

Direct seeding

One to two seeds of jatropha (sprouted) per hole can be sown directly in the main field. The main field should be ploughed and spots receiving the seeds may be marked and enriched with FYM and DAP. Seedlings grown by this method will take time for establishment and will have slow growth in the initial period. However, the development of a good taproot provides strength to withstand wind damage. There should be enough moisture in the soil to support germination. Damage by birds and rodents to germinating seeds results in reduced plant stand. Frequent weeding is required to prevent the seedlings from competing weed and shade. Similarly soil working around the seedlings will boost growth and will improve the moisture and rainfall infiltration.

Vegetative propagation (transplanting of pre-rooted cuttings/grafting)

Jatropha can be multiplied by raising cuttings in a raised bed and later transplanting in the main field. Cuttings of 2–3 cm thickness from the lower portion of the shoot and having a length of 25–30 cm may be prepared in the March, when plants shed most of their leaves. These cuttings are planted in nursery beds at a spacing of 30 × 30 cm or in poly-bags. Pretreatment of stem cuttings with 300 ppm IBA (indole butyric acid) solution for 5 min is desirable. Sprouting starts within 7 days of planting. Plants propagated by cuttings will normally produce seed within 1 year of planting and growth is rapid. However, it has been observed that seedlings raised from seeds have better root systems compared to cuttings. The plants raised through cuttings will be true to type and will have similar characteristics of the mother plant but are prone to damage by strong winds as they lack proper taproots.

15.3.2 Fertilization

Though jatropha is well adapted to marginal lands, it responds to fertilizer application. Nutrient management is one such aspect, which may play a pivotal role in economic cultivation on marginal lands. Information available on nutrient requirements of jatropha is scarce and that on wastelands is non-existent. The work to evaluate the effect of nitrogen and phosphorus on growth and productivity of 4-year-old jatropha planted on wastelands at ICRISAT showed response to fertilizer application. In the on-station study, a 4-year-old jatropha plantation recorded 1290 kg ha^{-1} seed yield in the unfertilized control plots. The seed productivity varied from 1320 to 1610 kg ha^{-1} with nitrogen and phosphorus fertilizers applied. The highest productivity was recorded at 80 kg N and 20 kg P_2O_5 ha^{-1}.

Nutrients requirement for targeted jatropha seed yield based on nutrient budgeting

Detailed nutrient concentrations in different plant parts along with amount of plant materials returned to soil and seed removed from the fields for the purpose of oil extraction enables us to look at the nutrient requirements to achieve the targeted crop yields. Our results showed that 1 t jatropha seed removed 22 kg N, 5 kg P and 8 kg K ha^{-1}. The productivity level of 3 t seeds ha^{-1} will remove 66 kg N, 15 kg P and 24 kg K ha^{-1} year^{-1}. In order to sustain the productivity of the field at this level it is necessary to return equivalent quantities of N, P and K in simple arithmetic terms.

However, nutrient availability in soil from different sources of nutrients varied significantly depending on the C:N ratio, biological activity along with soil moisture and temperature. For mineral fertilizers, recovery of nutrients, for example, in the case of N is about 40–45%, whereas from organic matter of wider C:N ratio such as fallen leaves with 40:1 C:N ratio, N and other nutrient availability will be far less (<20%) during the first year of application. Leaf fall data and return of nutrients through leaf fall indicated that a 3-year-old plantation returned 21 kg N, 2 kg P and 23 kg K ha^{-1}. Applying the efficiency of release and uptake of N from the fallen leaf material it may be able to provide about 5 kg N ha^{-1} at best. It means that for achieving target yield of 3 t ha^{-1} seeds, one will need to meet the demand of additional 60 kg plant available N per ha and applying the basis of even 50% recovery from the chemical fertilizers one will need to apply about 120 kg N ha^{-1}. This will also be the case for phosphorus, which will need application of 14 kg P, i.e. 38 kg P_2O_5 ha^{-1}. These studies clearly indicate that sustainable jatropha production will definitely need application of plant nutrients in sufficient quantities. Leaf fall and application of oilseed cake after extraction of oil will only partially meet plant demand for nutrients such as N and P. Regular applications of other nutrients such as K, Zn, B and S may not be needed, as removal of these nutrients through seeds is low and return through leaf fall is considerable. However, detailed studies on nutrient release patterns from the fallen leaf material as well as oilseed cake are needed to arrive at the right nutrient dose applications to sustain seed yields.

15.3.3 Growth regulators

Jatropha is a monoecious shrub exhibiting protandry, where the flowers are unisexual, and male and female flowers are produced in the same inflorescence and male flowers open first followed by female flowers. Inflorescences produce a central female flower surrounded by a group of male flowers. Generally, there are 1–5 female flowers and 25–90 male flowers per inflorescence. The average male to female flower ratio is 29:1. As it is well documented in cucurbits that the use of growth regulators has the ability to manipulate sex expression, studies were conducted at ICRISAT during 2009 to evaluate the effect of growth regulators such as naphthalene acetic acid (NAA), gibberellic acid (GA), chlorocholine chloride (CCC) and ethrel on the male and flower ratio and yield of jatropha planted in the year 2004. The results revealed that yields were in general low due to unfavourable weather, but application of either CCC or GA at 90 and 25 ppm, respectively, at the time of flower initiation improved the flowering characteristics and crop yields (Table 15.5).

15.3.4 Pruning

Jatropha produces flowers in racemose inflorescences with dichasial cyme pattern in new growth branches, and hence the number of new branchlets determines the number of inflorescences. Therefore, pruning is essential to increase the number of fruiting branchlets, which is carried out by nipping the terminal bud to induce secondary and tertiary branches.

A study was carried out to assess the effects of pruning on canopy characteristics of jatropha in which half of the plants in the block plantation were pruned at 45 cm in the first year and 75 cm in the second year of growth during the dry season (February–March) when the plants are dormant, and the remaining half were grown without pruning. During the third year, the top one-third of the secondary and tertiary branches of plants were nipped off under the pruning treatment. The effect of pruning was observed on plant height, stem girth at 10 cm above the ground, number of branches, crown area, and volume index during the third year (Fig. 15.5). The results showed that pruning significantly ($p < 0.05$) influenced plant height (224 cm), stem girth (36 cm) and crown index (29) compared to the non-pruned plants, where the plant height, stem girth and crown index were 197 cm, 27 cm and 15, respectively. Similarly, the number of branches per plant was also more (43) in the pruned compared to the non-pruned (26) plants.

Table 15.5. Effects of application of plant growth regulators on jatropha productivity at ICRISAT, Patancheru, India during 2009 (ICRISAT, 2010).

Treatments	No. of inflorescences	Pod no.	Pod weight (g)	Seed no.	Seed weight (g)
Control	10.2	70.9	162.0	173.0	102.9
NAA at flower initiation (FI)	8.2	55.5	108.5	122.3	71.9
NAA at 1 month after FI	12.5	7.0	13.9	18.8	8.1
NAA at FI and 1 month after FI	9.8	6.3	18.3	19.0	10.2
GA at FI	10.0	158.0	384.5	400.8	247.5
GA at 1 month after FI	10.5	26.5	50.6	58.3	28.9
GA at FI and 1 month after FI	6.8	9.8	17.6	22.3	9.5
CCC at FI	9.8	152.0	366.6	415.5	249.2
CCC at 1 month after FI	10.7	12.3	27.3	30.5	16.2
CCC at FI and 1 month after FI	11.0	128.8	276.0	320.5	188.9
Ethrel at FI	10.3	35.5	107.9	120.2	71.3
Ethrel at 1 month after FI	7.2	94.0	207.6	219.8	134.1
Ethrel at FI and 1 month after FI	15.8	21.0	38.4	46.2	22.2

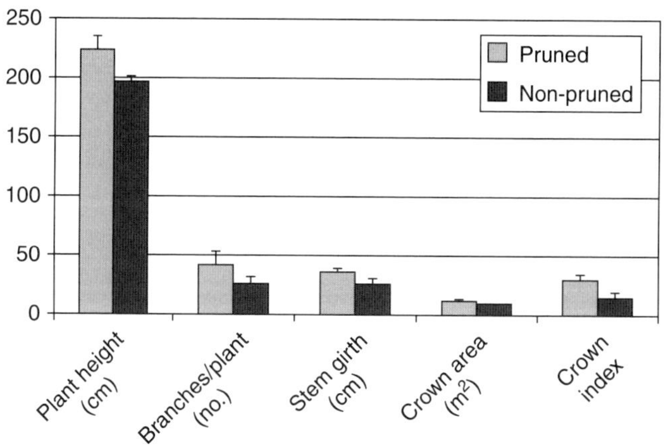

Fig. 15.5. Growth characteristics of jatropha plants due to summer pruning during the third year of establishment (Wani *et al.*, 2009).

Jatropha is well adapted to drought conditions, but shows profuse growth under irrigated conditions compared to no irrigation. As the fruits are borne on new branches, the enhanced growth and number of branches has a direct relation to fruit/seed yield.

15.3.5 Insect pests and diseases

Jatropha has been believed to be less susceptible to insect pests and diseases merely based on a few observations from isolated plants. Insects such as leaf-eating beetles, thrips, leafhoppers, grasshoppers, caterpillars and leaf miner feed on foliage. Shoot/stem borer and bark-eating caterpillars damage the stem. Blue bugs and green stink bugs suck on fruits while capsule borer damages the fruits (Wani *et al.*, 2006a). All the above pests may be controlled by spraying endosulfan at the rate of 3 ml l^{-1} water or any other pesticide recommended for that particular pest. Galls are formed due to the attack of mites and can be controlled by spraying dicofol at the rate of 5 ml l^{-1} water or wettable sulfur at the rate of 3 g l^{-1} water.

Diseases such as root rot, damping off, powdery mildew and leaf spots and cassava mosaic virus are frequently reported when the crop is raised as plantations (Wani *et al.*, 2006b; Divakara *et al.*, 2010). Black rot was observed during 2009 and 2010 in plantations in several Indian states including Andhra Pradesh, Assam, Chattisgarh and Madhya Pradesh (Srinivasa Rao *et al.*, 2011). Affected plants (Fig. 15.6) showed drying along with shrivelling, discoloration of the stem with sticky reddish brown exudation at the base of plants. Black lesions on the stem under the bark and cambium layer were also observed. The symptoms spread to leaves and petioles as 1–3 mm diameter black spots and shrivelling and gummosis of hard wood, finally leading to death of the infected plant. The casual fungus was identified as *Botryosphaeria dothidea*. Spraying affected plants at the early stage of symptoms with bavistin (Carbendazim 50% WP) at the rate of 2 g l^{-1} controlled the spread of the symptoms and led to recovery of plants with new leaf growth after the rains (Fig. 15.7). Studies on regional incidences of insect pests and diseases are needed for development of an effective package of practices for economic cultivation.

15.3.6 Benefits through intercropping in jatropha during gestation period

Because jatropha takes a minimum 3–4 years before producing economic yields, intercrops provide additional income to the farmers during the gestation period. The feasibility of growing drought-tolerant field crops as intercrops with jatropha was studied on 3-year-old plantations on vertisols at the ICRISAT farm, Patancheru, India. Intercrops such as sorghum, pearl millet, pigeonpea, chickpea, sunflower, safflower, soybean and mung bean were successfully cultivated and evaluated in jatropha plantations during the rainy season and post-rainy seasons. Sorghum, pearl millet, soybean and chickpea yielded more than 1 t ha^{-1}. The intercrops' productivity in terms of grain yield varied from 0.29 t ha^{-1} in green gram to

Fig. 15.6. *Jatropha curcas* plants infected with *Botryosphaeria dothidea*: (a) exudation of reddish brown gummy substance from the late infected stem and (b) black lesions over the stem under the bark.

Fig. 15.7. Initiation of new foliage from affected branches/ground after fungicide spraying in Velchal village plantation, Rangareddy district, during July 2009 (Srinivasa Rao *et al.*, 2011).

1.5 t ha^{-1} in sorghum. Total economic value from additional income through grains and fodder from these crops varied from Indian Rs 5355 to Rs 20,430 per hectare.

15.3.7 Seed yield

Productivity of jatropha depends on precipitation rates, soil moisture availability, soil characteristics including fertility (Francis *et al.*, 2005; Kumar and Sharma, 2008; da Schio, 2010; Jingura *et al.*, 2011), genetics (Kaushik *et al.*, 2007; Sunil *et al.*, 2008; Divakara *et al.*, 2010), plant age (Carels, 2009) and various management factors such as pruning, fertilization and disease control (Wani *et al.*, 2006a; Kaushik *et al.*, 2007; Achten *et al.*, 2008; Behera *et al.*, 2010; Ghosh *et al.*, 2011; Jingura *et al.*, 2011). Annual yield levels at 2–3 t dry seeds ha^{-1} have been proposed as achievable in semi-arid areas and on wastelands, while 5 t ha^{-1} can be obtained with good management on good soils receiving 900–1200 mm average annual rainfall (Foidl *et al.*, 1996; Francis *et al.*, 2005; Carels, 2009). Jongschaap *et al.* (2007) reported potential jatropha yields as high as 7.8 t dry seed ha^{-1} $year^{-1}$. The decorticated seeds yield about 28–40% oil (Wani *et al.*, 2006a; Divakara *et al.*, 2010), which can be transesterified and used for producing biodiesel (Jain and Sharma, 2010; Koh and Ghazi, 2011). Jatropha has not yet undergone breeding programmes with selection and improvement. The productivity varies greatly from plant to plant and environmental factors are reported to have a dominating role over genetics in determining seed size, weight and oil content (Kaushik *et al.*, 2007).

15.4 Factors Controlling Up-scaling of Jatropha Cultivation

15.4.1 Farmers' interest in crop

Lack of interest shown by farmers to jatropha cultivation is one of the main problems to its large-scale adoption. The success of jatropha seed oil to be used for biodiesel production lies in the sustainable and economically viable production of seeds at the field level. Farmers do not yet consider jatropha cultivation sufficiently profitable due to the wrong perspective being adopted, i.e. it is considered as a cash crop grown on good soil, whereas in practice it is to be grown on lands that cannot be used for growing food or other commercial crops. Therefore, large-scale cultivation of jatropha has been put into question and concerns have been raised regarding its success. At the start, in a rush to push the crop, some of the plantings were set on productive lands. But when returns from oilseed yields from these plantings were compared to usual cash crops, jatropha fared poorly. This resulted in a poor first impression and caused a setback to the programme as well. Here, clarity has to be kept in mind that promotion of jatropha is for growing on lands unsuitable for agriculture and where hardy plants only can be grown, thus tapping unused land resources and bringing in additional benefit to farmers.

Among other issues challenging up-scaling are current low yields, lack of high-yielding cultivars, high harvesting costs, diseases and pests in block plantations, water balance changes and off-site impacts. Over and above all, the market for the seed is not secured and farmers are not very sure where to sell the seed. On the other hand, in the absence of seed-collection and oil-extraction infrastructure, it will be difficult to persuade entrepreneurs to install transesterification plants.

Farmers also are hesitant in view of: (i) lack of confidence due to the delay in notifying, publicizing and explaining the government biodiesel policy; (ii) in the absence of long-term purchase contracts, there are no buy-back arrangements or purchase centres for jatropha; (iii) lack of availability of certified seeds with higher yield and oil content; and (iv) no announcement of incentives and other benefits to farmers. The government needs to take confidence-building measures and address farmers' concerns. Financial assistance should be given to organizations in developing a large-scale awareness/training programme for farmers.

15.4.2 Coherent national policy on biofuels

A country-wide policy to facilitate production of feedstock and processing is essential to bring the jatropha biodiesel programme to fruition. State by state policy differences create confusion and limited progress. In India, a petrol blending programme mandated 5% ethanol blending of petrol, initially for selected states and union territories and in 2006 extended to the whole country (Ministry of Petroleum & Natural Gas, 2008). Realizing the urgency, the Government of India (2009) has formulated and approved a National Policy on Biofuel along with setting up an empowered National Biofuel Coordination Committee, headed by the Prime Minister and a Biofuel Steering Committee headed by a Cabinet Secretary. Under the approved policy, the country aims to increase blending of biofuels with gasoline and diesel to 20% by the year 2017 (Achten *et al.*, 2010b). The policy focuses on indigenous production of biodiesel in waste/degraded/marginal lands. It incorporates the announcements on Minimum Support Price (MSP), with the provision of periodic revision for biodiesel oilseeds to provide a fair price to the growers, which would be based on the actual cost of production. In the case of biodiesel, the MSP could be linked to the prevailing retail diesel price. The National Biofuel Policy envisages that biofuels, namely, biodiesel and bioethanol, may be brought under the ambit of 'Declared Goods' by the Government to ensure unrestricted movement of biofuels within and outside the states. It is also stated in the policy that no taxes and duties would be levied on biodiesel.

15.4.3 Availability of sufficient degraded land for planting

Historically, while studying the bioenergy potential the availability of suitable land is the primary concern and focus. The recent Government of India assessments have identified 16% of the geographical area (>350 Mha) as wasteland (GOI, 2010). Keeping in mind the 400 million poor in India, and the fact that

70% of the poor in India are small/marginal farmers and landless labourers (GOI, 2010), it is essential that wasteland development programmes are undertaken to generate the needed socio-economic benefits for poor farmers and labourers.

The National Mission on Biofuels in India has identified about 13.4 Mha for jatropha (and pongamia) plantations in the immediate future and it covers poor, marginal, degraded, fallow, waste and other lands such as along the canals, roads, railway tracks, on farm and property bunds in the arid and semi-arid areas. Once success is achieved on potential lands, it should be possible to include lands with low fertility soils, which can be brought under jatropha plantation in an economically feasible manner to rehabilitate them. By adopting knowledge-based and pro-poor strategy, non-edible oils can be used for biodiesel production. Pro-poor biopower strategy leads to a win-win-win situation – improve livelihoods, protect the environment and have energy from renewable sources (ICRISAT, 2007). Other developing countries also have land area unfit for crop production and jatropha production may provide a suitable option.

15.4.4 Employment generation and social mainstreaming

The biofuels sector has the potential to successfully rehabilitate degraded lands and so improve the livelihoods of the rural people by providing employment and additional sources of income (Wani *et al.*, 2006a). Biofuel plantation activity on a commercial scale provides employment at village level through plantation, agronomic management, seed collection and through markets for fertilizer, pesticides, fuel and industrial raw materials for soap/cosmetics, etc. Developing nations are looking towards biofuels to help reduce their spiralling foreign oil import costs, and to mitigate pollution and global warming. The carbon credits gained through the fuel switch can also be a source of income to the rural people as these can be traded with the other developing or developed regions/countries.

A large part of the population in developing countries, mostly in rural areas, does not have access to energy services. There is an increasing gap between supply and demand, in addition to a continuous deterioration in quality of power and a low level of access to electricity. Lack of access to affordable energy services among the rural poor seriously affects their chances of benefiting from economic development and improved living standards. Under such circumstances decentralized power generation using biofuels is the need of the hour. Access to modern decentralized small-scale energy technologies, particularly renewable biofuels, is an important element for effective poverty-alleviation policies. A programme that develops energy from raw material grown in rural areas will go a long way in providing energy security to the rural people for their domestic use, farm production activities (e.g. irrigation, etc.) and engagement in livelihood activities (e.g. small processing mills), leading to improved incomes, better living and their social mainstreaming.

Another advantage of jatropha cultivation is that the former use of land for grazing can be continued. This means that nobody in the village will lose their customary right due to jatropha plantations. Grazing in jatropha plantations may raise concerns about the potential intoxication of livestock. Toxicity in jatropha is due to the presence of toxalbumin of nomecurcin (toxin protein), which irritates the gastrointestinal mucosa and also haemoagglutinates and causes nausea, vomiting, intense abdominal pain and diarrhoea with bloody stools (Ribeiro and Matavel, 2009). However, such incidence in a study village has not been reported, probably because animals quickly perceive the danger of consuming jatropha.

An additional benefit to the community from jatropha cultivation is higher groundwater tables, which improves access to water for domestic and agricultural use. Achten *et al.* (2010a) thoroughly discussed the benefits of jatropha cultivation in wastelands at a local scale. After oil extraction, seed cake could not be used for animal feed due to its toxic content but it could potentially be used as fertilizer and biopesticide/insecticide and molluscicide simultaneously (Rug *et al.*, 1997). Moreover seed cake could be used for biogas production through anaerobic digestion before using it as a soil amendment (Staubmann *et al.*, 1997).

The calculations show that the 3–5 years that it takes to cultivate 1 ha of jatropha has production costs between US$740 and US$920. As jatropha has such a long period without any financial return, there must be provisions to ensure the economic stability of the farmers. Flexible and long-term financing options could facilitate the expansion of agrofuel production. The assumed bank loan covering the initial expenditures could be broadened, providing different financing options for farmers, as well as increasing overall access to financial support and reducing the interest rate for repayment.

15.5 Genetic Improvement of Jatropha

Very little is known about the jatropha genome and not many systematic efforts for improvement of this crop have been undertaken. Improved varieties with desirable traits for specific growing conditions are not available, which makes growing jatropha a risky business (Wani *et al.*, 2006a; Jongschaap *et al.*, 2007). Chromosomes are of very small size (bivalent length 1–3.67 μm) with most species having $2n = 22$ and base number of $x = 11$ (Soontornchainaksaeng and Jenjittikul, 2003). It is an attractive candidate for genome sequencing with genome size (1C) of 416 Mbp (Carvalho *et al.*, 2008). Because jatropha is an introduced plant to many countries of Asia, Africa and Latin America, genetic variability has not been much explored for genetic improvement.

Genetic improvement of jatropha is suggested through mutation techniques. Selection of germplasm and identification of pistillate variants (Hegde, 2003) as followed for castor (*Ricinus communis* L.) belonging to the same family is also recommended (Sujatha, 2006; Sujatha *et al.*, 2008). Development of pistillate plants through mutation and interspecific hybridization techniques is time consuming. Alien gene transfer through interspecific hybridization and biotechnological interventions to bring the change in the desired traits and/or to identify divergent parents, which can later be exploited by heterosis, is also suggested. Meanwhile, physiological manipulation of sexuality, applying gametocides to enhance the M/F ratio and altering the expression levels of enzymes in the triacylglycerol-biosynthetic (Kennedy) pathway (King *et al.*, 2009) may also be considered to increase yield.

The objective for genetic up-gradation of the crop should aim at a higher number of female flowers or pistillate plants, high seed yield with high oil content, early maturity, resistance to pests and diseases, drought tolerance/resistance, reduced plant height and high natural ramification of branches. Jatropha is often a cross-pollinated crop and can be improved through mass selection, recurrent selection, mutation breeding, heterosis breeding and interspecific hybridization or biotechnological interventions to bring the change in the desired traits (Divakara *et al.*, 2010).

15.5.1 Tapping germplasm for potential productivity

Genetic diversity in plant species is a gift to mankind as it forms the basis for selection and further improvement for increasing the yield potential as well as other desired traits. However, comprehensive work on a jatropha germplasm collection, characterization and evaluation for growth, morphology, seed characteristics and yield traits is still in its infancy. The fact that jatropha has adapted itself to a wide range of edaphic and ecological conditions suggests the existence of a considerable amount of genetic variability to be exploited for potential realization (Rao *et al.*, 2008). Detailed and systematic provenance trials have not been carried out, and material from the centre of origin has not been sufficiently screened. The genetic background of jatropha grown in Africa and Asia is unclear and G × E interaction amongst the provenances needs to be taken into account while considering the reported variations (Namkoong *et al.*, 1988). Priority should be given to assess intra- and inter-accessional variability in the available germplasm, selection of pure lines and then multiplication.

Existence of natural hybrid complexes is reported in the genus *Jatropha*, such as the *J. curcas–canascens* complex in Mexico (Dehgan and Webster, 1978), *J. integerrima–hastata* complex in Cuba and West Indian islands (Pax, 1910) and *J. curcas–gossypifolia* (*J. tanjorensis*) in India (Prabakaran and Sujatha, 1999). Hence, germplasm exhibiting gross morphological differences should be the subject of pollen studies and lines with pollen abnormality or poor seed set should be investigated in detail before drawing conclusions about the distinctness (Divakara *et al.*, 2010). Makkar *et al.* (1997) reported large variations in contents of crude protein, crude fat, neutral detergent fibre and ash on 18 different provenances of jatropha from countries in West and East Africa, the Americas and Asia. Wani *et al.* (2006a) recorded variation in Indian accessions for oil content (27.8–38.4%) and 100-seed weight (44–77 g). Kaushik *et al.* (2007) explored the variability in Haryana, India accessions to find wide variation in 100-seed weight (49–69 g) and oil content (28–39%). Similarly, Rao *et al.* (2008) found wide variation in 100-seed weight (57–79 g) and oil content (30–37%) for Andhra Pradesh, India, accessions.The study conducted at ICRISAT on progeny trial evaluation comprising 99 jatropha accessions collected from different agro-eco regions of India and planted during 2006 showed great variability in plant height (105–274 cm), collar diameter (5.1–18.9 cm), number of branches (9.7–66.3), crown area (0.5–6.3 m^2), volume index (8056–85,708 cm^3), seed length (11.6–19.3 mm), seed width (9.5–11.7 mm), seed thickness (7.6–11.0 mm), 100-seed weight (38.9–67.1 g) and seed oil content (27.5–40.5) (Table 15.6). ICJC06116 recorded the highest 100-seed weight of 67.1 g. Genotype ICJC06004 recorded the highest plant height (274 cm) and ICJC06087 the highest volume index (85,708 cm^3) and collar diameter (18.9 cm). The number of branches and crown area were highest in ICJC06115 (66.3) and ICJC06010 (6.3 m^2), respectively. Highest seed length was seen in genotype ICJC06082 (19.3 cm) while genotype ICJC06055 showed the maximum seed width (11.7 cm) and thickness (11.0 cm). Highest oil content for seeds was 40.5% in genotype ICJC06019. As a whole, 44% of the genotypes have exhibited seed oil content in the range of 35.1–40%.

In India, the National Oilseeds and Vegetable Oils Development (NOVOD) Board (Ministry of Agriculture, Government of India) has identified 1855 candidate plus trees (CPTs) of jatropha and has over 5000 accessions

Table 15.6. Performance of selected *Jatropha curcas* genotypes for growth and seed traits at ICRISAT during 2008 (planted 2006) (ICRISAT, 2010).

Genotypes	Plant height (cm)	Collar diameter (cm)	Number of branches	Crown area (m^2)	Volume index (cm^3)	Seed length (mm)	Seed breadth (mm)	Seed thickness (mm)	100-seed weight (g)	Oil content (%)
ICJC06004	274	16.0	52.1	5.3	76,077	17.9	10.9	9.1	53.9	31.2
ICJC06010	226	15.0	54.5	6.3	51,563	17.9	10.9	8.6	51.2	37.8
ICJC06019	228	12.8	41.4	2.4	38,653	16.6	11.4	10.7	52.3	40.5
ICJC06055	135.9	5.8	14.9	0.8	8,972.8	16.7	11.7	11.0	46.1	34.9
ICJC06066	168	6.4	23.4	0.8	10,433	16.9	10.6	7.6	54.8	32.1
ICJC06082	207.5	14.0	32.4	1.5	41,756.3	19.3	10.3	9.3	53.3	35.0
ICJC06087	233.2	18.9	51.5	3.5	85,708	18.2	11.3	9.0	56.8	37.3
ICJC06091	202	18.3	51.0	3.4	68,345	17.6	11.5	10.3	64.3	39.1
ICJC06115	213	14.7	66.3	3.2	48,761	17.3	11.4	10.7	66.7	34.8
ICJC06116	205	17.6	61.9	3.5	64,976	19.1	11.5	8.9	67.1	29.4
ICJC06117	201	15.3	62.2	3.7	50,474	17.7	11.6	10.8	58.2	34.7
ICJC06120	198	10.6	48.5	2.3	23,850	18.7	11.0	8.6	66.6	34.4
Mean (of 99)	204	11.8	37.3	2.0	33,935	17.4	11.1	9.1	52.7	34.7
SEM	14.1	1.3	5.8	0.5	8,288.9	0.2	0.2	0.1	3.4	0.3
CD (5%)	43.6	3.9	17.8	1.3	25,664	0.6	0.5	0.4	11.2	0.8

collected with a network of 40 institutions. The oil content varies from 26.0% to 42.7% (Punia, 2007). Department of Biotechnology (DBT), Government of India, launched a micro-mission on the production of quality planting material with 30–40% oil and 3–5 t ha^{-1} seed yield; under this mission, 1500 accessions have been collected (Swarup, 2006). Kaushik *et al.* (2007) analysed 1000 samples of jatropha seeds representing 12 states of India for oil content and kernel seed-coat ratio and reported that a collection from Uttaranchal recorded the highest percentage of high-oil-yielding plants (73%). Most of the jatropha varieties spread worldwide (e.g. the variety 'Cape Verde', a variety from Nicaragua and a non-toxic variety from Mexico) were developed from good performing selections from the existing natural population and were not the result of targeted crossings (Heller, 1996). The first variety, 'SDAUJ I' ('Chatrapati'), has been identified for commercial cultivation in the semi-arid and arid regions of Gujarat and Rajasthan in India (http://www.icar.org.in/pr/10052006.htm). The Indian Council of Agricultural Research (ICAR) launched a network project in which 28 seed sources were identified as the best among 496 seed sources (Paramathma *et al.*, 2009).

15.5.2 Assessment of variability in jatropha accessions using molecular markers

Microsatellite markers are available for a few *Euphorbiaceae* members and are being developed for castor (Lespinasse *et al.*, 2000; Okogbenin *et al.*, 2006; http://castorbean.tigr.org). As the developmental costs for microsatellite markers are high, cross-taxa utility of molecular markers from *Hevea* and cassava could be assessed as done in other plant species (Rossetto, 2001). Divakara *et al.* (2010) reported that the majority of the studies are confined to characterization of accessions available in India; the exceptions are Basha and Sujatha (2007) with one non-toxic accession from Mexico, Montes *et al.* (2008) with accessions from 30 countries and Sun *et al.* (2008) with accessions from south China (Table 15.7). Regardless of the number of accessions used, the robustness of the primer and number of marker data points, all accessions from India clustered together. In general, diversity analysis with local germplasm revealed a narrow genetic base in India (Basha and Sujatha, 2007; Ganesh Ram *et al.*, 2008) and south China (Sun *et al.*, 2008), indicating the need for widening the genetic base of jatropha through introduction of accessions with broader geographical background and creation of variation through mutation and hybridization techniques. In contrary to the above studies, AFLP (amplified fragment length polymorphism)-based molecular characterization of jatropha accessions from Andhra Pradesh were found to be diverse as these were scattered in different groups, showed the occurrence of a higher number of unique/rare fragments and had greater variation in percentage oil content (Tatikonda *et al.*, 2009). Forty-eight accessions, collected from six different states of India, were used with seven AFLP primer combinations that generated a total of 770 fragments with an average of 110 fragments per primer combination. A total of 680 (88%) fragments showed polymorphism in the germplasm analysed, of which 59 (8.7%) fragments were unique (accession specific) and 108 (15.9%) fragments were rare (present in less than 10% accessions). In order

Table 15.7. Characterization of accessions using molecular markers (Divakara *et al.*, 2010).

Species/ accessions	Primers	Number	References
142	AFLP	–	DBTIndia, 2007
5	RAPD	18	Ganesh Ram *et al.*, 2008
22	RAPD DAMD	7 4	Ranade *et al.*, 2008
13	RAPD ISSR	20 14	Gupta *et al.*, 2008
7	RAPD AFLP	52 27	Pamidimarri *et al.*, 2008
20	RAPD AFLP	– –	Reddy *et al.*, 2007
43	RAPD ISSR	400 100	Basha and Sujatha, 2007
225	AFLP	–	Montes *et al.*, 2008
58	SSR AFLP	30 7	Sun *et al.*, 2008

to assess the discriminatory power of the seven primer combinations used, a variety of marker attributes such as polymorphism information content (PIC), marker index (MI) and resolving power (RP) values were calculated. Although the PIC values ranged from 0.20 (E-ACA/M-CAA) to 0.34 (E-ACT/M-CTT) with an average of 0.26 per primer combination and the MI values were observed in the range of 17.60 (E-ACA/M-CAA) to 32.30 (E-ACT/MCTT) with an average of 25.13 per primer combination, the RP was recognized as the real attribute for AFLP to determine the discriminatory power of the primer combination. The RP values for different primer combinations varied from 23.11 (E-ACA/M-CAA) to 46.82 (E-ACT/M-CTT) with an average of 35.21.

However, careful understanding of the phylogeny and use of an adequate number of molecular markers are essential prerequisites for drawing valid inferences about the genetic affinities (Sujatha *et al.*, 2008). Basha and Sujatha (2009) characterized jatropha species occurring in India using nuclear and organelle-specific primers, which revealed high interspecific genetic variation (98.5% polymorphism). Further characterization of both natural and artificially produced hybrids using chloroplast-specific markers revealed maternal inheritance of the markers (Basha and Sujatha, 2009). In support, genetic variation studies using RAPD (random amplified polymorphism DNA), AFLP and combinatorial tubulin-based polymorphism (cTBP) indicated higher possibilities of improving *J. curcas* by interspecific breeding (http://precedings.nature.com/documents/2782/version/1). Hence, molecular diversity estimates combined with the datasets on other agronomic traits will be very useful for selecting the appropriate accessions.

15.5.3 Jatropha genome project

The jatropha genome is approximately 400 million bp in size, similar to the size of the rice genome based on the sequencing of the genome, using both traditional Sanger sequencing and next-generation sequencing. Work in progress on the annotation of the genome to identify genes of interest helps in discovering genetic variations using marker-assisted breeding and provides information on factors controlling oil synthesis, maximizing yield, biotic and abiotic stress tolerance and low-curcin variants (http://greenbio.checkbiotech.org/news/first_Jatropha_genome_completed_synthetic_genomics_inc_and_asiati 511 c_centre_genome_tech).

Divakara *et al.* (2010) indicated that *J. curcas* is still in its infancy with respect to genetic improvement for increasing seed and oil yield per unit area. Genetic improvement activities so far have involved only collection and selection in local germplasm and hence initiatives to evaluate global germplasm is necessary, to study the pattern of diversity for various traits. Genetic improvement using conventional breeding approaches has to be initiated at more places and integrated with the latest biotechnological techniques for reducing time and increasing efficiency of breeding. Potential of the new varieties developed has to be further tested for their performance, through multilocation trials. Development of techniques, such as somoclonal variants, mutations, doubled haploids and gene transfer, that support plant-breeding activities should be emphasized. Similar to other crops, heritability of seed traits is the most common predictor of genetic gains for different breeding methods in jatropha. Application of information from the jatropha genome project can be used in marker-assisted breeding to accelerate and complement conventional breeding.

15.4 Jatropha Products

Biodiesel was probably the first of the alternative liquid fuels to become known to the public. The advantage in biodiesel is that it can be used in existing vehicles with little or no modification required when used as blended. During the last few decades researchers tried many edible and non-edible oils in compression ignition for different utilities. There are energy plants available that will produce a higher yield in kWh per area, but the simplicity of having a fuel that is fully compatible with present fuel and engine technology makes jatropha biodiesel very attractive. Jatropha biodiesel has good fuel properties, comparable to or even better than petroleum diesel (Rao *et al.*, 2008). The cetane number (an indication of its fuel-burning efficiency) is 51–52 for biodiesel from jatropha oil, which is higher than the cetane number of most petroleum diesels. It has 10% built-in oxygen content that helps it to burn fully. The esters of the long-chain fatty acids of biodiesel are excellent lubricants for the fuel injection system. It has a higher flash point than diesel, making it a safer fuel. But there are a few technical issues that need to be resolved. Biodiesel has a high viscosity at low temperatures, leading to flow problems at these temperatures.

Straight vegetable oils from jatropha are also used directly in high rpm diesel engines for power generation as well as pumping water (D'Silva *et al.*, 2004). The use of blended fossil fuels with biofuels results in reduction of unburnt hydrocarbons by about 30%, carbon monoxide by about 20% and particulate matter by about 25%. Moreover, sulfur content in the emissions from the use of blended fuels is almost negligible (Francis *et al.*, 2005). ICRISAT (2008) has assessed the performance of the vehicles (eight TATA Mobile 207Di, two Nissan Diesel and one Toyota Qualis) since May 2007 run on blended fuel at 10% biodiesel. The detailed examination of the engines by automobile engineers at ICRISAT indicated no abnormal wear or tear of the engine as well as any changes in the kilometreage with the blended fuel. No starting trouble or pick-up problem was observed in any vehicle. There was no abnormal smoke or other specific complaints from the users while driving the vehicle.

Jatropha oilcake contains all the macro- and micronutrients and it is an excellent organic fertilizer unlike inorganic fertilizers that supply one or two nutrients. A mass of 4 kg of jatropha seed gives about 3 kg of cake, which is generally sold at Indian Rs. 3 to 7 per kilogram depending upon the demand and supply scenario. The cake is mostly used for fertilizing plantation or commercial crops. An analysis of jatropha cake indicated the presence of all the essential elements required for plant growth, and was particularly found to be rich in nitrogen and sulfur.

Similar studies carried out in Zimbabwe have revealed that application of jatropha cake at the rate of 0, 0.25, 0.5 and 1.0 kg m^{-2} resulted in cabbage yield of 16.8, 23.6, 22.8 and 35.8 kg, respectively, and the crop was free from pests and diseases. Tasosa *et al.* (2001) have recorded a significant difference in growth rates of tomato and total above-ground dry matter with increased application rates of jatropha and castor cakes. Henning (2009) is of the opinion that jatropha plants can reduce the soil and water erosion when planted as a living fence and the cake obtained after oil extraction can help in building the organic matter content of the soils of Sahelian countries. Substitution of fertilizers with oilseed cake is likely to improve the fertility of the soils in the long run and the soils will overcome the widely observed deficiency of several nutrients such as N, P, Z, B, S, etc. Further, it will reduce the dependence of farmers on external input (fertilizers). Recycling of the cake serves the interest of farmers as well as government as the huge subsidy paid on fertilizers to industries in most countries can be reduced. The amount saved on subsidy of fertilizers can be used for encouraging biodiesel plantations in rural areas for ensuring energy, food and livelihood security (Osman *et al.*, 2009).

The biodiesel processing produces glycerol of about 88% purity, which is about 12% in volume to the biodiesel produced. At present there is limited market for glycerol but research efforts are underway to find alternative uses.

15.4 Economic Assessment

The key determinant of the financial analysis to estimate the economic viability of jatropha production is the calculation of the net present value (NPV; difference between the present value of cash inflows and the present value of cash outflows). The investment in jatropha seed production is judged as economically viable when the NPV is above zero. Different interest rates (6–16%) are used to simulate investment alternatives, which are also applied to the NPV calculations. The results showed that for the low-cost scenario, the production of jatropha seed to be used for JPPO (jatropha pure plant oil) becomes economically viable at rates above a crude-oil price of US$75 per barrel (interest rate 6%). For JME (methylated ester) the NPV (interest rate 6%) for jatropha seed production first becomes positive at crude-oil prices above US$90 per barrel. At an interest rate of 16% and under the low cost assumptions, the use of jatropha seeds for JPPO needs a crude oil price greater than US$105 per barrel and JME needs a crude oil price above US$120 per barrel. For the average cost scenario, jatropha seed production becomes economically viable for JPPO at crude-oil prices above US$100 (interest rate 6%) whereas JME requires crude-oil prices of at least US$115 (interest rate 6%). The economic viability of jatropha seed production under wasteland conditions can differ considerably depending on the applied interest rate and production costs. As shown in Figs 15.8 and 15.9 even at a crude-oil price of US$150 per barrel the high-cost scenario reaches no positive NPV for both jatropha fuel options.

Acknowledgements

The support from National Oilseeds and Vegetable Oils Development, Government of India is duly acknowledged. The authors are also grateful to Mr Ch. Srinivas Rao for his help in data collection/compilation.

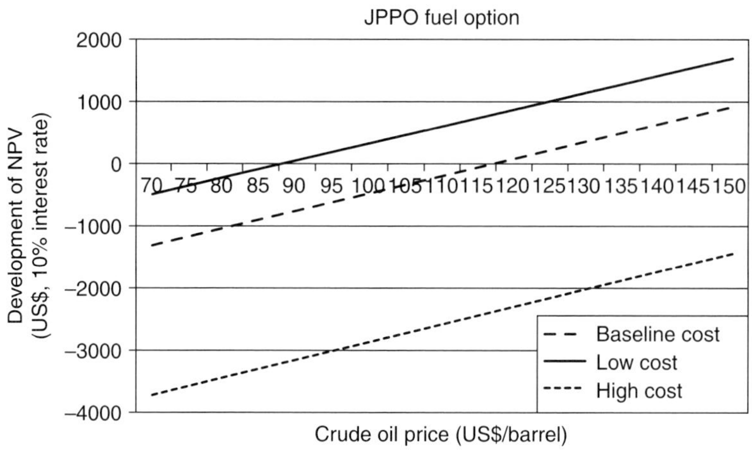

Fig. 15.8. Development of NPV for jatropha seed production according to the JPPO (Jatropha Pure Plant Oil) fuel option and different crude-oil price levels (Grass, 2010).

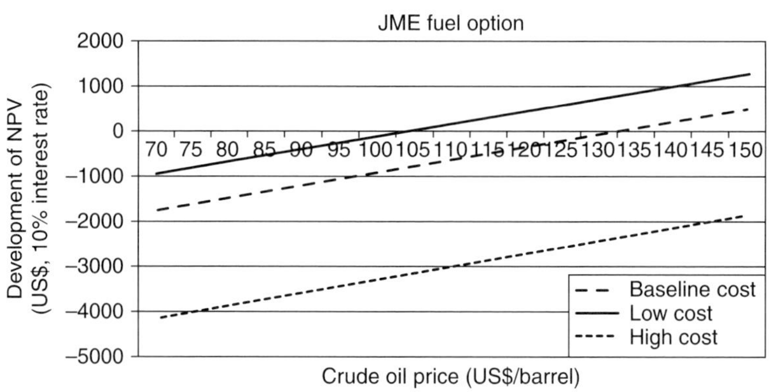

Fig. 15.9. Development of NPV for jatropha seed production according to the JME (Jatropha Methyl Ester) fuel option and different crude-oil price levels (Grass, 2010).

References

Achten, W.M.J., Verchot, L., Franken, Y.J., Mathijs, E., Singh, V.P. and Aerts, R. (2008) Review: Jatropha biodiesel production and use. *Biomass and Bioenergy* 32, 1063–84.

Achten, W.M.J., Almeida, J., Fobelets, V., Bolle, E., Mathijs, E., Singh, V.P., Tewari, D.N., Verchot, L.V. and Muys, B. (2010a) Life cycle assessment of Jatropha biodiesel as transportation fuel in rural India. *Applied Energy* 87, 3652–3660.

Achten, W.M.J., Maes, W.H., Aerts, R., Verchot, L., Trabucco, A., Mathijs, E., Singh, V.P. and Muys, B. (2010b) Jatropha: From global hype to local opportunity. *Journal of Arid Environments* 74, 164–165.

Basha, S.D. and Sujatha, M. (2007) Inter and intra-population variability of *Jatropha curcas* (L.) characterized by RAPD and ISSR markers and development of population specific SCAR markers. *Euphytica* 156, 375–386.

Basha, S.D. and Sujatha, M. (2009) Genetic analysis of Jatropha species and interspecific hybrids of *Jatropha curcas* using nuclear and organelle specific markers. *Euphytica* 168, 197–214.

Behera, S.K., Srivastava, P., Tripathi, R., Singh, J.P. and Singh, N. (2010) Evaluation of plant performance of *Jatropha curcas L.* under different agro-practices for optimizing biomass – a case study. *Biomass and Bioenergy* 34, 30–41.

Bhattacharyya, T., Chandran, P., Ray, S.K., Pal, D.K., Venugopalan, M.V., Mandal, C. and Wani, S.P. (2007) Changes in levels of carbon in soils over years of two important food production zones of India. *Current Science* 93, 1854–1863.

Bhattacharyya, T., Ray, S.K., Pal, D.K., Chandran, P., Mandal, C. and Wani, S.P. (2009) Soil carbon stocks in India – issues and priorities. *Journal of Indian Society of Soil Science* 57, 461–468.

Borchert, R. (1994) Soil and stem water storage determine phenology and distribution of tropical dry forest trees. *Ecology* 75, 1437–1449.

Carels, N. (2009) *Jatropha curcas*: a review. *Advances in Botanical Research* 50, 39–86.

Carvalho, C.R., Clarindo, W.R., Praça, M.M., Araújo, F.S. and Carels, N. (2008) Genome size, base composition and karyotype of *Jatropha curcas* L., an important biofuel plant. *Plant Science* 174, 613–617.

da Schio, B. (2010) *Jatropha curcas* L., a potential bioenergy crop. On field research in Belize. MSc dissertation, Padua University, Italy and Wageningen University and Research Centre, Plant Research International, the Netherlands.

DBTIndia (2007) Jatropha Germplasm Characterization for Biodiesel production. Available at: http://dbtindia.nic.in/Energy_bioscience/ABSTRACTS.pdf (accessed 15 November 2012).

Dehgan, B. and Webster, G.L. (1978) Three new species of *Jatropha* (Euphorbiaceae) from Western Mexico. *Madrono* 25, 30–39.

Divakara, B.N., Upadhyaya, H.D., Wani, S.P. and Gowda, C.L.L. (2010) Biology and genetic improvement of *Jatropha curcas* L.: a review. *Applied Energy* 87, 732–742.

D'Silva, E., Wani, S.P. and Nagnath, B. (2004) The making of new Powerguda: community empowerment and new technologies transform a problem village in Andhra Pradesh. *Global Theme on Agroecosystems* Report No. 11, Patancheru, Andhra Pradesh, India, 28 pp.

Foidl, N., Foidl, G., Sanchez, M., Mittelbach, M. and Hackle, S. (1996) *Jatropha curcas* L. as a source for the production of biofuel in Nicaragua. *Bioresource Technology* 58, 77–82.

Francis, G., Edinger, R. and Becker, K. (2005) A concept for simultaneous wasteland reclamation, fuel production, and socio-economic development in degraded areas in India: need, potential and perspectives of *Jatropha* plantations. *Natural Resources Forum* 29, 12–24.

Ganesh Ram, S., Parthiban, K.T., Kumar, R.S., Thiruvengadam, V. and Paramathma, M. (2008) Genetic diversity among Jatropha species as revealed by RAPD markers. *Genetic Resources and Crop Evolution* 55, 803–809.

Garg, K.K., Karlberg, L., Wani, S.P. and Berndes, G. (2011) Jatropha production on watersheds in India: opportunities and trade-offs for soil and water management at the watershed scale. *Biofuels, Bioproducts and Biorefining* 5, 410–430.

Ghosh, A., Chikara, J. and Chaudhary, D.R. (2011) Diminution of economic yield as affected by pruning and chemical manipulation of *Jatropha curcas* L. *Biomass and Bioenergy* 35, 1021–1029.

Government of India (2003) *Report of the Committee on Development of Bio-fuel*. Planning Commission, Government of India, New Delhi, 214 pp.

Government of India (2009) *National Policy on Biofuels*. Ministry of New & Renewable Energy, New Delhi.

Government of India (2010) Wasteland Atlas of India. Government of India (GOI), Ministry of Rural Development, Department of Land Resources, New Delhi, India. Available at: http://www.dolr.nic.in/wasteland_atlas.htm (accessed 9 April 2013).

Grass, M. (2010) Opportunities and constraints for agrofuels in developing countries: case studies on economic viability and employment effects of Jatropha production. PhD dissertation, University of Hohenheim, Germany.
Gupta, S., Srivastava, M., Mishra, G.P., Naik, P.K., Chauhan, R.S., Tiwari, S.K., Kumar, M. and Singh, R. (2008) Analogy of ISSR and RAPD markers for comparative analysis of genetic diversity among different *Jatropha curcas* genotypes. *African Journal of Biotechnology* 7(23), 4230–4243.
Hegde, D.M. (2003) Tree oilseeds for effective utilization of wastelands. In: *Compendium of Lecture Notes of Winter School on Wasteland Development in Rainfed Areas*. Central Research Institute for Dryland Agriculture, Hyderabad, India, pp. 111–119.
Heller, J. (1996) *Physic nut* Jatropha curcas *L. Promoting the conservation and use of underutilized and neglected crops*. Institute of Plant Genetic and Crop Plant Research Gatersleben/International Plant Genetic Resource Institute, Rome, Italy.
Henning, R. (2009) The Jatropha Book, The Jatropha System, an integrated approach of rural development. Available at: http://www.jatropha.de (accessed 30 March 2012).
Holl, M., Gush, M.B., Hallowes, J. and Versfeld, D.B. (2007) Jatropha curcas *in South Africa: an Assessment of its Water Use and Biophysical Potential*. Water Research Commission, Pretoria, South Africa.
Huante, P. and Rincon, E. (1998) Responses to light changes in tropical deciduous woody seedlings with contrasting growth rates. *Oecologia* 113, 53–66.
ICRISAT (International Crops Research Institute for the Semi-Arid Tropics) (2007) *Biopower Strategy*. ICRISAT, Patancheru, Andhra Pradesh, India.
ICRISAT (International Crops Research Institute for the Semi-Arid Tropics) (2008) *Supporting the Farmers' Activities in the Value-Chain of Biofuels*. Annual Project Report (October 2007–March 2008), ICRISAT, Patancheru, Andhra Pradesh, India.
ICRISAT (International Crops Research Institute for the Semi-Arid Tropics) (2010) *Harnessing the potential of water-use efficient bio-energy crops for enhancing livelihood opportunities of smallholder farmers in Asia, Africa and Latin America*. Annual Project Report (2009-10), ICRISAT, Patancheru, Andhra Pradesh, India.
Jain, S. and Sharma, M.P. (2010) Biodiesel production from *Jatropha curcas* oil. *Renewable and Sustainable Energy Reviews* 14, 3140–3147.
Jingura, R.M., Matengaifa, R., Musademba, D. and Musiyiwa, K. (2011) Characterisation of land types and agro-ecological conditions for production of Jatropha as a feedstock for biofuels in Zimbabwe. *Biomass and Bioenergy* 35, 2080–2086.
Jongschaap, R.E.E., Corré, W.J., Bindraban, P.S. and Brandenburg, W.A. (2007) *Claims and facts on* Jatropha curcas *L.* Plant Research International, Wageningen, the Netherlands.
Kaushik, N., Kumar, K., Kumar, S., Kaushik, N. and Roy, S. (2007) Genetic variability and divergence studies in seed traits and oil content of Jatropha (*J. curcas* L.) accessions. *Biomass and Bioenergy* 31, 497–502.
Kesava Rao, A.V.R., Wani, S.P., Singh, P., Srinivas, K. and Srinivasa Rao, C. (2012) Water requirement and use by *Jatropha curcas* in a semi-arid tropical location. *Biomass and Bioenergy*. Available online 21 January 2012, ISSN 0961-9534, 10.1016/ j.biombioe.2012.01.013.
King, A.J., He, W., Cuevas, J.A., Freudenberger, M., Ramiaramanana, D. and Graham, I.A. (2009) Potential of *Jatropha curcas* as a source of renewable oil and animal feed. *Journal of Experimental Botany* 60, 2897–2905.
Koh, M.Y. and Ghazi, T.I.D. (2011) Review of biodiesel production from *Jatropha curcas* L. oil. *Renewable and Sustainable Energy Reviews* 15, 2240–2251.
Krishnamurthy, L., Zaman-Allah, M., Marimuthu, S., Wani, S.P. and Kesava Rao, A.V.R. (2012) Root growth in Jatropha and its implications for drought adaptation, *Biomass and Bioenergy*, doi:10.1016/j. biombioe.2012.01.015.
Kumar, A. and Sharma, S. (2008) An evaluation of multipurpose oil seed crop for industrial uses (*Jatropha curcas* L.): a review. *Industrial Crops and Products* 28, 1–10.
Lespinasse, D., Rodier-Goud, M., Grivet, L., Leconte, A., Legnate, H. and Seguin, M.A. (2000) Saturated genetic linkage map of rubber tree (*Hevea* spp.) based on RFLP, AFLP, microsatellite and isozyme markers. *Theoretical Applied Genetics* 100, 127–138.
Maes, M.H., Achten, W.M.J. and Muys, B. (2009) Use of inadequate data and meteorological errors lead to an overestimation of water footprint of *Jatropha curcas*. *Proceedings of the National Academy of Sciences USA* 106, E91.DOI:10.1073/pnas.0906788106.
Makkar, H.P.S., Becker, K., Sporer, F. and Wink, M. (1997) Studies on nutritive potential and toxic constituents of different provenances of *Jatropha curcas*. *The Journal of Agricultural and Food Chemistry* 45, 3152–3157.

Ministry of Petroleum & Natural Gas, Government of India (2008) Petroleum Planning & Analysis Cell. Retail Selling Prices of Diesel – Major Price Revisions. Available at: http://www.ppac.org.in/OPM/Price_revision_other%20cities_HSD_oct.htm (accessed 30 March 2012).

Montes, L.R., Azurdia, C., Jongschaap, R.E.E., van Loo, E.N., Barillas, E., Visser, R. and Mejia, L. (2008) *Global Evaluation of Genetic Variability in* Jatropha curcas. Poster edn, Wageningen University Plant Breeding Research, Wageningen, the Netherlands.

Namkoong, G., Kang, H.C. and Brouard, J.S. (1988) Tree Breeding: Principles and Strategies. *Monographs on Theoretical and Applied Genetics*, vol. 11. Springer Verlag, New York.

Okogbenin, E., Marin, J. and Fregene, M. (2006) An SSR-based molecular genetic map of cassava. *Euphytica* 147, 433–440.

Osman, M., Wani, S.P., Balloli, S.S., Sreedevi, T.K., Srinivasarao, Ch. and D'Silva, E. (2009) Pongamia seed cake as a valuable source of plant nutrients for sustainable agriculture. *Indian Journal of Fertilisers* 2, 29–31.

Pamidimarri, D.V.N.S., Pandya, N., Reddy, M.P. and Radhakrishnan, T. (2008) Comparative study of interspecific genetic divergence and phylogenic analysis of genus Jatropha by RAPD and AFLP: Genetic divergence and phylogenic analysis of genus *Jatropha*. *Molecular Biology Reports* 36(5), 901–907.

Paramathma, M., Hegde, D.M., Parthiban, K.T., Mukta, N., Guptha, V.K. and Abraham, Y. (2009) Scope of tree borne oilseeds as source of vegetable oil and biofuels. In: Singh, H., Reddy, B.N. and Murthy, I. (eds) *Souvenir of the National Symposium on Vegetable Oils Scenario: approaches to meet the growing demands*. Directorate of Oilseeds Research, Hyderabad, India, pp. 47–49.

Parikh, J. (2005) Growing our own oils. *Biofuels India* 3, 7.

Passioura, J.B. (1982) The role of root system characteristics in the drought resistance of crop plants. In: *Drought Resistance of Crops with Emphasis on Rice*. International Rice Research Institute, Los Baños, the Philippines, pp. 71–82.

Pax, F. (1910) Euphorbiaceae–Jatropheae. In: Engler, A. (ed.) *Das Pflanzenreich IV*, vol. 147(42). Verlag von Wilhelm Engelmann, Leipzig, Germany.

Prabakaran, A.J. and Sujatha, M. (1999) *Jatropha tanjorensis* Ellis and Saroja, a natural interspecific hybrid occurring in Tamil Nadu, India. *Genetic Resources and Crop Evolution* 46, 213–218.

Punia, M.S. (2007) Current status of research and development on jatropha (*Jatropha curcas*) for sustainable biofuel production in India. In: *USDA Global Conference on Agricultural Biofuels: Research and Economics*, 20–22 August, Minneapolis, Minnesota.

Ranade, S.A., Srivastava, A.P., Rana, T.S., Srivastava, J. and Tuli, R. (2008) Easy assessment of diversity in *Jatropha curcas* L. plants using two single-primer amplification reaction (SPAR) methods. *Biomass Bioenergy*, doi: 10.1016/j.biombioe.2007.11.006.

Rao, Y.V.H., Voleti, R.S., Hariharan, V.S. and Raju, A.V.S. (2008) Jatropha oil methyl ester and its blends used as an alternative fuel in diesel engine. *International Journal of Agricultural and Biological Engineering* 1, 32–38.

Reddy, B.V.S., Ramesh, S., Ashok Kumar, A., Wani, S.P., Ortiz, R. and Ceballos, H. (2008) Biofuel crops research for energy security and rural development in developing countries. *Bioenergy Research* 1, 248–258.

Reddy M.P., Chikara, J., Patolia, J.S. and Ghosh, A. (2007) Genetic improvement of *Jatropha curcas* adaptability and oil yield. In: *Expert seminar on* Jatropha curcas *L. agronomy and genetics*, 26–28 March, Wageningen, the Netherlands. FACT Foundation, Wageningen, the Netherlands.

Ribeiro, D. and Matavel, N. (2009) *Jatropha! A socio-economic pitfall for Mozambique*. Justica Ambental (JA) and Uniao Nacional de Camponess (UNAC), Mozambique.

Rossetto, M. (2001) Sourcing of SSR markers from related plant species. In: Henry, R.J. (ed.) *Plant Genotyping; The DNA fingerprinting of plants*. CAB International, Wallingford, UK, pp. 211–224.

Rug, M., Sporer, F., Wink, M., Liu, S.Y., Henning, R. and Ruppel, A. (1997) Molluscicidal properties of *J. curcas* against vector snails of the human parasites *Schistosoma mansoni* and *S. japonicum*. In: Gubitz, G.M., Mittelbach, M. and Trabi, M. (eds) Biofuels and industrial products from *Jatropha curcas. Proceedings from the Symposium 'Jatropha 97'*. Dbv-Verlag, Graz, Austria, pp. 227–232.

Soontornchainaksaeng, P. and Jenjittikul, T. (2003) Karyology of Jatropha (Euphorbiaceae) in Thailand. *Thai Forest Bulletin (Botany)* 31, 105–12.

Srinivasa Rao, C., Kumari, M.P., Wani, S.P. and Marimuthu, S. (2011) Occurrence of black rot in *Jatropha curcas* L. plantations in India caused by *Botryosphaeria dothidea*. *Current Science* 100, 1547–1549.

Staubmann, R., Foidl, G., Foidl, N., Gubitz, G.M., Lafferty, R.M. and Valencia, A.V.M. (1997) Production of biogas from *J. curcas* seeds press cake. In: Gubitz, G.M., Mittelbach, M. and Trabi, M.M. (eds) Biofuels

and industrial products from *Jatropha curcas*. *Proceedings of the symposium 'Jatropha 97'*, 23–27 February 1997, Graz, Austria, pp. 123–131.
Sujatha, M. (2006) Genetic improvement of *Jatropha curcas* L. possibilities and prospects. *Indian Journal of Agroforestry* 8, 58–65.
Sujatha, M., Reddy, T.P. and Mahasi, M.J. (2008) Role of biotechnological interventions in the improvement of castor (*Ricinus communis L.*) and *Jatropha curcas* L. *Biotechnology Advances* 26, 424–435.
Sun, Q.-B., Lin-Feng, L., Yong, L., Wu Guo-Jiang, W. and Xue-Jun, G. (2008) SSR and AFLP markers reveal low genetic diversity in the biofuel plant *Jatropha curcas* in China. *Crop Science* 48, 1865–1871.
Sunil, N., Varaprasad, K.S., Sivaraj, N., Kumar, T.S., Abraham, B. and Prasad, R.B.N. (2008) Assessing *Jatropha curcas* L. germplasm in situ – a case study. *Biomass and Bioenergy* 32, 198–202.
Susanna, P. (2009) Assessing environmental impacts of rehabilitated degraded uplands in watershed with *Jatropha* plantation. MSc thesis (Submitted to JNTU, Kukatpally, Hyderabad, India), ICRISAT, Patancheru, India.
Swarup, R. (2006) Quality planting material and seed standards in Jatropha. In: Singh, B., Swaminathan, R. and Ponraj, V. (eds) *Proceedings of the Biodiesel Conference Toward Energy Independence – focus on Jatropha*, 9–10 June 2006, Hyderabad, India, pp. 129–135.
Tasosa, A., Chiduza, C., Robertson, I. and Manyowa, N. (2001) A comparative evaluation of the fertilizer value of castor and *Jatropha* press-cakes on the yield of tomato. *Crop Research, Haryana Agricultural University* 21, 66–71.
Tatikonda, L., Wani, S.P., Kannan, S., Beerelli, N., Sreedevi, T.K., Hoisington, D.A., Pratibha, D. and Varshney, R.K. (2009) AFLP-based molecular characterization of an elite germplasm collection on *Jatropha curcas* L., a biofuel plant. *Plant Science*, doi: 10.1016/j.plantsci.2009.01.006.
Tiwari, A.K., Kumar, A. and Raheman, H. (2007) Biodiesel production from Jatropha (*Jatropha curcas*) with high free fatty acids: an optimized process. *Biomass and Bioenergy* 31, 569–75.
Trabucco, A., Achten, W.M.J., Bowe, C., Aerts, R., Orshoven, J.V., Norgrove, L. and Muys, B. (2010) Global mapping of *Jatropha curcas* yield based on response of fitness to present and future climate. *GCB Bioenergy* 2, 139–151.
Wani, S.P. and Chaliganti, R. (2007) *Rural Livelihoods and Urban Environment: An Assessment of the Bio-fuel Programme for the emerging Megacity of Hyderabad*. Research Report 14, Humbolt-Universität zu Berlin, Berlin.
Wani, S.P. and Sreedevi, T.K. (2007) Strategy for rehabilitation of degraded lands and improved livelihoods through biodiesel plantations. In: *Proceedings of 4th International Biofuels Conference*, 1–2 February 2007, New Delhi, pp. 50–64.
Wani, S.P., Pathak, P., Jangawad, L.S., Eswaran, H. and Singh, P. (2003) Improved management of vertisols in the semiarid tropics for increased productivity and soil carbon sequestration. *Soil Use and Management* 19, 217–222.
Wani, S.P., Osman, M., D'Silva, E. and Sreedevi, T.K. (2006a) Improved livelihoods and environmental protection through biodiesel plantations in Asia. *Asian Biotechnology and Development Review* 8, 11–29.
Wani, S.P., Sreedevi, T.K. and Reddy, B.V.S. (2006b) Biofuels: Status, Issues and Approaches for Harnessing the Potential. Presentation made at CII Godrej Center, Hyderabad, India, 29–30 June 2006.
Wani, S.P., Sreedevi, T.K., Marimuthu, S., KesavaRao, A.V.R. and Vineela, C. (2009) Harnessing the potential of Jatropha and Pongamia plantations for improving livelihoods and rehabilitating degraded lands. In: *Proceedings of 6th International Biofuels Conference* (4–5 March, 2009, New Delhi, India). Winrock International, India, pp. 256–272..
Zhu, H.H., Wu, J.S., Huang, D.Y., Zhu, Q.H., Liu, S.L., Su, Y.R., Wei, W.X., Syers, J.K. and Li, Y. (2010) Improving fertility and productivity of a highly-weathered upland soil in subtropical China by incorporating rice straw. *Plant and Soil* 331, 427–437.

16 Oilseed Brassicas

G.S. Bañuelos,[1] K.S. Dhillon[2] and S.S. Banga[3]
[1]*USDA-ARS, Water Management Research Unit, California, USA;*
[2]*Department of Soil Science, Punjab Agricultural University, India (Retired);* [3]*Punjab Agricultural University, India*

16.1 Introduction

Oilseed brassicas, also known by the trade name of rapeseed-mustard, comprise *Brassica napus, B. juncea, B. carinata* and three ecotypes of *B. rapa*. Their current global production exceeds 54 Mt, making them the second most valuable source of vegetable oil in the world (Graner *et al.*, 2003; Dubuis *et al.*, 2005). Worldwide consumption of vegetable oils has increased by >50% during the past decade, and their prices have doubled since 2000 (Lu *et al.*, 2011). The leading brassica producers in 2009 were the European Union (EU), China, Canada and India (Fig. 16.1) (FEDIOL, 2006).

Winter-type *B. napus* predominates in Europe, parts of China and eastern USA, while spring-type *B. napus* is cultivated in Canada, northern Europe and China. The spring form of *B. rapa* is grown in Canada, northern Europe, China and India. Winter-type *B. rapa* has largely been replaced by more productive winter-type *B. napus* and spring crops in its traditional production areas. Only spring types exist for *B. juncea* and it is the leading brassica oilseed in the Indian subcontinent, and is now been actively considered as an option in drier areas of Canada, Australia and even in northern USA. The transition of these crops from high erucic to low erucic rapeseed, and the consequent explosive growth as oilseed crops began from Canada in 1968, with commercial release of a single low cultivar 'Oro' followed by several other single low cultivars, and the first canola cultivar 'Tower' in 1974. Almost all rapeseed produced in Australia, Canada and Europe and to a very large extent in China, is now canola. The cultivation of canola rapeseed-mustard has just begun in India.

Besides their pre-eminence as edible oils, brassica oilseeds are being seriously viewed as renewable sources for biodiesel production. The characteristics of oil also make it a favourable feedstock for biodiesel production (Carr, 1995). The higher oil content of canola results in a higher oil yield per hectare compared to competing oil crops due to higher productivity and oil content. Because of its low saturated fat content, biodiesel produced from canola oil exhibits better cold flow properties. Cold weather performance is a major quality factor for biodiesel and it is directly related to the level of unsaturation of the feedstock oil. A large number of canola-based biodiesel plants have been developed in northern America and the EU. Canada's largest biodiesel plant with an installed capacity of 265 Ml (70 million gallons) is being built in Alberta, which will use canola as a feedstock.

16.2 Taxonomy of the Genus

Crop brassicas are economically the most important members of family *Brassicaceae* and

©CAB International 2013. *Biofuel Crops: Production, Physiology and Genetics* (ed. B.P. Singh)

the subtribe *Brassicinae*. A number of taxonomic treatises of the family *Brassicaceae* are available. These include: Tournefort (1700), Linnaeus (1753), De Candolle (1821), Hooker (1862), Baillon (1871), Prantl (1891), Schulz (1919, 1936) and Beilstein *et al.* (2006). Schulz's by far has been the most comprehensive and authoritative account. A recent molecular account of the family was provided by Beilstein *et al.* (2006). *Brassiceae* is one among 19 tribes recognized by Schulz in the family *Brassicaceae* and is further subdivided into seven to nine subtribes (Gómez-Campo, 1980, 1999). *Brassica* is the major genus in this subtribe. Several members of related subtribes, such as *Raphaninae* and *Moricandiinae*, also show strong affinity with *Brassica*. This is despite the demonstrated distinctness of these species on the basis of chloroplast-DNA (cp-DNA) and restriction sites (Warwick and Black, 1991; Pradhan *et al.*, 1992; Warwick *et al.*, 1992). The cytogenetic investigations (Prakash and Hinata, 1980) have demonstrated that the three digenomic species, *B. carinata* ($2n = 34$; BBCC), *B. juncea* ($2n = 36$; AABB) and *B. napus* ($2n = 38$; AACC), have originated through crosses between any two of the three elementary species *B. nigra* ($2n = 16$; BB), *B. oleracea* ($2n = 18$; CC) and *B. rapa* ($2n = 20$; AA) (Fig. 16.2). It is also now known that *B. nigra, B. rapa* and *B. oleracea/B. rapa* are the cytoplasm donor species for *B. carinata, B. juncea* and *B. napus*, respectively (Banga *et al.*, 1983; Erickson, *et al.*, 1983). Major taxonomic forms are presented in Table 16.1.

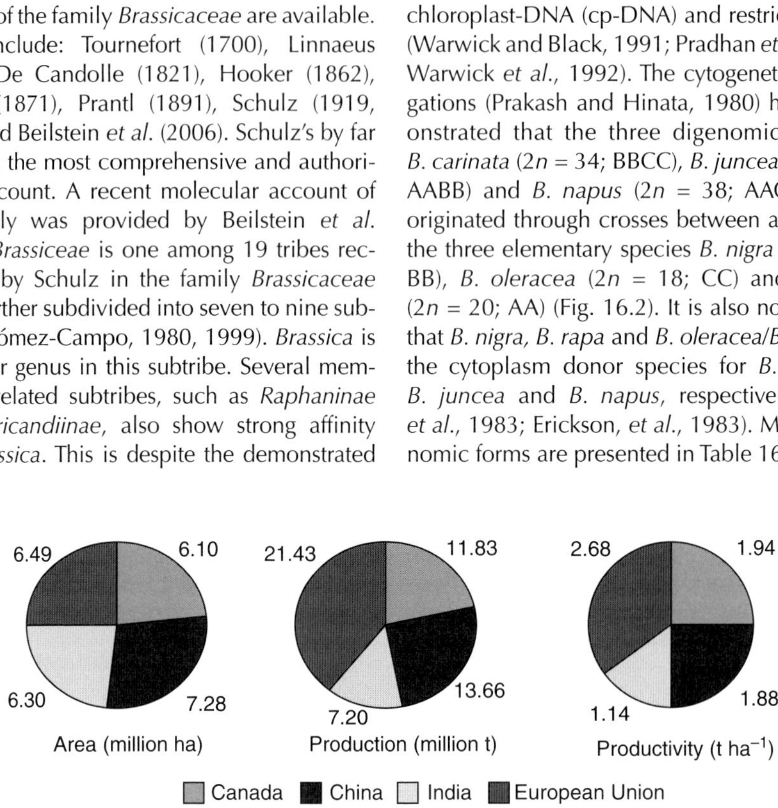

Fig. 16.1. Area, production and productivity of rapeseed-mustard crops in the world.

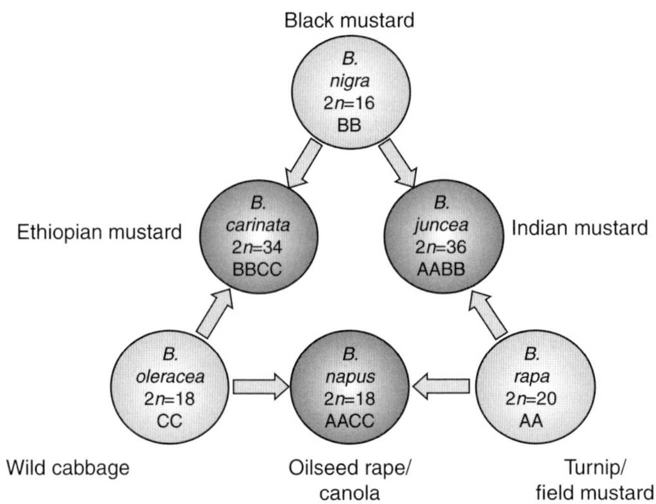

Fig. 16.2. Triangle of U, showing the affiliations among different *Brassica* species (U, 1935), as cited by Ahuja *et al.* (2010).

Table 16.1. Taxonomic forms of crop *Brassica* species.

Botanical name	Common name	Usage
B. nigra	Black mustard	Condiment (seed)
B. oleracea		
var. *acephala*	Kale	Vegetable fodder (leaves)
var. *capitata*	Cabbage	Vegetable (head)
var. *sabauda*	Savoy cabbage	Vegetable (head)
var. *gemmifera*	Brussels sprouts	Vegetable (terminal buds)
var. *gongylodes*	Kohlrabi	Vegetable, fodder (stem)
var. *italica*	Broccoli	Vegetable (inflorescence)
var. *fruticosa*	Branching bush kale	Fodder (leaves)
var. *alboglabra*	Chinese kale	Vegetable (stem, leaves)
B. rapa		
ssp. *oleifera*	Turnip rape	Oilseed
var. brown sarson	Brown sarson	Oilseed
var. yellow sarson	Yellow sarson	Oilseed
var. *toria*	Toria	Oilseed
ssp. *rapifera*	Turnip	Fodder, vegetable (root)
sp. *chinesis*	Pak choi	Vegetable (leaves)
ssp. *pekinesis*	Chinese cabbage	Vegetable, fodder (head)
ssp. *nipposinica*	–	Vegetable (leaves)
ssp. *pamchinensis*	–	Vegetable (leaves)
B. carinata	Ethiopian mustard	Vegetable oilseed
B. juncea	Mustard	Oilseed, vegetable
B. napus		
ssp. *oleifera*	Rapeseed	Oilseed
ssp. *rapifera*	Rutabaga, swede	Fodder

16.3 Ecology and History as a Crop

Brassicas were among the earliest crops to have been domesticated. *Brassica rapa* and *B. juncea* were mentioned in ancient Sanskrit literature dating to 1500 BCE, as well as in Greek, Roman and Chinese writings of 500 to 200 BCE (Downey and Röbbelen, 1989). The Chinese language equivalent of rapeseed was first recorded *c.*2500 years ago, and the oldest archaeological finds may date back as far as *c.*5000 BCE (Yan *et al.*, 1999). In India, seeds of *B. juncea* were found in archaeological sites as early as 2300 BCE (Prakash and Hinata, 1980). In Europe, domestication is believed to have occurred in the early middle ages and commercial plantings of rapeseed were carried out in the Netherlands as early as the 16th century. At that time, rapeseed oil was used primarily as oil for lamps. Later it became used as a lubricant for steam engines. Historically, *B. rapa* appears to have the widest distribution of brassica oilseeds. At least 2000 years ago, it was distributed from northern Europe to China and Korea, with the primary centre of diversity in the Himalayan region (Hedge, 1976).

16.3.1 *Brassica rapa*

Wild *B. rapa* subsp. *oleifera* (inc. var. *sylvestris*) is arguably the originator species of var. *rapa* L. (cultivated turnip) and var. *silvestris* (Lam.) Briggs (turnip-rape). It is native throughout Europe, Russia, Central Asia and the Near East (Prakash and Hinata, 1980), with Europe proposed as one centre of origin. The Asian and Near Eastern type may have originated from an independent centre of origin in Afghanistan, which then moved eastward following domestication. Prakash and Hinata (1980) believed that oleiferous *B. rapa* developed in two places giving rise to two different races, one European and the other Asian. Recent evidence (through analyses of chloroplast and mitochondrial DNA) suggests that *B. montana* (n = 9) might be closely related to the prototype that gave rise to both cytoplasms of *B. rapa* and *B. oleracea* (Song and Osborn, 1992). Large morphological variation observed in *B. rapa* might have resulted due to the long history of breeding for different traits along with natural selection for adaptation in different geographical regions. Oleiferous and turnip forms developed in Europe, while in eastern Asia and west Asia these evolved into leafy and oleiferous forms. Various types of leafy vegetables of *B. rapa*, such as Chinese cabbage (*B. rapa* ssp. *pekinensis*), non-heading pak choi (*B. rapa* ssp. *chinensis*) and mizuna, are found widely in East Asia, particularly in China, Korea and Japan. The oleiferous form of *B. rapa* (*B. rapa* ssp. *oleifera*) has been historically cultivated for production of vegetable oils in China, Canada, India and in northern Europe and is the third brassica oilseed crop. The Indian form developed into mainly oleiferous forms such as yellow sarson, brown sarson and toria type that are grown in large areas in eastern and north-eastern parts of India during the winter season (Singh, 1958).

16.3.2 *Brassica oleracea*

This crop was domesticated as early as 2000 BCE initially as primitive kales or cabbages. These are now believed to be the first cultivated forms of *B. oleracea* (Chiang *et al.*, 1993), especially by Celts whose name, 'Bresic', for cabbage may be the likely progenitor of '*Brassica*'. Heading cabbages and other leafy types probably had a common origin from the ancestral kales and non-heading cabbages (Herve, 2003) and they were further developed in Portugal, Spain and France. Wild forms of *B. oleracea* are found on the Atlantic coasts of Europe, northern France and England. Related wild species are endemic to the Mediterranean basin. Because the cultivated *B. oleracea* crops were grown in close proximity to wild relatives, the flow of genes from wild to cultivated types likely occurred. This coupled with mutation, human selection and adaptation, must have contributed to current morphogenetic variation in this crop.

16.3.3 *Brassica nigra*

Brassica nigra or black mustard was originally domesticated in Asia Minor or Iran, and has been cultivated since ancient times. It was the primary condiment mustard all over Europe but subsequently gave way to brown mustard or *B. juncea*. It is native to the Mediterranean region, and archaeological finds show that it has occurred as a weed in association with the cultivation of wheat and barley during domestication. Historically, black mustard appeared to have been harvested mainly from wild populations with only sporadic commercial cultivation.

16.3.4 *Brassica carinata*

This is an amphiploid derived from natural hybridization between *B. oleracea* and *B. nigra* in the Ethiopian highlands. Despite significant efforts, it has not been possible to locate any wild forms (Mizushima and Tsunoda, 1967). However, both the species are commonly observed growing close to each other in cultivation or as escapes. This crop has been domesticated for multiple uses such as oil, condiments, medicinal and vegetables (Astley, 1982; Riley and Belayneh, 1982).

16.3.5 *Brassica juncea*

This is a natural amphiploid derived from crosses between *B. rapa* and *B. nigra*. The origins of *B. juncea* are unclear, but it must have originated in areas like the Middle East and neighbouring regions, where distributions of *B. nigra* and *B. rapa* overlap (Prakash and Hinata, 1980). The concept of a separate origin of foliage types (Chinese) and oleiferous types (Indian) was mainly based on the investigations of Vaughan *et al.* (1963) and Vaughan (1977) on seed glucosinolate profiles. Vaughan (1977) also suggested that the Indian race was closer to the *B. rapa* progenitor, and the oriental race was closer to the *B. nigra* progenitor, and perhaps this indicates that *B. juncea* evolved in more than one region. Spect and Diederichsen (2001) considered the place of origin somewhere between eastern Europe and China, where there is parental sympatry. *Brassica juncea* has been divided into four subspecies, with different morphology, quality characteristics and uses (Spect and Diederichsen, 2001). These are: (i) ssp. *integrifolia*, used as a leaf vegetable in Asia; (ii) sp. *juncea*, cultivated mainly for its seed, occasionally as fodder; (iii) ssp. *napiformis*, used as a root-tuber vegetable; and (iv) ssp. *taisai*, stalks and leaves of which are used as vegetables in China.

16.3.6 *Brassica napus*

Brassica napus is thought to have multiple origins resulting from independent natural hybridization events between *B. oleracea* and *B. rapa*. In Europe, the winter form has predominantly become a common yellow crucifer of wild areas. In the British Isles, for example, it has been naturalized wherever oilseed rape is grown, and it is a relatively recent introduction into Canada, the USA, China and India.

16.4 Genetic Architecture

16.4.1 Morphological traits

A large number of genes influencing gross morphological characters of oilseed brassicas have been characterized (Banga, 1996; Séguin-Swartz *et al.*, 1997). The majority of the characters show monogenic recessive inheritance with an occasional character being influenced by a few modifiers. Seed colour is almost always associated with two or more genes. Because of wide adaptation of the crop, the genes for vernalization and photoperiodic response have been investigated. Spring habit is found to be dominant over winter habit and requirement for vernalization is controlled by a major and minor gene in *B. napus*. Acclimatization ability is regulated by genes with very small additive effects.

16.4.2 Yield and its components

Reports abound for the type of genetic variability for various quantitative traits. Primacy of additive genetic variance with some dominance and occasional epistasis has been demonstrated (Singh and Singh, 1972; Lefort-Buson and Datté, 1985). Despite these and several other studies, genetic architecture and determinants of yield have remained ambiguous (Banga and Banga, 2009). One of the most intractable problems is the interaction between genes and the environment. In an elaborate study, 85 quantitative trait loci (QTL) for seed yield along with 785 QTL for eight yield-associated traits from ten natural environments and two related populations of rapeseed were studied (Shi *et al.*, 2009). A trait-by-trait meta-analysis revealed 401 consensus QTL, of which 82.5% were clustered and integrated into 111 pleiotropic unique QTL by meta-analysis, and 47 of which were relevant for seed yield. These studies provided molecular basis to complexity of yield architecture, pleiotropy, synthesis, variability and plasticity of yield QTL. The idea of estimating indicator QTL for yield QTL and identifying potential candidate genes for yield provides an advance in methodology for complex traits. Experiments with different species have not revealed any predictors of heterosis based both on methods of molecular as well as genetic diversity (Bansal *et al.*, 2012). Generalization of such inferences has not been useful due to polyploidy, high genotypic specificity, heterogeneity of parental populations, lack of distinct gene pools with diverse gene frequencies and high genotype × experiment interactions. Although much research has been conducted, the genetic architecture of heterosis remains ambiguous. To unravel the genetic architecture of heterosis, a reconstructed F_2 population was produced by random intercross among 202 lines of a double haploid population in rapeseed (*Brassica napus* L.). Hundreds of QTL and epistatic interactions were identified for the 15 yield-correlated traits, involving numerous variable loci with moderate effect, genome-wide distribution and obvious hotspots. All kinds of mode-of-inheritance of QTL (additive, A; partial-dominant, PD; full-dominant, D; over-dominant, OD) and epistatic interactions (additive × additive, AA; additive dominant/dominant additive, AD/DA; dominant × dominant, DD) were observed and epistasis, especially AA epistasis, seemed to be the major genetic basis of heterosis in rapeseed

(Shi *et al.*, 2011). Consistent with the low correlation between marker heterozygosity and mid-parent heterosis/hybrid performance, a considerable proportion of dominant and DD epistatic effects were negative, indicating heterozygosity was not always advantageous for heterosis/hybrid performance.

16.5 Biotic Stresses

Seedling resistance to blackleg in winter rapeseed is controlled by a single dominant gene *Lml* (Delwiche, 1980). A second dominant gene (*Lm2*) was also identified in another cultivar. The cotyledon resistance in two spring oilseed rape lines (R 83-14 and 83-17) is governed by two genes with dominant alleles, designated *B1-1* and *B1-2*. Presence of both genes conferred a high level of resistance. Adult plant resistance in *B. napus*, introgressed from *B. juncea*, is controlled by three major genes exhibiting a complex interaction (Pang and Halloran, 1996). Genes for blackleg resistance appear to be located on one (Chevre *et al.*, 1996) or three B-genome chromosomes (Zhu *et al.*, 1993) in *B. juncea* and *B. carinata*. Delourme *et al.* (2004) mapped *Rlm1, Rlm3, Rlm4, Rlm7* and *Rlm9* on linkage group N7 and other researchers also identified *LEM1, LmR1* and *cRLMm, cRLMrb, cRLMj* and *aRLMj* on the same linkage group (Zhu and Rimmer, 2003; Rimmer, 2006). Additionally, *Rlm2* in *B. napus* was mapped on linkage group N10 (Delourme *et al.*, 2006a) and also on this linkage group, two of three blackleg resistance genes *LepR1, LepR2* and *LepR3* that were introduced from wild *B. rapa* var. *sylvestris* were mapped.

Multiple QTLs could be identified across four mapping populations, accounting for 13–33% of phenotypic variance (V(p)). QTL-linked marker data were considered suitable for implementation in breeding (Kaur *et al.*, 2009). Differential interactions between the resistant lines and a pathogen isolate were discovered with two resistance genes *BLMR1* and *BLMR2* identified through linkage analysis of five genome-specific molecular markers. *BLMR1* provides resistance through the hypersensitive response that protects inoculated cotyledons from becoming infected. Unlike *BLMR1, BLMR2* slows down the development of individual infection loci. *BLMR1* and *BLMR2* segregated independently in two large F3BC2 populations. Fine mapping of *BLMR1* was performed with 12 genome-specific molecular markers. The closest marker with a genetic distance of 0.13 cM to *BLMR1* was identified (Long *et al.*, 2011). For white rust, monogenic dominant resistance to *Albugo candida* (race 2) was reported in *B. nigra, B. rapa, B. juncea* and *B. carinata*. It was also possible to map locus AcB1-A4.1 to linkage group A4, and locus AcB1-A5.1 to linkage group A5 in *B. juncea*. In both the cases, closely linked flanking markers were developed based on synteny between *Arabidopsis* and *B. juncea* (Panjabi-Massand *et al.*, 2010). Two or possibly three independent dominant genes have also been suggested for inheritance of white rust resistance in *B. napus*. High heritability and nuclear gene control is suggested for *Sclerotinia* resistance in *B. napus*, while only variation for alternaria blight resistance has been demonstrated (Saharan, 1993). *Brassica* species with the B-genome appear to show greater tolerance to various insects attacking oilseed brassicas.

16.6 Crop Breeding Objectives

The crop brassicas harbour large intraspecific variability for the extent of cross-pollination. Out-crossing rates in predominantly self-pollinating crops of *B. juncea* (Labana and Banga, 1984) and *B. napus* (Olsson, 1955) vary from 7.6 to 18.1% and up to 16.1%, respectively, while *B. rapa* var. *toria* and 'brown sarson' are largely cross-pollinating due to self-incompatibility. Yellow sarson is, however, self-pollinating. The conventional crop improvement objectives focus largely on attempts to produce, protect and tailor the biomass to suit edible and industrial requirements. The goal of developing canola forms has been accomplished for *B. rapa, B. napus* and *B. juncea* but remains an important objective in *B. carinata*. Further modifications in the acid composition for specialized product applications such as biofuels are currently being sought. Brassicas are also confronted with several biotic stresses such as diseases (blackleg, sclerotinia rot, alternaria blight, white rust, etc.) and pests (aphids, beetles, etc.). As these crops are grown in a wide array of climate and cropping systems, they will require general or specific adaptation to specific situations. Varieties with varying maturity duration are required to escape frost (Canada) or late-season drought (southern Australia) or to fit in multiple cropping sequences (India, China). Breeding programmes are also concerned with cultivar's suitability for existing or emerging management practices, e.g. herbicide resistance or mechanical harvesting (resistance to pod shattering).

16.7 Breeding Methodologies

Success for improving yields depends on the ability to substitute bottleneck loci with more effective alleles (Banga and Banga, 2009). In the self-pollinating brassicas the yield improvement in conventional programmes has been achieved by various types of pedigree methods (Downey and Rimmer, 1993). Handling of segregating populations from biparental or multiple crosses as bulk populations before extraction of pure lines is a common practice. Visual selection in F_2 on single plant basis for seed yield and oil content has proved ineffective. The inefficiency of selection was attributed to large G × E interaction rather than preponderance of dominance genetic variance (Brown, 1995). Delaying selection until F_3 or later generations or through manipulation of simple characters or combination of characters has proved to be more effective. However, the influence of environment on yield components relationship (Thurling, 1993) modulates the response to selection. Dihaploid technique has helped to shorten the breeding cycle in *B. napus*, although it is not successful in *B. rapa* due to inbreeding depression. It may be desirable to first use incompatible types of *B. rapa*, whereby cultivars are genetically heterogeneous and yield improvement could be achieved through mass selection. Effectiveness of recurrent selection to improve various traits has been well documented in several cross-pollinating crops but no systematic attempt has been made to exploit this technique in *B. rapa*. Recurrent selection in self-pollinated brassicas can be facilitated by genetic male sterility. To maximize genetic advance, selection must be carried out in the target environment using criteria that combine yield with stability. Cultivars having resistance to blackleg have been bred in *B. napus* using indigenous variability or through introgression from *B. juncea* or *B. sylvestris*. The simple backcross method has been used to incorporate genes for white rust resistance in *B. juncea*. The introduction of chitinase gene into *B. napus* (Grison *et al.*, 1996) under the control of a constitutive promoter has been shown to increase field resistance to three fungal pathogens.

A high level (35–100%) of heterosis has been reported in the entire oilseed brassica (Banga, 1996; Banga and Banga, 2009). Heterosis is generally higher in crosses involving parents with diverse phenologies or geographic origin. Attempts at the production of F_1 hybrids have been simulated by the availability of large number of cytoplasmic male sterility (CMS) sources. These CMS systems are *nap, polima ogura, tournefortii, axyrrhina, siifolia, catholica, sinapis, trachystoma moricandia* and *lyratus*. Hybrids based on *polima* or *ogura* and/or transgenic *barnase-barstar* system have been commercialized in *B. napus* and now cover very large areas (>50%) in Canada, China and Europe. In India, hybrids have been commercialized in *B. juncea*, and are based on either *ogura*, INS 126 or *mori* CMS systems. Average yield heterosis over the best inbred parents, however, continues to be below 15%. In Canada, hybrids have averaged 11% higher yields over open-pollinated (OP) varieties during the past decade.

Due to ease of estimation and economic importance, emphasis so far has been on improving oil content, mostly at the cost of protein. This has helped in raising oil content to 38–44% in current cultivars. Further increase is possible from a switch over to yellow-seeded types, which are available in *B. rapa, B. juncea* and *B. carinata*. Attempts to breed stable, pure yellow-seeded cultivars are still continuing in *B. napus*. Oil quality in oilseed brassicas has been successfully modified using selection, hybridization, induced mutagenesis and transformation. Half-seed technique is routinely applied in a backcross programme to incorporate gene-preventing synthesis of erucic acid (Morice, 1974). Frequently, it is desirable to delay first backcross to F_2 generation as a cycle of recombination, because this provides the possibility of breaking certain undesirable linkages, as well as for improving yield and other agronomic traits (Röbbelen and Nitsch, 1975). Maternal influence on the heredity of glucosinolates complicates breeding for low glucosinolates using the backcross method. Mutagenesis or selective hybridization has helped in reducing linolenic acid content (<3%) and enhancing oleic acid levels (≈80%).

Extensive and exploitable primary, secondary and tertiary gene pools are available (Prakash *et al.*, 2009). Immunosuppressants, hormones, *in vitro* fertilization, embryo culture and protoplast fusion have helped in overcoming sexual barriers. Chromosome addition lines such as *B. rapa, B. oleracea, B. rapa, B. nigra, B. napus, Diplotaxis erucoides, B. napus–B. nigra* and *B. rapa–B. alboglabra* have been produced (Delourme *et al.*, 1989; Jahier *et al.*, 1989; Chen *et al.*, 1997), and C-genome chromosome addition lines in *B. juncea* have also been reported (Banga, 1988). It has also been possible to introgress wild alleles for tolerance to various stresses in the cultivated brassicas (Garg *et al.*, 2010; Warwick, 2010; Kumar *et al.*, 2011). Exploitation of alien cytoplasmic variability has been very effective in development of a large number of CMS lines (Prakash *et al.*, 2009).

Molecular markers are now increasingly being used to map agronomically important genes in oilseed brassicas. Several of the advance breeding programmes use them for marker assisted selection. The complete sequencing of *Arabidopsis thaliana, rapa* and *oleracea*, as well as the development of saturated linkage maps and the availability of comparative genetic and physical maps between brassicas and *Arabidopsis* have enormous potential for application in gene identification and breeding (http://www.brassica.info). Markers associated with linolenic acid and erucic acid (Jourdren *et al.*, 1996) have already been identified. These are likely to significantly improve selection efficiency. In addition, marker assisted selection during backcrossing can help to eliminate linkage drag and hasten the recovery of recurrent genotypes. High level of glucosinolates in rapeseed meal impairs palatability and acceptance by animals and can also lead to goitrogenic hypertrophy. One spring cultivar ('Bronowski') was identified as a low glucosinolate form and together with 'Liho' provided the alleles to create the first double-zero spring variety ('Tower') exhibiting low erucic acid and low glucosinolate content (8–15 μmol g^{-1}) in seeds. Four seed glucosinolate content QTLs located on chromosomes A9, C2, C7 and C9 have been recurrently mapped in different *B. napus* studies (Uzunova *et al.*, 1995; Howell *et al.*, 2003; Sharpe and Lydiate, 2003; Zhao and Meng, 2003; Basunanda *et al.*, 2007). Breeding efforts to further reduce glucosinolate content are ongoing and targeted to reducing specific types of glucosinolates. Interestingly, a recent study associated gene-linked SSR markers to seed glucosinolate content in *B. napus* by using comparative genomics with *Arabidopsis* (Hasan *et al.*, 2008). By this means, it will be possible to apply the vast biochemical knowledge developed in this model species to select molecular markers with putative linkage to genes active on specific branches of the glucosinolate biosynthetic pathway and thus facilitate targeting the reduction of specific glucosinolate types (Hasan *et al.*, 2008).

Ramchiary *et al.* (2007) provided an insight into the complexity of this trait in *B. juncea*, showing involvement of context-dependent loci (effects of QTL alleles differing in magnitude and direction in different genetic backgrounds) in the expression of the trait. Low glucosinolate trait was introgressed from canola-quality east European line 'Heera' to a major Indian cultivar 'Varuna'. Six 'true' QTLs, four major (*J2Gsl1, J3Gsl2, J9Gsl3* and *J16Gsl4*) and two minor QTLs (*J17Gsl5* and *J3Gsl6*) were identified, and their suitability in the marker-assisted introgression to the Indian germplasm was validated. The above-mentioned study has led to the identification of 'true' QTLs for marker-assisted introgression of the low glucosinolate trait from an east European donor to Indian cultivars. Bisht *et al.* (2009) undertook fine mapping of the QTL identified earlier by Ramchiary *et al.* (2007) for efficient marker-assisted introgression of low glucosinolate trait into the germplasm of the Indian gene pool. Based on the DNA sequences from *A. thaliana* and *B. oleracea* for the different genes involved in the aliphatic glucosinolate biosynthesis, candidate genes were cloned from high and low glucosinolate *B. juncea* lines 'Varuna' and 'Heera', respectively. Seventeen paralogues belonging to six gene families were mapped in *B. juncea*. Candidate gene *BjuA.GSL-ELONG.a* mapped to the QTL interval of *J2Gsl1, BjuA.GSLELONG.c, BjuA.GSL-ELONG.d* and *BjuA.Myb28.a* mapped to QTL interval of *J3Gsl2, BjuA.GSL-ALK.a* mapped to QTL interval of *J3Gsl6* and *BjuB.Myb28.a* mapped to QTL interval of *J17Gsl5*. These are likely to significantly improve selection efficiency. In addition, marker-assisted selection during backcrossing can help to eliminate linkage drag and hasten the recovery of recurrent genotype. Mapping and marker development for resistance genes to biotic stress factors represent major goals for brassica breeding. Map-based cloning of genes contributing to resistance against blackleg was a major achievement (Rimmer, 2006). Fine mapping of loci involved in seedling resistance to blackleg in two canola cultivars has been described (Mayerhofer *et al.*, 2005). Both loci localized to the same position on *B. napus* chromosome N07, and a collinear chromosome region was able to be identified in *Arabidopsis*. Candidate genes for blackleg resistance were also identified in *Arabidopsis* (Staal *et al.*, 2006). QTL-linked markers associated with resistance against *Verticillium longisporum* were introduced from *B. oleracea* (Rygulla *et al.*, 2007). QTL involved in resistance against clubroot disease has also been reported (Werner *et al.*, 2008).

16.8 Approaches to Enhance Oil Production

Increasing seed oil content is of great importance, particularly given the increasing market demands for renewable fuels in recent years (Durrett *et al.*, 2008; Lu *et al.*, 2011). Classical plant breeding, plant breeding aided by genomics, recombinant DNA and synthetic biology techniques have been used to enhance oil content and modify its fatty acid composition. Genetic sources are very important in improving oilseed brassicas for oil content. There is a wide range (35–45%) in oil content among accessions in the germplasm collections. However, genes affecting oil content and oil biosynthesis are strongly influenced by the environmental factors and crop agronomy followed (Röbbelen and Thies, 1980). A large number of QTLs for oil content in seeds have been identified in different mapping populations, and all of them exhibit relatively minor effects (Ecke *et al.*, 1995; Burns *et al.*, 2003; Zhao *et al.*, 2005; Delourme *et al.*, 2006; Qiu *et al.*, 2006; Zhao *et al.*, 2006). Three of these QTLs, located in linkage groups A1, A8 and C3, were consistently revealed across different populations (Delourme *et al.*, 2006b). Undoubtedly, the development of a consensus genetic map for the species will allow better comparison of QTLs and unifying results across different mapping populations. The QTLs located on linkage groups A8 and C3 coincide with major QTLs for erucic acid content in *B. napus* seeds (Howell *et al.*, 1996; Thormann *et al.*, 1996; Burns *et al.*, 2003).

Leaves and pods are known to be the principal photosynthate sources at different stages. At the early reproductive stage, leaves are the principal site of CO_2 assimilation. Photosynthates produced in leaves are transported to flowers and pods (Pechan and Morgan, 1985). Leaves establish the sink potential in terms of structures such as the number of pods and seeds. During pod filling, the contribution of leaves decreases as senescence approaches (Allen *et al.*, 1971) and pods take over the supply of assimilates for seed filling (Tayo and Morgan, 1979). From the conventional plant breeding perspective, photosynthetic activity in the silique may be a useful trait for the selection of varieties with high oil contents, such as the apetalous and erectophile pod lines that attract more photosynthetically active radiation (Fray *et al.*, 1996). Increased plant photosynthesis may also help to maximize lipid synthesis at the whole-plant level. Increased assimilate transport and/or cell-wall biosynthesis lead to increased carbon fixation and consequently influence seed metabolism and gene expression. Moreover, since a number of genes are involved in increased photosynthesis, there are many potential targets for improving oil content by genetic engineering (Reynolds *et al.*, 2009). The study by Hua *et al.* (2011) has emphasized that photosynthetic activity in the silique may be a useful trait for the selection of varieties with high oil contents, such as the apetalous and erectophile pod lines that attract more photosynthetically active radiation (Fray *et al.*, 1996). Several recent studies have reported the importance of photosynthesis for oil accumulation in green seeds. The light reactions of photosynthesis in the embryo can provide both ATP and reductant, and the Rubisco bypass allows more efficient lipid synthesis (Ruuska *et al.*, 2004; Schwender *et al.*, 2004, 2006; Goffman *et al.*, 2005). In addition to genetic options, available results have implications for cultivation practices, as high oil content requires efficient photosynthesis during the later stages of pod development. In support of this, seeds grown in higher-light environments tend to have higher oil contents (Li *et al.*, 2006). Thus, improvement of plant photosynthesis has the potential to maximize lipid synthesis at the whole-plant level.

Several attempts at engineering or designing nutritional or industrial fatty acids have been reviewed extensively (Nesi *et al.*, 2008; Weselake *et al.*, 2009). The majority of the genes encoding various steps of the fatty acid biosynthesis have been identified and functionally characterized, though the precise contribution of individual activities to flux through pathways and ultimate levels of triacylglycerol is unknown (Napier and Grahm, 2010). Through the isolation and characterization of *Arabidopsis* mutants it could be established that the final enzyme in the Kennedy pathway, acyl-CoA: diacylglycerol acyltransferases (DGAT), is a key activity in determining the levels of triacylglycerols in seed oils (Venegas-Calerón *et al.*, 2009). Quantitative genetic approaches in maize have also provided evidence for the importance of DGAT to seed oil levels (Zheng *et al.*, 2008).

In brassicas, silique wall is the predominant synthesizer of sucrose during seed filling. Local and tissue-specific photosynthetic activity in the silique wall (a maternal tissue) has been found to be responsible for the different seed oil contents. This effect was mimicked by *in planta* manipulation of silique wall photosynthesis. Furthermore, the starch content and expression of the important lipid synthesis regulatory gene *WRINKLED1* in developing seeds were linked with silique wall photosynthetic activity. Photosynthesis-related genes were over-represented in both total silique wall expressed genes and genes that were differentially expressed between genotypes. Further, oil content correlated closely with BnRBCS1A expression levels and Rubisco activities in the silique wall, but not in the leaf (Wei *et al.*, 2011). *WRINKLED1* participates in the interaction between the silique wall and the seed. Sucrose from silique walls is transported into developing seeds, and is then converted into hexose UDP-Glc and Fru. In the cytosol and the plastid of the embryo cells, hexose is further converted into acetyl-CoA, a precursor of FA biosynthesis (Baud *et al.*, 2002; Hill *et al.*, 2003; Schwender *et al.*, 2003). Using isolated plastids from developing rapeseed seeds, it was demonstrated that primary cytosolic metabolites, including Glc-6-P, malate, phosphoenol pyruvate and pyruvate, were imported into the plastid and then converted into acetyl-CoA through the glycolytic pathway to initiate *de novo* synthesis of fatty acids, of which pyruvate and Glc-6-P caused the highest rate of fatty acid biosynthesis (Rawsthorne, 2002). Apparently, increased carbon flux is necessary for a higher rate of fatty acid biosynthesis. Significant efforts have been made to manipulate key fatty acid synthetic genes in various species using transgenic technology. Studies with transgenes encoding key enzymes or enzyme subunits resulted in altered levels of lipids at varying degrees (Zou *et al.*, 1997; Jako *et al.*, 2001), and in some cases the oil content was reduced (Dehesh *et al.*, 2001). It is somewhat expected that the manipulation of single genes may not be an efficient approach because fatty acid synthesis is a highly coordinated process involved in not only fatty acid synthetic genes but also carbon metabolism genes, especially the control of carbon flux from glycolysis to fatty acid biosynthesis (Cahoon *et al.*, 2007; Weselake *et al.*, 2009). Therefore, genetic manipulations of key regulatory genes likely represented by transcription fatty acid sectors or protein kinases should be able to overcome the bottlenecks of the entire pathway of fatty acid synthesis (Ruuska *et al.*, 2002; Cahoon *et al.*, 2007; Mu *et al.*, 2008).

Expression of the yeast gene coding for cytosolic glycerol-3-phosphate dehydrogenase (*gpd*1) in transgenic oilseed rape under the control of the seed-specific napin promoter led to a twofold increase in glycerol-3-phosphate dehydrogenase activity and a three- to fourfold increase in the level of glycerol-3-phosphate in developing seeds (Vigeolas *et al.*, 2007). This precipitated a 40% increase in the final lipid content of the seed, with the protein content remaining substantially unchanged. Moreover, this was accompanied by a decrease in the glycolytic intermediate dihydroxyacetone phosphate, the direct precursor of glycerol-3-phosphate dehydrogenase. The levels of sucrose and various metabolites in the pathway from sucrose to fatty acids remained unaltered. The results show that glycerol-3-phosphate supply co-limits oil accumulation in developing seeds. MicroRNAs and siRNAs are important regulators of plant development and seed formation. A comparison of small RNA expression profiles of *B. napus* seeds at early embryonic developmental stages in a high oil content and a low oil content *B. napus* cultivar, both grown in two environments, led to the identification of a total of 50 conserved miRNAs and 11 new miRNAs, together with some new miRNA targets. It was considered possible that some miRNAs may be involved in the regulation of *B. napus* seed oil production (Zhao *et al.*, 2012). In *Arabidopsis*, several genes encoding embryo-specific transcription fatty acids factors have been shown to play important roles in the regulation of fatty acid biosynthesis. Best-understood examples are *LEAFY COTYLEDON1* (*AtLEC1*) and *LEC1-LIKE* (*AtL1L*).

16.9 Growing Brassicas

16.9.1 Climate

The growth and development of brassica species has a significant relationship with the type of climate prevailing at a particular location. The optimum temperature for seed germination ranges between 25 and 30°C. Seed germination is adversely affected at temperatures below 3°C and above 40°C. The crop requires higher temperatures during the vegetative growth period and cooler temperatures during the reproductive stage. Vigorous vegetative growth is expected between 30 and 32°C. Temperatures ranging between 20 and 22°C are highly conducive for a stronger reproductive phase. Frosty and cloudy weather restricts the pollinating activity of bees and other insects, thereby resulting in poor yield (Kumar, 1992).

16.9.2 Seedbed preparation under good soil conditions

Description of agronomic practices has been explained in great detail by Pouzet (1995). He clearly described choice of cultivar, soil preparation, sowing, fertilizer, irrigation, harvesting and storage associated with successfully growing brassicas, however, the range of crop management and cultural practices is too extensive to summarize in this chapter. The most suitable soil for cultivation of rapeseed and mustard is deep light-textured soil (loam or sandy loam), but these crops, particularly mustard, could also grow well on heavy-textured soils. Canola can be grown on most soil types, but it is best suited to clay loam soils that do not crust. Soil with good surface drainage is essential because it cannot tolerate standing water or waterlogged conditions. Canola is less tolerant to drought than small-grain crops (Berglund *et al.*, 2007). High clay content in the soil can be detrimental to root growth of the crop and it also affects water and subsequent uptake. The crop prefers near neutral soil pH (between 6.0 and 6.5), but can grow reasonably well in soil with lower pH values. The growth cycle of *Brassica* seed can range from less than 3 months to more than 1 year.

The objective of seedbed preparation is to encourage strong seedling establishment and good root development with plants spaced uniformly. Sometimes it is difficult to prepare soil adequately for the small *Brassica* seeds and it is tempting to compensate for poor preparation by increasing the seed rate. For winter planting (the practice used in central California to take advantage of winter rains), early sowing encourages establishment of strong, vigorous seedlings and plant population has little effect on final yields. Consequently, seed rate is of less significance in relation to seed yield, but it may have an effect on stem elongation. Competition between plants is low with late sowing and total-accumulated dry matter will be related to plant population at emergence. Winter hardiness will be poor until plants attain a minimum level of vigour. Physiological development is favoured by early sowing, however, one must be careful that seedbeds are not too dry as it leads to delayed or irregular emergence and seedlings may be more vulnerable to attack by insects, e.g. flea beetle or aphids. For spring planting, a soil temperature of at least 6°C is needed for germination; seed may rot at lower temperatures. The spring growing season is shorter than that for winter planting. Thus, it is recommended that plant densities be increased to ensure that dry matter accumulation before flowering is sufficient to achieve a high seed yield.

Growing brassicas on brassica stubble usually results in reduced yield compared to brassicas on cereal stubble. In addition, the brassica crop will be susceptible to disease and insect build-up. Brassica crops, e.g. canola and mustard, are excellent break crops for cereals. However, when pre-emergent herbicides are used in the brassica crops, it is recommended that oats and small-seeded grasses be sown next. Factors affecting crop establishment include the previous crop, crop residue, soil type, fertility, moisture availability, weeds, pests and diseases. Residues from the preceding crop, e.g. cereal straw ranging from 3 to 10 t ha^{-1}, may be unfavourable for the establishment of brassica seed. They may contribute to a highly porous soil texture and inhibit soil–root contact. A summer fallow before drilling winter brassica seeds allow weed seeds to germinate. Direct drilling is becoming increasingly popular with spring plantings as a means of conserving soil moisture. Direct drilling is often favoured as it is quicker and reduces the costs of labour and machinery. However, ploughing is better in soils with low organic matter content or with a poor structure. In such soils, root penetration and drainage are poor without deep cultivation and direct drilling is liable to make crops more susceptible to drought and to enhance nutritional deficiencies.

16.9.3 Growing brassicas under poor soil conditions

Brassica oilseeds are tolerant of a wide range of soils. They are considered moderately tolerant to salt and sodium, and can tolerate moderate salinity up to levels of 5 to 6 dS m^{-1}. In central California (USA), only poor quality soils/or waters are used by USDA-ARS (United States Department of Agriculture – Agricultural Research Service) researchers for growing brassica crops for phytomangement of Se (plant removal of Se via accumulation and volatilization). Generally, these field soils and groundwaters have a sodium sulfate-dominated salinity between 6 and 10 dS m^{-1} (EC: electrical conductivity), pH of 7.9, and concentrations of water-extractable Se and B ranging from 0.13 to 0.50 mg l^{-1} and 5 to 10 mg l^{-1}, respectively. The predominant soil type in the growing areas is generally classified as an Oxalis silty clay form (fine montmorillonitric, thermic Pachec Naploxeral with a well-developed salinity profile).

The most important effect of soil salinity, e.g. sodium, on brassica production is that it increases crusting and reduces seed emergence. Brassica has a small seed and at germination the cotyledons are pushed through the soil to the surface, making emergence more difficult in crusted soils. On soils with moderate salinity, the use of continuous cropping lengthens the rotation to the maximum possible. In this regard, crops use the water rather than allowing it to move deeply into the soil or accumulate near the soil surface and contribute to salt movement. In such soils, brassica seeds should be planted shallow and early so that seeds may germinate when surface salt levels are temporarily lowered. Importantly, subsurface drains can sometimes be used to drain water away and lower the water table, but this is very expensive and labour intensive. If soil salinity has already reached the stage where cereal and oilseed crops cannot be grown, then extensive salinity management will need to be undertaken to reclaim saline soil. When soil salinity is too high there is an increase in number of days from seeding to initial emergence. With severe salinity, after emergence begins, there is a rapid number of new plants but over time the percentage that survive drops off. Salinity under both moderate and severe conditions reduces average plant height, shoot and root biomass, number of leaves, leaf area, evapotranspiration, and seed yield at harvest. Lower seed yields are to be expected when growing under poor soil conditions. This expected result needs to be considered when growing a brassica crop for the purpose of phytomanagement of Se, e.g. in central California, or for the economic production of biofuel (described later).

16.9.4 Seed rate, spacing and sowing time

To ensure a good crop stand, 4–6 kg seed ha^{-1} is sufficient for mustard. Seed is placed in rows 30 cm apart and plant to plant distance of 10–15 cm, whereas rapeseed requires a row to row distance of 45 cm and plant to plant distance of 10 cm. Late-sown crops become vulnerable to white rust and aphid attack resulting in significant yield losses. In the case of canola, seed requirement varies from 6 to 9 kg ha^{-1} depending upon the seed size. Seed is placed in rows 15–18 cm apart with a grain-drill and seeding depth should not exceed 2.5 cm. Canola varieties differ widely in their growth habit and can be grown both as spring and winter annuals. In the USA, the spring type is best adapted to North Dakota conditions and should be planted from late April to early May. Winter canola can be seeded in cold, nearly frozen soil in the autumn like winter wheat. The seeds remain dormant after planting until early spring.

16.9.5 Irrigation

The crop performs well when the stored moisture in the soil profile is over 175 mm before sowing. Consumptive use of water in brassicas varies from 200 to 300 mm. In India, brassica crops are generally sown after heavy pre-sowing irrigation (10–12 cm) and thus may require first irrigation only after 3–4 weeks of sowing. In a deep alluvial well-drained sandy-loam soil, optimum yield of mustard (*Brassica juncea*) sown after pre-sowing irrigation needed only one post-sowing irrigation (80 mm) at 3–4 weeks after sowing (Prihar *et al.*, 1981). For obtaining higher grain yield, some studies based on phenological stages have recommended two irrigations, i.e. one at rosette stage 20–30 days-after-sowing (DAS) and another at 50–60 DAS. Best results were obtained when irrigation was applied at irrigation water:cumulative pan evaporation ratio of 0.6 (Rai *et al.*, 1993). In central California, non-saline (< 1 dS m^{-1}) canal water is initially applied to canola or mustard by a sprinkler system for a maximum of 30 days to initiate germination (Bañuelos, 2002). If additional water is needed during a winter planting, then irrigation scheduling is based on the weather data provided by the local California Irrigation Management Information System (CIMIS) (Howell *et al.*, 1984) and on general crop coefficients reported by Allen *et al.* (1998). Minimal furrow irrigation is then used with poor quality water, e.g. moderate levels of salinity. In central California, growing brassica under rain-fed conditions is highly desirable in such a water-limited region. Using minimum water, especially good quality water, is of the utmost importance for growing brassica crops for biofuel production in an agriculture-rich area like central California on a sustained basis. Ample water is simply not available in a reliable and consistent manner to sustain growing non-edible crops for biofuel production.

16.9.6 Fertilizer requirements

There is considerable variation in fertilizer use by *Brassica* species. To determine fertilizer requirements of brassicas, soil tests, cropping history of the field and plant nutrient status should be taken into account. Oil-bearing brassicas are considered to be heavy nutrient feeders removing 80–130 kg N, 17–25 kg P_2O_5, 70–130 kg K_2O and 25–45 kg S ha^{-1} (Sengupta and Das, 2003). In India, the fertilizer recommendation of N and P_2O_5 is in the ratio of 2:1 and application of K_2O is not mandatory in all the states. For optimum grain yields from medium fertility soils in the state of Punjab, brassicas require 100 kg N, 30 kg P_2O_5, 10 kg S and 10 kg Zn ha^{-1} under irrigated conditions and 40 kg N, 20 kg P_2O_5, 10 kg S and 10 kg Zn ha^{-1} under rain-fed conditions. At sowing time, a full dose of the phosphatic fertilizer and half of the nitrogenous fertilizer is drilled into the seedbed. The remaining half of the nitrogen fertilizer should be uniformly broadcast just before the first irrigation. Nitrogen and sulfur are the key elements for high canola yields in North Dakota (USA). Fertilizers containing N and K should not be placed in direct contact with canola seeds but should be broadcast or applied in a band at least 2.5 cm away from the seed. To produce 2280 kg grain ha^{-1} from a soil with medium fertility level, canola requires total availability of 170 kg N, 40 kg P_2O_5, 52 kg K_2O and 25–35 kg S ha^{-1}. Brassica seed yields can be increased further by increasing fertilizer use efficiency through green manuring. Figure 16.3 shows that the incorporation of green manure such as chopped guara (*Cyamopsis tetragonoloba*) shoots resulted in a significant increase in grain yield of mustard (*B. juncea* L.) over and above that obtained with different levels of inorganic N added alone (Dhillon *et al.*, 1984).

Brassica crops can accumulate large amounts of N. The N level in the plants is highest in the early seedling stage when young leaves constitute the majority of the plant's dry matter. As the plant grows into flowering stages, the overall N level declines due to stem material and leaf loss. By maturity, brassica straw contains just 0.5 to 1.5% N, while the seed contains between 4 and 6% N. Brassica responds well to fertilizer N on N-deficient soils. Generally, a spring planting of brassica accumulates 50–60 kg N for every tonne seed produced and winter planting accumulates 70 kg N. Management of N fertility is complex and requires substantial attention, especially because responses to N vary with soil and climate conditions. Researchers in central California growing brassica for biofuel generally apply less than 125 kg N ha^{-1} and do not exceed applying 150 kg N ha^{-1} (Bañuelos, 2002). In this

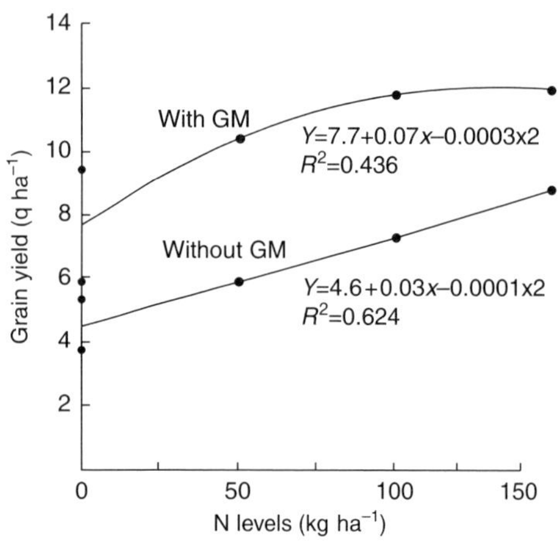

Fig. 16.3. Supplementing nitrogen requirement of *Brassica juncea* through green manuring (GM).

region, when using poor quality soil, a minimal amount of N fertilizer is applied since applying additional salts is not prudent on already saline soils. Excess applications of N also have an impact on the production of oil and subsequent biofuel production and may lead to reductions in seed yield and consequent oil production. Importantly, there is controversy about the balance between the energy expended on production (carbon output) compared with that released when utilized as a fuel. As N fertilizer accounts for one-third of total energy input (PROLEA, 1993), high inputs for biofuel production may be difficult to justify in the overall attempt to reduce carbon fingerprint. In regards to other sources of fertilizer, the amount of phosphate should not exceed 60 kg ha^{-1}. Because of its low mobility, phosphate should be incorporated into the soil. Generally speaking, the oil content of the brassica seed is, however, not affected by P fertilizers (Thomas, 1984) and there is a minimal effect on glucosinolate production (sulfonated oxime thioesters, which exist in the *Brassica* genus; Fenwick *et al.*, 1983).

Sulfur occupies a special place in the nutrition of oilseed crops, being a part of essential amino acids such as cysteine and methionine and having a strong influence on glucosinolate production in the seed (discussed later). Sulfur requirement of rapeseed-mustard crops is nearly three times more than that of cereals. Experiments conducted in the Indo-Gangetic plains of South Asia have revealed that increase in grain yield of rapeseed mustard crops varied from 83 to 839 kg ha^{-1} with application of 30 kg S ha^{-1} (Khurana *et al.*, 2008). In this regard, the maximum amount of S to apply should not exceed 60 kg ha^{-1}. Although in central California soluble concentrations of S (natural-occurring) can range from 2 to 3000 mg l^{-1} in the soil (Bañuelos, 2002), additional S applications are not needed and generally are never applied when growing brassica crops on the Se-laden soils.

16.9.7 Harvesting stage and postharvest care

Two major problems faced by growers to determine the stage of harvesting are: (i) the shattering of silique when over-ripe; and (ii) the lowering of oil content if the grains are not physiologically mature. The crop is ready for harvesting when the majority of the siliquae turn yellow. There is no risk of seed shattering at this stage and the seeds have attained physiological maturity to give potential oil yield. The harvested crop is then stacked for 7–10 days before threshing. After threshing and cleaning of the seed, it should be sun-dried for 3–4 days to bring down the moisture level to 8%, considered optimum moisture for storage.

Swathing canola or mustard at the optimum stage of ripening reduces green seed problems, shattering losses and maintains quality. Brassica crops are probably very near to the optimum time for swathing when the plants consist only of stems, stem branches and pods. Swathing can begin in Argentine canola at 15% seed colour change and at 20–25% seed colour change in Polish canola. Brassica should be allowed to cure and ripen for 10–14 days in the swath before combining. Very tight truck boxes and storage bins are required for storage and handling of their seeds. The optimum moisture level should be less than 8.5% at the time of storage of brassica seeds. Non-shattering mustard seeds are presently being used by Bañuelos (unpublished) in poor quality soils in central California. This characteristic is of special importance when growing brassicas in poor quality soils/water, since environmental stress tends to promote early flowering and subsequent seeding of plants. Seed shattering results in decreased yields and unwanted brassica volunteers in the subsequent crop on the same field site.

16.9.8 Weed control

Uncontrolled weed growth in mustard can remove 15 kg N, 4 kg P_2O_5 and 18.6 kg K_2O ha^{-1}, thus utilizing a major portion of fertilizers applied (Kaneria and Patel, 1993). Besides conventional hand weeding, a number of herbicides have proved effective in controlling annual grass and broad-leaved weeds (Table 16.2).

An effective weed control programme for brassica crops should include cultural, mechanical and chemical methods. Special attention should be given to control wild mustard, which is a serious seed contaminant in canola and can cause price discounts or rejection in the market. Among the herbicides, Trifluralin and Sonalan (ethalfluralin) are suitable for pre-plant incorporation in soil to control annual grasses and some broadleaf weeds, such as pigweed, common lamb's-quarters and kochia. Clopyralid is suitable for post-emergence control of weeds, which should be at the 2–6 leaf stage and prior to bolting. Sethoxydim (Poast), quizalofop (Assure II or Targa) and clethodim are also recommended for post-emergence grass control in canola (Berglund *et al.*, 2007). The development of Round-Up (Isopropprifamin salts of glyphosate)-resistant canola varieties may be of additional value in improving seed yields of canola.

16.9.9 Plant protection measures

Brassica crops suffer significant damages if not protected from the attack of insect pests and fungal diseases. The number of insect pests and

Table 16.2. Herbicides found suitable for controlling weeds in brassicas under Indian conditions.

Crops	Herbicides	Dose (ha^{-1})	Method and time of application
Mustard	Trifluralin (Treflan 48 EC)	1560 ml	Pre-plant incorporation
	Isoproturon (75 WP)	1000 g	Pre-emergence within 2 days of sowing or post-emergence 25–30 days after sowing
Rapeseed	Isoproturon (75 WP)	750 ml	Post-emergence before first irrigation
	Fluchloralin (Basalin 45 EC)	1625 ml	Pre-plant incorporation
	Trifluralin (Treflan 48 EC)	1560 ml	Pre-plant incorporation
	Atrazine (Atrataf 50 WP)	800 g	Pre-emergence within 2 days of sowing or post-emergence 25–30 days after sowing but 2–3 days before first irrigation

diseases attacking the brassicas keep on changing due to the spread of multiple cropping and change in cropping pattern in different agroclimatic regions worldwide. While some of the pests are described as major pests in one region, the same become minor in other parts of the world. Economically important insects and diseases are discussed below.

Insect pests

MUSTARD APHID (*LIPAPHIS ERYSIMI* KALTENBACH). The mustard aphid is considered a serious pest of cruciferous oilseeds causing damage to crops either directly by feeding or by transmitting plant viral diseases (Dawson *et al.*, 1990). Depending upon the severity of aphid infestation and crop stage, damage to cruciferous crops may range from 10 to 90%. More than 83% yield loss in case of *B. juncea* has been recorded in India and other tropical regions of the world (Sekhon and Åhman, 1993; Aslam and Ahmad, 2001). The insects breed parthenogenetically, grow very fast and become full-fed in 7–10 days. Rate of reproduction is higher during the early vegetative stage of mustard plants when the developmental period is the shortest and production of winged morphs is the lowest (Agarwala and Datta, 1999). They feed on plant tissue by piercing with their needle-like mouthparts (stylets), sucking water and nutrients from the phloem vascular system of the plant. Both the nymphs as well as adults suck cell-sap in huge quantities from leaves, stems, inflorescence and developing pods and thereby greatly reduce the vitality of plants and reduce oil content in grains (Mossler, 2005). The insect excreta on the plant provides a congenial environment for the growth of sooty mould.

PAINTED BUG (*BAGRADA HILARIS* BURMEISTER). This is a serious pest of cruciferous crops and is widely distributed in countries such as Myanmar, Sri Lanka, India, Iraq, Arabia and East Africa. It attacks the crop at two growth stages, during germination and at maturity. Both the nymphs as well as adults suck cell-sap from the foliage and developing pods, which subsequently wilt and dry up. Severe attack at seedling stage may ruin the whole crop. Adult bugs excrete a resinous material (i.e. glutinous), which spoils the pods. The insect population can be significantly reduced by applying first irrigation early, i.e. after 3–4 weeks of sowing. Clean cultivation and quick threshing of the harvested crop may also help in controlling the incidence of painted bug. The chemical control measures are reported in Table 16.3.

MUSTARD SAWFLY (*ATHALIA LUGENS* KLUG). This is widely distributed in Indonesia, Taiwan, Myanmar and the Indian subcontinent. Insect larvae are dark green in colour and the adults are orange yellow with black markings on the body and smoky wings. The insect larvae attack

Table 16.3. Effectiveness of different insecticides against common insect pests of mustard and rapeseed under Indian conditions.

	Effectiveness against				
Insecticides	Mustard aphid	Painted bug	Mustard sawfly	Leaf miner	Cabbage caterpillar
For foliar spray					
Metasystox 25 EC (oxydemeton methyl)	Yes			Yes	
Rogor 30 EC (dimethoate)	Yes			Yes	
Thiodane 35 EC (endosulfan)	Yes	Yes	Yes		Yes
Ekalux 25 EC (quinalphos)	Yes				
Malathion 50 EC (malathion)	Yes	Yes	Yes		Yes
Nuvan 85 SL (dichlorvos)					Yes
For soil application					
Thimet/Phorolik10 G (phorate) at 10 kg ha^{-1}	Yes			Yes	
Furadan 3 G (carbofuran) at 33 kg ha^{-1}	Yes				

fresh growth, biting holes into the leaves and thereby skeletonize the leaves. The older plants when attacked do not bear seed.

LEAF MINER (*CHROMATOMYIA HORTICOLA* GOUREAU). Insect larvae feed on leaves and may heavily damage the crop growth in some years.

HAIRY CATERPILLAR (*SPILOSOMA OBLIQUA* WALKER); CABBAGE CATERPILLAR (*PIERIS BRASSICAE* LINNAEUS). The caterpillars feed on leaves, young shoots and green pods. When young they feed gregariously but the grown-up caterpillars migrate from one field to another. They can be easily controlled in the gregarious stage by picking and destroying the infested leaves. When grown, the caterpillars can be controlled by the application of insecticides (Table 16.3).

DIAMONDBACK MOTH (*PLUTELLA XYLOSTELLA*). This was first noticed as a pest in South Africa in the early 1900s and is now considered as the most damaging insect pest of brassica crops worldwide with crop losses approaching 90% in some cases (Charleston and Kfir, 2000). Diamondback moth is highly mobile and can attack plants at all stages of growth. The insect population influxes from southern Canada can arrive in western parts of the USA approximately 1 month before even the vast monoculture canola crops are planted in the region (Sarfraz *et al.*, 2011). Similarly, the severity of infestation can vary considerably from year to year and may depend on the numbers and seasonal timing of migrating moths in North Dakota, USA (Knodel, 2010). Insect larvae feed on the leaves, buds, flowers, seed pods, green outer layer of the stems and, occasionally, on the developing seeds. Leaf feeding by young larvae results in a characteristic windowpane effect with small, irregular-shaped holes. Mature larvae may eat the entire leaf, leaving only the veins.

A number of natural factors can help in controlling the diamondback moth population. Heavy rainfall can wash/drown larvae of the first or second generations. Humid conditions can favour the development of lethal fungi such as *Entomophthorales*, which play an important role in controlling moth populations. Cool, windy weather reduces adult activity and females often die before they lay all of their eggs. In western Canada, three species of parasitic wasps, *Diadegma insulare* (Cresson), *Microplitis plutellae* (Muesebeck) and *Diadromus subtilicornis* (Gravenhorst), feed on the larvae of diamondback moth. Several predaceous arthropods, such as ground beetles, true bugs, lacewing larvae and spiders also prey on the diamondback moth larvae.

Chemical control is recommended when larvae count reaches approximately 2–3 larvae per plant of canola at the advanced pod stage. Application of insecticides such as bifenthrin (38–49 g AI ha^{-1}), deltamethrin (10 g AI ha^{-1}), gamma-cyhalothrin (9–17 g AI ha^{-1}) or zeta-cypermethrin (28 g AI ha^{-1}) has proved highly effective in controlling diamondback moth outbreaks in North Dakota (Knodel, 2010).

FLEA BEETLES (*PHYLLOTRETA CRUCIFERAE*). The crucifer flea beetle has become the dominant pest of canola in North America and the pollen beetles of oilseed rape in northern Europe. Adult flea beetles emerge in the spring and feed on the cotyledons and true leaves and can quickly devastate the canola field. The intensity of attack is influenced by weather conditions. Sunny, warm, dry conditions increase the feeding activity of beetles as compared to cool, damp weather. Damaged plants typically have a 'shot-holed' appearance when the tissues around the feeding sites in the cotyledons and leaves die. Crop damage due to flea beetles may exceed US$300 million annually in North America (Knodel and Olsen, 2002). Chemical control of flea beetles is recommended when damage to cotyledons and first true leaves is more than 25%. Some of the recommended insecticides are: Furadan CR-10 (carbofuran) at 2.85 kg ha^{-1}, Methyl parathion 4E, 5 (methyl parathion) at 0.6 kg ha^{-1}, Thiodan 3EC (endosulfan) at 1.0 kg ha^{-1} and Javelin WG (*Bacillus thuringiensis* ssp. *kurstaki*) at 0.14–1.7 kg ha^{-1}. Number of insecticide sprays should not exceed two per season.

Diseases

Of the 22 diseases affecting brassica crops, only a few may cause heavy damage to the crop. Some of the diseases considered critical for the growth of brassicas are discussed below.

ALTERNARIA BLIGHT (*ALTERNARIA BRASSICAE*). Brown to blackish spots with concentric rings appear on leaves, pods and stems. In cases of severe attack, the upper parts of stem and pod wither. The losses are heavier in yellow sarson followed by brown sarson and mustard. Low light intensity, 25°C temperature and 90% relative humidity are the optimum conditions that encourage germination of fungal spores. As a precautionary measure, the diseased debris from the previous crop must be removed from the field and destroyed.

DOWNY MILDEW (*PERONOSPORA PARASITICA* PERS.). The causal organism is a composite species consisting of many smaller morphological and biological species. The species occurring on *B. campestris* was often called *Peronospora brassicae*. The fungus attacking crucifers has recently been named as *Hyaloperonospora parasitica* (Constantinescu and Fatehi, 2002). The pathogen is primarily soil-borne through oospores lying in diseased plant debris as well as seeds. There is a possibility that *Peronospora* is capable of moving systematically from shoots to roots and the hyphae emerged from root tissues could infect neighbouring plants leading to new systemic infections. Appearance of small purplish brown or light green spots on the lower side of the leaves indicates the presence of downy mildew disease. All the aerial parts of the plants are affected by the disease, but it appears more severe on foliage, seed-bearing parts and stem. Temperatures ranging from 11.5 to 12.5°C and relative humidity of more than 75% are favourable for the growth and spread of disease. As a precautionary measure it is prudent to destroy any diseased debris from the previous crop.

WHITE BLISTERS OR WHITE RUST OF CRUCIFERS (*ALBUGO CANDIDA* KUNZ). The causal organism is an obligate or biotrophic parasite and perpetuates through oospores lying in the soil in diseased plant debris. Moist and cool weather favours disease development. A thin film of water is necessary for oospores' germination, which is the maximum at 12–14°C and in darkness. All the plant parts, except roots, are attacked by this disease. Prominent white creamy yellow scattered pustules appear on the lower side of leaves followed by swelling of affected parts. When the host epidermis ruptures, the exposed sori reveal a white powdery mass consisting of spores of the fungus. The axis of the inflorescence and the flower stalk may get enormously thick. Flowers are malformed and become sterile. Petals may become sepal-like green and stamens are transformed into leaf-like structures. Combined infection of leaves and inflorescence may cause up to 60% yield loss in *B. juncea*. In a late-sown crop, the yield loss due to floral infection alone may vary from 23 to 54.5%. This disease often appears in association with downy mildew (*P. brassicae*) and then the damage is enormous and requires prompt control measures.

PHYLLODY (*MESORHIZOBIUM LOTI*). Phyllody may be defined as the development of floral parts into leafy structures as a result of virus or phytoplasma infection. Evidence suggests that the phytoplasma down-regulates a gene involved in petal formation, instead causing leaves or leaf-like structures to form. Phyllody causes the affected plant to become partially or entirely sterile. In phyllody-infected plants, numerous shoots arise and the plants become bushy or broom shaped. The best way to control phyllody is to avoid early sowing and to rogue out the affected plants.

The crop damage due to the diseases discussed above, except phyllody, can be minimized by applying 3–4 foliar sprays at 15 day intervals with 625 g of Blitox or Indofil M-45 at about a 75-day-old crop.

BLACKLEG DISEASE (*LEPTOSPHAERIA MACULANS*). This occurs in two forms, i.e. mild or weakly virulent and a virulent or more aggressive form; the latter constitutes a serious threat to canola production in North America. The fungus overwinters in infected crop stubble and is also seed-borne. Round to irregular white to grey spots develop on cotyledons and leaves of canola. Basal stem cankers are formed that lead to premature ripening and lodging of infected plants. Black spots may also become visible on pods and seeds may be infected and shrivelled. Blackleg fungus is spread by wind-borne spores and infected seed. Varieties are available that have good tolerance or resistance to this disease.

Generally, the Argentine varieties are relatively more tolerant to blackleg than Polish varieties. Quadris fungicide is recommended for control of blackleg, applied at a rate of 430 g ha^{-1} at the 2–4-leaf stage of canola.

SCLEROTINIA STEM ROT (*SCLEROTINIA SCLEROTIORUM*). Recently, this has become the most serious disease of canola in North America with losses reaching as high as 50% in severely infected fields. The disease spreads fast especially during wet weather preceding flowering, and the first symptoms appear near the end of flowering stage. Hard, black bodies known as sclerotia produced in infected stems ultimately fall to the soil at harvest and may survive for several years in soil. Timely application of fungicides helps in controlling the disease effectively. Quadris is the most suitable fungicide, which should be applied at 10–25% bloom, or 3–7 days after initiation of flowering. Other fungicides, namely Ronilan EG, Thiophanate Methyl (1) or Endura, are also recommended for use in the suppression and control of sclerotinia in canola.

16.10 Growing Brassica Crops for Biofuel Production

The *Brassicaceae* family is a rich source of feedstocks for biodiesel production as evidenced by *Synapis alba* L., *B. carinata* L., *Camelina sativa* L. and *Raphanus sativus* L. oils. Canola (or rapeseed, *B. napus* L.) has already gained widespread acceptance as a common commodity feedstock for biodiesel production (Körbitz, 1995; McDonnel *et al.*, 2000; Oregon Environmental Council, 2007). Recent interest in the production of biofuels from agricultural feedstocks has resulted in considerable controversy (Jenkins *et al.*, 2009; Dale *et al.*, 2010). On one hand, biofuels offer partial relief from societal demand for petroleum, and their combustion products may contribute less to global climate change than fossil fuels. On the other hand, widespread production of biofuels with food or feed staple crops on fertile soils may result in food and animal feed scarcity.

Global increase in demand for biodiesel feedstock has encouraged production of canola (*B. napus* L.). Nevertheless, efforts are also being made to increase the production of other oilseed crops for biofuel production. Multilocation field experiments have been conducted to determine the oil yield potential of other crops related to *napus* canola in the semi-arid, short-season environment of the Canadian prairies (Blackshaw *et al.*, 2011). Oilseed crops evaluated included: *rapa* canola (*B. rapa* L.), mustard (*B. juncea* L.), Ethiopian mustard (*B. carinata* L.), oriental mustard (*B. juncea* L.), yellow mustard (*Sinapis alba* L.), camelina (*Camelina sativa* L.), flax (*Linum usitatissimum* L.) and soybean (*Glycine max* L.). The number of site-years (out of a total of 9) that crops attained similar or greater yields compared to *napus* canola were camelina (6), oriental mustard (5), mustard (3), flax (3), soybean (3), *rapa* canola (2), yellow mustard (2) and Ethiopian mustard (1). The ranking of seed oil concentration was: *napus* canola = *rapa* canola = mustard = flax > camelina = oriental mustard > Ethiopian mustard > yellow mustard > soybean. Considering the grain yield and oil concentration, the alternative oilseed crops exhibiting high potential for biodiesel feedstock were found to be camelina, flax, *rapa* canola and oriental mustard. Values reported by BioEnergy (2006) show one example of oil yields (l ha^{-1}) by different crops as follows: brassica rape (>1000) > sunflower (912) > safflower (757), camelina (560) and mustard (556). Oils from all these crops are easily converted to biodiesel and quality analyses indicated that all crops would be suitable for biodiesel feedstock with the addition of antioxidants that are routinely utilized in biodiesel fuels. Comparison of important characteristics and composition of wild mustard oil methyl esters with those of soybean oil and biodiesel standards (Table 16.4) revealed that wild mustard oil is an acceptable feedstock for biodiesel production.

Biodiesel possesses a number of technical advantages as well as disadvantages over petrodiesel. Important advantages are: origin from renewable and domestic feedstocks, displacement of imported petroleum, inherent lubricity, no sulfur, superior flash point and biodegradability, reduced toxicity and a reduction in many regulated exhaust emissions (Jham *et al.*, 2009). The disadvantages of biodiesel include: high

Table 16.4. Properties of methyl esters of wild mustard (WMME) and soybean oil (SME) in comparison to biodiesel specifications from American (ASTM D6751) and European Union (EN 14214) standards[a] (Jham *et al.*, 2009).

Parameters	WMME	SME	ASTM D6751	EN 14214
MW calc (g mol^{-1})	327.87	292.33	–	–
Acid value (mg KOH g^{-1})	0.06 (0.05)[b]	0.01 (0.01)	<0.50	<0.50
Gardner colour	10	1	–	–
Free glycerol (mass %)	0.002	0.010	<0.020	<0.020
Total glycerol (mass %)	0.224	0.156	<0.240	<0.250
Cloud point (°C)	4 (1)	0 (1)	–	–
Pour point (°C)	–21 (1)	–3 (1)	–	–
Cold filter plugging point (°C)	–3 (1)	–4 (1)	–	Variable
Oxidative stability				
Induction period, 110°C (h)	4.8 (0.1)	5.0 (0.1)	>3	>6
Onset temperature (°C)	170.2 (1.2)	171.6 (1.5)	–	–
Kinematic viscosity, 40°C (mm^2 s^{-1})	5.33 (0.01)	4.12 (0.01)	1.9–6.0	3.5–5.0
Specific gravity				
25°C	0.874 (0.001)	0.919 (0.001)	–	–
40°C	0.863 (0.001)	0.910 (0.001)	–	–
Wear scar, HFRR 60°C (μm)	151 (7)	135 (5)	520*	460**
Sulfur (ppm)	11	1	<15	<10
Phosphorus (mass %)	0.0002	0.0000	<0.001	<0.001
Derived cetane number	61.1	54.0	>47	>51
Iodine value (g I_2 100 g^{-1})	112	131	–	<120

[a]As specified in petrodiesel standards *ASTM D975 and **EN 590.
[b]Values in parentheses are standard deviations from the reported means (*n* = 3; *n* = 2 for lubricity).

cost of feedstock acquisition, lower volumetric energy content, inferior low temperature operability, inferior oxidative and storage stability, and higher NO_x exhaust emissions. Bhardwaj and Hamama (2008) evaluated the quality of winter-hardy rapeseed germplasm relative to biodiesel production. The investigations regarding the oil and fatty acid composition of winter-hardy rapeseed accessions grown in this region revealed that oil of *B. napus* group contained the following fatty acids in descending order: C22:1, C18:1, C18:2, C18:3, C20:1, C16:0, C20:2, C16:1, C18:0, C20:0, C22.2, C14:0, C22:0. The contents of fatty acids in *B. rapa* oil had a similar distribution except that it had a higher content of C18:2 than C18:1. Based on desirability of low viscosity and higher cetane numbers, it was concluded that *B. napus* oil may be superior to *B. rapa* oil for biodiesel manufacturing. Selection index (a calculated value based on closeness of fatty acid profiles of *B. napus* and *B. rapa* accessions to an optimal fatty acid profile of soybean oil) also indicated that *napus* accessions may be more close to the target fatty acid profile than *rapa* accessions.

As per American and European standard specifications, a standardized definition for biodiesel includes fatty acid methyl esters (FAME) from any kind of feedstock, including recycled. The primary feedstock for biodiesel is triglycerides. These triglycerides comprise three long fatty acid chains attached to a glycerol molecule, the glycerol being only a very small part of the triglyceride molecule. The remainder of the triglyceride comprises long hydrocarbon chains. With this basic structure any oil can be transformed into biodiesel. Important fuel properties of biodiesel that are influenced by the fatty acid profile include: ignition quality, heat of combustion, cold flow, oxidative stability, exhaust emissions, viscosity and lubricity. Superior cold-temperature flow characteristics require a fuel with low saturated fatty acids, whereas increasing oxidative stability and reducing NO_2 emissions requires decreasing the amounts of unsaturated and polyunsaturated fatty acids. Additionally, the ignition quality (as measured by cetane number) is also adversely affected by increased unsaturation. Despite this, unsaturated fats are considered superior for cold-weather biodiesel production and use. However, they are more prone to oxidation and rancidification and will need some oxidative stabilizer to be used safely as fuel. In view of these contrasting requirements, there is no fatty acid profile that will provide a fuel for which all these parameters are optimal. Excellent compromise can be reached by considering a fuel high in the mono-unsaturated fatty acids, such as oleate or palmitoleate, and low in both saturated and polyunsaturated fatty acids. Canola oil and olive oil are typically mono-unsaturated and have very few issues with oxidative stability. They also tend to have a low gel point.

Biodiesel (BD) 20 biofuel (20% plant-based and 80% petrol diesel) can be directly used for powering diesel engines on irrigation booster pumps and on-site tractors. Results have indicated that engines run on a 20% blend of mustard oil showed a closer performance to pure diesel (Anbumani and Singh, 2010). All the parameters tested, viz. total fuel consumption, specific energy consumption, specific fuel consumption, brake thermal efficiency and cylindrical peak pressure were improved. In addition to this, use of esterified mustard oil at 20% blends (BD 20) compared well with the important fuel properties as per ASTM specifications of biodiesel. Its use led to an improvement in engine performance and some emission characteristics without requiring any modifications in the engine. Preliminary emissions analyses on using California-produced BD 20 biofuels produced from brassica crops in irrigation booster pumps showed positive effects on lowering air pollutant emissions of particulate matter, CO, total hydrocarbons, and exhibited questionable results on NO_x emissions (Bañuelos, unpublished data). Reduced combustion and consumption of diesel, in addition to lowering emissions of specific pollutants, is a positive secondary effect for air quality upon using BD 20 biofuels in central California, which has one of the most contaminated valley basins in the USA. Importantly, the USA plans to increase first generation biofuel production to 1.3×10^{11} l by 2017, whereas Europe proposes that 5.75% of all petrol and diesel for transportation purposes (calculated on the basis of energy content) should be placed on the market (McDermott, 2006).

Bañuelos and Dhillon (2011) and Dhillon and Dhillon (2009) grew brassica plants as a primary part of a plant-based technology to manage the movement of selenium into water tables, surface- and tail-water runoff and drainage effluent in central California and in north-western India. Briefly, the impact of Se on the biological ecosystems in arid and semi-arid regions of the world with alkaline seleniferous soils derived from marine-type sediments caused great alarm throughout the western USA, as well parts of China (Zhu *et al.*, 2009) and India (Dhillon and Dhillon, 1991). In central California, the bioaccumulation of soluble Se in the avian food chain and the resulting high mortality and reproduction failure in migratory aquatic birds led to a well-publicized environmental disaster (Ohlendorf and Hothem, 1995). Because of selenium's chemical and physical relationship with sulfur, and *Brassica*'s affinity for sulfur, brassica crops (e.g. mustard, canola) are suitable for managing soluble Se in the soil and water via accumulation and volatilization of Se. This process is called 'phytomanagement'. Using brassica species for phytomanagment of Se, USDA-ARS scientists, in cooperation with Red Rock Ranch in California, have been investigating the production of biodiesel from oil extracted from canola and mustard plants grown under selenium-laden soil and water conditions. In general, brasssica yield has ranged from 1 to 2 tons of seed per acre (2.2–4.4 t ha^{-1}) when grown with soil and water containing an average sulfate-dominated salinity of 6–7 dS m^{-1} and 150 parts per billion (ppb) soluble Se under field conditions. Initial efforts have resulted in the capability to process 1 t *Brassica* seed h^{-1} using an Insta Pro 1500 Oil Press and a 2000 RC Extruder (Insta Pro International, Des Moines, Iowa, USA). With an ideal oil extraction efficiency of almost 90%, the expelled oil is remarkably stable due to the high levels of natural antioxidants and low levels of phospholipids and peroxides. All recovered oil is stored in airtight tanks allowing for natural separation of canola and mustard oils and any remaining residual particles. After transesterification of the brassica oil, biodiesel is then usable for operating on-site diesel-powered equipment. The resulting Se-enriched seed meals from either canola or mustard after pressing and extruding brassica seeds are now available for other uses (e.g. animal feed, bioherbicide), as described below.

16.11 By-products from Brassica Seed Meal

The economic value obtained from producing biofuel and potentially other by-products from brassicas will determine the economic viability and sustainability of brassica feedstocks. Production strategies are needed to guarantee the supply of seed needed by oil processors to make the production for bioenergy profitable without subsidies on a long-term basis. Moreover, economic studies are needed to identify the impact that shifts into brassica production may exert on crop food prices and crop and livestock profitability when the crop is grown in agricultural regions. Bañuelos (2009) has used brassica crops for over 15 years on large farm experiments on Se-laden soils. He has clearly realized that in order for the brassica species to play an important role in remediation strategies in California and in other parts of the world (Dhillon and Dhillon, 2003), viable economic uses must be developed for their harvested plant products.

Producing only biofuel from brassica crops grown on unproductive soils of poor quality is also not economically viable when we consider typical production costs incurred on soils in central California. At the current cost of diesel in California, e.g. US\$1.07 l^{-1}, an estimated US\$1027 ha^{-1} could grossly be earned from selling the BD 100 biofuels produced from brassica crops. For example, if we assume oil crushing/processing costs associated with pressing the seed in central California, e.g. circa US\$150 day^{-1}, it is imperative that more revenue needs to be generated in order to make it feasible to produce biofuel by producing other bio-based products from the seed. In this regard, after pressing and extracting oil from seed, the residual seed meal is one of the most widely used protein supplements in the animal feed trade, and generally competes as a second tier product to soya meal. Nutritionally, Se, nutrient and vitamin concentrations in canola are adequate when compared to other reported values for grass and legume forage (NRC, 1985). Importantly, the Se concentrations in the canola meal produced from central California are generally less than 2 mg kg^{-1} DM (Bañuelos, 2006), which allows its safe use as part of the daily feed ration for lactating cows (Bañuelos *et al.*, 2010) (note: high soluble S concentrations in California soils prevents an excessive accumulation of Se by the plant). Other vital animal forage components within the Se-biofortified meal such as acid detergent fibre (ADF), neutral detergent fibre (NDF) and non-fibrous carbohydrates (NFC), are slightly lower for canola meal than typical values for good quality legume, grass and legume grass hays and silages (NRC, 1985); however, the crude protein (CP), crude fat (CF), soluble protein (SP) and total digestible nutrient (TDN) values are slightly better than legume hays. The low glucosinolate content in the canola meal (<21 μmol g^{-1}) is an important feed quality of the canola seed meal.

Selenium phytomanagement strategies using brassica plants in seleniferous soils can easily produce Se-biofortified meals because of the accumulation of Se in the protein-rich seed. Importantly, this meal product can attain higher market values when sold as Se-enriched seed meal. The Se-enriched seed meal could be easily sold for a price between US\$200 and US\$300 t^{-1}. This Se-enriched animal feed, e.g. canola meal, produced can be of special importance for providing Se as an essential trace element to central California's large livestock and dairy industries. Moreover, the bioavailability of organic Se (C–Se–C forms) supplied within the Se-enriched seed meal and their subsequent more efficient absorption by animal tissues than inorganic forms of Se (Agobossamey *et al.*, 1998), makes this product valuable for animal nutrition. Providing animals with Se-enriched meal is a practical and economical strategy for utilizing seed meals produced from plants grown on Se-rich field sites. In contrast to canola seed meal, residual mustard seed meal also possesses bioactive properties from its higher glucosinolate content (e.g. 300 μmol g^{-1}), which makes it potentially useful for biofumigation strategies (Stapleton and Bañuelos, 2009). The glucosinolates can hydrolyse into isothiocyanates (ITC) with potentially broad-spectrum biocidal activity.

Creating an alternative biofumigant product from mustard seed meal with its high glucosinolate content may have practical use, especially in organic agriculture. Prices from such products, e.g. biofumigants, are currently selling for US$20 for 4 kg bags for use in organic agriculture. These niche-markets are available for increasing the economic viability of growing mustard in poor quality, e.g. westside soils of central California.

16.12 Conclusion

Among the oilseed crops, the majority of the biodiesel in the world is produced from rapeseed, primarily by European producers. Canola's oil yield has been substantially optimized, while other brassica species, e.g. mustard, have likely not reached their optimum value. While further research and breeding is essential for increasing seed yield under a variety of growing conditions, the most realistic approach will be to develop a range of brassica crops that can be grown in unproductive soils and with a minimum of good quality water. Because growing crops for biofuel production can be controversial, energy inputs, i.e. fertilizers, equipment activities, greenhouse gas emissions, compared to energy outputs, and the food versus fuel debate, must always be considered when growing brassica or other biofuel crops under good quality growing conditions. Successful and sustainable production of brassica should not displace annual agricultural crops or utilize good quality waters that would normally be used to grow agricultural food crops, especially in water-limited regions of the world. The development of new varieties of brassica that yield high biofuel returns with minimal carbon debts, or that do not require good land for production, should be encouraged. Competing land uses should always be subjected to comprehensive evaluation, so as to optimize socio-economic and ecological benefits when producing brassica feedstock as a renewable source of bioenergy.

References

Agarwala, B.K. and Datta, N. (1999) Life history response of the mustard aphid *Lipaphis erysimi* to phenological changes in its host. *Journal of Bioscience* 24, 224–231.

Agobossamey, Y.R., Petit, H.V., Seoane, J.R. and St Laurent, G.J. (1998) Performance of lambs fed either hay or silage supplemented with canola or fish meals. *Canadian Journal of Animal Science* 78, 135–141.

Ahuja, I., Rohloff, J. and Bones, A.M. (2010) Defence mechanisms of *Brassicaceae*: implications for plant-insect interactions and potential for integrated pest management. A review. *Agronomy for Sustainable Development* 30, 311–348.

Allen, E.J., Morgan, D.G. and Ridgman, W.I. (1971) A physiological analysis of the growth of oilseed rape. *Journal of Agricultural Science, Cambridge* 77, 339–341.

Allen, R.G., Pereira, L.S., Raes, D. and Smith, M. (1998) *FAO Irrigation and Drainage Paper 56*. FAO, Rome, Italy.

Anbumani, K. and Singh, A.P. (2010) Performance of mustard and neem oil blends with diesel fuel in C. I. engine. *ARPN Journal of Engineering and Applied Sciences* 5, 14–20.

Aslam, M. and Ahmad, M. (2001) Effectiveness of some insecticides against mustard aphid, *Lipaphis erysimi* (Kalt.) (Aphididae: Homoptera) on three different crops. *Journal of Research Science* 12, 19–25.

Astley, D. (1982) Collecting in Ethiopia. *Cruciferae Newsletter* 7, 3–4.

Baillon, H.E. (1871) *Histoire des plantes. Tome 3, Partie 4, Monographie des crucifères.* L. Hachette, Paris, France.

Banga, S.S. (1988) C-genome chromosome substitution lines in *Brassica juncea* L. Coss. *Genetica* 77, 81–84.

Banga, S.S. (1996) Breeding and genetics. In: Chopra, V.L. and Shyam Prakash (eds) *Oilseeds and Vegetable Brassicas: Indian perspective*. Oxford & IBH, New Delhi, India, pp. 50–76.

Banga, S.S. and Banga, S.K. (2009) Crop improvement strategies in rapeseed-mustard. In: Hegde, D.M. (ed.) *Vegetable Oils Scenario: Approaches to meet the growing demands.* ISOR, Hyderabad, India, pp. 13–35.

Banga, S.S., Banga, S.K. and Labana, K.S. (1983) Nucleocytoplasmic interactions in *Brassica. Proceedings of the 6th International Rapeseed Conference*, Paris, France, pp. 602–606.

Bansal, P., Banga, S.K. and Banga, S.S. (2012) Heterosis investigated in terms of polyploidy and genetic diversity using designed *Brassica juncea* amphiploid and its progenitor diploid species. *PLoS ONE* 7(2), e29607.doi:10.1371/journal.pone0029607.
Bañuelos, G.S. (2002) Irrigation of broccoli and canola with boron-and selenium-laden effluent. *Journal of Environmental Quality* 31, 1802–1808.
Bañuelos, G.S. (2006) Phyto-products may be essential for sustainability and implementation of phytoremediation. *Journal of Environmental Pollution* 144, 19–23.
Bañuelos, G.S. (2009) Phytoremediation of selenium-contaminated soil and water produces biofortified products and new agricultural by products. In: Bañuelos, G.S. and Lin, Z.Q. (eds) *Development and Uses of Biofortified Agricultural Products*. CRC Press, Boca Raton, Florida, pp. 57–70.
Bañuelos, G.S. and Dhillon, K.S. (2011) Developing a sustainable phytomanagement strategy for excessive selenium in Western United States and India. *International Journal of Phytoremediation* 13, 208–222.
Bañuelos, G.S., Da Roche, J. and Robison, J. (2010) Developing selenium-enriched animal feed and biofuel from canola planted for managing Se-laden drainage waters in the westside of central California. *International Journal of Phytoremediation* 12, 243–254.
Basunanda, P., Spiller, T.H., Hasan, M., Gehringer, A., Schondelmaier, J., Lühs, W., Friedt, W. and Snowdon, R.J. (2007) Marker-assisted increase of genetic diversity in a double-low seed quality winter oilseed rape genetic background. *Plant Breeding* 126, 581–587.
Baud, S., Boutin, J.P., Miquel, M., Lepiniec, L. and Rochat, C. (2002) An integrated overview of seed development in *Arabidopsis thaliana* ecotype WS. *Plant Physiology and Biochemistry* 40, 151–160.
Beilstein, M.A., Al-Shehbaz, I.A. and Kellogg, E.A. (2006) Brassicaceae phylogeny and trichome evolution. *American Journal of Botany* 93, 607–619.
Berglund, D., McKay, K. and Knodel, J. (2007) Canola Production. Available at: http://www.ag.ndsu.edu/pubs/plantsci/crops/a686w.htm (accessed 1 January 2011).
Bhardwaj, H.L. and Hamama, A.A. (2008) Oil quality of winter hardy rapeseed gemplasm relative to biodiesel production. *World Journal of Agricultural Sciences* 4, 1–6.
BioEnergy (2006) Oil Seed Crops. Available at: http://www.bioenergy.wa.gov/OilSeed.aspx (accessed 1 January 2012).
Bisht, N.C., Gupta, V., Ramchiary, N., Sodhi, Y.S., Mukhopadhyay, A., Arumugam, N., Pental, D. and Pradhan, A.K. (2009) Fine mapping of loci involved with glucosinolate biosynthesis in oilseed mustard (*Brassica juncea*) using genomic information from allied species. *Theoretical and Applied Genetics* 118, 413–421.
Blackshaw, R.E., Johnson, E.N., Gan, Y., May, W.E., McAndrew, D.W., Barthet, V., McDonald, T. and Wispinski, D. (2011) Alternative oilseed crops for biodiesel feedstock on the Canadian prairies. *Canadian Journal of Plant Science* 91, 889–896.
Brown, J. (1995) Early generation selection in spring canola (*Brassica napus* L.). *Proceedings of the 6th International Rapeseed Conference*, Cambridge, UK, 3, 699–701.
Burns, M.J., Barnes, S.R., Bowman, J.G., Clarke, M.H.E., Werner, C.P. and Kearsey, M.J. (2003) QTL analysis of an intervarietal set of substitution lines in *Brassica napus*: (i) Seed oil content and fatty acid composition. *Heredity* 90, 39–48.
Cahoon, E.B., Shockey, J.M., Dietrich, C.R., Gidda, S.K., Mullen, R.T. and Dyer, J.M. (2007) Engineering oilseeds for sustainable production of industrial and nutritional feedstocks: solving bottlenecks in fatty acid flux. *Current Opinion in Plant Biology* 10, 236–244.
Carr, R.A. (1995) Processing the seed and oil. In: Kimber, D. and McGregor, D.I. (eds) *Brassica Oilseeds – Production and Utilization*. CAB International, Wallingford, UK, pp. 267–290.
Charleston, D.S. and Kfir, R. (2000) The possibility of using Indian mustard, *Brassica juncea*, as a trap crop for the diamondback moth *Plutella xylostella* in South Africa. *Crop Protection* 19, 455–460.
Chen, B.Y., Jørgensen, R.B., Cheng, B.F. and Heneen, W.K. (1997) Identification and chromosomal assignment of RAPD markers linked with a gene for seed colour in a *Brassica campestris-alboglabra* addition line. *Hereditas* 126, 133–138.
Chevre, A.M., Barret, P., Eber, F., Dupuy, H., Brun, P., Tanguy, X. and Renard, M. (1996) Selection of stable *Brassica napus–B. juncea* recombinant lines resistant to black leg (*Leptosphaeria maculans*). 1. Identification of molecular markers, chromosomal and genomic origin of the introgression. *Theoretical Applied Genetics* 95, 1104–1111.
Chiang, M.S., Chong, C., Landry, R.S. and Crete, R. (1993) Cabbage *Brassica oleracea* subsp. *capitata* L. In: Kalloo, G. and Bergh, B.O. (eds) *Genetic Improvement of Vegetable Crops*. Pergamon Press, Oxford, UK, pp. 113–155.

Constantinescu, O. and Fatehi, J. (2002) Peronospora-like fungi (Christa, Peronosporales) parasitic on Brassicaceae and related hosts. *Nova Hedwigia* 74, 291–301.
Dale, V.H., Kline, K.L., Wiens, J. and Fargione, J. (2010) Biofuels: implications for land use and biodiversity. Available at: http://www.esa.org/biofuelsreports (accessed 1 January 2012).
Dawson, G.W., Griffiths, D.C., Merritt, L.A., Mudd, A., Pickett, J.A., Wadhams, L.J. and Woodcock, C.M. (1990) Aphid semichemical – a review, and recent advances on the sex pheromone. *Journal of Chemical Ecology* 16, 3019–3030.
De Candolle, A.P. (1821) Cruciferae. *Systema Naturale* 2, 139–700.
Dehesh, K., Tai, H., Edwards, P., Byrne, J. and Jaworski, J.G. (2001) Overexpression of 3-ketoacyl-acyl-carrier protein synthase IIIs in plants reduces the rate of lipid synthesis. *Plant Physiology* 125, 1103–1114.
Delourme, R., Eber, F., and Chevere, A.M. (1989) Intergeneric hybridization of *Diplotaxis erucoides* with *Brassica napus L.* Cyogenetic analysis of F1 and BC1 progeny. *Euphytica* 41, 113–122.
Delourme, R., Pilet-Nayel, M.L., Archipiano, M., Horvais, R., Tanguy, X., Rouxel, T., Brun, H., Renard, M. and Balesdent, M.H. (2004) A cluster of major specific resistance genes to *Leptosphaeria maculans* in *Brassica napus*. *Phytopathology* 94, 578–583.
Delourme, R., Chèvre, A.M., Brun, H., Rouxel, T., Balesdent, M.H., Dias, J.S., Salisbury, P., Renard, M. and Rimmer, S.R. (2006a) Major gene and polygenic resistance to *Leptosphaeria maculans* in oilseed rape (*Brassica napus*). *European Journal of Plant Pathology* 114, 41–52.
Delourme, R., Falentin, C., Huteau, V., Clouet, V., Horvais, R., Gandon, B., Specel, S., Hanneton, L., Dheu, J.E., Deschamps, M., Margale, E., Vincourt, P. and Renard, M. (2006b) Genetic control of oil content in oilseed rape (*Brassica napus* L.). *Theoretical and Applied Genetics* 113, 1331–1345.
Delwiche, P.A. (1980) Genetic aspects of blackleg (*Leptosphaeria maculans*) resistance in rapeseed (*Brassica napus*). PhD thesis, University of Wisconsin, Madison, Wisconsin.
Dhillon, K.S. and Dhillon, S.K. (1991) Selenium toxicity in soils, plants and animals in some parts of Punjab, India. *International Journal of Environmental Studies* 37, 15–24.
Dhillon, K.S. and Dhillon, S.K. (2003) Distribution and management of seleniferous soils. *Advances in Agronomy* 79, 119–184.
Dhillon, K.S., Pasricha, N.S., Bahl, G.S. and Baddesha, H.S. (1984) Supplementation of nitrogen through green manuring in raya (*Brassica juncea* L.). *Bulletin of the Indian Society Soil Science* 13, 304–309.
Dhillon, S.K. and Dhillon, K.S. (2009) Phytoremediation of selenium-contaminated soils: the efficiency of different cropping systems. *Soil Use and Management* 25, 441–453.
Downey, R.K. and Rimmer, S.R. (1993) Agronomic improvement in oilseed *Brassicas*. *Advances in Agronomy* 50, 1–65.
Downey, R.K. and Röbbelen, G. (1989) *Brassica* species. In: Röbbelen, G., Downey, R.K. and Ashri, A. (eds) *Oil Crops of the World*. McGraw-Hill, New York, pp. 339–362.
Dubuis, P.H., Marazzi, C., Stadler, E. and Mauch, F. (2005) Sulphur deficiency causes a reduction in antimicrobial potential and leads to increased disease susceptibility of oilseed rape. *Journal of Phytopathology* 153, 27–36.
Durrett, T.P., Benning, C. and Ohlrogge, J. (2008) Plant triacylglycerols as feedstocks for the production of biofuels. *Plant Journal* 54, 593–607.
Ecke, W., Uzunova, M. and Weissleder, K. (1995) Mapping the genome of rapeseed (*Brassica napus L.*). II. Localization of genes controlling erucic acid synthesis and seed oil content. *Theoretical and Applied Genetics* 91, 972–977.
Erickson, I.R., Straus, N.A., and Beversdrof, W.D. (1983) Restriction patterns reveal origins of chloroplast genomes in *Brassica* amphiploids. *Theoretical and Applied Genetics* 65, 202–206.
FEDIOL (2006) Rapeseed Facts. Available at: http://www.soyatech.com/rapeseed_facts.htm (accessed 11 December 2011).
Fenwick, G.R., Heaney, R.K. and Mullin, W.J. (1983) Glucosinolates and their breakdown products in food and food plants. *CRC Critical Reviews in Food Science and Nutrition* 18, 123–201.
Fray, J., Evans, E.J., Lydiate, D.J. and Arthur, A.E. (1996) Physiological assessment of apetalous flowers and erectophile pods in oilseed rape (*Brassica napus*). *Journal of Agricultural Sciences* 127, 193–200.
Garg, H.C., Atri, P.S., Sandhu, B., Kaur, M., Renton, S.K., Banga, H., Singh, C., Singh, M.J., Barbetti, and Banga, S.S. (2010) High level of resistance to *Sclerotinia sclerotiorum* in introgression lines derived from hybridization between wild crucifers and the crop *Brassica* species *Brassica napus* and *Brassica juncea*. *Field Crop Research* 117, 51–58.
Goffman, F.D., Alonso, A.P., Schwender, J., Shachar-Hill, Y. and Ohlrogge, J.B. (2005) Light enables a very high efficiency of carbon storage in developing embryos of rapeseed. *Plant Physiology* 138, 2269–2279.

Gómez-Campo, C. (1980) Morphology and morphotaxonomy of the tribe Brassiceae. In: Tsunoda, S., Hinata, K. and Gómez-Campo, C. (eds) *Brassica Crops and Wild Allies: Biology and Breeding*. Japan Scientific Society Press, Tokyo, Japan, pp. 3–31.

Gómez-Campo, C. (1999) Seedless and seeded beaks in the tribe Brassiceae (Cruciferae). *Cruciferae Newsletter* 21, 11–12.

Graner, G., Hamberg, M. and Meijer, J. (2003) Screening of oxylipins for control of oilseed rape (*Brassica napus*) fungal pathogens. *Phytochemistry* 63, 89–95.

Grison, R., Grezes-Besset, B., Schneider, M., Lucante, N., Olsen, L., Leguay, J.J. and Toppan, A. (1996) Field tolerance to fungal pathotypes of *Brassica napus* constitutively expressing a chimeric chitinase gene. *Nature Biotechnology* 14, 643-646

Hasan, V., Friedt, W., Pons-Kühnemann, J., Freitag, N.M., Link, K. and Snowdon, R.J. (2008) Association of gene-linked SSR markers to seed glucosinolate content in oilseed rape (*Brassica napus* ssp. *napus*). *Theoretical and Applied Genetics* 116, 1035–1049.

Hedge, I.C. (1976) A systematic and geographical survey of the Old World Cruciferae. In: Vaughan, J.G., MacLeod, A.J. and Jones, M.G. (eds) *The Biology and Chemistry of the Cruciferae*. Academic Press, London, pp. 1–45.

Herve, Y. (2003) Choux. In: Pitrat, M. and Foury, C. (eds) *History de legumes, des origins a l'oree du XXI siecle*. INRA, Paris, France, pp. 222–234.

Hill, L.M., Morley-Smith, E.R. and Rawsthorne, S. (2003) Metabolism of sugars in the endosperm of developing seeds of oilseed rape. *Plant Physiology* 131, 228–236.

Hooker, J.D. (1862) In: Bentham, G. and Hooker, J.D. (eds) *Genera Plantarum*, Vol. 1. Lovell Reed, London, pp. 57–102.

Howell, B.C., Barker, G.C., Jones, G.H., Kearsey, M.J., King, G.J., Kop, E.P., Ryder, C.D., Teakle, G.R., Vicente, J.G. and Armstrong, S.J. (2003) Integration of the cytogenetic and genetic linkage maps of *Brassica oleracea*. *Genetics* 161, 1225–1234.

Howell, P.M., Marshall, D.F. and Lydiate, D.J. (1996) Towards developing intervarietal substitution lines in *Brassica napus* using marker-assisted selection. *Genome* 39, 348–358.

Howell, T.A., Phene, C.J., Miller, R.J. and Meek, D.W. (1984) Evaporation from screened class A pans in a semi-arid climate. *Agricultural and Forest Meteorology* 29, 111–124.

Hua, W., Li, R.J., Zhan, G.M., Liu, J., Li, J., Wang, X.F., Liu, G.H. and Wang, H.Z. (2011) Maternal control of seed oil content in *Brassica napus*: the role of silique wall photosynthesis. *Plant Journal* doi: 10.1111/j.1365-313X.2011.04802.x.

Jahier, J., Chèvre, A.M., Tanguy, A.M. and Eber, F. (1989) Extraction of disomic addition lines of *Brassica napus – B. nigra*. *Genome* 32, 408–413.

Jako, C., Kumar, A., Wei, Y., Zou, J., Barton, D.L., Giblin, E.M., Covello, P.S. and Taylor, D.C. (2001) Seed-specific over-expression of an Arabidopsis cDNA encoding a diacylglycerol acyltransferase enhances seed oil content and seed weight. *Plant Physiology* 126, 861–874.

Jenkins, B.M., Williams, R.B., Gildart, M.C., Kaffka, S.R., Hartsough, B. and Demster, P.G. (2009) Sustainable use of California biomass resources can help meet state and national bioenergy targets. *California Agriculture* 63(40), 168–177.

Jham, G.N., Moser, B.R., Shah, S.N., Holser, R.A., Dhingra, O.D., Vaughn, S.F., Berhow, M.A.,Winkler-Moser, J.K., Isbell, T.A., Holloway, R.K., Walter, E.L., Natalino, R., Anderson, J.C. and Stelly, D.M. (2009) Wild Brazilian mustard (*Brassica juncea* L.) seed oil methyl esters as biodiesel fuel. *Journal of the American Oil Chemists' Society* 86, 917–926.

Jourdren, C., Barret, P., Horvais, R., Foisset, N., Delourme, R. and Renard, M. (1996) Identification of RAPD markers linked to the loci controlling erucic acid level in rapeseed. *Molecular Breeding* 2, 61–71.

Kaneria, B.B. and Patel, Z.G. (1993) Effect of weed management practices and nitrogen on mustard yield and nutrient losses through weeds. *Journal of Oilseeds* 10, 246–250.

Kaur, S., Cogan, N.O., Ye, G., Baillie, R.C., Hand, M.L., Ling, A.E., McGearey, A.K., Kaur, J., Hopkins, C.J., Todorovic, M., Mountford, H., Edwards, D., Batley, J., Burton, W., Salisbury, P., Gororo, N., Marcroft, S., Kearney, G., Smith, K.F., Forster, J.W. and Spangenberg, G.C. (2009) Genetic map construction and QTL mapping of resistance to blackleg (*Leptosphaeria maculans*) disease in Australian canola (*Brassica napus* L.) cultivars. *Theoretical and Applied Genetics* 120, 71–83.

Khurana, M.P.S., Sadana, U.S. and Bijay-Singh (2008) Sulfur nutrition of crops in the Indo-Gangetic of South Asia. In: Jez, J. (ed.) *Sulfur: a Missing Link between Soils, Crops, and Nutrition*. Agronomy Monograph Number 50. American Society of Agronomy, Crop Science Society of America, Soil Science Society of America, Madison, Wisconsin, pp. 11–24.

Knodel, J.J. (2010) North Dakota Field Crop Insect Management Guide. Available at: http://www.ag.ndsu.edu/pubs/plantsci/pests/e1143ww1htm (accessed 1 December 2011).

Knodel, J.J. and Olsen, D.L. (2002) Crucifer Flea Beetle – Biology and Integrated Pest Management in Canola. Available at: http://www.ag.ndsu.edu/pubs/plantsci/pests/e1234.pdf (accessed 1 December 2011).

Körbitz, W. (1995) Utilization of oil as a biodiesel fuel. In: Kimber, N. and McGregor, D.I. (eds) *Brassica Oilseeds – Production and Instillation*. CAB International, Wallingford, UK, pp. 353–371.

Kumar, D. (1992) Production technology for yield enhancement of Indian mustard under irrigated conditions. In: Kumar, D. and Rai, M. (eds) *Advances in Oilseed Research: Rapeseeds and Mustard*, Vol. 1. Scientific Publishers, Jodhpur, Rajasthan, India, pp. 71–95.

Kumar, S.C., Atri, M.K., Sangha and Banga, S.S. (2011) Screening of wild crucifers for resistance to mustard aphid, *Lipaphis erysimi* (Kaltenbach) and attempt at introgression of resistance gene(s) from *Brassica fruticulosa* to *Brassica juncea*. *Euphytica* 179, 461–470.

Labana, K.S. and Banga, S.K. (1984) Floral biology in Indian mustard (*Brassica juncea* L. Coss). *Genetica Agraria* 38, 131–138.

Lefort-Buson, M. and Datté, Y. (1985) Etude del' heterosis chezle conzla oleagineux d'hiver (*Brassica napus* L.) I. Comparison de deux populations, Fune homozygote et Fautre heterozygote. *Agronomie* 5, 101–110.

Li, R.J., Wang, H.Z., Mao, H., Lu, Y.T. and Hua, W. (2006) Identification of differentially expressed genes in seeds of two near-isogenic *Brassica napus* lines with different oil content. *Planta* 224, 952–962.

Linnaeus, C. (1753) *Species. Plantarum* II. Stockholm, Sweden.

Long, Y., Wang, Z., Sun, Z., Dilantha Fernando, W.G., McVetty, P.B.V. and Li, G. (2011) Identification of two blackleg resistance genes and fine mapping of one of these two genes in a *Brassica napus* canola cultivar 'Surpass 400'. *Theoretical and Applied Genetics* 122, 1223–1231.

Lu, C., Napier, J.A., Clemente, T.E. and Cahoon, E.B. (2011) New frontiers in oilseed biotechnology: meeting the global demand for vegetable oils for food, feed, biofuel, and industrial applications. *Current Opinion in Biotechnology* 22, 252–259.

Mayerhofer, R., Wilde, K., Mayerhofer, M., Lydiate, D., Bansal, V., Good, A. and Parkin, I. (2005) Complexities of chromosome landing in a highly duplicated genome: towards map based cloning of a gene controlling blackleg resistance in *Brassica napus*. *Genetics* 171, 1977–1988.

McDermott, W.E. (2006) Biofuels in Europe. *McDermott Newsletter*, 17 October 2006.

McDonnell, K.P., Ward, S.M., McNully, P.B. and Howard Hildige, R. (2000) Results of engine and vehicle testing of semi refined rapeseed oil. *Transactions of the American Society of Agricultural Engineers* 43, 1309–1316.

Mizushima, U. and Tsunoda, S. (1967) A plant exploration in *Brassica* and allied genera. *Tohoku Journal of Agricultural Research* 17, 249–276.

Morice, J. (1974) Selection d'une variete de colza, sans acids erucique et sans glucosinolates. In: Giestten, W. (ed.) *Proceedings of the 4th International Rapeseed Conference*, Giessen, Germany, pp. 31–47.

Mossler, M.A. (2005) *Florida Crop/Pest Management Profile: Specialty* Brassicas *(Arrugula, Bok Choy, Chinese Broccoli, Chinese Mustard, Napa)*. Pesticide Information Office, Gainesville, Florida.

Mu, J., Tan, H., Zheng, Q., Fu, F., Liang, Y., Zhang, J., Yang, X., Wang, T., Chong, K., Wang, X-J. and Zuo, J. (2008) LEAFY COTYLEDON1 is a key regulator of fatty acid biosynthesis in Arabidopsis. *Plant Physiology* 148, 1042–1054.

Napier, J.A. and Grahm, I.A. (2010) Tailoring plant lipid composition: designer oilseeds come of age. *Current Opinion in Plant Biology* 13, 330–337.

National Research Council (NRC) (1985) *Mineral Tolerances of Domestic Animals*. National Academic Press, Washington, DC.

Nesi, N., Delourme, R., Brégeon, M., Falentin, C. and Renard, M. (2008) Genetic and molecular approaches to improve nutritional value of *Brassica napus* L. seed. *Biology* 331, 763–771.

Ohlendorf, H.M. and Hothem, R.L. (1995) Agricultural drain water affects on wildlife in central California. In: Hoffman, D.F., Rattner, B.A., Burton, G.A. Jr and Cairns, J. (eds) *Handbook of Ecotoxicology*. Lewis Press, Boca Raton, Florida, pp. 577–595.

Olsson, G. (1955) Investigators of the degree of cross pollination in white mustard and rape. *Sver-Usaedesfoeren Tidskr* 62, 311–322.

Oregon Environmental Council (2007) Biofuels for Oregon. Available at: http://www.oeconline.org/our-work/economy/biofuels4oregon (accessed 1 January 2012).

Pang, E.C.K. and Halloran, G.M. (1996) The genetics of adult plant blackleg (*Leptosphaeria maculans*) resistance from *B. juncea* in *B. napus*. *Theoretical and Applied Genetics* 92, 382–387.

Panjabi-Massand, P., Yadava, S.K., Sharma, P., Kaur, A., Kumar, A., Arumugam, N., Sodhi, Y.S., Mukhopadhyay, A., Gupta, V., Pradhan, A.K. and Pental, D. (2010) Molecular mapping reveals two independent loci conferring resistance to *Albugo candida* in the east European germplasm of oilseed mustard *Brassica juncea*. *Theoretical and Applied Genetics* 121, 137–145.
Pechan, P.A. and Morgan, D.G. (1985) Defoliation and its effects on pod and seed development in oil seed rape (*Brassica napus* L.). *Journal of Experimental Botany* 36, 458–468.
Pouzet, A. (1995) Agronomy. In: Kimber, P. and McGregor, D.I. (eds) *Brassica Oilseeds – Production and Utilization*. CAB International, Wallingford, UK, pp. 92-95.
Pradhan, A.K., Prakash, S., Mukhopadhyay, A. and Pental, D. (1992) Phylogeny of *Brassica* and allied genera based on variation in chloroplast and mitochondrial DNA patterns: molecular and taxonomical classifications are incongruous. *Theoretical and Applied Genetics* 85, 331–340.
Prakash, S. and Hinata, K. (1980) Taxonomy, cytogenetics and origin of crop *Brassicas*, a review. *Opera Botanica* 55, 1–57.
Prakash, S., Bhat, S.R., Quirós, C.F., Kirti, P.B. and Chopra, V.L. (2009) *Brassica* and its close allies: cytogenetics and evolution. *Plant Breeding Review* 31, 21–187.
Prantl, K. (1891) Cruciferae. In: Engler, A. and Prantl, K. (eds) *Die Natürlichen Pflanzenfamilien*. Wilhelm Englmann, Leipzig, Germany, pp. 145–208.
Prihar, S.S., Sandhu, K.S., Khera, K.L. and Sandhu, B.S. (1981) Effects of irrigation schedules on yield of mustard (*Brassica juncea* L.). *Experimental Agriculture* 17, 105–111.
PROLEA (1993) Ecobilan du Diester: Evaluation comparie der filiires Gazdi et Diester. PROLEA, Paris, France.
Qiu, D., Morgan, C., Shi, J., Long, Y., Liu, J., Li, R., Zhuang, X., Wang, Y., Tan, X., Dietrich, E., Weihmann, T., Everett, C., Vanstraelen, S., Beckett, P., Fraser, F., Trick, M., Barnes, S., Wilmer, J., Schmidt, R., Li, J., Li, D., Meng, J. and Bancroft, I. (2006) A comparative linkage map of oilseed rape and its use for QTL analysis of seed oil and erucic acid content. *Theoretical and Applied Genetics* 114, 67–80.
Rai, B., Singh, R.B., Singh, Y., Singh, R. and Singh, R.K. (1993) Technology for increasing rapeseed and mustard production in eastern Uttar Pradesh. *Indian Farming* 43, 3–6.
Ramchiary, N., Bisht, N.C., Gupta, V., Mukhopadhyay, A., Arumugam, N., Sodhi, Y.S., Pental, D. and Pradhan, A.K. (2007) QTL analysis reveals context-dependent loci for seed glucosinolate trait in the oilseed *Brassica juncea*: importance of recurrent selection backcross scheme for the identification of 'true' QTL. *Theoretical and Applied Genetics* 116, 77–85.
Rawsthorne, S. (2002) Carbon flux and fatty acid synthesis in plants. *Progress in Lipid Research* 41, 182–196.
Reynolds, M., Foulkes, M.J., Slafer, G.A., Berry, P., Parry, M.A., Snape, J.W. and Angus, W.J. (2009) Raising yield potential in wheat. *Journal of Experimental Botany* 60, 1899–1918.
Riley, K.W. and Belayneh, H. (1982) Report from an oilcrop collection trip in Ethiopia. *Cruciferae Newsletter* 7, 5–6.
Rimmer, S.R. (2006) Resistance genes to *Leptosphaeria maculans* in *Brassica napus*. *Canadian Journal of Plant Pathology* 28, S288–S297.
Röbbelen, G. and Nitsch, A. (1975) Genetical and physiological investigations on mutants for polyenoic fatty acids in rapeseed *B. napus* L. *Journal of Plant Breeding* 75, 93–105.
Röbbelen, G. and Thies, W. (1980) Biosynthesis of seed oil and breeding for improved oil quality of rapeseed. In: Tsunoda, S., Hinata, K. and Gomez-Campo, C. (eds) *Brassica Crops and Wild Allies*. Japan Science Society Press, Tokyo, Japan, pp. 253–283.
Ruuska, S.A., Girke, T., Benning, C. and Ohlrogge, J.B. (2002) Contrapuntal networks of gene expression during Arabidopsis seed filling. *Plant Cell* 14, 1191–1206.
Ruuska, S.A., Schwender, J. and Ohlrogge, J.B. (2004) The capacity of green oilseeds to utilize photosynthesis to drive biosynthetic processes. *Plant Physiology* 136, 2700–2709.
Rygulla, W., Snowdon, R.J., Friedt, W., Happstadius, I., Cheung, W.Y. and Chen, D. (2007) Identification of quantitative trait loci for resistance against *Verticillium longisporum* in oilseed rape (*Brassica napus*). *Phytopathology* 98, 215–221.
Saharan, G.S. (1993) Disease resistance. In: Labana, K.S., Banga, S.S. and Banga, S.K. (eds) *Breeding Oilseed Brassicas*. Springer-Verlag, Berlin, Germany, pp. 181–219.
Sarfraz, R.M., Dosdall, L.M., Keddie, A.B. and Myers, J.H. (2011) Larval survival, host plant preferences and developmental responses of the diamondback moth *Plutella xylostella* (Lepidoptera: Plutellidae) on wild Brassicaceous species. *Entomological Science* 14, 20–30.
Schulz, O.E. (1919) *Cruciferae-Brassiceae*. Part I: *Brassicinae and Raphaninae*. Heft 68–70. In: Engler, A. (ed.) *Das Pflanzenreich*. Wilhelm Engelmann, Leipzig, Germany, pp. 1–290.

Schulz, O.E. (1936) Cruciferae. In: Engler, A. and Prantl, A.P. (eds) *Die Natürlichen Pflanzenfamilien*. Wilhelm Engelmann, Leipzig, Germany, pp. 17b, 227–658.
Schwender, J., Ohlrogge, J.B. and Shachar-Hill, Y. (2003) A flux model of glycolysis and the oxidative pentose-phosphate pathway in developing *Brassica napus* embryos. *Journal of Biological Chemistry* 278, 29442–29453.
Schwender, J., Goffman, F., Ohlrogge, J.B. and Shachar-Hill, Y. (2004) Rubisco without the Calvin cycle improves the carbon efficiency of developing green seeds. *Nature* 432, 779–782.
Schwender, J., Shachar-Hill, Y. and Ohlrogge, J.B. (2006) Mitochondrial metabolism in developing embryos of *Brassica napus*. *Journal of Biological Chemistry* 281, 34040–34047.
Séguin-Swartz, G., Warwick, S.I. and Scarth, R. (1997) *Cruciferae: Compendium of Trait Genetics*. Agriculture and Food Canada, 177 pp.
Sekhon, B.S. and Åhman, I. (1993) Insect resistance with special reference to mustard aphid. In: Labana, K.S., Banga, S.S. and Banga, S.K. (eds) *Breeding Oilseed Brassicas*. Springer-Verlag, Berlin, Germany, pp. 206–217.
Sengupta, K. and Das, P.K. (2003) *Cultivated Annual Oil Seed Crops of India*, 1st edn. Naya Udyog, 206. Bidhan Sarani, Kolkata, India, pp. 31–84.
Sharpe, A.G. and Lydiate, D.J. (2003) Mapping the mosaic of ancestral genotypes in a cultivar of oilseed rape (*Brassica napus*) selected via pedigree breeding. *Genome* 46, 461–468.
Shi, J., Li, R., Qiu, D., Jiang, C., Long, Y., Morgan, C., Bancroft, I., Zhao, J. and Meng, J. (2009) Unraveling the complex trait of crop yield with quantitative trait loci mapping in *Brassica napus*. *Genetics* 182, 851–861.
Shi, J., Li, R., Zou, J., Long, Y. and Meng, J. (2011) A dynamic and complex network regulates the heterosis of yield-correlated traits in rapeseed (*Brassica napus* L.). *PLoS ONE* 6(7), e21645.
Singh, D. (1958) *Rape and Mustard*. Indian Central Oilseeds Committee Examiner, Bombay.
Singh, S.P. and Singh, D.P. (1972) Inheritance of yield and other agronomic characters in Indian mustard (*Brassica juncea*). *Canadian Journal of Genetics and Cytolology* 14, 227-233.
Song, K.M. and Osborn, T.C. (1992) Polyphyletic origins of *Brassica napus*: new evidence based on organelle and nuclear RFLP analyses. *Genome* 35, 992–1001.
Spect, C.E. and Diederichsen, A. (2001) *Brassica*. In: Hanelt, P. (ed.) *Mansfeld's Encyclopedia of Agricultural and Horticultural Crops*, vol. 3. Heidelberg, New York, pp. 1453–1456.
Staal, J., Kaliff, M., Bohman, S. and Dixelius, C. (2006) Transgressive segregation reveals two Arabidopsis TIR–NB–LRR resistance genes effective against *Leptosphaeria maculans*, causal agent of blackleg disease. *Plant Journal* 46, 218–230.
Stapleton, J.J. and Bañuelos, G.S. (2009) Biomass crops can be used for biological disinfestation and remediation of soils and water. *California Agriculture* 63, 41–46.
Tayo, T.O. and Morgan, D.G. (1979) Factors influencing flower and pod development in oil-seed rape (*Brassica napus* L.). *Journal of Agricultural Science* 92, 363–373.
Thomas, P. (1984) *Canola Growers Manual*. Canola Council of Canada Publication, Winnipeg, Canada.
Thormann, C.E., Romero, J., Mantet, J. and Osborn, T.C. (1996) Mapping loci controlling the concentrations of erucic and linolenic acids in seed oil of *Brassica napus* L. *Theoretical and Applied Genetics* 93, 282–286.
Thurling, N. (1993) Physiological constraints and their genetic manipulation. In: Labana, K.S., Banga, S.S. and Banga, S.K. (eds) *Breeding Oilseed Brassicas*. Springer-Verlag, Berlin, Germany, pp. 44–68.
Tournefort, J.P. de (1700) *Institutiones rei herbariae editio altera*, 1, 219–227. Paris.
U, N. (1935) Genomic analysis in *Brassica* with special reference to the experimental formation of *B. napus* and peculiar mode of fertilization. *Japan Journal of Botany* 7, 389–452.
Uzunova, M., Ecke, W., Weissleder, K. and Röbbelen, G. (1995) Mapping the genome of rapeseed (*Brassica napus* L.). I. Construction of an RFLP linkage map and localization of QTLs for seed glucosinolate content. *Theoretical and Applied Genetics* 90, 194–204.
Vaughan, J.G. (1977) A multidisciplinary study of the taxonomy and origin of *Brassica* crops. *BioScience* 27, 35–40.
Vaughan, J.G., Hemingway, J.S. and Schofield, H.J. (1963) Contributions to a study of variations in *Brassica juncea* Coss et. Czern. *Botany Journal of the Linnean Society* 58, 435–447.
Venegas-Calerón, M., Sayanova, O. and Napier, J.A. (2009) An alternative to fish oils: metabolic engineering of oil-seed crops to produce omega-3 long chain polyunsaturated fatty acids. *Progress Lipid Research* 49, 108–119.
Vigeolas, H., Waldeck, P., Zank, T. and Geigenberger, P. (2007) Increasing seed oil content in oil-seed rape (*Brassica napus* L.) by over-expression of a yeast glycerol-3-phosphate dehydrogenase under the control of a seed-specific promoter. *Plant Biotechnology Journal* 5, 431–441.

Warwick, S.I. (2010) *Brassicaceae* in Agriculture. In: Schmidt, R. and Bancroft, I. (eds) *Genetics and Genomics of the Brassicaceae*. Genetics and Genomics: Crops and Models 9, DOI 10.1007/978-1-4419-7118-0_2, C Springer Science and Business Media, pp. 1–67.

Warwick, S.I. and Black, L.D. (1991) Molecular systematics of *Brassica* and allied genera (subtribe *Brassicinae, Brassiceae*) – chloroplast genome and cytodeme congruence. *Theoretical and Applied Genetics* 82, 81–92.

Warwick, S.I., Black, L.D. and Aguinagalde, I. (1992) Molecular systematics of *Brassica* and allied genera (subtribe Brassicinae, Brassiceae) – chloroplast DNA variation in the genus *Diplotaxis. Theoretical and Applied Genetics* 83, 839–850.

Wei, H., Rong-Jun, L., Gao-Miao, Z., Jing, L., Jun, L., Xin-Fa, W., Gui-Hua, L. and Wang, H.Z. (2011) Maternal control of seed oil content in *Brassica napus*: the role of silique wall photosynthesis. *Plant Journal*, doi: 10.1111/j.1365-313X.2011.04802

Werner, S., Diederichsen, E., Frauen, M., Schondelmaier, J. and Jung, C. (2008) Genetic mapping of clubroot resistance genes in oilseed rape. *Theoretical and Applied Genetics* 116, 363–372.

Weselake, R.J., Taylor, D.C., Rahman, M.H., Shah, S., Laroche, A., McVetty, P.B. and Harwood, J.L. (2009) Increasing the flow of carbon into seed oil. *Biotechnology Advances* 27, 866–878.

Yan, Z., Tian, Z., Huang, B., Huang, R. and Meng, J. (1999) Production of somatic hybrids between *Brassica oleracea* and the C3–C4 intermediate species *Moricandia nitens*. *Theoretical Applied Genetics* 99, 1281–1286.

Zhao, J. and Meng, J. (2003) Detection of loci controlling seed glucosinolate content and their association with *Sclerotinia* resistance in *Brassica napus*. *Plant Breeding* 122, 19–23.

Zhao, J., Becker, H., Zhang, D., Zhang, Y. and Ecke, W. (2005) Oil content in an European x Chinese rapeseed population: QTL with additive and epistatic effects and their genotype-environment interactions. *Crop Science* 45, 51–59.

Zhao, J., Becker, H., Zhang, D., Zhang, Y. and Ecke, W. (2006) Conditional QTL mapping of oil content in rapeseed with respect to protein content and traits related to plant development and grain yield. *Theoretical and Applied Genetics* 133, 33–38.

Zhao, Y.T., Wang, M., Fu, S.X., Yang, W.C., Qi, C.K. and Wang, X.J. (2012) Small RNA profiling in two *Brassica napus* cultivars identifies microRNAs with oil production and developmental correlated expressions and new small RNA classes. *Plant Physiology* DOI:10.1104/pp.111.187666.

Zheng, P., Allen, W.B., Roesler, K., Williams, M.E., Zhang, S., Li, J., Glassman, K., Ranch, J., Nubel, D. and Solawetz, W. (2008) A phenylalanine in DGAT is a key determinant of oil content and composition in maize. *Nature Genetics* 40, 367–372.

Zhu, B. and Rimmer, S.R. (2003) Inheritance of resistance to *L. maculans* in two accessions of *B. napus*. *Canadian Journal of Plant Pathology* 3, 98–103.

Zhu, J.S., Struss, D. and Röbbelen, G. (1993) Studies on resistance to *Polima lingam* in *Brassica napus-B. nigra* addition lines. *Plant Breeding* 111, 192–197.

Zhu, Y.G., Pilon-Simts, E.A.H., Zhao, F.J., Williams, P.N. and Meharg, A.A. (2009) Selenium in higher plants: understanding mechanisms for biofortification and phytoremediation. *Trends in Plant Science* 14, 436–442.

Zou, J., Katavic, V., Giblin, E.M., Barton, D.L., MacKenzie, S.L., Keller, W.A., Hu, X. and Taylor, D.C. (1997) Modification of seed oil content and acyl composition in the Brassicaceae by expression of a yeast sn-2 acyltransferase gene. *Plant Cell* 9, 909–923.

17 Camelina (*Camelina sativa*)

C. Eynck[1] and K.C. Falk[2]
[1]*Linnaeus Plant Sciences, Inc., Saskatchewan, Canada;*
[2]*Agriculture and Agri-Food Canada, Saskatchewan, Canada*

17.1 History of Camelina Cultivation

Camelina (*Camelina sativa* L. Crantz) is an ancient oilseed that belongs to the *Brassicaceae* (mustard) family. It is also known as gold-of-pleasure, false flax, large-seeded false flax, wild flax (UK), linseed dodder, Dutch flax, German sesame, or Siberian oilseed (Putnam *et al.*, 1993; Zubr, 1997). While some of these names hint at the plant's resemblance to flax (*Linum usitatissimum*), the latter three denotations provide an indication of the geographical origin of this species. Thus, archaeological records indicate that the south-eastern Europe–south-western Asian steppe regions are most likely the centre of origin of camelina (Knörzer, 1978; Zohary and Hopf, 2000).

Observations from mostly carbonized seeds and capsules suggest that the evolution of camelina as a crop was initiated when the wild species *Camelina microcarpa* and *C. alyssum* (synonym: *linicola, macrocarpa, dentata*) were domesticated in the late Neolithic in south-eastern Europe (Zinger, 1909; Knörzer, 1978; Putnam *et al.*, 1993). With the expansion of agriculture, camelina very early invaded cultivated land as a weed in flax (Schwanitz, 1967). But soon the nutritional value of the oleiferous seeds was discovered and camelina was grown as a crop in monoculture. In the Bronze Age (1500–400 BCE), camelina was well established in the south-eastern part of Europe and it became a commonly grown crop in several parts of the European mainland and Scandinavia during the Iron Age (400 BCE–CE 500) (Plessers *et al.*, 1962; Knörzer, 1978; Bouby, 1998). Seeds of *C. alyssum* were a substantial part of the human diet, together with flax and other cereals (Hjelmquist, 1979). The species that is now known as *Camelina sativa* L. Crantz probably emerged during prehistoric times from plantings of *C. microcarpa* and/or *C. alyssum* (Knörzer, 1978). It is also not clear how or when the event of speciation of *C. sativa* took place.

Genetic analyses provide an alternative approach to address the question of the origin of a species. A recent study (Ghamkhar *et al.*, 2010) used AFLP fingerprinting to investigate the genetic diversity in a previously not accessible set of *C. sativa* accessions from the Russian-Ukrainian area, together with accessions from central and western Europe, China and Australia. In accordance with archaeological records, it identified the Russian-Ukrainian region as a genetic diversity hotspot and potential centre of origin of the species.

The importance of camelina as a food crop declined during the Middle Ages and it was grown only sporadically until modern times (Knörzer, 1978; Hjelmquist, 1979). Aside from its limited utilization as edible oil, camelina oil was the lamp oil of choice until the discovery of

©CAB International 2013. *Biofuel Crops: Production, Physiology and Genetics* (ed. B.P. Singh)

gas and electricity; later it was used in the production of soft soaps and coating material. Prior to the 1930s, the crop was still grown on sandy soils in France (Guillaumin, 1946), Belgium, Holland, the Balkan region and in Russia from Caucasus to Siberia (Wacker, 1934; Plessers *et al.*, 1962; Zubr, 1997). The cultivation of camelina declined further with the introduction of oilseed rape. According to Crowley (1999), the lower cost of hydrogenating rape oil and the lack of knowledge of potential large-scale applications for camelina oil were the major reasons for the lack of interest in the crop.

With the colonization of the New World by European settlers, camelina was introduced for a second time into a new environment as a weed in flax (Putnam *et al.*, 1993). In his voluminous encyclopaedia on the fauna of the central and southern USA, Porcher (1863) described the positive agronomic attributes of camelina, praising its low fertilizer requirements, high yield, early maturity, frost and drought tolerance and resistance to certain insect pests. In the middle of the 20th century, a 30-year research programme was started at the University of Minnesota, whose results added significantly to the body of knowledge on cultivation and potential uses of camelina in the USA (Robinson, 1987). The results of this comprehensive study showed that camelina was similar to flax, rape, mustard and canola in maturity, yield and requirements for harvesting machinery and that its cultivation was accompanied with lower production costs than any of the other oilseed crops. Trials conducted in Canada (Plessers *et al.*, 1962) obtained results similar to those in the USA and led to the suggestion that camelina should be given serious consideration as a potential oilseed crop for northern latitudes.

However, despite numerous favourable accounts of camelina as a low-input oilseed crop, until recently camelina has not been cultivated except for small isolated plantings (Gugel and Falk, 2006). Similar to the situation in Europe, the lack of a market for the oil has hindered the establishment of camelina as a crop in North America. However, with an increasing demand for vegetable oils and rising petroleum prices this may change. Camelina has recently gained renewed interest as an oilseed crop in Europe (Zubr, 1997; Hebard, 1998; Leonard, 1998; Bonjean and Le Goffic, 1999), North America (Putnam *et al.*, 1993) and Australia (Francis and Campbell, 2003), primarily because of the exceptional level of the ω-3 essential fatty acid α-linolenic acid in its oil, favourable agronomic attributes, its potential as a low-input source of biofuels and as a feedstock for the oleochemical industry (Fröhlich and Rice, 2005). Currently, it is being grown as a feedstock for biofuels in several northern US states (Fröhlich and Rice, 2005; Moser and Vaughn, 2010), such as Montana, North Dakota and Washington.

17.2 Biology

Camelina is a herbaceous, fast-growing annual with both spring and winter forms (Putnam *et al.*, 1993). Plants are taprooted and can attain heights of between 30 and 110 cm (Putnam *et al.*, 1993; Francis and Warwick 2009; Berti *et al.*, 2011). Initially, a rosette is formed; winter-type camelina as well as spring-type camelina sown in the autumn in climates with mild winters remain dormant in this stage over winter (Crowley, 1999). In spring or otherwise 1 week to 10 days after a rosette is established, a single erect, either smooth or sparsely haired, stem is initiated (Putnam *et al.*, 1993; Francis and Warwick, 2009). Depending on the genotype, stems are more or less heavily branched and become woody as they mature. Leaves are arrow- to lance-shaped with smooth or lobed edges. They are typically 2–8 cm long and 2–10 mm wide and can be either smooth or feature a few, primarily dichotomic, hairs (Putnam *et al.*, 1993; Francis and Warwick, 2009). Lacking a petiole, the leaves are usually clasping and arranged in an alternate manner.

From 4 to 6 weeks after germination, depending on the prevailing weather conditions, camelina plants start to flower, with a flowering period of about 2 weeks (Akk and Ilumäe, 2005). Inflorescences are arranged as racemes with the flowers in terminal clusters (Francis and Warwick, 2009). Flowers are small and pale yellow to greenish yellow in colour with a diameter of 5–7 mm (Fig.17.1). The organization of the flower is similar to that of other cruciferous crop plants but features delicate, lance-shaped petals. Camelina is primarily

Fig. 17.1. (a) Raceme, (b) flower and (c) seed pod of *C. sativa* (photo courtesy Prakash Venglat, Genome Prairie).

self-pollinating (Plessers *et al.*, 1962; Zubr, 1997; Mulligan, 2002). To the best of our knowledge the only study that investigated the rate of out-crossing in camelina is a field study conducted in Alberta, reporting rates of at maximum 0.28% at 20 cm distance (Walsh *et al.*, 2012). However, cross-pollination mediated by insects, i.e. bees, has been suggested by several authors (Blamey and Grey-Wilson, 1989; Corbet *et al.*, 1991).

The fruits of camelina are smooth pear-shaped siliques 7–9 mm in length. At first sight, they resemble the seed capsules of flax, a character that led to denotations such as false flax or wild flax in the European vernacular. Siliques usually contain 5–15 golden, brown or reddish brown seeds; however, up to 25 seeds per silique have been reported (Schuster and Friedt, 1998). Seeds are oblong-oval and rough, having a deeply ridged surface (Putnam *et al.*, 1993; Francis and Warwick, 2009). They are small with 1000-seed weights averaging about 1.0 g and ranging from 0.3 to 2.3 g, depending on genotype and growing conditions during seed development (Gugel, personal communication; Plessers *et al.*, 1962; Zubr, 1997; Schuster and Friedt, 1998; Gugel and Falk, 2006; Vollmann *et al.*, 2007).

Seed oil contents of 300–460 g kg^{-1} seed have been reported (Robinson, 1987; Putnam *et al.*, 1993; Budin *et al.*, 1995; Angelini *et al.*, 1997; Zubr and Matthäus, 2002; Zubr, 2003; Gugel and Falk, 2006; Vollmann *et al.*, 2007) and a considerable environmental variation in oil content and its composition has repeatedly been demonstrated (Zubr, 2003; Vollmann *et al.*, 2007). In northerly locations the oil content of camelina is usually higher than in more southerly locations (Plessers *et al.*, 1962; Gugel and Falk, 2006), which is probably due to the fact that the predominantly cool and moist weather in northern areas is conducive to proper seed filling and maturation.

Camelina oil is largely composed of unsaturated fatty acids with 25–42% α-linolenic acid (C18:3), 13–21% linoleic acid (C18:2), 14–20% oleic acid (C18:1), 12–18% gondoic (eicosenoic) acid (C20:1) and 2–4% erucic acid (C22:1) (Budin *et al.*, 1995; Zubr and Matthäus, 2002; Vollmann *et al.*, 2007).

This high content of unsaturated fatty acids usually is suggestive of a low degree of oxidative stability, but the oil is more stable than expected due to its high content of phenolics and tocopherols (vitamin E), which act as antioxidants (Budin *et al.*, 1995; Abramovič and Abram, 2005). Therefore, camelina oil is more stable to oxidative degradation than other unsaturated oils, such as fish or flax oil.

Crude, cold-pressed camelina seed oil destined for human consumption possesses a high purity, i.e. homogeneity, relative to crude oils from brassicas; therefore, the only treatment necessary after pressing of the seed is filtration of the oil. Deodorization, neutralization, degumming and bleaching, which are necessary for processing crude canola oil into edible oil, are not necessary (C. Rosengren, Three Farmers, personal communication). Also, according to Zubr (1997), these processes can have adverse effects on the quality of the oil. However, camelina oil is prone to photo-oxidation and should therefore be stored in the dark (Abramovič and Abram, 2005).

Compared to other vegetable oils, camelina oil is unusually high in cholesterol (45 mg $100g^{-1}$ compared to 10 mg $100g^{-1}$) (Shukla *et al.*, 2002). Antinutritive components, such as glucosinolates, sinapine, inositol phosphate and condensed tannins, have been identified, even though only at low levels (Matthäus and Angelini, 2005).

17.3 Agronomics

17.3.1 Time of planting

Camelina can be grown as a spring annual, a winter annual or a biennial winter crop, the latter requiring vernalization. Even though biennial forms do exist, they are not commonly grown and camelina seeded in the autumn in regions of the northern USA or southern Canada are generally spring types that overwinter. This is possible because the plants are extremely frost tolerant in the seedling and rosette stage (Putnam *et al.*, 1993; Angelini *et al.*, 1997). Thus, in northern France, winter annual camelina has been reported to tolerate several days with temperatures of −10 to −14°C without snow cover (Bonjean and Le Goffic, 1999). Even in Saskatchewan with its harsh winters, spring camelina has been successfully sown in autumn. Alternatively, it can be sown on frozen ground in late November to early December. Since the seedlings are able to cope with several cycles of freeze–thaw (Putnam *et al.*, 1993; Angelini *et al.*, 1997), the seeds start to germinate early in spring at temperatures near 0°C, well before the emergence of weeds, a feature that increases the competitiveness of the crop (Putnam *et al.*, 1993; Zubr, 1997). Autumn or winter seeding of spring camelina is particularly advantageous in dryland regions where it maximizes the amount of soil moisture initially available for the crop by taking advantage of rainfall during the cooler spring months (Pilgeram *et al.*, 2007). However, in areas with wet mild winters, as found in several parts of Europe, winter-sowing has proven to be unsuccessful particularly without the use of herbicides and fungicides due to high levels of weed competition, diseases and excessive lodging early in the season, which result in difficult harvesting conditions and consequently low yields (Crowley, 1999).

Aside from the research mentioned above, the effect of seeding date on the performance of camelina has not been thoroughly studied, although the choice of a proper seeding date may be critical for the success of camelina production in a certain region. Currently, most annual camelina, when seeded in spring, is seeded at the same time as spring-type brassica oilseeds. Camelina at different developmental stages is shown in Fig. 17.2.

17.3.2 Seeding rates

Agegnehu and Honermeier (1997) found that camelina showed the highest yields at a seeding rate of 400 seeds m^{-2}. At a seeding rate of 800 seeds m^{-2} yield components such as branches/plant, pods/plant, seeds/pod and seed weight/plant decreased. Similarly, Bramm (1993), Graf and Vetter (1994) and Schuster and Friedt (1995) suggested that sowing 400 seeds m^{-2} is most effective for maximum seed production. In trials studying the effect of seeding rate (5, 8 and 11 kg ha^{-1}) on the yield of camelina in Ireland,

Fig. 17.2. Some developmental stages of *C. sativa* according to the BBCH scale developed by Martinelli and Galasso (2011): (a) rosette stage (BBCH 15); (b) bolting (BBCH 50); (c) nursery plots in early flowering (BBCH 62); (d) racemes of *C. sativa* at full flowering (BBCH 65); (e) seed pod development (BBCH 70–79) (photographs courtesy C. Eynck).

the different seeding rates did not result in significant yield differences. This observation led the authors to suggest that a lower seeding rate of 5 kg or 300 seeds m^{-2} is sufficient to achieve adequate plant stands and consequently satisfactory yields (Crowley and Fröhlich, 1998). However, this recommendation was made on the assumption that the germination rate was 90% and the 1000 seed-weight was 1.5 g on average. Other authors likewise reported a seeding rate of 300 seeds m^{-2} to be sufficient (Pearson and Walker, 1999; Francis and Campbell, 2003). In contrast, other researchers recommended seeding rates of up to 600 seeds m^{-2} (Zubr, 2003; Urbaniak *et al.*, 2008). The most recent study on seeding rates was a field trial conducted in Saskatchewan. It determined optimum plant densities to be between 114 and 172 plants m^{-2}, which corresponded to a seeding rate of 316 to 437 seeds m^{-2} (average germination in this study was 42%). This study also revealed that higher plant stands hastened the maturity of camelina plants by as much as 7 days. The authors recommended a seeding rate of 550–700 seeds m^{-2} since this rate would provide targeted plant densities, even if only 25% of the seeds emerged. These rates correspond to 5.5 to 7.0 kg ha^{-1}, depending on 1000-seed weight, soil conditions, weed pressure, etc. (Johnson *et al.*, 2010).

17.3.3 Nitrogen and water requirements

A number of studies have looked at nitrogen (N) requirements of camelina, with quite variable results. For example, camelina grown in trials in Minnesota responded well to a N rate of 90 kg ha^{-1} (Budin *et al.*, 1995), while studies in Montana showed that between 80 and 100 kg ha^{-1} combined soil and fertilizer NO_3-N is required for optimum production of camelina (Jackson, 2008). Results from Denmark (Zubr, 2003), as well as Saskatchewan (Johnson *et al.*, 2010), indicated that the optimum N supply is attained at about 100 kg ha^{-1}. Researchers in Ireland investigated the effect of applied N on yield, oil content and the level of diseases in camelina. From the results of this 2-year study, the authors inferred that the optimum economic response to N application was achieved at 75 kg ha^{-1}. While the incidence of disease (*Botrytis cinerea*) was found to intensify with increasing N rates, the oil content was not significantly affected (Crowley and Fröhlich, 1998). In contrast, studies conducted in Poland established that the amount of applied N had a direct effect on oil content and fatty acid composition and therefore, seed quality (Zadernowski *et al.*, 1999). Similar results were obtained from research conducted in Romania where N levels of 100 kg ha^{-1} led to a yield increase of 58% with a slight decrease in oil content (Bugnarug and Borcean, 2000). Likewise, in an Austrian study, increasing the N rate from 60 to 120 kg ha^{-1} N resulted in a 30% yield increase and a significant decline in oil content (Agegnehu and Honermeier, 1997). Cultivars tested in a Canadian study responded to increasing N with increasing plant height, seed yield, seed protein and total plant nitrogen and a decrease in oil content. At higher levels of N supplementation (>60–80kg ha^{-1} N), significant changes in these parameters did not occur (Urbaniak *et al.*, 2008). In trials in western Canada, a significant yield response was observed only at one out of three sites (Day and Chalmers, 2007). In contrast, camelina yields responded positively to high levels of N (140 kg ha^{-1}) at one site out of two in another Irish study (Crowley, 1999). On the other hand, higher levels of N on the second, more fertile site led to greater lodging and disease, consequently reducing yields. However, over-application of N in camelina is most likely not to occur, because that would remove one of the reported economic advantages of growing camelina – namely that it is a low input crop.

It is well documented that camelina is drought and heat tolerant (Porcher, 1863; Plessers *et al.*, 1962; Robinson, 1987; Putnam *et al.*, 1993; Angelini *et al.*, 1997; Zubr, 1997; Blackshaw *et al.*, 2011). There are two consecutive studies that investigated the water use of a camelina crop in the irrigated semi-arid environment of Arizona, USA, in order to provide camelina growers in arid regions with practical tools for managing irrigation. The first study revealed that the average seasonal water consumption of camelina ranged from 333 to 423 mm (French *et al.*, 2009). A subsequent study by

Hunsaker *et al.* (2011) investigated the effect of irrigation water on yields and found that varying total water amounts did not have a significant effect on yield. In this study, the total evapotranspiration for camelina, which equates to the minimum water requirement of the crop, was 332 to 371 mm, which is markedly less than that typically needed by grain and vegetable crops (600–655 mm) (Hunsaker *et al.*, 2011) and that needed by *Lesquerella fendleri* (668 mm), an industrial oilseed crop that is native to the semi-arid south-western USA (Hunsaker *et al.*, 1998). These data strongly support the fact that camelina merits serious consideration as a cropping alternative in regions where long-standing droughts, urbanization and strong competition for water supplies hinder successful production of crops such as wheat or cotton (French *et al.*, 2009).

17.3.4 Yield potential

Since shattering is not an issue in camelina (Zubr, 1997), it can be either swathed or directly combined. Combines used to harvest other brassica oilseeds can be used to harvest camelina provided that a 3 mm lower sieve is fitted (Crowley and Fröhlich, 1998) (Fig. 17.3). In addition, combines and grain trailers should be carefully sealed to prevent the loss of seed.

Camelina is a crop that has not undergone intensive breeding relative to other oilseeds; however, it shows a high yield potential. This high yield potential was first documented in the New World by Porcher in 1863. In the middle of the last century, Plessers *et al.* (1962) evaluated a number of camelina lines from Europe and Asia for agronomic performance in comparison to flax and rape, in Ottawa, Ontario and Fort Vermillion, Alberta. The authors found that camelina matured 3 to 4 weeks earlier than flax in northern Alberta and that the yields ranged from 1357 to 1963 kg ha^{-1} (average of 1650 kg ha^{-1}), which was 2.5 times more than the average yield of flax and 50% higher than the yield of summer rapeseed. As mentioned earlier, due to its earliness and high yield potential the authors suggested that camelina should be considered as an oilseed crop for northerly regions. More recently, field trials conducted over a period of 9 years in Minnesota showed that camelina had a yield potential similar to that of other cruciferous oilseeds such as *Brassica napus*, *Sinapis alba* and *Crambe abyssinica* (Putnam *et al.*, 1993; Budin *et al.*, 1995). In this study, seed yields averaged 1150 kg ha^{-1} and ranged from 600 to 1700 kg ha^{-1}. In field trials conducted by Agegnehu and Honermeier (1997) in Austria, yields averaged 1440 kg ha^{-1} and ranged from 820 to 1950 kg ha^{-1}. However, in a previous study by the same authors, yields of camelina were as

Fig. 17.3. Harvesting of camelina with a plot combine in Saskatoon, Saskatchewan.

high as 3100 kg ha^{-1} in one treatment (loamy soil with 120 kg ha^{-1} applied N) (Honermeier and Agegnehu, 1994). Another Austrian study reported yields of up to 3250 kg ha^{-1} (Vollmann *et al.*, 1996).

Gugel and Falk (2006) studied the agronomic performance of several camelina accessions in western Canada in comparison to *B. napus*, *B. rapa* and *B. juncea*. They found that the camelina accessions matured relatively early and that most accessions had seed yields competitive with those of the brassicas. The yield of camelina ranged from 1177 kg ha^{-1} under drought conditions in Saskatchewan to 3012 kg ha^{-1} in northern Alberta. In contrast, the yield of *B. napus* (AC Excel), *B. rapa* (AC Parkland), and *B. juncea* (AC Vulcan) ranged from 650 to 2419 kg ha^{-1}, 620 to 3246 kg ha^{-1} and 1023 to 3700 kg ha^{-1}, respectively, at those locations. In field trials in Australia camelina yielded on average 1700 kg ha^{-1}, out-yielding canola at some locations (Francis and Campbell, 2003). Zubr (1997) again reported yields of 2600 kg ha^{-1} and 3300 kg ha^{-1} for spring and winter varieties, respectively. A more recent study comparing various field crops at multiple locations in western Canadian provinces was conducted by Blackshaw *et al.* (2011). The study was conducted in Alberta, Saskatchewan and Manitoba with the objective of determining the oil yield potential of various crops relative to that of *B. napus* canola in the semi-arid, short-season environment of the Canadian prairies. The number of site-years (out of a total of 9) that crops attained similar or significantly greater yields compared to canola were camelina (6), oriental mustard (5), *B. juncea* canola (3), flax (*Linum usitatissimum*) (3), soybean (*Glycine max*) (3), *B. rapa* canola (2), yellow mustard (*S. alba*) (2) and Ethiopian mustard (*B. carinata*) (1). Notably, camelina performed well at many locations and attained greater yields than *B. napus* canola under higher than normal rainfall as well as lower than normal rainfall (Blackshaw *et al.*, 2011). Pavlista *et al.* (2012) conducted a study on the growth patterns of *B. napus*, *B. juncea* (brown mustard) and *C. sativa* in the High Plains of western Nebraska. The results showed that the growth curves of canola, brown mustard and camelina differed considerably. Thus, canola tended to accumulate more biomass than the other two crops, while brown mustard tended to be larger. However, all three crops did not differ significantly with regard to yield, which suggests that the harvest index (weight of a harvested product as a percentage of the total weight of a crop) of camelina is superior to that of canola and brown mustard.

17.3.5 Pests and diseases

Camelina is tolerant to many insect pests such as the crucifer flea beetle (*Phyllotreta cruciferae* Goeze (Coleoptera: Chrysomelidae)) (Pachagounder *et al.*, 1998; Henderson *et al.*, 2004) and the cabbage root fly (*Delia brassicae* Wiedemann (Diptera: Anthomyiidae)) (Finch, 1978). Camelina is also resistant to blackleg (*Leptosphaeria maculans* (Desmaz.) (Ces. & de Not.)) (Salisbury, 1987) and alternaria blight (*Alternaria brassicae* (Berk.) Sacc.) (Conn *et al.*, 1988; Narasimhulu *et al.*, 1994), which has been attributed to the production of the phytoalexin camalexin. However, camelina is susceptible to aster yellows disease (*Candidatus* Phytoplasma asteris) (Fig. 17.4), clubroot (*Plasmodiophora brassicae* Woronin), white rust (*Albugo candida*), sclerotinia stem rot (*Sclerotinia sclerotiorum* Lib. De Bary), brown girdling root rot (*Rhizoctonia solani* Kühn) and downy mildew (*Hyaloperonospora camelinae* Gäum.). Genotypes resistant to brown girdling root rot (Conn *et al.*, 1988), aster yellows (Olivier, unpublished data), downy mildew (Vollmann *et al.*, 2001) and sclerotinia stem rot (Eynck *et al.*, unpublished data) (reviewed in Séguin-Swartz *et al.*, 2009) have been identified, a necessary step in developing cultivars resistant to these diseases.

17.3.6 Weed control

Possibly the most intractable factor that limits the adoption of camelina as an oilseed crop is its sensitivity to broadleaf herbicides. While the grass herbicide Assure II received Minor Use registration in 2010, camelina has not shown tolerance to any of the post-emergence broadleaf herbicides tested so far, including herbicides of the imidazolinone, pyridine,

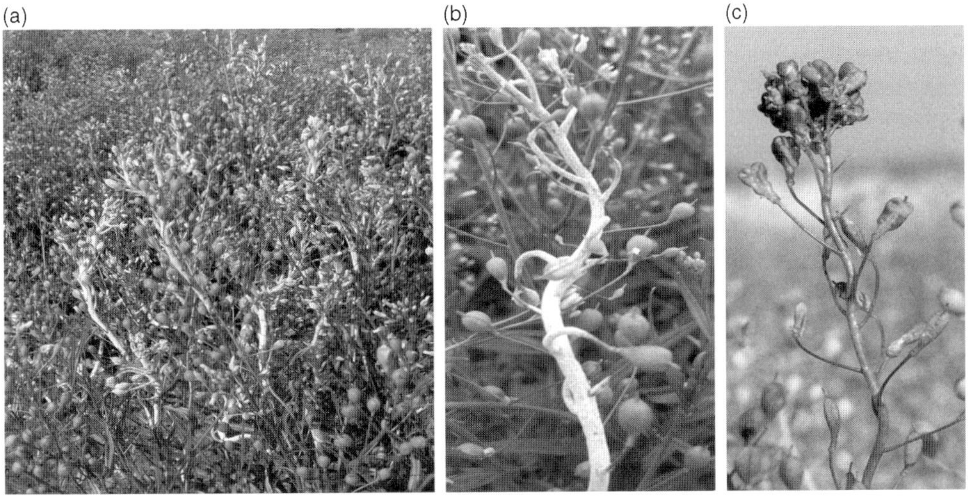

Fig. 17.4. (a) Camelina research plot with several plants exhibiting downy mildew (*Hyaloperonospora camelinae*) symptoms (photo courtesy Doug MacDonald, Dalhousie University). (b) Close-up of downy mildew on camelina. Symptoms include greyish white mycelial growth on leaf surfaces, stems and siliques. Severely infected plants may be malformed (photo courtesy C. Eynck). (c) Aster yellows, caused by *Candidatus* Phytoplasma asteris. Stems, leaves and siliques of plants exhibiting aster yellows symptoms turn greenish yellow or red, often with distorted inflorescences. Further phyllody (development of floral parts into leafy structures) is observed. Infected plants often show small, flattened siliques containing small and misshapen seeds (photo courtesy Richard Gugel, PGRC).

aromatic acid, nitrile, phenoxy, triazolopyrimidine, organophosphorous, anilide, cyclopropylisoxazole or triazinylsulfonylurea classes, plus clomazone and bentazone (Johnson *et al.*, 2010). However, camelina tolerates the pre-emergence application of trifluralin, which is currently the only chemical broadleaf weed control option available. Recently, a camelina line showing some tolerance to acetolactate synthase (ALS) inhibitor herbicides was created through mutagenesis using ethyl methane sulfonate (EMS) at Washington State University, Pullman, Washington, USA. Although this line is not completely tolerant to ALS-inhibitor herbicides (it still shows stunting symptoms after direct application of the recommended field rate), it exhibits its tolerance to herbicide carry-over and may therefore help to create an ALS-inhibitor tolerant camelina cultivar that can be incorporated into current crop rotations that use ALS-inhibitor herbicides (Walsh *et al.*, 2011).

Interestingly, a considerable suppressive effect on weeds was observed when camelina was used in mixed cropping systems. Saucke and Ackermann (2006) explored the use of camelina in an intercrop with peas and found that there were significantly fewer weeds in the intercrop in comparison to either monocrop. In his studies on intercropping of wheat, lupines and peas with camelina, Paulsen (2007) likewise noted an increased weed control. The author noted that if a substantial market was available for camelina oil and meal, the economics of intercropping would be favourable.

17.3.7 Invasive potential of camelina

However disadvantageous the high sensitivity to broadleaf herbicides may be for the establishment of camelina as a widely grown crop, this putative shortfall may be somewhat advantageous. For example, in situations where camelina is accidently introduced, control should be relatively simple. Also, in our experience camelina does not appear to have any persistence issues, i.e. dormancy or volunteering, which is a characteristic of recurring annual weeds. Robinson (1987) for instance stated that volunteer camelina in following crops has never been a problem, even after 30 years of growing camelina on experimental fields. Despite this, camelina is listed in the Canadian Federal Weed Seeds Order as a secondary noxious weed (class 3) (http://laws-lois.justice.gc.ca/eng/regulations/SOR-2005-220/FullText.html?term=camelina). However, the Canadian Food Inspection Agency (CFIA) has indicated that the process of removing camelina from the Weed Seed Order has been initiated (S. Vyas, CFIA, personal communication).

17.4 Breeding

Although the use of camelina by humans traces back to prehistoric times, until recently there have been few efforts to improve the crop through systematic breeding (Vollmann *et al.*, 2005). Thus, to the best of our knowledge, the first record of camelina breeding in Europe was published in the late 1980s when Seehuber *et al.* (1987) proposed the single-seed-descent method as an approach to achieve substantial yield gains in a short time. Since then, a comprehensive body of knowledge on agronomy, seed quality, as well as breeding has been created by Josef Zubr at the Royal Veterinary and Agricultural University in Frederiksberg, Denmark and Johann Vollmann at the University of Natural Resources and Life Sciences in Vienna, Austria. Zubr developed a number of cultivars adapted to northern Europe, e.g. the cultivar 'Hoga' (Crowley and Fröhlich, 1998; Fröhlich and Rice, 2005). Vollmann developed a camelina cultivar that is probably the best known and most successful in North America: 'Calena' (Vollmann, personal communication). This cultivar shows excellent yield potential and high oil content. However, it is highly susceptible to downy mildew (*Hyaloperonospora camelinae*). Other cultivars that were developed by European breeding companies and introduced to North America include 'Lindo' and 'Ligena' (DSV Deutsche Saatveredelung, Germany) and 'Celine' (Limagrain, France). Currently, there is no active camelina breeding programme in Europe, except for that of the Agricultural Foundation of Trade in Finland.

With the increasing interest in camelina in recent years, several breeding programmes have been initiated in the USA and in Canada. To date only a few cultivars have been released such as 'Blaine Creek' and 'Suneson' by Montana State University, Bozeman and 'Cheyenne' from Blue Sun Biodiesel, Colorado, USA.

If camelina is to be widely grown in North America, a concerted effort to develop adapted cultivars/germplasms needs to occur. The major breeding objectives for camelina include, but are not limited to, developing adapted early-maturing strains with superior seed yield, high seed oil and meal protein contents as well as resistance/tolerance to biotic stresses such as disease and insect pests.

Genetic diversity or variation is a key factor in the development of new genetic material and it appears from early work in the USA and research from Gugel and Falk (2006) that adapted germplasms are available to at least begin the process of cultivar development.

A high degree of genetic diversity was reported by Ghamkhar *et al.* (2010), who used amplified fragment length polymorphism (AFLP) fingerprinting to analyse the genetic diversity among 53 camelina accessions. In contrast, Vollmann *et al.* (2005), who applied a set of 15 random amplified polymorphic DNA (RAPD) markers, observed only little genetic diversity among 41 camelina accessions from a range of geographic locations. In the same material, however, they found considerable phenotypic variation in agronomic as well as seed quality traits, which per se reflects genetic distinctiveness between genotypes. In agreement with this observation, several previous publications have described significant genetic variation in several seed quality parameters (Budin *et al.*, 1995; Schuster and Friedt, 1998; Zubr and Matthäus, 2002; Zubr, 2003). Since the two estimates of diversity in the study of Vollmann *et al.* (2005) sampled different segments of the genome, i.e. regions coding for seed characteristics on the one hand and phenotypically neutral genomic regions flagged by discrete markers on the other, and the number of markers was generally low, it is not surprising that the two approaches did not yield similar results (Vollmann *et al.*, 2005). However, the considerable variation in agronomic and seed quality characters observed suggests that sufficient variation is present in current camelina germplasm to allow good progress toward increased seed oil content and yield, two of the main selection objectives in breeding oilseed crops. The fact that the oil content, a highly heritable trait in camelina (Vollmann *et al.*, 1996), and seed yield were positively correlated (Seehuber, 1984; Vollmann *et al.*, 1996, 2007; Gehringer *et al.*, 2006) is a helpful factor in the selection of desirable genotypes.

Another breeding objective that deserves attention is the improvement of seed size. As mentioned earlier, most cultivars are relatively small seeded when compared to other oilseeds such as canola or flax. Camelina's current seed size may in fact hinder its adoption in modern agriculture since small-seeded crops are generally more difficult to manage with large farm equipment (both seeding and harvest). Larger seeded forms would also result in better seedling emergence and thus better establishment of the crop under dry seeding conditions. Also, postharvest considerations may benefit from a larger seed since oil extraction is generally more efficient with large-seeded cultivars (Vollmann *et al.*, 1996). However, Vollmann *et al.* (1996, 2007) concluded from their observations that an improvement of the 1000-seed weight of camelina above 1.5 g may be of low immediate value due to concomitant reductions of both oil content and seed yield. Seed oil and meal protein content are correlated with seed size since the hull would make up a smaller (or larger) proportion of the total seed depending on the oil or protein content. In sunflower *(Helianthus annuus* L.), for example, Putt (1943) found a small negative correlation between seed size and oil content. He also found a strong positive correlation between seed yield and seed size. Ollson (1960) found a strong negative correlation between seed size and oil content in *B. napus*. However, it is interesting to note that Ollson was able to select for both large seed and high seed oil content as evidenced by the release of the cv. 'Tanka', which combines both traits.

In a study on material exhibiting a smaller range of seed sizes, Seehuber (1984) did not observe a significant correlation between seed size and oil content in camelina. Plessers *et al.* (1962) in their early performance evaluations of camelina noted that the cultivar with the largest seeds also possessed the lowest oil content. At the same time, however, the authors considered the combination of large seed and high oil content in one line quite possible. Our own preliminary observations likewise suggest that certain populations derived from crosses of large-seeded lines and lines with high oil content in fact bear individuals that combine both features.

17.4.1 Marker assisted selection, mapping and genomics

Very little work has been done on the creation of a genetic map and the identification of quantitative trait loci (QTLs) associated with agronomic or quality characteristics in camelina. The only study published thus far reported on the development of a recombinant inbred line (RIL) mapping population that was used to construct a linkage map from 157 AFLP markers and three brassica simple sequence repeat (SSR) markers on a total of 20 linkage groups. QTLs for oil content, seed yield, plant height, 1000-seed weight, oleic, linoleic, linolenic, eicosenoic (gondoic) and erucic acid were identified. Interestingly, the major QTL for oil content co-localized with a QTL for seed yield and might therefore be a promising target for the simultaneous improvement of seed yield and oil content (Gehringer *et al.*, 2006). Another interesting observation from this study was that a large proportion of the homologous SSR primers amplified more than one locus and that several AFLP markers showed a skewed segregation pattern. This led the authors to conclude that the camelina genome may possess a polyploid or duplicated structure. This assumption is supported by results of a study on intron-length polymorphisms in the β-tubulin gene family in camelina (Galasso *et al.*, 2011). The authors found an exceptionally high number of β-tubulins and a general correspondence between *Arabidopsis* and camelina β-tubulin genes of 1:2, which resembles the situation between barley and the closely related hexaploid wheat with 5 and 15 α-tubulin gene sequences, respectively (Farajalla and Gulick, 2007). Hutcheon *et al.* (2010) further suggested a hexaploid, more precisely, an allohexaploid structure of the camelina genome; a hypothesis that is supported by the observation that camelina shows diploid inheritance (Gehringer *et al.*, 2006; Lu and Kang, 2008). According to Gehringer *et al.* (2006), polyploidy might be one reason for the conflicting results with regard to chromosome numbers. Thus, chromosome counts in camelina have been reported as $n = 6$ or 14, or $2n = 12$, 26 or 40, with $2n = 40$ being the most common number (Warwick *et al.*, 1999; Mulligan, 2002; Gehringer *et al.*, 2006). Among the members of the tribe *Brassiceae*, a base chromosome number as high as $n = 20$ has only been found among known alloploids or autoploids (Warwick *et al.*, 2000).

The finding that camelina is most likely an allohexaploid is not only most interesting from a scientific point of view, but also of major importance from a practical standpoint since the manipulation of agronomic properties, with the help of marker-assisted breeding, mutagenesis or transgenesis requires knowledge about the duplication status of the genome (Hutcheon *et al.*, 2010).

17.4.2 Bioplatform development

Another area of interest for plant breeders and plant molecular biologists is the development of camelina as a dedicated non-food oilseed crop or bioplatform crop that can be used for metabolic engineering of novel oil compositions for industrial applications. Conventional selection has led to the identification of lines with an increased content of linolenic acid (Seehuber, 1984; Zubr and Matthäus, 2002; Zubr, 2003; Vollmann *et al.*, 2005, 2007), as has EMS-based mutagenesis (Büchsenschütz-Nothdurft *et al.*, 1998). However, none of these lines exhibited linolenic acid concentrations high enough to make camelina oil, which does not currently fit any particular use, into valuable industrial oil. While the variability inherent in camelina has yet to be fully exploited using conventional breeding methods, transgenesis is a promising option for developing specific oil profiles. Recently, an *Agrobacterium*-based floral infiltration method for the genetic transformation of camelina was developed (Lu and Kang, 2008). This protocol was first developed for the model plant *Arabidopsis thaliana*, which is interestingly the closest relative to camelina among the *Brassicaceae* family (Flannery *et al.*, 2006) and whose genome is therefore an ideal reference for the development of genetic and genomic tools in camelina (Hutcheon *et al.*, 2010). Largely due to the availability of this transformation protocol, camelina is likely to gain increasing interest as a platform for producing novel, high value industrial oils. For instance, Lu and Kang (2008) demonstrated the accumulation of the hydroxy fatty acid ricinoleic acid in camelina seeds through the seed-specific expression of a castor fatty acid hydroxylase gene as an economic alternative to castor oil. At this point in time, no genetically modified camelina cultivars are grown in Canada. However, first risk assessment studies under field conditions with genetically modified camelina lines, featuring an altered fatty acid profile, are currently being done by Linnaeus Plant Sciences, Inc., in collaboration with the laboratory of Linda Hall, University of Alberta (Edmonton, Alberta, Canada).

17.5 Camelina Oil as a Feedstock for Biofuels

17.5.1 Straight vegetable oil (SVO)

First results on the use of neat camelina straight vegetable oil (SVO) as a biofuel were described by Bernardo *et al.* (2003), who used cold-pressed and filtered camelina oil directly in a car engine to test its applicability as a fuel for unmodified diesel engines in comparison to mineral diesel fuel. The results of the study revealed that camelina oil showed an increased maximum power output with 43.25 kW compared to 38.50 kW for the mineral fuel and greater fuel consumption with 12.57 km l^{-1} compared to 14.03 km l^{-1}. Smoke opacity and carbon monoxide production were 50% less with camelina oil than with mineral diesel fuel.

Nitric oxide emissions were higher (6%) for the camelina oil at higher numbers of revolutions, but similar for the two fuels at less than 3500 rpm. Little nitrogen dioxide (NO_2) was produced by combustion of either fuel, and both fuels produced similar amounts of carbon dioxide (CO_2). Oxygen (O_2) levels in the exhaust were also similar. Being highly unsaturated, camelina oil was shown to be susceptible to oxidation. Accordingly, accelerated thermal aging tests (heating to 170°C) revealed a significant reduction in the degree of unsaturation and a corresponding increase in viscosity.

Next to mineral diesel fuel, Paulsen *et al.* (2011) compared the fuel characteristics of neat camelina oil with those of cold-pressed rape SVO and a mixture of both SVOs containing 700 $dm^3\ m^{-3}$ rape SVO and 300 $dm^3\ m^{-3}$ camelina SVO in several diesel engine types. Oil quality, burning behaviour and exhaust emission data revealed that the SVOs and their mixtures are suitable for use as fuels in diesel engines. Notwithstanding, the authors did not recommend the use of pure camelina SVO due to high Conradson Carbon Residue (CCR) values and low oxidation resistance. High CCR values may result in the formation of carbonaceous deposits in certain parts of the engine; furthermore, low oxidation resistance limits storage time. Although these characteristics are negative if camelina is to be used as a fuel, the remedy is relatively simple. First, injection nozzles can be cooled to prevent carbon deposition and the oil can be supplemented with antioxidant additives. Addition of the latter would help to avoid auto-oxidation, the polymerization process and the formation of lacquer-like films and consequently increase the duration of biofuel storage time (Paulsen *et al.*, 2011).

17.5.2 Biodiesel

Camelina oil has been successfully converted into biodiesel (fatty acid methyl and ethyl esters, FAME and FAEE) by a number of methods, including transesterification with alkaline catalysts (Rice, 1995; Fröhlich and Rice, 2005; Moser and Vaughn, 2010; Wu and Leung, 2011), heterogeneous metal oxide catalysts, both with and without microwave irradiation (Patil and Deng, 2009), and at uncatalysed sub- and supercritical conditions employing co-solvents with methanol (Patil *et al.*, 2010). However, the method most often reported is transesterification in the presence of a homogeneous alkali catalyst and excess methanol at elevated temperatures (Moser, 2010). Critical parameters that determine biodiesel yield from camelina oil are: (i) nature and concentration of the catalyst; (ii) reaction time; and (iii) reaction temperature; as well as (iv) the mole ratio between camelina oil glyceride esters and alcohol, i.e. either methanol or ethanol.

Two studies investigated in detail the ester- and fuel-specific characteristics of biodiesel obtained through homogeneous alkali-based transesterification from camelina oil in the context of European and American specifications for standard biodiesel. An Irish study (Fröhlich and Rice, 2005) evaluated the production of biodiesel-grade methyl ester from camelina oil and compared its properties with the limits in ÖNORM C 1191 (1997), which was the first national standard for fatty acid methyl esters as engine fuels in Europe and which was later replaced by the European standard EN 14214 (European Committee for Standardization, 2001). The study indicated that the ester-specific properties of camelina biodiesel were within the limits of ÖNORM C 1191, except for two parameters. First, the iodine value (IV), which is a measure for the degree of unsaturation in an oil sample, was far above the limit of 120 owing to the high levels of polyunsaturated acids in camelina oil. Second, the ash levels were higher than allowed in some cases, but this was found to be due to insufficient washing in this particular experiment.

The fuel-specific properties of camelina biodiesel were also within the specifications of ÖNORM C 1191; however, the authors state that the cold filter plug point (CFPP, temperature at which wax crystals form in a fuel to the extent that they plug filters) of −4°C may cause problems under low temperatures. It is worth noting that the average CFPP of camelina methyl ester was similar to that of rapeseed oil, which is to be expected since both oils possess about the same amount of saturated fatty acid esters which can crystallize due to their high (>30°C) melting points.

However, the addition of admixtures improved the CFPP of camelina biodiesel to up to −12°C and the pour point (lowest temperature at which an oil will flow) from −8°C to −19°C. Likewise, the low-temperature properties of camelina methyl ester improved when mixed with diesel oil proportional to the amount of diesel oil in the blend. In vehicle trials, no evidence of any adverse effect of the high iodine value such as an increase in viscosity near the end of an oil-change interval or deterioration of the fuel was observed. This finding is in agreement with that of Prankl *et al.* (1999), who performed comprehensive trials with car and tractor engines in Austria. Further, the metal wear levels in the lubricating oil were identified to be normal in the study of Fröhlich and Rice (2005). As is common for biodiesel from other feedstocks, the authors documented a slight but not statistically significant reduction in fuel economy (Fröhlich and Rice, 2005).

Moser and Vaughn (2010) conducted a thorough examination of the fuel properties of camelina methyl and ethyl esters prepared from base-catalysed transesterification, along with an evaluation of camelina methyl ester as a blend component in ultra-low-sulfur diesel fuel. They also conducted a comparison of camelina methyl ester with canola and palm methyl esters, along with soybean methyl and ethyl esters. Parameters included temperature operability, oxidative stability, cetane number (CN, measurement of the combustion quality of diesel fuel during compression ignition), acid value (AV, measure of acid in the fuel), sulfur and phosphorus content, lubricity, iodine value, surface tension and kinematic viscosity. Again, the relatively high polyunsaturated and trienoic acid (C18:3, C20:3) contents of camelina oil resulted in an IV substantially higher than that of other oils. Although the American biodiesel standard ASTM D6751 (B 100 standard of the American Society for Testing and Materials) does not include an IV limit, the European biodiesel standard EN 14214, as mentioned above, limits maximum IV values to 120. Thus, in order for camelina oil to be acceptable as a feedstock in areas where the European standard is applied, blending with feedstocks with higher saturated fat content would be necessary.

Similar to the results of Fröhlich and Rice (2005) and also Freedman *et al.* (1984) who worked with other vegetable oils, Moser and Vaughn (2010) found that the yield of camelina methyl and ethyl ester after transesterification were not quantitative. This was most likely due to the free fatty acid content in camelina oil as measured by the AV. Free fatty acids react with homogeneous basic catalysts such as sodium hydroxide and methoxide to form soap and water (or methanol in the case of methoxide), thus irreversibly quenching the catalyst and reducing biodiesel yields (Lotero *et al.*, 2005). Although the AVs of crude camelina methyl and ethyl esters were higher than those of refined, bleached and deodorized canola, palm and soybean methyl esters, they were still below the maximum prescribed limits according to ASTM D6751 and EN 14214:2003.

Other parameters, such as kinematic viscosity, derived cetane number (CN), lubricity, sulfur and phosphorus content, as well as surface tension of camelina methyl and ethyl esters, were satisfactory according to ASTM D6751 and EN 14214:2003. However, as has been observed in previous studies (Rice, 1995; Fröhlich and Rice, 2005), camelina biodiesel performed poorly under cold temperatures, similar to biodiesel prepared from canola and soybean oils, but significantly better than that derived from palm oil. According to Moser and Vaughn (2010), strategies to improve the low temperature operability (cloud point (temperature below which the waxes in biodiesel solidify, giving the biodiesel a cloudy appearance), pour point and cold filter plug point) of camelina alkyl esters may include the use of additives, blending with biodiesel produced from other feedstocks, crystallization fractionation, transesterification with long- or branched-chain alcohols and blending with petrodiesel.

According to the ASTM D7467 (B6 to B20) standard, the oxidative stability of camelina methyl and ethyl esters were unsatisfactory. Therefore the addition of antioxidant additives or blending with more oxidatively stable feedstocks would be necessary. Otherwise, camelina/petrodiesel blends were indistinguishable from soybean/petrodiesel blends with regard to low temperature performance, kinematic viscosity, lubricity and surface tension.

In summary, the results of both studies indicate that camelina methyl ester and ethyl ester are acceptable as biodiesel fuels provided the oxidative stability is improved through the addition of feedstocks with a lower IV or supplementation with antioxidants. Overall, the properties of camelina alkyl esters, both pure and mixed with petrodiesel, were similar to other commonly used biodiesel fuels, such as soybean, canola and palm methyl esters (Fröhlich and Rice, 2005; Moser and Vaughn, 2010).

17.5.3 Jet fuel

The first step in the production of renewable jet fuel from vegetable oils, such as camelina oil, is hydrodeoxygenation, also referred to as hydrotreatment. The reaction product, renewable diesel, consists of linear C15–C18 paraffins and freezes at temperatures encountered at the high altitudes reached by most aircraft. Therefore, in order to improve the cold flow properties of the fuel, isomerization to branched paraffins is crucial. According to ASTM D1655 (American standard for aviation turbine fuels), jet fuels must have freezing points below –40°C. Concurrently, paraffins are transformed to shorter-chain hydrocarbons by a process called selective catalytic cracking, which lowers the boiling range of jet fuel in comparison to that of diesel fuel. The resulting product is known as hydroprocessed renewable jet (HRJ) fuel. As a last step, aromatics are introduced, most commonly by blending the renewable hydrocarbons resulting from the reactions described above with conventional petroleum jet fuel.

Corporan *et al.* (2011) compared camelina-derived jet fuel with other alternative jet fuels. Parameters such as oxidative thermal stability, elastomer swell compatibility and combustion emissions were compared to conventional JP-8 fuels. Interestingly, and in contrast to the results often reported for camelina biodiesel, camelina-derived jet fuel exhibited superior thermal oxidative stability compared to JP-8. Furthermore, the data from this study showed that the use of camelina-derived jet fuel produced significantly lower soot and moderately lower unburned hydrocarbons and carbon monoxide than JP-8 fuel. However, elastomer swelling was significantly lower than for JP-8, which may lead to fuel leaks in aircraft systems. In summary, it was demonstrated that camelina-derived jet fuel possessed very similar properties to conventional fuels including excellent physical, chemical and combustion characteristics for use in aviation turbine engines (Corporan *et al.*, 2011).

Based on the promising performance of camelina-derived jet fuel in preliminary tests, several US companies have initiated the production of renewable jet fuel using camelina oil. Starting in 2009, The US Air Force (USAF) has successfully test-flown different fighter jets on a blend of standard JP-8 jet and camelina-derived renewable jet fuel. Furthermore, camelina derived jet fuel was used by the Thunderbirds, the USAF's official aerial demonstration team, to emphasize the military's commitment to sustainable technology. Not only has the US military shown interest in jet fuel from camelina oil, commercial airlines such as KLM Royal Dutch and Japan Airlines have performed successful test flights with fuel blends containing camelina-derived renewable jet fuel. The popularity of camelina oil as a feedstock for renewable jet fuel was exemplified in the summer of 2011 when two freight planes powered by camelina/JP-8 blend landed after completion of a transatlantic flight at the Paris Air Show.

17.5.4 Renewable diesel from camelina meal

Camelina meal, a by-product of the oil extraction process, can be converted to high carbon, high-energy liquid fuel intermediates by a process called fast pyrolysis. Fast pyrolysis is the thermal conversion of biomass by rapid heating to temperatures between 450 and 600°C, in the absence of oxygen. Such fuel intermediates can be used as feedstocks for the production of renewable diesel (C15–C18 paraffins produced via hydrotreatment, see above) or as a source of aromatic hydrocarbon compounds that can be utilized in the formulation of renewable jet fuel (Boateng *et al.*, 2010).

17.5.5 Life cycle analysis of camelina-derived biofuels

A life cycle analysis (LCA), also known as life cycle assessment or ecobalance, is a method for determining all costs associated with the use of a fuel or product. LCA would therefore consider raw materials, production, transportation and final use, including the depletion of resources and the release of polluting and harmful substances and their impacts, both at the local and global scale (http://www.csa.com; www.fluorocarbons.org).

A recent LCA of camelina-derived biodiesel, renewable diesel and jet fuel concluded that the life cycle greenhouse gas emissions (water vapour, carbon dioxide, methane, nitrous oxide (N_2O) and ozone), the cumulative energy demand ('entire demand, valued as primary energy, which arises in connection with the production, use and disposal of an economic good' (VDI, 1997)) and the fossil energy demand of all three camelina-derived fuel types was substantially lower than for petroleum-derived fuel (Shonnard *et al.*, 2010). According to this study, specifically the life cycle greenhouse gas emissions were reduced by 75–80% for camelina-derived biodiesel, renewable diesel and jet fuel. These findings are of significant importance because they show that all three camelina-derived fuel types qualify as advanced biofuels according to the Renewable Fuels Standard (RFS2) created by the US Energy Independence and Security Act (2007) and administered by the US Environmental Protection Agency (EPA) (http://www.epa.gov).

Another life cycle analysis compared the life cycle energy balance and greenhouse gas emissions of biodiesel derived from camelina grown under different scenarios, with those of biodiesel derived from the traditional biodiesel crops soybean and canola, as well as with those of petrodiesel (Krohn and Fripp, 2012). Without the consideration of land-use changes, the life cycle fossil energy demand and life cycle emissions of camelina were less than that of canola and considerably less than that of petroleum diesel, but higher than that of soybean. Soybean is a legume and thus able to fix N with the help of symbiotic bacteria, which results in very low N requirements and consequently low fossil energy consumption and greenhouse gas emissions. However, when land-use changes were accounted for in the LCA, the results indicate that camelina biodiesel had lower fossil energy demands and emitted fewer greenhouse gases than biodiesel from soybean or canola (Krohn and Fripp, 2012).

17.5.6 Political, social and environmental aspects

Over the past 20 years, there has been a growing awareness of the negative political, social and environmental effects of fossil fuels and considerable effort is being directed towards the development of alternative energy sources that are both domestic and renewable (Krohn and Fripp, 2012). The major crops currently grown for biofuel production are soybean, maize and canola. However, since all three are traditional food crops, they have two substantial drawbacks. First, the demand for biofuels may lead to direct or indirect changes in land use, such as the conversion of forested areas, with large amounts of stored carbon, into agricultural land. This conversion or change in land use results in the release of sequestered carbon into the atmosphere creating a so-called 'carbon debt'. Paying off this carbon debt through the use of biofuel may take years, if not hundreds of years (Searchinger *et al.*, 2008; Krohn and Fripp, 2012). The second disadvantage to using soybean, maize or canola as a biofuel is the dilemma of 'food versus fuel'. With an ever-growing world population, and thus a steadily increasing demand for food, it seems a moral imperative to reserve arable land and crop production for food (Krohn and Fripp, 2012).

Since camelina is on the one hand not traditionally a food crop in North America and on the other hand grows well under many biotic and abiotic stresses, it may offer producers and processors an alternative that circumvents the issues discussed above. For example, its heat tolerance, low water requirements and ability to mature in short growing seasons render it well suited to fill fallow periods in dryland wheat farming or as a double crop with short season soybeans or sunflowers (Carlson, 2009). Furthermore, camelina is so hardy that it can be grown on land that does not support the economic production of other crops. Thus, camelina could be successfully grown on soils reclaimed from mine spoil in the Czech Republic (Petrikova *et al.*, 1996) and in the Westlands Water District in Western Fresno County, California, where thousands of hectares of once-irrigated farmland had been abandoned due to the lack of water and salt build-up (http://westernfarmpress.com/markets/camelina-offers-hope-california-biofuel-crop).

All these scenarios neither involve land-use changes (LUC) nor take arable land out of food production, which makes camelina one of the most promising alternative oilseeds of the future for the sustainable production of biofuels.

17.6 Camelina meal

Camelina meal, or ground camelina oil cake or press cake, is a by-product of the oil extraction process. It consists of approximately 10% residual oil, 45% crude protein (similar to canola meal (Frame *et al.*, 2007)), 13% fibres, 5% minerals, and minor amounts of other substances, such as vitamins (Zubr, 1993, 1997). As already discussed for seed oil, the percentage of individual components in the meal depends on the plant genotype and is further affected by the growing conditions of the crop (Agegnehu and Honermeier, 1997; Angelini *et al.*, 1997; Zubr, 2003).

The protein in camelina meal is of high nutritional value owing to several essential and conditionally essential amino acids, such as arginine, cysteine, glycine, lysine, methionine and threonine (Zubr, 1997). Further, camelina meal contains relatively small amounts of antinutritive compounds such as sinapine (1.7–4.2 mg g^{-1} meal), condensed tannins (1.0–2.4 mg g^{-1} meal) (Zubr and Matthäus, 2002) and glucosinolates (13–36 μmol g^{-1} meal) (Zubr, 1997; Schuster and Friedt, 1998). The latter have been identified as 9-methyl-sulfinyl-nonyl glucosinolate, 10-methyl-sulfinyl-decyl glucosinolate (glucocamelinine) and 11-methyl-sulfinyl-undecyl glucosinolate (Lange *et al.*, 1995; Schuster and Friedt, 1998). Whereas the first two glucosinolates have also been identified in other cruciferous plants, such as alpine rock-cress (*Arabis alpina*) and shepherd's purse (*Capsella bursa-pastoris*), 11-methyl-sulfinyl-undecyl glucosinolate has only been detected in camelina (Daxenbichler *et al.*, 1991). Antinutritives present in camelina, such as glucosinolates, are not well understood.

Because of its high nutritional value, the meal was traditionally used as a fodder component (Ollech, 1884; Knörzer, 1978) and with the rising interest in camelina oil as an industrial feedstock, methods of using the meal as a fat- and protein-rich ingredient in today's feed rations are being explored again (e.g. Zubr, 1993; Peiretti *et al.*, 2007; Cherian *et al.*, 2009). Frame *et al.* (2007), for example, investigated the use of camelina meal as a feed ingredient in turkey poult starter diets. The results of this study indicated that increasing amounts of camelina meal (>10%) were accompanied by a corresponding decrease in weight gain and a poorer feed conversion when compared to conventional feedstuffs. Based on this, the authors concluded that camelina meal may be useful only as a minor ingredient in turkey diets and that its percentage should not exceed 5% of the total diet.

A Polish study on dairy ewes revealed that supplementing the diet with small amounts (3 and 6%) of camelina meal resulted in an improved fat composition of the milk, with elevated concentrations of mono- and polyunsaturated fatty acids. Further, the milk from ewes fed camelina meal featured lower atherogenic (causing atherosclerosis) and thrombogenic (causing thrombosis) indices, suggesting potential beneficial effects on consumer health (Szumacher-Strabel *et al.*, 2010).

A feeding experiment with lactating cows found that moderate amounts of camelina expeller supplemented to a base diet consisting of red clover had no adverse effects on silage dry matter intake, nutrient digestion, or milk production; however, the milk fat composition was altered. Thus, as had been observed for dairy ewes (Szumacher-Strabel *et al.*, 2010), a decrease in saturated fatty acids and an increase in unsaturated fatty acids were observed (Halmemies-Beauchet-Filleau *et al.*, 2011). Another feeding experiment with dairy cows revealed that the addition of camelina seed and meal to the diet led to a decrease in total milk fat yield and softer, more spreadable butter, owing to a higher content of unsaturated fatty acids in the milk (Hurtaud and Peyraud, 2007). In an experiment on the effect of different diets on the reproductive performance of heifers, Moriel *et al.* (2011) found that heifers fed camelina meal as part of their ration had significantly higher first-service pregnancy rates after timed artificial insemination than heifers fed a control ration.

Based on the results obtained thus far, the US Food and Drug Administration (USDA) approved the use of camelina meal in feed rations given to beef cattle and broiler chickens, provided that the percentage is not higher than 10% (Moser, 2010). In Canada, camelina meal does not have the 'Generally Regarded as Safe' (GRAS) status and therefore cannot be used as a feed ingredient.

17.7 Concluding Remarks

Camelina sativa, also known as false flax or gold-of-pleasure, is a cruciferous oilseed belonging to the *Brassicaceae* (mustard) family. It is native to Europe and naturalized in North America where it grows well under a wide range of climatic and soil conditions.

The renewed interest in camelina is due to its numerous valuable agronomic attributes and its seed quality. Thus, camelina matures early, has a high yield potential, good drought and heat tolerance and resistance to many common pests and diseases. Its rather unique fatty acid profile is largely unsaturated (>90%) and the seed meal is relatively low in glucosinolates when compared with other crucifers. The meal further has a favourable balance of amino acids, making it a potentially valuable feed for poultry, swine and ruminants.

Although breeding and agronomic research on camelina has been conducted over the past 50 years, relative to other field crops this work has not been exhaustive. Therefore, it is likely that the full potential of this crop remains largely unexplored. Future breeding efforts should therefore focus on seed yield, oil and protein content, seed size, broadleaf herbicide tolerance and disease resistance. Another area of interest for plant breeders and plant molecular biologists is the development of camelina as a dedicated non-food oilseed crop or bioplatform crop that can be used for metabolic engineering of novel oil compositions for industrial applications.

Based on our work, and that of others, the prospects of developing improved germplasms/cultivars of camelina are promising.

Camelina oil has been successfully converted into biodiesel, renewable diesel and renewable jet fuel with similar physical, chemical and combustion characteristics to conventional petroleum-derived fuel. Life cycle assessments comparing camelina-derived biofuels with conventional fuel have shown that camelina-derived biofuels have lower energy demands and dramatically reduced greenhouse gas emissions (75–80%) compared to conventional fuel. Due to its drought tolerance and short-season nature, camelina is well suited to fill fallow periods in dryland wheat farming, as a double crop with short season soybeans or sunflowers or as a crop on land that does not support the economic production of other crops. These scenarios neither involve land-use changes (LUC) nor take arable land out of food production, which makes camelina one of the most promising alternative oilseeds of the future for the sustainable production of biofuels.

References

Abramovič, H. and Abram, V. (2005) Physio-chemical properties, composition and oxidative stability of *Camelina sativa* oil. *Food Technology and Biotechnology* 43, 63–70.

Agegnehu, M. and Honermeier, B. (1997) Effects of seeding rates and nitrogen fertilization on seed yield, seed quality and yield components of false flax (*Camelina sativa* Crtz.). *Die Bodenkultur* 48, 15–20.

Akk, E. and Ilumäe, E. (2005) Possibilities of growing *Camelina sativa* in ecological cultivation. In: *Proceedings of the seminar: Environmental friendly food production systems: requirements for plant breeding and seed* production ((ENVIRFOOD) FP-2003-SSA-1-007003), 31 May–3 June 2005, Talsi, Lithuania. State Stende Plant Breeding Station, Talsi, pp. 120–122.

Angelini, L.G., Moscheni, E., Colonna, G., Belloni, P. and Bonari, E. (1997) Variation in agronomic characteristics and seed oil composition of new oilseed crops in central Italy. *Industrial Crops and Products* 6, 313–323.

Bernardo, A., Howard-Hildige, R., O'Connell, A., Nichol, R., Ryan, J. *et al.* (2003) Camelina oil as a fuel for diesel transport engines. *Industrial Crops and Products* 17, 191–197.

Berti, M., Wilckens, R., Fischer, S., Solis, A. and Johnson, B. (2011) Seeding date influence on camelina seed yield, yield components, and oil content in Chile. *Industrial Crops and Products* 34, 1358–1365.

Blackshaw, R.E., Johnson, E.N., Gan, Y., May, W.E., McAndrew, D.W. *et al.* (2011) Alternative oilseed crops for biodiesel feedstock on the Canadian prairies. *Canadian Journal of Plant Science* 91, 889–896.

Blamey, M. and Grey-Wilson, C. (1989) *The Illustrated Flora of Britain and Northern Europe*. Hodder & Stoughton, Toronto, Ontario.

Boateng, A.A., Mullen, C.A. and Goldberg, N.M. (2010) Producing stable pyrolysis liquids from the oil-seed presscakes of mustard family plants: pennycress (*Thlaspi arvense* L.) and camelina (*Camelina sativa*). *Energy and Fuels* 24, 6624–6632.

Bonjean, A. and Le Goffic, F. (1999) La cameline – *Camelina sativa* (L.) Crantz: une opportunité pour l'agriculture et l'industrie européenne. (Camelina – *Camelina sativa* (L.) Crantz: an opportunity for European agriculture and industry.) OCL 6, 28–34. (In French, English summary.)

Bouby, L. (1998) Two early finds of gold-of-pleasure (*Camelina* sp.) in middle Neolithic and Chalcolithic sites in western France. *Antiquity* 72, 391–398.

Bramm, A. (1993) Produktionsmöglichkeiten pflanzlicher Öle und ihre Verwendung im technischen Bereich. *Landbauforschung Völkenrode* 43, 112–120. (In German.)

Büchsenschütz-Nothdurft, A., Schuster, A. and Friedt, W. (1998) Breeding for modified fatty acid composition via experimental mutagenesis in *Camelina sativa* (L.) Crantz. *Industrial Crops and Products* 7, 291–295.

Budin, J.T., Breene, W.M. and Putnam, D.H. (1995) Some compositional properties of camelina (*Camelina sativa* L. Crantz) seeds and oils. *JAOCS* 72, 309–315.

Bugnarug, C. and Borcean, I. (2000) A study on the effect of fertilizers on the crop and oil content of *Camelina sativa* L. *Lucrari Stiintifice – Agricultura, Universitatea de Stiinte Agricole si medicina Veterinara a Banatului Timisoara* (Banat's University of Agricultural Sciences and Veterinary Medicine) 32, 541–544.

Carlson, A.S. (2009) Plant oils as feedstock alternatives to petroleum – a short survey of potential oil crop platforms. *Biochimie* 91, 665–670.

Cherian, G., Campbell, A. and Parker, T. (2009) Egg quality and lipid composition of eggs from hens fed *Camelina sativa*. *Journal of Applied Poultry Research* 18, 143–150.

Conn, K.L., Tewari, J.P. and Dahiya, J.S. (1988) Resistance to *Alternaria brassicae* and phytoalexin-elicitation in rapeseed and other crucifers. *Plant Science* 56, 21–25.

Corbet, S.A., Williams, I.H. and Osborne, J.L. (1991) Bees and the pollination of crops and wild flowers in the European community. *Bee World* 72, 47–59.

Corporan, E., Edwards, T., Shafer, L., DeWitt, M.J., Klingshirn, C. *et al.* (2011) Chemical, thermal stability, seal swell, and emissions studies of alternative jet fuels. *Energy and Fuels* 25, 955–966.

Crowley, J.G. (1999) Evaluation of *Camelina sativa* as an alternative oilseed crop. *Teagasc publication* 1 84170 049 5. Crops Research Centre, Oak Park, Carlow, Ireland.

Crowley, J.G. and Fröhlich, A. (1998) Factors affecting the composition and use of camelina. *Teagasc publication* 1 901138 66 6. Crops Research Centre, Oak Park, Carlow, Ireland.

Daxenbichler, M.E., Spencer, G.F., Carlson, D.G., Rose, G.B., Brinker, A.M. *et al.* (1991) Glucosinolate composition of seeds from 297 species of wild plants. *Phytochemistry* 30, 2623–2638.

Day, S. and Chalmer, S. (2007) *Westman Agricultural Diversification Organization Annual Report 2007*. Manitoba Agriculture, Food and Rural Initiatives, Manitoba.

European Committee for Standardization (2001) Automotive Fuels – Fatty Acid Methyl Esters (FAME) for Diesel Engines: Requirements and Test Methods. Draft prEN 14214. CEN-CENELEC Management Centre, Avenue Marnix 17, B-1000 Brussels.

Farajalla, M.R. and Gulick, P. (2007) The α-tubulin gene family in wheat (*Triticum aestivum* L.) and differential gene expression during cold acclimation. *Genome* 50, 502–510.

Finch, S. (1978) Volatile plant chemicals and their effect on host plant finding by the cabbage root fly (*Delia brassicae*). *Entomologia Experimentalis et Applicata* 25, 150–159.

Flannery, M.L., Mitchell, F.J., Coyne, S., Kavanagh, T.A., Burke, J.I. *et al.* (2006) Plastid genome characterization in Brassica and Brassicaceae using a new set of nine SSRs. *Theoretical and Applied Genetics* 113, 1221–1231.

Frame, D.D., Palmer, M. and Peterson, B. (2007) Use of *Camelina sativa* in the diets of young turkeys. *Journal of Applied Poultry Research* 16, 381–386.
Francis, A. and Warwick, S.I. (2009) The biology of Canadian weeds. 142. *Camelina alyssum* (Mill.) Thell.; *C. microcarpa* Andrz. ex DC; *C. sativa* (L.) Crantz. *Canadian Journal of Plant Science* 89, 791–810.
Francis, C.M. and Campbell, M.C. (2003) *New high quality oil seed crops for temperate and tropical Australia*. Publication 03/045. Available from: Rural Industries Research and Development Corporation, 15 National Circuit, Barton, ACT 6000, Australia.
Freedman, B., Pryde, E.H. and Mounts, T.L. (1984) Variables affecting the yields of fatty esters from transesterified vegetable oils. *JAOCS* 61, 1638–1643.
French, A.N., Hunsaker, D., Thorp, K. and Clarke, T. (2009) Evapotranspiration over a camelina crop at Maricopa, Arizona. *Industrial Crops and Products* 29, 289–300.
Fröhlich, A. and Rice, B. (2005) Evaluation of *Camelina sativa* oil as a feedstock for biodiesel production. *Industrial Crops and Products* 21, 25–31.
Galasso, I., Manca, A., Braglia, L., Martinelli, T., Morello, L. *et al.* (2011) h-TBP: an approach based on intron-length polymorphism for the rapid isolation and characterization of the multiple members of the β-tubulin gene family in *Camelina sativa* (L.) Crantz. *Molecular Breeding* 28, 635–645.
Gehringer, A., Friedt, W., Lühs, W. and Snowdon, R.J. (2006) Genetic mapping of agronomic traits in false flax (*Camelina sativa* subsp. *sativa*). *Genome* 49, 1555–1563.
Ghamkhar, K., Croser, J., Aryamanesh, N., Campbell, M., Kon'kova, N. *et al.* (2010) Camelina (*Camelina sativa* (L.) Crantz) as an alternative oilseed: molecular and ecogeographic analyses. *Genome* 53, 558–567.
Graf, T. and Vetter, A. (1994) Leindotter. In: Ergebnisse Thüringer Feldversuche 1992–1993, *Thüringer Ministerium für Landwirtschaft und Forsten*, 211–214. (In German.)
Gugel, R.K. and Falk, K.C. (2006) Agronomic and seed quality evaluation of *Camelina sativa* in western Canada. *Canadian Journal of Plant Science* 86, 1047–1058.
Guillaumin, A. (1946) *Les plantes cultivées*. Librairie Science et Vie, Paris. (In French.)
Halmemies-Beauchet-Filleau, A., Kokkonen, T., Lampi, A.-M., Toivonen, V., Shingfield, K.J. *et al.* (2011) Effect of plant oils and camelina expeller on milk fatty acid composition in lactating cows fed diets based on red clover silage. *Journal of Dairy Science* 94, 4413–4430.
Hebard, A. (1998) *Camelina sativa* – a pleasurable experience or another false hope? *Lipid Technology* 10, 81–83.
Henderson, A.E., Hallett, R.H. and Soroka, J.J. (2004) Prefeeding behaviour of the crucifer flea beetle, *Phyllotetra cruciferae*, on host and nonhost crucifers. *Journal of Insect Behavior* 17, 17–39.
Hjelmquist, H. (1979) Beiträge zur Kenntnis der prähistorischen Nutzpflanzen in Schweden. *Opera Bot.* 47, 34–57. (In German.)
Honermeier, B. and Agegnehu, M. (1994) Zur Anbaueignung von Sommerleindotter (*Camelina sativa* Crtz.). *Mitteilungen der Gesellschaft für Pflanzenbauwissenschaften* 7, 331–334 (in German).
Hunsaker, D.J., Nakayama, F., Dierig, D. and Alexander, W. (1998) Lesquerella seed production: water requirement and management. *Industrial Crops and Products* 8, 167–182.
Hunsaker, D.J., French, A.N., Clarke, T.R. and El-Shikha, D.M. (2011) Water use, crop coefficients, and irrigation management criteria for camelina production in arid regions. *Irrigation Science* 29, 27–43.
Hurtaud, C. and Peyraud, J.L. (2007) Effects of feeding camelina (seeds or meal) on milk fatty acid composition and butter spreadability. *Journal of Dairy Science* 90, 5134–5145.
Hutcheon, C., Ditt, R.F., Beilstein, M., Comai, L. and Schroeder, J. (2010) Polyploid genome of *Camelina sativa* revealed by isolation of fatty acid synthesis genes. *BMC Plant Biology* 10, 233.
Jackson, G.D. (2008) Response of camelina to nitrogen, phosphorus, and sulfur. Montana State University Extension Service *Fertilizer Facts* 49, 2 pp.
Johnson, E.N., Falk, K., Klein-Gebbinck, H., Lewis, L., Malhi, S. *et al.* (2010) Agronomy of *Camelina sativa* and *Brassica carinata*. Final Report. Available at AAFC Scott Research Farm, Box 10, Scott, Saskatchewan, Canada.
Knörzer, K.H. (1978) Entwicklung und Ausbreitung des Leindotters (*Camelina sativa* s.l.) (Evolution and spreading of Gold of Pleasure (*Camelina sativa* s.l.).) *Berichte der Deutschen Botanischen Gesellschaft* 91, 187–195. (In German, English summary.)
Krohn, B.J. and Fripp, M. (2012) A life cycle assessment of biodiesel derived from the 'niche filling' energy crop camelina in the USA. *Applied Energy* 92, 92–98.
Lange, R., Schumann, W., Petrzika, M., Buschund, H. and Marquard, R. (1995) Glucosinolate in Leindottersamen. *Fat Science Technology* 97(4), 146–152. (In German.)
Leonard, E.C. (1998) Camelina oil: α-linolenic source. *Inform* 9, 830–838.

Lotero, E., Liu, Y., Lopez, D.E., Suwannakarn, K., Bruce, D.A. *et al.* (2005) Synthesis of biodiesel via acid catalysis. *Industrial & Engineering Chemistry Research* 44, 5353–5363.
Lu, C. and Kang, J. (2008) Generation of transgenic plants of a potential oilseed crop *Camelina sativa* by *Agrobacterium*-mediated transformation. *Plant Cell Reports* 27, 273–278.
Martineeli, T. and Galasso, I. (2011) Phenological growth stages of *Camelina sativa* according to the extended BBCH scale. *Annals of Applied Biology* 158, 87–94.
Matthäus, B. and Angelini, L.G. (2005) Anti-nutritive constituents in oilseed crops from Italy. *Industrial Crops and Products* 21, 89–99.
Moriel, P., Nayigihugu, V., Cappellozza, B.I., Gonçalves, E.P., Krall, J.M. *et al.* (2011) Camelina meal and crude glycerine as feed supplements for developing replacement beef heifers. *Journal of Animal Science* 89, 4314–4324.
Moser, B.R. (2010) Camelina (*Camelina sativa* L.) oils as a biofuels feedstock: golden opportunity or false hope? *Lipid Technology* 22, 270–273.
Moser, B.R. and Vaughn, S.F. (2010) Evaluation of alkyl esters from *Camelina sativa* oil as biodiesel and as blend components in ultra low-sulfur diesel fuel. *Bioresource Technology* 101, 646–653.
Mulligan, G.A. (2002) Weedy introduced mustards (*Brassicaceae*) of Canada. *Canadian Field Naturalist* 116, 623–631.
Narasimhulu, S.B., Kirti, P.B., Bhatt, S.R. and Chopra, V.L. (1994) Intergeneric protoplast fusion between *Brassica carinata* and *Camelina sativa*. *Plant Cell Reports* 13, 657–660.
Ollech, V. (1884) *Die Rückstande der Ölfabrikation als Futtermittel für die landwirtschaftlichen Nutztiere*. Karl Scholtze, Leipzig, Germany, 12 pp. (In German.)
Ollson, G. (1960) Some relations between number of seeds per pod, seed size and oil content and the effects of selection for these characters in Brassica and Sinapis. *Hereditas* 46, 29–70.
Pachagounder, P., Lamb, R.J. and Bodnaryk, R.P. (1998) Resistance to the flea beetle *Phyllotreta cruciferae* (Coleoptera: Chrysomelidae) in false flax, *Camelina sativa* (Brassicaceae). *Canadian Entomologist* 130, 235–240.
Patil, P.D. and Deng, S. (2009) Transesterification of *Camelina sativa* oil using heterogenous metal oxide catalysts. *Energy Fuels* 23, 4619–4624.
Patil, P.D., Gude, V.G. and Deng, S. (2010) Transesterification of *Camelina sativa* oil using supercritical and subcritical methanol with cosolvents. *Energy and Fuels* 24, 746–751.
Paulsen, H.M. (2007) Mischfruchtanbausysteme mit Ölpflanzen im ökologischen Landbau. 1. Ertragsstruktur des Mischfruchtanbaus von Leguminosen oder Sommerweizen mit Leindotter (*Camelina sativa* L. Crantz) (Organic mixed cropping systems with oilseeds. 1. Yields of mixed cropping systems of legumes or spring wheat with false flax (*Camelina sativa* L. Crantz.)). *Landbauforschung Völkenrode* 57, 107–117. (In German, English summary.)
Paulsen, H.M., Wichmann, V., Shuemann, U. and Richter, B. (2011) Use of straight vegetable oil mixtures of rape and camelina as on farm fuels in agriculture. *Biomass and Bioenergy* 35, 4015–4024.
Pavlista, A.D., Baltensperger, D.D., Isbell, T.A. and Hergert, G.W. (2012) Comparative growth of spring-planted canola, brown mustard and camelina. *Industrial Crops and Products* 36, 9–13.
Pearson, N. and Walker, K.C. (1999) The performance of *Camelina sativa* in the UK. *Aspects of Applied Biology* 56, 249–255.
Peiretti, P.G., Mussa, P.P., Prola, L. and Meineri, G. (2007) Use of different levels of flase flax (*Camelina sativa* L.) seed in diets of fattening rabbits. *Livestock Production Science* 107, 192–198.
Petrikova, V., Vana, J. and Ustjak, S. (1996) Production and utilization of industrial and energy crops on reclaimed soils. *Metod Zemědělské Praxi* 17, 1–24. (In Czech, English summary.)
Pilgeram, A.L., Sands, D.C., Boss, D., Dale, N., Wichman, D. *et al.* (2007) *Camelina sativa*, a Montana omega-3 fatty acid and fuel crop. In: Janick, J. and Whipkey, A. (eds) *Issues in New Crops and New Uses*. ASHS Press, Alexandria, pp. 129–131.
Plessers, A.G., McGregor, W.G., Carson, R.B. and Nakoneshny, W. (1962) Species trials with oilseed plants. II. Camelina. *Canadian Journal of Plant Science* 42, 452–459.
Porcher, F.P. (1863) Resources of the Southern Fields and Forests, Medical, Economical, and Agricultural. Being also a Medical Botany of the Confederate States; with Practical Information on the Useful Properties of the Trees, Plants, and Shrubs. Steam- Power Press of Evans and Cogswell, No. 3 Broad Street, Charleston, VA. xxv + 601 pp., ill. (Rare Book Collection, University of North Carolina at Chapel Hill). (Online) Available at: http://docsouth.unc.edu/imls/porcher/porcher (accessed 23 March 2012).
Prankl, H., Krammer, K. and Worgetter, M. (1999) Fleet test with Camelina methyl ester. Final report, ALTENER Contract XVII/4.1030/Z/96-013, pp. 44–64.

Putnam, D.H., Budin, J.T., Field, L.A. and Breene, W.M. (1993) Camelina: a promising low input oilseed. In: Janick, J. and Simon, J.E. (eds) *New Crops*. Wiley, New York, pp. 314–322.
Putt, E.D. (1943) Association of seed yield and oil content with other characters in the sunflower. *Scientific Agriculture* 23, 377–383.
Rice, B. (1995) Promotion of the use of vegetable oil as a diesel engine fuel extender/replacement in Ireland. Altener Contract No. XVII/4, 1030/93–12, 21 pp.
Robinson, R.G. (1987) Camelina: a useful research crop and a potential oilseed crop. University of Minnesota Agricultural Experiment Station Bulletin 579-1987 (Item No. AD-SB-3275), pp. 1–12.
Salisbury, P.A. (1987) Blackleg resistance in weedy crucifers. *Eucarpia Cruciferae Newsletter* 12, 90.
Saucke, H. and Ackermann, K. (2006) Weed suppression in mixed cropped grain peas and false flax (*Camelina sativa*). *Weed Research* 46, 453–461.
Schuster, A. and Friedt, W. (1995) *Camelina sativa*: old face – new prospects. *Eucarpia Cruciferae Newsletter* 17, 6–7.
Schuster, A. and Friedt, W. (1998) Glucosinolate content and composition as parameters of quality of *Camelina* seed. *Industrial Crops and Products* 7, 297–302.
Schwanitz, F. (1967) *Die Evolution der Kulturpflanzen*. Munich, 463 pp. (In German.)
Searchinger, T., Heimlich, R., Houghton, R., Dong, F., Elobeid, A. *et al.* (2008) Use of OS croplands for biofuels increases greenhouse gases through emissions from land-use change. *Science* 319, 1238.
Seehuber, R. (1984) Genotypische Variabilität in Ertrags- und Qualitätsmerkmalen bei Mohn und Leindotter. (Genotypic variation for yield and quality traits in poppy and false flax.) *Fette Seifen Anstrichm* 5, 177–180 (In German, English summary.)
Seehuber, R., Vollmann, J. and Dambroth, M. (1987) Anwendung der Single-Seed-Descent Methode bei Leindotter (*Camelina sativa* (l.) Crantz) zur Erhöhung des Ertragsniveaus. *Landbauforschung Völkenrode* 37, 132–136. (In German.)
Séguin-Swartz, G., Eynck, C., Gugel, R.K., Strelkov, S.E., Olivier, C.Y. *et al.* (2009) Diseases of *Camelina sativa* (false flax). *Canadian Journal of Plant Pathology* 31, 375–386.
Shonnard, D.R., Williams, L. and Kalnes, T.N. (2010) Camelina-derived jet fuel and diesel: sustainable advanced fuels. *Environmental Progress and Sustainable Energy* 29, 382–392.
Shukla, V.K.S., Dutta, P.C. and Artz, W.E. (2002) Camelina oil and its unusual cholesterol content. *JAOCS* 79, 965–969.
Szumacher-Strabel, M., Cieślak, A., Zmora, P., Pers-Kamczyc, E., Bieliñska, S. *et al.* (2010) *Camelina sativa* cake improved unsaturated fatty acids in ewe's milk. *Journal of the Science of Food and Agriculture* 91, 2031–2037.
Urbaniak, S.D., Caldwell, C.D., Zheljazkov, V.D., Lada, R. and Luan, L. (2008) The effect of cultivar and applied nitrogen on the performance of *Camelina sativa* L. in the maritime provinces of Canada. *Canadian Journal of Plant Science* 88, 111–119.
VDI (1997) Richtlinie 4600. *Kumulierter Energieaufwand – Begriffe, Definitionen, Berechnungsmethoden (Cumulative energy demand – terms, definitions, methods of calculation)*. Verein Deutscher Ingenieure. Association of German Engineers, Düsseldorf, Germany.
Vollmann, J., Damboeck, A., Eckl, A., Schrems, H. and Ruckenbauer, P. (1996) Improvement of *Camelina sativa*, an underexploited oilseed. In: Janik, J. (ed.) *Progress in New Crops*. ASHS Press, Alexandria, Virginia, pp. 357–362.
Vollmann, J., Steinkellner, S. and Glauninger, J. (2001) Variation in resistance of camelina (*Camelina sativa* (L.) Crtz.) to downy mildew (*Peronospora camelinae* Gäum.). *Journal of Phytopathology* 149, 129–133.
Vollmann, J., Grausgruber, H., Stift, G., Dryzhyruk, V. and Lelley, T. (2005) Genetic diversity in camelina germplasm as revealed by seed quality characteristics and RAPD polymorphism. *Plant Breeding* 124, 446–453.
Vollmann, J., Moritz, T., Kargl, C., Baumgartner, S. and Wagentristl, H. (2007) Agronomic evaluation of camelina genotypes selected for seed quality characteristics. *Industrial Crops and Products* 26, 270–277.
Wacker, J. (1934) *Die Ölfrüchte. Anbau, Pflege und Verwertung*. Landwirtschaftliche Hefte 32/33, Berlin. (In German.)
Walsh, D.T., Babiker, E.M., Burke, I.C. and Hulbert, S.H. (2011) Camelina mutants resistant to actolactate synthase inhibitor herbicides. *Molecular Breeding*, Online First™, 27 December 2011.
Walsh, K.D., Puttick, D.M., Hills, M.J., Yang, R.-C., Topinka, K.C. and Hall, L.M. (2012) First report of outcrossing rates in camelina (*Camelina sativa* (L.) Crantz), a potential platform for bioindustrial oils. *Canadian Journal of Plant Science* 92, 681–685.

Warwick, S.I., Francis, A. and Mulligan, G.A. (1999) Brassicaceae of Canada. AAFC(Online). Available at: http://www.scib.gc.ca/spp_pages/brass/index_e.php (accessed 5 March 2012).

Warwick, S.I., Francis, A. and La Fleche, J. (2000) Guide to the Wild Germplasm of Brassica and Allied Crops (tribe Brassiceae, Brassicaceae) (online), 2nd edn. Available at: http://www.brassica.info/resources/crucifer_genetics/guidewild.htm (accessed 22 May 2012).

Wu, X. and Leung, D.Y.C. (2011) Optimization of biodiesel production from camelina oil using orthogonal experiment. *Applied Energy* 88, 3615–3624.

Zadernowski, R., Budzynski, W., Nowak-Polakowska, H., Rashed, A.A. and Jankowski, K. (1999) Effect of fertilization on the composition of lipids from false flax (*Camelina sativa* L. Cr.) and crambe (*Crambe abissinica* Hochst.). *Rosliny Oleiste* 20, 503–510.

Zinger, N. (1909) Über die im Lein als Unkraut auftretenden Camelina- und Spergula-Arten und ihre Abstammung. *Zeitschr. ind. Abst. und Vererbungsl.* 2, 340–349. (In German.)

Zohary, D. and Hopf, M. (2000) *Domestication of Plants in the Old World.* Oxford University Press, New York.

Zubr, J. (1993) New source of protein for laying hens. *Food Compounder* 23–25.

Zubr, J. (1997) Oil-seed crop: *Camelina sativa. Industrial Crops and Products* 6, 113–119.

Zubr, J. (2003) Qualitative variation of *Camelina sativa* seed from different locations. *Industrial Crops and Products* 17, 161–169.

Zubr, J. and Matthäus, B. (2002) Effects of growth conditions on fatty acids and tocopherols in *Camelina sativa* oil. *Industrial Crops and Products* 15, 155–162.

18 Oil Palm

Somashekhar M. Punnuri and Bharat P. Singh
Fort Valley State University, Georgia, USA

18.1 Introduction

The majority of crops contain some form of oil, but only crops with oil content high enough to be extracted for food or industry are labelled as oil crops. There are approximately 40 plant species with the potential to be used as oilseed but, at present, only a handful of crops supply most of the world's edible/non-edible oil market (Salunkhe *et al.*, 1992). International trade in vegetable oil is mainly dominated by oil palm (*Elaeis guineensis* Jacq.), soybean (*Glycine max* L. Merrill), sunflower (*Helianthus annuus* L.), groundnut (*Arachis hypogaea* L.), cotton (*Gossypium hirsutum* L.), rapeseed (*Brassica napus* L.) and coconut (copra, *Cocos nucifera* L.). In these crops, the major energy reservoir is stored in the form of oil and the fruit or seed parts of the plants are the source of most oils. Based on the fatty acid composition, oils are utilized for different purposes. A relatively small amount of oil is also extracted from different parts of the plants, such as glands in flowers, seeds, leaves, gum, bark, wood and roots of a number of plants for a distinctive aroma, flavour or other properties. These are often referred to as essential oils. Crops such as cotton and maize (*Zea mays*), the main purpose of whose production is other than oil, nevertheless are also significant commercial sources of oil.

The oil palm is produced in 43 countries around the world. The leading producing countries are Indonesia and Malaysia, which together account for 87% of the total world palm oil production, estimated at 52.7 Mt (Fig. 18.1) (FAO, 2010; FAS-USDA, 2012). Thailand has emerged recently as an important producer of oil palm. The major oil palm exporting countries are Indonesia and Malaysia, driven by the consumer markets in China and India (Murphy, 2007; Gunstone, 2011) for cooking oil and elsewhere for biofuel. Palm oil accounts for 23% of the global oil supply (Basiron, 2012).

Oil palm is a major contributor to world oil production. It is also the highest yielding oil crop with an average oil yield of around 3–4 t ha^{-1} (Murphy, 2003; Wahid *et al.*, 2005). Oil is processed from fruit or seed by expeller pressing alone or by expeller pressing followed by solvent extraction. The residual oil in the leftover oilcake/meal from expeller pressing is in the vicinity of 4–6% compared to 1–2% in the combined process (Salunkhe *et al.*, 1992). Oilcake/meal is a rich source of protein and used as animal feed if free of antinutritional factors, or otherwise applied as manure for soil improvement.

18.2 Palm Oil for Biodiesel

Biodiesel is chemically a fatty acid alkyl ester that can be produced from vegetable oil such

©CAB International 2013. *Biofuel Crops: Production, Physiology and Genetics* (ed. B.P. Singh)

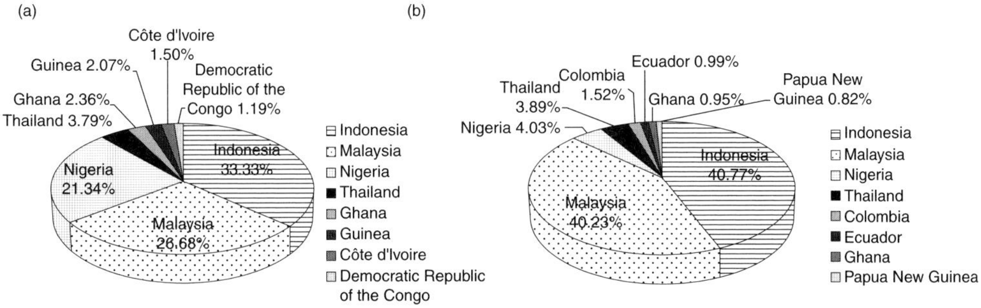

Fig. 18.1. Worldwide oil palm hectarage and production. (a) Oil palm area harvested (FAO, 2010) and (b) oil palm production (tons) expressed as percentages shown in the top eight of a total of 43 countries (FAO, 2010).

as palm oil by an esterification of triglyceride with methanol under a catalyst reaction (Marchetti *et al.*, 2007). Therefore, palm oil diesel (POD) is chemically a methyl ester of palm oil (fatty acid), which is also known by the acronym FAME (fatty acid methyl esters).

There is a burgeoning demand for fuel with ever-increasing population, urbanization and economic prosperity (Tan *et al.*, 2009). The transportation sector is the major driving force for future demands for oil (World Oil Outlook, 2011). Biodiesel is an alternative to conventional diesel, which is an economically feasible and environmentally acceptable fuel compared to fossil-derived petroleum fuel. A study on potential biodiesel-producing countries has shown that Malaysia is top among all countries in terms of its capacity to supply biodiesel. Among these countries, the top five countries (Malaysia, Indonesia, Argentina, the USA and Brazil) account for 80% of the total biodiesel produced worldwide, utilizing mainly palm oil and soybean oil as feedstock (Johnston and Hollaway, 2007). World biodiesel production, which is no longer in an exponential phase of growth, was estimated at 23.2 Mt in 2012 (Oil World, 2012) (more information on biodiesel can be found at http://www.afdc.energy.gov/afdc/fuels/biodiesel.html).

18.2.1 Why palm oil?

Palm oil is more affordable when compared to other oils and has a well-established and reliable supply chain (Lam *et al.*, 2009; Wahid, 2009). It produces nearly 5612 l biodiesel ha^{-1} (600 gallons of biodiesel per acre), which is 10–12 times more compared to other vegetable oils (Pahl, 2005; Gardner *et al.*, 2010). As a result, palm oil has taken over other oil crops in terms of demand, consumption and realistic biodiesel yield potential (Fig. 18.2) (Johnston *et al.*, 2009). Palm oil exhibits several physicochemical characters such as higher cetane number, viscosity, calorific energy released and engine efficiency compared to regular diesel, adding to its value for biodiesel production (Kalam and Masjuki, 2002; Nagi *et al.*, 2008). Palm diesel produces lower sulfur dioxide and other greenhouse gases such as CO_2 and has better lubricating properties than conventional fuel (May *et al.*, 2005). Several of the other properties such as carbon residue, flash point, pour point, oxygen content and blending also make it a useful biofuel. Its ignition properties are acceptable to the diesel engine and safe for transportation. There exists promise of the enhancement in the biofuel feedstock value of oil palm with genetic engineering. An ideal biodiesel composition should consist of a high degree of mono-unsaturated fatty acids such as oleic acids and palmitoleic acids, although palm oil contains more oleic acid and less palmitoleic acid, but transgenic palm oil can be genetically manipulated to produce a suitable feedstock possessing a high amount of these acids at the expense of palmitic acid (Parveez *et al.*, 2009; Pinzi *et al.*, 2009).

In general, palm oil as an alternative to fossil fuel has positive impacts on the environment. It is renewable, non-toxic and easily

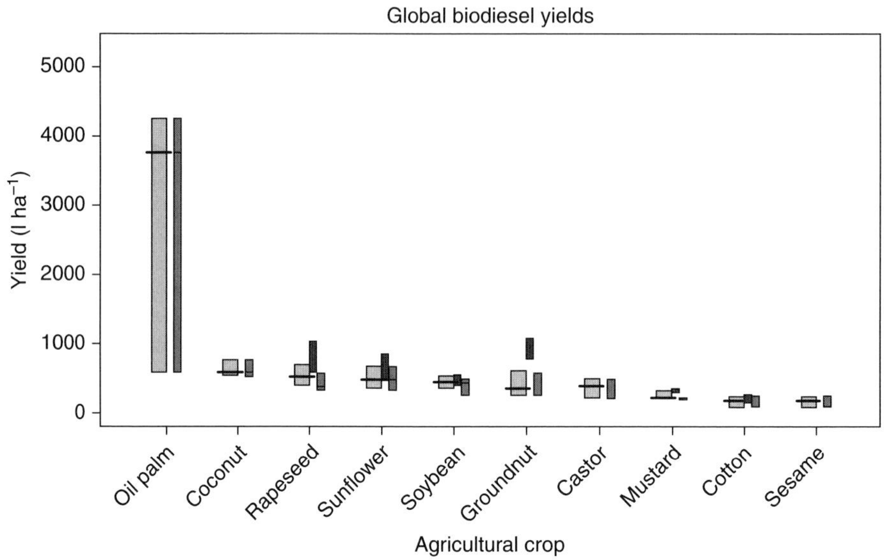

Fig. 18.2. Oil palm leads among ten edible oils in terms of biodiesel potential yield (Johnston *et al.*, 2009, used with permission from IOP Publishing). Light grey bars, biodiesel yield values for entire globe; medium grey bars, values for developed countries; dark grey bars, values for developing countries. Black horizontal lines represent median yields.

biodegradable. It reduces global warming and related effects. It helps the agrarian economy. It improves the socio-economic status of oil palm growers. The economies of developing countries exporting palm oil will also obtain a much needed boost. It creates new jobs across the globe, including in countries producing and exporting palm oil. It also enhances the energy security of producing and importing countries by diversification of liquid fuel source.

18.3 Origin and Distribution

Oil palm (*Elaeis guineensis* Jacq.) originated in Africa with its natural habitat spreading from Sierra Leone and Liberia in West Africa and dispersing through the diffused belts of the Democratic Republic of Congo and Angola in the south-west of Africa (Corley and Tinker, 2007). The source of genetic diversity for this species exists in south-eastern Nigeria and western Cameroon. The recent molecular genetic diversity studies also point to the fact that Nigeria might be the centre of diversity for wild oil palm (Maizura *et al.*, 2006). The oil value of the plant was realized by the native African population very early and thus it was cultivated and domesticated. Oil palm was exported from the African continent to other parts of the world by European traders engaged in the slave trade during the 15th century (Corley and Tinker, 2007).

The earliest international trade in palm oil and kernels started between West Africa and Europe during the colonial period of the 19th century. The demand for volume and higher quality of palm oil had experienced a considerable escalation by the beginning of the 20th century, leading to the establishment of commercial plantations. The first oil palm plantation was established in Indonesia in 1911 out of four seedlings planted in the Botanical Garden of Bogor (Indonesia) in 1848 (Rajanaidu *et al.*, 1979). The same population was used for establishing and propagating another plantation in Malaysia in 1917 (Hartley, 1988; Corley and Tinker, 2007). Major expansion in oil palm plantations took place during 1970s due to the burgeoning demand for the oil. Large numbers of plantations were established in tropical countries of South-east Asia such as Indonesia, Malaysia, Thailand and Papua New Guinea, tropical countries of South America such as Colombia, Ecuador and Costa Rica and African countries such as Ghana, Côte d'Ivoire and Cameroon (Poku, 2002). Oil palm plantations are also found in the Philippines, Solomon Islands, China, India and Sri Lanka. Currently, oil palm plantations are in another expansion phase due to the demand created for oil in the production of biodiesel.

18.4 Botanical Description

A complete and elaborate description of oil palm botany can be found in several sources (Hartley, 1988; Ataga and van der Vossen, 2007; Corley and Tinker, 2007), hence a brief summary of the botanical description is presented in this chapter. The oil palm (*Elaeis guineensis* Jacq.) belongs to the family *Arecaceae* and the genus *Elaeis* comprising three species, i.e. *Elaeis oleifera, Elaeis odora* and *Elaeis guineensis*. Out of the three species, the African oil palm, *E. guineensis* Jacq. and tropical American oil palm, *E. oleifera,* are usually found under cultivation. Of these two, African oil palm has favourable traits for domestication and hence is widely cultivated throughout the world. It is a branchless tree that can attain heights of up to 10 to 18 m. The tree has an extensive fasciculate adventitious root system in the upper soil. Petioles holding the leaves are 1–2 m long with spines. In young plants, petioles emerge from the trunk making it rough, but they are later shed leaving the trunk with only scars. In a mature tree, the trunk is swollen at the bottom and cylindrical and erect in the remainder. Leaf growth is slow initially at the rate of two leaves per month and mature leaves are pinnately compound, arranged spirally in sheaths, up to 8 m long. A whorl of more than 50 leaves, which also comprise the top inflorescence of the tree, forms a leaf crown with a circumference of 10–16 m. Leaflets number around 200–300 per leaf, formed at the intersection of petiole and rachis, and are 35–65 cm × 2–4 cm in size (Ataga and van der Vossen, 2007).

Oil palm is a monoecious crop with an axillary inflorescence tightly enclosed in spindle-shaped or ovate bracts that split before anthesis. Mature inflorescences are abundant within 3 years of planting. Although it is common to find both male and female flowers, only one of them develops into a mature inflorescence. The male inflorescence is highly influenced by environmental conditions such as high temperature and drought (Williams and Thomas, 1970; Purseglove, 1972). Male and female flowers are produced alternately every 6 months (Hartley, 1988; Corley and Donough, 1995; Adam *et al.*, 2005). The inflorescence emerges from the base of the petiole and consists of peduncle and spadix. The male inflorescence is ovoid containing 700–1200 closely packed flowers, while the female inflorescence is globose with 10–25 spirally arranged flowers and a terminal spine. Male flowers are comprised of six stamens and female flowers have a superior, three-celled ovary and sessile three-lobed, creamy white stigma, but only one carpel out of three develops into a seed. The female flower with moist separated stigma lobes exhibits receptivity for pollen for 36–48 h after initial opening (Latiff, 2000). Although wind and insects are possible means of dispersal, pollination is primarily carried out by insects such as weevils, *Elaeidobius kamerunicus* in Cameroon and *Thrips hawaiiensis* in Peninsular Malaysia (Syed, 1979; Hartley, 1988).

The female infructescence (fruit bunch; the fruiting stage of the inflorescence) contains a large number (around 1500–3000) of tightly packed ovoid fruits. Fruit maturity takes 6 months after pollination and individual fruits weigh 10–20 g. An infructescence, fresh fruit bunch (FFB), can weigh between 10 and 40 kg (Hartley, 1988). The number of fruit bunches is influenced by the ratio of male to female flowers, which is controlled by genetic and environmental factors (Corley and Donough, 1995; Adam *et al.*, 2005). Fruit is typically a sessile drupe, which has a single seed (kernel) surrounded by a soft oily pulp. Oil is extracted from both the pulp of the fruit (palm oil, edible and non-edible) and the kernel (kernel oil, mainly for soap manufacture). For every 100 kg of fruit, typically 20–22 kg of palm oil and 1.6 kg of palm kernel oil can be extracted. The three most commonly found fruit types are *dura, pisifera* and *tenera. Pisifera* has more desired characters such as more pericarp and thinner or shell-less outer shell, but is a female sterile. Therefore, it is crossed with *dura,* which has a thick shell and smaller pericarp. The resulting hybrid is the *tenera* type, producing more oil content.

18.5 Soil and Climate

Oil palm is adapted to humid tropical lowlands. For optimal growth and development of the tree, a large loamy to loamy-clay soil area with high light intensity and enough soil water content are needed. A well-distributed rainfall of 1800–2000 mm with minimum of 1600 mm year^{-1}, high air humidity, 5–7 h direct sunlight day^{-1} and a temperature range of 22–33°C constitute a congenial oil palm climate. Long dry spells or average monthly temperatures below 18°C or above 35°C will drastically hamper the palm cultivation.

Oil palms can grow on a wide range of soils and are adapted to pH from 4 to 7. It prefers soils that are deep and possess high available moisture holding capacity, contain organic carbon above 1.5% in the topsoil (upper soil) and cation exchange capacity that exceeds 100 mmol kg^{-1} (Ataga and van der Vossen, 2007). Oil palm trees can tolerate short periods of flooding but are affected by prolonged waterlogging.

18.6 Cultivation

The cultivation of oil palm is most suitable within 10° latitude of the equator as this is congenial for better growth and yield. Freshly harvested oil palm seeds are dormant. Dormancy is often overcome by induced heat treatment of 39–40°C for 2–3 months with subsequent cooling and drying. The germinating seed develops into a seedling with plumule, radicle and adventitious roots within 4 weeks. Seedlings are planted in the nursery using cylindrical heavy-duty polyethylene containers. Raising seedlings in a nursery for 1 year is a common practice for better establishment. Initially, during the first 2–3 months, seedlings are kept in shade. Field transplanting is carried out when plants are big enough, around 12–14 months old. Culling is practised in the main nursery to remove diseased and abnormal plants (Corley and Tinker, 2007).

Oil palms are generally planted on flat or gently sloping land. Soils with poor permeability may require the construction of a drainage system to avoid waterlogging. If forest land is to be cleared for the new planting, boundaries for the plantation are first laid out, trees cut within the demarcated and surrounding area to clear the land for planting and create a firebreak from the rest of the forested area. Fire is set to kill the underbrush. For planting in existing cleared land, soil is prepared by disk ploughing followed by harrowing. Palms that have reached heights of 6–9 m are difficult to harvest and are killed by lethal injection of herbicides or bulldozed. Planting is commonly done at the start of the rainy season. Plants are spaced 9 m apart in a triangular pattern, making an optimum population and a standard density of 143 plants ha^{-1} (Corley and Tinker, 2007). Mulch is spread around the seedlings to prevent weed germination and soil drying. Wire netting is placed around the seedling for about 18 months to protect it from foraging by animals. To protect bare soil and incorporate nitrogen into it, legumes are planted, often in a mixture such as *Pueraria phaseoloides* (Roxb., Benth.), *Centrosema pubescens* Benth. and *Calopogonium mucunoides* Desv.

The area surrounding the tree is kept clean manually or by applying herbicides. During the first 4 years, the open land area is either planted to legumes or intercropped with annual food crops. As the canopy closes in the mature plantations, the under-canopy is replaced by natural vegetation of grasses and ferns. Regular slashing is carried out to keep the under-canopy growth under control. Leaf number is maintained around 40 per tree for optimum yield. The excess leaves are cut and spread in the spacing between tress to provide ground cover.

Irrigation is generally not practised as most oil palm plantations are located in geographical regions with high rainfall. However, irrigation is practised to maintain high yields in southern India and Colombia, where it is not uncommon to encounter long dry spells. Fertilizer recommendations for oil palms during the first 3 years of planting in Nigeria are 0.1–0.6 kg N, 0.3–1.4 kg P and 0.4–1.2 kg K per palm per year, with rates increasing from field planting to 3 years (Ataga and van der Vossen, 2007). In subsequent years, the fertilizer needs vary with the soil type, age of the palm, land use pattern, history of the land before planting and number of years under production.

Potassium fertilizer is crucial for fruiting trees. Fertility trials and foliar analysis can be used to determine fertilizer needs of individual plantations. Some of the fertilizer needs can be substituted by applying composted waste comprising of empty bunches, fibre and sludge from palm oil mills.

Palm trees start yielding fruit only after 2–3 years of establishment in the main field. Fruits are harvested after 2.5 years in South-east Asia and after 3–3.5 years in West Africa. Harvesting of ripened fruit bunches is year-round, carried out at 10–15 day intervals. Fruit bunches are considered ready for harvest as fruit abscission starts. In plantations, trees are kept to a height manageable for harvest by chisels or hooked knives attached to poles. However, smallholders in Africa have tall oil palm trees, which have to be climbed with a climbing rope for harvest. Loose fruits on the ground are collected to add to the total yield. Cultivation of oil palm is very labour intensive. It is estimated that one field worker per 4 ha is required for proper management. The most labour-intensive part of the operation is harvesting. Development of a mechanical harvester has not been successful, because of the difficult problem of distinguishing ripe and unripe fruit bunches and selective harvesting without harming the unripe fruit.

The yield of oil palm is highest during the initial 4–5 years of production. The average economic and productive lifespan of oil palm is about 25 years. Well-managed plantations in South-east Asia commonly produce yearly bunch yields of 24–32 t ha^{-1}. Calculating by the usual factory oil extraction rate of 22% (*tenera* type), this amounts to oil yields of 5.2–7.1 Mt ha^{-1}. Average yield per ha reported for 2005 was 2.8 t palm oil and 0.7 t palm kernels (45% oil and 55% meal) (Ataga and van der Vossen, 2007).

18.7 Pests and Diseases

Oil palm insect pests in South-east Asia differ from those in Africa. Most commonly observed in Africa are leaf miner (*Coelaenomenodera lameensis* (syn. *Coelaenomenodera elaeidis*)), rhinoceros beetle (*Oryctes monoceros*) and palm weevil (*Rhynchophorus phoenicis*). Those observed in Asia include bagworms (*Cremastopsyche pendula, Metisa plana, Mahasena corbetti*), nettle and slug caterpillars (*Darna trima* and *Setora nitens*) and rhinoceros beetle (*Oryctes rhinoceros*) (Hartley, 1988; Ataga and vander-Vossen, 2007). There are several strategies recommended along with integrated pest management practices for different regions, including pheromone-based mass trapping, application of narrow spectrum insecticides and use of biological control agents. Rodents can cause considerable damage to maturing fruit bunches. Baits are placed near the trees for rat population control. Nest boxes are placed in plantations for barn owls (*Tyto alba*), which prey on as many as 2000 rodents per year.

Several fungal diseases attack young seedlings and adult plants. The prominent ones at seedling stage are seedling blight caused by *Curvularia* spp., corticium leaf rot caused by *Corticium solani*, cylindrocladium caused by *Cylindrocladium macrosporum* manifested as brown lesions, anthracnose caused by *Botryodiplodia* spp., *Glomerella* spp. and *Melanconium* spp. and blast, a root disease caused by *Rhizoctonia lamellifera* (syn. *Macrophomina phaseolina*) and *Pythium* spp. (Gascon *et al.*, 1989; Ataga and van der Vossen, 2007). Another important disease is cercospora leaf spot caused by *Cercospora elaeidis*, which affects young seedlings only in Africa and is prevalent during the wet season (Turner, 1981). Cultural practices that minimize fungal spread and fungicides are the usual methods for controlling these diseases. Young palm trees 2–4 years old can also suffer from a physiological disorder known as crown disease, which is inherited by a single recessive gene causing distortion of leaves.

The most important disease of adult palm trees in South-east Asia is basal stem rot (caused by *Ganoderma* sp.) (Hartley, 1988; Gascon *et al.*, 1989; Corley and Tinker, 2007). The disease has the potential of causing losses especially when a prior coconut or oil palm field is replanted. Fungal infection takes place through contact with decaying roots and stems. Sanitation by removing all debris prior to replanting and destroying diseased oil palm trees in the vicinity are measures used to keep the disease under control. The most notable among other fungal diseases confined to Africa is vascular wilt caused by *Fusarium oxysporum* f. sp. *elaedis*, for which interspecific hybrids have been developed as resistant cultivars.

18.8 Oil Extraction for Biodiesel Production

The oleochemicals from palm oil are used in producing biodiesel through the transesterfication process. The most commonly used grade of palm oil for biodiesel production is RBDPO (Refined Bleached Deodorized Palm Oil).

After the fruit harvest, the extraction of crude palm oil is done through several steps that are common to both large and small processing industries. Fruits are separated from bunches, crushed and macerated to extract crude palm oil through hydraulic pressing. Further downstream processing is done to separate the crude palm oil from fibres, free fatty acids, phospholipids and pigments. The crude palm oil as such cannot be used directly in biodiesel production as it needs a high amount of catalyst, several additional biodiesel plants and is also associated with emulsion problems. Therefore, a transesterification reaction is a common route to produce biodiesel from triglycerides using an acidic or basic catalyst. The catalyst can either be homogeneous or heterogeneous based on the requirement. A multitude of studies have been carried out to test different catalysts in order to optimize the transesterfication reaction and increase the efficiency of conversion (Alkabbashi *et al*., 2009; Sharmila, 2010;Yee *et al*., 2010; Taufiq-Yap *et al*., 2011). A general prototype as described in Yee *et al*. (2010) has been utilized for carrying out transesterfication reactions in order to determine the optimum temperature and best possible catalyst for efficient conversion of palm oil to palm oil diesel. Recently, transesterfication has been carried out without using a catalyst by changing the temperature to supercritical alcohol conditions (higher temperature and pressure), which is known as the supercritical fluids (SCF) reaction (Tan and Lee, 2011).

18.9 Physiology of Oil Palm Tree

18.9.1 Growth and development

The canopy architecture of oil palm is ideal for maximum light interception. The leaves are produced at the apical meristem and the new leaves point vertically upwards, creating gaps for sunlight to penetrate to the lower canopy. In this configuration, top leaves are able to utilize their full leaf surface to intercept sunlight and achieve light saturation. The photosynthetic rate of leaves at light saturation is approximately 25 μmol m^{-2} s^{-1}, which is a relatively high value for a perennial tree crop (Dufrene and Saugier, 1993). The angle of leaf rachis to the stem increases from top to bottom, which ensures that while the top canopy efficiently utilizes the sun irradiance, it also permits light to infiltrate to the lower canopy. The gradually increasing horizontal orientation of the leaves permits full ground coverage from one end to the other of the plant periphery, so that much of the incident irradiance gets absorbed before reaching ground level. The architecture of each individual leaf is also ideal for light dispersion. The leaflets are narrow (~6 cm) and long (up to 1.3 m) and number 200–300 per mature leaf (Corley and Tinker, 2007), meaning that while a high leaf area is maintained, the leaflets also provide gaps for downward light travel. To further enhance the photosynthetic surface, the linear leaflets are arranged alternately on each side of the leaf stalk, essentially forming a light sieve.

The oil palm canopy structure allows it to fully cover the ground surface at an early age for maximum light interception and to achieve a high leaf area index (LAI) of functional leaves for maximum photosynthesis. Squire and Corley (1987) stated that oil palm had an LAI of 6 (high value) and covered the ground surface fully at the age of 5–6 years when planted at a density of 130–150 plants ha^{-1}. By year 10, the plant canopy intercepted 96% of photosynthetically-active radiation. The oil palm also has a self-thinning system to get rid of old, photosynthetically inefficient leaves, leaving the crown with 30–50 leaves. In plantations, the lower leaves are pruned off earlier to better sustain the active leaves.

In annual oilseed crops, during the reproductive period, plants prioritize the translocation of assimilates to seeds over vegetative organs. But for oil palm, it has been taken as corroborated fact that the reverse is true, i.e. vegetative parts are favoured over inflorescences and fruits (Corley *et al.*, 1971). The answer to this perceived discrepancy may lie in the difference in the morphologic make-up between oil palm and annual seed crops. Annual crops in most cases complete the formation of the axial or apical morphological receptacle for the formation of reproductive organs and thus further expenditure of dry matter to vegetative organ formation does not contribute to the propagation of its life cycle. On the other hand, in oil palm, there exists a bud in the axil of every leaf, which develops into an inflorescence. It is understood, then, that in the absence of a new leaf, there would not be any new inflorescence. Furthermore, as the leaves are borne apically, the stem extension is a prerequisite for the formation of new leaves. Similarly, in the case of oil palm, the shifting priority from reproductive growth to vegetative growth may also be for the same purpose of ensuring the best possible chance for life cycle preservation by the formation of new leaves where flower primordia can develop.

Similar to annual seed crops, trunks of oil palm also serve as the storage organ for assimilates for the fruit when the current rate of photosynthesis is not sufficient to supply the developmental needs of actively developing fruits. Henson *et al.* (1999) found that from 54 to 82% of the carbohydrate in the trunk biomass was non-structural soluble sugars and on average the available carbohydrate reserve in an 8-year-old palm tree was sufficient for the production of 20 kg of dry fruit bunch. Thus, instead of using the surplus carbohydrate for luxury growth, the plant clearly showed a preference to preserve it in favour of reproductive growth. The report by Breure *et al.* (1990) points to the fact that up to 65% of the carbohydrate assimilated by the palm tree was partitioned to the fruit.

Oil palm is a tropical perennial crop that produces fruits throughout the year. This permits the plant to fully exploit resources available to it to produce much higher yields than seasonal yielders. Physiological age is considered an important factor in governing the source–sink relationship in whole-plant and the varied responses of phytomers (Jones, 1997).

18.9.2 Effect of environment on oil palm physiology

Photoperiod influences oil palm phenological rhythms in the alternating male and female inflorescence ratio, which eventually affects the fruit bunch (Borchert *et al.*, 2005; Legros *et al.*, 2009). Water stress significantly reduces female flowers by way of abortion and hence affects fruit yields. Therefore, environment plays a major role other than genetic content in dictating physiological phenomena such as sex differentiation and sex determination (Corley, 1976; Adam *et al.*, 2011).

Oil palm belongs to the group of C_3 plants and as such should benefit from the predicted increase in the atmospheric level of CO_2. Carbon dioxide enrichment may be of practical significance in the raising of seedlings. Higher CO_2 enhances net assimilation rate (NAR) to divert the extra photoassimilates towards developing organs (Eric *et al.*, 2000; Ibrahim *et al.*, 2010). Ibrahim *et al.* (2010) observed increased seedling growth rate and reduction in nursery acclimatization period by 27 weeks in a CO_2-enriched environment in comparison to conventional nursery management.

The differential day and night time temperature requirement in plants is a well-known phenomenon, but in oil palm the existence of such preference does not appear to be strong and it has been difficult to demarcate day and night temperature effects (Corley and Tinker, 2007). According to Hartley (1988), as climates of West Africa and Malaysia were similar except for lower mean night temperature in Africa, the yield advantage of Malaysian palm may result from the prevalence of a narrower gap between day and night temperatures. In trees, the seasonal temperature variation to trigger some important physiological phenomenon is not uncommon, but no such consequential triggering seems to be needed for oil palm. Goh (2000) found that oil palm grew best in geographical regions where temperature variation during different times of the year was minimal. Only time-seasonal temperature becomes of consequence when it goes below the minimal threshold during certain period during the year. Hartley (1988) reported a winter mean temperature of approximately 18°C in Tela, Honduras, which made fruit harvest seasonal (June–December) compared to year-round in Malaysia.

Oil palm has an advantage over other crops because of its ability to go through reasonable drought periods without wilting (Corley and Tinker, 2007) and as a result continue photosynthesis and maintain growth uninterrupted. The plant can withstand drought because its leaves are endowed with multiple features for this purpose. The upper surface of the leaf has a thick cuticle to prevent water loss. The stomata are present on the abaxial (bottom) rather than adaxial (top) portion of the leaf. As the bottom portion of the leaf is shaded, it is cooler than the top and thus transpires less than the top leaf surface, preserving plant water. The construction of stomata is also favourable to the preservation of water as guard cells are thick walled and the opening ridges close tightly throughout the entire length so as not to permit escape of water vapour from the leaf to the atmosphere.

However, Ochs and Daniel (1976) found oil palm yield was adversely affected when water deficit reached 100 mm. The extent of damage depended on whether the water deficit existed as unpredictable isolated events or occurred regularly at certain periods of the year (Caliman and Southworth, 1998). For single droughts, the relationship between water deficit and FFB yield was rather weak, but regular seasonal drought affected yield through influencing floral initiation, sex differentiation and abortion rate.

18.10 Biochemistry of Oil Synthesis

Palm oils contain triacylglycerols (TGAs), which is the raw material used in biodiesel production. Hence, a comprehensive analysis of TGA biosynthesis and its engineering in oil palm to increase unsaturated fatty acids will enhance the supply of sustainable lipid feedstock for biofuel production (Durrett *et al.*, 2008). Triglyceride chemically consists of one molecule of glycerol as a backbone to which three molecules of fatty acids are attached. Different fatty acids having different properties, such as chain length, degree of saturation and polarity, will make a different composition for TGAs. The complete biosynthetic pathway for triacylglycerols, oil deposition and fruit ripening in oil palm was compiled and reviewed by Sambanthamurthi *et al.* (2000). The oil palm produces ripe fruits 16–22 weeks after pollination during which time the amount of TAGs increases in the mesocarp. The mesocarp of the fruit contains major fatty acids such as palmitic acid (16:0) and oleic acid (18:1) and of these, palmitic acid (44%) dominates in the matured fruits depending on the different characteristic fruit types (Osagie and Bafor, 1990; Sambanthamurthi *et al.*, 2000). Palm kernel oil mainly contains lauric acid.

Like other plants, the basic mechanisms to synthesize fatty acid through fatty acid synthase (FAS) complex in the plastids and triacylglycerol (TAG) synthesis through the Kennedy pathway in the endoplasmic reticulum (ER) remain the same in oil palm (Fig. 18.3; Durrett *et al.*, 2008), but its ability to convert most of the carbon fixed through photosynthesis into oil droplets through quantitative regulation is not completely understood (Ohlrogge and Jaworski, 1997). Oil palm's carbon source is diverted to mesocarp, where TAGs are stored in the cytoplasm as lipid oil droplets amounting to 90% of the tissue dry weight (Osagie and Bafor, 1990; Ohlrogge and Chapman, 2011). Lipid droplets (LDs) contain no oleosins in oil palm and are synthesized by mechanisms unique to oil palm (Chapman et al., 2012). Several studies have helped explain the metabolic dynamics of carbon flux to lipid biosynthesis (Bao and Ohlrogge, 1999; Ramli *et al.*, 2002a, b, 2005). Furthermore, recently a similar and more advanced study has also highlighted the regulatory aspects of carbon flux control and fatty acid synthesis, which controls the entire process of lipid synthesis (Ramli *et al.*, 2009). Plastids play a critical role in increasing fatty acid synthesis (Voelker, 2011). The photosynthates

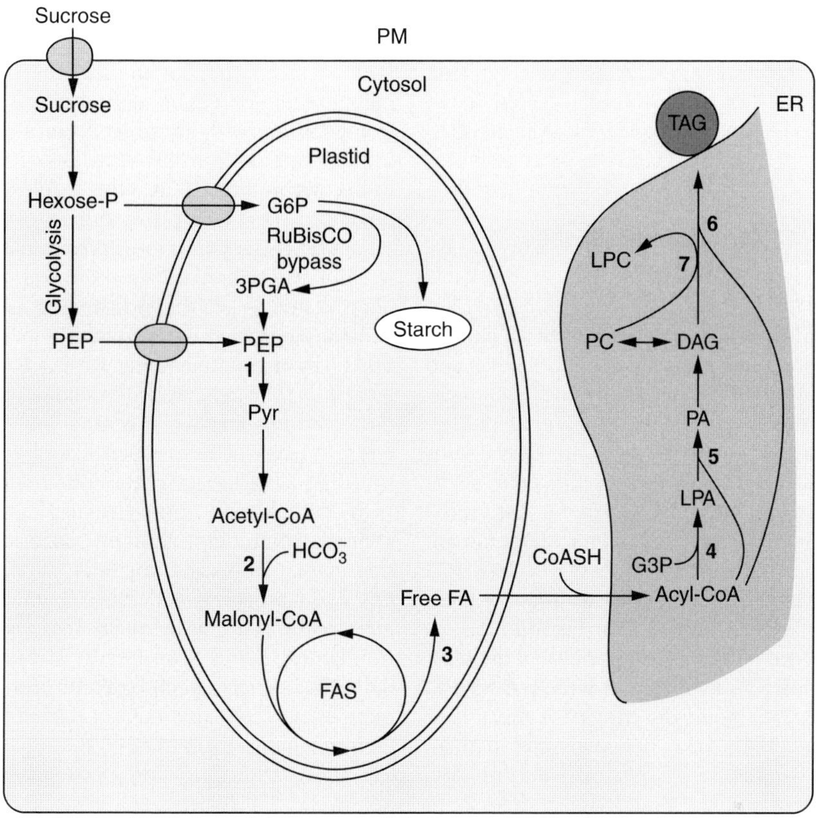

Fig. 18.3. Overview of FA and TAG synthesis in plants (Durrett *et al.*, 2008, used with permission from John Wiley & Sons). For simplicity, not all possible reactions are shown and multiple reactions may be represented by single arrows. Some pathways, for example, the Rubisco bypass pathway, are not present in every species. Bold numbers represent key enzymes: 1, plastidic pyruvate kinase; 2, acetyl CoA carboxylase; 3, acyl ACP thioesterases FatA and FatB; 4, glycerol-3-phosphate acyl transferase; 5, lyso-phosphatidic acid acyltransferase; 6, diacylglycerol acyl transferase; 7, lyso-phosphatidylcholine acyl transferase. 3PGA, 3-phosphoglycerate; DAG, diacylglycerol; ER, endoplasmic reticulum; FAS, fatty acid synthesis; G3P, glycerol-3-phosphate; G6P, glucose-6-phosphate; LPA, lyso-phosphatidic acid; LPC, lyso-phosphatidylcholine; PA, phosphatidic acid; PC, phosphatidylcholine; PEP, phosphoenolpyruvate; PM, plasma membrane; Pyr, pyruvate; TAG, triacylglycerol.

and intermediates such as hexoses are channelled into plastids for massive amounts of fatty acid production, which are further sequestered into TAGs as oil bodies (Huang, 1992; Dyer and Mullen, 2008; Voelker, 2011). Several of the rate-limiting enzymes in biosynthetic pathways along with compartmentalization of fatty acid synthesis and TAG synthesis could be associated with specialized production of oil droplets in oil palm (Coleman and Lee, 2004; Li-Beisson *et al.*, 2010). Synthesis and supply of fatty acids was the regulating factor in TAG biosynthesis in developing embryos of seeds (Bao and Ohlrogge, 1999). A recent understanding of TAG synthesis through transcriptome and metabolome analysis have given detailed insight into the regulatory mechanism involved in oil palm during fruit maturation (Bourgis *et al.*, 2011; Tranbarger *et al.*, 2011). Focks and Bening (1998) observed that the *WRI1* mutant possessed significantly reduced ability (up to 80% reduction) to synthesize oil, indicating this gene being a key transcription factor involved in production of oil droplets. Furthermore, *WRI1* was shown to be up-regulated in date palm mesocarp, which contains less oil (Bourgis *et al.*, 2011). Therefore, *WRI1* can be best regarded as a 'master regulator' of all the key enzymes involved in conversion of sucrose to FA (Chapman and Ohlrogge, 2012).

18.11 Genetic Stocks and Breeding

Oil palm has a diploid genome with 16 pairs of homologous linkage groups ($2n = 2x = 32$) and $2C = 3.4 \times 10^9$ bp based on flow cytometry analysis (Rival *et al.*, 1997; Maria *et al.*, 1995). The plant is monoecious with distinguishing female and male flowers. Oil palm shell-thickness has been shown to be controlled by a single Mendelian bi-allelic codominant gene (*Sh*) (Beirnaert and Vanderweyen, 1941; Hartley, 1988). Shell thickness indirectly gives a measure of the amount of oil that can be extracted from the mesocarp (pulp) of the fruit. The commercial hybrid most widely used and which renders good oil content is the *tenera* type, an intermediate between, and derived from, two common parents, *dura* and *pisifera*. Attributes of *dura* are a thick shell and lesser or smaller pericarp. In contrast, *pisifera* has more pericarp with thinner outer shell or shell-less and is a sterile female.

Oil palm germplasm collections are maintained at several research institutions across the world. However, these collections have a narrow genetic base, often with a base population derived from common parents, *deli dura* and *pisifera*. The Malaysian Palm Oil Board (MPOB) maintains a large oil palm germplasm collection. The Nigerian Institute for Oil Palm Research (NIFOR) has collected accessions from different parts of Africa and as well from America. The National Centre for Agricultural Research (CNRA) in Côte d'Ivoire also maintains a small set of collections.

Breeding programmes in oil palm have been carried out in three major institutes, FELDA (Federal Land Development Authority) Malaysia, CIRAD (French Agricultural Research Centre for International Development) and MPOB. The initiative taken by the department of agriculture, Malaysia, led to the establishment of a *dura* × *pisifera* breeding population in 1953 (Kushairi *et al.*, 1999). Furthermore, Malaysia also imported AVROS1 *tenera* × *pisifera* (T × P) seeds from Indonesia. Therefore, oil palm breeding programmes witnessed a boom in production with introduction of high yielding *tenera* type from D × P planting material in the 1960s and 1970s, which demonstrated an improvement in oil to bunch of 26% in *tenera* type compared to its predecessor, 16% in *dura* type (Sambanthamurthi *et al.*, 2009). Several studies were conducted to assess the genetic parameters for oil yield (Soh and Tan, 1983; Jalani and Rajanaidu, 1992; Okwuagwu and Tai, 1995). A study to characterize 20 different traits for their breeding values in different crosses of Deli-AVROS *dura* × *pisifera* breeding populations, revealed moderate to high heritabilities (broad sense) and good general combining abilities, indicating the importance of genetic material (Bilal, 2004). Oil palm breeding activities were also accelerated in Malaysia with the introduction of 12 Planting Series from PS1 in 1994 to PS12 in 2006, where each specific series was targeted for one particular commercial trait (Sambanthamurthi *et al.*, 2009).

Conventional breeding is credited with achieving the present-day fresh fruit bunch yield and oil content of *tenera* type oil palms. Extensive breeding success was achieved through various breeding schemes like mass selection (MS), reciprocal recurrent selection (RRS), modified recurrent selection (MRS) and pedigree selection; which subsequently have provided an increase of 1% genetic gain (Rival and Jaligot, 2010). Notable among these methods is MRS, which proved successful in improving oil yield from 0.18 t ha^{-1} to 4–5 t ha^{-1} (Jalani and Ariffin, 1994).

18.12 Tissue Culture

Clonal propagation is essential to increase the oil palm yield potential and to meet the growing demands of planting material required by oil palm growers worldwide. Micropropagation of oil palm through leaf and embryo explants has been routinely practised for production of healthy and uniform plantlets. Although the importance of tissue culture was realized in the

1960s and 1970s, *in vitro* technology was not routinely followed due to problems such as flower abnormalities and fruit bunch failure (Corley *et al.*, 1986). More than 3.5 million ramets are produced every year worldwide and Malaysia alone accounted for the production of 2.53 million ramets (Kushairi *et al.*, 2010). The supply of a huge amount of planting material is a prerequisite for sustainable palm oil production, which is facilitated by ramets produced through tissue culture laboratories. Tissue culture is expected to meet the future demand of 40 million ramets by 2017 (Kushairi *et al.*, 2010).

Oil palm tissue-culture techniques have been developed over several decades and are used as a tool in genetic engineering and transformation of elite material to improve oil yield. Clonal hybrids are better compared to sexual hybrids as they are genetically uniform and high oil yielding. There are more than 20 private and public tissue-culture laboratories around the world producing good quality plantlets. Of these, more than a dozen are operating in Malaysia, including the one owned by MPOB. Among these, Advanced Agriecological Research (AAR) is the top producer of oil palm ramets in the world. An initiative by MPOB in collaboration with a private firm, TSH Bio-Tech, has produced a high quality ramet, 'WAKUBA', which proved to be better performing with oil yield of 10 t ha^{-1} and an increased oil extraction rate (OER) of 26% (TSH, 2012). The commercial application of tissue culture can be best harnessed for the production of homozygous double-haploid (DH) lines for better consistency and germplasm improvement. Tissue culture studies have also helped in building EST resources, development of markers and annotation of genes associated with callogenesis and embryogenesis (Low *et al.*, 2008; Thawaro and Te-chato, 2009; Tranbarger *et al.*, 2012).

18.13 Molecular Markers

18.13.1 Use of molecular markers in genetic diversity determination

Genetic diversity of oil palm has been narrowed due to the limited number of founder lines used in the breeding population. It is, however, well emphasized that broadening the genetic base would help increase the oil palm yield (Meunier, 1976; Hardon *et al.*, 1985; Hartley, 1988; Mayes *et al.*, 1997). It was observed that natural sources of variation existing in oil palm are low and hence have led many studies to exploit the different sources of variation to increase the oil palm genetic diversity. The role of molecular markers and their utility in oil palm breeding has been elaborated and demonstrated in several diversity studies (Jones and Hughes, 1988; Jones, 1989; Jack and Mayes, 1993; Jack *et al.*, 1995; Mayes *et al.*, 2000, 2008; Hayati *et al.*, 2004). The studies on germplasm resources have facilitated the increase in the genetic base for incorporation into palm breeding programmes for high oil yield and yield components. Molecular markers such as RFLPs (restriction fragment length polymorphisms), RAPDs (randomly amplified polymorphic DNA), SSR (simple sequence repeats), PCR-RFLP (RFLP converted to PCR marker), AFLPs (amplified fragment length polymorphisms) have aided in accelerating these breeding programmes, which otherwise would have taken a long time in making selections. Molecular markers have facilitated overcoming the long breeding cycle, which usually takes a minimum of 10 years. The diversity studies were initially conducted with isozyme markers and RFLP markers before the advent of PCR-based markers (Baudouin, 1992; Jack and Mayes, 1993; Shah *et al.*, 1993; Mayes *et al.*, 1996). More recently, RAPD markers have been used in the determination of genetic variants through somaclonal variation and in differentiating between a mother palm and its progenies (Shah *et al.*, 1994; Rival *et al.*, 1998). The genetic variability of breeding material coming from different countries (Deli, Côte d'Ivoire, Zaire, Cameroon) was also assessed using isozyme and AFLP markers, which proved successful in removing off-types from reciprocal recurrent selection breeding programmes (Purba *et al.*, 2000). RFLP markers employed in linkage mapping were exploited as probes in deciphering the diversity and also in elucidating the genetic structure and architecture of the breeding population (Mayes *et al.*, 2000). Furthermore, genetic divergence studies between two different species of oil palm (African and

American) with AFLP markers proved to be powerful compared to RFLP markers in delivering equivalent genetic information and more divergence (Barcelos *et al.*, 2002). Microsatellites (commonly known as simple sequence repeats; SSRs) were developed and employed in an array of studies such as diversity, fingerprinting, phylogenetic and cross-transferability across palm taxa (Billotte *et al.*, 2001; Ting *et al.*, 2010; Zaki *et al.*, 2012). Microsatellites were also employed in a study conducted by Maizura *et al.* (2006), which suggests that the origin of palm diversity might be in Nigeria and, therefore, harnessing the diversity alleles from the Nigerian oil palm population could increase the genetic base for palm breeding programmes. Development and use of these markers laid the foundation for linkage and QTL studies.

SNP markers are a recent choice in oil palm breeding and have renewed interest because of their potential to detect single base/point mutations and the abundant polymorphic information they reveal. They are applicable in genetic mapping, conventional pedigree-based mapping and association mapping, diversity analysis, fingerprinting, marker-assisted selection, population genetic and phylogenetic studies of oil palm. These markers are known to occur at certain frequency (>1%) in the genome and are characterized by single base pair differences in homologous sequences. The expression studies have accumulated huge EST databases, which have been deposited at the palm genes website (http://palmoils.mpob.gov.my/palm-genes.html). 'Palm genes' is the first of its kind in making genomic resources available to the public. Oil palm EST sequences are used in the identification of SNPs as they represent gene-rich regions and exon regions (Singh *et al.*, 2008a, 2009a). Riju *et al.* (2007) used 576 ESTs from EST libraries to mine SNPs present in the seven different types of tissues. These results found a high frequency of 2.91 SNPs and indels 100 bp^{-1} in mesocarp tissue among all tissues. Recent efforts have also established an SNP database (http://riju.byethost31.com/oilpalm/index.php) available to the palm scientific community (Riju and Arunachalam, 2009). The efforts from these studies have yielded applicable SNPs linked to several monogenic traits such as fruit colour and shell genes. A total of 63 SNP markers labelled as VALOPS1 have been identified at MPOB and were tested in several populations for their accuracy (85–90%) to predict fruit forms and colour (Cheng-Li *et al.*, 2011). These resources will further enhance breeding applications such as genomic selection in improving oil palm productivity.

18.13.2 Mapping and quantitative trait loci (QTLs) identification

Out of the above mentioned institutes focusing on oil palm development, FELDA, CIRAD and MPOB have also pioneered the development of genomic resources for constructing high density linkage maps and germplasm diversity studies. Linkage mapping further helps in identifying QTLs for yield and oil quality.

The first framework map of oil palm was constructed with 97 RFLP loci in an F_2 mapping population of 98 palms, which spanned over the genome covering a total genetic distance of 860 cM (Mayes *et al.*, 1997). Subsequently, a linkage map was established using the pseudo-test cross mapping strategy, which comprised 121 loci obtained from 73 RAPD markers with a major objective of identifying markers linked to alleles controlling shell thickness (Moretzsohn *et al.*, 2000). Map information produced by Mayes *et al.* (1997) and Jack *et al.* (1998) was further extended with addition of a few more markers having better resolution (Rance *et al.*, 2001). A high density linkage map was constructed with 944 genetic markers comprised of SSRs and AFLPs, which represented for the first time a complete set of 16 homologous chromosomes of oil palm and was further used as reference map for other oil palm linkage studies (Billotte *et al.*, 2005). A map containing gene-specific markers was also developed from 123 cDNA-RFLP markers genotyped in 136 selfed progeny palms of T128 (Singh *et al.*, 2008b). An oil palm genetic map for an interspecific cross consisting of a Colombian *E. oleifera* (UP1026) and a Nigerian *E. guineensis* (T128) was constructed with 252 markers of AFLPs, RFLPs and SSRs (Singh *et al.*, 2009b). Recently, an integrated map consisting of 571 SSRs, AFLPs and

PCR-RFLPs was produced using a FELDA high-yielding commercial oil palm mapping population (DA41) along with its parents. This integrated map gives a good coverage of the genome with high number of informative markers (Seng *et al.*, 2011).

These maps were the initial efforts for leveraging QTL studies related to several important agronomic traits in oil palm. The first QTL study to take advantage of the available map was the efforts by Rance *et al.* (2001). This work was an improvement of oil palm map resolution over the works of Jack *et al.* (1995), Mayes *et al.* (1997) and Jack *et al.* (1998), with an additional 49 marker loci. The improved map was used to identify QTLs for fruit yield components and several other vegetative growth parameters. A total of 147 of 153 RFLP informative markers were used in constructing the linkage map, which yielded 22 linkage groups with map distance of 852 cM. QTLs were identified for 11 out of 13 traits included in the study, which accounted for phenotypic variation of 8.2–44.0% from single marker and interval mapping analysis. Moretzsohn *et al.* (2000) also used their own map in identifying two closely linked RAPD markers (R11–1282 and T19–1046) on either side of the locus controlling shell thickness. This work also attempted to identify QTLs for several traits, but did not reveal any significant associations between markers and traits. Singh *et al.* (2008b) identified a QTL for oil to wet mesocarp (O/WM) on linkage group 1 of their map at a distance of 11 cM from one of the gene-specific markers. This QTL was detected with LOD score of 3.54 and with phenotypic variation of 29.8% for the trait, (O/WM). A specific QTL study for iodine value and fatty acid composition in oil palm was conducted in an interspecific cross population of 118 palms, which exhibited 19.7–56.6% of variance associated with loci emanating from multiple regions of the genome. More recently, Billotte *et al.* (2010) identified 76 QTLs associated with 24 quantitative traits in a multi-parent population. This work involved detection of QTLs in multi-parent linkage mapping involving within-family and across-family analysis.

Currently, work is in progress to saturate the map developed in FELDA's oil palm breeding programme and to identify QTLs for yield components using a larger DA41 mapping population, which is genetically closer to commercial planting material (Seng *et al.*, 2011).

18.13.3 Marker-assisted selection

Marker-assisted selection (MAS) facilitates conventional palm breeding by associating a trait with marker location thereby increasing selection efficiency and gain. It is very effective in reducing the number of generations required for a breeding cycle and generation time (8–10 years) as selection can be made in the nursery before transplanting to the field. QTL mapping and linkage studies have given several potential markers for making selection of plants for higher oil yield, broader genetic base and disease-resistant clones. MAS is useful when phenotype is distinct and a tight linkage exists with the marker, in screening late-expressed traits such as sterility, fatty acid composition and shell thickness. MAS can expedite rapid selection in reciprocal recurrent selection and facilitate introgression of previously identified QTLs. It is anticipated that MAS will become more common with decreasing marker assay costs (Xie and Xu, 1998; Rance *et al.*, 2001), which is already becoming a reality (discussed in Section 18.13.5).

18.13.4 Oil palm genomics, transcriptomics and metabolomics

The sequencing projects for different species have been increasing in an exponential manner since 2000. As of now, approximately 3173 projects have been completed, 10,479 are ongoing and at least 1918 targeted for future (source: http://www.genomesonline.org/cgi-bin/GOLD/index.cgi). This has been possible due to the rapid development in sequencing technologies at a cheaper cost.

The first draft sequence of the oil palm genome was completed in 2008 by Synthetic Genomics Inc (USA) and the Asiatic Centre for Genome Technology (ACGT) Sdn. Bhd. (Malaysia). In 2009, a consortium headed by The Malaysian Palm Oil Board, and along with several other private companies from the USA, Malaysia, South

Korea and Australia, sequenced three oil palm genomes of two oil palm species, *E. oleifera and E. guineensis*, which included *pisifera* and *dura* palms. Among the private firms, St Louis-based Orion Genomics co-led the sequencing project with supporting firms such as MOgene LC and The Genome Center at Washington University (USA), Macrogen Inc. (South Korea) and GeneWorks Pty Ltd (Australia). This project has rendered the complete annotated genetic code of three oil palm genomes with an enormous transcriptome sequence (Wahid, 2009).

In May 2009, Sime Darby Bhd, a Malaysia-based oil company together with Synamatix Sdn. Bhd., a bioinformatics company, completed sequencing the genome with 30 times coverage and with genome coverage of 93.8%.

The physical size of the oil palm genome is approximately 1.8 billion base pairs of DNA, which is two-thirds the size of the maize genome and four times the size of the rice genome. A complete annotation of the palm genome would help in understanding the key genes affecting oil yield, drought resistance, disease resistance, evolution of the three genomes, comparative mapping and synteny studies. It will also augment efforts in increasing genomic resources required for molecular breeding.

Oil palm transcriptomics has just begun to unravel the genetic basis of expressed traits, more specifically oil bodies produced in the mesocarp of oil palm. Some of the metabolomics studies have made remarkable progress in understanding the key genes underlying oil synthesis. The mystery behind the synthesis of oil in oil palm is no more a hidden secret as several transcriptomic studies have started shedding light on the regulatorial aspect of genes involved in the biochemical pathway (Voelker, 2011). The review by Voelker (2011) has compiled several important works on FAs biosynthesis mechanism including transcriptomic studies on oil palm and date palm (Bourgis *et al.*, 2011; Tranbarger *et al.*, 2011), which delineated on the carbon partitioning to TAGs and their associated metabolic pathway and biochemistry of acyl lipid synthesis. These works highlight why mesocarp of oil palm is able to divert all its sugars to triacylglycerol (TAG) synthesis (90%) compared to date palm which contains no TAGs with sugars accumulated to a maximum of 50% of dry weight. It is clear from the comparative transcriptomic studies by Bourgis *et al.* (2011) that oil palm has a unique mechanism of carbon partitioning, where hexose and intermediaries derived from sucrose are routed to plastids. Glycolysis and FA synthesis is highly up-regulated in plastids leading to a large flux of FAs. Further, their assembly in endoplasmic reticulum (ER) leads to production of TAGs, which are accumulated as oil droplets in the cytoplasm (Voelker, 2011). Several other ongoing transcriptomics and metabolics studies involving systems biology will broaden our knowledge of important traits such as sex differentiation, roles of MADS-box genes, mesocarp development and fruit ripening (Tranbarger *et al.*, 2012).

18.13.5 Application of next-generation sequencing, newer mapping approaches and genomic selection

Next-generation sequencing (NGS) technologies, such as Roche/454, Illumina, AB SOLiD and Helicos heliscope, will hasten the current ongoing breeding programme in oil palm with affordable costs and longer read lengths. These high-throughput technologies will increase the amount of genetic data for oil palm to build the genomic resources for development of markers such as SNPs and SSRs (genomic and genic), insertion deletion sites (indels), conserved orthologue sets (COS), which compares genes between species and genomes, and conserved intron spanning primers (CISP) (Paux *et al.*, 2011). The markers developed will hasten the marker-assisted selection, gene introgression and wide crosses. Availability of the complete reference genome sequence would further bolster genomics-assisted technologies for identifying candidate genes underlying a functional variation in the oil palm genome.

There are various newer mapping approaches using next-generation technologies to identify QTLs, which involve the scoring of a large number of genotypes in a short time. Parental genomes of a mapping population can be resequenced to identify the variants in the genome. Further, these variants can be referenced to the known genome sequence to identify functional variants. Newer approaches used for mapping include Rad-seq, restriction-site associated DNA (Miller *et al.*, 2007), genotyping-by-sequencing (Elshire *et al.*, 2011), DNA chip and RNA seq. These methods have not been yet applied in oil palm and at this time they remain to be tested for their applicability to produce reduced representation libraries from a small size of mapping population available in this crop.

Development of a linkage map is tedious and time consuming in outcrossed and monoecious crops such as oil palm. Unlike in a bi-parental crossed mapping population, it is very difficult to develop a mapping population in oil palm as it takes several years to produce mapping progenies. Recently, a new form of marker-assisted selection known as genomic selection has been introduced, which predicts breeding values for a large set of markers screened across the genome (Meuwissen *et al.*, 2001). Genomic selection can be superior to marker-assisted recurrent selection and phenotypic selection in oil palm breeding and can lower cost per unit gain from 26 to 65% based on marker costs (Wong and Bernardo, 2008). It is also easier to trace the pedigree of a population with genomic selection.

With high-throughput machines, genotyping has become much easier for genome-wide association mapping studies. Variations at the sequence level are associated with observed variation for a trait in natural populations, improving the precision and efficiency for predicting phenotypes from genotypes. By screening a large number of individuals, it would be possible to compare sequence variation across species and genomes of palm. It will also be easier to exploit genomic selection for identification of candidate genes and their deployment through genomics-assisted breeding.

18.14 Sustainability of Oil Palm Production

Whenever new lands are cleared for agriculture, it is done at the expense of previous vegetation and the animals occupying that land space. The newly established agricultural ecosystems are usually narrower, supporting fewer life species and excluding a multitude of original inhabitants. Destruction of the prairie in the USA, elimination of tropical forests in South-east Asia or clearing of Amazon forests for agriculture all had adverse ecological outcomes, but since tropical ecosystems support larger and diverse life forms, consequences are more apparent and severe in their case. Deforestation has impacted many species that are rare and threatened.

Almost all recent deforestation concerns have been centred in developing countries. It is an ironic fact that rapid human population growth has taken place in the developing countries where agriculture remains the primary occupation. The current agricultural hectarage in these countries is not enough to provide agriculture-based livelihood to a large new workforce, nor have the countries been successful in finding alternative livelihoods. Therefore, encroachment by human beings into the tropical forests has been inevitable. However, the manner in which these forest clearings have been carried out with indiscriminate destruction of natural fauna and flora and near extinction of several rare species of animals and plants is inexcusable. Establishment of large tree plantations during the same time frame without environmental guidelines has exacerbated the problem further. But a substantial portion of blame must also be shared by the lumber interests, which convinced the local population for forest clearing with the promise of profitable plantations and left land unattended after clear felling to the vagaries of nature. It was fortunate that environmental groups raised the alarm over these destructive occurrences. The worldwide attention to the negative consequences of plantation cropping in South-east Asia had its impact. As these plantation crops are grown mainly for export, the producing countries are more amenable to put into place measures to ensure environmentally friendly production practices to ward against restrictions to their commerce by importing countries. Several certification programmes have been recently initiated to guarantee eco-friendly production systems with due safeguard for environmental concerns.

When the deforestation is considered in particular reference to oil palm, after a reasonable analysis of the situation, it becomes apparent that the damage claims are exaggerated. Most of the land brought under oil palm plantations is previously cleared land to establish plantation crops in Asia and pastures in Latin America, and crop shift resulted from greater profitability of oil palm. Plantation crops such as oil palm originally constituted part of a flourishing forest system with its own ecosystem of thriving fauna and flora, thus in the managed planting environment they should qualify as 'planted forests' similar to managed tree farming in developed countries. The oil palm planted forests should be able to support a wide variety of fauna and flora because of the tropical environment, albeit not the same as in the natural forest. Mielke (2001) pointed out that going by the FAO projections of deforestation to the year 2020 and assuming ambitious expansion calculations in hectarage without any gain in per unit yield, the oil palm share of the deforestation will amount to approximately 2.6% of expected forest loss. Clearly, deforestation is a problem of much wider scope and calls for comprehensive rather than piece-meal solutions.

The biofuel-producing countries have identified eroded and poor fertility soils unfit for food crop production and the focus is to confine expansion of biofuel feedstock crops to such lands. A sufficient knowledge base is also available to implement measures such as contouring to prevent future erosion of soil and use of legumes and crop waste to enrich the soil to ensure long-term sustainability of oil palm production systems. The emphasis has shifted from bringing more hectarage under oil palm to increased yield to get the existing hectarage under the trees to produce more, and the results have been promising. The prospect of nutrient recycling is also very promising for oil palm as oil contains very low amounts of nutrients and with the emphasis of returning dead biomass, effluents and wastes from processing plants on to the plantation, much of the mineral nutrients can be returned for reuse and the added biomass can support soil microbial populations effective in fixing or increasing the availability of soil nutrients to the plant.

The perennial and evergreen nature of oil palm makes it more environmentally friendly than annual oil crops. Once planted the soil is left undisturbed to preserve soil texture and structure, prevent loss of soil through disturbance and allow the soil to reach microbial equilibrium. The plant can make use of solar radiation year-round to assimilate more photosynthate in a smaller land area compared to plants with an annual growth habit. Oil palm is the highest producing oil crop. The capability to produce high yields of oil palm in itself is a sustainable trait as a high yielding crop can produce similar yields on far less land than needed for lower yielding crops. It should always be borne in mind that no one feedstock can meet the needs of the biofuel industry. The hectarage, location and the feedstock species has to be decided using the best decision-making tools available.

18.15 Conclusion

Oil palm is considered to support multiple sectors of society and economy. It is a multifaceted crop because of the value-added by-products obtained from it, such as the fibre and left-over biomass, which can be used in generating heat energy. The left over palm kernel after oil extraction can be used in making feeds for broilers.

Palm oil is a preferred biodiesel feedstock because of the high yield and year-round production ability of the oil palm tree. However, safeguards and precautions are in order to ensure its production in an ecofriendly manner and regulatory mechanisms are also essential to avoid the food versus fuel conflict.

There is ample opportunity to increase oil palm yields especially through genetic improvement. The present ongoing research needs to concentrate on several important traits to scale-up the production. For example, monogenic controlled traits such as shell thickness and fruit colour can be focused to develop SNP markers through the next-generation sequencing systems, which has not been utilized to the full extent in this crop. This will also increase the genomic resource for genomics-assisted selection. As two palms (oil palm and date palm) have been completely sequenced, this will make for easy access to harness benefits out of genomic studies. Although oil palm functional genomics is still in its infancy, it should see accelerated outcomes with the present trend in transcriptomics and metabolomics studies, aided by the rapidly evolving automated technologies.

References

Adam, H., Jouannic, S., Escoute, J., Duval, Y., Verdeil, J.-L. and Tregear, J.W. (2005) Reproductive developmental complexity in the African oil palm (*Elaeis guineensis*, Arecaceae). *American Journal of Botany* 92, 1836–1852.

Adam, H., Collin, M., Richaud, F., Beulé, T., Cros, D. *et al.* (2011) Environmental regulation of sex determination in oil palm: current knowledge and insights from other species. *Annals of Botany* 108, 1529–1537.

Alkabbashi, A.N., Alam, M.Z., Mirghani, M.E.S. and Al-Fusaiel, A.M.A. (2009) Biodiesel production from crude palm oil by transesterification process. *Journal of Applied Sciences* 9, 3166–3170.

Ataga, C.D. and van der Vossen, H.A.M. (2007) *Elaeis guineensis* Jacq. In: vander Vossen, H.A.M. and Mkamilo, G.S. (eds) *Vegetable Oils/Oléagineux.* PROTA 14: Vegetable oils/Oléagineux (CD-Rom). PROTA, Wageningen, the Netherlands.

Bao, X. and Ohlrogge, J. (1999) Supply of fatty acid is one limiting factor in the accumulation of triacylglycerol in developing embryos. *Plant Physiology* 120, 1057–1062.

Barcelos, E., Amblard, P., Berthaud, J. and Seguin, M. (2002) Genetic diversity and relationship in American and African oil palm as revealed by RFLP and AFLP molecular markers. *Pesquisa Agropecuária Brasileira* 37, 1105–1114.

Basiron, Y. (2012) Where lies the future for Malaysian palm oil and rubber industries. Presentation by YB Tan Sri Datuk Dr Yusof Basiron, CEO of Malaysian Palm Oil Council (MPOC) for seminar at IKMAS, Universiti Kebangsaan Malaysia, Bangi, 19 April, 2012. Available at: http://www.mpoc.org.my/upload/The-Future-of-Malaysian-Palm-Oil-and-Rubber-Industry.pdf (accessed 1 July 2012).

Baudouin, L. (1992) Use of molecular markers for oil palm breeding. I. Protein markers. *Oleagineaux* 47, 681–691.

Beirnaert, A. and Vanderweyen, R. (1941) *Contribution B l'ktude gknktique et biomktrique des variktks d'* Elaeis guineensis *Jacq*. Publications de l'institut national pour l'ktude agronomique du Congo Belge, skrie scientifique, No. 27.

Bilal, M. (2004) Genetics of the Deli-Avros breeding populations of the oil palm (*Elaeis guineensis Jacq.*). MSc thesis, University Malaysia Pahang.

Billotte, N., Risterucci, A.M., Barcelos, E., Noyer, J.L., Amblard, P. *et al.* (2001) Development, characterization, and across-taxa utility of oil palm (*Elaeis guineensis* Jacq.) microsatellite markers. *Genome* 44, 413–425.

Billotte, N., Marseillac, N., Risterucci, A.M., Adon, B., Brottier, P. *et al.* (2005) Microsatellite-based high density linkage map in oil palm (*Elaeis guineensis* Jacq.). *Theoretical and Applied Genetics* 110, 754–765.

Billotte, N., Jourjon, M.F., Marseillac, N., Berger, A., Flori, A. *et al.* (2010) QTL detection by multi-parent linkage mapping in oil palm (*Elaeis guineensis* Jacq.). *Theoretical and Applied Genetics* 120, 1673–1687.

Borchert, R., Renner, S.S., Calle, Z., Navarrete, D., Tye, A. *et al.* (2005) Photoperiodic induction of synchronous flowering near the Equator. *Nature* 433, 627–629.

Bourgis, F., Kilaru, A., Cao, X., Ngando-Ebongue, G.-F., Drira, N. *et al.* (2011) Comparative transcriptome and metabolite analysis of oil palm and date palm mesocarp that differ dramatically in carbon partitioning. *Proceedings of the National Academy of Sciences of the United States of America* 108, 12527–12532.

Breure, C.J., Menendez, T. and Powell, M.S. (1990) The effect of planting density on the yield components of oil palm (*Eleais guineensis* Jacq.). *Experimental Agriculture* 26, 117–124.

Caliman, J.P. and Southworth, A. (1998) Effect of drought and haze on the performance of oil palm. In: Jatmika, A. *et al.* (eds) *Proceedings 1998 International Oil Palm Conference, 'Commodity of the Past, Today and the Future'.* Indonesian Oil Palm Research Institute, Medan, Indonesia, pp. 250–274.

Chapman, K.D. and Ohlrogge, J.B. (2012) Compartmentation of triacylglycerol accumulation in plants. *Journal of Biological Chemistry* 287, 2288–2294.

Chapman, K.D., Dyer, J.M. and Mullen, R.T. (2012) Biogenesis and functions of lipid droplets in plants. *Journal of Lipid Research* 53, 215–226.

Cheng-Li, L.O., Ti, L.L.E. and Singh, R. (2011) SNP markers for genetic studies and prediction of monogenic traits in oil palm. *MPOB information series* MPOB TT No. 482.

Coleman, R.A. and Lee, D.P. (2004) Enzymes of triacylglycerol synthesis and their regulation. *Progress in Lipid Research* 43, 134–176.

Corley, R.H.V. (1976) Inflorescence abortion and sex differentiation. In: Corley, R.H.V., Hardon, J.J. and Wood, B.J. (eds) *Developments in Crop Science 1: Oil Palm Research.* Elsevier, Amsterdam, pp. 37–54.

Corley, R.H.V. and Donough, C.R. (1995) Effects of defoliation on sex differentiation in oil palm clones. *Experimental Agriculture* 31, 177–189.
Corley, R.H.V. and Tinker, P.B. (2007) *The Oil Palm*, 4th edn. Blackwell Publishing, Oxford, UK, 592 pp.
Corley, R.H.V., Gray, B.S. and Ng, S.K. (1971) Productivity of the oil palm (*Eleais guineensis Jacq.*) in Malaysia. *Experimental Agriculture* 7, 129–136.
Corley, R.H.V., Lee, C.H., Law, I.H. and Cundall, E. (1986) Abnormal flower development in oil palm clones. In: *Proceedings 1987 International Oil Palm Conference*. Palm Oil Research Institute Malaysia, Kuala Lumpur, pp. 173–185.
Dufrene, E. and Saugier, B. (1993) Gas exchange of oil palm in relation to light, vapour pressure deficit, temperature and leaf age. *Functional Ecology* 7, 97–104.
Durrett, T.P., Benning, C. and Ohlrogge, J. (2008) Plant triacylglycerols as feedstocks for the production of biofuels. *The Plant Journal* 54, 593–607.
Dyer, J.M. and Mullen, R.T. (2008) Engineering plant oils as high-value industrial feedstocks for biorefining: the need for underpinning cell biology research. *Physiologia Plantarum* 132, 11–22.
Elshire, R.J., Glaubitz, J.C., Sun, Q., Poland, J.A., Kawamoto, K. *et al.* (2011) A robust, simple genotyping-by-sequencing (GBS) approach for high diversity species. *PLoS ONE 6: e19379.*
Eric, L.S., Donald, R.O. and Evan, H.D. (2000) Diurnal regulation of photosynthesis in understory saplings. *New Phytologist* 145, 39–49.
FAO (2010) FAO Crop Statistics. Available at: http://www.fao.org/corp/statistics/en (accessed 15 May 2012).
FAS-USDA (2012) Foreign Agricultural Service, US Department of Agriculture. Available at: http://www.fas.usda.gov/oilseeds/Current (accessed 1 July 2012).
Focks, N. and Benning, C. (1998) wrinkled1: A novel, lowseed- oil mutant of Arabidopsis with a deficiency in the seed-specific regulation of carbohydrate metabolism. *Plant Physiology* 118, 91–101.
Gardner, C.J., Quinn, N.W.T., Gerpen, J.V. and Simonpietri, J. (2010) Oilseed and Algal Oils as Biofuel Feedstocks. Available at: http://www.swcs.org/documents/resources/Chapter_8__Gardner__Oilseed_B191CC8A1C9DB.pdf (accessed 1 July 2012).
Gascon, J.P., Noiret, J.M. and Meunier, J. (1989) Oil palm. In: Röbbelen, G., Downey, R.K. and Ashri, A. (eds) *Oil Crops of the World*. McGraw-Hill Publishing, New York, pp. 475–493.
Goh, K.J. (2000) Climatic requirements of the oil palm for high yields. In: Goh, K.J. (ed.) *Managing Oil Palms for High Yields: Agronomic Principles*. Malaysian Society of Soil Science and Param Agricultural Surveys, Kula Lumpur, Malaysia, pp. 1–17.
Gunstone, F.D. (2011) Production and trade of vegetable oils. In: Gunstone, F.D. (ed.) *Vegetable Oils in Food Technology: Composition, Properties and Uses*, 2nd edn. Blackwell Publishing, Oxford, UK, pp. 1–24.
Hardon. J.J., Rao, V. and Rajanaidu, N. (1985) A review of oil palm breeding. In: Russel, G.E. (ed) *Progress in Plant Breeding*. Butterworths, London, pp. 139–163.
Hartley, C.W.S. (1988) *The Oil Palm*, 3rd edn. Longman, London.
Hayati, A., Wickneswari, R., Maizura, I. and Rajanaidu, N. (2004) Genetic diversity of oil palm (*Elaeis guineensis* Jacq.) germplasm collections from Africa: implications for improvement and conservation of genetic resources. *Theoretical and Applied Genetics* 108, 1274–1284.
Henson, I.E., Change, K.C., Mustakim, S.N.A., Chai, S.H., Hasnuddin, M.Y. and Zakaria, A. (1999) The oil palm trunk as a carbohydrate reserve. *Journal of Oil Palm Research* 11, 98–113.
Huang, A.H.C. (1992) Oil bodies and oleosins in seeds. *Annual Reviews of Plant Physiology and Plant Molecular Biology* 43, 177–200.
Ibrahim, M., Jaafar, H., Harun, M. and Yusop, M. (2010) Changes in growth and photosynthetic patterns of oil palm (*Elaeis guineensis* Jacq.) seedlings exposed to short-term CO_2 enrichment in a closed top chamber. *Acta Physiologiae Plantarum* 32, 305–313.
Jack, P.L. and Mayes, S. (1993) Use of molecular markers for oil palm breeding. II. Use of DNA markers (RFLPs). *Oleagineux* 48, 1–8.
Jack, P.L., Dimitrejevic, T.A.F. and Mayes, S. (1995) Assessment of nuclear, mitochondria1 and chloroplast RFLP markers in oil palm (*Elaeis guineensis* Jacq.). *Theoretical and Applied Genetics* 90, 643–649.
Jack, P.L., James, C., Price, Z., Groves, L., Corley, R.H.V., Nelson, S. and Rao, V. (1998) Application of DNA markers in oil palm breeding. In: *Proceedings of International Oil Palm Conference*, Indonesian Oil Palm Research Institute, Medan, Indonesia, pp. 315–324.
Jalani, B.S. and Ariffin, D. (1994) Quality practices in R&D in the plantation sector. In: *Proceedings of the Palm Oil Quality Conference*, Palm Oil Research Institute of Malaysia, pp. 42–52.
Jalani, B.S. and Rajanaidu, N. (1992) Oil palm improvement and productivity. In: *The Impact of Biological Research on Agricultural Productivity*. Proceedings SABRAO International Symposium Taichung, Taiwan, 10–13 March, 1992.

Johnston, M. and Hollaway, T. (2007) A global comparison of national biodiesel production potentials. *Environmental Science and Technology* 41, 7967–7973.
Johnston, M., Foley, J.A., Holloway, T., Kucharik, C. and Monfreda, C. (2009) Resetting global expectations from agricultural biofuels. *Environmental Research Letters* 4: doi:10.1088/1748-9326/4/1/014004.
Jones, L.H. (1989) Prospects for biotechnology in oil palm (*Elaeis guineensis* Jacq.) and coconut (*Cocos nucifera*) improvement. *Biotechnology and Genetic Engineering Reviews* 7, 281–296.
Jones, L.H. (1997) The effects of leaf pruning and other stresses on sex determination in the oil palm and their representation by a computer simulation. *Journal of Theoretical Biology* 187, 241–260.
Jones, L.H. and Hughes, W.A. (1988) Oil Palm (*Elaeis guineensis* Jacq.). In: Bajaj, Y.P.S. (ed.) *Biotechnology in Agriculture and Forestry, 5: Trees II*. Springer-Verlag, Berlin, Germany, pp. 176–202.
Kalam, M.A. and Masjuki, H.H. (2002) Biodiesel from palmoil – an analysis of its properties and potential. *Biomass and Bioenergy* 23, 471–479.
Kushairi, A., Rajanaidu, N., Jalani, B.S., Rafii, M.Y. and Mohd, D.A. (1999) PORIM Oil palm planting materials. *PORIM Bulletin* 38, 1–13.
Kushairi, A., Tarmizi, A.H., Zamzuri, I., Ong-Abdullah, M., Samsul Kamal, R. *et al.* (2010) Production, Performance and Advances in Oil Palm Tissue Culture. Paper presented at ISOPB International Seminar on Advances in Oil Palm Tissue Culture, Yogyakarta Indonesia, 29 May 2010.
Lam, K.M., Tan, K.T., Lee, K.T. and Mohamed, A.R. (2009) Malaysian palm oil: surviving the food versus fuel dispute for a sustainable future. *Renewable and Sustainable Energy Reviews* 13, 1456–1464.
Latiff, A. (2000) The biology of the genus *Elaeis*. In: Basiron, Y., Jalani, B.S. and Chan, K.W. (eds) *Advances in Oil Palm Research*, Vol. 1. Malaysian Palm Oil Board, pp. 19–38.
Legros, S., Mialet-Serra, I., Caliman, J.-P., Siregar, F.A., Clement-Vidal, A. *et al.* (2009) Phenology, growth and physiological adjustments of oil palm (*Elaeis guineensis*) to sink limitation induced by fruit pruning. *Annals of Botany* 104, 1183–1194.
Li-Beisson, Y., Shorrosh, B., Beisson, F., Andersson, M.X., Arondel, V. *et al.* (2010) Acyl lipid metabolism. In: Last, R. (ed.) *The Arabidopsis Book*. American Society of Plant Biologists, Rockville, Maryland.
Low, E.-T., Alias, H., Boon, S.-H., Shariff, E., Tan, C.-Y. *et al.* (2008) Oil palm (*Elaeis guineensis* Jacq.) tissue culture ESTs: Identifying genes associated with callogenesis and embryogenesis. *BMC Plant Biology* 8, 62.
Maizura, I., Rajanaidu, N., Zakri, A. and Cheah, S. (2006) Assessment of genetic diversity in oil palm (*Elaeis guineensis* Jacq.) using restriction fragment length polymorphism (RFLP). *Genetic Resources and Crop Evolution* 53, 187–195.
Marchetti, J.M., Miguel, V.U. and Errazu, A.F. (2007) Possible methods for biodiesel production. *Renewable and Sustainable Energy Reviews* 11, 1300–1311.
Maria, M., Clyde, M.M. and Cheah, S.C. (1995) Cytological analysis of *Elaeis guineensis* (tenera) chromosomes. *Elaeis* 7, 122–134.
May, C.Y., Liang, Y.C., Foon, C.S., Ngan, M.A., Hook, C.C. *et al.* (2005) Key fuel properties of palm oil alkyl esters. *Fuel* 84, 1717–1720.
Mayes, S., James, C.M., Horner, S.F., Jack, P.L. and Corley, R.H.V. (1996) The application of restriction fragment length polymorphism for the genetic fingerprinting of oil palm (*Elaeis guineensis* Jacq.). *Molecular Breeding* 2, 175–180.
Mayes, S., Jack, P.L., Marshall, D.F. and Corley, R.H.V. (1997) Construction of a RFLP genetic linkage map for oil palm (*Elaeis guineensis* Jacq.). *Genome* 40, 116–122.
Mayes, S., Jack, P.L. and Corley, R.H.V. (2000) The use of molecular markers to investigate the genetic structure of an oil palm breeding programme. *Heredity* 85, 288–293.
Mayes, S., Hafeez, F., Price, Z., MacDonald, D., Billotte, N. *et al.* (2008) Molecular research in oil palm, the key oil crop for the future. In: Moore, P.H. and Ming, R. (eds) *Genomics of Tropical Crop Plants*. Springer, New York, pp. 371–404.
Meunier, J. (1976) Les prospections de Palmacées. Une nécessité pour l'amélioration des palmiers oléagineux. *Oléagineux* 31, 153–157.
Meuwissen, T.H.E., Hayes, B.J. and Goddard, M.E. (2001) Prediction of total genetic value using genome-wide dense marker maps. *Genetics* 157, 1819–1829.
Mielke, S. (2001) Can production of oils and fats keep pace with future production? Paper presented at the ISF International Congress, Berlin, 20 September 2001.
Miller, M., Dunham, J., Amores, A., Cresko, W. and Johnson, E. (2007) Rapid and cost effective polymorphism identification and genotyping using restriction site associated DNA (RAD) markers. *Genome Research* 17, 240–248.
Moretzsohn, M.C., Nunes, C.D.M., Ferreira, M.E. and Grattapaglia, D. (2000) RAPD linkage mapping of the shell thickness locus in oil palm (*Elaeis guineensis* Jacq.). *Theoretical and Applied Genetics* 100, 63–70.

Murphy, D.J. (2003) Working to improve the oil palm crop. *International News on Fats, Oils and Related Materials* 14, 670–671.

Murphy, D.J. (2007) Future prospects for oil palm in the 21st century: biological and related challenges. *European Journal of Lipid Science and Technology* 109, 296–306.

Nagi, J., Ahmedand, S.K. and Nagi, F. (2008) Palm biodiesel an alternative green renewable energy for the energy demands of the future. *International Conference on Construction and Building Technology* 7, 79–94.

Ochs, R. and Daniel, C. (1976) Research on techniques adapted to dry regions. In: Corley, R.H.V., Hardon, J.J. and Wood, B.J. (eds) *Oil Palm Research*. Elsevier, Amsterdam, pp. 315–330.

Ohlrogge, J.B. and Chapman, K. (2011) The seeds of green energy – expanding the contribution of plant oils as biofuels. *The Biochemist* 33, 34–38.

Ohlrogge, J.B. and Jaworski, J.G. (1997) Regulation of fatty acid biosynthesis. *Annual Review of Plant Physiology and* Plant *Molecular Biology* 48, 109–136.

Oil World (2012) *Oil World Reports*. ISTA Mielke GmbH, Hamburg, Germany.

Okwuagwu, C.O. and Tai, G.C.C. (1995) Estimation of variance components and heritability of bunch yield and yield components in the oil palm (*Elaeis guineensis* Jacq.). *Plant Breeding* 114, 463–465.

Osagie, A.U. and Bafor, M.E. (1990) Triacylglycerols of oil palm (*Elaeis guineensis* var. *dura*) mesocarp during fruit maturation. *Biochemistry and Cell Biology* 68, 313–317.

Pahl, G. (2005) *Biodiesel: Growing a New Energy Economy*. Chelsea Green Publishing Company, White River Junction, Vermont, 281 pp.

Parveez, G.K.A., Abrizah, O., Umi Salamah, R., Masani, A.M.Y., Hashim, A.T. *et al.* (2009) Improving value of oil palm using genetic engineering. In: Hou, C.T. (ed.) *Biocatalysis and Molecular Engineering*. John Wiley and Sons, Hoboken, New Jersey.

Paux, E., Sourdille, P., Mackay, I. and Feuillet, C. (2011) Sequence-based marker development in wheat: advances and applications to breeding. *Biotechnology Advances* doi:10.1016/j.biotechadv.2011.09.015.

Pinzi, S., Garcia, I.L., Lopez-Gimenez, F.J., Luque de Castro, M.D., Dorado, G. *et al.* (2009) The ideal vegetable oil-based biodiesel composition: a review of social, economical and technical implications. *Energy and Fuels* 23, 2325–2341.

Poku, K. (2002) Small-scale palm oil processing in Africa. *FAO Agricultural Services Bulletin* 148. FAO, Rome. Available at: ftp://ftp.fao.org/docrep/fao/005/y4355E/y4355E00.htm (accessed 8 June 2012).

Purba, A.R., Noyer, J.L., Baudouin, L., Perrier, X., Hamon, S. and Lagoda, P.J.L. (2000) A new aspect of genetic diversity of Indonesian oil palm (*Elaeis guineensis* Jacq.) revealed by isoenzyme and AFLP markers and its consequences for breeding. *Theoretical and Applied Genetics* 101, 956–961.

Purseglove, J.W. (1972) *Tropical Crops. Monocotyledons*. Longman, London, 607 pp.

Rajanaidu, N., Arasu, N.T. and Obasola, C.O. (1979) Collection of oil palm (*Elaeis guineensis*) genetic material in Nigeria II. Phenotypic variation of natural populations. *MARDI Research Bulletin* 7, 1–27.

Ramli, U.S., Baker, D.S., Quant, P.A. and Harwood, J.L. (2002a) Control mechanisms operating for lipid biosynthesis differ in oil palm (*Elaeis guineensis* Jacq.) and olive (*Olea europaea* L.) callus cultures. *Biochemistry Journal* 364, 385–391.

Ramli, U.S., Baker, D.S., Quant, P.A. and Harwood, J.L. (2002b) Control analysis of lipid biosynthesis in tissue cultures from oil crops shows that flux control is shared between fatty acid synthesis and lipid assembly. *Biochemistry Journal* 364, 393–401.

Ramli, U.S., Salas, J.J., Quant, P.A. and Harwood, J.L. (2005) Metabolic control analysis reveals an important role for diacylglycerol acyltransferase in olive but not in oil palm lipid accumulation. *FEBS Journal* 272, 5764–5770.

Ramli, U.S., Salas, J.J., Quant, P.A. and Harwood, J.L. (2009) Use of metabolic control analysis to give quantitative information on control of lipid biosynthesis in the important oil crop, *Elaeis guineensis* (oil palm). *New Phytologist* 184, 330–339.

Rance, K.A., Mayes, S., Price, Z., Jack, P.L. and Corley, H.R.V. (2001) Quantitative trait loci for yield components in oil palm (*Elaeis guineensis* Jacq.). *Theoretical and Applied Genetics* 103, 1302–1310.

Riju, A. and Arunachalam, V. (2009) Interspecific differences in single nucleotide polymorphisms (SNPs) and indels in expressed sequence tag libraries of oil palm *Elaeis guineensis* and *E. oleifera*. *Nature Proceedings*. Available at: http://hdl.handle.net/10101/npre.2009.3593.1 (accessed 2 July 2012).

Riju, A., Chandraseker, A. and Arunachalam, V. (2007) Mining for single nucleotide poly-morphisms and insertions/deletions in expressed sequence tag libraries of oil palm. *Bioinformation* 2, 128–131.

Rival, A. and Jaligot, E. (2010) Oil palm biotechnologies are definitely out of infancy. *Oléagineux corps gras lipides* 17, 368–374.

Rival, A., Bertrand, L., Beulé, T., Combes, M.C., Trouslot, P. and Lashermes, P. (1998) Suitability of RAPD analysis for the detection of somaclonal variants in oil palm (*Elaeis guineensis* Jacq). *Plant Breeding* 117, 73–76.

Rival, A., Beulé, T., Barre, P., Hamon, S., Duval, Y. and Noirot, M. (1997) Comparative flow cytometric estimation of nuclear DNA content in oil palm (*Elaeis guineensis* Jacq.) tissue cultures and seed-derived plants. *Plant Cell Reports* 16, 884–887.

Salunkhe, D.K., Chavan, J.K., Adsule, R.N. and Kadam, S.S. (1992) *World Oilseeds: Chemistry, Technology, and Utilization*. Van Nostrand Reinhold, New York, 554 pp.

Sambanthamurthi, R., Sundram, K. and Tan, Y.A. (2000) Chemistry and biochemistry of palm oil. *Progress in Lipid Research* 39, 507–558.

Sambanthamurthi, R., Singh, R., Parveez, G.K.A., Meilina, O.A. and Kushairi, A. (2009) Opportunities for the oil palm via breeding and biotechnology. In: Jain, S.M. and Priyadarshan, P.M. (eds) *Breeding Plantation Tree Crops: Tropical Species*. Springer, New York.

Seng, T.-Y., Mohamed, Saad, S.H., Chin, C.-W., Ting, N.-C., Harminder Singh, R.S. *et al.* (2011) Genetic linkage map of a high yielding FELDA Deli×Yangambi oil palm cross. *PLoS ONE* 6:e26593.

Shah, F.H., Zuliaha, S. and Omar, R. (1993) Polymorphism of rDNA locus of oil palm: a preliminary investigation. *Proceedings of Biochemistry Society Symposium* 17, 81–84.

Shah, F.H., Rashid, O., Simons, A.J. and Dunsdon, A. (1994) The utility of RAPD markers for the determination of genetic variation in oil palm (Efaeirguineemir). *Theoritical and Applied Genetics* 89, 713–718.

Sharmila, A. (2010) Development of heterogeneous base catalyst for transesterfication reaction. MSc thesis, University of Malaysia, Pahang, Malaysia.

Singh, R., Zaki, N., Ting, N.-C., Rosli, R., Tan, S.-G. *et al.* (2008a) Exploiting an oil palm EST database for the development of gene-derived SSR markers and their exploitation for assessment of genetic diversity. *Biologia* 63, 227–235.

Singh, R., Tan, S.G., Panandam, J.M., Rahman, R.A. and Cheah, S.C. (2008b) Identification of cDNA-RFLP markers and their use for molecular mapping in oil palm (*Elaeis guineensis*). *Journal of Molecular Biology and Biotechnology* 16, 53–63.

Singh, R., Ti, L.L.E.I., Cheng Li, L.O., Lan, C.P., Abdullah, M.O. and Ghulam Kadir, A.P. (2009a) Exploitation of Gene Rich Sequences for Improved Oil Palm Productivity. Available at: http://isopb.org/docs/2009/P3%20Rajinder%20Paper_Exploitation%20of%20gene%20rich%20sequencies%20for%20improved%20oil%20palm%20productivity%20Rajinde.pdf (accessed 2 July 2012).

Singh, R., Tan, S., Panandam, J., Rahman, R., Ooi, L. *et al.* (2009b) Mapping quantitative trait loci (QTLs) for fatty acid composition in an interspecific cross of oil palm. *BMC Plant Biology* 9, 114.

Soh, A.C. and Tan, S.T. (1983) Estimation of genetic variance, heritability and combining ability in oil palm breeding. In: Yap T.C., Graham, K.M. and Sukaimi, J. (eds) *Crop Improvement Research*. Bangi, Selangor, Malaya, pp. 379–388.

Squire, G. and Corley, R.H.V. (1987) Oil Palm. In: Sethuraj, M.R. and Raghavendra, A.S. (eds) *Tree Crop Physiology*. Elsevier Science, Amsterdam, pp. 141–167.

Syed, R.A. (1979) Studies on oil palm pollination by insects. *Bulletin of Entomological Research* 69, 213–224.

Tan, K.T. and Lee, K.T. (2011) A review on supercritical fluids (SCF) technology in sustainable biodiesel production: potential and challenges. *Renewable and Sustainable Energy Reviews* 15, 2452–2456.

Tan, K.T., Lee, K.T., Mohamed, A.R. and Bhatia, S. (2009) Palm oil: addressing issues and towards sustainable development. *Renewable and Sustainable Energy* 13, 420–427.

Taufiq-Yap, Y.H., Abdullah, Y.H. and Basrim, N.F. (2011) Biodiesel production via transesterification of palm oil using $NaOH/Al_2O_3$ catalysts. *Sains Malaysiana* 40, 587–594.

Thawaro, S. and Te-chato, S. (2009) Application of molecular markers in the hybrid verification and assessment of somaclonal variation from oil palm propagated in vitro. *ScienceAsia* 35, 142–149. Available at: http://www.scienceasia.org/2009.35.n2/scias35_142.pdf (accessed 2 July 2012).

Ting, N.-C., Zaki, N., Rosli, R., Low, E.-T., Ithnin, M. *et al.* (2010) SSR mining in oil palm EST database: application in oil palm germplasm diversity studies. *Journal of Genetics* 89, 135–145.

Tranbarger, T.J., Dussert, S., Joët, T., Argout, X., Summo, M. *et al.* (2011) Regulatory mechanisms underlying oil palm fruit mesocarp maturation, ripening, and functional specialization in lipid and carotenoid metabolism. *Plant Physiology* 156, 564–584.

Tranbarger, T.J., Roongsattham, P., Jantasuriyarat, C., Dussert, S., Joët, T. *et al.* (2012) Exploring the oil palm transcriptome: mechanisms underlying reproductive development. *Plant and Animal Genome Conference XX*, 14–18 January 2012, San Diego, California.

TSH (2012) TSH Resources Berhad. Available at: http://www.tsh.com.my/web/images/stories/pdf/re9-tsh_brochure.pdf (accessed 2 July 2012).
Turner, P.D. (1981) *Oil Palm Diseases and Disorders*. Oxford University Press, Kuala Lumpur.
Voelker, T. (2011) Secrets of palm oil biosynthesis revealed. *Proceedings of the National Academy of Sciences* 108, 12193–12194.
Wahid, M.B. (2009) Sequencing the oil palm genome. In: *The beginning, The International Palm Oil Congress, Biotechnology Sustainability Conference* Kuala Lumpur, Malaysia, 9–12 November 2009, 1, 3.
Wahid, M.B., Abdullah, S.N.A. and Henson, I.E. (2005) Oil palm-achievements and potential. *Plant Production Science* 8, 288–297.
Williams, C.N. and Thomas, R.L. (1970) Observations on sex differentiation in the oil palm, *Elaeis guineensis* L. *Annals of Botany* 34, 937–963.
Wong, C.K. and Bernardo, R. (2008) Genomewide selection in oil palm: increasing selection gain per unit time and cost with small populations. *Theoretical and Applied Genetics* 116, 815–824.
World Oil Outlook (2011) The OPEC Secretariat. Available at: http://www.opec.org/opec_web/static_files_project/media/downloads/publications/WOO_2011.pdf (accessed 2 July 2012).
Xie, C.Q. and Xu, S.Z. (1998) Efficiency of multistage marker-assisted selection in the improvement of multiple quantitative traits. *Heredity* 80, 489–498.
Yee, K.F., Kansedo, J. and Lee, K.T. (2010) Biodiesel production from palm oil via heterogeneous transesterification: optimization study. *Chemical Engineering Communications* 197, 1597–1611.
Zaki, N.M., Singh, R., Rosli, R. and Ismail, I. (2012) *Elaeis oleifera* genomic-SSR markers: exploitation in oil palm germplasm diversity and cross-amplification in Arecaceae. *International Journal of Molecular Sciences* 13, 4069–4088.

19 Willow

Martin Weih
Department of Crop Production Ecology, Swedish University of Agricultural Sciences, Sweden

19.1 Introduction

Various willow species (genus *Salix*) are part of the natural flora especially in temperate-arctic regions, and products from naturally grown willows have been used traditionally in many countries of the world for centuries, e.g. for basket-making, fencing and small-scale husbandry. Willows have recently been undergoing a renaissance as biomass crops for bioenergy and bioproducts, for production of wood and for environmental purposes (Verwijst, 2001; Ball *et al.*, 2005). Especially when grown for energy purposes, intensively managed plantations of fast-growing trees on agricultural land are gaining increasing interest in many countries of the world, due to their efficient and sustainable land use in combination with an increasing demand for renewable energy sources (Volk *et al.*, 2004; Weih, 2004; Karp and Shield, 2008) and the additional possibilities for environmental control (phytoremediation, Crites, 1984; Aronsson and Perttu, 2001). For example in Sweden, willow cultivation as a source of biomass for energy purposes has been developed within the framework of a National Energy Forestry Programme (Sirén *et al.*, 1987). Based on these early activities, appropriate production systems were developed and commercial biomass plantations were established in Sweden. Thus in Sweden, the culture of willow has been most important during recent years and in contrast to most other countries in the world, complete biomass chains for the commercial production of biomass from willow short-rotation plantations are in place.

Life cycle analysis of willow biomass production on agricultural land based on commercial clones and cultivation practices in Sweden revealed high energy efficiency of the production system, because the produced biomass yields 20 times more energy compared to the energy input, a figure which is more than 3 times higher than the energy balance (i.e. energy output/energy input) of many annual energy crops such as wheat and rape (Keoleian and Volk, 2005; Börjesson, 2006). In addition, plantations of fast-growing willows require a reduced chemical input when compared to conventional arable crops, can provide an alternative use of agricultural land and can enhance the local environment through increased biodiversity. Thus, agricultural land cropped with fast-growing willows can become an important supplier of energy with marginal contribution to the increasing quantity of greenhouse gases in the atmosphere. These plantations can sequester large amounts of carbon (Grogan and Matthews, 2002; Baum *et al.*, 2009a) and may have a great potential to contribute to carbon-managed future economies in terms of fossil fuel replacement and carbon sequestration (Hall and House, 1994; Tuskan and Walsh, 2001; Styles and Jones, 2007).

©CAB International 2013. *Biofuel Crops: Production, Physiology and Genetics* (ed. B.P. Singh)

About 450,000 ha indigenous willow stands and 133,000 ha willow plantations (Fig. 19.1) are currently grown worldwide, in comparison to about 75 million ha of indigenous or planted poplar stands grown worldwide (FAO, 2008). Thus, the area of willow grown for biomass is still rather small, but may grow rapidly in the future due to a generally increasing need for biomass and some specific advantages of willows compared to other trees grown for biomass (e.g. poplars and aspens), such as ease of vegetative propagation, short cutting cycles, and possibly support of more biodiversity in some landscapes. In this chapter, an overview of the biology and taxonomy of willows will be followed by a short presentation of production systems and breeding issues with respect to willow use as biomass crop, and finally a discussion of environmental and landscape design issues including the possibility for environmental control by means of phytoremediation.

19.2 The Species

The genus *Salix* Linnaeus (willow, osier) belongs, together with *Populus*, to the family *Salicaceae* s.str., order *Salicales*, class *Magnoliopsida*, subclass *Dilleniidae*. There are about 330 to 500 species of *Salix* worldwide, predominately in temperate and arctic zones, but also in subtropical and tropical zones (Argus, 1997). Most species of *Salix* are native to China (*c*.275 species), the former Soviet

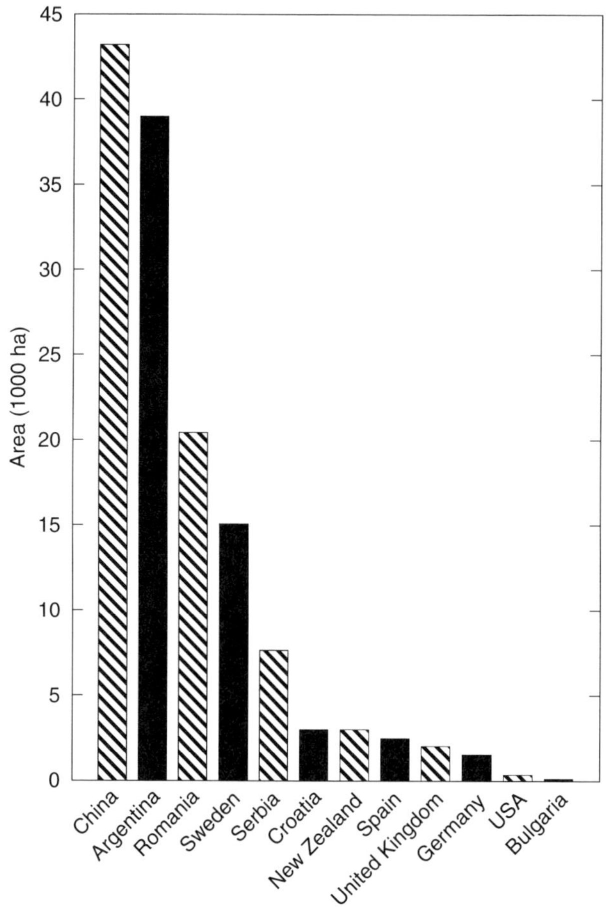

Fig. 19.1. Areas of planted willow (*Salix* spp.) worldwide, including agroforestry plantations, in 2007 (FAO, 2008).

Union (*c*.120 species), North America (*c*.100 species) and Europe (*c*.65 species) (Argus, 1999; Zhenfu *et al*., 1999). *Salix* is a difficult genus to identify at the species level due to considerable individual variability and polymorphism, interspecific hybridization and polyploidy. In addition, non-concurrent phenology for flowers and leaves frequently makes the observation of morphological characters at one time very difficult. Several taxonomical treatments on willows are available, with extensive classifications and rather stable nomenclature (see Kuzovkina *et al*., 2008, and references therein).

Broad physiological and ecological amplitudes of willows and adaptations to a wide range of climates make them suitable for cultivation in many regions of the world. In addition, extensive exchange of *Salix* species between continents has occurred through human activities especially since colonial times and thereafter. Willows occur as different morphological growth forms – tree, shrub or prostrate – with a variety of heights from 25 m tall trees to ground-covering species of only a few centimetres in height. The subgenus *Salix* is comprised of mostly tree-type species sharing many common characteristics with *Populus*. Willows have separate generative and vegetative buds; generative or flower buds, which contain embryonic flowers, occur mostly on the upper portion of the branch, and vegetative or leaf buds occur below the generative buds. The spring and autumn leaf phenology comprises important characteristics determining the shoot growth and biomass production potential of willows used for biomass production (Weih, 2009a). Many willows have characteristics typical of early successional plants colonizing newly opened habitats such as recently flooded, burned, logged, or glaciated areas. These characteristics include: the annual production of abundant seeds with an effective system of seed dispersal enabling long-distance travel; an ability to survive in extreme water fluctuations where root exposure, sediment deposition and prolonged periods of partial inundation occur; and an ability to develop new adventitious roots on buried stems. Thus, the vegetative propagation of broken branches dispersed by water is a common feature of colonization for many willow species, and forms the basis for the efficient vegetative propagation by means of stem cuttings used in willow culture for biomass production. Another feature of many willows that has been exploited in cultivation is their ability to re-sprout from stumps or stools even after repeated cutting or harvesting.

Willows are C_3 plants, but can achieve a considerably greater photosynthetic rate and light-use efficiency than most other trees of the temperate zone, when well supplied with water and nutrients. Some physiological characteristics important for the remarkable biomass production capacity of many *Salix* species include high light-use efficiency when well supplied with water and nutrients, an ability to partition a great fraction of the assimilated biomass to stem parts, rather than roots, and high frost tolerance during winter, making these species suitable as energy crops also at high latitudes (Cannell *et al*., 1987, 1988). Those characteristics combined are part of an explanation for the ability of many *Salix* species to rapidly reach their maximum annual increment and to attain high biomass production levels after only a few years of growth, to vigorously re-establish after harvesting of above-ground biomass and to tolerate high planting densities, which makes many willows promising biomass crops (Keoleian and Volk, 2005). Further important characteristics for the use of willows especially for phytoremediation applications include the tolerance of some species to high concentrations of heavy metals and the ability to accumulate significant amounts of metal in plant tissues, coupled with an efficient nutrient uptake system. Another characteristic of willows relevant for biomass production is the capacity of many *Salix* species to host two types of mycorrhizas (vesicular arbuscular mycorrhiza and ectomycorrhiza) on the same plant, which possibly promotes nutrient transfer and growth of these willows and also could have potentially favourable consequences for carbon mitigation (Rooney *et al*., 2009).

19.3 Production System and Yields

The ease of establishment from unrooted stem cuttings is one of the most well-known attributes of willow cultivation. Based on that characteristic, the crop is usually vegetatively propagated from dormant shoots while seed propagation is limited to breeding efforts. Pieces of shoots are typically harvested from 1-year-old shoots during the dormant season either as 18–25 cm cuttings or 1–2 m long whips and cold stored. Cuttings are normally planted in the ground either with mechanical planters or manually (e.g. Larsson *et al.*, 2007). Optimum establishment of cuttings is achieved by planting in early spring, providing moist conditions at early stages of rooting, long photoperiods, and protection from competition from other vegetation (i.e. weeds).

Willow biomass production is today mostly based on an intensive production system known as short rotation coppice (SRC) systems (Sage, 1998). It involves the cultivation of high-yielding cultivars of willow on agricultural land at high densities of 10,000 to 20,000 plants ha^{-1}. The cuttings are usually planted in a twin row formation to allow specialized harvest machinery to harvest two rows at a time. The distance between the rows usually alternates at 75 and 150 cm and the distance between the plants within rows is around 60 cm, which gives a plantation density of approximately 15,000 cuttings ha^{-1} (Christersson *et al.*, 1993; Larsson *et al.*, 2007). Other plantation schemes are also in use, but it is important to adjust the plantation scheme to the harvest machinery if the plantation is to be run commercially. The SRC cultivation practice is based on intensive site preparation prior to planting: weed control, double-row mechanical planting, mineral nutrient fertilizer inputs at the beginning of each cutting cycle if needed, and multiple harvests on 2- to 5-year cutting cycles (Fig. 19.2; Larsson *et al.*, 2007). Because of economic limitations, irrigation is not typically used for large-scale production operations. Especially in temperate regions and when using old plant material, fertilization with mineral nutrients has frequently been reported not to enhance tree growth significantly (Makeschin, 1999; Larsson *et al.*, 2007). However, under the short growing season in cold-temperate regions and when using modern hybrid willows with greater fertilizer response compared to older plant material (Weih, 2001), the use of commercial nitrogen fertilizer or sludge is recommended in commercial plantations (Weih, 2004; Larsson *et al.*, 2007). The need for nitrogen fertilization varies depending on the plant material, climate and soil conditions, and

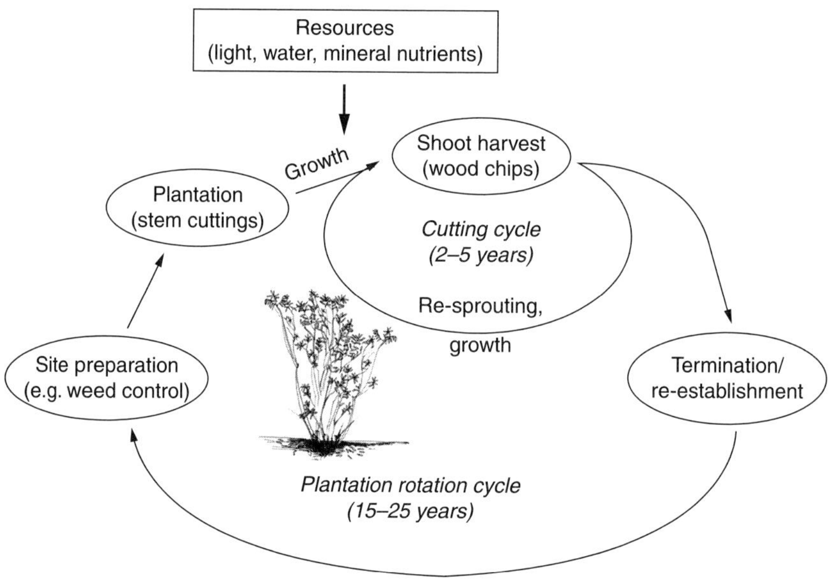

Fig. 19.2. Overview of the actions and processes in a typical *Salix* short rotation coppice (SRC) plantation grown for energy purposes.

fertilization rates of at least between 60 and 100 kg N ha^{-1} $year^{-1}$ are frequently recommended in cold-temperate regions and when using modern willow hybrids.

Biomass harvesting takes place in winter, when growth has ceased and leaves are shed. Harvests are usually performed using specialized and cost-effective machinery that cuts the shoots and directly processes them into wood chips, which are loaded into a trailer either driven parallel to the harvester or connected directly to the harvester, and transported to the end-user. The most commonly used harvesting machines are modified maize harvesters with headers adjusted specifically for use with willow.

The plantation consists of densely planted (from cuttings), high-yielding varieties of willow. The above-ground shoots are harvested during winter on the basis of 2- to 5-year cutting cycles. The shoots are typically harvested and processed to wood chips on site by using specialized machinery. The rootstock or stools remain in the ground after harvest with new shoots re-sprouting the following spring. A plantation could be viable for more than 20 years before re-establishment becomes necessary (plantation rotation cycle).

Combustion of willow wood chips can be done by many different methods depending on the end use, which can be electricity generation through dedicated biomass power stations or co-firing in coal-fired power plants, fueling of heating plants as well as combined heat and power (CHP) plants. The calorific value of willow biomass is similar to the biomass from conventional forest trees (Hakkila, 1989), and 1 t of dry *Salix* wood chips yield approximately 4 to 5 MWh energy when used in a CHP plant, which is equivalent to *c*.0.5 t oil. Wood fuel quality is usually higher in biomass from willow plantations with greater stem diameter and thereby lower bark proportion, which is influenced by willow variety and also management actions that affect stem diameter, e.g. length of cutting cycles (Adler *et al*., 2005, 2008). Biomass production rates of more than 20 t ha^{-1} $year^{-1}$ were reported after 3 to 5 years of growth in fertilized experimental willow stands grown in Sweden and Canada (Christersson, 1986; Labrecque and Teodorescu, 2003). However, such high production figures have been achieved only in small-scale experimental plantations. Biomass yields in commercial willow plantations grown in southern and central Sweden are usually in the range of 5 to 10 t ha^{-1} $year^{-1}$ or lower, with great variation between growers (Mola-Yudego and Aronsson, 2008). In the near future, the mean yields in commercial willow plantations are expected to significantly increase, due to the more frequent use of improved hybrid varieties and better management of the plantations (Mola-Yudego, 2011).

A willow coppice plantation can be grown for more than five cutting cycles before it needs to be re-established. However, biomass production tends to decline from the third cutting cycle due to, for example, density-dependent shoot mortality (Nordh, 2005), and replanting with fresh and modern plant material after four or five harvests is probably the more economically viable option in many cases. The termination and re-establishment of a *Salix* energy crop plantation is not particularly complicated, because the roots in a plantation are relatively shallow and degrade easily.

The economy of willow biomass cultivation for bioenergy has been much debated and greatly depends on market prices for biomass, energy and agricultural policies. In general, willow energy-crop cultivation has the potential to offer farmers a modestly profitable alternative to declining returns from conventional land uses in many European regions, but it currently seems unlikely that this energy crop will be adopted by a great number of European farmers without active support mechanisms and long-term stability of the status of energy crops in common agricultural policy (Rosenqvist and Dawson, 2005; Ericsson *et al*., 2006; Styles *et al*., 2008). A strongly positive effect on the economy of willow culture can be achieved by including environmental applications, e.g. the use of willows for biomass production and vegetation filter purposes (Rosenqvist and Ness, 2004). Thus, the development of willow production systems for biomass provides a basis for using willows also for other environmental applications. Today, short-rotation willow coppice is cultivated not only to produce biomass for energy, but also to filter waste products through contaminant uptake from the soil and water, and is adapted for environmental applications, such as phytoremediation and riparian buffers.

19.4 Genetics and Breeding

Most species of *Salix* are diploid ($2n = 38$), but many species are polyploid (Argus, 1999). Interspecific hybridization is common both naturally occurring and induced through cross-pollination, and many willow hybrids exhibit hybrid vigour (heterosis), i.e. the offspring outperform the parents. Families of species hybrids can yield both male and female individuals, of which many are fertile (Mosseler and Zsuffa, 1989).

There is wide genetic variation within the genus *Salix*. Low levels of domestication, a short generation time, a relatively small genome size and close taxonomical relationship to *Populus*, a model organism for tree genetics and physiology, have made *Salix* attractive for genetic studies and breeding. However, a major challenge in breeding of perennial crops, such as *Salix* grown as SRC, is the production of plant material with high productivity over many years. The cyclic nature of perennial production systems, with harvests every 2–5 years for a period of up to 25 years, also stresses the need for specific breeding goals compared to the breeding of annual agricultural crops. Some key requirements for perennial crops include high yield along with natural resistances to pathogens and pests, and with the ability to withstand abiotic stresses such as drought and frost, which are of particular importance, given the length of time that perennial crops are in the ground. Nevertheless, the great potential to increase yields and resistance or tolerance traits of willows through traditional breeding and hybridization has resulted in many plant breeding programmes that annually introduce new willow varieties (for an overview by country see Kuzovkina *et al.*, 2008). For example, breeding programmes for biomass *Salix* were launched in Europe and in North America (Gullberg, 1993; Lindegaard and Barker, 1997; Smart *et al.*, 2005). The early breeding programmes were based on traditional crossings and breeding methods, whereas modern breeding programmes increasingly attempt to utilize molecular methods and marker assisted breeding. Specific breeding objectives for biomass willows focus not only on biomass production, but also growth habit (to adapt the crop to commercial harvesting machinery), water use efficiency and drought tolerance, frost and disease resistance, and also biomass quality characters.

Within the molecular breeding approaches, molecular markers such as amplified fragment length polymorphism (AFLP), restriction fragment length polymorphism (RFLP) and microsatellites have been used for the construction of genetic linkage maps and for the study of the genetic architecture in *Salix* (Bradshaw *et al.*, 1994; Barker *et al.*, 2003; Trybush *et al.*, 2006). Also, mapping populations and genetic linkage maps have been developed for different *Salix* species or hybrids (Hanley *et al.*, 2002; Tsarouhas *et al.*, 2002; Barcaccia *et al.*, 2003; Rönnberg-Wästljung *et al.*, 2003; Berlin *et al.*, 2010). An approach to investigate the genetic basis of phenotypic traits is to use quantitative trait loci (QTL) analysis, in which areas on the chromosomes containing loci with genes for traits that show continuous (quantitative) variation are identified (e.g. Lander and Botstein, 1989). The QTL analysis can complement classical breeding programmes, and in *Salix*, QTL affecting growth traits such as shoot height and diameter have been mapped, as well as QTL affecting frost tolerance and phenology, water-use efficiency and drought tolerance (Tsarouhas *et al.*, 2002, 2003, 2004; Rönnberg-Wästljung *et al.*, 2005, 2006; Weih *et al.*, 2006). Many years of breeding have resulted in the accumulation of considerable information on the inheritance of traits and the creation of a genetic map to facilitate future breeding efforts for desired traits (Hanley *et al.*, 2002; Rönnberg-Wästljung *et al.*, 2003). The relatedness between the willow and poplar genomes has also been demonstrated (Hanley *et al.*, 2006; Berlin *et al.*, 2010). With the large variation found in the mapping populations, the increasing number of tools available and the rapidly increasing genomics information being released from *Populus*, there is a great potential to gain maximized benefit in the breeding of complex crop characters important for the long-term performance of this perennial crop. Moreover, breeding programmes for future perennial biomass crops, among them *Salix*, could greatly benefit from the integration of ecological information affecting long-term productivity, e.g. ecophysiological growth determinants at stand level and the biological control of pests (Weih *et al.*, 2008).

19.5 Environmental Issues

The current areas of willow biomass plantations are relatively small compared to other agricultural crops, but the increasing demand for bioenergy from biomass could result in major land-use changes over short timescales (Dauber *et al.*, 2010), and increased areas of willow biomass plantations. A potential shift from currently-grown agricultural crops to willow SRC, or the increased cultivation of willows on agricultural land currently not under use, will have negative and positive implications on many environmental issues (Dimitriou *et al.*, 2011). The following sections summarize the potential implications on soil and water, biodiversity and landscape issues.

19.5.1 Soil and water

Intensively managed plantations of willows grown for bioenergy imply the risk of negative effects on the environment in terms of, e.g. herbicide application, deteriorated soil properties and nutrient leakage. However, when compared to conventional arable crops, the perennial character of the willows grown as SRC commonly leads to reduced herbicide application, because herbicide treatment is required only during the establishment of the crop. Further, afforestation with fast-growing trees (e.g. willows) on former arable land has often positive effects on soil properties due to long-term non-tillage management, increased litter amounts and increased carbon sequestration in the topsoil (Baum *et al.*, 2009a). The perennial nature of tree plantations also leads to less nutrient leakage compared to conventional arable crops provided that the rate and timing of fertilization is adjusted to the nutrient delivery and nutrient holding capacity of the soil in relation to the uptake capacity of the trees (Ericsson *et al.*, 1992). If willow SRC plantations are managed according to those principles, they are unlikely to endanger the environment by means of groundwater contamination (Aronsson *et al.*, 2000). Although there is a risk for greater water use in willow SRC compared to annual agricultural crops, water quality is likely to be improved (Dimitriou *et al.*, 2009). In addition, based on the large water and nutrient quantities taken up by fast-growing willows, *Salix* plantations can be used as recipients for municipal wastewater and industrial sludge and simultaneous biomass production (multifunctional biomass plantations) (Perttu, 1993; Aronsson and Perttu, 2001; Isebrands and Karnosky, 2001; Labrecque and Teodorescu, 2003). A major problem occurs with this system in regions with extended dormant periods during winter, where only very small quantities of nutrients are taken up by plants. This sort of wastewater treatment can therefore be accomplished in winter-cold regions only if the wastewater is stored in ponds or lagoons during the winter period (Dimitriou and Aronsson, 2005).

19.5.2 Biodiversity

Studies on biodiversity in plantations of fast-growing trees often arrive at contradictory conclusions, especially when different kinds of organisms are considered (Hartley, 2002). Thus, the landscape context (forest or agricultural, Hanowski *et al.*, 1997; Weih *et al.*, 2003) along with the land-use type that the plantation replaces (Christian *et al.*, 1994) and spatial scale aspects (e.g. size or shape of plantation; Christian *et al.*, 1994; Berg, 2002) influence the impact of establishment of tree plantations on biodiversity. In addition, many animals use multiple habitats and therefore depend on certain habitat combinations (With *et al.*, 1997; Law and Dickman, 1998) and the diversity of land-use types in a given landscape can have large impacts on biodiversity.

Willow SRC plantations have been concluded to frequently increase vascular-plant diversity compared to managed coniferous forests and different kinds of farmland (Gustafsson, 1987; Augustson *et al.*, 2006; Baum *et al.*, 2009b).

Similar to the observations on floras, fauna diversity (birds and mammals) is frequently found to be higher in willow stands compared to agricultural croplands (Christian *et al.*, 1998; Berg, 2002). Thus, the more extensive management of tree plantations compared to intensively managed cereal crops can improve habitat quality for many organisms including plants, birds and arachnids (Berg, 2002; Dhondt and Wrege, 2003; Weih *et al.*, 2003). In addition, plantations of fast-growing trees appear to have a potential as important habitats for gamebirds (Sage and Robertson, 1994). The arrangement of tree plantations within the landscape context have a strong influence on their effect on biodiversity, and there are many ways to arrange and shape willow SRC plantations in such a way that landscape biodiversity is affected positively (Weih, 2009b). For example, mixtures of different willow varieties can be established either by planting them in different blocks across the field or random mixtures within each row (Volk *et al.*, 2004). These mixtures increase structural and functional diversity and can contribute to reduce the impact of pests and diseases (Ramstedt, 1999; McCracken and Dawson, 2001). Finally, a few metres-wide strips along the edges of tree plantations could be left without trees to support the development of a rich native flora and fauna (Weih, 2009b). It hence appears that intensively managed willow SRC plantations, if not too large in size and if planned creatively, can increase the abundance and diversity of many organisms, particularly in landscapes dominated by either conventional agricultural fields or managed coniferous forests, by increasing structural diversity (Baum *et al.*, 2009b). However, willow plantations should be avoided close to open habitats of high conservation values such as certain types of wet meadows and buffer zones along streams, because such sites often have a conservation value linked to openness of the habitat (Berg, 2002). With the above restrictions in mind, small-scale willow plantations (or subdivided larger plantations) grown on agricultural land may have the potential to positively affect biodiversity in many farmland regions.

There is a risk that biomass willows, as these are introduced organisms with improved characteristics, might invade local floras and threaten biodiversity. For the varieties used so far this does not seem to be a problem, because most varieties used for biomass production have been around for many years or decades without showing signs of invasiveness. This is possibly related to the fact that these cultivars, when grown in non-native environments, are poor competitors in the juvenile plant stage and require weed control in order to establish properly. If new varieties are produced in the future with characteristics making them stronger competitors, e.g. to reduce the costs for weed control, the risk of threatening local floras might greatly increase.

19.5.3 Landscape

Concerns are frequently raised regarding the impact of willow plantations on the landscape (Skärbäck and Becht, 2005): will the open agricultural landscape be 'darkened' by a new type of forest? How will the new crop be adapted to the traditional field crops of agricultural landscapes? Will we soon experience a total transformation of our ingrained cultural heritage? Studies regarding the effects of willow plantations on visual landscape perception often arrive at the conclusion that the more open the landscape is, the larger can the size of the plantations be, and particularly small plantations placed in open agricultural landscape can enrich landscape perception (Rode, 2005; Skärbäck and Becht, 2005). Thus, if used creatively as part of active landscape analysis and design, willow plantations can improve the visual and recreational values of a landscape (Rode, 2005; Blomqvist, 2007).

19.5.4 Synergies and conflicts of interests

Localization, shape and management of commercial willow plantations are partly conflicting with the primary interests of nature conservation and landscape design, while no serious conflicts of interest appear to exist between the interests of commercial growers, nature conservationists and landscape designers, if plantations are established outside areas of high nature conservation and cultural heritage values (Weih, 2009b). Synergies between commercial and nature conservational interests could be achieved by localization of plantations near to native woods, which favour biodiversity and also shelter the plantation from extremes in climate (e.g. frost) and, thereby, improve long-term health and productivity of the plantation. No serious conflicts of interests are expected between nature conservation and landscape design principles of shaping and managing tree plantations (Weih, 2009b). Possibly synergies could be achieved through the establishment of edge habitats consisting of attractive (flowering) plants that have high conservational value for plants and insects. In conclusion, there are potential negative impacts of commercial willow plantations on environment and landscape, but these may be controlled and minimized by application of a creative and integrated attitude to ensure that establishment and management of willow plantations operate in an optimal way to secure the positive impacts on rural economic activity, nature conservation (biodiversity) and cultural heritage.

References

Adler, A., Verwijst, T. and Aronsson, P. (2005) Estimation and relevance of bark proportion in a willow stand. *Biomass and Bioenergy* 29, 102–113.

Adler, A., Dimitriou, I., Aronsson, P., Verwijst, T. and Weih, M. (2008) Wood fuel quality of two *Salix viminalis* stands fertilised with sludge, ash and sludge-ash mixtures. *Biomass and Bioenergy* 32, 914–925.

Argus, G.W. (1997) Infrageneric classification of *Salix* (Salicaceae) in the New World. *Systematic Botany Monographs* 52.

Argus, G.W. (1999) Classification of *Salix* in the New World. Version 5:5. July 1999. *Botanical Electronic News* (BEN) 227. Available at: http://www.ou.edu/cas/botany-micro/ben227.html (accessed 18 March 2012).

Aronsson, P. and Perttu, K. (2001) Willow vegetation filters for wastewater treatment and soil remediation combined with biomass production. *Forestry Chronicle* 77, 293–299.

Aronsson, P.G., Bergström, L.F. and Elowson, S.N.E. (2000) Long-term influence of intensively cultured short-rotation willow coppice on nitrogen concentrations in groundwater. *Journal of Environmental Management* 58, 135–145.

Augustson, Å., Lind, A. and Weih, M. (2006) Floristic diversity in willow biomass plantations. *Svensk Botanisk Tidskrift* 100, 52–58. (in Swedish, with English summary.)

Ball, J., Carle, J. and Del Lungo, A. (2005) Contribution of poplars and willows to sustainable forestry and rural development. *Unasylva* 221, 3–9.

Barcaccia, G., Meneghetti, S., Albertini, E., Triest, L. and Lucchin, M. (2003) Linkage mapping in tetraploid willows: Segregation of molecular markers and estimation of linkage phases support an allotetraploid stucture for *Salix alba* x *Salix fragilis* interspecific hybrids. *Heredity* 90, 169–180.

Barker, J.H.A., Pahlich, A., Trybush, S., Edwards, K.J. and Karp, A. (2003) Microsatellite markers for diverse *Salix* species. *Molecular Ecology Notes* 3, 4–6.

Baum, C., Leinweber, P., Weih, M., Lamersdorf, N. and Dimitriou, I. (2009a) Effects of short rotation coppice with willows and poplar on soil ecology. *Landbauforschung (Volkenrode)* 59, 183–196.

Baum, S., Weih, M., Busch, G., Kroiher, F. and Bolte, A. (2009b) The impact of Short Rotation Coppice plantations on phytodiversity. *Landbauforschung (Volkenrode)* 59, 163–170.

Berg, Å. (2002) Breeding birds in short-rotation coppices on farmland in central Sweden – the importance of *Salix* height and adjacent habitats. *Agricultural Ecosystems and Environment* 90, 265–276.

Berlin S., Lagercranz, U., Von Arnold, S., Öst, T. and Rönnberg-Wästljung, A.C. (2010) High-density linkage mapping and evolution of paralogs and orthologs in *Salix* and *Populus*. *BMC Genomics* 11, 129. doi:10.1186/1471-2164-11-129.

Blomqvist, L. (2007) Attractive landscape and biofuel production – a possible combination? Licentiate thesis, Swedish University of Agricultural Sciences, Department of Bioenergy, Report No. 15. Uppsala, Sweden.

Börjesson, P. (2006) *Life cycle assessment of willow production*. Lund University, Dept. of Technology and Society, Lund, Sweden: Report no. 60, 21 pp. (In Swedish, with English abstract.)

Bradshaw, H.D., Villar, M., Watson, B.D., Otto, K.G., Stewart, S. and Stettler, R.F. (1994) Molecular genetics of growth and development in *Populus*. III. A genetic linkage map of hybrid poplar composed of RFLP, STS, and RAPD markers. *Theoretical and Applied Genetics* 89, 167–178.

Cannell, M.G.R., Milne, R., Sheppard, L.J. and Unsworth, M.H. (1987) Radiation interception and productivity of willow. *Journal of Applied Ecology* 24, 261–278.
Cannell, M.G.R., Sheppard, L.J. and Milne, R. (1988) Light use efficiency and woody biomass production of poplar and willow. *Forestry* 61, 125–136.
Christersson, L. (1986) High technology biomass production by *Salix* clones on a sandy soil in southern Sweden. *Tree Physiology* 2, 261–277.
Christersson, L., Sennerby-Forsse, L. and Zsuffa, L. (1993) The role and significance of woody biomass plantations in Swedish agriculture. *Forestry Chronicle* 69, 687–693.
Christian, D.P., Niemi, G.J., Hanowski, J.M. and Collins, P. (1994) Perspectives on biomass energy tree plantations and changes in habitat for biological organisms. *Biomass and Bioenergy* 6, 31–39.
Christian, D.P., Hoffman, W., Hanowski, J.M., Niemi, G.J. and Beyea, J. (1998) Bird and mammal diversity on woody biomass plantations in North America. *Biomass and Bioenergy* 14, 395–402.
Crites, R.W. (1984) Land use of wastewater and sludge. *Environmental Science and Technology* 18, 140–147.
Dauber, J., Jones, M.B. and Stout, J.C. (2010) The impact of biomass crop cultivation on temperate biodiversity. *Global Change Biology Bioenergy* 2, 289–309.
Dhondt, A.A. and Wrege, P.H. (2003) Avian biodiversity studies in short-rotation woody crops. Final report prepared for the US Dept of Energy under cooperative agreement No. DE-FC36-96GO10132. Cornell University Laboratory of Ornithology, Ithaca, New York.
Dimitriou, I. and Aronsson, P. (2005) Willows for energy and phytoremediation in Sweden. *Unasylva* 56, 47–50.
Dimitriou, I., Busch, G., Jacobs, S., Schmidt-Walter, P. and Lamersdorf, N. (2009) A review of Short Rotation Coppice cultivation on water issues. *Landbauforschung (Volkenrode)* 59, 197–206.
Dimitriou, I., Baum, C., Baum, S., Busch, G., Schulz, U. *et al.* (2011) Quantifying environmental effects of Short Rotation Coppice (SRC) on biodiversity, soil and water. *IEA Bioenergy Task* 43, 2011:01. 34 pp.
Ericsson, K., Rosenqvist, H., Ganko, E., Pisarek, M. and Nilsson, L. (2006) An agro-economic analysis of willow cultivation in Poland. *Biomass and Bioenergy* 30, 16–27.
Ericsson, T., Rytter, L. and Linder, S. (1992) Nutritional dynamics and requirements of short rotation forests. In: Mitchell, C.P., Ford-Robertson, J.B., Hinckley, T. and Sennerby-Forsse, L. (eds) *Ecophysiology of Short Rotation Forest Crops*. Elsevier, London, pp. 35–65.
FAO (2008) Synthesis of Country Progress Reports received, prepared for the 23rd Session of the International Poplar Commission, jointly hosted by FAO and by the Beijing Forestry University, the State Forest Administration of China and the Chinese Academy of Forestry; Beijing, China, 27–30 October 2008. International Poplar Commission, Working, Paper IPC/6. Forest Management Division, FAO, Rome. Available at: http://www.fao.org/docrep/011/k3380e/k3380e00.htm (accessed 18 March 2012).
Grogan, P. and Matthews, R. (2002) A modelling analysis of the potential for soil carbon sequestration under short rotation coppice willow bioenergy plantations. *Soil Use and Management* 18, 175–183.
Gullberg, U. (1993) Towards making willows pilot species for coppicing production. *Forestry Chronicle* 69, 721–726.
Gustafsson, L. (1987) Plant conservation aspects of energy forestry – a new type of land use in Sweden. *Forest Ecology and Management* 21, 141–161.
Hakkila, P. (1989) *Utilization of Residual Forest Biomass*. Springer-Verlag, New York.
Hall, D.O. and House, J.I. (1994) Trees and biomass energy: carbon storage and/or fossil fuel substitution? *Biomass and Bioenergy* 6, 11–30.
Hanley, S., Barker, J.H.A., Van Ooijen, J.W., Aldam, C., Harris, S. *et al.* (2002) A genetic linkage map of willow (*Salix viminalis*) based on AFLP and microsatellite markers. *Theoretical and Applied Genetics* 105, 1087–1096.
Hanley, S., Mallott, M.D. and Karp, A. (2006) Alignment of a *Salix* linkage map to the *Populus* genomic sequence reveals macrosynteny between willow and poplar genomes. *Tree Genetics and Genomes* 3, 35–48.
Hanowski, J.M., Niemi, G.J. and Christian, D.C. (1997) Influence of within-plantation heterogeneity and surrounding landscape composition on avian communities in hybrid poplar plantations. *Conservation Biology* 11, 936–944.
Hartley, M.J. (2002) Rationale and methods for conserving biodiversity in plantation forests. *Forest Ecology and Management* 155, 81–95.
Isebrands, J.G. and Karnosky, D.F. (2001) Environmental benefits of poplar culture. In: Dickmann, D.I., Isebrands, J.G., Eckenwalder, J.E. and Richardson, J. (eds) *Poplar Culture in North America. Part A.* NRC Research Press, National Research Council of Canada, Ottawa, Canada, pp. 207–218.

Karp, A. and Shield, I. (2008) Bioenergy from plants and the sustainable yield challenge. *New Phytologist* 179, 15–32.

Keoleian, G.A. and Volk, T.A. (2005) Renewable energy from willow biomass crops: life cycle energy, environmental and economic performance. *Critical Reviews in Plant Sciences* 24, 385–406.

Kuzovkina, Y., Weih, M., Abalos Romero, M., Belyaeva, I., Charles, J. *et al.* (2008) *Salix*: botany and global horticulture. *Horticultural Reviews* 34, 447–489.

Labrecque, M. and Teodorescu, T.I. (2003) High biomass yield achieved by *Salix* clones in SRIC following two 3-year coppice rotations on abandoned farmland in southern Quebec, Canada. *Biomass and Bioenergy* 25, 135–146.

Lander, E.S. and Botstein, D. (1989) Mapping Mendelian factors underlying quantitartive traits using RFLP linkage maps. *Genetics* 121, 185–199.

Larsson, S., Nordh, N.E., Farrell, J. and Tweddle, P. (2007) *Manual for SRC Willow Growers*. Lantmännen Agroenergi AB (www.agroenergi.se), Örebro, Sweden, 18 pp.

Law, B.S. and Dickman, C.R. (1998) The use of mosaic habitats by vertebrate fauna: implications for conservation and management. *Biodiversity Conservation* 7, 323–333.

Lindegaard, K.N. and Barker, J.H.A. (1997) Breeding willows for biomass. *Aspects of Applied Biology* 49, 155–162.

Makeschin, F. (1999) Short rotation forestry in Central and Northern Europe – introduction and conclusions. *Forest Ecology and Management* 121, 1–7.

McCracken, A.R. and Dawson, W.M. (2001) Disease effects in mixed varietal plantations of willow. *Aspects of Applied Biology* 65, 255–262.

Mola-Yudego, B. (2011) Trends and productivity improvements from commercial willow plantations in Sweden during the period 1986–2000. *Biomass and Bioenergy* 35, 446–453.

Mola-Yudego, B. and Aronsson, P. (2008) Yield models for commercial willow biomass plantations in Sweden. *Biomass and Bioenergy* 32, 829–837.

Mosseler, A. and Zsuffa, L. (1989) Sex expression and sex ratios in intra- and interspecific hybrid families of *Salix* L. *Silvae Genetica* 38, 12–17.

Nordh, N.E. (2005) Long-term changes in stand structure and biomass production in short rotation willow coppice. ActaUniversitatis Agriculturae Suecia 120. PhD dissertation. Swedish University of Agricultural Sciences, Uppsala, Sweden.

Perttu, K.L. (1993) Biomass production and nutrient removal from municipal wastes using willow vegetation filters. *Journal of Sustainable Forestry* 1, 57–70.

Ramstedt, M. (1999) Rust disease on willows – virulence variation and resistance breeding strategies. *Forest Ecology and Management* 121, 101–111.

Rode, M. (2005) Energetische Nutzung von Biomasse und der Naturschutz. *Natur und Landschaft* 80, 403–412 (in German, with English summary).

Rönnberg-Wästljung, A.C., Tsarouhas, V., Semerikov, V. and Lagercranz, U. (2003) A genetic linkage map of a tetraploid *Salix viminalis* x *S. dasyclados* hybrid based on AFLP markers. *Forest Genetics* 10, 185–194.

Rönnberg-Wästljung, A.C., Glynn, C. and Weih, M. (2005) QTL analyses of drought tolerance and growth for a *Salix dasyclados* x *S. viminalis* hybrid in contrasting water regimes. *Theoretical and Applied Genetics* 110, 537–549.

Rönnberg-Wästljung, A.C., Åhman, I., Glynn, C. and Widenfalk, O. (2006) Quantitative trait loci for resistance to herbivores in willow: field experiments with varying soils and climate. *Entomologia Experimentalis et Applicata* 118, 163–174.

Rooney, D.C., Killham, K., Bending, G.D., Baggs, E., Weih, M. *et al.* (2009) Mycorrhizas and biomass crops: opportunities for future sustainable development. *Trends in Plant Science* 14, 542–549.

Rosenqvist, H. and Dawson, M. (2005) Economics of willow growing in Northern Ireland. *Biomass and Bioenergy* 28, 7–14.

Rosenqvist, H. and Ness, B. (2004) An economic analysis of leachate purification through willow coppice vegetation filters. *Bioresource Technology* 94, 321–329.

Sage, R.B. (1998) Short rotation coppice for energy – towards ecological guidelines. *Biomass and Bioenergy* 15, 39–47.

Sage, R.B. and Robertson, P.A. (1994) Wildlife and game potential of short rotation coppice in the UK. *Biomass and Bioenergy* 6, 41–48.

Sirén, G., Sennerby-Forsse, L. and Ledin, S. (1987) Energy plantations – short rotation forestry in Sweden. In: Hall, D.O. and Overend, R.P. (eds) *Biomass*. Wiley and Sons, London, pp. 119–143.

Skärbäck, E. and Becht, P. (2005) Landscape perspective on energy forests. *Biomass and Bioenergy* 28, 151–159.

Smart, L.B., Volk, T.A., Lin, J., Kopp, R.F., Phillips, I.S., Cameron, K.D., White, E.H. and Abrahamson, L.P. (2005) Genetic improvement of shrub willow (*Salix* spp.) crops for bioenergy and environmental applications in the United States. *Unasylva* 221, 51–55.

Styles, D. and Jones, M.B. (2007) Energy crops in Ireland: Quantifying the potential life-cycle greenhouse gas reductions of energy-crop electricity. *Biomass and Bioenergy* 31, 759–772.

Styles, D., Thorne, F. and Jones, M.B. (2008) Energy crops in Ireland: an economic comparison of willow and *Miscanthus* production with conventional farming systems. *Biomass and Bioenergy* 32, 407–421.

Trybush, S., Hanley, S., Cho, K.H., Jahodova, S., Grimmer, M. *et al.* (2006) Getting the most out of fluorescent amplified fragment length polymorphism. *Canadian Journal of Botany* 84, 1347–1354.

Tsarouhas, V., Gullberg, U. and Lagercranz, U. (2002) An AFLP and RFLP linkage map and quantitative trait locus (QTL) analysis of growth traits in Salix. *Theoretical and Applied Genetics* 105, 277–288.

Tsarouhas, V., Gullberg, U. and Lagercrantz, U. (2003) Mapping of quantitative trait loci controlling timing of bud flush in *Salix*. *Hereditas* 138, 172–178.

Tsarouhas, V., Gullberg, U. and Lagercrantz, U. (2004) Mapping of quantitative trait loci (QTLs) affecting autumn freezing resistance and phenology in *Salix*. *Theoretical and Applied Genetics* 108, 1335–1342.

Tuskan, G.A. and Walsh, M.E. (2001) Short-rotation woody crop systems, atmospheric carbon dioxide and carbon management: a U.S. case study. *Forestry Chronicle* 77, 259–264.

Verwijst, T. (2001) Willows: an underestimated resource for environment and society. *Forestry Chronicle* 77, 281–285.

Volk, T.A., Verwijst, T., Tharakan, P.J., Abrahamson, L.P. and White, E.H. (2004) Growing fuel: a sustainability assessment of willow biomass crops. *Frontiers in Ecology and Environment* 2, 411–418.

Weih, M. (2001) Evidence for increased sensitivity to nutrient and water stress in a fast-growing hybrid willow compared with a natural willow clone. *Tree Physiology* 21, 1141–1148.

Weih, M. (2004) Intensive short rotation forestry in boreal climates: present and future perspectives. *Canadian Journal of Forest Research* 34, 1369–1378.

Weih, M. (2009a) Genetic and environmental variation in spring and autumn phenology of biomass willows (*Salix* spp.): effects on shoot growth and nitrogen economy. *Tree Physiology* 29, 1479–1490.

Weih, M. (2009b) Willow short rotation coppice commercially grown on agricultural land in Sweden – possibilities for improvement of biodiversity and landscape design. IEA Technical Review No. 4, 36 pp. Available at: http://www.shortrotationcrops.org (accessed 18 March 2012).

Weih, M., Karacic, A., Munkert, H., Verwijst, T. and Diekmann, M. (2003) Influence of young poplar stands on floristic diversity in agricultural landscapes. *Basic and Applied Ecology* 4, 149–156.

Weih, M., Rönnberg-Wästljung, A.C. and Glynn, C. (2006) Genetic basis of phenotypic correlations among growth traits in hybrid willow (*Salix dasyclados* x *S. viminalis*) grown under two water regimes. *New Phytologist* 170, 467–477.

Weih, M., Didon, U.M.E., Rönnberg-Wästljung, A.-C. and Björkman, C. (2008) Integrated agricultural research and crop breeding: Allelopathic weed control in cereals and long-term productivity in perennial biomass crops. *Agricultural Systems* 97, 99–107.

With, K.A., Gardner, L.H. and Turner, M.G. (1997) Landscape connectivity and population distribution in heterogeneous environments. *Oikos* 78, 151–169.

Zhenfu, F., Shidong, Z. and Skvortsov, A.K. (1999) Salicaceae. In: Zheng-yi, W. and Raven, P.H. (eds) *Flora of China*. Missouri Botanical Garden Press, St Louis, Missouri, pp. 139–274.

20 Pines

C. Dana Nelson,[1] Gary F. Peter,[2] Steven E. McKeand,[3] Eric J. Jokela,[2] Robert B. Rummer,[4] Leslie H. Groom[5] and Kurt H. Johnsen[6]

[1]*USDA Forest Service, Southern Research Station, Southern Institute of Forest Genetics, Mississippi, USA;* [2]*School of Forest Resources and Conservation, University of Florida, USA;* [3]*Department of Forestry and Environmental Resources, North Carolina State University, USA;* [4]*USDA Forest Service, Southern Research Station, Alabama, USA;* [5]*USDA Forest Service, Southern Research Station, Louisiana, USA;* [6]*USDA Forest Service, Southern Research Station, North Carolina, USA*

20.1 Introduction

The southern pines (yellow or hard pines, genus *Pinus*, subgenus *Pinus*, section *Pinus*, subsection *Australes*) occupy an immense land-base in the south-eastern region of the USA (Little and Critchfield, 1969). In addition, they are planted and managed for wood production on millions of hectares worldwide, including in China, Brazil, Argentina and Australia. The taxonomic subsection *Australes* consists of 11 species, ranging from relatively minor to major in terms of land base occupied and management opportunities. For example, Table-mountain pine (*Pinus pungens*) sporadically occupies higher elevation sites in the southern Appalachian Mountains and due to declining habitat is considered a species of concern for conservation (Erickson *et al.*, 2012). In contrast, loblolly pine (*Pinus taeda*) has a large native range in the USA with an even larger managed land-base (across the lower and upper Coastal Plains and the Piedmont physiographic region) as a result of extensive planting and intensive silviculture in response to the wood products industry. In addition to loblolly pine, three other southern pine species are considered major due to their large native ranges in the USA: shortleaf pine (*Pinus echinata*), longleaf pine (*Pinus palustris*) and slash pine (*Pinus elliottii* var. *elliottii*). Of note, the former two are also considered to be species of concern for conservation (Erickson *et al.*, 2012) due to long-standing land management practices that have favoured loblolly pine. Similarly to loblolly pine, although on a smaller scale, slash pine has been widely planted and managed for wood and fibre production. Because of this, slash pine could be an important component of southern pine production for bioenergy purposes, sharing many similar features in this respect to loblolly pine. However, for the purpose of this chapter we will focus our discussion on loblolly pine and consider the general features and properties in bioenergy production, genetics and breeding for bioenergy traits, silvicultural practices for bioenergy production, tree harvesting and chip processing, bioenergy opportunities and challenges, and sustainability of bioenergy production systems. Socio-economic analyses and their implications are critical for the whole system but are beyond the scope of this chapter.

20.2 Southern Pines in Bioenergy Production

With the implementation of global carbon reduction goals, concerns over the stability of

©CAB International 2013. *Biofuel Crops: Production, Physiology and Genetics* (ed. B.P. Singh)

supply and cost of crude oil, renewable forms of energy from biomass have received significant increases in commercial interest. In 2007, the US federal government passed the Energy Independence and Security Act, which raised the renewable fuel standard (RFS). The RFS now calls for 36 billion gallons of renewable fuels, 21 billion of which are required to be obtained from cellulosic ethanol and other biofuels, by 2022. In addition, the law requires that the renewable fuels be produced with at least 20% lower life cycle greenhouse gas emissions relative to gasoline and diesel from crude oil. Current US production of cellulosic based biofuels is still in its infancy with a number of companies, including Abengoa in Kansas, BlueFire and KiOR in Mississippi, Dupont in Iowa, GEVO in Minnesota, Mascoma in Michigan, POET in Iowa and ZeaChem in Oregon, building or planning to build intermediate scale facilities. This indicates indicating that additional feedstocks will be required including forest trees such as the southern pines, both as purpose-grown crops and residues from other harvest operations.

The southern pines are a proven sustainable source of biomass for renewable chemicals, materials and bioenergy. They grow on over 37 Mha in the southern USA and supply ~18% of the global supply of industrial roundwood (Prestemon and Abt, 2002). The US forest products industry already generates 77% of all industrial biomass energy by burning wood waste and lignin at high thermal efficiencies in wood processing facilities. Even though southern US forests produce more industrial roundwood than any other single country, annual wood growth exceeds harvest rates. An important reason growth exceeds removals is the more than tenfold greater productivity of planted compared with naturally regenerated pines (Fox *et al.*, 2007). In the region, there are 13 Mha of loblolly and slash pine plantations, and nutrients and competition are managed to substantially increase yields (Conner and Hartsell, 2002; Munsell and Fox, 2010). Since the mid-1980s, virtually all of these plantations have been established with genetically improved planting stock.

Commercial interest in using southern pine for bioenergy and biofuel production is strong because key questions related to scale, cost and sustainable supplies are readily answered. An extensive, robust supply chain for roundwood has been developed for the pulp and paper industry, and this supply chain operates year round. Excellent inventories of available standing biomass and reliable predictions of yield for planted pines enable siting of facilities at locations with adequate immediate and future supplies of biomass. Because harvests of the higher density woody (i.e. lignocellulosic) pine biomass can occur year round, the logistics of using wood for bioenergy and biofuel are simpler than perennial grasses and crop residues that need to be compressed and stored. Production, harvesting and transportation costs are understood, and delivered southern pine wood costs have been more stable than other agricultural commodities. Finally, net energy yields from southern pine to ethanol are estimated to be higher than maize starch, cane sugar and sweet sorghum (Evans and Cohen, 2009). This higher net energy is due to the substantially lower energy inputs used in growing, harvesting and transporting planted pine compared with herbaceous grass crops and the higher energy obtained from lignin. Thus, overall southern pines are a very attractive immediate and future source of biomass for bioenergy production.

The expansive and productive southern pine forests have attracted substantial new interest as a source of biomass for standalone facilities to produce bioenergy and biofuel. For example, five large commercial facilities that produce wood pellets for biopower have been built, and a number of woody biomass to electrical power or biofuel facilities are planned or are being built in the region. Compared with biopower, production of biofuels from lignocellulosic biomass is more complex and technically challenging, in part because cost and environmental metrics will need to be met simultaneously. The central technical challenge for converting lignocellulosic biomass to liquid fuel is that sugars, which account for the majority of the carbon, have high oxygen content whereas fuels are carbon rich and have no or very low oxygen content. Currently a large number of approaches, broadly categorized into biochemical and thermochemical methods to convert lignocellulosic biomass to biofuels, are being researched and developed. Several of the most promising approaches and their potential for implementation with the southern pines are discussed below in detail.

Clearly, the issue of sustainability of southern pine energy-wood (i.e. stemwood and/or residues harvested for bioenergy uses) production systems needs careful consideration, in particular the implications of whole-tree harvesting on shorter rotations. Can sites continue producing biomass at the same or increasing levels over multiple rotations? Can sites continue to provide ecosystem services at socially and culturally acceptable levels? Over the last several decades, the regions' forestry sector has gained much insight into these questions through silviculture research and land management experience, and the answers seem favourable for sustainable energy-wood production. However, bioenergy policy affecting economic and regulatory considerations will require continued monitoring and study of their impact on sustainability issues.

20.3 Genetics and Breeding for Bioenergy Traits

In recent years, about 1 billion loblolly and slash pine seedlings have been planted each year in the US south, and virtually every one of these seedlings has come from intensive tree-breeding programmes (McKeand *et al.*, 2003). The cooperative tree improvement programmes in the southern USA have been responsible for the vast majority of tree breeding with southern pines for the last 50+ years. These cooperative programmes are: the Cooperative Forest Genetics Research Program (CFGRP) at the University of Florida (http://www.sfrc.ufl.edu/cfgrp), the NC State University Cooperative Tree Improvement Program (http://treeimprovement.org) and the Western Gulf Forest Tree Improvement Program at the Texas Forest Service and Texas A&M University (http://www.ars-grin.gov/misc/wgftip).

Breeding strategies have emphasized population improvement for broad adaptability and value improvement. The traits of interest for improvement have been volume production, stem straightness, and resistance to fusiform rust (caused by the fungus *Cronartium quercuum* f. sp. *fusiforme*). In some programmes, wood density has been modified, but relatively little effort has been focused on changing wood properties in most populations. The focus on volume and stem quality versus wood properties is in large part a function of the wood market in the southern USA. Currently, landowners realize most of their financial benefit when saw-timber and poles are harvested as compared to pulpwood, so most breeding and deployment emphasis has been on growth and stem quality traits, and value improvements have been impressive (Vergara *et al.*, 2004, 2007; McKeand *et al.*, 2006a). Thus, pine breeders will continue emphasis on selecting high-yielding varieties that grow across a range of different sites, which will have positive impacts on energy yields per hectare per year.

The harvest index of southern pine plantations is already high, and for energy-wood plantings is expected to be higher as whole trees are likely to be harvested to maximize biomass yields and reduce harvesting costs. Given this possibility, altering carbon allocation between stems and branches may be of limited value, whereas increasing carbon allocation to the stem over roots could improve biomass yields. Little is known about the genetic control of allocation between shoots and roots in southern pine, but it is clear that fertilization significantly increases total carbon accumulation in shoots and roots (Retzlaff *et al.*, 2001). Biomass yield is also influenced by wood density; consequently increasing juvenile wood density in fast-growing pine trees offers the potential for raising yields in energy-wood plantations. However, achieving increases in wood density could be difficult by traditional breeding and selection in some populations, because of the negative genetic correlation between wood density and growth (e.g. McKinley *et al.*, 1982; Belonger *et al.*, 1996; Atwood *et al.*, 2002). It may be possible to increase wood density and growth simultaneously through traditional breeding in many populations where the traits are independent, or there is a negligible correlation (Zobel and van Buijtenen, 1989; Gräns, 2012).

In addition to improving biomass yield, breeding for wood properties can also enhance bioenergy yields from southern pine. For example, in extractive free wood, the heating value is linearly correlated with Klason lignin content, increasing by ~1% for each 1% increase in lignin content (White, 1987). Research demonstrates that wood chemical composition is under weak to moderate genetic control, and substantial genetic variation exists for selecting germplasm with altered wood properties (Sykes *et al.*, 2006). Natural and breeding populations of southern pine have wood lignin contents that range from 25 to 35%, thus the heating value of extractive free wood can differ by as much as 10% (G.F. Peter, unpublished). With biochemical methods for conversion to liquid fuels, the chemical composition and structure of the lignocellulosic biomass dramatically affects the efficiency of saccharification and fermentation and total yield of available sugars (Himmel *et al.*, 2006). Thus, identifying varieties with low and high lignin should both increase energy and fuel yields, depending on the conversion technology. The limited evidence available suggests that wood chemistry is not genetically correlated with growth, as has been reported for angiosperm (i.e. hardwood) trees (Novaes *et al.*, 2010).

In addition to breeding, genetic engineering provides an important alternative approach for altering wood chemical properties for bioenergy and biofuels. The pathways for synthesis of all major wood chemicals, including cellulose, galactoglucomannan, arabinoglucuronoxylan, lignin, lipids, sterols and terpenes are largely conserved with other land plants. For example, in loblolly pine, the three isoforms of the catalytic subunits of cellulose synthase involved in secondary wall synthesis in xylem are well conserved with woody and herbaceous angiosperm species (Nairn and Haselkorn, 2005). Enzymes in the pathway that catalyse synthesis of coniferyl alcohol from phenylalanine are well conserved with those in angiosperms (Peter and Neale, 2004). Pines synthesize and accumulate mono- and diterpenes in the wood via the conserved 2-C-methyl-D-erythritol 4-phosphate/1-deoxy-D-xylulose 5-phosphate pathway (Zulak and Bohlmann, 2010), suggesting that fundamental knowledge from angiosperm plants can be applied to genetic engineering of wood chemical composition in pine. For example, introducing two genes that mediate syringyl alcohol synthesis in angiosperms into pine should lead to syringyl lignin formation (Li *et al.*, 2003).

In angiosperms, syringyl lignin is more readily extracted during pulping and *Populus* trees with more syringyl relative to guaiacyl lignin saccharify better, requiring milder pretreatment (Studer *et al.*, 2011). Large increases in wood heating value can be achieved by increasing wood extractive content. Extractives are composed of lipids, fatty acids, sterols and terpenes, all hydrocarbon-rich compounds with similar heating values as crude oil. Loblolly and slash pine synthesize and accumulate substantial amounts of terpenes; in mature trees these can be up to 20% of the dry weight of wood (Stubbs *et al.*, 1984). Thus, increasing wood terpene content in trees grown for short rotations will also increase energy yields per hectare per year. While pine terpenes are valuable chemicals and have been recovered at commercial scales for a long time, the interest for liquid fuel production is new. For example, pinenes can be efficiently dimerized to produce a compound with similar properties as jet fuel from petroleum (Harvey *et al.*, 2010). Thus, altering wood properties offers an excellent opportunity to improve energy yields by improving conversion efficiencies.

An important question is whether energy-wood markets will value these traits sufficiently to justify these efforts given the long generation intervals and rotation times for pines. Breeding for these additional traits requires economic justification. If more traits are included in a selection and breeding programme, a reduction in genetic gain will occur in other traits. For example, if higher wood density is desired and is given the same weight as volume in a selection index, then the potential gain in volume is reduced in half assuming the two traits are independent. If there is a negative correlation, then reduction in volume would be greater than 50%, suggesting that genetic engineering approaches may be favoured for altering wood property traits in pine clonal varieties optimized for energy-wood yield for bioenergy production.

The most likely scenario, at least in the short run, for genetics having an impact on bioenergy production is that specific existing families or clonal varieties with desirable wood properties will be identified by researchers, and these genotypes can be operationally deployed or selectively harvested if they have already been planted. In current deployment populations, there are hundreds of different families that could be screened and utilized immediately (McKeand *et al.*, 2003). Even though the parents of these families have been highly selected for growth, stem quality and disease resistance, there will assuredly be high degrees of genetic variation for almost any wood-property trait. Breeders can screen families in deployment populations for desired traits, and if there are economic incentives for landowners to plant them, these families will be utilized (Byram *et al.*, 2005; Peter *et al.*, 2007). An additional option could develop for landowners if valuable varieties of loblolly or slash pines are identified. Ten years ago, 59% of the loblolly pine plantations and 43% of the slash pine plantations were established as single-family blocks (McKeand *et al.*, 2003); these percentages are substantially higher (perhaps up to 80% for both species) today. Most large landowners know the genetic identity of their plantations, so if valuable varieties are found, selective harvest of these varieties would be possible.

Loblolly and slash pine are essentially undomesticated species, having been through only three cycles of selection and breeding, and large increases in tree growth are still obtainable through traditional breeding methods. These gains in growth will likely be achieved much faster with implementation of molecular marker-based selection (Nelson and Johnsen, 2008; Resende *et al.*, 2012; Zapata-Valenzuela *et al.*, 2012). The likelihood of utilizing genetic differences in bioenergy traits in southern pines will depend more on economic and market forces rather than genetic factors. If traits are economically important, then breeders and tree-improvement foresters will take advantage of the information to breed, engineer and deploy specific varieties for energy-wood plantations.

20.4 Silvicultural Practices for Bioenergy Production

20.4.1 Site selection

Of the southern pines, loblolly has the most extensive natural and managed range and, therefore, site selection is critical relative to evaluating its performance and suitability as a biofuel species. It can be found growing in a variety of habitats that vary based on physiography, geology, soils and climate. Throughout its natural range, loblolly pine occurs within multiple physiographic regions such as the Coastal Plain (Atlantic and Gulf), Piedmont, Ozark Plateaus and Quachita Mountains, Ridge and Valley, and the Appalachian and Interior Low Plateaus (Morris and Campbell, 1991; Schultz, 1997). Major soil orders among these regions include: Alfisols, Entisols, Histosols, Inceptisols, Mollisols, Spodosols and Ultisols.

Site classification systems used across the southern USA for species deployment decisions tend to be multi-faceted and include elements related to soil type (e.g. series, drainage class, depth and characteristics of the subsoil–argillic horizon), site quality (actual and potential), climate and disease hazard rating (e.g. fusiform rust). As loblolly pine tends to be a nutrient-demanding species compared to other southern pines, it grows best on high quality sites (soils) that are fertile, moderately acidic, have imperfect to poor surface drainage, a thick medium-textured surface layer and fine-textured subsoil. Poorest growth is often associated with shallow, eroded, or very wet or waterlogged soils.

The Coastal Plain region accounts for the majority (75%) of the managed loblolly pine plantations (Shultz, 1997). This physiographic region is subdivided into the lower, middle and upper Coastal Plain and the topography can vary from level to gently rolling to hilly and undulating. As such, the soils can range from very poorly to excessively drained and also differ markedly in fertility based on drainage class, soil texture, parent materials and historical land use patterns.

Soil groupings, based on easily recognizable features, have been used successfully in the southern USA by foresters and natural resource specialists to identify sites where available nutrient supplies are low, or where other site factors (e.g. moisture availability) influence growth and species performance. To understand the general distribution of forest soils and their fertility in the region, it may be helpful to consider the major land areas of the Coastal Plain region, and a soil classification system developed in the 1980s by the Cooperative Research in Forest Fertilization (CRIFF) programme at the University of Florida (Fisher, 1981; Fig. 20.1). The eight CRIFF soil groups (A–H) are defined using drainage, texture and depth of the subsurface soil horizons. Table 20.1 defines the nature of each soil group in relation to the major land areas.

The CRIFF soil classification system is still widely used today as a basis for stratifying forestland for species deployment decisions (Fox, 2004; Fig. 20.2) and prescribing silvicultural treatments. Suitable sites for loblolly pine plantations are commonly found on CRIFF A, B, C, E and F group soils (Fig. 20.1 and Fig. 20.2). For example, wet mineral flats in the lower Coastal Plain are characterized as poorly to very poorly drained, fine-textured soils that developed from slack-water deposits (e.g. CRIFF A; Paleaquults, Haplaquults). An argillic (Bt) horizon commonly occurs at depths less than 50 cm. These soils are inherently phosphorus deficient, but can produce some of the most productive stands of loblolly pine when bedded and fertilized. Similarly, Coastal Plain flatwoods sites found in northern Florida and southern Georgia are typified by poorly to somewhat poorly drained soils that developed from coarse-textured marine sands (e.g. CRIFF C; Alaquods, Haplaquods). Loblolly pine tends to grow best on those soils having well-developed clayey (argillic) subsoil, especially when combined with silvicultural treatments that alleviate nutrient deficiencies. In the upper Coastal Plain and Piedmont regions, where topographic relief has contributed to a fairly dissected and eroded landscape, loblolly pine growth rates can range from fair to excellent. As most soils in this region (CRIFF E and F groups) were once farmed for cotton,

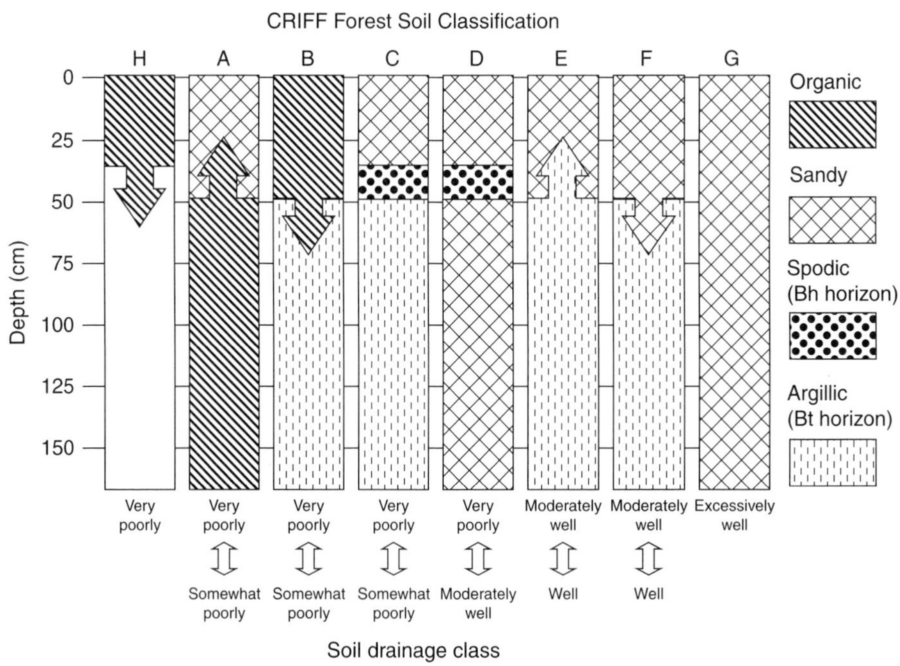

Fig. 20.1. Cooperative Research in Forest Fertilization (CRIFF) soil groups for the southern USA (adapted from Jokela *et al.*, 1991a).

Table 20.1. Definitions of CRIFF soil groups within the Coastal Plain region of the southern USA.

CRIFF soil group	Major land area	Soil drainage class	Important features
A	Savannahs	Very poor to somewhat poor	Sand to loamy sand surface layer less than 50 cm thick, with a finer textured soil (argillic) horizon below
B	Savannahs	Very poor to somewhat poor	Sand to loamy sand surface layer greater than 50 cm thick, with a finer textured soil horizon (argillic) below
C	Flatwoods	Poor to somewhat poor	Spodic horizon below the surface layer. Sandy loam or finer textured soil horizon (argillic) below the spodic horizon
D	Flatwoods	Poor to somewhat poor	Spodic horizon below the surface layer. Sand to loamy sand soil horizon below the spodic horizon (no argillic horizon)
E	Uplands	Moderate to well	Sand to loamy sand surface layer less than 50 cm thick, with a finer textured soil horizon (argillic) below
F	Uplands	Moderate to well	Sand to loamy sand surface layer greater than 50 cm thick, with a finer textured soil horizon (argillic) below
G	Sandhills	Excessive	Sand to loamy sand surface layer at least 100 cm thick (no argillic horizon)
H	Depressions	Very poor	High in decomposing plant residues, often an organic soil

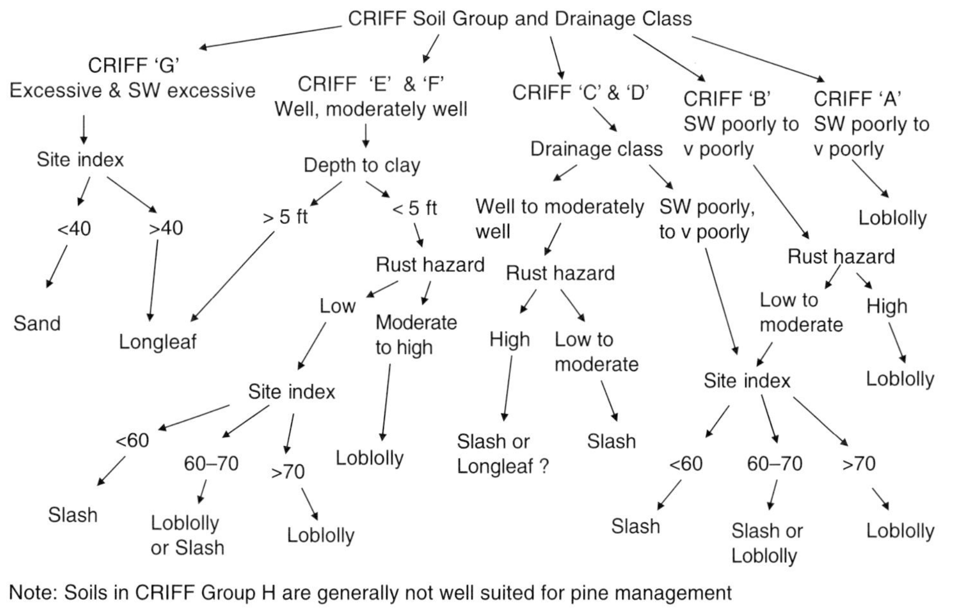

Fig. 20.2. Species deployment decision key for southern pines in the Coastal Plain of Georgia, Florida and Alabama (from Fox, 2004). SW, somewhat.

erosion processes have led to losses of the fertile topsoil, resulting in exposure of subsoil horizons that have lower fertility and higher mechanical resistance to root penetration and development. Conversely, many loblolly pine stands in this region have benefitted from the 'old-field' effect, where residual soil fertility levels that resulted from recent or former agricultural practices have contributed to highly productive stands of loblolly pine.

Site quality, expressed as site index (i.e. average height of dominant and co-dominant trees at index age of 25 years), can be variable both among and within soil groups, as well as physiographic regions. Research has clearly demonstrated that site quality is not a 'fixed' attribute, but can be modified with site-specific silvicultural treatments (Munsell and Fox, 2010). For example, Jokela *et al.* (2010) reported that site index for loblolly pine growing on a somewhat poorly drained Ultic Alaquod (CRIFF C group soil) in northern Florida increased from 22.5 to 26.5 m when fertilizer and understorey competition control treatments were prescribed. These changes in site index were also expressed in volume yields at 25 years, with the treated stands having 2.2-fold more standing volume than the untreated controls (367 versus 165 m^3 ha^{-1}). With today's genetics and silvicultural technologies, it is possible on some sites to achieve site index levels of about 30 m.

20.4.2 Site preparation

Mechanical treatments

The successful establishment of loblolly pine stands for energy-wood production will require some form of site preparation, be it mechanical (e.g. bedding, shearing, spot raking, subsoiling, disking, combination ploughing) or chemical (herbicides). Prescribed fire may also be used. The choice of site preparation methods depends on many factors, including: management objective, physiographic region, past management practices, harvesting method, soil characteristics, site quality, understorey competition levels, tract size, special needs considerations and economics. For stand establishment, some combination of treatments will be used to facilitate the control of understorey competition levels, reduce harvest debris (slash) that impedes planting operations, improve surface soil drainage on problematic sites and ameliorate soil physical properties (e.g. reduce surface soil compaction (skid trails and ramp areas) and soil strength). The collective impact of these site preparation treatments can increase future harvest yields, reduce rotation lengths and increase economic returns to the landowner by improving seedling survival, altering site resource availability (e.g. through competition control) and encouraging greater root development for water and nutrient uptake (Morris and Lowery, 1988; Allen *et al.*, 1990; Lowery and Gjerstad, 1991).

Following the harvest of a previously established loblolly pine plantation, non-marketable stems may be sheared and the residual logging slash (branches, foliage, roots, stumps) spot-raked into small piles to facilitate other site preparation activities and future planting operations. The amount of logging slash left on site can vary, but estimates from Mississippi suggest an average of about 34.5 dry t ha^{-1} (Schultz, 1997). Bentley and Johnson (2008) examined logging utilization on softwood harvests in Alabama and estimated that about 12% of total softwood volume was left behind as logging residues. These slash piles may be left in place, burned, or alternatively chipped on site and collected when managing for biofuels production. Regardless, careful supervision of the raking/collection operation is warranted to avoid displacement of surface soil into the piles and the possible subsequent reductions in future site productivity (Morris *et al.*, 1983).

On poorly and very poorly drained soils, characteristic of many lower Coastal Plain sites, bedding is commonly used to increase pine survival and growth because high water tables lead to anaerobic conditions. Gent *et al.* (1986), for example, reported that bedding increased height growth by 1 to 3 m. Bedding is also a form of surface soil tillage that mixes organic debris (forest floor) into the mineral soil, which can lead to increased rates of N and P mineralization (nutrient supply) and reduce levels of woody competition. Single- or double-pass bedding operations may be used depending upon the site, with the first pass commonly conducted in the spring and the second bed pass conducted in mid- to late summer. Typically, the beds will need to settle before planting to avoid having seedling root contact with unsettled air pockets, which can lead to increased mortality. Contour bedding is also essential on sites with slopes to reduce the risks of erosion and sediment transport. As is the case with any silvicultural operation, including site preparation activities, users are encouraged to follow the best management practices (BMPs) guidelines for their state (an example BMPs document for the state of Florida can be found at: http://www.floridaforestservice.com/publications/silvicultural_bmp_manual2011.pdf).

Other mechanical site preparation treatments used prior to planting may include roller drum chopping, disking and subsoiling. Chopping breaks up existing woody vegetation and crushes the logging slash into smaller pieces to enhance burning and the decomposition process. It does little, however, to control the development of re-invading woody and herbaceous competition, and is less frequently used today than in the past. Subsoiling may be used on soils having high mechanical resistance (strength) or on soils with cemented horizons that reduce root penetration and effective occupancy of the lower solum.

Chemical treatments

The use of chemical site preparation continues to grow as a cost-effective tool for establishing southern pine plantations. The primary purpose of this treatment is to reduce sprouting and re-establishment of woody and herbaceous forms of competition and to increase pine growth. Today, the availability and efficacy of both foliar and soil active compounds (including generic formulations), coupled with a competitive cost structure and resultant growth benefits to the pines, have made their use commonplace, either singly or in combination with other mechanical site preparation (e.g. bedding) techniques.

The selection of herbicides to be used for site preparation will be dependent on the suite of species to be controlled, soil conditions (texture), costs and site-specific factors related to environmental sensitivity and health and safety (Nelson and Cantrell, 2002). Common site preparation herbicides (pre-plant) for loblolly pine plantations include: glyphosate, hexazinone, imazapyr, triclopyr and metsulfuron. Tank mixes that combine chemicals are commonly used to increase treatment efficacy. Application methods for chemical site preparation vary and may include broadcast or banded applications using tractors/skidders mounted with boom or boomless sprayers, aerial systems, basal stem treatments, cut stump applications, hack and squirt (cut stem) treatments and backpacked foliar sprays.

The US Environmental Protection Agency, responsible for registering all herbicides, classifies them as general or restricted-use. In the latter case, purchase and use requires a certified, licensed applicator. For some compounds (e.g. triclopyr), a wait period may be necessary before planting occurs to avoid mortality losses. Users are encouraged to consult both a professional applicator and the herbicide label for determining the appropriate rates, carrier (e.g. water, oil), volume of spray per hectare, timing, and health and safety concerns. For example, on upland sites a late summer/early autumn application of glyphosate and imazapyr (Chopper®) may be used for broad spectrum control, whereas imazapyr and triclopyr (Garlon®) may be more efficacious on lower Coastal Plain flatwoods sites that contain gallberry (*Ilex glabra*) and saw-palmetto (*Serenoa repens*).

Depending upon the degree of competition control at establishment and the growth rates of the pines, additional herbicide (pine release) treatments may be required during the first 2 to 5 years. Common pine release herbicides may include, among others, imazapyr, hexazinone, sulfometuron methyl and glyphosate. Broadcast and banded applications are commonly conducted, depending upon the herbicide being used and site-specific factors. Tank mixes may also be used to increase the effectiveness of the treatments. In a regional experiment conducted across the southern USA, that included a common study design, Miller *et al.* (1991) examined and compared loblolly pine growth responses across 14 sites to four types of competition control (total control, herbaceous control only, woody control only and no control). After the first 5 years, diameter growth was more responsive than height to the treatments. For example, pine volume associated with the total control treatment was fourfold greater than the no control treatment. The results suggested that herbaceous competition control during the early stages of stand development was critical, with stand volumes being increased on average by 171% across sites compared to 67% with woody control. Subsequent research has also demonstrated that woody competition can become a significant factor decreasing pine growth in older stands (Miller *et al.*, 2003).

20.4.3 Planting stock and initial spacing

Loblolly pine stands being managed for a biofuels objective will generally utilize the same types of planting stock as those used for meeting traditional forestry objectives. The vast majority of seedlings planted will be of bare-root origin (1-0 stock, i.e. 1 year in the seedbed and 0 years in the transplant bed) that are genetically improved (as discussed earlier) for growth, form, disease resistance (e.g. fusiform rust, pitch canker), or possibly chemical composition. Containerized seedlings may also be used in special cases, but they are about twice as expensive as bare-root stock. In general, as the level of genetic improvement increases (e.g. open-pollinated, control-pollinated, clonal varieties) so too does the cost of seedlings. Care must be exercised when selecting the proper planting stock for a site. For example, on sites having a high fusiform rust hazard rating, deployment of fusiform rust-resistant seedlings are recommended as they will decrease the likelihood of associated mortality and growth losses. By comparison, on low rust-hazard sites, seedlings that are genetically improved for growth may be preferentially selected as they can significantly increase biomass yields. Seedling deployment decisions, therefore, should consider the trade-offs between genetic improvement for disease resistance versus growth (e.g. Vergara *et al.*, 2007).

Proper care of seedlings (lifting, storing and transporting), optimal planting season and weather conditions are central elements necessary for establishing a successful loblolly pine stand (Wakeley, 1954, 1969; Harrington and Howell, 1998). Seedlings should be transported from the nursery to the planting site in refrigerated vans that maintain proper temperature and humidity. Most guidelines used to 'time' the planting operations in the southern USA concentrate on the period between early November and mid-February. Several southern states also use a weather classification system based on temperature, relative humidity and wind speed to determine normal, marginal and critical planting conditions (Long, 1991). Both machine- and hand-planting operations are used with loblolly pine, and the choice is often based on topography, residual harvesting debris, physiographic region, prior experience and costs. For example, hand planting may be preferable on steep slopes and those sites with broken topography, whereas machine planting may be used on gentle slopes and 'clean' sites that will not result in a poor planting job with a high rate of skips. Survival surveys are normally conducted after the first growing season, as seedling losses and stand failure will likely occur at that time (Matney and Hodges, 1991). Seedling failures are not common, but must be addressed early in the life of the stand to avoid yield reductions. Growth and yield simulation models, coupled with financial analyses, provide a more realistic basis for evaluating the trade-offs associated with mortality losses than establishing an arbitrary number of seedlings surviving per unit area (Matney and Hodges, 1991).

Decisions on initial planting density will vary depending upon ownership objectives. Some landowners may prefer to establish stands with a dedicated biofuels objective, while others may choose an initial planting density that produces a mixture of traditional forest products (pulpwood, chip-n-saw, sawtimber), including biomass. Stands planted at close spacings tend to produce more total biomass, but the yields are distributed over smaller diameter trees that have less value. For example, Zhao *et al.* (2012) reported stand-level biomass accumulation for loblolly pine grown in a region-wide culture-density experiment that included six different initial spacings. After 12 growing seasons, both cultural intensity (operational versus intensive) and planting density significantly affected total above-ground biomass accumulation, but their interaction was not significant. Stands managed under the intensive culture regime accumulated more total biomass than the operational treatment. Stands planted at a 3.66 × 3.66 m (747 trees ha^{-1}) spacing had significantly less total biomass than all other spacings, but no significant differences in total biomass were found among the remaining spacings, which ranged from 2.44 × 2.74 m (1496 trees ha^{-1}) to 1.83 × 1.22 m (4479 trees ha^{-1}). These results suggest that planting seedlings on a 2.4 × 2.4 m (1736 trees ha^{-1}) to 3.0 × 3.0 m (or equivalent spacing; 1111 trees ha^{-1}) spacing will generally provide flexibility for meeting most landowner objectives with loblolly pine (Smith and Strub, 1991).

Munsell and Fox (2010) conducted a feasibility analysis for increasing woody biomass production from pine plantations in the southern USA. They examined the yields and merchandized product values associated with a 24-year rotation that included two silvicultural treatments (moderate and intensive), two planting densities (1235 trees ha^{-1} versus 1853 trees ha^{-1}), and thinning compared to a dedicated biomass crop managed on three 8-year rotations. Their results suggested that pine plantations managed intensively for a mixture of forest products, or high-density plantings managed for a dedicated biomass supply, could be profitable at current prices in the southern USA. However, biomass stumpage prices and markets would likely need to increase substantially before pine plantations would be managed solely for biomass production on cutover sites.

Alternative spacing arrangements, that also include planting seedlings of varying genetic improvement, have been proposed as a means to provide landowners with an opportunity to meet multiple timber product objectives, including biomass and saw-timber. Referred to as 'FlexStand™ systems' (ArborGen, Inc.), they may include closely spaced, dual rows of genetically improved, open-pollinated seedlings (for biomass) planted between rows of control pollinated or clonal seedlings (for saw-timber). As the stand grows and develops, the alternate biomass rows would be removed first, leaving the more valuable, controlled pollinated/clonal planting stock to meet future saw-timber objectives.

20.4.4 Thinning and fertilization

Density management through intermediate cuttings (thinning) may or may not be a component of a biofuels silvicultural system used with loblolly pine. For example, across the region, loblolly pine begins to exhibit self-thinning mortality at a basal area of about 30 m^2 ha^{-1} when established at traditional spacings (e.g. 1.8 × 3.6 m or 1543 trees ha^{-1}) used for meeting integrated forest product objectives (Jokela *et al.*, 2004; Fig. 20.3). Thus, unthinned stands could either be harvested or thinned prior to reaching that level of stand density. Maximum fibre production typically occurs at the culmination of mean annual increment (MAI), which in unthinned, intensively managed loblolly pine stands may be as early as age 13 years (Martin and Jokela, 2004). It follows that the culmination of MAI in unthinned stands planted at closer initial

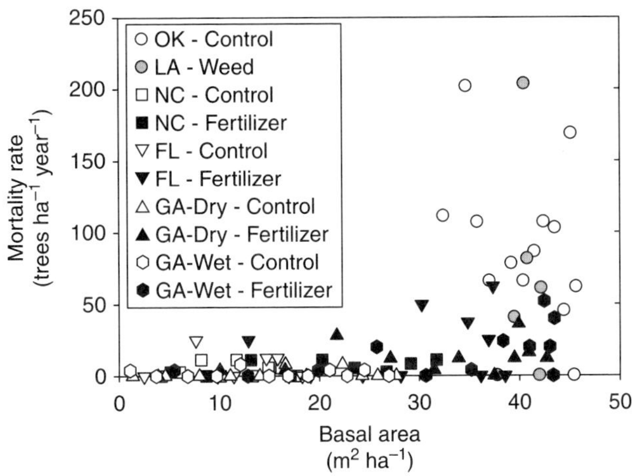

Fig. 20.3. Mortality rate of unthinned loblolly pine stands in relation to stand basal area (adapted from Jokela *et al.*, 2004).

spacings could occur even earlier (e.g. 8 years; Munsell and Fox, 2010). Density management diagrams have been developed for loblolly pine to aid thinning decisions related to timing and the desired upper and lower limits of residual stocking (Dean and Baldwin, 1993, 1996). Depending upon the site, initial planting density and local market conditions, landowners managing loblolly pine stands for an integrated products objective may use their first thinning entry (between ages 10 and 15 years) to supply fibre to a biofuels market.

Soils supporting loblolly pine stands in the southern USA tend to be infertile and nutrient additions are typically required to achieve optimum rates of production (Pritchett and Comerford, 1982; Jokela *et al.*, 1991a; Albaugh *et al.*, 2007; Fox *et al.*, 2007). Early site occupancy and the development of a large and functioning leaf area (leaf area index (LAI)) represents an important strategy for enhancing pine productivity, and correcting nutrient deficiencies through fertilizer additions is an important silvicultural tool for achieving that objective (Colbert *et al.*, 1990). Phosphorus plus N, and P alone, are the nutrient elements that tend to be the most widely limiting and applied to loblolly pine stands. Applications of N alone are not generally recommended in young stands because it often stimulates competing vegetation. In some cases, K and other nutrients may limit loblolly pine growth after N and P demands have been met (Jokela, 2004; Kyle *et al.*, 2005). For example, micronutrient deficiencies have been documented (Mn, Cu) in southern pine stands that were managed intensively using N + P fertilization and understorey competition control treatments (Jokela *et al.*, 1991b; Vogel and Jokela, 2011).

With loblolly pine, fertilization treatments typically occur at or near time of planting and at mid-rotation. On poorly drained, P-deficient clayey soils (e.g. CRIFF A group soils), P applications typically occur in year 1 (40–50 kg ha^{-1} elemental P) and volume responses may average 2.8 to 3.5 m^3 ha^{-1} year^{-1} for 20 years or longer (Pritchett and Comerford, 1982; Jokela *et al.*, 1991a). Responses to P alone have also been documented on upland, well-drained Coastal Plain sites (CRIFF E and F group soils; Allen and Lein, 1998; Leggett and Kelting, 2006). Similarly, in young loblolly pine stands growing on lower Coastal Plain sites (CRIFF A, B, C, D group soils), applications of N + P (e.g. 40 kg ha^{-1} N and 45 kg ha^{-1} P) may be applied alone or in conjunction with herbaceous weed-control treatments within the first 5 years of establishment. Nitrogen and P fertilizer sources that are most commonly used on such sites include diammonium phosphate (DAP; 18-46-0), monoammonium phosphate (MAP; 11-52-0) and urea (45-0-0).

Mid-rotation fertilizer applications are commonly applied at the time of crown closure, when growth demands exceed soil supply for N and P, and nutrient deficiencies restrict loblolly pine leaf area development and stand growth (Allen *et al.*, 1990; Fox *et al.*, 2007). A combination of N and P is most commonly applied over either element alone because it provides the greatest probability and magnitude of growth response (Fox *et al.*, 2007; Fig. 20.4).

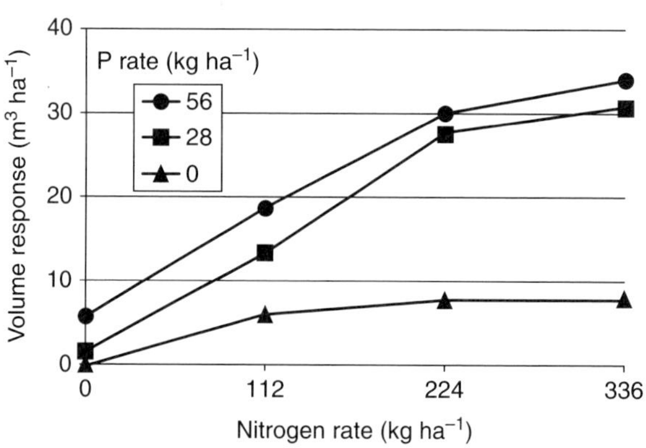

Fig. 20.4. Eight-year volume responses of loblolly pine to N and P fertilization (age 9–16 years) in the southern USA (adapted from Fox *et al.*, 2007).

Application rates range from 150 to 200 kg N ha^{-1} plus 25 kg P ha^{-1} (Jokela, 2004) using DAP and urea as fertilizer sources. Growth responses may last from 6 to 10 years and average about 1.5 m^3 ha^{-1} $year^{-1}$ (Fox *et al.*, 2007).

A number of diagnostic tools have been used, with varying degrees of success, for identifying loblolly pine stands that will be responsive to fertilizer additions, including soil groups, foliar analyses, soil analyses and LAI assessments. For example, critical concentrations of foliar N and P in loblolly pine stands are 1.2% and 0.11%, respectively (Allen, 1987; Jokela *et al.*, 1991a). Stands that have foliar nutrient concentrations that fall below these levels would have a greater likelihood of responding to fertilizer additions, although quantitative relationships that predict the magnitudes of response are still limited. Guidelines for identifying P deficiencies based on soil analyses have been developed and responsive soils would be characterized by having Mehlich-3 extractable P levels below 6 ppm. Similarly, as LAI in loblolly pine stands is highly correlated with potential productivity (Vose and Allen, 1988; Colbert *et al.*, 1990; Albaugh *et al.*, 1998), mid-rotation stands may be identified as responsive to N and P fertilizer additions if projected LAI levels in fully stocked stands are less than 3.5 (Fox *et al.*, 2007).

20.4.5 Treatment interactions

Understanding the importance and magnitude of interactions among silvicultural treatments may enhance the growth potential of loblolly pine stands being managed for biofuels objectives. For example, Jokela *et al.* (2000) reported on ten regional loblolly pine experiments that were established using a common study design that were treated at time of planting (i.e. control, fertilizer only (223 kg ha^{-1} DAP), herbaceous weed control only (hexazionone and sulfometuron methyl) and fertilizer + herbaceous weed control). The percentage of loblolly pine sites exhibiting significant volume responses at age 5 years were 60% for fertilizer only, 40% for herbaceous weed control only and 100% for the fertilizer + herbaceous weed control treatment. In the majority of tests, the fertilizer and herbaceous weed control treatments were additive in nature, suggesting that each treatment was independent relative to its effects on volume growth. However, when both treatments equally contribute to increased soil nutrient supply, less than additive responses may be expected (Jokela *et al.*, 2000; Albaugh *et al.*, 2003). After 8 years, the volume responses for the fertilizer + herbaceous weed control treatment over the untreated control averaged 100% on CRIFF A group soils (100 m^3 ha^{-1} versus 50 m^3 ha^{-1}) and 52% on CRIFF C and D group soils (48 m^3 ha^{-1} versus 32 m^3 ha^{-1}). The early treatment responses declined between ages 5 and 8 years, especially for the herbaceous weed control treatment, suggesting that additional silvicultural inputs, in the form of mid-rotation fertilizer applications, would be necessary to sustain the early growth benefits.

Genotype-by-environment interactions are less common in loblolly pine stands, especially with open-pollinated families that originated from first- or second-generation seed orchards (McKeand *et al.*, 2006b). That is, families have generally exhibited stable performance (few rank changes) across sites within a climatic zone. Less information is available, however, regarding genotype-by-silviculture treatment interactions (McKeand *et al.*, 1997, 2000), but these interactions could become more important as less genetically diverse full-sib families and clones are deployed using more intensive silvicultural regimes (McKeand *et al.*, 2006b). For example, Roth *et al.* (2007) used full-sib family block plot experiments that included two levels of planting density and silvicultural treatment intensity, and reported significant genotype-by-location and genotype-by-silvicultural treatment intensity interactions for loblolly pine at age 5 years. The nature of the interactions and instability of family performance (volume accumulation) were mainly the result of scale effects, where certain families either outperformed or underperformed their peers with increasing intensity of silvicultural treatments.

20.4.6 Yields and rotation ages

Loblolly pine grows faster than slash pine when given adequate nutrients, producing larger diameter stems at younger ages. For example, at the end of 15 years, loblolly and slash pine on average accumulate 136 and 118 t carbon ha^{-1}, respectively (Gonzalez-Benecke *et al.*, 2010, 2011), with about 70% of the carbon in the stem, branches and needles, and over 85% of this above-ground carbon in the stem and branch wood (Johnsen *et al.*, 2004; Gonzales-Benecke *et al.*, 2011).

The expected yields of loblolly pine stands being managed for a biofuels objective will depend upon a number of factors, including site quality, silvicultural management intensity, genetics, planting density and rotation ages. Numerous growth and yield models have been developed for loblolly pine (e.g. PTAEDA4.0; FASTLOB) and most are based on traditional forest product objectives (Clutter, 1963; Sullivan and Clutter, 1972; Burkhart *et al.*, 1985; Baldwin and Feduccia, 1987; Matney and Farrar, 1992; Baldwin *et al.*, 2001; Burkhart, 2008). In addition, many studies have documented biomass accumulation (dry weight) of loblolly pine stands under varying scenarios of management intensity (Colbert *et al.*, 1990; Albaugh *et al.*, 1998; Jokela and Martin, 2000; Adegbidi *et al.*, 2002; Samuelson *et al.*, 2004; Aspinwall *et al.*, 2011). Zhao *et al.* (2012) reported in 12-year-old loblolly pine stands that total above-ground biomass accumulation averaged about 150 dry t ha^{-1} under intensive management compared to 120 dry t ha^{-1} under operational management (Samuelson *et al.*, 2008). Munsell and Fox (2010) modelled biomass yields for loblolly pine stands planted under varying management regimes (biomass, traditional, integrated, traditional + thinning and integrated + thinning), planting densities and site index. Short rotation (8 year) cumulative biomass estimates over a 24-year period for stands planted at 1235 stems ha^{-1} ranged from 215 to 336 wet t ha^{-1} for site index 19.8 m to 25.9 m, respectively. When planting density was increased to 1835 stem ha^{-1} for the same levels of site index, wet cumulative biomass yields ranged from 240 to 493 t ha^{-1}.

Other management options being considered when using loblolly pine to meet a biofuels objective involve an agroforestry intercropping system, where switchgrass (*Panicum virgatum*) is interplanted with loblolly pine. The concept recognizes that the forest has the potential to produce a biofuels feedstock (switchgrass) while still maintaining a supply of traditional forest products. The loblolly pine stands would be established using a relatively low planting density (740 trees ha^{-1}; 6 m between planting rows and 2.25 m within), with switchgrass interplanted between the pine rows. Once the perennial switchgrass stand has been established using herbicides, disk harrowing and seeding procedures, it could be harvested annually. The pines would be harvested occasionally when they reach merchantable sizes for meeting traditional forest products objectives. At present, however, few published studies have quantified the expected combined yields of switchgrass and loblolly pine for biofuels production (Blazier, 2009).

20.5 Tree Harvesting and Chip Processing for Bioenergy

Well-developed harvesting technologies exist to recover forest products from loblolly pine plantations. Harvesting systems have evolved from manual chainsaw and shortwood pulpwood to current highly mechanized operations in response to changing technology, costs and product specifications. Product markets encourage silvicultural and harvesting operations that minimize total cost per tonne. Landowners are incentivized to grow a stand that will maximize their economic return. The management decisions affect stocking, tree size and spacing, and rotation age. At the same time product markets define merchantability specifications including piece size, wood or fibre quality, market pricing and delivery form. The harvesting operation fits between these two with the cost and equipment requirements determined by multiple factors. The interplay of these constraints has defined the most common economically viable plantation management model with a single mid-rotation thinning treatment recovering primarily pulpwood products and a final harvest with saw-timber, oriented strand board or other higher-valued products. In most cases energy-wood is simply another potential product and is harvested using conventional operational technology. A few specialized pine energy harvesting systems have been developed and are described below.

20.5.1 Harvesting systems

Pine harvesting systems (Fig. 20.5) can be classified into two basic types, single product and multi-product. Whether in a thinning or final harvest, the simplest harvesting system takes everything to a common delivery point (single product). Nearly all pine felling is accomplished by feller-bunchers using high-speed disk sawheads (Baker and Greene, 2008). Bunches of felled trees are skidded from the stand to a roadside landing by grapple skidders. At the landing a knuckleboom log loader de-limbs and tops the trees to meet product specifications using a stationary de-limber. Finally, the loader places tree-length material on log trailers for transport to the mill.

A single product harvest operation may be used when there are very few product classes in the management treatment. For example, an early thinning with smaller diameter trees may have few product options other than energy-wood or pulpwood. Similarly, a final regeneration cut on a short rotation pine plantation may have uniform small diameter trees that are only energy-wood material. Taking everything to one market delivery point simplifies operations and maximizes the productivity of in-woods activities. As an example there are pellet mills in the southern USA that purchase tree-length pine for feedstock. This gives the pellet mill control of the debarking and chipping process to ensure product quality. The bark is used in the mill for process heat while the stemwood ends up in the pellets.

While less common, in-woods chipping may also be employed in a single-product system. Smaller diameter and shorter trees (early thinning or energy-wood) are harder to load effectively as tree-length material and may be better utilized by roadside chipping. In this system everything is skidded to the roadside. If the product specification requires low bark

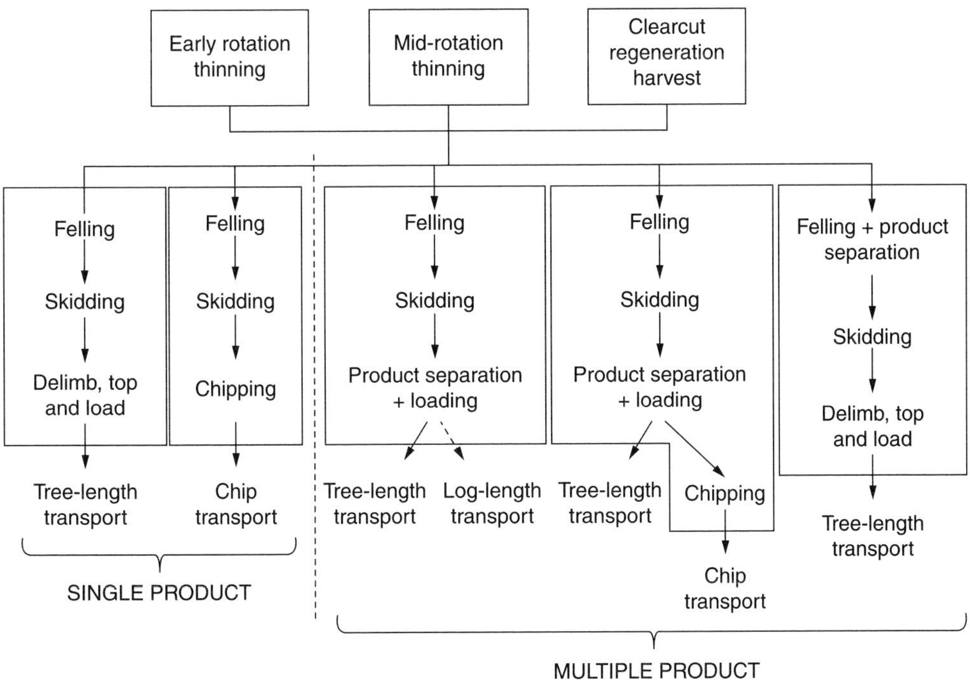

Fig. 20.5. Alternative harvesting systems for plantation pine and energy-wood.

content, the stems are passed through a flail debarker before chipping (clean chips). If product specifications are not constrained by bark content, it is faster and cheaper to process the whole tree through the chipper (whole-tree chips).

Multi-product harvesting systems are used when there is more than one product class in the stand. The system becomes more complex by introducing a sorting and product separation function at the roadside. Merchandizing may be as simple as sorting trees by product classes and loading separate trailers. More complex merchandizing involves bucking trees to meet different market specifications using a slasher saw or processor. Cass *et al* (2009) examined economic trade-offs in product sorting and found that the value differential among product classes was critical to determine the optimum number of product sorts for a given harvest situation. Product sorting adds extra handling, adding cost that must be more than offset by an increase in total value. In a region-wide study of logging contractors deHoop *et al.* (2002) found that more than half of logging contractors used between four and seven product sorts.

One version of the multi-product operation includes sorting and chipping in addition to conventional tree-length or log-length products. Westbrook *et al.* (2007) evaluated the effect of adding a small chipper to concurrently recover residues as fuel chips on a conventional southern pine harvesting operation. They found that residue utilization recovered an additional 15% volume per hectare and reduced site preparation costs by 30%. Another key finding was that the stand product mix had to include a sufficient volume of chippable material to make it worth the additional expense of bringing in another machine.

Another version of a multi-product harvesting system detailed in Fig. 20.5 is product separation at the felling stage. This may occur in a single-pass operation with the feller-buncher cutting and stacking trees in different bunches by product class. A plantation with significant hardwood ingrowth might be an example where this method would be appropriate to separate pine pulpwood from mixed whole-tree energy-wood. This system can also be implemented as a two-pass system with felling and skidding of the energy-wood prior to felling and skidding larger material. Watson *et al.* (1986) compared one- and two-pass systems and identified advantages and disadvantages of each.

While harvesting systems for thinning or final cuts utilize the same functional technology, there are some significant differences. Thinning systems require operation in a residual stand with due consideration for impacts that could affect mortality or wood quality. Stem scars on residual trees can lead to log degrade or pathogen entry. Rutting, soil disturbance and compaction can affect root systems and lead to growth and vigour impacts. Best practices to minimize impacts in thinning include selection of smaller or more manoeuvrable equipment, careful operation to reduce residual stem damage, and more attention to soil conditions and weather to avoid rutting. Equipment manoeuvrability is particularly important if the thinning pattern is a combination of row and selection harvest.

Thinning is also more expensive than final clear-cutting. On a cost per tonne basis felling usually represents the greatest single functional cost, followed by skidding and then loading. If chipping is included in woods operations it is the largest cost component. The single most important factor affecting felling productivity and cost is piece size (Visser and Stampfer, 2003). Cutting time for modern high-speed feller-bunchers is relatively unaffected by diameter over a reasonable range of tree sizes (constant time per tree). However, because piece volume is not linearly related to diameter, the volume cost curve for felling is strongly non-linear (Fig. 20.6). In the example a 10 cm diameter tree is half the diameter of a 20 cm tree but the cost per volume harvested is three times greater. The exponentially increasing costs of smaller trees establish a minimum economically viable thinning size.

20.5.2 Pine energy-wood options

Energy-wood markets introduce some new challenges and opportunities for pine harvesting. Obviously there is the potential to utilize more of the total stand volume if energy specifications accept material not otherwise

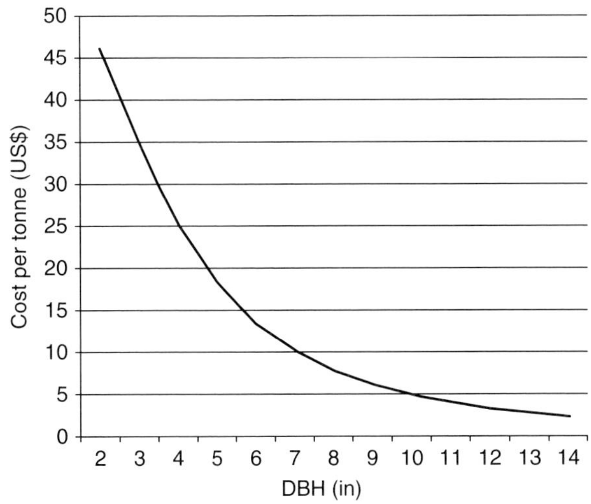

Fig. 20.6. Example felling cost function illustrating the effect of piece size on cost per tonne. DBH, diameter at breast height.

merchantable. Small diameter material, cull logs and defects and understorey hardwood encroachment are all potential energy feedstocks that can be recovered from pine stands. Logging residues like these are often cited as available bioenergy feedstock because they are considered 'waste'. Logging utilization studies (e.g. Bentley and Johnson, 2008) typically find that at least 10% of total softwood volume is left behind in conventional harvests. At an annual softwood harvest of about 170 million m^3, southern pine harvesting has a potential residue volume of about 14.5 Mt green.

Not all logging residues are technically recoverable. Current systems can recover residues from roadside processing by adding a portable grinder. These materials are often high in bark, foliage and ash and may only be suitable for direct combustion. Meadows *et al.* (2011) tested an innovative residue bundler as an alternative to chipping or grinding. They found that this system cost an additional US$13 per tonne to collect residues in bundle form. The bundles still need to be transported and ground for final utilization. Other equipment has been proposed for residue collection, including terrain chippers.

Pine understorey is another potentially available energy feedstock. This may include natural pine regeneration and hardwood ingrowth. Several versions of swath baling machines have been developed that can mow and bale understorey material up to about 15 cm in diameter. For example, do Canto *et al.* (2011) evaluated the performance of a swath baling machine in a pine stand in southern Georgia. The system showed the potential to treat about 0.5 ha h^{-1} but was only able to recover about 10 green t ha^{-1}. The economic viability of such a system is highly dependent on the volume of understorey material to be removed. Koch (1980) envisioned purpose-grown pine energywood with strips of dense, seeded loblolly alternating with rows of planted trees. His system proposed using a swath cutting system to harvest the seeded pine energy-wood strips at age 4 years. This concept may be viable with the new swath baling machines.

As discussed above, loblolly pine can also be grown as a single product stand for energywood (Munsell and Fox, 2010; Zhao *et al.*, 2012), where proposed stands would have more than 1700 trees ha^{-1} and would be harvested in a single clearcut at 8 to 14 years of age, eliminating thinning and product sorting thereby simplifying harvesting operations. Klepac *et al.* (2011) describe an ongoing test of a harvesting system optimized for high-density pine plantations. The biomass feller-buncher uses a high-speed shear to reduce costs. The new grapple skidder design increases grapple size to obtain optimal payloads. Initial results show that the specialized harvesting system will be able to collect bioenergy feedstock more efficiently than current conventional pine harvesting systems.

20.5.3 Technology to manage feedstock specifications

Energy-wood offers the potential to utilize a wider range of tree sizes. However, the conversion process can impose demanding constraints that do not apply to conventional forest products. Net heat value for example may be important if the feedstock is going directly to a thermal combustion process. While most wood has a similar higher heating value (btu per dry kg) the inherent moisture content in the wood reduces the actual heat value. Field moisture content is often about 50% (wet basis). The net heat content, if this material is delivered to a combustion process, is less than half of the potential heat value of the material because of the moisture. Thus, one specification may be moisture content. Harvesting operations that can deliver drier material would be producing feedstock with a higher market value. Currently the only demonstrated method to reduce moisture content is field drying – leaving felled material on the ground for some period of time prior to chipping or loading and transport. Cutshall *et al.* (2011) tested drying periods up to 8 weeks. They found that the moisture content of chips was reduced from 53% to 39% between chipping green and chipping dried material. This represents a 40% btu increase in the delivered final product.

Mitchell (2006) describes other bioenergy product specifications that may be important depending on conversion process. Particle size affects process variables in many types of conversion. Smaller pieces digest faster, burn faster, or densify more easily. Typical pulp chip specifications are optimized for pulp digesters but may not be optimal for bioenergy use. Thus new in-woods equipment has been developed to create 'micro-chips' that are more appropriate for bioenergy use. Micro-chips are about half the size of conventional pulp chips.

Another critical bioenergy specification may be a limit on ash content. Some uses may even specify the mineralogical composition of the ash such as percentage silica. Manipulating ash usually requires separation of bark and foliage from stemwood using a mechanical debarker. Stemwood has the lowest ash content of these three components and the most stringent ash specification would require total debarking. Less exacting specifications may be met by simply de-limbing to separate the limbs and foliage. Logging residues are inherently high in ash content and they are difficult to process. Currently, the only viable technology for cleaning logging residues is post-processing chipped material (Dooley, 2012). Chip beneficiation is one such process in which materials are separated sequentially using screening, flotation and other methods.

20.6 Bioenergy Opportunities and Challenges

Bioenergy in its most basic definition is the release of energy derived from a biological feedstock. In the case of pine as a feedstock, the available woody biomass can be defined in several ways depending on the location within the pine tree. The main stem of the pine tree has several features that make it ideal as a bioenergy feedstock. The near cylindrical shape of the main stem in conjunction with the large percentage of overall biomass tree volume make this portion of the tree the most economical to harvest and transport (Harrill and Han, 2012). The chemical composition of stemwood is of particular interest as this typically accounts for over 60% of the weight of the harvested biomass. In loblolly pine, wood chemical composition varies with cambial meristem age: juvenile wood has slightly higher lignin and lower cellulose contents than mature wood (Sykes *et al.*, 2006; Yeh *et al.*, 2006). Differences in lignin content with tree height are relatively small (Yeh *et al.*, 2006). The greatest difference in wood chemical composition is between mature stems and severe compression wood, which forms on the underside of branches (knots) and bent stems. Severe compression wood contains ~33% more lignin and ~17% less cellulose than mature stemwood (Table 20.2).

Table 20.2. Average percentage chemical composition of loblolly and slash pine tissue.

	Juvenile wood	Mature wood	Compression wood	Bark
Cellulose	44	46	38	20
Hemicellulose	24	24	23	10
Lignin	29	27	36	50
Hydrocarbon extractives	3	3	3	12

Economics play a defining role in the final end-product of the stem material. Typically, large-diameter pine stems are converted to structural materials such as lumber, leaving the smaller-diameter stems for uses such as composites and bioenergy feedstock.

Besides transportation benefits, another advantage of the main stem is the ease with which it can be debarked. This usually results in better mill flexibility and efficiency. The bark, being high in ash content but also possessing higher heats of combustion (Howard, 1973), is ideally suited as a combustion feedstock. The resultant debarked pine stemwood is comprised of approximately 50% carbon as well as the lowest ash portion of the tree, typically between 0.1 and 0.5% (Panshin and de Zeeuw, 1980), thus making it a desirable feedstock for thermochemical processes.

One of the primary sources of available woody pine biomass is residuals, which generally encompass logging residues and pre-commercial thinnings with recent estimates of available forest harvest residuals around 54 Mt dry annually (White, 2010). Characterization of this feedstock is difficult due to the inconsistent proportion of bark, needles and clear wood percentages that individually determine the physical, chemical and thermal properties of the feedstock. Baker *et al.* (2012) examined chip properties from conventional southern pine harvests and found that ash content was generally less than 1% by weight, whereas logging residues have been found to have higher ash content because they may be contaminated with additional ash from soil contact. Although ash percentage from forest residues is markedly different from the corresponding clear-wood values, other important property differences are bulk density, mineral composition and heating value (Das *et al.*, 2011).

20.6.1 Thermochemical conversion

Thermochemical conversion processes are used to breakdown lignocellulosic biomass, releasing heat that can be captured to generate power and smaller carbon-containing molecules that can be polymerized into liquid fuels. A myriad of methods is available and well known (Fig. 20.7), but only three are of practical significance for the southern pines: combustion, gasification and pyrolysis. In the following we discuss these three methods and follow this up with a discussion on biochemical approaches to releasing and capturing energy from lignocellulosic biomass.

Combustion

The combustion of lignocellulosic biomass (wood and woody residues) is the single largest form of energy production in the world, accounting for up to 97% of the world's non-industrial energy production. In essence, combustion of woody biomass is the exothermic oxidation in the presence of air of the principal wood constituents, with carbon being oxidized to carbon dioxide and hydrogen oxidized to water. This series of chemical reaction results can be useful in the production of two types of energy outputs: heat and fuelling steam turbines and steam engines both towards electricity generation. The heat produced from combustion is directly related to the bulk density of the material in conjunction with the

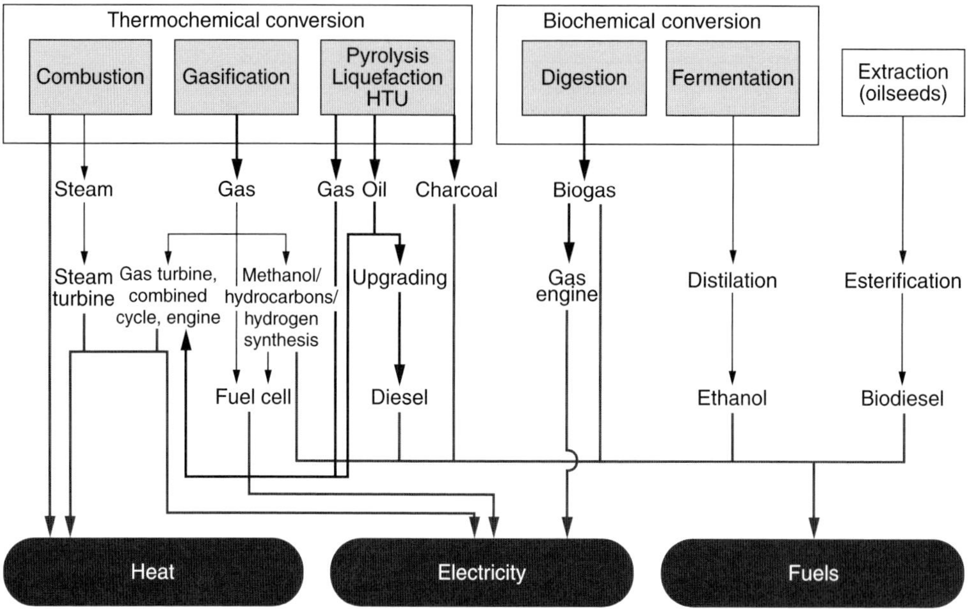

Fig. 20.7. Various thermochemical and biochemical conversion technologies for woody biomass and their various end-product scenarios (from Faaji, 2006).

energy content of the woody biomass. The energy content of woody biomass on a dry, ash-free basis falls within the range of 17 to 21 MJ kg^{-1} (McKendry, 2002a). The energy content of southern pine wood, bark and residuals falls within the upper end of the range at 19 to 21 MJ kg^{-1} (Phanphanich and Mani, 2009).

Southern pine fuel characteristics that govern the efficacy of the woody biomass feedstock are bulk density (which in turn is affected by growth features such as variants, species, geographic distribution, silviculture and age at harvest), material properties such as moisture content and ash content, particle size, and processing variables such as whole-tree chipping and debarking. Geographic distribution can also be a critical factor in overall ring density in the southern pines. For example, loblolly pine ring density is higher in the Gulf and South Atlantic regions than in the Piedmont and Hilly regions (Clark and Daniels, 2004; Jordan *et al.*, 2008).

Gasification

Gasification is a three-step process resulting in the partial oxidation of biomass at temperatures in a restricted oxygen environment into a gaseous fuel. The woody biomass undergoes pyrolysis, combustion and reduction reactions resulting in a mixture of hydrogen, carbon dioxide, carbon monoxide, low molecular weight hydrocarbons (primarily methane) and nitrogen. The resulting gas mixture is commonly referred to as synthesis gas or syngas. The production of syngas generally occurs between 800 and 1200°C. Gasifier designs are typically a function of size. Smaller throughput gasifiers generally are fixed bed reactors that limit the mobility of the raw material. Gas flow within these systems is commonly referred to as updraught or downdraught due to the nature of syngas directional movement. Larger gasification systems require a greater flexibility associated with fuel types and are most typically circulating fluidized bed reactors, allowing for a more dynamic movement of particles during gasification.

The composition and scrubbing of the syngas dictates its final end use. For example, in its gaseous state syngas can be used to power gas turbines or as fuel for diesel or modified internal combustion engines. Syngas produced from woodchips via a fixed bed gasifier was used to power automobile engines in Sweden as far back as the 1920s (McKendry, 2002b). Engines run from syngas typically are used to turn generators that provide electrical power to remote locations or to supplement existing power grids. Although also showing potential for use in fuel cells, technologies are developing that convert the gaseous syngas into various chemical compounds as well as liquid transportation fuels. The Fischer-Tropsch conversion pathway was patented in 1930 (Fischer and Tropsch, 1930) but has always been limited in its use because of economic barriers. Development of catalysts with increased surface area and activation, along with techniques such as the Mobil process (Gujar *et al.*, 2009), has sparked an increase in research efforts.

Due to the heterogeneity of woody biomass as well as the complexity of a typical gasification system, evaluating the effect of woody biomass on syngas composition can be problematic. The carbon conversion reactions during biomass gasification can shift appreciably as a result of different operating conditions and thus determine syngas composition (Deglise and Magne, 1987; Devi *et al.*, 2003). Operating parameters that affect syngas composition are temperature profile within the gasification chamber (Pletka *et al.*, 2001), residence time (Babu and Chaurasia, 2003) and air supply (Chen *et al.*, 2003).

Determination of the effect various woody feedstocks have on syngas composition can be difficult because the differences can be masked by slight fluctuations in operating parameters. Geyer and Walawender (2000) showed that various *Populus* clone feedstocks had no effect on syngas composition. Evaluating at a wider range of hardwood feedstocks, Walawender and Geyer (1988) showed a greater separation of syngas composition although the differences were still small. Elder and Groom (2011) evaluated mixed hardwoods and southern pine mix in a downdraught gasifier and did see appreciable differences in the syngas composition. They evaluated the feedstocks at several flow rates and found that for all flow rates, the hardwood mix produced less carbon monoxide and more hydrogen than the southern pine mix. The carbon monoxide to hydrogen ratio is a critical factor in the conversion to a liquid transportation fuel.

It should be pointed out that although the tree species may have an effect on syngas composition, moisture content of the woody biomass has a greater effect on syngas composition. Higher moisture contents reduce thermal efficiency and thus lower reactor temperatures. The results are lower carbon monoxide levels, increased hydrogen production, lower syngas heating values and higher tar values (Wei *et al.*, 2009).

Pyrolysis

Pyrolysis as it relates to woody biomass is broadly defined as the thermal decomposition of biomass in a non- or low-oxygen environment. Unlike gasification in which there is one desired end-product (syngas), the pyrolysis process produces a liquid tar, a hydrocarbon-rich gas mixture and a solid carbon-rich char. The proportions of these components are primarily attributable to the processing method. The production of char is favoured by low processing temperatures and long residence times. Conversely, moderate temperatures and short residence times optimize the liquid fraction, with high temperatures and long residence times favouring the production of gases (Bridgwater, 2003).

Pyrolysis processes are most often categorized as conventional or fast pyrolysis (Mohan *et al.*, 2006). Pyrolyses are varied in design and are often conducted at temperatures somewhere between the extremes of conventional and fast pyrolysis. Conventional pyrolysis is typically between 400 and 600°C with residence times from 5 to 30 min (Bridgwater, 2003). Fast pyrolysis operates in the same general temperature range (400–550°C) but relies on shorter residence times (<2 s) and increased heat transfer rates thus requiring ground feedstocks (Mohan *et al.*, 2006). Pyrolysis is continuing to evolve with the development of processes such as vacuum pyrolysis (Garcia-Perez *et al.*, 2006), flash pyrolysis (Demirbas, 2002) and vacuum flash pyrolysis (Goyal *et al.*, 2008).

No matter the method of pyrolysis, the end result is a crude liquid referred to as pyrolysis oil, bio-oil, bio-crude-oil and others. The crude pyrolysis liquid is usually dark brown and is composed of a very complex mixture of oxygenated hydrocarbons with an appreciable proportion of water from both the original moisture and reaction product. Solid char and dissolved alkali metals from ash are often present in what is actually a micro-emulsion (Bridgwater, 2003). Although the pyrolytic liquid fraction is composed of a wide spectrum of organic substances (Bridgwater, 1994), the fraction actually consists of two phases: an aqueous phase containing a wide variety of organo-oxygen low molecular weight compounds and higher molecular weight compounds comprising the non-aqueous phase containing insoluble organics (Wenzl, 1970).

The pyrolysis oil can be burned, converted to diesel or other liquid transportation fuels, or used as a chemical feedstock for a suite of commodity products such as adhesives and fertilizers as well as more specialized chemicals such as acetic acid and levoglucosan (Bridgwater, 2003). All chemicals are attractive possibilities due to their much higher added value compared to fuels and energy products, and lead to the possibility of a biorefinery concept in which the optimum combinations of fuels and chemicals are produced.

Evaluating the link between woody biomass species and pyrolysis oil can be challenging due to heterogeneous variables such as juvenility, bark percentage and location within a tree (Table 20.2). Wood that has had the bark removed generally results in a pyrolysis oil yield of 72 to 80% on a weight basis (Mohan *et al.*, 2006), depending on the relative amounts of cellulose and lignin. Pyrolysis oil yields tend to favour lower lignin contents (Mohan *et al.*, 2006). Typically, the lignin content of hardwoods is in the range of 21 ± 3% as opposed to the 27 ± 2% range associated with *Pinus* spp. (Pettersen, 1984). Bark is also high in lignin compared to wood, generally yielding 60–65% pyrolysis oil (Mohan *et al.*, 2006).

20.6.2 Biochemical conversion

For southern pine, thermochemical methods have been favoured over biochemical conversion technologies because of the relatively lower wood sugar levels compared with hardwood (angiosperm) tree species. However, despite lower sugar content, efficient methods for chemical and biochemical conversion of pine (softwood trees) wood to ethanol have been demonstrated, and cost modelling supports the potential of using these conversion technologies at commercial scales (Frederick *et al.*, 2008; Zhu *et al.*, 2010). Concentrated acid hydrolysis to isolate sugars is being commercialized, with a large production facility being built in Mississippi that will use southern pine wood.

Biochemical conversion relies on recovering the sugars present in cellulose and hemicelluloses for fermentation by microbes to produce fuel and other valuable chemicals. A major impediment to cost-effective production of sugars is the inherent recalcitrance of the plant lignocellulosic cell walls to degradation (Himmel *et al.*, 2006). Biochemical conversion methods involve biomass size reduction, pretreatment, hydrolysis/saccharification, microbial fermentation and distillation. While wood can be efficiently chipped, higher amounts of energy are needed for further size reduction and pretreatment compared with lignocellulosic biomass from grasses and crop residues, largely due to the higher density and lignin content in pine wood. Two main approaches have been investigated for solubilizing sugars from lignocellulosic biomass. Concentrated acid at low temperatures directly hydrolyses the lignocellulosic biomass into simple sugars, which are separated from the lignin and acid prior to fermentation (Taherzadeh and Keikhosro, 2007). Acid hydrolysis works with minimal changes in protocol with lignocellulosic biomass, including southern pine, with near theoretical yields of sugars (Galbe and Zacchi, 2002). The other main approach is to pretreat the biomass to increase the surface area and accessibility to enzymatic degradation, thereby increasing the efficiency of saccharification and final sugar yields.

A large number of pretreatment methods have been developed and tested with grass and woody biomass (Mosier *et al.*, 2006). Steam explosion has been tested extensively and is a low-cost effective pretreatment method for grass and hardwood biomass; however, with pine wood addition of dilute acid is required and results in more condensation and enrichment of lignin and yields of sugars are adversely affected (Shevchenko *et al.*, 2001). With wood from pine, modified chemical pulping methods appear to be more effective pretreatment approaches. For example, the SPORL pretreatment method with a furfural removal coupled with enzymatic hydrolysis and yeast fermentation works effectively with pine wood, yielding 304 l per dry tonne (Zhu *et al.*, 2010). A significant advantage of the SPORL pretreatment method is that it starts with wood chips and partial removal of lignin dramatically reduces the energy of fiberization (Zhu *et al.*, 2009), thereby increasing the surface area and accessibility of the cellulose and hemicelluloses to enzymatic degradation.

20.6.3 Improving pines for bioenergy

For the bioenergy and biofuel market to expand with southern pine plantations as a dedicated source of biomass, continued developments that decrease delivered pine biomass cost, increase the conversion efficiency to bioenergy and biofuels and improve value are needed. Total energy yield depends on the efficiency of the growing and harvesting system, as well as the efficiency of converting biomass to usable energy. Energy yield per hectare per year is especially important for biofuels as the efficiency of conversion of different sources of biomass vary for each processing method and fuel synthesis approach. Thus, energy yield on a land area and time basis can be increased by improving growth and biomass yields, carbon allocation to harvested components, and partitioning of carbon into compounds that enhance the yield of energy and fuels from biomass.

Simulated returns of three 8-year biomass rotations with a 24-year mixed product rotation indicate that with intensive management and a typical planting density, a stumpage price of US$11 per wet tonne would return the same value to the landowner as the longer mixed traditional product rotation (Munsell and Fox, 2010). This suggests that increases in biomass yield of short rotation, intensively managed stands will bring additional value to the landowner and lower delivered wood costs while helping to maintain sustainable supplies, and increase in energy per hectare per year. Increasing stem volume has been a principal focus of tree improvement programmes, and genetically improved loblolly and slash pine breeding populations grow faster and yield substantially more wood than unimproved pines (Li *et al.*, 1999; Vergara *et al.*, 2004). As discussed earlier, continued focus on breeding and selecting for faster growth will benefit both traditional and bioenergy/biofuel markets.

20.7 Sustainability in Southern Pine Bioenergy Systems

We have demonstrated that the southern pines (especially loblolly and slash pines) are strong candidates for providing feedstocks for bioenergy products in the southern USA. The goal of sustainable use of native forests is complex and controversial (Nambiar, 1996). However, as rotations become shorter and above-ground organic matter removal becomes more complete, pine silviculture for bioenergy products becomes more akin to agriculture. Similar to increases in agricultural productivity via the Green Revolution, improved silviculture has already, in fact, increased productivity of loblolly pine plantations in the southern USA more than fourfold (Munsell and Fox, 2010). In addition, there is a reliable history of successful multi-rotation tree plantations across an array of environments worldwide (Nambiar, 1996). Where pine plantations in the southern USA use native species, intensively managed tree plantations across the rest of the world more often than not utilize exotic species. As in agriculture, one question is if short rotation intensive culture for bioenergy production is sustainable in the long term. First, of course, we need to define sustainability. One basic goal of sustainable short rotation intensive culture should be that the trend in productivity is non-declining over successive rotations (Nambiar, 1996). However, we should also endeavour to increase productivity over time by applying further advances in forest genetics and silviculture tailored to specific site needs.

There is a deep history of successful plantation forestry around the world (Nambiar, 1996). However, the concept of sustainable use of forests is still complex and controversial. In his treatise on sustainable productivity, Powers (1999) asserted that obvious declines in productivity in plantations using conventional logging are rare and can be attributed to poor management. However, clear declines have been observed following whole-tree harvesting (WTH) operations as would occur in short rotation bioenergy plantations. Sitka spruce (*Picea sitchensis*) trees following WTH were 18% shorter than controls after 9 years (Proe and Dutch, 1994). Trees from the WTH plots had substantially less N, P and K than controls. On a nitrogen-limited site following WTH, 24-year-old Scots pine (*Pinus sylvestris*) trees had 20% less wood and bark biomass than trees from the conventionally harvested treatment (Egnell and Valinger, 2003). Similarly, 31 years after treatments were imposed, basal area of Norway spruce (*Picea abies*) was 25% less than trees from the conventionally harvested plots (Egnell, 2011). This reduction was due to a temporary reduction in growth in years 8 to 12 when foliar N was also observed to be low. He suggested N fertilization would have ameliorated the response. Like the cases above, yields from radiata pine (*Pinus radiata*) plantations (not subjected to WTH) on sites low in nutrients declined over time (Keeves, 1966). However, research on *P. radiata* silviculture in Australia has resulted in subsequent plantations being highly productive (Nambiar, 1996).

In the late 1980s and early 1990s, the US Forest Service initiated a series of studies called the Long-Term Soil Productivity Studies (LTSP), which among other factors examined WTH across an array of forest types. Generally, after 10 years, WTH has had no impact on site productivity (Powers *et al.*, 2005) similar to 15-year results of Johnson and Todd (1998) in a mixed oak stand in Tennessee, and specifically, WTH did not impact 10 year productivity of loblolly pine stands in North Carolina and Louisiana (Sanchez *et al.*, 2006). In the loblolly pine examples, productivity declines were negligible even when WTH was combined with total surface organic removal.

Again, we should not only strive to sustain basic site productivity but to maintain, and even increase, high levels of production achievable via intensive forestry (Fox, 2000). Earlier in this chapter we focused on the importance of nutrient management in southern pine intensive forestry. Soil management is seen increasingly as the basis for sustainable forest productivity (Nambiar, 1996; Powers *et al.*, 2005). Loblolly pine grown in factorial combination of nutrient and water treatments on a droughty site indicated the overwhelming importance of soil nutrition in loblolly pine productivity (Albaugh *et al.*, 2004). Given the state of knowledge of nutrient supply and demand in loblolly pine, fertilization will clearly be required to sustain high productivity on most sites in the southern USA (Eisenbies *et al.*, 2009). We also discussed the importance of tree improvement for growth and pest resistance and its requirement in continuing to increase productivity in loblolly pine plantations.

It has been claimed that intensively managed monoculture plantations are at greater risk for catastrophic loss than low intensity, multi-species plantations, however evidence to this effect is minimal (Nambiar, 1996). Risks, for southern pine bioenergy plantation monocultures exist, but we consider them minor based on the vast degree of confirmation from traditional and the more intensive southern pine culture of recent decades. In fact, risks from abiotic agents such as hurricanes (Johnsen *et al.*, 2009) become lower, and matching pest-resistant genotypes to site become easier as rotation ages become shorter.

Short rotation intensive culture for bioenergy production will result in more mechanical entries into plantations potentially increasing soil bulk density and thus decreasing soil porosity (Powers *et al.*, 2005). After two decades, topsoil removal with moderate and substantial compaction reduced stand volumes by 8 and 42%, respectively (Murphy *et al.*, 2004), in radiata pine stands. However, the effects of topsoil removal and compaction were confounded and could not be isolated. In the LTSP experiments, soil compaction was included as a discrete treatment so compaction responses could be isolated; forest productivity responses depended on other factors. Production declined with compaction on compacted clays and increased on sands (Powers *et al.*, 2005). In their further work on mechanical harvesting impacts on productivity, Eisenbies *et al.* (2006, 2007) found that productivity was not reduced by soil physical disturbances. These results suggest that bedding (wet sites) and other tillage such as subsoiling or ripping (dry sites) may become more important as harvesting operations on a site become more frequent (Edwards *et al.*, 2006) such as will occur in bioenergy plantations.

Is there an indicator that can be simply measured to track sustainability and provide warnings as to obstructions to its continuity? Such indicators will likely be soil based. One such indicator suggested by Nambiar (1996) and expounded upon by Sanchez (1998) is the quantification of soil organic matter as it is so entwined with nutrient and water retention. However, loblolly pine plantations do not appear to be impacted by severe alterations of soil organic matter. In the LTSP experiments, after 10 years, soil C concentrations were reduced from the removal of all surface organic matter but there were negligible effects on growth rates. In the WTH-only plots soil organic matter was not impacted (Powers *et al.*, 2005), similar to results observed across four disparate hardwood and pine forests in the southern USA (Johnson *et al.*, 2002). The most likely reason is that following harvest, fine and coarse root decomposition inputs C into the soil. Data from the North Carolina loblolly pine LTSP installation shows a temporary pulse of soil C concentration over the 10-year period (Johnsen *et al.*, 2004). Such soil C inputs will even be higher in loblolly pine if taproot decomposition is considered (Ludovici *et al.*, 2002; Maier and Johnsen, 2010; Maier *et al.*, 2012).

Forest productivity has been shown to be directly related to the fertility rating (FR) as encompassed in the model 3-PG (Physiological Principles Predicting Growth, Landsberg and Waring (1997)), which is a simple process-based, stand-level model of growth. The FR ranges from 0 to 1 and for the most part has been estimated by statistical 'fitting' from representative calibration stands and testing the calibrated model on adjacent stands (Landsberg *et al.*, 2001) or stands of the same species subjected to similar management (Sands and Landsberg, 2002). Progress has been made in relating FR to soil texture (Sampson *et al.*, 2008). Thus, FR might provide a basis for tracking southern pine bioenergy plantation sustainability if it can be related to a set of simply measured soil properties. Therefore, continued research in this area might prove particularly fruitful. For now, however, sustainability will need to be tracked via subsampling productivity of bioenergy operations, stratified perhaps by soil type, over time.

Thus, it is very likely that southern pine energy plantations' productivity can be sustained and even increased over time. The goal will be to direct environmental resources such as light, water and nutrients to the crop trees via the judicious use of herbicides, pesticides, fertilizer and site preparation and to do so in a manner that takes into account the impacts of increasing vegetation and nutrient removals and, perhaps, the impact of more intensive mechanical harvesting. Continued genetic improvements in growth rate, pest and pathogen resistance will enhance productivity and value. A plethora of research (Jokela *et al.*, 2004; Carter and Foster 2006; Vance *et al.*, 2010) can be directed toward these efforts, and the infrastructure is in place to research issues that might arise or are particular to the use of southern pine as a bioenergy crop for feedstock production.

20.8 Summary and Conclusions

The southern pines, particularly loblolly and slash pines, offer many advantages as lignocellulosic feedstocks for bioenergy and biofuel production. Silvicultural practices and genetic improvements in planting stocks have been implemented broadly on millions of hectares of land across the southern USA where food crop production is not viable. Furthermore, southern pine energy-wood plantation productivity can be sustained over time and even increased with continuing genetic and silvicultural improvements. The emerging energy-wood market product specifications offer the potential to utilize smaller trees than traditional markets (i.e. pulpwood, saw-timber) opening new opportunities for earlier harvests (i.e. thinning or short rotations), closer tree-to-tree spacing and thus better site utilization. However, the higher costs associated with handling greater numbers of small trees needs to be overcome to improve the economic feasibility of delivering southern pine biomass for transportation fuels that are inherently of limited value. Altering wood chemical composition through genetics offers the opportunity to enhance energy yields and provide higher value chemical feedstocks that will improve the economics of southern pine energy-wood production.

References

Adegbidi, H.G., Jokela, E.J., Comerford, N.B. and Barros, N.F. (2002) Biomass development of intensively managed loblolly pine plantations growing on spodosols in the southeastern USA. *Forest Ecology and Management* 167, 91–102.

Albaugh, T.J., Allen, H.L., Dougherty, P.M., Kress, L.W. and King, J.S. (1998) Leaf-area and above- and belowground growth responses of loblolly pine to nutrient and water additions. *Forest Science* 44, 317–328.

Albaugh, T.J., Allen, H.L., Zutter, B.R. and Quicke, H.E. (2003) Vegetation control and fertilization in midrotation *Pinus taeda* stands in southeastern United States. *Annals Forest Science* 60, 619–624.

Albaugh, T.A., Allen, H.L., Dougherty, P.M. and Johnsen, K.H. (2004) Long term growth responses of loblolly pine to optimal nutrient and water resource availability. *Forest Ecology and Management* 192, 3–19.

Albaugh, T.J., Allen, H.L. and Fox, T.R. (2007) Historical patterns of forest fertilization in the southeastern United States. *Southern Journal of Applied Forestry* 31(3), 129–137.

Allen, H.L. (1987) Fertilizers: adding nutrients for enhanced forest productivity. *Journal of Forestry* 85, 37–46.

Allen, H.L. and Lein, S. (1998) Effects of site preparation, early fertilization, and weed control on 14-year old loblolly pine. In: *Proceedings Southern Weed Science Society* 51, 104–110.

Allen, H.L., Dougherty, P.M. and Campbell, R.G. (1990) Manipulation of water and nutrients – practice and opportunity in southern US pine forests. *Forest Ecology and Management* 30, 437–453.

Aspinwall, M.J., King, J.S., McKeand, S.E. and Bullock, B.P. (2011) Genetic effects on stand-level uniformity, and above- and belowground biomass production in juvenile loblolly pine. *Forest Ecology and Management* 262, 609–619.

Atwood, R.A., White, T.L. and Huber, D.A. (2002) Genetic parameters and gains for growth and wood properties in Florida source loblolly pine in the southeastern United States. *Canadian Journal of Forest Research* 32, 1025–1038.

Babu, B.V. and Chaurasia, A.S. (2003) Modeling, simulation, and estimation of optimum parameters in pyrolysis of biomass. *Energy Conversion and Management* 44(13), 2135–2158.

Baker, S. and Greene, D. (2008) Changes in Georgia's logging workforce, 1987-2007. *Southern Journal of Applied Forestry* 32(2), 60–68.

Baker, S., Greene, D. and Wilson, A. (2012) Fuel characteristics of woods – run whole tree southern pine chips. *Biomass and Bioenergy* 37, 67–72.

Baldwin, V.C., Jr and Feduccia, D.P. (1987) Loblolly pine growth and yield prediction for managed West Gulf plantations. *Research Paper* SO-236. US Department of Agriculture, Forest Service, Southern Forest Experiment Station, New Orleans, Louisiana, 27 pp.

Baldwin, V.C., Jr, Burkhart, H.E., Westfall, J.A. and Peterson, K.D. (2001) Linking growth and yield and process models to estimate impact of environmental changes on growth of loblolly pine. *Forest Science* 47, 77–82.

Belonger, P.J., McKeand, S.E. and Jett, J.B. (1996) Genetic and environmental effects on biomass production and wood density in loblolly pine. In: Dieters, M.J., Matheson, A.C., Nikles, D.G., Harwood, C.E. and Walker, S.M. (eds) *Tree Improvement for Sustainable Tropical Forestry*. Proc. QFRI-IUFRO Conference, Caloundra, Queensland, Australia, 27 October–1 November 1996, pp. 307–310.

Bentley, J.W. and Johnson, T.G. (2008) Alabama harvest and utilization study. Resource Bulletin SRS-141. US Department of Agriculture Forest Service, Southern Research Station, Asheville, North Carolina, 24 pp.

Blazier, M. (2009) Perfect pair for biofuel: switchgrass and trees. Louisiana Agriculture. Available at: http://text.lsuagcenter.com/en/communications/publications/agmag/Archive/2009/fall/Perfect+Pair+for+Biofuel+Switchgrass+and+Trees.htm (accessed 1 June 2012).

Bridgwater, A.V. (1994) Catalysis in thermal biomass conversion. *Applied Catalysis A: General* 116 (1–2), 5–47.

Bridgwater, A.V. (2003) Renewable fuels and chemicals by thermal processing of biomass. *Chemical Engineering Journal* 91(2–3), 87–102.

Burkhart, H.E. (2008) Modelling growth and yield for intensively managed forests. *Forest Science* 24, 173–180.

Burkhart, H.E., Cloeren, D.C. and Amateis, R.L. (1985) Yield relationships in unthinned loblolly pine plantations on cutover, site-prepared lands. *Southern Journal of Applied Forestry* 9, 84–91.

Byram, T.D., Myszewski, J.H., Gwaze, D.P. and Lowe, W.J. (2005) Improving wood quality in the western gulf forest tree improvement program: the problem of multiple breeding objectives. *Tree Genetics and Genomes* 1, 85–92.

Carter, M.C. and Foster, D.C. (2006) Milestones and millstones: a retrospective on 50 years of research to improve productivity in loblolly pine plantations. *Forest Ecology and Management* 227, 137–144.

Cass, R.D., Baker, S. and Greene, D. (2009) Costs and productivity impacts of product sorting on conventional ground-based timber harvesting operations. *Forest Products Journal* 59(11/12), 108–114.

Chen, G., Andries, J., Luo, Z. and Spliethoff, H. (2003) Biomass pyrolysis/gasification for product gas: the overall investigation of parametric effects. *Energy Conversion and Management* 44(11), 1875–1884.

Clark, A. and Daniels, R. (2004) Modeling the effect of physiogeographic region on wood properties of planted loblolly pine in the southern United States. In: *Connection between Forest Resource and Wood Quality: Modeling Approaches and Simulation Software*, Fourth Workshop IUFRO Working Party S5.01-04, 8–15 September 2002, Harrison Hot Springs, British Columbia. INRA-Centre de Recherches de Nancy, France, pp. 54–60.

Clutter, J.L. (1963) Compatible growth and yield models for loblolly pine. *Forest Science* 9, 354–371.

Colbert, S.R., Jokela, E.J. and Neary, D.G. (1990) Effects of annual fertilization and sustained weed control on dry matter partitioning, leaf area, and growth efficiency of juvenile loblolly and slash pine. *Forest Science* 36, 995–1014.

Conner, R.C. and Hartsell, A.J. (2002) Forest area and conditions. In: Wear, D.N. and Greis, J.G. (eds) *The Southern Forest Resource Assessment*. USDA Forest Service, General Technical Report SRS-53, Asheville, North Carolina, 635 pp.

Cutshall, J., Greene, D., Baker, S. and Mitchell, D. (2011) Transpirational drying effects on energy and ash content from wholetree chipping. In: *Proceedings of the 34th Council on Forest Engineering meeting*, Quebec City, Quebec, 9 pp.

Das, K.C., Singh, K., Bibens, B., Hilten, R., Baker, S.A., Greene, W.D. and Peterson, J.D. (2011) Pyrolysis characteristics of forest residues obtained from different harvesting methods. *Applied Engineering in Agriculture* 27(1), 107–113.

Dean, T.J. and Baldwin, V.C., Jr (1993) Using a density management diagram to develop thinning schedules for loblolly pine plantations. *Research Paper* SO-275. US Department of Agriculture, Forest Service, Southern Research Station, New Orleans, Louisiana, 7 pp.

Dean, T.J. and Baldwin, V.C. (1996) Growth in loblolly pine plantations as a function of stand density and canopy properties. *Forest Ecology and Management* 82, 49–58.

Deglise, X. and Magne, P. (1987) Pyrolysis and industrial charcoal. In: *Biomass: Regenerable Energy*. John Wiley and Sons, London, pp. 221–235.

deHoop, C.F., Egan, A.F., Greene, W.D. and Mayo, J.H. (2002) Surveys of the logging contractor population – eight southern states and Maine. *Working Paper 55*. Louisiana Forest Products Development Center, Louisiana State University, 13 pp.

Demirbas, A. (2002) Analysis of liquid products from biomass via flash pyrolysis. *Energy Sources* 24(2), 337–345.

Devi, L., Ptasinski, K.J. and Janssen, F.J.J.G. (2003) A review of the primary measures for tar elimination in biomass gasification processes. *Biomass and Bioenergy* 24, 125–140.

do Canto, J.L., Klepac, J., Rummer, B., Savoie, P. and Seixas, F. (2011) Evaluation of two round baling systems for harvesting understory biomass. *Biomass and Bioenergy* 35, 2163–2170.

Dooley, J. (2012) Beneficiation of chipped and shredded woody biomass. *Abstract at International Wood Composites Symposium*, 11–14 April 2012. Washington State University, Seattle. Pullman, Washington, USA.

Edwards, S.L., Ezell, A.W. and Demarais, S. (2006) A comparison of planted loblolly pine (*Pinus taeda*) growth in areas receiving different levels of establishment regime intensity. *Journal of Sustainable Forestry* 23, 1–16.

Egnell, G. (2011) Is the productivity decline in Norway spruce following whole-tree harvesting in the final felling in boreal Sweden permanent or temporary? *Forest Ecology and Management* 261, 148–153.

Egnell, G. and Valinger, E. (2003) Survival, growth and growth allocation of planted scots pine trees after different levels of biomass removal in clear felling. *Forest Ecology and Management* 177, 65–74.

Eisenbies, M.H., Burger, J.A., Aust, W.M., Patterson, S.C. and Fox, T.R. (2006) Assessing change in soil-site productivity on intensively managed loblolly pine plantations. *Soil Science Society of America Journal* 70, 130–140.

Eisenbies, M.H., Burger, J.A., Aust, W.M. and Patterson, S.C. (2007) Changes in site productivity and recovery of soil properties following wet- and dry-weather harvesting disturbance in the Atlantic Coastal Plain for a stand of age 10 years. *Canadian Journal of Forest Research* 37, 1336–1348.

Eisenbies, M.H., Vance, E.D., Aust, W.M. and Seiler, J.R. (2009) Intensive utilization of harvest residues in southern pine plantations: quantities available and implications for nutrient budgets and sustainable site productivity. *Bioenergy Research* 2, 90–98.

Elder, T. and Groom, L.H. (2011) Pilot-scale gasification of woody biomass. *Biomass and Bioenergy* 35, 3522–3528.

Erickson, V., Aubry, C., Berrang, P., Blush, T., Bower, A., Crane, B., DeSpain, T., Gwaze, D., Hamlin, J., Horning, M., Johnson, R., Mahalovich, M., Maldonado, M., Sniezko, R. and St Clair, B. (2012) *Genetic Resource Management and Climate Change: Genetic Options for Adapting National Forests to Climate Change.* USDA Forest Service, Forest Management, Washington, DC, 24 pp.

Evans, J.M. and Cohen, M.J. (2009) Regional water resource implications of bioethanol production in the southeastern United States. *Global Change Biology* 15, 2261–2273.

Faaji, A. (2006) Modern biomass conversion technologies. *Mitigation and Adaption Strategies for Global Change* 11, 343–375.

Fischer, F. and Tropsch, H. (1930) Process for the production of paraffin-hydrocarbons with more than one carbon atom. *US Patent Number* 1,746,464. 3 pp.

Fisher, R.F. (1981) Soils interpretations for silviculture in the southeastern Coastal Plain. In: Barnett, J. (ed.) *Proceedings of the First Biennial Southern Silvicultural Research Conference,* Atlanta, Georgia, 6–7 November 1980. USDA Forest Service General Technical Report SO-GTR-34, pp. 323–330.

Fox, T.R. (2000) Sustained productivity in intensively managed forest plantations. *Forest Ecology and Management* 138, 187–202.

Fox, T.R. (2004) Species deployment strategies for the southern pines: site specific management practices for the flatwoods of Georgia and Florida. In: Dickens, E.D., Barnett, J.P., Hubbard, W.G. and Jokela, E.J. (eds) *Slash Pine: still growing and growing! Proceedings of the Slash Pine Symposium, General Technical Report 76,* US Department of Agriculture, Forest Service, Southern Research Station, Asheville, North Carolina, pp. 50–55.

Fox, T.R., Allen, H.L., Albaugh, T.J., Rubilar, R. and Carlson, C.A. (2007) Tree nutrition and forest fertilization of pine plantations in the southern United States. *Southern Journal of Applied Forestry* 31, 5–11.

Frederick, W.J. Jr, Lien, S.J., Courchene, C.E., DeMartini, N.A., Ragauskas, A.J. and Iisa, K. (2008) Production of ethanol from carbohydrates from loblolly pine: a technical and economic assessment. *Bioresource Technology* 99(11), 5051–5057.

Galbe, M. and Zacchi, G. (2002) A review of the production of ethanol from softwood. *Applied Microbiology and Biotechnology* 59, 618–628.

Garcia-Perez, M., Lappas, P., Hughes, P., Dell, L., Chaala, A. Kretschmer, D. and Roy, C. (2006) Evaporation and combustion characteristics of biomass vacuum pyrolysis oils. *IFRF Combustion Journal,* Volume (2006), 200601, 28 pp.

Gent, J.A., Allen, H.L., Campbell, R.G. and Wells, C.G. (1986) Magnitude, duration, and economic analysis of loblolly pine growth response following bedding and phosphorus fertilization. *Southern Journal of Applied Forestry* 10, 124–128.

Geyer, W.A. and Walawender, W.P. (2000) Biomass and gasification properties of young *Populus* clones. *Wood and Fiber Science* 32(3), 375–384.

Gonzalez-Benecke, C.A., Martin, T.A., Cropper, W.P. Jr and Bracho, R. (2010) Forest management effects on *in situ* and *ex situ* slash pine forest carbon balance. *Forest Ecology and Management* 260, 795–805.

Gonzalez-Benecke, C.A., Martin, T.A., Jokela, E.J. and De La Torre, R. (2011) A flexible hybrid model of life cycle carbon balance for loblolly pine (*Pinus taeda* L.) management systems. *Forests* 2, 749–776.

Goyal, H.B., Seal, D. and Saxena, R.C. (2008) Bio-fuels from thermochemical conversion of renewable resources: a review. *Renewable and Sustainable Energy Reviews* 12(2), 504–517.

Gräns, O.A.D. (2012) Effects of genetics and silviculture and their interactions on loblolly pine and Norway spruce growth and wood properties. PhD dissertation, NC State University, Raleigh, North Carolina, 181 pp. (Available at: http://repository.lib.ncsu.edu/ir/bitstream/1840.16/7616/1/etd.pdf)

Gujar, A.C., Guda, V.K., Nolan, M., Yan, Q., Toghiani, H. and White, M.G. (2009) Reactions of methanol and higher alcohols over H-ZSM-5. *Applied Catalysis A: General* 363(1–2), 115–121.

Harrill, H. and Han, H.-S. (2012) Productivity and cost of integrated harvesting of wood chips and sawlogs in stand conversion. *International Journal Forestry Research,* Volume (2012), 893079, 10 pp.

Harrington, T.B. and Howell, K.D. (1998) Planting cost, survival, and growth one to three years after establishing loblolly pine seedlings with straight, deformed, or pruned taproots. *New Forests* 15, 193–204.

Harvey, B.G., Wright, M.E. and Quintana, R.L. (2010) High-density renewable fuels based on selective dimerization of pinenes. *Energy Fuels* 24, 267–273.

Himmel, M.E., Ding, S.-Y., Johnson, D.K., Adney, W.S., Nimlos, M.R., Brady, J.W. and Foust, T.D. (2006) Biomass recalcitrance: engineering plants and enzymes for biofuels production. *Science* 315, 804–807.

Howard, E.T. (1973) Heat of combustion of various southern pine materials. *Wood Science* 5(3), 194–197.

Johnsen, K.H., Teskey, R., Samuelson, L., Butnor, J., Maier, C., Sampson, D. and McKeand, S. (2004) Carbon sequestration in loblolly pine plantations: methods, limitations and research needs for estimating storage pools. In: Rausche, M.H.M. and Johnsen K.H. (eds) *Southern Forest Science: Past, Present, and Future.* General Technical Report SRS-75, USDA Forest Service, Southern Research Station, Asheville, North Carolina, 394 pp.

Johnsen, K.H., Butnor, J.R., Kush, J.S., Schmidtling, R.C. and Nelson, C.D. (2009) Longleaf pine displays less wind damage than loblolly pine. *Southern Journal of Applied Forestry* 33, 178–181.

Johnson, D.W. and Todd, D.E. (1998) Harvesting effects on long-term changes in nutrient pools of mixed oak forest. *Soil Science Society of America Journal* 62, 1725–1735.

Johnson, D.W., Knoepp, J.D., Swank, W.T., Shan, J., Morris, L.A., Van Lear, D.H. and Kapeluck, P.P. (2002) Effects of forest management on soil carbon: results from some long-term resampling studies. *Environmental Pollution* 116, S201–S208.

Jokela, E.J. (2004) Nutrient management of southern pines. In: Dickens, E.D., Barnett, J.P., Hubbard, W.G. and Jokela, E.J. (eds) Slash Pine: still growing and growing! *Proceeding of the Slash Pine Symposium, USDA Forest Service General Technical Report* SRS-76, pp. 27–35.

Jokela, E.J. and Martin, T.A. (2000) Effects of ontogeny and soil nutrient supply on production, allocation and leaf area efficiency in loblolly and slash pine stands. *Canadian Journal Forest Research* 30, 1511–1524.

Jokela, E.J., Allen, H.L. and McFee, W.W. (1991a) Fertilization of southern pines at establishment. In: Duryea, M.L. and Dougherty, P.M. (eds) *Forest Regeneration Manual.* Kluwer Academic Publishers, Dordrecht, the Netherlands, pp. 263–277.

Jokela, E.J., McFee, W.W. and Stone, E.L. (1991b) Micronutrient deficiency in slash pine: response and persistence of added manganese. *Soil Science Society of America Journal* 55, 492–496.

Jokela, E.J., Wilson, D. and Allen, J.E. (2000) Early growth responses of slash and loblolly pine following fertilization and herbaceous weed control treatments at establishment. *Southern Journal of Applied Forestry* 24, 23–30.

Jokela, E.J., Dougherty, P.M. and Martin, T.A. (2004) Production dynamics of intensively managed loblolly pine stands in the southern United States: a synthesis of seven long- term experiments. *Forest Ecology and Management* 192, 117–130.

Jokela, E.J., Martin, T.A. and Vogel, J.G. (2010) Twenty-five years of intensive forest management with southern pines: important lessons learned. *Journal of Forestry* 108, 338–347.

Jordan, L., Clark III, A., Schimleck, L.R., Hall, D.B. and Daniels, R.F. (2008) Regional variation in wood specific gravity of planted loblolly pine in the United States. *Canadian Journal of Forest Research* 38, 698–710.

Keeves, A. (1966) Some evidence of loss of productivity with successive rotations of *Pinus radiata* in the southeast of South Australia. *Australian Forestry* 30, 51–63.

Klepac, J., Rummer, B. and Thompson, J. (2011) Harvesting small trees for bioenergy. In: *Proceedings of the 34th Council on Forest Engineering Meeting*. June 2011, Quebec City, Quebec, 11 pp.

Koch, P. (1980) Concept for southern pine plantation operation in the year 2020. *Journal of Forestry* 78(2), 78–82.

Kyle, K.H., Andrews. L.J., Fox, T.R., Aust, W.M., Burger, J.A. and Hansen, G.H. (2005) Long-term impact of drainage, bedding, and fertilization on growth of loblolly pine (*Pinus taeda* L.) in the Coastal Plain of Virginia. *Southern Journal of Applied Forestry* 29, 205–214.

Landsberg, J.J. and Waring, R.H. (1997) A generalized model of forest productivity using simplified concepts of radiation-use efficiency, carbon balance and partitioning. *Forest Ecology and Management* 95, 209–228.

Landsberg, J.J., Johnsen, K.H., Albaugh, T.J., Allen, H.L. and McKeand, S.E. (2001) Applying 3-PG, a simple process-based model designed to produce practical results, to data from loblolly pine experiments. *Forest Science* 47, 43–51.

Leggett, Z.H. and Kelting, D.L. (2006) Fertilization effects on carbon pools in loblolly pine plantations on two upland sites. *Soil Science Society of America Journal* 70, 279–286.

Li, B., McKeand, S. and Weir, R. (1999) Tree improvement and sustainable forestry – impact of two linking growth and yield and process models to estimate impact of environmental changes on growth of loblolly pine. *Forest Science* 47, 77–82.

Li, L., Zhou, Y., Cheng, X., Sun, J., Marita, J.M., Ralph, J. and Chiang, V.L. (2003) Combinatorial modification of multiple lignin traits in trees through multigene cotransformation. *The Proceedings of the National Academy of Sciences* USA 100, 4939–4944.

Little, E.L. Jr and Critchfield, W.B. (1969) *Subdivisions of the Genus* Pinus *(Pines)*. US Dept. of Agriculture Forest Service Miscellaneous Publication No. 1144, 51 pp.

Long, A.J. (1991) Proper planting improves performance. In: Duryea, M.L. and Dougherty, P.M. (eds) *Forest Regeneration Manual*. Kluwer Academic Publishers, Dordrecht, the Netherlands, pp. 303–320.

Lowery, R.F. and Gjerstad, G.H. (1991) Chemical and mechanical site preparation. In: Duryea, M.L. and Dougherty, P.M. (eds) *Forest Regeneration Manual*. Kluwer Academic Publishers, Dordrecht, the Netherlands, pp. 251–261.

Ludovici, K.H., Zarnoch, S.J. and Richtor, D.D. (2002) Modeling *in-situ* pine root decomposition using data from a 60-year chronosequence. *Canadian Journal of Forest Research* 32, 1675–1684.

Maier, C.A. and Johnsen, K.H. (2010) Quantifying carbon sequestration in forest plantations by modeling the dynamics of above- and below-ground C pools. *Proceedings of the 14th Biennial Southern Silviculture Research Conference,* General Technical Report SRS–121, US Department of Agriculture, Forest Service, Southern Research Station, Asheville, North Carolina, pp. 3–8.

Maier, C.A., Johnsen, K.H., Dougherty, P., McInnis, D., Anderson, P. and Patterson, S. (2012) Effect of harvest residue management on tree productivity and carbon pools during early stand development in a loblolly pine plantation. *Forest Science* 58, 430–445.

Martin, T.A. and Jokela, E.J. (2004) Stand development and production dynamics of loblolly pine growing on spodosols in north-central Florida USA. *Forest Ecology and Management* 192, 39–58.

Matney, T.G. and Farrar, R.M. Jr (1992) A thinned/unthinned loblolly pine growth and yield simulator for planted cutover site-prepared land in the mid-gulf South. *Southern Journal of Applied Forestry* 16, 70–75.

Matney, T.G. and Hodges, J.D. (1991) Evaluating regeneration success. In: Duryea, M.L. and Dougherty, P.M. (eds) *Forest Regeneration Manual*. Kluwer Academic Publishers, Dordrecht, the Netherlands, pp. 321–331.

McKeand, S.E., Crook, R.P. and Allen, H.L. (1997) Genotypic stability effects on predicted family responses to silvicultural treatments in loblolly pine. *Southern Journal of Applied Forestry* 21, 84–89.

McKeand, S.E., Grissom, J.E., Handest, J.A., O'Malley, D.M. and Allen, H.L. (2000) Responsiveness of diverse provenances of loblolly pine to fertilization – age 4 results. *Journal of Sustainable Forestry* 10, 87–94.

McKeand, S., Mullin, T., Byram, T. and White, T. (2003) Deployment of genetically improved loblolly and slash pine in the South. *Journal of Forestry* 101(3), 32–37.

McKeand, S.E., Abt, R.C., Allen, H.L., Li, B. and Catts, G.P. (2006a) What are the best loblolly pine genotypes worth to landowners? *Journal of Forestry* 104, 352–358.

McKeand, S.E., Jokela, E.J., Huber, D.A., Byram, T.D., Allen, H., Lee Li, B. and Mullin, T.J. (2006b) Performance of improved genotypes of loblolly pine across different soils, climates, and silvicultural inputs. *Forest Ecology and Management* 227, 178–184.

McKendry, P. (2002a) Energy production from biomass (Part 1): overview of biomass. *Bioresource Technology* 83, 37–46.

McKendry, P. (2002b) Energy production from biomass (Part 3): gasification technologies. *Bioresource Technology* 83, 55–63.

McKinley, C.R., Lowe, W.J. and van Buijtenen, J.P. (1982) Genetic improvement of wood specific gravity in loblolly pine (*P. taeda*) and its relation to other traits. In: *Proceedings TAPPI R&D Division Conference*, Asheville, North Carolina, pp. 153–158.

Meadows, S., Gallagher, T. and Mitchell, D. (2011) A new slash bundling concept for use in a southern US logging system. *Forest Products Journal* 61, 210–215.

Miller, J.H., Zutter, B.R., Zedaker, S.M., Edward, M.B. and Haywood, J.D. (1991) A regional study on the influence of woody and herbaceous competition on early loblolly pine growth. *Southern Journal of Applied Forestry* 15, 169–179.

Miller, J.H., Zutter, B.R., Zedaker, S.M., Edwards, M.B. and Newbold, R.A. (2003) Growth and yield relative to competition for loblolly pine plantations to midrotation – a southeastern United States regional study. *Southern Journal of Applied Forestry* 27, 1–16.

Mitchell, D. (2006) Perspectives on woody biomass fuel value and specifications in Alabama. *American Society of Agricultural and Biosystems Engineers Paper* 06-8050, St Joseph, Michigan, 7 pp.

Mohan, D., Pittman, C.U. Jr and Steele, P.H. (2006) Pyrolysis of wood/biomass for bio-oil: a critical review. *Energy and Fuels* 20, 848–889.

Morris, L.A. and Campbell, R.G. (1991) Soil and site potential. In: Duryea, M.L. and Dougherty, P.M. (eds) *Forest Regeneration Manual*. Kluwer Academic Publishers, Dordrecht, the Netherlands, pp. 183–206.
Morris, L.A. and Lowery, R.F. (1988) Influence of site preparation on soil conditions affecting seedling establishment and early growth. *Southern Journal of Applied Forestry* 12, 170–178.
Morris, L.A., Pritchett, W.L. and Swindel, B.F. (1983) Displacement of nutrients into windrows during site preparation of a flatwoods forest soil. *Soil Science Society of America Journal* 47, 951–954.
Mosier, N., Wyman, C., Dale, B., Elander, R., Lee, Y.Y., Holtzapple, M. and Ladish, M. (2006) Features of promising technologies for pretreatment of lignocellulosic biomass. *Bioresource Technology* 96, 673–686.
Munsell, J.F. and Fox, T.R. (2010) An analysis of the feasibility for increasing woody biomass production from pine plantations in the southern United States. *Biomass and Bioenergy* 34, 1631–1642.
Murphy, G., Firth, J.G. and Skinner, M.F. (2004) Long-term impacts of forest harvesting related soil disturbance on log product yields and economic potential in a New Zealand forest. *Silvae Fennica* 38, 279–286.
Nairn, C.J. and Haselkorn, T. (2005) Three loblolly pine CesA genes expressed in developing xylem are orthologous to secondary cell wall CesA genes of angiosperms. *New Phytologist* 166, 907–915.
Nambiar, E.K.S. (1996) Sustained productivity of forests is a continuing challenge to forest science. *Soil Science Society of America Journal* 60, 1629–1642.
Nelson, C.D. and Johnsen, K.H. (2008) Genomic and physiological approaches to advancing forest tree improvement. *Tree Physiology* 28, 1135–1143.
Nelson, L.R. and Cantrell, R.L. (2002) *Herbicide prescription manual for southern pine management*. Clemson University Extension Document EC 659, 53 pp.
Novaes, E., Kirst, M., Chiang, V., Winter-Sederoff, H. and Sederoff, R. (2010) Lignin and biomass: a negative correlation for wood formation and lignin content in trees. *Plant Physiology* 154, 555–561.
Panshin, A.J. and de Zeeuw, C. (1980) *Textbook of Wood Technology*. McGraw-Hill, New York, 722 pp.
Peter, G.F. and Neale, D.B. (2004) Molecular basis for the evolution of xylem lignification. *Current Opinion in Plant Biology* 7, 737–742.
Peter, G.F., White, D.E.W., De la Torre, R., Singh, R. and Newman, D. (2007) The value of forest biotechnology: a cost modeling study with loblolly pine and kraft linerboard in the southeastern USA. *International Journal of Biotechnology* 9, 415–435.
Pettersen, R.C. (1984) The Chemical Composition of Wood. In: Rowell, R.M. (ed.) *The Chemistry of Solid Wood*. The American Chemical Society, Washington, DC, pp. 57–126.
Phanphanich, M. and Mani, S. (2009) Drying of pine residues. *BioResources* 5(1), 108–121.
Pletka, R., Brown, R.C. and Smeenk, J. (2001) Indirectly heated biomass gasification using a latent heat ballast: 1. Experimental evaluations. *Biomass Bioenergy* 20(4), 237–325.
Powers, R.F. (1999) On the sustainable productivity of planted forests. *New Forests* 17, 263–306.
Powers, R.F., Scott, D.A., Sanchez, F.G., Volset, R.A., Page-Dumroese, D., Elioff, J.D. and Stone, D.M. (2005) The North American long-term soil productivity experiment: Findings from the first decade of research. *Forest Ecology and Management* 220, 31–50.
Prestemon, J.P. and Abt, R.C. (2002) Timber products supply and demand. In: Wear, D.N. and Greis, J.G. (eds) *The Southern Forest Resource Assessment*. USDA Forest Service, General Technical Report SRS-53, Asheville, North Carolina, pp. 299–325.
Pritchett, W.L. and Comerford, N.B. (1982) Long-term response of phosphorus fertilization on selected southern Coastal Plain soils. *Soil Science Society of America Journal* 46, 640–644.
Proe, M.F. and Dutch, J. (1994) Impact of whole-tree harvesting on second rotation growth of Sitka Spruce: the first 10 years. *Forest Ecology and Management* 66, 39–54.
Resende, M.F.R., Munoz, P., Acosta, J.J., Peter, G.F., Davis, J.M., Grattapaglia, D., Resnede, M.D.V. and Kirst, M. (2012) Accelerating the domestication of trees using genomic selection: accuracy of prediction models across ages and environments. *New Phytologist* 193, 617–624.
Retzlaff, W.A., Handest, J.A., O'Malley, D.M., McKeand, S.E. and Topa, M.A. (2001) Whole-tree biomass and carbon allocation of juvenile trees of loblolly pine (Pinus taeda): influence of genetics and fertilization. *Canadian Journal of Forest Research* 31, 960–970.
Roth, B.E., Jokela, E.J., Martin, T.A., Huber, D.A. and White, T.L. (2007) Genotype × environment interactions in selected loblolly and slash pine plantations in the southeastern United States. *Forest Ecology and Management* 238, 175–188.
Sampson, D.A., Wynne, R.H. and Seiler, J.R. (2008) Edaphic and climatic effects on forest stand development, net primary production, and net ecosystem productivity simulated for Coastal Plain loblolly pine in Virginia. *Journal of Geophysical Research* 113, G01003, doi:10.1029/2006JG000270.

Samuelson, L.J., Johnsen, K. and Stokes, T. (2004) Production, allocation and stemwood growth efficiency of *Pinus taeda* L. stands in response to 6 years of intensive management. *Forest Ecology and Management* 192, 59–70.

Samuelson, L.J., Butnor, J., Maier, C., Stokes, T.A., Johnsen, K. and Kane, M. (2008) Growth and physiology of loblolly pine in response to long-term intensive management: defining growth potential in the southeastern United States. *Canadian Journal of Forest Research* 38, 721–732.

Sanchez, F.G. (1998) Soil organic matter and soil productivity: Searching for the missing link. In: Mickler, R. and Fox, S. (eds) *The Productivity and Sustainability of Southern Forest Ecosystems in a Changing Environment*. Springer-Verlag, New York, pp. 543–556.

Sanchez, F.G., Scott, D.A. and Ludovici, K.H. (2006) Negligible effects of severe organic matter removal and soil compaction on loblolly pine growth over 10 years. *Forest Ecology and Management* 227, 145–154.

Sands, P.J. and Landsberg, J.J. (2002) Parameterization of 3-PG for plantation grown *Eucalyptus globulus*. *Forest Ecology and Management* 163, 273–292.

Schultz, R.P. (1997) *Loblolly Pine: the ecology and culture of loblolly pine (*Pinus taeda *L.)*. Agriculture Handbook 713, US Department of Agriculture, Forest Service, Washington, DC, 493 pp.

Shevchenko, S.M., Chang, K., Dick, D.G., Gregg, D.J. and Saddler, J.N. (2001) Structure and properties of lignin in softwoods after SO_2-catalyzed steam explosion and enzymatic hydrolysis. *Cellulose Chemical Technology* 35, 487–502.

Smith, W.D. and Strub, M.R. (1991) Initial spacing: how many trees to plant. In: Duryea, M.L. and Dougherty, P.M. (eds) *Forest Regeneration Manual*. Kluwer Academic Publishers, Dordrecht, the Netherlands, pp. 281–289.

Stubbs, J., Roberts, D.R. and Outcalt, K.W. (1984) *Chemical Stimulation of Lightwood in Southern Pines*. USDA Forest Service General Technical Report SE-25, 54 pp.

Studer, M.H., DeMartini, J.D., Davis, M.F., Sykes, R.W., Davison, B., Keller, M., Tuskan, G.A. and Wyman, C.E. (2011) Lignin content in natural *Populus* variants affects sugar release. *Proceedings of the National Academy of Sciences USA* 108, 6300–6305.

Sullivan, A.D. and Clutter, J.L. (1972) A simultaneous growth and yield model for loblolly pine. *Forest Science* 18, 76–86.

Sykes, R., Li, B., Isik, F., Kadla, J. and Chang, H.-M. (2006) Genetic variation and genotype by environment interaction of juvenile wood properties in loblolly pine (*Pinus taeda* L.). *Annals of Forest Science* 63, 897–904.

Taherzadeh, M.J. and Keikhosro, K. (2007) Acid-based hydrolysis processes for ethanol from lignocellulosic materials: a review. *BioResources* 2, 472–499.

Vance, E.D., Maquire, D.A. and Zalesny, R.S. Jr (2010) Research strategies for increasing productivity of intensively managed forest plantations. *Journal of Forestry* 108, 183–192.

Vergara, R., White, T.L., Huber, D.A., Shiver, B.D. and Rockwood, D.L. (2004) Estimated realized gains for first-generation slash pine (*Pinus elliottii* var. *elliottii*) tree improvement in the southeastern United States. *Canadian Journal of Forest Research* 34, 2587–2600.

Vergara, R., White, T.L., Huber, D.A. and Schmidt, R.A. (2007) Realized genetic gains of rust resistant selections of slash pine (*Pinus elliottii* var. *elliottii*) planted in high rust hazard sites. *Silvae Genetica* 56, 231–242.

Visser, R. and Stampfer, K. (2003) Tree-length system evaluation of second thinning in a loblolly pine plantation. *Southern Journal of Applied Forestry* 27(2), 77–82.

Vogel, J.G. and Jokela, E.J. (2011) An examination of micronutrient limitations in two young managed southern pine stands planted on Florida spodosols. *Soil Science Society of America Journal* 75, 1117–1124.

Vose, J.M. and Allen, H.L. (1988) Leaf area, stemwood growth, and nutrition relationships in loblolly pine. *Forest Science* 34, 547–563.

Wakeley, P.C. (1954) *Planting the Southern Pines*. USDA Agricultural Monograph No. 18, Government Printing Office, Washington, DC, 233 pp.

Wakeley, P.C. (1969) Results of southern pine planting experiments established in the middle twenties. *Journal of Forestry* 67, 237–241.

Walawender, W.P. and Geyer, W.A. (1988) Influence of tree species and wood deterioration on downdraft gasifier performance. *Biomass* 17, 51–64.

Watson, W.F., Stokes, B.J. and Savelle, I.W. (1986) Comparisons of two methods of harvesting biomass for energy. *Forest Products Journal* 36(4), 63–68.

Wei, L., Thomasson, J.A., Bricka, R.M., Sui, R., Wooten, J.R. and Columbus, E.P. (2009) Syn-gas quality evaluation for biomass gasification with a downdraft gasifier. *Transactions of the American Society of Agricultural and Biological Engineers* 52(1), 21–37.

Wenzl, H.F.J. (1970) Further destructive processing of wood. In: *The Chemical Technology of Wood*. Academic Press, New York, pp. 253–270.

Westbrook, M.D., Greene, W.D. and Izlar, R.L. (2007) Utilizing forest biomass by adding a small chipper to a tree-length southern pine harvesting operation. *Southern Journal of Applied Forestry* 31(4), 165–169.

White, E.M. (2010) *USDA Forest Service, Pacific. Woody Biomass for Bioenergy and Biofuels in the United States – A briefing paper*. Northwest Research Station General Technical Report PNW-GTR-825, 45 pp.

White, R.H. (1987) Effect of lignin content and extractives on higher heating value of wood. *Wood and Fiber Science* 19, 446–452.

Yeh, T.-F., Braun, J.L., Goldfarb, B., Chang H.-M. and Kadla, J.F. (2006) Morphological and chemical variations between juvenile wood, mature wood, and compression wood in loblolly pine (*Pinus taeda* L.). *Holzforschung* 60, 1–8.

Zapata-Valenzuela, J., Isik, F., Maltecca, C., Wegrzyn, J., Neale, D., McKeand, S. and Whetten, R. (2012) SNP markers trace familial linkages in a cloned population of *Pinus taeda* – prospects for genomic selection. *Tree Genetics and Genomes* DOI 10.1007/s11295-012-0516-5.

Zhao, D., Kane, M., Borders, B., Subedi, S. and Akers, M. (2012) Effects of cultural intensity and planting density on stand-level aboveground biomass production and allocation of 12-year old loblolly pine plantations in the Upper Coastal Plain and Piedmont of the southeastern United States. *Canadian Journal of Forest Research* 42, 111–122.

Zhu, J.Y., Pan, X.J., Wang, G.S. and Gleinser, R. (2009) Sulfite pretreatment (SPORL) for robust enzymatic saccharification of spruce and red pine. *Bioresource Technology* 100, 2411–2418.

Zhu, J.Y., Zhu, W., O'Bryan, P.J., Dien, B.S., Tian, S., Gleisner, R. and Pan, X.J. (2010) Ethanol production from SPORL-pretreated lodgepole pine: preliminary evaluation of mass balance and process energy efficiency. *Applied Microbiology and Biotechnology* 86(5), 1355–1365.

Zobel, B.J. and van Buijtenen, J.P. (1989) *Wood Variation, Its Causes and Control*. Springer-Verlag, Berlin, 363 pp.

Zulak, K.G. and Bohlmann, J.J. (2010) Terpenoid biosynthesis and specialized vascular cells of conifer defense. *Integrated Plant Biology* 52, 86–97.

21 Eucalyptus and Bamboo

Masazumi Kayama
Kyushu Research Centre, Forestry and Forest Products Research Institute, Japan

21.1 Eucalyptus

21.1.1 Introduction

The genus *Eucalyptus* contains woody species in the myrtle family, *Myrtaceae*. There are over 800 species in the genus, most of them distributed in Australia (Coppen, 2002). Eucalyptus is also found in Indonesia, Papua New Guinea and the southern Philippines, however, there are few species in these countries. Eucalyptus is often the dominant tree in forest and woodland in Australia. Eucalyptus forest (closed forest) is mostly distributed along the eastern and south-eastern coasts of Australia and gives way to woodland in areas where rainfall is below 900 mm (Florence, 1996). In the part of Australia with rainfall below 400 mm and dominated by shrubland, eucalyptus trees are not widespread and are absent from regions where annual rainfall is below 200 mm.

In Australia, forest fire has occurred with high frequency since 20 million years ago (Kemp, 1981). Eucalyptus forest has high flammability because of the oil content in its leaf (Florence, 1996). On the other hand, eucalyptus has high adaptation capacity against forest fire (Florence, 1996). The seed of eucalyptus is covered with a woody capsule. The capsule is released sporadically at the end of a dry period. In the event of forest fire, the capsule is not burned, and seeds in the capsule are released simultaneously. The uniform stand of eucalyptus is regenerated after forest fire.

Eucalyptus was introduced to other countries in the 19th century for alleviation of malaria (Bennett, 2011). At present, it is planted in various continents and countries, including Asia (Fig. 21.1), South America, Africa, Spain and Portugal (Eldridge *et al.*, 1994). Eucalyptus is fast-growing and it is possible to obtain economic return within a short period (Coppen, 2002). Eucalyptus are utilized for multipurpose wood materials, and processed into pulp, fuelwood, essential oil, etc. (FAO, 1979). Honey produced from the nectar of the tree is also an important product. Moreover, when the culm of eucalyptus is felled, coppice shoots grow from the stump. Eucalyptus plantations can be utilized for short-rotation silviculture by coppicing (Diaz-Balteiro and Rodriguez, 2006). The ability of eucalyptus to coppice also makes it a useful biofuel crop.

21.1.2 Morphology, physiology and biochemistry

Eucalyptus has a dimorphic root system for adaptation to water deficit. The widely spreading

©CAB International 2013. *Biofuel Crops: Production, Physiology and Genetics* (ed. B.P. Singh)

Fig. 21.1. A plantation of eucalyptus (*Eucalyptus camaldulensis*) in Thailand (photograph provided by Kenzo Tanaka).

lateral root system below the soil surface accompanied by a deeply penetrating taproot enables the plant to fully utilize soil resources (Whitehead and Beadle, 2004). The deep taproot system permits uptake of groundwater well below 8 m (Dye, 1996). Roots of *E. marginata* in south-western Australia have been found to extend down to a depth of 20 m (Carbon *et al.*, 1980). The bark of eucalyptus has various forms. Many eucalypts have a smooth bark, and their outer bark is shed at a particular time of the year (Coppen, 2002). Meanwhile, a few eucalyptus, such as *E. macrorhyncha* and *E. calophylla*, have a rough bark, and dead bark accumulates year after year (Coppen, 2002). The seed morphology of eucalyptus is variable. Shape, size, colour and surface ornamentation are inherited traits of the taxonomic groups (Boland *et al.*, 1991).

One of the most peculiar morphological traits of eucalyptus is that essential oil is contained in its leaves. The essential oil is a marketable commodity and sold commonly as eucalyptus oil. There are numerous oil glands in the leaves of eucalyptus (Coppen, 2002). The range of oil yield of eucalyptus is from a trace to 8.6% (Coppen, 2002).

The photosynthetic capacity of eucalyptus is relatively high. In particular, *E. grandis* shows high photosynthetic capacity, and maximum photosynthetic rate at light saturation is 33 $\mu mol\ m^{-2}\ s^{-1}$ (Clearwater and Meinzer, 2001). Photosynthetic rate is dependent on nutrients in the leaf, and shows a positive correlation with leaf nitrogen and phosphorus (Clearwater and Meinzer, 2001; Whitehead and Beadle, 2004). Moreover, eucalyptus has a high tolerance to water deficit. When eucalyptus is grown on an arid site, water use efficiency shows a higher value compared with a moist site (Whitehead and Beadle, 2004); this may be because the hydraulic properties assist in the maintenance of homeostatic adjustment.

The woods of several species of eucalyptus contain 45–59% cellulose and 16–27% lignin (Clarke *et al.*, 1997). In particular, *E. smithii* shows a high ratio of cellulose and low ratio of lignin (Table 21.1; Clarke *et al.*, 1997). The wood of *E. globulus* shows a low content of lignin (18.7%) compared with other eucalyptus (Rencoret *et al.*, 2007). Compared with other woody species, wood of eucalyptus shows a high ratio of cellulose and a low ratio of lignin (Sjöström, 1993). High cellulose and low lignin content makes eucalyptus desirable for biofuel feedstock.

21.1.3 Plant breeding

Eucalyptus has hermaphrodite flowers and is pollinated by insects or birds (Griffin, 1989). Eucalypts reproduce by mixed mating systems, with both out-crossing and inbreeding (Moran and Bell, 1983). Breeding methods of eucalyptus are based on pollination techniques introduced by Eldridge *et al.* (1994). General pollination methods are divided into open pollination and controlled pollination. In the open pollination system, trees are pollinated by natural mechanisms and accelerated out-crossing in a seed orchard. However, open-pollinated seeds of eucalyptus show significant loss of productivity because of inbreeding depression (Horsley *et al.*, 2010). Alternatively, controlled pollination is followed to select for high-yielding and high-wood-quality offspring (see Eldridge *et al.*, 1994). Pure pollen is collected from mature flower buds isolated by bagging. Collected pollen is kept under low humidity, and stored in a freezer (−18°C) for preservation over an extended period. Before pollination, flower buds of the recipient parents are emasculated to avoid pollen contamination. After emasculation, the flower should be enclosed in a bag. For applying the pollen, the bag is opened at one end and closed after the operation. The bags are kept on until 2–4 weeks after pollination.

Eucalyptus is one of the woody species in which a considerable amount of breeding has been carried out. Interspecific hybrids of eucalyptus have been used in forestry for decades (Potts and Dungey, 2004). The main hybrids used commercially are *E. grandis* × *urophylla*, *E. grandis* × *camaldulensis*, etc. (Potts and Dungey, 2004). For example, wood of *E. grandis* × *urophylla* shows high quality, such as high density and thick fibre wall

Table 21.1. Chemical analysis of nine species of eucalyptus planted on a plantation in South Africa (Clarke *et al.*, 1997).

	Cellulose (%)		Lignin (%)		Pentosan (%)	
Species	Range	Mean	Range	Mean	Range	Mean
E. grandis	46.8–56.9	51.5	19.4–25.2	22.3	16.8–20.0	18.6
E. saligna	46.8–55.6	51.5	20.1–25.6	23.0	15.9–20.0	17.9
E. dunnii	47.3–55.7	51.7	18.3–24.2	20.9	17.0–22.0	20.0
E. macarthurii	45.4–51.1	48.2	17.8–25.7	22.0	17.2–21.6	18.9
E. nitens	46.3–53.9	50.4	17.8–24.7	20.0	20.1–23.9	22.0
E. smithii	49.6–58.7	55.0	15.7–23.0	19.9	17.2–21.9	19.6
E. fastigata	46.0–53.1	48.8	21.5–26.5	23.8	18.1–22.3	20.2
E. fraxinoides	49.8–54.8	52.4	17.4–22.9	20.7	19.7–23.2	21.2
E. oreades	49.8–54.0	52.0	19.7–20.4	20.1	19.7–20.4	20.1

(Quihó *et al.*, 2006). Eucalyptus cultivars with cold and salt tolerance have also been developed (Hendrati *et al.*, 2011; Navarro *et al.*, 2011).

21.1.4 Cultural practices

Eucalyptus grows well on a site where the soil has a good depth, structure and water storage capacity (Florence, 1996). Shallow and podzolic soils require intensive amelioration treatment (Florence, 1996). High compaction soil that decreases porosity, aeration and water movement affects root growth and productivity (Williamson, 1990). The soil compaction has a long-lasting effect, is difficult to ameliorate, and the cumulative effect is low harvest (Florence, 1996). Thus, a site with highly compacted soil should be avoided for planting of eucalyptus. Before planting, tillage is important to improve water and root penetration; a rotary hoe is suitable for this purpose. In dry areas, planting in grooves is preferable because water is concentrated near the plant (Coppen, 2002).

Eucalyptus does not flower and seed uniformly every year: 2–4 years may elapse between two good seed harvests. Therefore seed collection is usually made during the year of heavy flowering and seeding to lower the cost of collection and to obtain seeds of better quality (Florence, 1996). It is important to collect enough seeds to meet current planting needs as well as for subsequent years when scant flowering and seeding may make seed collection more costly and difficult. Collected seeds should be stored in the dark in sealed containers at a temperature of 1–4°C (FAO, 1979). Seeds of eucalyptus are sown in trays and the germinated seedlings pricked out into containers. After pricking out, the containers should be kept under shade for about 2 days and then stored in bays until field planting. In temperate regions, eucalyptus seeds are normally sown in midsummer and planted out during autumn–winter rains (Florence, 1996). In subtropical regions (summer-rainfall area), tree planting is best after substantial rains have fallen in spring (Florence, 1996).

Direct seeding is also used widely in the temperate zone. In contrast, direct seeding is not common practice in the subtropical zone because of high competition with other species during germination and seedling growth. The detail of direct seeding has been described by Florence (1996). Lockett (1991) recommended sowing rates in the range of 0.3–0.8 kg ha^{-1}, and this range should be changed according to the suitability of sites (unfavourable sites should have an increased sowing rate).

In general, eucalyptus has a tolerance to water deficits by maintaining turgor through osmotic adjustment and tissue elasticity (White *et al.*, 2000; Ngugi *et al.*, 2003). However, establishment of commercial eucalyptus plantations in the subtropical region is marginal at 700–1000 mm rainfall (Ngugi *et al.*, 2004). In this region, production is limited by low water availability and high summer vapour pressure deficit (Ngugi *et al.*, 2004) and irrigation may improve growth. Also, irrigation should be applied in low rainfall (<600 mm) regions; however, in many cases irrigation water is saline (Theiveyanathan *et al.*, 2004) in these areas. Theiveyanathan *et al.* (2004) has provided an irrigation scheduling model for saline water, which is a useful watering management tool to avoid salinity problems on eucalyptus plantations.

In Australian forest soils, phosphorus is the most critical nutrient, and forest stands exhibit symptoms of phosphorus deficiency (Florence, 1996). Growths of many species of eucalyptus are enhanced by fertilization (Schönau and Herbert, 1989). FAO (1979) introduced eucalyptus fertilization schemes in various countries. For example, in Australia (New South Wales) it is 50 g N and P (15:30) per plant, and this combination produces the best growth of *E. grandis* (Florence, 1996). When fertilizers are used in the field, they are applied a few weeks to 3 months after planting (FAO, 1979). Of the other nutrients, boron deficiency is exhibited in African countries (e.g. Zambia and Zimbabwe; Schönau and Herbert, 1989). Up to 60 g of fertilizer per plant is applied in these countries (FAO, 1979).

Eucalyptus is sensitive to grass competition, especially during the first year in a plantation (FAO, 1979). Thus, the weeds that compete with the seedlings of eucalypts must be controlled during the first year, and weeding should be continued until canopy closure. Herbicides can be applied before or after planting. Detail about the application of herbicides is covered by Florence (1996).

One of the most serious pests of eucalyptus is longhorned borer (*Phorocantha semipunctata*), which is native to Australia and also found in many other countries (Florence, 1996). Various species of eucalyptus, such as *E. globulus, E. grandis, E. saligna*, etc. are susceptive to longhorned borer (Duffy, 1963). The beetle attacks damaged trees that have suffered from drought stress. Eggs of this beetle are laid in a crack or depressions in the bark of the trunk or branches, and upon hatching larvae tunnel inside the tree extending into the heartwood (Duffy, 1963). Currently there is no known way to control the longhorned borer, and in order to prevent the spread infested trees have to be destroyed.

Short-rotation taking about 5–9 years is proposed for eucalyptus plantations in subtropical regions (Hawaii, Whitesell *et al.*, 1992; Brazil, Diaz-Balteiro and Rodriguez, 2006). In temperate regions (e.g. Spain), the period of short rotation is extended to 13–18 years (Diaz-Balteiro and Rodriguez, 2006). Silvicultural methods are divided into clearfelling and selection-type harvesting. The clearfelling method consists of removing all of the existing growing stock. This method is effective for even-aged and mature eucalyptus stands. In the selection method, silvicultural practice is to maintain an uneven-aged forest through the periodic harvest. This is an efficient method to manage wood production of eucalypts. One type of selection harvesting depends on diameter-limit cutting, in which standing trees are divided into diameter classes, and thick diameter trees are cut. For example, in the Queensland Forest Service, all trees with diameters over 100 cm are removed for market. Trees whose diameters are 40–99 cm are retained as acceptable trees, and several conditions are established for acceptance. Another type of selective clearing is known as shelterwood-type cutting. In this system, 40–50% of the canopy is removed during the first cut to allow more sunlight to reach the remaining trees for increased growth and for the sunlight to reach the plantation floor to provide conditions for the new tree crop to germinate and be ready to replace the existing one when they are taken out.

Mechanical felling is also practised in eucalyptus plantations. This method is suitable for clearfelling on a short rotation type management system (Hillis and Brown, 1984; Whitesell *et al.*, 1992). The trees are cut by feller-buncher or harvester, and hauled by grapple skidders or forwarder (Hillis and Brown, 1984).

21.1.5 Transport, postharvest handling and feedstock preparation

Hauled trees of eucalyptus are transported by roundwood trucks. Immediate chipping of eucalyptus trees after hauling is effective to cut down the cost of transportation (whole-tree system, Spinelli *et al.*, 2009). Preservation of eucalyptus wood postharvest is detailed by Hillis and Brown (1984). Harvested sapwood of eucalyptus is susceptible to decay and termite attack because it contains starch, however, preservative treatment is not difficult. Several chemicals are available for preservation of eucalyptus sapwood. In contrast, heartwood of eucalyptus is difficult to penetrate with preservatives (Hillis and Brown, 1984). For the preservation of heartwood, pressured treatment or diffusion treatment is essential. When eucalyptus is processed for biofuel, the tree is chipped into small fragments, and milled to pass an 8 mm screen (Romanì *et al.*, 2010).

21.1.6 Uses

The use of eucalyptus for biofuel is in the evolutionary stage. Charcoal is notable among the by-products from the use of eucalyptus for biofuel. Charcoal of eucalyptus has a high calorific value compared with its wood, and is considered as a high quality fuel (FAO, 1979). Eucalyptus is utilized as wood material, pulp, fuelwood, essential oil, etc. Tannin is produced from the bark (FAO, 1979). Eucalyptus oil is contained in the leaves and thus collected at the time of harvest. Eucalyptus oil is a valuable multipurpose product, which increases the economic value of eucalyptus. The oil contains various terpenes, the main component of which is 1,8-cineol (Coppen, 2002), which has anti-inflammatory activity (Juergens *et al.*, 2003; Batish *et al.*, 2008) and is utilized as an ingredient of mouthwashes and cough suppressants. *Eucalyptus polybractea* contains over 90% of 1,8-cineol in its oil, and is exploited as a commercial source (Coppen, 2002). Eucalyptus oil also has antifungal, antimicrobial, herbicidal and acaricidal properties (Batich *et al.*, 2008).

21.1.7 Sustainability related issues

Eucalyptus has a high capacity for regeneration by the growth of coppice shoots (FAO, 1979). This trait can be utilized for sustainable silviculture of eucalyptus. Diaz-Balteiro and Rodriguez (2006) introduced optimal rotations of eucalyptus plantations. Under their scheme, coppice wood of eucalyptus can be harvested at least three times. The optimal periods between harvesting are about 6–8 years in Brazil and 14–16 years in Spain. In the case of Brazil, eucalyptus wood can be harvested for 18–24 years with a short rotation. Eucalyptus is considered a highly sustainable woody material source because it can be harvested without replanting for long periods.

21.1.8 Status of current crop use for biofuel, constraints and future prospects

Charcoal is the energy product from eucalyptus that is being utilized at present. Commercial production of bioethanol from eucalyptus has not yet started. Total production cost of bioethanol from eucalyptus is cheaper than that from straw, poplar and switchgrass (Pandey *et al.*, 2011). Thus, eucalyptus is considered one of the most suitable bioethanol feedstocks. For processing eucalyptus into bioethanol, Gonzalez *et al.* (2011) proposed 'dilute acid prohydrolysis followed by enzymatic hydrolysis process model'. This model was developed by NRCL (National Renewable Energy Laboratory, USA) and is available for commercial production of bioethanol from eucalyptus. Based on the trial calculation by Gonzalez *et al.* (2011), one of the constraining factors is the cost of enzymes. As eucalyptus feedstock contains a high ratio of carbohydrate (cellulose and hemicellulose) and low ratio of lignin, its use will discount the production cost of bioethanol.

21.2 Bamboo

21.2.1 Introduction

Bamboo is a vernacular term for woody species of the family *Poaceae*. Bamboo consists of 75 genera and 1250 species (Tewari, 1993). The main distribution of bamboo is in Asia, and 14 Mha of the earth's surface is covered by bamboo forests. Bamboo is distributed across East Asia, from 51°N latitude at Sakhalin, south to northern Australia and west to India. Bamboos are also distributed in African (from Sudan at 6°N to Mozambique at 22°S) and American (from eastern part of USA at 40°N to Chile at 47°S) regions (Yang and Hui, 2010).

Bamboos are considered as one of the fastest growing plants and their culms can grow to 3–30 m within a few months (Liese and Weiner, 1995). Culms that emerge early in the season can fully develop during the warm summer in temperate climates and during the wet season in tropical climates (Kleinhenz and Midmore, 2001). In one of the most peculiar traits of bamboo, the pattern of flowering varies with species. The majority of bamboos display gregarious flowering and an entire clump at one location will produce flowers and then die back over the course of 2–3 years. The period until flowering in bamboo is about 30–40 years, with one example of 70 years (Tewari, 1993). The observation of flowering is quite rare, and the physiology of flowering has not been elucidated. In contrast, there are a few bamboos that flower annually or frequently. For example, *Bambusa atra* distributed in New Guinea, Moluccas and Sangihe Island flowers continuously, and does not die back after flowering (Dransfield and Widjaja, 1995).

Bamboos have been used for centuries in eastern, south-eastern and southern Asian countries. They are processed into multipurpose materials, such as pulp, handicrafts, furniture, construction, food, medicine, etc. (Tewari, 1993). In China, there is a history for utilization of bamboos dating back to 5000 years ago. There is a record that the first planting of bamboo was in the Han Dynasty (206 BCE–220 CE; Guan, 1994). In Japan, *Phyllostachys pubescens* was introduced from China for food (new shoots) and planted at Kyoto in 1650 (Suzuki, 1996). Thus, utilization of bamboo is part of the culture in Asian countries.

Culms of new bamboo reach full expansion in a single growing season, and can be harvested within 1 year. Moreover, chemical properties for organs of bamboos are similar to other woody species (Scurlock *et al.*, 2000). Therefore, bamboos can be potential feedstock for biofuel production.

21.2.2 Morphology, physiology and biochemistry

On the morphological traits, bamboos are divided into monopodial and sympodial bamboos (Fig. 21.2; Kleinhenz and Midmore, 2001). The monopodial bamboos have long leptomorph rhizome systems, and lateral buds on nodes of the rhizome produce new shoots or rhizome. The sympodial bamboos have wide patchmorph rhizome systems, and lateral buds on nodes of the rhizome produce only rhizome (Fig. 21.3). In sympodial bamboos, new culms develop from buds on elongated culm necks. Monopodial bamboos are native to temperate climates, whereas sympodial bamboos are from tropical climates with a pronounced dry season (Kleinhenz and Midmore, 2001). The difference of rhizome is regarded as adaptation to the climatic condition, and sympodial bamboos have high adaptation capacity for dehydration during dry seasons (Farrelly, 1984). Cells of bamboo culm have many parenchyma cells and fibre cells in vascular bundle sheaths, and this trait is different from other woody

Fig. 21.2. (a) A monopodial bamboo forest (*Phyllostachys pubescens*) in Japan (author's photograph) and (b) a sympodial bamboo forest (*Gigantochloa hasskarliana*) in Thailand (photograph provided by Hiroshi Tanaka).

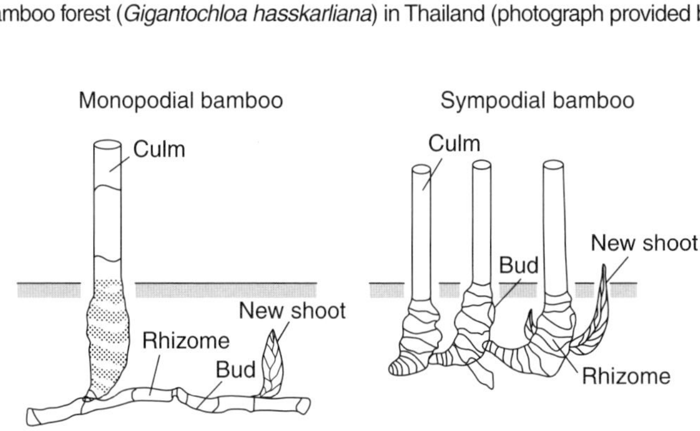

Fig. 21.3. The rhizome systems of two types of bamboos (illustrations provided by Masatoshi Watanabe, Japan Bamboo Society).

species (Abe and Yano, 2010). Parenchyma and fibre cells have a similar trend in cellulose crystallinity and tensile strength (Abe and Yano, 2010).

Younger leaves expanded in the current year show high concentrations of nutrients (Zhou and Wu, 1997) and high photosynthetic rates (Huang *et al.*, 1989). After full expansion, the photosynthetic rate of bamboo declines with aging (Huang *et al.*, 1989). There is a fundamental difference between monopodial and sympodial bamboos on the longevity of leaves. The longevity of leaves of monopodial bamboo is no more than 2 years (Li *et al.*, 1998) whereas for those of sympodial bamboo it is approximately 6 years (Pearson *et al.*, 1994).

Cellulose and lignin contents in organs of bamboo are in the same range compared with other woody species (Table 21.2). In particular, *Bambusa balcooa* shows a high ratio of cellulose content (Bhattacharya *et al.*, 2010). Fatty acid content of several bamboo culms have been analysed by Peng *et al.* (2009). The content of lipophilic extractive is 22.7 g kg^{-1} for *Phyllostachys bambusoides*, and this value is quite high compared with other woody species (Freire *et al.*, 2006). Lipophilic extractive can be changed into biodiesel. Thus, bamboo wood is quite suitable for production of biofuel because of its quality of feedstock and quite high productivity.

21.2.3 Plant breeding

Unfortunately, the majority of bamboos take a long period before flowering, and as a result, acquisitions of pollen and seed of bamboo are difficult. There are only a few cases where hybrid bamboos have been generated (Lin *et al.*, 2010). Tissue culture methods have been developed for the genetic improvement of bamboo (Gielis *et al.*, 2002). Gielis *et al.* (2002) report that *in vitro* flowering can be induced in controlled tissue culture. In any case, the current status of technology of breeding bamboos is not developed enough to be employed towards enhancing productivity of bamboo culms. Recently, though, a new hybrid bamboo 'Bamboo Sur' has been introduced by Van de Ven (Electric Energy Publications Inc., 2010). 'Bamboo Sur' is the product of biotechnology, and has thick-walled culms. Under intensive cultivation with a population of 2471 plants ha^{-1} (1000 plants per acre), the biomass obtained was 75 t ha^{-1} in year 2.

21.2.4 Cultural practices

Bamboo productivity is affected by physical properties of the soil, such as slope of land, texture, bulk density, moisture-holding capacity and temperature (Kleinhenz and Midmore, 2001). The productivity of bamboo is better on a flatland site (Hassan *et al.*, 1988; Virtucio *et al.*, 1994). Virtucio *et al.* (1994) suggested that bamboo should not be planted on sloping land because of low productivity and a risk of soil erosion. Soil texture affects productivity of bamboo. Bamboo growth parameters are negatively related to sand and clay content, and positively related to silt content (Chung and Ramm, 1990). Tillage during the non-shoot season is important for better growth of culms and rhizomes (Ling, 1984; Wu, 1984).

Table 21.2. Wet chemical analysis of several species of bamboo (mean).

Species	Cellulose (%)	Hemicellulose (%)	Lignin (%)	Reference
Bambusa balcooa	58.1	No data	25.0	1
Dendrocalamus asper	40.7	26.5	27.1	2
Phyllostachys bambusoides	43.3	27.7	26.2	3
Phyllostachys bissetii	42.8	27.5	27.7	3
Phyllostachys nigra	43.4	26.9	27.1	3
Phyllostachys pubescens	47.0	24.4	22.8	4

1, Bhattacharya *et al.* (2010); 2, Leenakul and Tippayawong (2010); 3, Scurlock *et al.* (2000); 4, Li *et al.* (2007). In Scurlock *et al.* (2000), glucan and other polysaccharides were interpreted as cellulose and hemicellulose, respectively.

The detail of the traditional methods for planting bamboo is given by Dransfield and Widjaja (1995). In general, bamboos flower at long intervals and collection of viable seeds is quite rare (Tewari, 1993). When seeds are available, propagation by seed can be established. The viability of seeds last only 2 months, however, seeds can be stored under cool conditions. Seeds should be sown on moist medium, and will germinate readily. After germination, seedlings should be transported into individual containers. When seedlings grow to 0.75–1 m, saplings can be planted out in the field. The season of planting is best from November to February in temperate regions (Yang and Hui, 2010) and during the rainy season in tropical regions (Tewari, 1993).

The other common methods for propagation of bamboo are from cuttings from rhizome or culms (for details, see Dransfield and Widjaja, 1995). For rhizome cuttings, 1- to 2-year-old culms should be selected. The cut rhizomes should be at least 0.3 m in size, and are separated from the mother plant and planted directly. For propagation by culm cuttings, whole culm or culm segments should be planted. Complete culms should be buried to stimulate the alternative buds. Culm segments with nodes should be collected from culms at the age of 6 months to 2 years. One internode of a culm will form two young plants very rapidly. The branches are pruned to a length of 10–20 cm and foliage is removed. Cuttings are planted horizontally at a depth of 5–10 cm. Recently, a method of micropropagation of bamboo (tissue culture) has been developed (Zamora, 2001) and seedlings of bamboo can now be produced by micropropagation of new shoots and branches.

Compared to other tree species, bamboo has high sensitivity to reduction of soil moisture (Dierick *et al.*, 2010). Bamboo suffers easily from water stress, and decrease of xylem water potential is remarkable for a species that shows high xylem hydraulic conductivity (Saha *et al.*, 2009). As a result, distribution and growth of bamboo is restricted by precipitation (Biswas, 1988). Thus, water supply is quite important for cultivation of bamboo. The productivity of sympodial bamboos in particular can be maximized with year-round irrigation (Kleinhenz and Midmore, 2001). In contrast, monopodial bamboos distributed in temperate regions do not need irrigation in winter (Kleinhenz and Midmore, 2001).

Fertilization is the most limiting factor of bamboo productivity on infertile sites. Hong (1987) recognized the beneficial effect of fertilizer application on bamboo production on poor soils in China. Widjaja (1991) stated that fertilization was the most important management technique in Indonesia. Moreover, productivity of bamboo forests under minimum management could increase by fertilizer application (Qiu *et al.*, 1992). In China, Yang and Hui (2010) recommend 103.5 kg N, 63 kg P and 40.5 kg K ha^{-1} as the optimal rate of fertilization.

Newly planted areas should be weeded because young bamboo cannot compete well with weeds (Dransfield and Widjaja, 1995). Direct-sowed plantings need continuous weeding until the seedlings are well established (Tewari, 1993). Also, weeding and scarifying have a positive effect on the elimination of the shoot-boring noctuid *Oligia vulgaris*, because it can wipe out overwintering eggs (Yang and Hui, 2010).

Over 800 species of pests of bamboo have been identified (Wang *et al.*, 1996). The most serious pest for decrease of productivity of bamboo (*Phyllostachys pubescens*) is the shoot-boring noctuid (*Oligia vulgaris*) (Wang *et al.*, 1996). *O. vulgaris* is a night-flying moth and is distributed in China, India and Japan (Wang *et al.*, 1996). The damage is caused by larvae, which bore inside new shoots and cause the death of damaged shoots (Yang and Hui, 2010). Effective control of *O. vulgaris* is by spraying insecticides in the winter season (Wang *et al.*, 1996). Insecticide solutions can be 1:5000 water using dimethoate or 80% dichlorvos (Yang and Hui, 2010). Culm borers (e.g. *Cyrtotrachelus dux, C. longimanus*) are distributed widely in Asia (Wang *et al.*, 1996), and affect culms because their larvae bore long tunnels that pass internally through several internodes (Singh and Bhandari, 1988). Infection is promoted under overcrowding and unhygienic conditions (Tewari, 1993). Culm borers can be managed in plantations by silvicultural practices, such as thinning and harvesting at regular intervals (Tewari, 1993). Weak and bent culms should be removed regularly (Singh and Bhandari, 1988).

Harvesting of bamboo culms in tropical regions should be carried out at the beginning of the dry season (Dransfield and Widjaja, 1995), because culms are attacked by borers in the monsoon (Tewari, 1993). Moreover, starch content is the lowest during the dry season (Sulthoni, 1987). The 1–2-year-old culms are suitable for pulp production, including fuel materials. The 3-year-old culms are mostly suitable for building materials. There are two methods of harvesting culms: clearfelling and selective felling. In clearfelling, the entire clump is cut down, and production of culms will be smaller in the next year. It is not advisable to use this method when aiming at a staggered harvest of culms. The selective felling method is generally used for exploiting and maintaining bamboo clumps. Correct felling intensity is essential for the maintenance of the rhizome. In Malaysia, about 70% of harvesting intensity per clump is recommended, and about 30% of the culms are left aside (Ahmad and Haron, 1994). A sugarcane harvester can be adapted to clearfelling bamboo with a uniform culm width (Gielis, 2000). In contrast, a harvester is unfit for selective felling and for large-diameter bamboo. Ito *et al.* (2010) described mechanical selective felling of bamboo using chainsaws that are specialized for bamboo. Felled bamboo is hauled by forwarder.

21.2.5 Transport and postharvest handling

In general, bamboo culms are transported by truck. However, bamboo culm is hollow, and the cost of transportation is expensive compared with other woody species (Ito *et al.*, 2010). High transportation cost becomes a serious problem when culms are used for biofuel feedstock as a vast amount of raw material is needed to keep the biorefineries running. Immediate chipping of bamboo culms after harvesting is effective to cut down the cost of transportation (Ito *et al.*, 2010). When bamboo culms are to be processed into biofuel, culms are chipped into small fragments (length: 5–8 cm, width: 2–3 cm, Yamashita *et al.*, 2010).

Dransfield and Widjaja (1995) describe in detail postharvest handling of bamboo. Bamboo culms become degraded during storage due to attack by rotting fungi and insects (beetles, borers, termites). If bamboo culms are to be preserved for a long duration, treatments by chemical or non-chemical methods are available. Several chemicals are used for the chemical method; however, the toxicity of the chemicals is of concern. Non-chemical methods are traditional in Asia, and include curing, smoking, etc.

21.2.6 Uses

Bamboos are used for a multitude of purposes, such as pulp, handicraft, furniture, construction, food, medicine, etc. Bamboo culms can be used for making charcoal, and waste from bamboo processing can be made into charcoal briquettes (Guan, 2004). Bamboo charcoal has a high calorific value and is a good alternative for firewood for heating and cooking purposes (Ganesh, 2003). Moreover, bamboo oil and vinegar can be extracted during processing for bamboo charcoal (Ganesh, 2003; Guan, 2004). Bamboo vinegar contains approximately 200 chemical components; the main one being acetic acid (Mu *et al.*, 2003). Bamboo charcoal has many uses other than for fuel, such as a purifying agent of water and air, deodorant, dehumidifying agent, etc. (Guan, 2004). Recently, bamboo charcoal has found application in stabilization of toxic inorganic and organic compounds (e.g. Hua *et al.*, 2009; Ma *et al.*, 2010). Bamboo vinegar has various efficacy, such as to diminish inflammation, for sterilization, promotion of crop growth, etc. (Mu *et al.*, 2003; Guan, 2004).

21.2.8 Sustainability related issues

Bamboo is one of the most sustainable woody resources in the world. In general, woody plants take decades for harvest. In contrast, new buds of bamboo are regenerated rapidly from the rhizome after clearcutting of culms, buds show marvellous expanding ability, and culms of new bamboo reach full expansion in a single growing season (Kleinhenz and Midmore, 2001). Bamboo culms can be harvested within 1 year, and take 4 years for full maturation. The chemical properties of bamboo culms are almost same as other woody species (Scurlock *et al*., 2000). Therefore, the period for silviculture of bamboos is quite short, an attribute to sustainable production of their culms.

21.2.9 Status of current use of the crop for biofuel, constraints and future prospects

At present, utilization of bamboo charcoal is quite prevalent in Asian countries. However, technology for the conversion of bamboo into other fuels, such as bioethanol, is delayed compared with other lignocellulosic biofuel crops. There are several reports of bamboo saccharification by enzymatic hydrolysis (Sathitsuksanoh *et al*., 2010; Yamashita *et al*., 2010). If these technologies are developed, commercial production of bioethanol from bamboo can be realized. One of the constraining factors for production of bioethanol from bamboo is cost of transportation; this problem can probably be solved by immediate chipping. Moreover, enzymes used in saccharification of bamboo are expensive (Sathitsuksanoh *et al*., 2010); this cost needs to come down for the commercial production of bioethanol from bamboo to become economical.

References

Abe, K. and Yano, H. (2010) Comparison of the characteristics of cellulose microfibril aggregates isolatetd from fiber and parenchyma cells of Moso bamboo (*Phyllostachys pubescens*). *Cellulose* 17, 271–277.

Ahmad, I.H. and Haron, N.H. (1994) The sustainability of bamboo supply in peninsular Malaysia. In: Thammincha, S., Anantachote, A., Rao, Y.S. and Muraille, B. (eds) *Bamboo in Asia and the Pacific, Proceedings of the 4th International Bamboo Workshop*. International Development Research Centre, FAO, Canada, pp. 87–91.

Batish, D.R., Singh, H.P., Kohli, R.K. and Kaur, S. (2008) Eucalyptus essential oil as a natural pesticide. *Forest Ecology and Management* 256, 2166–2174.

Bennett, B.M. (2011) A global history of Australian trees. *Journal of the History of Biology* 44, 125–145.

Bhattacharya, S., Ghosh, J.S., Sahoo, D.K., Dey, N. and Pal, A. (2010) Screening of superior fiber-quality-traits among wild accessions of *Bambusa balcooa*: efficient and non-invasive evaluation of fiber developmental stages. *Annals of Forest Science* 67, 611.

Biswas, S. (1988) Studies on bamboo distribution in north-eastern region of India. *Indian Forester* 114, 514–531.

Boland, D.J., Brophy, J.J. and House, A.P.N. (1991) *Eucalyptus Leaf Oils. Use, chemistry, distillation and marketing*. Inkata Press, Melbourne, Australia.

Carbon, B.A., Bartle, G.A., Murray, A.M. and McPherson, D.K. (1980) The distribution of root length and the limits to flow of soil water to roots in a dry sclerophyll forest. *Forest Science* 26, 656–664.

Chung, Y.G. and Ramm, C.W. (1990) Relationships between soil-site properties and bamboo (*Phyllostachys bambusoides*) growth. *Journal of Korean Forestry Society* 79, 16–20.

Clarke, C.R.E., Garbutt, D.C.F. and Pearce, J. (1997) Growth and wood properties of provenances and trees of nine eucalypt species. *Appita Journal* 50, 121–130.

Clearwater, M.J. and Meinzer, F.C. (2001) Relationships between hydraulic architecture and leaf photosynthetic capacity in nitrogen-fertilized *Eucalyptus grandis* trees. *Tree Physiology* 21, 683–690.

Coppen, J.W. (2002) *Eucalyptus, the Genus* Eucalyptus. Taylor and Francis, London.

Diaz-Balteiro, L. and Rodriguez, L.C.E. (2006) Optimal rotations on Eucalyptus plantation including carbon sequestration – A comparison of results in Brazil and Spain. *Forest Ecology and Management* 229, 247–258.

Dierick, D., Hölscher, D. and Schwendenmann, L. (2010) Water use characteristics of a bamboo species (*Bambusa blumeana*) in the Philippines. *Agricultural and Forest Meteorology* 150, 1568–1578.

Dransfield, S. and Widjaja, E.A. (1995) *Plant Resources of South-east Asia. No. 7, Bamboos*. Backhuys Publishers, Leiden, the Netherlands.

Duffy, E.A.J. (1963) *A Monograph of the Immature Stages of Australasian Timber Beetles (Cerambycidae)*. British Museum (Natural History), London.

Dye, P.J. (1996) Response of *Eucalyptus grandis* trees to soil water deficits. *Tree Physiology* 16, 233–238.
Eldridge, K., Davidson, J., Harwood, C. and Van Wyk, G. (1994) *Eucalypt Domestication and Breeding*. Oxford University Press, Oxford, UK.
Electric Energy Publications Inc. (2010) Bamboo Sur introduce a New Bamboo Hybrid that can be the solution for biomass power plants and worldwide deforestation. Available at: http://www.electricenergyonline.com/?page=show_news&rss=1&id=126838 (accessed 3 February 2010).
FAO (1979) *Eucalyptus for Planting*. FAO Forestry series 11, FAO, Rome, Italy.
Farrelly, D. (1984) *The Book of Bamboo*. Sierra Club Books, San Francisco, California.
Florence, R.G. (1996) *Ecology and Silviculture of Eucalypt Forests*. CSIRO Publishing, Collingwood, Australia.
Freire, C.S.R., Pinto, P.C.R., Santiago, A.S., Silvestre, A.J.D., Evtuguin, D.V. and Pascoal Neto, C. (2006) Comparative study of lipophilic extractives of hardwood and corresponding ECF bleached kraft pulps. *Bioresources* 1, 3–17.
Ganesh, A. (2003) *Bamboo Characterization for Thermochemical Conversion and Feasibility Study of Bamboo Based Gasification and Charcoal Making*. National Mission on Bamboo Applications, New Delhi, India.
Gielis, J. (2000) *Future Possibilities for Bamboo in European Agriculture*. Bamboo Thematic Network, Rijkevorsel, Belgium.
Gielis, J., Peeters, H., Gillis, K., Oprins, J. and Debergh, P.C. (2002) Tissue culture strategies for genetic improvement of bamboo. In: Van Huylebroeck, J., Van Bockstaele, E. and Debergh, P. (eds) *Proceedings of the Twentieth International Eucarpia Symposium, Section Ornamentals*. International Society for Horticultural Science, Leuven, Belgium, pp. 195–203.
Gonzalez, R., Treasure, T., Phillips, R., Jameel, H., Saloni, D., Abt, R. and Wright, J. (2011) Converting *Eucalyptus* biomass into ethanol: financial and sensitivity analysis in a co-current dilute acid process. Part II. *Biomass and Bioenergy* 35, 767–772.
Griffin, A.R. (1989) Strategies for the genetic improvement of yield in Eucalyptus. In: Pereira, J.S. and Landsberg, J.J. (eds) *Biomass Production by Fast-growing Trees*. Kluwer, Dordrecht, the Netherlands, pp. 247–265.
Guan, C.Y. (1994) Textual research on the history of bamboo garden in China. *Journal of Bamboo Research* 13, 53–63 (in Chinese).
Guan, M.J. (2004) *Manual for Bamboo Charcoal Production and Utilization*. Bamboo Engineering Research Centre, Nanjing, China.
Hassan, M.M., Alam, M.K. and Mazumder, A.H. (1988) Distribution of bamboo under the edaphic and climatic conditions of Bangladesh. *Indian Forester* 114, 505–513.
Hendrati, R., Byrne, M., Barbour, E. and Plummer, J. (2011) Effect of genetic relatedness among parents on gain in salt tolerance in progeny of crosses of *Eucalyptus occidentalis*. *Silvae Genetica* 60, 45–55.
Hillis, W.E. and Brown, A.G. (1984) *Eucalyptus for Wood Production*. CSIRO/Academic Press, Melbourne/Sydney, Australia.
Hong, S.S. (1987) A preliminary study on optimum amounts and properties of fertilizers for the growth of *Ph. pubescens*. *Journal of Bamboo Research* 6, 35–41 (in Chinese).
Horsley, T.N., Johnson, S.D. and Myburg, A.A. (2010) Comparison of different control-pollination techniques for small-flowered eucalypts. *New Forests* 39, 75–88.
Hua, L., Wu, W.X., Liu, Y.X., McBride, M.B. and Chen, Y.X. (2009) Reduction of nitrogen loss and Cu and Zn mobility during sludge composting with bamboo charcoal amendant. *Environmental Science and Pollution Research* 16, 1–9.
Huang, Q.M., Yang, D.D., Gao, A.X., Shen, Y.K., Qiu, G.X., Long, S.P., Beadle, C.L., Hall, D.O. and Scurlock, J.M.O. (1989) Research on photosynthesis of bamboo. *Journal of Bamboo Research* 8, 8–18 (in Chinese).
Ito, T., Murakami, M. and Taniyama, T. (2010) The key of bamboo utilization is cost of extraction. *Shinrin Kagaku* 58, 20–23 (in Japanese).
Juergens, U.R., Dethlefsen, U., Steinkamp, G., Gillissen, A., Repges, R. and Vetter, H. (2003) Anti-inflammatory activity of 1,8-cineol in bronchial asthma: a double-blind placebo-controlled trial. *Respiratory Medicine* 97, 250–256.
Kemp, E.M. (1981) Pre-quaternary fire in Australia. In: Gill, A.M., Groves, R.H. and Noble, I.R. (eds) *Fire and the Australia*. Australian Academy of Science, Canberra, Australia, pp. 3–21.
Kleinhenz, V. and Midmore, D.J. (2001) Aspects of bamboo agronomy. *Advances in Agronomy* 74, 99–145.
Leenakul, W. and Tippayawong, N. (2010) Dilute acid pretreatment of bamboo for fermentable sugar production. *Journal of Sustainable Energy and Environment* 1, 117–120.

Li, R., Werger, M.J.A., During, H.J. and Zhong, Z.C. (1998) Carbon and nutrient dynamics in relation to growth rhythm in the giant bamboo *Phyllostachys pubescens*. *Plant and Soil* 201, 113–123.

Li, X.B., Shupe, T.F., Peter, G.F., Hse, C.Y. and Eberhardt, T.L. (2007) Chemical changes with maturation of the bamboo species *Phyllostachys pubescens*. *Journal of Tropical Forest Science* 19, 6–12.

Liese, W. and Weiner, G. (1995) Ageing of bamboo culms. A review. *Wood Science and Technology* 30, 77–89.

Lin, X.C., Lou, Y.F., Liu, J., Peng, J.S., Liao, G.L. and Fang, W. (2010) Crossbreeding of *Phyllostachys* species (Poaceae) and identification of their hybrids using ISSR markers. *Genetics and Molecular Research* 9, 1398–1404.

Ling, J.X. (1984) Investigation on the bamboo shoot growth of *Lignania wenchouensis* and its management. *Journal of Bamboo Research* 3, 92–101 (in Chinese).

Lockett, E.J. (1991) *Eucalypt Seed and Sowing*. Native forest silviculture technical bulletin no. 1, Forestry Commission, Hobart, Australia.

Ma, J.W., Wang, F.Y., Huang, Z.H. and Wang, H. (2010) Simultaneous removal of 2,4-dichliriohenol and Cd from soils by electrokinetic remediation combined with activated bamboo charcoal. *Journal of Hazardous Materials* 176, 715–720.

Moran, G.F. and Bell, J.C. (1983) Eucalyptus. In: Tanksley, S.D. and Orton, T.J. (eds) *Isozymes in Plant Genetics and Breeding, part B*. Elsevier, Amsterdam, the Netherlands, pp. 423–441.

Mu, J., Uehara, T. and Furuno, T. (2003) Effect of bamboo vinegar on regulation of germination and radicle growth of seed plants. *Journal of Wood Science* 49, 262–270.

Navarro, M., Ayax, C., Martinez, Y., Laur, J., El Kayal, W., Marque, C. and Teulieres, C. (2011) Two *EguCBF1* genes overexpressed in *Eucalyptus* display a different impact on stress tolerance and plant development. *Plant Biotechnology Journal* 9, 50–63.

Ngugi, M.R., Doley, D., Hunt, M.A., Dart, P. and Ryan, P. (2003) Leaf water relations of *Eucalyptus cloeziana* and *Eucalyptus argopholia* in response to water deficit. *Tree Physiology* 23, 335–343.

Ngugi, M.R., Doley, D., Hunt, M.A., Ryan, P. and Dart, P. (2004) Physiological responses to water stress in *Eucalyptus cloeziana* and *E. argopholia* seedlings. *Trees* 18, 381–389.

Pandey, A., Larroche, C., Ricke, S.C., Dussap, C.-G. and Gnansounou, E. (2011) *Biofuels: alternative feedstocks and conservation processes*. Academic Press, San Diego, California.

Pearson, A.K., Pearson, O.P. and Gomez, I.A. (1994) Biology of the bamboo *Chusquea culeou* (*Poaceae: Bambusoideae*) in southern Argentina. *Vegetatio* 111, 93–126.

Peng, P., Yang, F.X., Xu, F., Fang, J.M., Sun, R.C. and Jones, G.L. (2009) Identification and characterization of lipophilic extractives in four species of bamboo culms. *Journal of Biobased Materials and Bioenergy* 3, 354–362.

Potts, B.M. and Dungey, H.S. (2004) Interspecific hybridization of Eucalyptus: key issues for breeder and geneticists. *New Forests* 27, 115–138.

Qiu, G.X., Shen, Y.K., Li, D.Y., Wang, Z.W., Huang, Q.M., Yang, D.D. and Gao, A.X. (1992) Bamboo in sub-tropical China. In: Long, S.P., Jones, M.B. and Roberts, M.J. (eds) *Primary Productivity of Grass Ecosystems of the Tropics and Sub-tropics*. Chapman and Hall, London, pp. 159–188.

Quihó, T., Miranda, I. and Pereira, H. (2006) Within-tree variation in wood fibre biometry and basic density of the urograndis eucalypt hybrid (*Eucalyptus grandis* × *E. urophylla*). *IAWA Journal* 27, 243–254.

Rencoret, J., Gutiérrez, A. and del Rio, J.C. (2007) Lipid and lignin composition from different eucalypt species. *Holzforschung* 64, 165–174.

Romaní, A., Garrote, G., Alonso, J.L. and Parajó, J.C. (2010) Bioethanol production from hydrothermally pretreated *Eucalyptus globlus* wood. *Bioresource Technology* 101, 8706–8712.

Saha, S., Holbrook, N.M., Montti, L., Goldstein, G. and Cardinot, G.K. (2009) Water relations of *Chusquea ramosissima* and *Merostachys claussenii* in Iguazu National Park, Argentina. *Plant Physiology* 149, 1992–1999.

Sathitsuksanoh, N., Zhu, Z., Ho, T.J., Bai, M.D. and Zhang, Y.H.P. (2010) Bamboo saccharification through cellulose solvent-based biomass pretreatment followed by enzymatic hydrolysis at ultra-low cellulose loadings. *Bioresource Technology* 101, 4926–4929.

Schönau, A.P.G. and Herbert, M.A. (1989) Fertilizing eucalypts at plantation establishment. *Forest Ecology and Management* 29, 221–239.

Scurlock, J.M.O., Dayton, D.C. and Hames, B. (2000) Bamboo: an overlooked biomass resource? *Biomass and Bioenergy* 19, 229–244.

Singh, P. and Bhandari, R.S. (1988) Pests of bamboos and their control in India. *Indian Forester* 114, 670–783.

Sjöström, E. (1993) *Wood Chemistry. Fundamentals and Applications*, 2nd edn. Academic Press, San Diego, California.

Spinelli, R., Ward, S.M. and Owende, P.M. (2009) A harvest and transport cost model for *Eucalyptus* spp. fast-growing short rotation plantations. *Biomass and Bioenergy* 33, 1265–1270.

Sulthoni, A. (1987) Traditional preservation of bamboo in Java, Indonesia. In: Rao, A.N., Dhanarajan, G. and Sastry, C.B. (eds) *Recent Research on Bamboos. Proceedings of the international bamboo workshop, 6–14 October 1985, Hangzhou, China*. China and International Development Research Centre, Canada, pp. 349–357.

Suzuki, S. (1996) *Illustrations of Japanese Bambusaceae* (revised edn). Jukai Shorin, Funabashi, Japan (in Japanese).

Tewari, D.N. (1993) *A monograph on Bamboo*. International Book Distributors, Dehra Dun, India.

Theiveyanathan, S., Benyon, R.G., Marcar, N.E., Myers, B.J., Polglase, P.J. and Falkiner, R.A. (2004) An irrigation-scheduling model for application of saline water to tree plantations. *Forest Ecology and Management* 193, 97–112.

Virtucio, F.D., Manipula, B.M. and Schlegel, F.M. (1994) Culm yield and biomass productivity of Laak (*Sphaerobambos philippinensis*). In: *Fourth International Bamboo Workshop on Bamboo in Asia and the Pacific, held at Chiangmai, Thailand, 27–30 November 1991*. FAO, Regional office for Asia and the Pacific, Bangkok, Thailand, pp. 27–30.

Wang, H.J., Varma, R.V. and Xu, T.S. (1996) *Insect Pests of Bamboos – an illustrated manual*. International Network for Bamboo and Rattan, Beijing, China.

White, D.A., Turner, N.C. and Galbraith, J.H. (2000) Leaf water relations and stomatal behavior of four allopatric Eucalyptus species planted in Mediterranean southwestern Australia. *Tree Physiology* 20, 1157–1165.

Whitehead, D. and Beadle, C.L. (2004) Physiological regulation of productivity and water use in *Eucalyptus*: a review. *Forest Ecology and Management* 193, 113–140.

Whitesell, C.D., DeBell, D.S., Schubert, T.H., Strand, R.F. and Crabb, T.B. (1992) *Short-rotation Management of Eucalyptus: guideline for plantations in Hawaii*. Pacific Southwest Research Station, Albany, California.

Widjaja, E.A. (1991) Socio-ecological observations of bamboo forests in Indonesia. *The Journal of the American Bamboo Society* 8, 125–135.

Williamson, J.R. (1990) *The Effects of Mechanized Forest Harvesting Operations on Soil Properties and Site Productivity*. Research Report No. 5, Tasmanian Forest Research Council Inc., Hobart, Australia.

Wu, B.S. (1984) A preliminary analysis of the relationship between the underground structure of a *Phyllostachys pubescens* forest and its yield. *Journal of Bamboo Research* 3, 49–58 (in Chinese).

Yamashita, Y., Shono, M., Sasaki, C. and Nakamura, Y. (2010) Alkaline peroxide pretreatment for efficient enzymatic saccharification of bamboo. *Carbohydrate Polymers* 79, 914–920.

Yang, Y.M. and Hui, C.M. (2010) *China's Bamboo*. International Network for Bamboo and Rattan, Beijing, China.

Zamora, A.B. (2001) *Micro-propagation of Bamboo*. International Network for Bamboo and Rattan, Beijing, China.

Zhou, Y.C. and Wu, B.S. (1997) Study of nutrient characteristics of the leaves of *Bambusa distegia*. *Journal of Bamboo Research* 16, 13–18 (in Chinese).

22 Phytoremediation Trees for Biofuel

D.L. Rockwood,[1] J.G. Isebrands[2] and P.J. Minogue[3]
[1]*School of Forest Resources and Conservation, University of Florida, USA;* [2]*Environmental Forestry Consultants, Wisconsin, USA;* [3]*North Florida Research and Education Center, University of Florida, USA*

22.1 Introduction

Many plants have phytoremediation potential (McIntyre, 2005). Over a hundred tree species worldwide could be grown as short rotation woody crops (SRWC) for biofuels (El Bassam, 1998) on nearly 10 Mha, with *Eucalyptus* spp. constituting about 38% of all SRWCs (FAO, 2003). The eucalypts which are proven woody biomass producers are largely unexplored for biofuels (Bartle, 2009). Grown on 18 Mha in tropical and temperate climates throughout Africa, South America, Asia and Australia and in more temperate regions of Europe, South America, North America and Australia, eucalyptus is a nearly universal woody biomass feedstock for biofuels. In the world's temperate regions, poplar (*Populus* spp.) and willow (*Salix* spp.) predominate as SRWCs.

SRWC systems involve short rotations of intensively cultured fast-growing trees planted at close spacing. Dendroremediation, i.e. phytoremediation using trees to remediate contaminated soil and groundwater (Rockwood *et al.*, 2004b), often uses non-hyperaccumulating, fast-growing *Populus, Salix* and *Eucalyptus* species. Due to long growing seasons and increased soil temperature, tropical and subtropical environments accelerate dendroremediation (Paquin *et al.*, 2004). In general, improvement of SRWC productivity depends on screening of species, density, fertilization, irrigation, stand management and harvesting techniques.

Broadly defined, dendroremediation may link SRWCs with agroforestry systems. Dendroremediation could include planting fast-growing trees for land reclamation and restoration of disturbed lands to improve soils, control invasive species and/or transition to native plants. Contaminated sites, landfills, wastewater treatment facilities, storm-water collection areas, agricultural lands and mined lands may provide planting opportunities. For example, thousands of sites worldwide were arsenic (As) contaminated by mining, wood treatment and pesticide uses.

Another opportunity is mixing an SRWC planted for dendroremediation and/or timber product with a crop specially suited for dendroremediation instead of agricultural production. A mixture involving a hyperaccumulating understorey plant that sequesters a contaminant from the upper soil profile and an SRWC that accumulates the same contaminant at a non-hazardous concentration but primarily from a lower soil profile could increase the efficiency and effectiveness of a dendroremediation system. For arsenic remediation in the south-eastern USA, for example, such a potential combination may involve a high but non-toxic-accumulating *Populus deltoides* clone with the Chinese brake fern (*Pteris vittata*).

©CAB International 2013. *Biofuel Crops: Production, Physiology and Genetics* (ed. B.P. Singh)

Overall dendroremediation needs could encourage agroforestry approaches combining SRWC and dendroremediation. The phytoremediation market was projected to be US$235–400 million in 2005 (Glass, 1999). Dendroremediation of degraded land using biomass fuel crops is feasible when former industrial land has little development potential and low environmental risks, agricultural land is contaminated through atmospheric depositions or waste disposal, and agricultural land is contaminated with fertilizer residue (Riddell-Black and Ferguson, 2002). Urban areas may create closed-loop renewable biomass energy systems by applying urban reclaimed water and yard wastes to bioenergy crops (e.g. Rockwood *et al.*, 2001).

This chapter reviews: (i) research on phytoremediating trees; (ii) commercial dendroremediation; (iii) the potential uses of phytoremediating trees as biofuel feedstocks; and (iv) when trees harvested from contaminated (high nutrient, heavy metal, chemical, etc.) sites could be directly used for biofuels and when pretreatment is necessary.

22.2 Research on Phytoremediating Trees (Dendroremediation)

SRWC dendroremediation research has expanded in several ways since Rockwood *et al.* (2004b) compiled listings of worldwide and US projects.

22.2.1 World

Sewage effluent greatly increased the growth of *Eucalyptus grandis* and *Pinus radiata* in Australia (Table 22.1). Wastewater sludge significantly increased willow biomass productivity and recycled undesirable residue (Hytönen, 1994; Labrecque *et al.*, 1997; Aronsson and Perttu, 2001).

Toxicity, uptake and transformation of atrazine by three poplars were assessed in Korea. Poplar cuttings exposed to lower concentrations of atrazine grew well, transpired at a constant rate, and took up, hydrolysed and de-alkylated atrazine to less toxic metabolites. Metabolism of atrazine occurred in roots, stems and leaves and became more complete with increased residence time in tissue. Phytoremediation is a viable approach to removing atrazine from contaminated water and should be considered for other contaminants.

Other species have been considered in India and China (Table 22.1). *Acacia farnesiana* was one of several species tolerant of biomethanated diluted spent wash in India that could then have multiple uses. *Jatropha curcas* could be suitable for phytostabilization of polymetallic acid mine tailings. The order of metal accumulation was roots>stems>leaves. Improved growth and decreases in phytoavailable soil Al (95%), Zn (~75%), and Cu (~65%) contents occurred at 0.5% lime. Of 36 species screened for heavy metal (As, Cu, Pb, Sn and Zn) tolerance in India, *Cyperus rotundus* was a potential

Table 22.1. Representative SRWC dendroremediation studies worldwide.

Country	Estab. date	Contaminant	Species	Reference
Australia	1991	Nutrients	*Eucalyptus, Pinus*	Myers *et al.* (1996)
Slovenia	1993	Heavy metals	Hybrid poplar	EPA (2000e)
Switzerland	1997	Heavy metals	Willow	Hammer *et al.* (2003)
Belgium	1999	Hydrocarbons	Willow	Vervaeke *et al.* (2003)
Korea		Atrazine	Poplars	Chang *et al.* (2005)
India		Biomethanated diluted spent wash	*Acacia farnesiana*, etc.	Sharma *et al.* (2011)
China		Polymetallic acid mine tailings	*Jatropha curcas*	Wu *et al.* (2011)
India		Heavy metals	Various	Ashraf *et al.* (2011)
Germany		Heavy metals	Willow	Meers *et al.* (2007)
Italy		As, Cd, Cu, Zn	*Populus, Salix*	Zacchini *et al.* (2009)
Switzerland	2010	Heavy metals	*Salix, Populus, Betula*	Evangelou *et al.* (2012)

Sn-hyperaccumulator, *Mimosa pudica* a metal accumulator, and *Acacia podalyriifolia* metal tolerant (Ashraf *et al.*, 2011). Roots had the highest metal concentration followed by leaves, shoots and flowers.

The potential of five *Salix* clones to extract and accumulate Cd, Cr, Cu, Ni, Pb and Zn was assessed in a 1-year open pot experiment with three different soil types in Germany (Table 22.1). Three of the five clones had the highest Cd and Zn concentrations and warrant further field validation. These clones could annually extract 5–27 kg ha^{-1} for Zn and 0.25–0.65 kg ha^{-1} for Cd, based on the soil type. Uptake of Cd and Zn could not be enhanced in one soil type, but could be increased by 50–100% in two other soil types. Soil treatment also had a distinct effect on Cu uptake.

Recent use of *in vitro* culture has expanded identification and testing of plants for phytoremediation. Three *Populus alba* clones were evaluated for As, Cd, Cu and Zn tolerance, accumulation and translocation through *in vitro* screening (Table 22.1). A fast-growing commercial clone without the highest metal concentration had the highest metal content, suggesting biomass production was key to evaluating phytoextraction capacity. Hydroponic screening of some poplar and willow clones evaluated their capability for Cd tolerance, accumulation and translocation (Zacchini *et al.*, 2009). Poplar clones varied widely whereas willow clones were more homogeneous in Cd accumulation and distribution. Willows had a much greater ability to tolerate Cd than poplars. The capacity of willows to translocate was double that of poplars, but total uptake by poplar clones was far higher.

The biomass production and metal uptake of willow (*Salix viminalis*), poplar (*Populus monviso*), birch (*Betula pendula*) and oak (*Quercus robur*) on five different soils with varying high concentrations of Cu, Zn, Cd and Pb were investigated in Switzerland (Table 22.1). Biomass was reduced in all species except oak primarily due to high Cd and Zn uptake. Willow leaves had up to 15 mg Cd kg^{-1} and 2000 mg Zn kg^{-1}. Only Pb was higher in wood than leaves, and birch had the highest accumulation of Pb with up to 135 mg kg^{-1} in wood and 78 mg kg^{-1} in leaves. Birch could be suitable for phytostabilization of soils with high Cd and Zn but low Pb, while poplars and willows could be used to stabilize soils with high Cu and Pb and low Zn and Cd.

Biomass production by heavy metal-accumulating willow varieties near Freiburg, Germany was compared to setting the land aside or switching from high value vegetable production to less-profitable cereals (Lewandowski *et al.*, 2006). Dendroremediation was economically beneficial for farmers but strongly dependent on the potential income from the remediated area, the duration of crop production after cleanup, and the length of phytoremediation.

22.2.2 USA

Hybrid poplar, cottonwood and willow are the most promising trees in the temperate USA, with *Eucalyptus* preeminent in semi-tropical and tropical regions. Over 20,000 ha of hybrid poplar plantations in the Pacific Northwest and Minnesota are grown for pulpwood, with ~25% of the total biomass used as fuel (Stanton *et al.*, 2002).

Dendroremediation has been implemented by many industries and communities. In Minnesota, electricity-generating companies are planting 20,000 ha to poplars. An SRWC pilot test project on a 50 ha clay settling area (CSA) near Lakeland, Florida, representative of ~160,000 ha of phosphate mined lands in central Florida, developed guidelines for establishing and managing SRWCs on CSAs and evaluated genetic, cultural and harvesting options for further improving the cost effectiveness of using cottonwood (*Populus deltoides, PD*), *Eucalyptus amplifolia* (*EA*) and *E. grandis* (*EG*) (Segrest *et al.*, 2001).

In Florida, several species have shown potential for nutrient, metal and hydrocarbon dendroremediation (Table 22.2). For example, bald cypress (*Taxodium distichum*) is promising for Cu remediation. SRWCs such as *PD* and *EG* have a high demand for water and nutrients. *PD* and *Eucalyptus* species have wastewater dendroremediation potential. Three of ten *PD* clones in a sewage effluent sprayfield at Tallahassee, Florida, were particularly productive and recommended to optimize biomass production and nutrient uptake.

Table 22.2. Representative SRWC dendroremediation studies in the USA.

Study location	Estab. date	Contaminant	Species	Reference
Tallahassee, FL	1992	Nutrients	*PD*	Rockwood *et al.* (1996)
Tampa, FL	1993	Nutrients	*EA, EC, EG*	Pisano and Rockwood (1997)
Edgewood, MD	1996	Chlorinated solvents	Hybrid poplar	EPA (2005)
New Gretna, NJ	1996	Chlorinated solvents	Hybrid poplar	EPA (2005)
East Palo Alto, CA	1997	As, VOCs, metals	Eucalyptus, tamarisk	EPA (2005)
Orlando, FL	1998	Nutrients	*PD, EG, EA, ...*	Rockwood *et al.* (2001)
Piketon, OH	1999	TCW	Hybrid poplar	EPA (2005)
Quincy, FL	2000	As	*PD*	Rockwood *et al.* (2001)
Archer, FL	2000	As	*PD*	Rockwood *et al.* (2001)
St Augustine, FL	2000	Toluene	*PD, EA*	Rockwood *et al.* (2004b)
Hill AFB, UT	2001	TCE	Fruit trees	Chard *et al.* (2003)
LaSalle, IL	2002	TCE PCE	Poplar, willow	Rockwood *et al.* (2004b)
Aiken, SC	2002	TCE	*PD*, loblolly pine	Stanhope *et al.* (2008)
Gary, IN		Hydrocarbons	*Populus, Salix*	Zalesny *et al.* (2005)
Tallahassee, FL	2004	Landfill gas	*EA, PD*	Abichou *et al.* (2012)
Tallahassee, FL	2008	Nutrients	*EA, EG, PD*	Minogue *et al.* (2012)
Hawaii	Various	Various	>100 species	Paquin *et al.* (2004)

CA, California; FL, Florida; IL, Illinois; IN, Indiana; MD, Maryland; NJ, New Jersey; OH, Ohio; SC, South Carolina; UT, Utah.

One of the earliest studies of the potential of *Eucalyptus* dendroremediation addressed stormwater cleanup in Tampa, Florida, in 1993 (Table 22.2). While three *Eucalyptus* species tolerated flooding in the stormwater retention treatment pond, *EG* was the most vigorous after 3.75 years. Although water use could be as high as 1600 mm annually, the exact amount of water and nutrients taken up by *EG* depends on climate, tree vigour and the timing and extent of stormwater applications. One *EG* clone appeared superior for stormwater dendroremediation in central Florida.

After 3 years, *EG* biomass yields were more than twice those of *PD* in response to application of sewage effluent, compost and/or mulch in a study at Orlando, Florida (Table 22.2). The superior productivity of *EG* has obvious potential environmental values, as well as economic implications. *EG* plantations can increase water loading and reduce nutrient leaching, even when compost amendments are applied. The trees can effectively increase the amount of effluent that can be applied. *EG* may reduce N and P leaching by up to 75% when water only is applied and 85% when mulch is added for weed control.

Dendroremediation may effectively address contamination associated with the widespread manufacturing and use of wood treated with preservatives such as chromated copper arsenate (CCA). Growth and arsenic uptake by *PD* at a CCA contaminated site in Archer, Florida, varied with season, plant tissue and clone (Table 22.2). Within tree arsenic concentration was highest in the leaves, next highest in branch bark, followed by stem bark, then branch wood and lowest in stemwood. Variation among clones was large for the tree components with the highest arsenic concentrations. Several high and low arsenic-concentrating *PD* clones were noted. Total arsenic uptake by *PD* will be a balance between biomass production and arsenic concentration. Annual harvesting of *PD* clones will probably result in the highest arsenic removal. To maximize arsenic removal, high density planting, e.g. 1 × 1 m, of high concentrating clone(s) should be practised.

A dendroremediation study at St Augustine, Florida, used highly contaminated dual phase extraction wells to pump toluene-contaminated groundwater for treatment and to contain the plume (Table 22.2). *PD* and *EA* were equally vigorous, but *EA* had higher survival, and an *EA* progeny was the most productive genotype. A study at LaSalle, Illinois, also provides evidence of the importance of various factors in establishing dendroremediation systems (Table 22.2).

Dendroremediation of chlorinated ethene-contaminated water was investigated at the Savannah River Site in Aiken, South Carolina (Table 22.2). Loblolly pine, hybrid poplar, coyote willow and sweetgum took up both perchloroethylene (PCE) and trichloroethylene (TCE). Cis-dichloroethylene (cDCE), a by-product from rhizosphere microbial activity, was detected in the soils as well as some tree tissues.

Populus and *Salix* species and species hybrids have been used successfully for dendroremediation of petroleum hydrocarbons in Gary, Indiana (Table 22.2). Of 20 poplar and two willow clones, different combinations of genotype and cutting length were recommended, with commercial poplar clones given priority. Specific clones should be selected based on site location and soils.

The potentials of *Populus* clones and *Eucalyptus* progenies as alternative landfill covers were evaluated at a Tallahassee, Florida landfill (Table 22.2). After 3 months, 32 local clones and 16 progenies varied in survival and tree height but still surpassed two non-adapted poplar clones by 0.3–1.0 m in height. Through 19 months, the local clones were taller than the progenies in both the lined and unlined portions of the study. At 45 months, however, the clones had only ~30% survival compared to ~90% for the progenies. By 84 months, while overall survival decreased slightly, survival of the clones was still 60% less than the progenies, and the *EA* progenies were notably larger and healthier. Long-term evaluation was necessary to confirm genetic superiority of *EA*.

Various progenies of *EG* and *EA* and clones of *PD* were evaluated for nitrogen and phosphorus uptake through 27 months at a Tallahassee, Florida, municipal waste sprayfield (Table 22.2). Tertiary treated wastewater containing 2.73 mg l^{-1} nitrate N and 0.30 mg l^{-1} total P (TP) was applied by sprinklers (93.8 m^3 ha^{-1} day^{-1}). *EA* and *EG* survived and grew very poorly after two successive exceptionally cold winters and were not evaluated for nutrient removal. Survival and growth of *PD* demonstrated suitability for dendroremediation, and selected clones were evaluated for biomass and nutrient content. Concentrations of both total N (TN) and TP were foliage > green branches > woody branches > stem. TN and TP uptakes were greatest for main stem (36% and 44%, respectively) and foliage (44% and 36%, respectively), suggesting total tree removal when trees are foliated for optimum N and P dendroremediation. Low biomass-producing clones generally had higher nutrient concentrations, but high biomass-producing clones removed more TN and TP. The ~789 kg ha^{-1} TN and 103 kg ha^{-1} TP removed by the highest biomass-producing *PD* clone represented 215% of N and 615% of P inputs.

Testing of over 100 plant species for remediation of explosive residues, hydrocarbons, pesticide residues, soil stabilization and slaughterhouse wastewater at seven chemically contaminated sites in Hawaii suggested that salt- and/or drought-tolerant plants are well suited for phytoremediation (Table 22.2).

Engineered plant-associated endophytic bacteria can significantly contribute to the efficiency of the phytoremediation of organic contaminants (Baraca *et al.*, 2009). Endophytic communities isolated from trees grown on groundwater contaminated with benzene, toluene, ethylbenzene and xylene (BTEX) or trichloroethylene (TCE) included many strains able to degrade BTEX compounds or resist TCE. Inoculating plants with endophytic bacteria engineered to optimally metabolize the contaminant can improve overall phytoremediation.

Dendroremediation can reduce pollution and decrease the impact of pesticides. Plants act directly on pesticides by taking up, accumulating, or detoxifying organic substances, or indirectly stimulating soil microbial activity to breakdown organic compounds. The efficiency of vegetative filter strips (VFS) to protect water from pesticide runoff contamination makes them a good farming practice. Trees such as poplar or willow are particularly useful in riparian zones or in phytoremediation processes due to large evapotranspiration capacities. Genetic engineering technologies broaden the scope of improving dendroremediation.

22.3 Commercial Dendroremediation

The following represent commercial-scale dendroremediation projects that have biofuel production potential.

22.3.1 World

Many opportunities exist worldwide for linking dendroremediation with tangible biomass economic opportunities such as bioenergy and many intangible opportunities (Licht and Isebrands, 2005). Landfill-leachate management using land-treatment systems has been practised in several countries: Sweden and Finland (Ettala, 1987; Hasselgren, 1998), Canada (Gordon *et al.*, 1989), the USA (Licht, 1994), Hong Kong (Wong and Leung, 1989) and the UK (Alker *et al.*, 2002). These projects suggest that using SRWCs for leachate management, reuse and recovery is cost effective compared to conventional methods.

In Northern Ireland, France, Sweden and Greece, sewage effluent improved willow yields, which in turn resulted in high nutrient removals from effluent, equivalent to the efficacy of conventional sewage tertiary treatment systems (Larsson, 2003). Wastewater application did not increase heavy metal content of the wood.

At Kågeröd, Sweden, wastewater irrigation greatly increased the growth of willows and the removal rate of N and P in the willow-soil system was higher than conventional nitrification/denitrification and phosphate chemical precipitation treatment processes (Hasselgren, 1998).

In Sao Paulo State, Brazil, wet and dry sewage sludges supplemented with K and B were large nutrient sources for *E. grandis* and significantly increased woody biomass production (da Silva *et al.*, 2011). Both sludges similarly enhanced the biological cycling of nutrients. Sewage sludge application in eucalypt plantations may be a valuable option and considerably reduce mineral fertilizer requirements. However, ammonia volatilization may cause seedling mortality if planting is soon after wet sewage sludge application.

While SRWCs and dendroremediation may often be justified independently, combining the two can increase their economic feasibility, and incorporating both into agroforestry systems can further enhance annual income possibilities. The generally favourable cost of dendroremediation with SRWCs, e.g. for remediating landfill-leachate, can be augmented by nearby fuelwood or other timber markets when the planting is sufficiently large to produce merchantable biomass quantities. Similarly, dendroremediating SRWCs in an agroforestry format will have periodic value, when planted on a large scale in a marketable area, that can add to the more frequent returns from the agronomic component. SRWCs in the riparian component of agroforestry systems dendroremediate and provide numerous other benefits.

As noted by Isebrands and Karnosky (2001), timber-belts, multi-row shelterbelts of SRWCs, can also provide the environmental benefits of windbreaks/shelterbelts while producing the economic benefits of timber products. Although substantial fuelwood is derived from converting forests to farmland in many areas, considerable amounts of fuelwood also are obtained from agroforestry practices on marginal and farm land (FAO, 1996).

Low intensity agroforestry systems suggest the prospects for combining SRWC and dendroremediation. Between 350,000 and 500,000 ha of woodland in Tanzania have been restored by dry-season fodder reserves that are farmer-led initiatives that retain standing vegetation during the rainy season (Kamwenda, 2002). This traditional practice of carefully managed trees dominated by *Acacia* spp. and understorey grasses, herbs and forbs is suitable for sustainable production of dry-season fodder, food, fuelwood and mitigation of land degradation.

The 'Integrated Oil Mallee' project in Western Australia grows SRWC eucalyptus mallee biomass in 4–5 m wide strips and harvests trees in 3–4 year cycles to provide multiple energy products (Table 22.3; Bartle *et al.*, 2007; Sudmeyer and Goodreid, 2007; Bartle and Abadi, 2010). Planting eucalypts for leaf oil may simultaneously provide a commercial incentive for vegetative restoration, sustainable control of groundwater and salinity, an environmentally benign substitute for widely used ozone-damaging solvent, speciality chemical products, biomass fuels and carbon sequestration. Large scale plantings are needed since dryland salinity extends over millions of hectares of arable land across several Australian states.

Bartle and Abadi (2010) suggest that steep agricultural lands in China afford opportunities to pair agroforestry and dendroremediation. Due to severe erosion on new agricultural areas, intercropping systems are being developed to control erosion and maintain land productivity.

22.3.2 USA

The USA has over 200 demonstration or commercial phytoremediation sites (Tables 22.2 and 22.4; http://www.clu-in.org/products/phyto/search/phyto_list.cfm). Field studies and commercial applications have identified the potential of SRWCs for dendroremediating nutrient, metal and hydrocarbon contamination. An early application of hybrid poplars was for trichloroethylene (TCE) dendroremediation at Oconee, Illinois, in 1989.

A poplar-based dendroremediation system was installed at former refinery and tank farm sites in Cabin Creek, West Virginia, to clean-up petroleum-contaminated soils and groundwater (Table 22.4). Benzene, toluene, ethylbenzene, xylene and gasoline organic concentrations decreased 81%, 90%, 67%, 78% and 82%, respectively, in the lower soil horizons and 34%, 84%, 12%, 19% and 59%, respectively, in groundwater. Tree roots dewatered soils and allowed the oxygen penetration necessary for rhizosphere bioremediation. Dendroremediation is a cost-effective strategy if human exposure and risk are not issues.

Poplars are ideal for phytoremediation because of their high water usage, fast growth and deep root systems. Superior-performing poplar clones exhibited great potential for phytoremediation applications where elevated water usage is critical. Extensive variation in responses to leachate has been observed among 25 poplar clones, along with the potential of using phyto-recurrent selection to choose superior genotypes (Zalesny *et al.*, 2007). The clones were irrigated with leachate during three cycles of phyto-recurrent selection to test genotypic differences. Eight clones were selected in cycles 1–3 and subsequently tested in an *in situ* landfill study in cycle 4.

Dendroremediation of a TCE and PCE contaminated site in Orlando, Florida, with chlorinated volatile organic compounds (CVOCs) in the soil and shallow groundwater was designed to enhance their natural attenuation and rhizodegradation and their uptake by phytodegradation and phytovolatilization processes (Table 22.4; Nzengung *et al.*, 2004), as well as to minimize their transport to a nearby lake. Whips of poplar clone DN34, *PD* clones and one local willow species (*Salix caroliniana*) were densely and deeply planted (0.6 m above water table) over the 1.2 ha plume. The *PD* whips averaged 5.5 m in height after 6 months (Rockwood *et al.*, 2004a). Locally adapted *PD* clones typically grew more than the widely planted DN34, which had proportionately more root biomass, though rooting was limited due to a higher water table. After 1 year, dendroremediation was already influencing the plume and uptake of the CVOCs in root, stem and leaf tissues. The trees were expected to evapotranspire 76 cm year^{-1} of groundwater.

Table 22.3. Representative commercial SRWC dendroremediation applications worldwide.

Location	Estab. date	Contaminant	Species	Species
Canada	1996	Hydrocarbons	Hybrid poplar	EPA (2000b)
Canada	1999	Hydrocarbons	Hybrid poplar	Dan *et al.* (2007)
Denmark	1999	Hydrocarbons	Poplar, willow	Trapp *et al.* (2001)
Australia	2004	Heavy metals	Poplar, willow	Laidlaw and Baker (2006)
Australia	2009	Salinity	Eucalyptus	Bartle (2009)

Table 22.4. Representative commercial SRWC dendroremediation applications in the USA.

Study location	Estab. date	Contaminant	Species	Reference
Oconee, IL	1989	TCE	Hybrid poplar	EPA (2005)
Martelle, IA	1993	Nitrates	Hybrid poplar	EPA (2005)
Grundy Center, IA	1993	Landfill	Hybrid poplar	EPC (2000d)
Anderson, SC	1993	Heavy metals	Hybrid poplar	EPA (2005)
Klamath Falls, OR	1994	Hydrocarbons	Hybrid poplar	Green and Hoffnagle (2004)
Lafayette, LA	1995	TCE	Hybrid poplar	EPA (2000c)
East Palo Alto, CA	1996	As	Eucalyptus, tamarisk	EPA (2005)
Woodburn, OR	1996	Nutrients	Hybrid poplar	EPA (2000a)
Wayne, MI	1997	TCE	Hybrid poplar	EPA (2005)
Courtland, CA	1997	Nitrate	Eucalyptus	EPA (2003)
Findlay, OH	1997	TCE	Hybrid poplar, willow	EPA (2005)
Argonne, IL	1999	TCE tritium	Hybrid poplar, willow	EPA (2005)
Cabin Creek, WV	1999	Petroleum hydrocarbons	Hybrid poplar	El-Gendy *et al.* (2009)
Rhinelander, WI	2002	Landfill	Poplars	Zalesny *et al.* (2006)
Orlando, FL	2002	TCE PCE	*PD*, poplar	Spriggs *et al.* (2004)
Gulfport, MS	2003	Hydrocarbons	*PD*	
Denham Springs, LA	2004	Heavy metals	Eucalyptus, poplar	EPA (2005)
Texas City, TX	2004	Heavy metals	Eucalyptus	EPA (2005)
Manhattan, KS	2005	Dioxane	Cottonwood	Christianson *et al.* (2007)
Panama City, FL	2006	As	*EA*, *PD*	Rockwood (pers. comm.)
Winter Garden, FL	2009	Nutrients	*EG*	Rockwood (pers. comm.)
Wood River, IL	2009	Petroleum hydrocarbons	Willow, poplar etc.	BP (2011)

CA, California; FL, Florida; IA, Iowa; IL, Illinois; KS, Kansas; LA, Louisiana; MI, Michigan; MS, Mississippi; OH, Ohio; OR, Oregon; SC, South Carolina; TX, Texas; WI, Wisconsin; WV, West Virginia.

Past use of TCE as a degreasing solvent has resulted in widespread groundwater contamination at Air Force bases (AFB) (Klein, 2011). TCE removal was estimated at 839 g year^{-1} at Travis and 18 g year^{-1} at Fairchild (50 and 5% of the mass of TCE in the groundwater, respectively) AFBs, with the majority from leaf and soil volatilization. Soil surface volatilization of TCE was greater inside the planted areas than outside the planted areas, indicating that the trees enhance this removal by this mechanism.

Combining SRWCs, dendroremediation and agroforestry can enhance biofuel possibilities in the USA as well. Rapidly growing poplar trees effectively lowered subsurface NO_3-N concentrations and stabilized degraded agricultural stream banks (O'Neill and Gordon, 1994). SRWCs in multispecies riparian buffer strips in central Iowa effectively removed nutrients, pesticides and sediments while stabilizing stream banks, providing wildlife habitat, affording commercial timber and other products and improving agroecosystem aesthetics (Schultz *et al.*, 1995). Bartle and Abadi (2010) suggest riparian buffers along the Mississippi River as a way to address major nutrient, sediment and agrochemical problems while producing copious biomass quantities.

Repeating Rockwood *et al.* (2004b) for installation guidelines for dendroremediation projects, we again recommend a phased, smaller start up that tests genotypes, cultural options, etc., to ensure that the appropriate genotypes of a species are matched with the necessary silvicultural options. SRWCs will reach their dendroremediation potential only when the system is properly established and maintained.

22.4 Biofuel Production Options for Trees

Genotypes derived from traditional tree improvement programmes can be very helpful as planting stock for dendroremediation systems. Such genotypes, typically genetically superior for growth, survival and pest incidence bring many of the characteristics needed for maximum dendroremediation.

Poplar breeding is conducted worldwide. In the USA, poplar breeding programmes such as in the north-central USA (Riemenschneider *et al.*, 2001) and the Pacific Northwest (Stanton *et al.*, 2002) produce superior clones that can be used regionally for dendroremediation. Most US clones used for dendroremediation are older clones produced in the north-eastern USA and Canada in the 1990s or clones naturalized from Europe (Dickmann *et al.*, 2001).

Willow breeding programmes are also active worldwide. New York has a willow-breeding programme for new hybrids for biomass production and dendroremediation (Kopp *et al.*, 2001). Tharakan *et al.* (2003) observed variation in key energy feedstock characteristics among willow and hybrid poplar clones. Clonal variation presents opportunities for manipulating feedstock quality through selection, breeding and plantation management.

Many eucalyptus breeding programmes in the world may be sources of genotypes for dendroremediation. For example, some *EG* clones developed for timber production in Florida have done well in various dendroremediation applications (Rockwood *et al.*, 1996).

The use of bacteria and mycorrhizae and genetic engineering has global relevance for further significant progress in remediation of contaminated environments. Selected bacteria and mycorrhizae can improve extraction and degradation (Capuana, 2011). In particular, genetic engineering has promise for improving characteristics of poplars and eucalyptus for dendroremediation. The genomes of these species have been mapped and afford tremendous opportunities for engineering traits that enhance dendroremediation (Grattapaglia, 2008; Grattapaglia and Resende, 2011).

Current poplars, willows and eucalypts however, unlike hyperaccumulators, may not require special treatment or disposal and are suitable for biofuels using most conversion processes, which have been variously classified as first-, second-, third- and even fourth-generation processes (Masarovicova *et al.*, 2009; Bartle and Abadi, 2010; da Silva Perez *et al.*, 2010). Trees are not well suited for the first-generation biofuels based on starch, sugar and oilseed.

As second-generation biofuels are based on lignocellulosic sources, dendroremediating trees are suitable feedstocks. Bioethanol may be obtained by hydrolysis of polysaccharides and biodiesel from syngas produced by gasification (da Silva Perez *et al.*, 2010). These processes are expected to have high energy output to input, while SRWC dendroremediation can provide ecosystem services.

The phosphoric lignocellulosic biorefinery under development and depicted in Fig. 22.1 exemplifies the types of processes that may use biomass from dendroremediation. Use of phosphoric acid for hydrolysis is expected to decrease costs and improve several environmental aspects. Overall, the process has fewer steps and lower costs and has the potential to produce high-value co-products.

SRWC willow, poplar and eucalyptus can be utilized whole to produce heat and electricity directly through combustion or indirectly through conversion for use as biofuels such as methanol and ethanol. *Eucalyptus* species can be readily converted biologically for biofuel and by-product production. Under mild pretreatment conditions, such as sulfuric acid pretreatment, over 80% cellulose conversion can be achieved, and over 30 g glucose can be obtained from 100 g of wood. Among and within Florida-grown *EG, EA* and *CT*, differences were noted for wood properties (Rockwood *et al.*, 2008), suggesting that deployment of favourable clones would be advantageous in producing energy products. The major chemicals in clones EG1 and EA4 are lignin, glucan and xylan. Glucan contents are about 40%, lower than other hardwood species. The main hemicellulose xylan (11%) can be easily removed in pretreatment and fermented. Lignin content (~34%) suggests that SRWC *Eucalyptus* species can be ideal for bioenergy production through thermoconversion.

(a) (b)

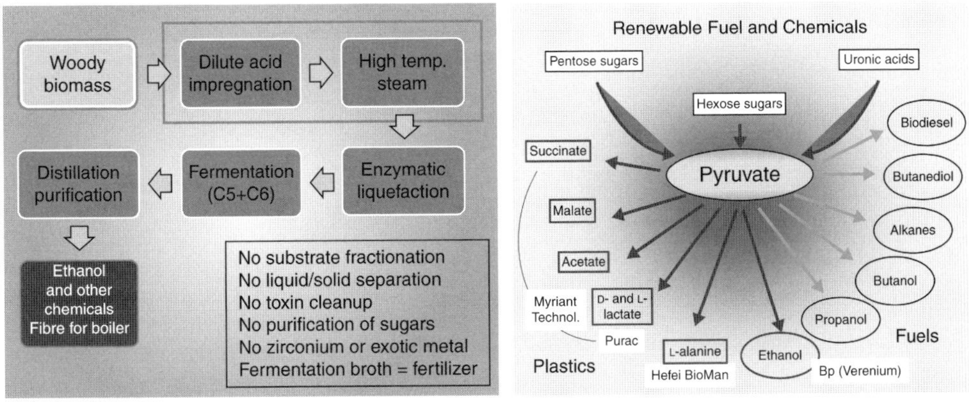

Fig. 22.1. (a) Conversion steps in a proposed phosphoric lignocellulosic biorefinery and (b) its potential co-products (L.O. Ingram, personal communication).

Gasification of biomass residues into syngas is expected to produce new value streams (Rockwood *et al.*, 2008). Syngas could be converted into biofuels, power, chemicals and other high value materials. Methanol can be produced from biomass by gasification, gas upgrading and eventual synthesis over a copper-zinc oxide catalyst. Methanol could be an attractive transportation fuel, provided effective methanol-powered fuel cells and fuel-cell/electric vehicles are developed.

For example, the 'Integrated Oil Mallee' project in Western Australia involves more than heat and power generation (Bartle *et al.*, 2007). Direct liquefaction of mallee biomass results in a lower cost and higher efficiency transportation of the liquid biofuel to a central user or processing facility. Biomass from SRWC eucalyptus mallee may provide pharmaceutical oils, activated carbon, heat and power, tradable renewable energy certificates and even a carbon credit. Large volumes of high-cineole eucalyptus oil could be produced from mallee eucalypts well below current market prices. Large industrial solvent markets are currently in transition following withdrawal of 1,1,1-trichloroethane as a result of international measures to control ozone depletion.

Third-generation biofuels will be based on genetically altered feedstocks (Masarovicova *et al.*, 2009). Given recent and anticipated progress in genetic mapping and understanding of gene function, genomic selection is expected to increase the productivity and convertibility of poplars, willows and eucalypts.

Fourth-generation biofuels will combine genetically optimized feedstocks with synthesized microbes (Masarovicova *et al.*, 2009). Due to the increased efficiencies of both feedstock production and conversion, fourth-generation biofuels promise to be 'carbon negative'.

22.5 Biofuel Uses of Untreated Dendroremediating Trees

The chemical composition of solid biofuels affects their thermal utilization (Obernberger *et al.*, 2006). Carbon, hydrogen and oxygen are the main components and are of special relevance for the gross calorific value, and H also affects net calorific value. Nitrogen content is responsible for NO_x formation and emissions, which are the main environmental factors of solid biofuel combustion. Chlorine and sulfur are responsible for deposit formation and corrosion. Furthermore, Cl causes HCl as well as PCDD/F and S SO_x emissions, which are involved in the formation of aerosols. Ash content influences the choice of combustion technology and influences deposit formation and fly ash emissions. Major ash forming elements (Al, Ca, Fe, K, Mg, Na, P, Si, Ti) are relevant for the ash melting behaviour, deposit formation and corrosion. Volatile elements such as Na and K are main constituents of aerosols. Volatile minor elements (As, Cd, Hg, Pb, Zn) play a major role in gaseous and especially aerosol emissions as well as in deposit formation, corrosion and ash utilization/disposal. Either partly or non-volatile minor elements (Ba, Co, Cr, Cu, Mo, Mn, V) influence ash utilization.

Incomplete combustion or charring of various biomass sources under low oxygen produces thermal energy, combustible gases and oils, and biochar in varying amounts depending on process conditions and biomass source. Use of biochar from fast-growing trees, grasses, crop residues and organic wastes as a soil amendment is advantageous for long term carbon sequestration and other agricultural benefits (Atkinson *et al*., 2010). While the largely aromatic carbon structure of biochar may provide a recalcitrant carbon pool in soil, Bliss *et al*. (2011) noted differences in recalcitrance among soils and biofuel crop sources, as the biochar of low phenol content eastern cottonwood (*Populus deltoides*) had significantly greater mineralization rates in both sandy and clayey forested soils than the biochar of high phenol content loblolly pine (*Pinus taeda*).

Metal tissue concentrations in fast-growing tree species grown on metal-contaminated soil are considerably lower compared to hyperaccumulators and even coal (Table 22.5). High above-ground biomass production may compensate for lower tissue contaminant concentrations in these species allowing gradual soil contaminant clean-up while the ability to substitute the biomass for coal in energy production is not compromised by elevated tissue metal concentrations. Arsenic concentration in *PD* was highest in the leaves, next highest in branch bark, followed by stem bark, then branch wood, and lowest in stemwood (Table 22.5). Position was significant for leaf concentration as leaves from the lowest part of the crown had more than twice the concentration of leaves from the upper crown. Bark from upper branches tended to have concentrations some 6 mg kg^{-1} less than upper leaves. Lower leaf concentrations of nearly 40 mg kg^{-1} decreased to less than half as much in the upper leaves. Branch bark concentrations increased from just more than 8 mg kg^{-1} in the lower crown to almost 11 mg kg^{-1} in the upper crown. Branch-wood, stem-bark and stem-wood concentrations changed little with crown or stem position. Variability over time for As concentration in leaves appears relatively small but seasonal. Arsenic is probably translocated from leaf tissues into stem and root tissues prior to dormancy. Variation among clones appears large for the tree components that have the highest As concentrations. A threefold difference was common between the clones with the lowest and highest concentrations in leaves. Based on cumulative results, several *PD* clones are notable. Negative correlations of tree size with concentration suggested a tendency for

Table 22.5. Representative values of Cu, Cr and As concentrations in tree tissues compared to coal and oil.

	Cu (mg kg^{-1})	Cr (mg kg^{-1})	As (mg kg^{-1})	Source
Salix leaves	10.44–15.8	0.21–2.79		Riddell-Black (1994)
Salix stems	5.86–8.15			Riddell-Black (1994)
Poplar leaves	10.3–15.1		2.9–6.3	Yu *et al.* (1996)
Poplar wood	3.0–3.9		2.3–2.9	Yu *et al.* (1996)
Poplar bark	5.3–6.6		2.3–3.1	Yu *et al.* (1996)
PD leaves			10.7–58.4	Ma and Rockwood (2005)
PD branch bark			3.0–23.7	Ma and Rockwood (2005)
PD branch wood			2.4–20.6	Ma and Rockwood (2005)
PD stem bark			3.7–10.4	Ma and Rockwood (2005)
PD stemwood			2.5–4.1	Ma and Rockwood (2005)
Salix ash	205	5.3		Sander (1997)
Coal (mean US)	19	15	15	Valkovic (1983)
Coal ash	3.7–349	3.4–437	0.5–279	Korcak (1998)
Used lube oil		0.23 mg l^{-1}		Johnson (2010)
Gas oil		0.41 mg l^{-1}		
#6 fuel oil		1.39 mg l^{-1}		
Various crude oils	23–195 μg kg^{-1}	<19–161 μg kg^{-1}	<10–37 μg kg^{-1}	Stigter *et al.* (2000)

larger trees to have lower As concentrations in virtually all tree components. Trees typically had As concentrations 1/1000 those of hyper-accumulating Chinese brake fern (*Pteris vittata*), with higher levels in leaves than stems.

Metal concentrations are important in wood used commercially (Evangelou *et al.*, 2012). High concentrations of inorganic compounds in gasified wood can produce gaseous species and char at the expense of bio-oil (Patwardhan *et al.*, 2010). Uncontaminated short rotation coppice willow wood typically contains 3000 mg kg^{-1} K and 5000 mg kg^{-1} Ca (Obernberger *et al.*, 2006). Because young trees combine high metal concentrations in bark with high bark to wood ratio, their metal uptakes are significantly higher (Obernberger *et al.*, 2006). Since older tree trunks have a lower concentration of metals than younger tree trunks, due to a lower bark to wood ratio (Adler *et al.*, 2008), willows grown as a few large stems would have a better wood fuel quality than many small stems. Thus for biomass for biofuel/bioenergy production, longer cutting cycles would improve wood fuel quality, and the wood could also be used for furniture, construction or industrial purposes. Because only birch grown on one soil exceeded concentrations of 50 mg kg^{-1} Cd, 90 mg kg^{-1} Pb, or 40 mg kg^{-1} Cu, the biomass from these contaminated soils could also be used for industrial or furniture products.

Jensen *et al.* (2009), studying growth and heavy metal uptake by willow (*Salix viminalis*) in field and growth-chamber trials, noted field uptakes 2–10 times higher than growth-chamber uptakes. Willow grew poorly on strongly polluted soils, with high concentrations of Cd (≥80 mg kg^{-1}) and Zn (≥3000 mg kg^{-1}) in leaves grown on strongly polluted soil with up to 18 mg Cd kg^{-1}, 1400 mg Cu kg^{-1}, 500 mg Pb kg^{-1} and 3300 mg Zn kg^{-1}. However, willow was promising on moderately polluted soils (2.5 mg Cd kg^{-1} and 400 mg Zn kg^{-1}), where it extracted 0.13% of total Cd and 0.29% of the total Zn per year.

Patwardhan *et al.* (2010), noting that existing petroleum refinery infrastructure is feasible for bio-oil production and upgrading processes, identified that bio-oil chemical composition depends on biomass composition and the process conditions of fast pyrolysis reactions. Concentrations of inorganic mineral salts (NaCl, KCl, $MgCl_2$, $CaCl_2$, $Ca(OH)_2$, $Ca(NO_3)_2$, $CaCO_3$ and $CaHPO_4$) affect the formation of low molecular weight species, furan ring derivatives and anhydro sugar. Mineral salts and higher temperatures accelerated reactions leading to the formation of low molecular weight species compared to those leading to anhydro sugars.

Sebastiani *et al.* (2004) determined the effects of non-hazardous levels of heavy metal-enriched organic waste on heavy metal accumulation of two poplar clones. Trees grown in treated soil differed in stem and root biomass allocation compared to controls. Of Zn, Cu, Cr and Cd in the industrial organic waste, only Zn, Cu and Cr concentrations differed consistently between clones or soil treatments, while Cd levels were always below detection limits. Only non-hazardous industrial biosolid levels may be environmentally sustainable for poplar plantations.

Other estimates of heavy metal concentrations include those of Eriksson and Ledin (1999) for Cd after long-term *Salix* cropping, Giachetti and Sebastiani (2006) for poplars grown with industrial wastes, Hasselgren (1999) for sewage sludge influence on short-rotation forestry, and Hermle *et al.* (2006) on the effects of metal-contaminated soil on young trees.

22.6 Biofuel Uses of Treated Dendroremediating Trees

While natural contaminant levels in dendroremediating trees are typically low, some constituents may need to be managed. Contaminant levels induced by genetic engineering or enhanced uptake will be more important.

For example, natural variation among 30 willow and seven hybrid poplar clones may affect their conversion into bioenergy via thermochemical and biochemical processes (Tharakan *et al.*, 2003). Substantial clonal variation occurred in the stem concentrations of N (2.9–5.0 g kg^{-1}), P (0.4–0.8 g kg^{-1}), K (1.2–2.4 g kg^{-1}), Na (0.09–0.20 g kg^{-1}), Ca (3.9–8.9 g kg^{-1}), Mg (0.2–0.6 g kg^{-1}), ash (13.2–27.2 g kg^{-1}) and bark percentage (3.6–8.1%). Less variation was documented for specific gravity (0.33–0.48 g cm^{-3}) and percentage moisture (49–56%). Bark had a higher concentration of inorganic elements and ash, relative to wood. Willow clones had a higher specific gravity, bark percentage and calcium concentration than hybrid poplar clones, which had a higher K concentration.

Eucalyptus may need Cl pretreatment before some conversion processes. In Brazil, Cl content in eucalyptus varies from 300 to 500 mg kg^{-1} dry wood and K content ranges from 500 to 1000 mg kg^{-1} dry wood (Pekkanen, 2010). In South Africa, Cl contents were up to 1100 mg kg^{-1} dry wood. Elevated amounts of Cl and K can cause fouling and corrosion.

The use of bacteria, mycorrhizae and genetic engineering in dendroremediation raises prospects that SRWC biomass could have elevated levels of heavy metals, etc. (Capuana, 2011). Metal hyperaccumulation and tolerance are genetic traits involving many potential cellular mechanisms in the detoxification of heavy metals and thus tolerance to metal stress (Sarma, 2011). Insights into the molecular basis of metal hyperaccumulation and metal hypertolerance can improve the performance of phytoremediation technology as well as plant adaptation to extreme metallic environments.

As concluded by Vangronsveld *et al.* (2009), in spite of growing interest and successful pilot and field-scale applications, more fundamental dendroremediation research is needed to better exploit plant diversity and to understand interactions between contaminants, soil, roots and microorganisms. Economic analyses would also build public, policy maker and stakeholder support for biofuel use of SRWC dendroremediation systems.

22.7 Conclusions

Effective dendroremediation by SRWCs depends on tree-contaminant interactions, on silvicultural and environmental factors and on the use of locally adapted trees. Even at the local level, genetic differences within SRWC species can be large and should be utilized to maximize biomass productivity and contaminant uptake. In general, dendroremediating SRWC species do not have contaminant concentrations that limit their biofuel use, but biotechnologies may increase contaminants to levels of concern. To meet biomass quantity needs for biofuel production, fast-growing SRWC dendroremediation requires large but highly concentrated areas, such as may be provided by agroforestry, riparian buffer and windbreak projects. Their cost efficient conversion to biofuels is dependent on conversion processes, which are currently numerous and undergoing refinement.

References

Abichou, T., Musagasa, J., Yuan, L., Chanton, J., Tawfiq, K., Rockwood, D. and Licht, L. (2012) Field performance of alternative landfill covers vegetated with cottonwood and eucalyptus trees. *International Journal of Phytoremediation* 14, 47–60.

Adler, A., Dimitriou, I., Aronsson, P., Verwijst, T. and Weiha, M. (2008) Wood fuel quality of two *Salix viminalis* stands fertilised with sludge, ash and sludge-ash mixtures. *Biomass Bioenergy* 32, 914–925.

Alker, G.R., Godley, A.R. and Hallett, J.E. (2002) Landfill leachate management using short rotation coppice final technical report. WRC Report No. CO 5126, 201 pp. Available at: http://www.wrcplc.co.uk/Final_Technical_Report_Leachate_& _SRC_CO5126.pdf (accessed 28 February 2012).

Aronsson, P. and Perttu, K. (2001) Willow vegetation filters for wastewater treatment and soil remediation combined with biomass production. *Forest Chronicles* 77, 293–299.

Ashraf, M.A., Maah, M.J. and Yussof, I. (2011) Heavy metals accumulation in plants growing in ex tin mining catchment. *International Journal of Environment Science and Technology* 8, 401–416.

Atkinson, C.J., Fitzgerald, J.D. and Hipps, N.A. (2010) Potential mechanisms for achieving agricultural benefits from biochar application to temperate soils: a review. *Plant Soil* 337, 1–18.

Baraca, T., Weyensa, N., Oeyena, L., Taghavib, S., Van der Lelieb, D., Dubinc, D., Spliet, M. and Vangronsveld, J. (2009) Field note: hydraulic containment of a BTEX plume using poplar trees. *International Journal of Phytoremediation* 11, 416–424.

Bartle, J. (2009) Integrated production systems. In: Nuberg, I., George, B. and Reid, R. (eds) *Agroforestry for Natural Resource Management*. CSIRO publishing, Collingwood, Victoria, Australia.

Bartle, J.R. and Abadi, A. (2010) Toward sustainable production of second generation bioenergy feedstocks. *Energy and Fuels* 24, 2–9.

Bartle, J.R., Olsen, G., Cooper, D. and Hobbs, R. (2007) Scale of biomass production from new woody crops for salinity control in dryland agriculture in Australia. *International Journal of Global Energy* 27, 115–137.

Bliss, C.M., Minogue, P.J., Mackowiak, C.L., Wright, D. and Comerford, N.B. (2011) Carbon mineralization influenced by biochar loading rates under different land uses and soil textures. *Proceedings of Soil Science Society of America Annual Meeting*, San Antonio, Texas, 16–19 October 2011.

British Petroleum (BP) (2011) Wood River Site, Ecological Revitalization Project Profiles. Available at: http://cluin.org/products/ecorev/er_details.cfm?Project_ID=108 (accessed 28 February 2012).

Capuana, M. (2011) Heavy metals and woody plants – biotechnologies for phytoremediation. *iForest* 4, 7–15. [online 2011-01-27] Available at: http://www.sisef.it/iforest/contents/?ifor0555-004 (accessed 28 February 2012).

Chang, S., Lee, S. and Je, C. (2005) Phytoremediation of atrazine by poplar trees: toxicity, uptake, and transformation. *Journal of Environmental Science and Health, Part B* 40. DOI:10.1080/03601230500227483.

Chard, B.K., Doucette, W.J., Chard, J.K., Bugbee, B. and Gorder, K. (2003) Greenhouse study to determine the uptake of trichloroethylene by fruit trees. *Proceedings of The Seventh International Symposium on In Situ and On-Site Bioremediation*, 2–5 June, Orlando, Florida. Available at: http://www.battelle.org/bookstore (accessed 28 February 2012).

Christianson, L., Barden, C., Davis, L. and Bussen, P. (2007) Use of a multi-year balance to assess a phytoremediation plantation's effectiveness. *Proceedings of The 4th International Phytotechnologies Conference*, 24–26 September, Denver, Colorado.

Dan, T., Foye, D. and Higgins, J. (2007) Facilitated degradation of xylenes and petroleum hydrocarbons in ground water using a hybrid poplar phytoremediation cover at a former bulk fuel loading facility in Southern Ontario. *Proceedings of The 4th International Phytotechnologies Conference*, 24–26 September, Denver, Colorado.

Da Silva Perez, D., Guillemain, A., Berthelot, A., N'Guyen-The, N., De Morogues, F. *et al.* (2010) Evaluation of forestry biomass quality for the production of second-generation biofuels. *Cellulose Chemistry and Technology* 44, 1–14.

Da Silva, P.H.M., Poggiani, F. and Laclau, J.P. (2011) Applying sewage sludge to *Eucalyptus grandis* plantations: effects on biomass production and nutrient cycling through litterfall. *Applied and Environmental Soil Science* 2011, 11.

Dickmann, D.I., Isebrands, J.G., Eckenwalder, J.E. and Richardson, J. (2001) *Poplar Culture in North America*. NRC Press, Ottawa, Canada, 397 pp.

El Bassam, N. (1998) *Energy Plant Species: their use and impact on environment and development*. James and James Science Publishers, London, 334 pp.

El-Gendy, A.S., Svingos, S., Brice, D., Garretson, J.H. and Schnoor, J. (2009) Assessments of the efficacy of a long-term application of a phytoremediation system using hybrid poplar trees at former oil tank farm sites. *Water Environment Research* 81, 486–498.

EPA (2000a) Remediation technologies development forum phytoremediation site profiles: Woodburn WWTP. Available at: http://www.rtdf.org (accessed 28 February 2012).

EPA (2000b) Remediation technologies development forum phytoremediation site profiles: former land treatment facility. Available at: http://www.rtdf.org (accessed 27 March 2012).

EPA (2000c) Remediation technologies development forum phytoremediation site profiles: Lafayette. Available at: http://www.rtdf.org (accessed 27 February 2012).

EPA (2000d) Remediation technologies development forum phytoremediation site profiles: Grundy County Landfill. Available at: http://www.rtdf.org (accessed 24 February 2012).

EPA (2000e) Remediation technologies development forum phytoremediation site profiles: Barje Landfill. Available at: http://www.rtdf.org (accessed 28 February 2012).

EPA (2003) Remediation technologies development forum phytoremediation site profiles: Courtland. Available at: http://www.rtdf.org (accessed 25 February 2012).

EPA (2005) Use of field-scale phytotechnology for chlorinated solvents, metals, explosives and propellants, and pesticides. EPA 542-R-05-002. Available at: http://www.epa.gov/tio/ download/remed/542-r-05-002.pdf (accessed 24 February 2012).

Eriksson, J. and Ledin, S. (1999) Changes in phytoavailability and concentration of cadmium in soil following long term Salix cropping. *Water Air Soil Pollution* 114, 171–184.

Ettala, M.O. (1987) Influence of irrigation with leachate on biomass production and evapotranspiration on a sanitary landfill. *Aqua Fennica* 17, 69–86.

Evangelou, M.W.H., Bürgi, A., Robinson, B.H., Günthardt-Goerg, M.S. and Schulin, R. (2012) Metal uptake and allocation in trees grown on contaminated land: implications for biomass production. *International Journal of Phytoremediation*. DOI:10.1080/15226514.2012.670317.

FAO (1996) Wood Energy News 11. Food and Agricultural Organization Regional Office for Asia, Bangkok, Thailand. Available at: http://www.rwedp.org/wen11-2.html (accessed 27 February 2012).

FAO (2003) State of the World's Forests. Food and Agricultural Organization of the United Nations, Rome. Available at: http://www.fao.org/docrep/005/y7581e/Y7581e00.htm (accessed 27 February 2012).
Giachetti, G. and Sebastiani, L. (2006) Metal accumulation in poplar plant grown with industrial wastes. *Chemosphere* 64, 446–454.
Glass, D.J. (1999) US and international markets for phytoremediation, 1999–2000. Available at: http://www.channel1.com/dglassassoc/INFO/phy99exc.htm (accessed 9 June 2003).
Gordon, A.M., McBride, R.A. and Fisken, A.J. (1989) The effects of landfill leachate spraying on foliar nutrient concentrations and leaf transpiration in a northern hardwood forest. *Canada Forestry* 6, 19–28.
Grattapaglia, D. (2008) Genomics of eucalyptus, a global tree for energy, paper and wood. In: Moore, P.H. and Ming, R. (eds) *Genomics of Tropical Crop Plants*. Springer, New York, pp. 259–298.
Grattapaglia, D. and Resende M.D.V. (2011) Genomic selection in forest tree breeding. *Tree Genetics and Genomes* 7(2), 241–255.
Green, C. and Hoffnagle, A. (2004) Phytoremediation field studies database for chlorinated solvents, pesticides, explosives, and metals. Available at: http://clu-in.org/download/ studentpapers / hoffnagle-phytoremediation.pdf (accessed 27 February 2012).
Hammer, D., Kayser, A. and Keller, C. (2003) Phytoextraction of Cd and Zn with *Salix viminalis* in field trials. *Soil Use and Management* 19, 187–192.
Hasselgren, K. (1998) Use of municipal waste products in energy forestry: highlights from 15 years of experience. *Biomass and Bioenergy* 15, 71–74.
Hasselgren, K. (1999) Utilization of sewage sludge in short-rotation energy forestry: a pilot study. *Waste Management and Research* 17, 251–262.
Hermle, S., Günthardt-Goerg, M.S. and Schulin, R. (2006) Effects of metal-contaminated soil on the performance of young trees growing in model ecosystems under field conditions. *Environmental Pollution* 144, 703–714.
Hytönen, J. (1994) Effect of fertilizer application rate on nutrient status and biomass production in short rotation plantations of willows on cut away peatland areas. *Suo* 45, 65–77.
Isebrands, J.G. and Karnosky, D.F. (2001) Environmental benefits of poplar culture. In: Dickmann, D.I., Isebrands, J.G., Eckenwalder, J.E. and Richardson, J. (eds) *Poplar Culture in North America. Part A*. NRC Research Press, National Research Council of Canada, Ottawa, Canada, pp. 207–218.
Jensen, J.K., Holm, P.E., Nejrup, J., Larsen, M.B. and Borggaar, O.K. (2009) The potential of willow for remediation of heavy metal polluted calcareous urban soils. *Environmental Pollution* 157, 931–937.
Johnson, D. (2010) Determination of metals in oils by ICP-OES. Available at: http://www.chem.agilent. com/ Library/ applications/Icpes-13.pdf (accessed 28 February 2012).
Kamwenda, G.J. (2002) Ngitili agrosilvipastoral systems in the United Republic of Tanzania. *Unasylva* 211, 53, 46–50.
Klein, H.A. (2011) Measuring the removal of trichloroethylene from phytoremediation sites at Travis and Fairchild air force bases. All Graduate Thesis and Dissertations. 860 pp. Available at: http://digitalcommons.usu.edu/etd/860 (accessed 27 February 2012).
Kopp, R.F., Smart, L.B., Maynard, C.A., Isebrands, J.G., Tuskan, G.A. and Abrahamson, L.P. (2001) The development of improved willow clones for eastern North America. *Forest Chronicles* 77, 287–292.
Korcak, R.F. (1998) Agricultural uses of coal combustion byproducts. In: Wright, R.J. *et al.* (eds) *Agricultural Uses of Municipal, Animal and Industrial Byproducts*. USDA-ARS, Conservation Research Report 44.
Labrecque, M., Tedodorescu, T.I. and Daigle, S. (1997) Biomass productivity and wood energy of *Salix* species after two years growth in SRIC fertilized with waste water sludge. *Biomass Bioenergy* 12, 409–417.
Laidlaw, W.S. and Baker, A.J.M. (2006) Developing a Willow SRC System to Phytoextract Heavy Metals from Biosolid Stockpiles at Melbourne Water's Western Treatment Plant. *Proceedings of Environmental Applications of Poplar and Willow Working Party, Northern Ireland, 18–20 May 2006.*
Larsson, S. (2003) Short-rotation willow biomass plantations irrigated and fertilised with wastewaters. European Commission DG VI, Agriculture, 48 pp. Available at: http://www2.gtz.de/Dokumente/oe44/ecosan/en-willow-biomass-plantations-irrigated-wastewaters-2003.pdf (accessed 28 February 2012).
Lewandowski, I., Schmidt, U., Londo, M. and Faaij, A. (2006) The economic value of the phytoremediation function – assessed by the example of cadmium remediation by willow (*Salix* spp). *Agricultural Systems* 89, 68–89.
Licht, L.A. (1994) *Populus* spp. (Poplar) capabilities and relationships to landfill water management. *Proceedings of Air and Waste Management Association. 87th Annual Meeting*, Cincinnati, Ohio, June 1994, Paper 94WA86.02, pp. 19–22.
Licht, L.A. and Isebrands, J.G. (2005) Linking phytoremediation pollutant removal to biomass economic opportunities. *Biomass Bioenergy* 28, 203–218.

Ma, L.Q. and Rockwood, D.L. (2005) Fern-tree combinations for maximizing arsenic phytoremediation in Florida. In: *School of Natural Resources and Environment Mini-Grant Programs Annual Report 2003–2004*, SNRE/University of Florida, pp. 64–70.

Masarovicova, E., Kralova, K. and Pesko, M. (2009) Cost and benefit of energetic plants – Challenges for environment friendly management. *Proceedings of ECOpole* 3, 259–264.

McIntyre, T.C. (2005) Plant and microorganism selection for phytoremediation of hydrocarbons and metals. *Water Encyclopedia* pp. 628–637.

Meers, E., Vandecasteele, B., Ruttens, A., Vangronsveld, J. and Tack, F.M.G. (2007) Potential of five willow species (*Salix* spp.) for phytoextraction of heavy metals. *Environmental and Experimental Botany* 60, 157–168.

Minogue, P.J., Miwa, M., Rockwood, D.L. and Mackowiak, C.L. (2012) Phytoremediation of nitrogen and phosphorus by eucalyptus and populus at a tertiary treated wastewater sprayfield. *International Journal of Phytoremediation* 14, 1–14.

Myers, B.J., Theiveyanathan, S., O'Brien, N.D. and Bond, W.J. (1996) Growth and water use of effluent-irrigated *Eucalyptus grandis* and *Pinus radiata* plantations. *Tree Physiology* 16, 211–219.

Nzengung, V.A., Spriggs, T., Tsangaris, S. and Nwokike, B. (2004) Phytoremediation of a chlorinated solvent plume in Orlando, Florida. In: Magar, V.S. and Kelley, M.E. (eds) *In Situ* and On-Site Bioremediation-2003. *Proceedings of the Seventh International In Situ and On-Site Bioremediation Symposium*. 2–5 June 2003, Orlando, Florida, Paper F-13. Battelle Press, Columbus, Ohio.

Obernberger, I., Brunner, T. and Barnthaler, G. (2006) Chemical properties of solid biofuels – significance and impact. *Biomass and Bioenergy* 30, 973–982.

O'Neill, G.J. and Gordon, A.M. (1994) The nitrogen filtering capacity of Carolina poplar in an artificial riparian zone. *Journal of Environmental Quality* 23, 1218–1823.

Paquin, D.G., Campbell, S. and Li, Q.X. (2004) Phytoremediation in subtropical Hawaii – a review of over 100 plant species. *Remediation Journal* 14, 127–139. doi: 10.1002/rem.20005.

Patwardhan, P.R., Satrio, J.A., Brown, R.C. and Shanks, B.H. (2010) Influence of inorganic salts on the primary pyrolysis products of cellulose. *Bioresource Technology* 101, 4646–4655.

Pekkanen, M.J. (2010) Chemical balances of a modern eucalyptus bleached kraft pulp mill. *Proceedings of XXI TECNICELPA Conference and Exhibition / VI CIADICYP*, Lisbon, Portugal, 12–15 October 2010.

Pisano, S.M. and Rockwood, D.L. (1997) Stormwater phytoremediation potential of Eucalyptus. In: *Proceedings of The 5th Biennial Stormwater Research Conference*, Tampa, Florida. Southwest Florida Water Management District, Brooksville, Florida, 5–7 November 1997, pp. 32–42.

Riddell-Black, D. (1994) Heavy metal uptake by fast growing willow species. In: Aronsson, P. and Perttu, K. (eds) *Willow Vegetation Filters for Municipal Wastewaters and Sludges. A biological purification system*. Swedish University of Agricultural Sciences, Uppsala, Sweden, pp. 145–151.

Riddell-Black, D. and Ferguson, R. (2002) Country Report United Kingdom, 71 pp. Appendix G in Johannesson, M. *et al.* (eds) *The Market Implication of Integrated Management for Heavy Metals Flows for Bioenergy use in the European Union*. Kalmar University, Department of Biology and Environmental Science, Kalmar, Sweden, 115 pp., ISBN 91-89584-13-9.

Riemenschneider, D.E., Berguson, W.E., Dickmann, D.I., Hall, R.B., Isebrands, J.G. *et al.* (2001) Poplar breeding and testing strategies in the north-central US: demonstration of potential yield and consideration of future research needs. *Forest Chronicles* 77, 245–253.

Rockwood, D.L., Pisano, S.M. and McConnell, M.V. (1996) Superior cottonwood and Eucalyptus clones for biomass production in waste bioremediation systems. *Proceedings of Bioenergy 96, 7th National Bioenergy Conference*, 15–20 September 1996, Nashville, Tennessee, pp. 254–261.

Rockwood, D.L., Ma, L.Q., Alker, G.R., Tu, C. and Cardellino, R.W. (2001) Phytoremediation of contaminated sites using woody biomass. Final Report to the Florida Center for Solid and Hazardous Waste Management, June 2001, 95 pp. Available at: http://www.floridacenter.org/publications/ publicationsnew.htm (accessed 27 February 2012).

Rockwood, D.L., Cardellino, R., Alker, G., Lin, C., Brown, N. *et al.* (2004a) Fast-growing trees for heavy metal and chlorinated solvent phytoremediation. In: Magar, V.S. and Kelley, M.E. (eds) *In Situ and On-Site Bioremediation: Proceedings of the Seventh International In Situ and On-Site Bioremediation Symposium*, Orlando, Florida, 2–5 June 2003, Paper F-12. Battelle Press, Columbus, Ohio.

Rockwood, D.L., Naidu, C.V., Carter, C.V., Rahmani, M., Spriggs, T. *et al.* (2004b) Short-rotation woody crops and phytoremediation: opportunities for agroforestry? In: Nair, P.K.R., Rao, M.R. and Buck, L.E. (eds) *A Compendium for the 1st World Congress of Agroforestry 2004: New Vistas in Agroforestry*. Kluwer Academic Publishers, Dordrecht, the Netherlands, pp. 51–63. Available at: http://www.springerlink.com/content/n8258x036171221r/fulltext.pdf (accessed 27 February 2012).

Rockwood, D.L., Rudie, A.W., Ralph, S.A., Zhu, J. and Winandy, J.E. (2008) Energy product options for *Eucalyptus* species grown as short rotation woody crops. *International Journal of Molecular Sciences* 9, 1361–1378. Available at: http://www.mdpi.org/ijms/papers/i9081361.pdf (accessed on 26 March 2012).
Sander, M.L. (1997) Biofuel ash use in *Salix* plantations, biomass production, nutrient uptake and heavy metal circulation. Doctoral thesis, Swedish University of Agricultural Sciences, Uppsala, Sweden.
Sarma, H. (2011) Metal hyperaccumulation in plants: a review focusing on phytoremediation technology. *Journal of Environmental Science and Technology* 4, 118–138.
Schultz, R.C., Collettii, J.P., Isenhart, T.M., Simpkins, W.W., Mize, C.W. and Thompson, M.L. (1995) Design and placement of a multispecies riparian buffer strip system. *Agroforestry Systems* 31, 117–132.
Sebastiani, L., Scebba, F. and Tognetti, R. (2004) Heavy metal accumulation and growth responses in poplar clones Eridano (*Populus deltoides x maximowiczii*) and I-214 (*P. x euramericana*) exposed to industrial waste. *Environmental and Experimental Botany* 52, 79–88.
Segrest, S.A., Rockwood, D.L., Stricker, J.A. and Alker, G.R. (2001) Partnering to cofire woody biomass in central Florida. In: Abstracts 5th Biomass Conference of the Americas, pp. 2. Available at: http://bioproducts-bioenergy.gov/pdfs/bcota/abstracts/4/z280.pdf (accessed 28 February 2012).
Sharma, K.P., Singh, P.K., Kumar, S., Sharma, S. and Kumar, R. (2011) Tolerance of some hardy plant species to biomethanated spent wash of distilleries. *Indian Journal of Biotechnology* 10, 97–112.
Spriggs, T., Tsangaris, S., Nzengung, V.A. and Nwokike, B. (2004) Phytoremediation of a chlorinated solvent plume in Orlando, Florida. In: Magar, V.S. and Kelley, M.E. (eds) *Situ and On-site Bioremediation, 2003: Proceedings of the Seventh International In Situ and On-site Bioremediation Symposium*. CD-ROM. Battelle Press, Columbus, Ohio.
Stanhope, A., Berry, C.J. and Brigmon, C.J. (2008) Field Note: Phytoremediation of chlorinated ethenes in seepline sediments: Tree selection. *International Journal of Phytoremediation* 10, 529–546.
Stanton, B., Eaton, J., Johnson, J., Rice, D., Schuette, B. and Moser, B. (2002) Hybrid poplar in the Pacific Northwest: the effects of market-driven management. *Journal of Forestry* 100, 28–33.
Stigter, J.B., de Haan, H.P.M., Guicherit, R., Dekkers, C.P.A. and Daane, M.L. (2000) Determination of cadmium, zinc, copper, chromium and arsenic in crude oil cargoes. *Environmental Pollution* 107, 451–464.
Sudmeyer, R.A. and Goodreid, A. (2007) Short-rotation woody crops: a prospective method for phytoremediation of agricultural land at risk of salinisation in southern Australia. *Ecological Engineering* 29, 350–361.
Tharakan, P.J., Volk, T.A., Abrahamson, L.P. and White, E.H. (2003) Energy feedstock characteristics of willow and hybrid poplar clones at harvest age. *Biomass Bioenergy* 25, 571–580.
Trapp, S., Koehler, A., Larsen, L.C., Zambrano, K.C. and Karlson, U. (2001) Phytotoxicity of fresh and weathered diesel and gasoline to willow and poplar trees. *Journal of Soils and Sediments* 1, 71–76.
Valkovic, V. (1983) *Trace Elements in Coal*, Vol. 1. CRC Press, Boca Raton, Florida.
Vangronsveld, J., Herzig, R., Weyens, N., Boulet, J., Adriaensen, K. *et al.* (2009) Phytoremediation of contaminated soils and groundwater: lessons from the field. *Environmental Science and Pollution Research* 16, 765–794.
Vervaeke, P., Luyssaert, S., Mertens, J., Meers, E., Tack, F.M.G. and Lust, N. (2003) Phytoremediation prospects of willow stands on contaminated sediments: a field trial. *Environmental Pollution* 126, 275–282.
Wong, M.H. and Leung, C.K. (1989) Landfill leachate as irrigation water for tree and vegetable crops. *Waste Management and Research* 7, 311–324.
Wu, Q., Wang, S., Thangavel, P., Li, Q., Zheng, H., Bai, J. and Qiu, R. (2011) Phytostabilization potential of *Jatropha curcas* L. in polymetallic acid mine tailings. *International Journal of Phytoremediation* 13, 788–804.
Yu, G., Wu, Y. and Wang, X. (1996). Transferring of heavy metals in and out of poplar tree before and after its leaf fallen. *Chinese Journal of Applied Ecology* 7, 201–206.
Zacchini, M., Pietrini, F., Mugnozza, G.S., Iori, V., Pietrosanti, L. and Massacci, A. (2009) Metal tolerance, accumulation and translocation in poplar and willow clones treated with Cadmium in hydroponics. *Water, Air, and Soil Pollution* 197, 23–34.
Zalesny, J.A., Zalesny, R.S. Jr, Wiese, A.H. and Hall, R.B. (2007) Choosing tree genotypes for phytoremediation of landfill leachate using phyto-recurrent selection. *International Journal of Phytoremediation* 9, 513–530.
Zalesny, R.S., Bauer, E.O., Hall, R.B., Zalesny, J.A., Kunzman, J. *et al.* (2005) Clonal variation in survival and growth of hybrid poplar and willow in an *in situ* trial on soils heavily contaminated with petroleum hydrocarbons. *International Journal of Phytoremediation* 7, 177–197.
Zalesny, R.S. Jr, Wiese, A.H., Bauer, E.O. and Riemenschneider, D.E. (2006) Sapflow of hybrid poplar (*Populus nigra* L. × *P. maximowiczii* A. Henry 'NM6') during phytoremediation of landfill leachate. *Biomass and Bioenergy* 30, 784–793.

23 Lignocellulosic Feedstock Preparation by Size Reduction and Pretreatment

K. Muthukumarappan
Department of Agricultural and Biosystems Engineering, South Dakota State University, USA

23.1 Introduction

About 60% of the world's ethanol is produced from maize and sugarcane in the USA and Brazil, respectively. Economic viability and grain supply limitation lead to the search for alternative feedstocks. In order to meet the ever-growing fuel demand, researchers around the world have been exploring alternative renewable resources for more than three decades. Because of the renewable nature, positive environmental impact and abundant supply, biomass such as agricultural residues, woody materials and native grasses appears to be an attractive feedstock. The US National Research Council has set a goal to provide at least 10% of liquid fuel by the year 2020 and to provide 50% of liquid fuel by the year 2050 from renewable resources (NRC, 2000). Further, the 2007 Energy Act mandates the production of 21 billion gallons of biofuels from non-maize starch materials by 2022.

Ethanol production from biomass is energy intensive and cost prohibitive compared to ethanol production from maize, because the accessibility of C6 and C5 sugars in biomass to hydrolytic enzymes is more difficult than the starch in grain (Gibbons *et al.*, 1986). The complex structure and recalcitrant nature of biomass necessitates an additional process called pretreatment. The primary purposes of pretreatment are to open up the biomass structure, to increase accessible surface area, to reduce the cellulose crystallinity and to increase the porosity, pore size and pore volume. Yang and Wyman (2008) reported that feedstock preparation such as size reduction and pretreatment are the key steps in converting biomass into biofuel because of their pervasive impact on the downstream processing. Numerous pretreatment methods such as physical, chemical, biological and their combinations can be found in the literature. This chapter presents a brief review and discussion on size reduction and traditional and novel pretreatments of different feedstocks.

23.2 Biomass Processing

The goal for cellulosic biofuel production envisions large biorefineries (approximately 2000–5000 dry t feedstock day^{-1}), which are fully integrated and containing all unit operations in a single location. However, such refineries may be impractical from a logistical standpoint. For example, a large-scale biorefinery would require one truck every 3 min (24 h day^{-1}), contracts with many farmers, and would need to raise hundreds of millions of dollars to begin operations. The difficulties with this

©CAB International 2013. *Biofuel Crops: Production, Physiology and Genetics* (ed. B.P. Singh)

approach are further limited by a short harvest season, requiring biomass to be stored for months to ensure year-round biofuel production. In contrast, a much smaller facility would require relatively much smaller investment and would be a simpler, more easily manageable operation. One approach to help resolve these problems is to create regional storage locations, storage facilities, or depots. These facilities are currently envisioned as transitory storage locations for square or round bales typically using standard farm equipment and sometimes including moderate processing, usually densification of biomass, before delivering biomass to the biorefinery.

The logistics of harvesting, handling, transporting and storing low bulk density feedstocks is a significant challenge to development of the bioeconomy (Klein, 2007). These issues are especially critical for herbaceous feedstocks, which may have low per hectare productivities. For example, biomass that yields 70 gallons (265 l) of ethanol per tonne, baled at a density of 120 kg m^{-3} would require over ten times the volume of material for a given volume of ethanol, compared with maize grain. Therefore, biomass densification at distributed locations (to minimize transport of feedstock bales) is critically needed, but conventional extrusion pelleting has proven too costly (Sokhansanj and Turhollow, 2004; Mani *et al.*, 2006). Various densification methods are available, such as pelletizing, briquetting, or cubing, which all have different capital and energy requirements. In general, a high bulk density (pelletization) requires higher energy costs, more unit operations, and more finely ground material than low bulk density material (cubing). Because biomass is naturally heterogeneous, different harvests of a particular species may have different physical properties that can impact biofuel production. Grinding, densifying, mixing and storing the biomass produces a consistent product that can be supplied to biorefineries or other markets while standardizing the supply system schedule and logistics.

Muthukumarappan (2012) proposed a solution based on a novel densification process being developed by Federal Machine, Inc., Fargo, North Dakota (USA). The ComPAKco gear mesh system compresses biomass through a tapering channel between adjacent gear teeth, forming high density comPAKed material termed as PAKS. The system operates at lower temperature, pressure and energy requirements than pelletizers. To provide binding properties, prior to densification the biomass will be pretreated with the Ammonia Fiber Expansion (AFEX) process (see Section 23.3). This pretreatment will also open the biomass structure, thereby increasing the efficiency of subsequent hydrolysis and fermentation.

Carolan *et al.* (2007) suggested the concept of distributed bioprocessing using a network of regional biomass pre-processing centres (RBPCs) to form a biomass supply chain to feed larger centralized biorefineries. Their study evaluated AFEX pretreatment followed by conventional pelletizing. An economic analysis based on AFEX process information that was several years old, showed that the RBPC would be profitable at a 12% internal rate of return at operational scales as low as 250 t day^{-1}. Recent AFEX process improvements will make the operation profitable at even smaller scales (Fig. 23.1).

The study above used pelletizing to provide a compact and highly flowable material for transport, storage and subsequent processing. Pelletizing is a traditional form of densification (Sokhansanj *et al.*, 2003) and binders are often added to the feedstock to form stable pellets (Tabil *et al.*, 1997). Unfortunately, traditional ring/die pelletizing systems are very expensive and energy intensive. A typical system might require a 400 HP motor to process 4–5 t h^{-1}. This system would cost in the range of US$300,000. Due to the high friction, the feedstock material could reach 180–200°C, potentially degrading the carbohydrates.

An alternative compaction device is being developed by Federal Machine (Fargo, North Dakota). The proof-of-concept device (Fig. 23.2) has been used to form PAKS of various biomass sources. Pilot-scale testing with various binding agents has produced compacted PAKS with a specific gravity of 1.13 (79 lb ft^{-3} or 1266 kg m^{-3}), that have stored for over 1 year with no apparent degradation. We estimate a full scale system would use a 180 HP motor and produce 7–8 t h^{-1} of compacted PAKS. The capital investment is estimated to be less than half of a comparable pelletizer, with total operating and capital costs of US$5–10 t^{-1}, versus up to

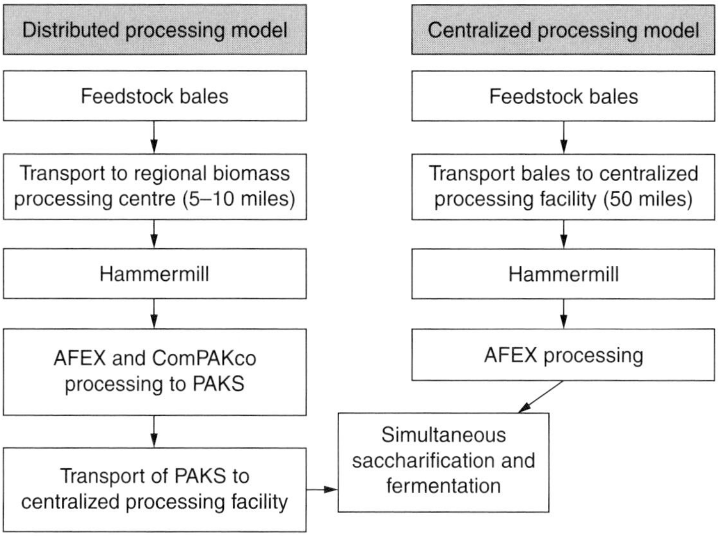

Fig. 23.1. Comparison of distributed versus centralized processing model for bioprocessing.

ComPAKco device ComPAKed biomass

Fig. 23.2. Proof-of-concept compaction device (Federal Machine, Inc., Fargo, North Dakota).

US$50–65 t^{-1} for conventional pelletization or cubing (Sokhansanj and Turhollow, 2004; Mani *et al.*, 2006). Moreover, this system operates at temperatures of ambient to 60°C, meaning less biomass degradation.

To eliminate external binders, Muthukumarappan (2012) used Ammonia Fiber Expansion (AFEX) to both open the structure of the biomass and provide an intrinsic binding capability. The AFEX process, developed at Michigan State University, heats biomass to 60–100°C in the presence of concentrated ammonia (Teymouri *et al.*, 2004). A rapid pressure drop then causes a physical disruption of the biomass structure, exposing cellulose and hemicellulose fibres, without the extreme sugar degradation common to many pretreatments. Other advantages of AFEX are listed below:

- Nearly all of the ammonia can be recovered and reused while the remaining serves as a nitrogen source for microbes in fermentation. There is no wash stream in the process.
- Dry matter recovery following the AFEX treatment is essentially 100%. AFEX is

basically a dry to dry process. Treated biomass is stable for long periods and can be fed at very high solids loadings in enzymatic hydrolysis or fermentation process.
- Cellulose and hemicellulose are well preserved in the AFEX process (little degradation).
- There is no need for neutralization prior to the enzymatic hydrolysis of AFEX-treated biomass. Enzymatic hydrolysis of AFEX-treated biomass produces clean sugar streams for subsequent fermentation.

AFEX treatment also transfers some lignin and hemicellulose oligomers to the surface, where it can act as a binding agent. Hard, dense and durable pellets have been made from AFEX-treated maize stover and switchgrass with a specific gravity of up to 1.40 (Muthukumarappan, 2012) without using any binders, whereas untreated maize stover pellets are much less dense and are not durable at all (Fig. 23.3).

The combined system created dense, durable biomass PAKS that will substantially lower transportation costs. The increased flowability of these PAKS will make possible automated loading and unloading of transport vehicles and storage systems, as well as transport through the processing facility. The material handling and storage equipment currently used for grains may well be adaptable for biomass PAKS. The PAKS also have reduced moisture content and water activity, and this should increase storage life. Prior analyses suggest that the integrated AFEX/ComPAKco system will be scalable for sizes that enable regional biomass processing centres (RBPCs) that minimize the distance that low density feedstock bales will be transported. This process will link AFEX pretreatment with a novel densification process to produce dense, solid biomass PAKS that retain their original composition, are more susceptible to hydrolysis and fermentation, and can use existing transportation/handling infrastructure. The RBPCs model will minimize transportation of low density feedstock bales. Densified PAKS will then be more efficiently transported to centralized processing facilities as shown in Fig. 23.1.

In preliminary trials, Muthukumarappan (2012) subjected maize stover and switchgrass to AFEX at Michigan State University (MSU). After pretreatment, samples were PAKed using the ComPAKco device. Figure 23.3 shows the pretreated and PAKed materials resulting from these optimized trials. South Dakota State University (USA) then subjected the AFEX pretreated and PAKed feedstocks to preliminary physical tests such as angle of repose (°), water activity (no units), true density (kg m^{-3}), bulk density (kg m^{-3}), porosity, water absorption index (no units) and water solubility index (%) and the results are shown in Table 23.1. Based on this initial work, it appears that the technology will perform as anticipated, and that further improvements can be made by optimizing these unit operations with respect to each

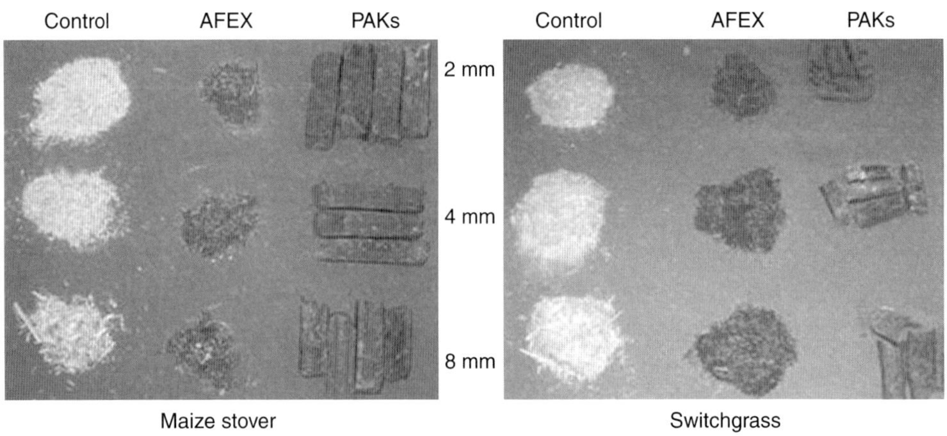

Fig. 23.3. Comparison of control, AFEX treated and AFEX-PAKed biomass samples.

Table 23.1. Physical properties of AFEX-treated versus AFEX-treated and PAKed switchgrass and maize stover.[a]

Biomass type	AoR (°)	Water activity	TD (kg m^{-3})	BD (kg m^{-3})	Porosity	WAI	WSI (%)
Switchgrass (AFEX and PAKed)	57.4[a]	0.575[c]	918[a]	547.2[a]	0.487	6.30[b]	6.74[a]
Switchgrass (AFEX only)	56.0[a]	0.787[a]	421[c]	411.9[b]	0.640[a]	6.17[b]	6.14[a]
Maize stover (AFEX and PAKed)	60.6[a]	0.451[b]	722[b]	549.2[a]	0.376[b]	7.14[a, b]	4.36[a]
Maize stover (AFEX only)	54.4[a]	0.672[b]	452[c]	447.7[b]	0.657[a]	8.03[a]	5.63[a]

[a]Similar superscripts for a given property are not significantly different at $\alpha = 0.05$.
AoR, angle of repose; TD, true density; BD, bulk density; WAI, water absorption index; WSI, water solubility index.

other, and in relation to storage characteristics, physical attributes, and hydrolysis and fermentation processing. Based on preliminary tests, the combined system will create dense, durable biomass PAKS that will substantially lower transportation costs. The increased flowability of these PAKS will make possible automated loading and unloading of transport vehicles and storage systems, as well as transport through the processing facility. Handling and storage equipment currently used for grains may well be adaptable for biomass PAKS. The PAKS also have reduced moisture content and water activity resulting in increased storage life.

23.3 Feedstock Preparation

Pretreatment is an important step for a practical cellulose conversion process. Unlike maize grain where the major carbohydrate is starch, biomass is composed of 35–50% cellulose, 20–35% hemicellulose and 10–25% lignin (Saha and Bothast, 1997). Because biomass is naturally resistant to breakdown of its component sugars, the pretreatment step is needed to open up the structure of the material and to make it accessible for enzymes to hydrolyse the cellulose and hemicellulose components at appreciable rates and acceptable yields (Wyman, 1994). Lignin, hemicellulose and the acetyl groups are the known chemical barriers. The physical factors of biomass, such as crystallinity (Fan *et al.*, 1980), surface area (Burns *et al.*, 1989; Lee and Yu, 1995) and degrees of polymerization (Puri, 1984) have also been known to influence the enzymatic hydrolysis. Among these factors, lignin has been considered as the major impeding factor. Pretreatment is required to alter the biomass macroscopic and microscopic size and structure as well as its submicroscopic chemical composition and structure so that hydrolysis of the carbohydrate fraction to monomeric sugars can be achieved more rapidly and with greater yields (Sun and Cheng, 2002; Mosier *et al.*, 2005a). Pretreatment affects the structure of biomass by solubilizing hemicellulose, reducing crystallinity and increasing the available surface area and pore volume of the substrate. An effective pretreatment must: (i) produce fibre, in which the cellulose fraction could be hydrolysed by cellulases effectively; (ii) preserve the utility of the hemicelluloses; and (iii) avoid the formation of inhibitory by-products (Laser *et al.*, 2002). An economical pretreatment should use inexpensive chemicals and require simple equipment and procedures (McGinnis *et al.*, 1983). Pretreatment has been considered as one of the most expensive processing steps in converting biomass to fermentable sugars with costs as high as $0.30 per gallon of ethanol produced (Mosier *et al.*, 2005b).

23.2.1 Size reduction

Biomass size reduction is a pretreatment step that changes the particle size, shape, total surface area and bulk density of the material. Size reduction of lignocellulosic biomass is also necessary to eliminate mass and heat transfer limitations during the hydrolysis reactions (Schell and Harwood, 1994). Common biomass size reduction devices or grinders are hammer mills, choppers, chippers, shredders and disk mills. Size reduction is energy inefficient as only from 0.06 to 1% of the energy is used in breaking and creating new surfaces (Mohsenin, 1986). Miu *et al.* (2006) suggested hammer mill, knife mill and disc mill as the proper equipment for biomass comminution. Hammer mills are widely used due to high size reduction ratios, good control of particle size range, with relatively good cubic shape of particles. Knife mills work successfully for shredding forages under various crops and machine conditions (Ige and Finner, 1976). Disk mills produce very small particles if input feed is provided by knife mills or hammer mills. Different types of size reduction equipment perform the comminution processes differently as one or more phenomena (impact, attrition, shear and compression) become dominant in the grinding/milling space. Ideal equipment has a combination of stress loadings that leads to efficient disintegration of biomass to a fineness that corresponds to optimal reacting conditions for microorganisms or chemical substances. In a hammer mill, a rotating drum with fixed or swinging blades uses shear, impact, attrition and compression to reduce material size. Shear is responsible for size reduction in the knife mill whereas attrition and shear are major forces in disk mills. Schell and Harwood (1994) studied the energy requirement for milling of lignocellulosic biomass and concluded that hammer milling used less energy than disk milling.

Vibratory ball milling has been found to be more effective in breaking down the cellulose crystallinity of spruce and aspen chips and improving the digestibility of the biomass than ordinary ball milling (Millet *et al.*, 1976). In general, ball milling significantly increases yield and selectivity of saccharification. Degree of crystallinity was found to decrease linearly with respect to time of ball milling (Sidiras and Koukios, 1989). The power requirement of mechanical comminution of agricultural materials depends on the final particle size and the biomass characteristics (Cadoche and Lopez, 1989). Costs and energy requirements for particle size reduction increase geometrically with decreasing particle size. The grinding process can account for one-third of the energy requirements of the entire process (van Walsum *et al.*, 1996). Milling has been suggested as a unique pretreatment method before cellulose hydrolysis, since the required equipment is less expensive than the equipment needed for other pretreatment methods such as steam explosion or ammonia fibre explosion process. These processes account for 6–20% of capital costs of the process whereas milling equipment accounts for about 1% only. However, milling has elevated energetic costs. Alvo and Belkacemi (1997) point out that milling of perennial grasses requires much less energy than milling of wood. These authors consider that milling as sole pretreatment method should not be discarded as an option because of two advantages: toxic products of degradation are not formed, and soluble carbohydrates of the initial biomass are not destroyed. Also many rural communities can acquire mills more easily than other expensive pretreatment systems. A simultaneous ball milling and enzymatic hydrolysis could improve the rate of saccharification and/or reduce the amount of enzyme required to attain total hydrolysis of the carbohydrate moieties (Mais *et al.*, 2002).

Biomass can be comminuted by a combination of chipping, grinding and milling to reduce cellulose crystallinity. The size of the materials is usually 10–30 mm after chipping and 0.2–2 mm after milling or grinding. According to US patent No 5677154, the ethanol production process needs a size of ground biomass between 1 and 6 mm. Biomass final size is the main criterion for selecting size reduction equipment if the target is the conversion of energy because of power consumption. Size reduction equipment uses a wide range of specific energy (10–50 kW Mg^{-1}) depending on the material and grinding mechanisms, i.e. shear, impact, attrition (Spinelli and Hartsough, 2001). Mani *et al.* (2004) recorded the specific energy consumption of switchgrass (27.6 kWh t^{-1}) and maize stover (11.0 kWh t^{-1}) at 3.2 mm screen size. Jannasch *et al.* (2005) reported a specific energy of 55.9 kWh t^{-1} for the hammer mill screen sizes of 5.6 and 2.8 mm for switchgrass. Currently a research team at the University of Tennessee is working on biomass (maize stover and switchgrass) size reduction using a hammer mill, disk mill and linear knife grind. They reported that the energy requirement of 1.69 ± 0.8, 0.48 ± 0.08 and 0.49 ± 0.04, 0.14 ± 0.02 kWh t^{-1} for the particle size of 50 and 101 mm of maize stalks and switchgrass, respectively. Specific energy consumption based on the new surface area generated for 50 and 101 mm particle size were 87.78 ± 24.5, 63.34 ± 12.3 and 21.50 ± 1.9, 15.53 ± 1.7 kN m^{-1} for maize stalk and switchgrass, respectively (Igathinathane *et al.*, 2006). High moisture maize stalk required greater stress and 1.3 times the energy than the low moisture material (Igathinathane *et al.*, 2007).

Different pretreatment processes require different levels of size reduction. Generally, grinding raw material to a particle size of the order of 1 mm is required before chemical pretreatment. Size reduction itself can be evaluated as a pretreatment tool. Before attempting to develop any new equipment, it is better to evaluate existing equipment from similar or different industries. For this reason we propose the evaluation of existing size reduction equipment from a sugarcane factory (shredder) as first size reduction and from the wheat flour industry as second size reduction equipment.

Sidiras and Koukios (1989) reported acid saccharification of more than 50% of barley straw with minimal glucose degradation. Total saccharification of 51.4% with enzyme activities of 5 FPU g^{-1} is reported for timothy (Alvo and Belkacemi, 1997). This saccharification yield is very low when considering industrial applications. Less heat buildup in the roller mill appeared as an advantage. It should be pointed out that the grain was perhaps more brittle than many biomass feedstocks. Tensile and shear properties of the biomass can influence the energy requirements for biomass size reduction. Size reduction studies show that mean shear strength is approximately one-fifth of the tensile strength (Igathinathane *et al.*, 2007). Size reduction equipment is more efficient by applying shear stress rather than tensile stress, because shear mechanisms may be considered the weakest mode of failure or natural fracture. As a result of better size reduction, i.e. uniform particle size, increase in surface area and porosity of biomass, we expect saccharification yield will be near to the industry's expectation.

23.2.2 Pretreatment

Pretreatment is an important step towards efficient hydrolysis and downstream process for ethanol production. It normally considers all processes performed on the lignocellulosic feedstocks to make them suitable for enzyme or acid digestion (for hydrolysis). Lignocellulose, although having abundant carbohydrate polysaccharides, does not yield sugar when subjected to hydrolysis as such in a natural state. The crystallinity of cellulose fibres and other components besides cellulose such as lignin and hemicelluloses restrict the accessibility of the hydrolyte into the hydrolysis site. Thus an effective pretreatment method needs to: (i) perform optimum delignification; (ii) disrupt fibres such that surfaces are exposed; (iii) have a low energy requirement; (iv) preserve the pentose fraction for hydrolysis; and (v) prevent the formation of products that are inhibitory for hydrolysis and fermentation (Sun and Cheng, 2002). Major pretreatment strategies can be broadly categorized into: (i) physical pretreatment (size reduction, milling), which we discussed in detail in the previous section; (ii) physiochemical (steam treatment, hydrothermolysis); and (iii) chemical (alkali, acid, oxidizing agents). Several pretreatment strategies have been discussed in detail (Sun and Cheng, 2002; Mosier *et al.*, 2005a; Kumar *et al.*, 2009).

Liquid hot-water treatment

Liquid hot-water treatment uses the high temperature water brought about by increased pressure in the reactor to remove primarily lignin and hemicelluloses and also removes cellulose to a lesser extent. Hot-water treatment can be in co-current, counter-current or fixed bed flow-through fashion. Liquid hot water cooks the biomass and removes much of the lignin such that the cellulose is susceptible to enzymatic attack. Further, there is no need of a greater degree of size reduction in this process. Liquid hot-water treatment also does not need complex washing techniques or conditioning of biomass before hydrolysis as no other chemicals are usually added (Zheng *et al.*, 2009). However, the pH of water lowers with increase in temperature, thus at high temperatures there is the chance of degradation of monomers from hemicelluloses to form products such as furfural and 5-hydroxymethyl furfural, which are inhibitory to the fermenting microorganism in later stages (Mosier *et al.*, 2005b).

Steam treatment

Steam treatment involves the use of high temperature steam (up to 240°C) to solubilize hemicelluloses, to make cellulose better utilized for hydrolysis and prevent the formation of inhibitors. Steam treatment may be just a simple treatment strategy where the steam is applied to the biomass in a large vessel to solubilize hemicelluloses and lignin or it could be the application of steam at high pressure and then quickly releasing the steam such that the so-called 'explosion effect' takes place. The latter strategy is also known as steam explosion (Hendriks and Zeeman, 2009). The effect of steam pretreatment is due to the acetic acid and other acid released during the treatment that hydrolyses the hemicelluloses and also further catalysing the glucose and xylose from acetic acid produced from the acetyl groups of hemicelluloses. Steam increases the temperature of the biomass rapidly and does not dilute the released sugars as with other similar treatment processes (Walch *et al.*, 1992; Mosier *et al.*, 2005b).

Acid pretreatment

Acid pretreatment has been used in various cellulosic materials to make them more amenable to hydrolysis. The treatment commonly uses sulfuric acid, hydrochloric acid, nitric acid, or phosphoric acid to remove the hemicelluloses from the lignocelluloses such that a high level of enzymatic digestibility is obtained. However, as concentrated acid removes the hemicelluloses from the lignocelluloses, it is expensive unless an effective recovery method is developed and also corrosive, thus hazardous and requiring expensive construction materials. In response, dilute acid has received considerable attention in pretreatment. Among the several acids, dilute sulfuric acid has been studied extensively as it is cheaper and has proved effective (Zheng *et al.*, 2009). Even then dilute sulfuric acid also comes with the disadvantage of corrosiveness, requirement of neutralization of biomass, formation of degradation compound and release of natural inhibitors of fermentation (Kumar *et al.*, 2009).

Ammonia Fiber Expansion treatment

Ammonia Fiber Expansion treatment (AFEX) is a pretreatment method in which the liquefied ammonia at normal temperature is used to treat the lignocelluloses and then at a later stage the pressure is dropped quickly so that the liquid changes to gas, bringing about the expansion of cellulosic fibres. It is similar to the steam explosion process where steam is rapidly vented out to bring about expansion of fibres. In AFEX, the combined chemical effect of cellulose decrystallization and the physical effect of increased surface area due to expansion helps in better enzymatic hydrolysis (Holtzapple *et al.*, 1991). Various kinds of biomass such as switchgrass, coastal bermudagrass, (Reshamwala *et al.*, 1995) and maize fibre (Hespell *et al.*, 1997) have been treated using AFEX. It has been found that there is no production of inhibitory compound from AFEX treatment as in dilute acid pretreatment. Agricultural waste and herbaceous biomass are well suited for AFEX treatment; however, it is less suitable for hardwood and even less suitable for softwood (McMillan, 1994; Sun and Cheng, 2002).

Alkaline pretreatment

Alkaline treatment involves the use of sodium hydroxide, liquid ammonia, aqueous ammonia, lime or other alkalis for pretreatment of lignocelluloses. Alkali treatment is found to cause swelling, decrease the polymerization and delignify the lignocelluloses and increase the surface area. In general, the alkaline pretreatment process has been used with lower temperature and longer periods, usually hours or days. Saponification of intermolecular ester bonds, which crosslink the xylan, hemicelluloses and other components such as lignin and other hemicelluloses, is thought to be the mechanism of the alkaline treatment. The removal of these crosslinks increases the porosity of the lignocelluloses (Tarkow and Feist, 1969; Sun and Cheng, 2002; Mosier *et al.*, 2005a).

Among various alkalis, sodium hydroxide is probably the most used by researchers. Sodium hydroxide treatment of rice straw was able to reduce lignin by 28.4% and hemicellulose by 36.8% (He *et al.*, 2008). Xu *et al.*

(2010) studied the effect of 2% sodium hydroxide for 96 h on switchgrass and found up to 85.8% lignin reduction. Silverstein *et al.* (2007) found that sodium hydroxide pretreatment (2% NaOH, 90 min, 121°C at 15 psi) of cotton stalk produced delignification of 65.6% and cellulose conversion of 60.8%. Aqueous ammonia, another pretreatment chemical, is also found to be an effective delignifying agent. Both the pretreatments have been used by soaking biomass or in flow-through reactors. Kim and Lee (2007) studied the effect of soaking (10–60 days) maize stover in aqueous ammonia and found that 55–74% of lignin was removed from biomass but nearly 100% glucan and 85% xylan was retained. They also found removal of 62% lignin and retention of 100% glucan and 85% xylan when treated at moderate temperature of 60°C with 15% ammonia for 12 h. In percolation reactors, 59–70% lignin removal and 48–57% xylan retention was found for maize stover, with digestibility of 95% with 60 FPU g^{-1} glucan (Kim *et al.*, 2006). A pilot-scale fermentation from aqueous ammonia-soaked switchgrass yielded about 73% ethanol of the theoretical yield in the 50 l reactor (Isci *et al.*, 2009). Similarly, various researchers have studied switchgrass, bagasse, maize husk and hybrid poplar (Iyer *et al.*, 1996, Kurakake *et al.*, 2001) and found similar yields.

Lime pretreatment involves use of $Ca(OH)_2$ solution in water in a concentration usually of about 1g g^{-1} dry biomass. Lime is a mild base, which reacts in similar way as other alkalis to delignify the lignocelluloses; however, the action of lime is much slower than ammonia and other strong bases. But the low cost and lower toxicity makes it an attractive pretreatment chemical. Usually, the temperature of treatment using lime is around room temperature, which may extend to 130°C in order to reduce the treatment time. High temperature often comes with a trade-off with energy requirement and the solubility of lime (as lime is less soluble at higher temperature) (Wyman *et al.*, 2005). Lime treatment has been found to reduce the lignin content by 33% and acetyl group by about 100% regardless of temperature (Chang *et al.*, 1997; Gandi *et al.*, 1997). Lime was found to improve the digestibility of bagasse, African millet and Indian millet at a rate of twice the control sample when placed in the rumen of a cannulated steer (Gandi *et al.*, 1997). After 4 weeks of treatment at 50°C with aeration, Kim and Holtzapple (2005) found that lime removed 87.5% lignin at maximum and 91.3% glucan and 51.8% of glucan and xylan were converted to glucose and xylose. Although lime treatment combined with high temperature is sufficient to improve the digestibility of low lignin biomass, it is not sufficient for high lignin biomass, thus supplementation with oxidation is needed. Poplar wood chips were able to give 7.5-fold the combined recovery of glucan and xylan than untreated biomass when aeration-treated with lime at 65°C for 4 weeks (Sierra *et al.*, 2011). Several other researchers have worked on wheat straw, rice straw, switchgrass and maize stover (Chang *et al.*, 1998; Kaar and Holtzapple, 2000; Saha and Cotta, 2007; Park *et al.*, 2010).

23.4 New Pretreatments to Improve Ethanol Production

Most of these pretreatments often involve high temperature and acidic conditions resulting in degradation and leading to the formation of potential fermentation inhibitors. Moreover, most of the conventional pretreatment methods are batch in nature with low throughput and solids loading rate, resulting in high energy consumption. Dilute acid pretreatment requires corrosion-resistant construction materials, neutralization of the pretreated feedstocks; hydrothermal pretreatment requires elevated temperature; chemicals are expensive and difficult to recover and reuse; and in addition safety becomes an important concern. An effective pretreatment must expose the fibre, thereby the cellulose fraction could be hydrolysed by cellulases without significant difficulty, preserve the utility of the hemicellulose, and avoid the formation of inhibitory by-products (Laser *et al.*, 2002). Inexpensive chemicals, simple equipment and procedures enable the pretreatment to be economical. Issues related to various pretreatments are given in Table 23.2.

Thermo-mechanical/extrusion pretreatment meets the above listed criteria and addresses the listed issues (Table 23.2) associated with other pretreatment methods.

Table 23.2. Merits and demerits of various pretreatment methods.

Method	Merits	Demerits
Size reduction	Increases surface area; decreases crystallinity	High energy, low yield
Hot water	No chemicals; removes hemicellulose	High pressure and high water consumption; lignin is not removed
Steam explosion	Effectively hydrolyses hemicellulose; promotes delignification	High equipment costs
Wet oxidation	Production of only fewer by-products; no detoxification step; mild temperature	High pressure; cost of oxygen
AFEX, ARP	No sugar degradation; inhibitor formation is very limited; gas neutralization is not necessary	Cost of ammonia, ammonia recovery system and safety
Acid	High cellulose digestibility; removal of hemicellulose and lignin	Downstream neutralization is required; compounds such as acetic acid and furfural are formed; high equipment costs
Alkali	High efficiency; pressure vessel is not required	Cost of chemicals (NaOH, O_2); long residence time

23.4.1 Extrusion pretreatment

Numerous pretreatments have been investigated in the past with various degrees of success and investigations are still in progress. The extrusion/thermo-mechanical process is a continuous viable pretreatment and easy to adapt to industry scale with flexibility in process modifications. Extrusion, another important mechanical treatment process, has been studied as a potential pretreatment strategy of lignocelluloses. Agricultural residues and dedicated energy crops, such as maize stover, prairie cordgrass, wheat straw, rice straw, rape straw, barley straw, wheat bran, soybean hull, switchgrass, miscanthus, big bluestem, and forest products, such as poplar, pine, Douglas fir and eucalyptus, were subjected to extrusion pretreatment with or without chemicals. Extrusion is a continuous cost-effective, fast and simple process, hence practical and useful for large scale operation with high throughput and adaptability for many different process modifications (high pressure, explosion, chemical addition and reactive extrusion). Extrusion does not produce any effluent thereby resulting in no effluent disposal cost and no solid loss, and no safety issues are involved. The unique advantages of extrusion pretreatment are high shear, rapid mixing, short residence time, excellent temperature and screw speed control, and flexibility in screw configuration to mention a few. Moderate temperature employed in extrusion pretreatment prevents the degradation and formation of potential fermentation inhibitors (de Vrije *et al.*, 2002). Extruder can accommodate a wide range of feedstock sizes especially larger sizes when compared to other leading pretreatment methods, thereby extruder leads to remarkable savings in terms of size reduction.

In the extrusion process, as the feedstock passes through an extruder barrel, high shear is exerted by the screw and high pressure and temperatures are developed (Lamsal *et al.*, 2010) causing defibration, fibrillation and shortening the fibres (change in aspect ratio). When the pretreated feedstock exits through the die section, some of the moisture present in the feedstock is flashed off into steam due to the sudden drop in the pressure resulting in expansion and a porous structure. Several researchers have attributed possible reasons for the increase in sugar recovery in extrusion pretreatment, including increases in surface area, specific surface area, pore size and pore quantity, a decrease in cellulose crystallinity, all of which facilitate the access of enzymes to cellulose. The developed mechanical high shear breaks down the structure of feedstock and facilitates contact between feedstock and chemicals (acid/alkali/cellulose affinity additives) added during extrusion due to effective mixing. In general, pretreatment efficiency (sugar recovery) depends on several extruder parameters such as compression ratio, screw configuration, screw speed and barrel temperature. This process is considered less expensive and has the possible advantage of acting as a continuous pretreatment process (Karunanithy and Muthukumarappan, 2011a, b). Karunanithy and Muthukumarappan (2011a, b) were able to pretreat various feedstocks using a single screw extruder and obtained about 82% and 90% of total sugar recovery from switchgrass and maize stover, respectively. They compared alkali soaking of different feedstocks with extrusion and they found the performance of extrusion (48–57%) was better than alkali soaking (35–39%). However, when alkali-soaked, feedstocks pretreated in an extruder increased the total sugar recovery to 86–95%. The details of the extruder and samples of control and extruded maize stover and switchgrass are presented in Fig. 23.4. The authors reported about 90% of the control passed through an 8 mm sieve and about 95% of optimum pretreated maize stover and switchgrass passed through 1.78 and 1.85 mm sieves, respectively. Recently, Chen *et al.* (2011) found the mean particle diameter of extrusion pretreated rice straw in the range of 400–500 μm depending upon the screw speed, whereas extrusion followed by hot water extraction resulted in a lower range of 90–300 μm. The above results provide evidence that the increase in sugar recovery from different feedstocks is due to decrease in fibre diameter.

23.4.2 Ultrasound pretreatment

Ultrasound treatment involves the use of sonic waves to disrupt lignocelluloses to make them more vulnerable to digestion. It is audible to the human ear and has a range above 2–20 MHz. The mechanism of ultrasound disruption is found to be an effect known as cavitation, where microscopic bubble formation, growth and violent collapse occurs in the medium due to the pressure difference created by the propagating waves. This cavitation causes the rupturing of the cell structures further facilitating the other sono-chemical effect (Feng *et al.*, 2008). Ultrasound has been found to remove 90% lignin from poplar in assistance with sodium hydroxide and various chemicals (Tong-Qi Yuan, 2011). Ultrasound treatment (40 KHz) of maize meal for 5 min at 60°C increased the glucose by 6.82% over the control (Nikolic *et al.*, 2010). Ultrasound has also been helpful in enhancing the rate and extent of saccharification when

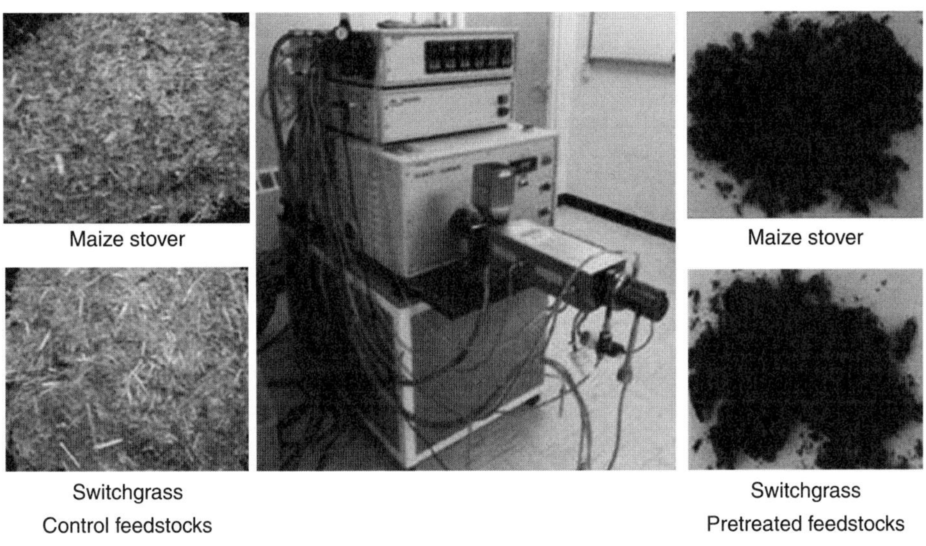

Fig. 23.4. Comparison of control and pretreated feedstocks in a single-screw extruder.

applied during the hydrolysis in bagasse (Rolz, 1986). Sodium hydroxide treatment followed by ultrasound treatment was found to increase the degradation efficiency of the waste sludge from 38.0 to 50.7%. It was also reported that the combined treatment of the alkali and ultrasound was better than each treatment alone (Jin *et al.*, 2009). Recently, Donepudi (2011) evaluated the effect of ultrasound in combination with acid or alkali treatment on sugar recovery from various feedstocks. In general, alkali and ultrasound combined treatment yielded higher total sugar recovery compared to acid and ultrasound combined treatment.

23.4.3 Ozone pretreatment

Ozone treatment involves the use of ozone gas to treat the lignocelluloses in order to delignify them. The process essentially removes lignin and hemicelluloses without affecting much of the cellulose. The process in general involves sparging of the ozone gas (generated from the ozone generator) on to biomass with some degree of moisture. The whole process takes place at normal temperature and pressure, however not all of the ozone sparged is utilized for the pretreatment, thus unused ozone escapes to the atmosphere. This underutilization of ozone adds to the cost of pretreatment. Several researchers have used ozone treatment to treat biomass including wheat and rye straw (García-Cubero *et al.*, 2009), high lignin fraction of maize stalks (Quesada *et al.*, 1999), sawdust, lumber and board waste (Sugimoto *et al.*, 2009) and have found success in recovering sugar from the treated biomass except cotton stalks (Silverstein *et al.*, 2007). Donepudi (2011) observed that acid and alkali pretreatment along with ozone pretreatment yielded higher sugar recovery than with ozone treatment alone for various feedstocks.

23.5 Future Considerations

Extrusion pretreatment of different feedstocks clearly identify it as one of the continuous pretreatments that is ready for scaling up. In general, the extrusion pretreatment alone is capable of producing a total sugar recovery in the range of 50–65%. Addition of alkali improved the sugar recovery to 80–90%. This indicates sequential pretreatment would be a viable option. Considering the cost economics of chemicals and effluent treatment, hot water/steam pretreatment in the front- (preferably) or back-end of extrusion would be explored for different feedstocks. Hot-water pretreatment at the front end would reduce the torque requirement remarkably during extrusion pretreatment, thereby saving energy costs. The other opportunities are in exploring different sequential pretreatments consisting of microwave, ultrasound or ozone at the front- or back-end of extrusion. In general, there could be about 20–30% increase in sugar recovery depending upon the pretreatment conditions. However, fermentation results of the extrusion pretreated feedstocks are missing. In order to prove extrusion pretreatment capability, extruded feedstock should be subjected to fermentation trials to check types of inhibitors and their levels and effects. Energy and economic analysis of the extrusion pretreatment in sequence with other novel pretreatments such as ultrasound and ozone pretreatments should be evaluated and documented. The provision of complete energy analyses of the extrusion pretreatment of different feedstocks would help the biorefinery industry.

Acknowledgement

The author would like to thank the Agricultural Experiment Station, South Dakota State University, Brookings, South Dakota for providing funds to complete this project.

References

Alvo, P. and Belkacemi, K. (1997) Enzymatic saccharification of milled timothy and alfalfa. *Bioresource Technology* 61, 185–198.

Burns, D.S., Ooshima, H. and Converse, A.O. (1989) Surface area of pretreated lignocellulosics as a function of the extent of enzymatic hydrolysis. *Applied Biochemistry and Biotechnology* 20/21, 79–94.

Cadoche, L. and Lopez, G.D. (1989) Assessment of size reduction as a preliminary step in the production of ethanol from lignocellulosic wastes. *Biological Wastes* 30, 153–157.

Carolan, J.E., Joshi, S.V. and Dale, B.E. (2007) Technical and financial feasibility of distributed bioprocessing using regional biomass pre-processing centers. *Journal of Agricultural Food and Industrial Organization* Special Issue 5, 10.

Chang, V., Burr, B. and Holtzapple, M. (1997) Lime pretreatment of switchgrass. *Applied Biochemistry and Biotechnology* 63–65, 3–19.

Chang, V., Nagwani, M. and Holtzapple, M. (1998) Lime pretreatment of crop residues bagasse and wheat straw. *Applied Biochemistry and Biotechnology* 74, 135–159.

Chen, W.H., Xu, Y.Y., Hwang, W.S. and Wang, J.B. (2011) Pretreatment of rice straw using an extrusion/extraction process at bench-scale for producing cellulosic ethanol. *Bioresource Technology* 102, 10451–10458.

de Vrije, T., de Haas, G.G., Tan, G.B., Keijsers, E.R.P. and Claassen, P.A.M. (2002) Pretreatment of miscanthus for hydrogen production by *Thermotoga elfii*. *International Journal of Hydrogen Energy* 27, 1381–1390.

Donepudi, A. (2011) Sugar recoveries from corn stover, prairie cord grass, switchgrass, and big bluestem using microwave, ultrasound, and ozone pretreatment technologies. Unpublished Masters thesis, South Dakota State University, Brookings, South Dakota.

Fan, L.T., Lee, Y.H. and Beardmore, D.H. (1980) Mechanism of the enzymatic hydrolysis of cellulose: effects of major structural features of cellulose on enzymatic hydrolysis. *Biotechnology and Bioengineering* 22, 177–199.

Feng, H., Yang, W. and Hielscher, T. (2008) Power ultrasound. *Food Science and Technology International* 14, 433–436.

Gandi, J., Holtzapple, M.T., Ferrer, A., Byers, F.M., Turner, N.D. *et al.* (1997) Lime treatment of agricultural residues to improve rumen digestibility. *Animal Feed Science and Technology* 68, 195–211.

García-Cubero, M.T., González-Benito, G., Indacoechea, I., Coca, M. and Bolado, S. (2009) Effect of ozonolysis pretreatment on enzymatic digestibility of wheat and rye straw. *Bioresource Technology* 100, 1608–1613.

Gibbons, W.R., Westby, C.A. and Dobbs, T.L. (1986) Intermediate-scale, semicontinuous solid phase fermentation process for production of fuel ethanol from sweet sorghum. *Applied Environment Microbiology* 51, 115–122.

He, Y., Pang, Y., Liu, Y., Li, X. and Wang, K. (2008) Physicochemical characterization of rice straw pretreated with sodium hydroxide in the solid state for enhancing biogas production. *Energy and Fuels* 22, 2775–2781.

Hendriks, A.T.W.M. and Zeeman, G. (2009) Pretreatments to enhance the digestibility of lignocellulosic biomass. *Bioresource Technology* 100, 10–18.

Hespell, R., O'Bryan, P., Moniruzzaman, M. and Bothast, R. (1997) Hydrolysis by commercial enzyme mixtures of AFEX-treated corn fiber and isolated xylans. *Applied Biochemistry and Biotechnology* 62, 87–97.

Holtzapple, M., Jun, J.-H., Ashok, G., Patibandla, S. and Dale, B. (1991) The ammonia freeze explosion (AFEX) process. *Applied Biochemistry and Biotechnology* 28–29, 59–74.

Igathinathane, C., Womac, A.R., Miu, P.I., Yu, M., Sokhansanj, S. *et al.* (2006) Linear knife grid application for biomass size reduction. *ASABE paper no. 066170*. ASABE, St Joseph, Michigan.

Igathinathane, C., Womac, A.R., Sokhansanj, S. and Narayan, S. (2007) Size reduction of wet and dry biomass by linear knife grid device. *ASABE paper no. 076045*. ASABE, St. Joseph, Michigan.

Ige, M.T. and Finner, M.F. (1976) Optimization of the performance of the cylinder type forage harvester cutterhead. *Transactions of the ASAE* 19, 455–460.

Isci, A., Himmelsbach, J., Strohl, J., Pometto, A., Raman, D. *et al.* (2009) Pilot-scale fermentation of aqueous-ammonia-soaked switchgrass. *Applied Biochemistry and Biotechnology* 157, 453–462.

Iyer, P., Wu, Z.-W., Kim, S. and Lee, Y. (1996) Ammonia recycled percolation process for pretreatment of herbaceous biomass. *Applied Biochemistry and Biotechnology* 57–58, 121–132.

Jannasch, R., Quan, Y. and Samson, R. (2005) A process and energy analysis of palletizing switchgrass. Resource Efficient Agricultural Production (REAP-Canada). Available at: http://www.reapcanada.com (accessed 2 January 2008).

Jin, Y., Li, H., Mahar, R.B., Wang, Z. and Nie, Y. (2009) Combined alkaline and ultrasonic pretreatment of sludge before aerobic digestion. *Journal of Environmental Sciences* 21, 279–284.

Kaar, W.E. and Holtzapple, M.T. (2000) Using lime pretreatment to facilitate the enzymic hydrolysis of corn stover. *Biomass and Bioenergy* 18, 189–199.

Karunanithy, C. and Muthukumarappan, K. (2011a) Optimization of alkali soaking and extrusion pretreatment of prairie cord grass for maximum sugar recovery by enzymatic hydrolysis. *Biochemical Engineering Journal* 54, 71–82.

Karunanithy, C. and Muthukumarappan, K. (2011b) Optimization of switchgrass and extruder parameters for enzymatic hydrolysis using response surface methodology. *Industrial Crops and Products* 33, 188–199.

Kim, S. and Holtzapple, M.T. (2005) Lime pretreatment and enzymatic hydrolysis of corn stover. *Bioresource Technology* 96, 1994–2006.

Kim, T. and Lee, Y. (2007) Pretreatment of corn stover by soaking in aqueous ammonia at moderate temperatures. *Applied Biochemistry and Biotechnology* 137–140, 81–92.

Kim, T., Lee, Y., Sunwoo, C. and Kim, J. (2006) Pretreatment of corn stover by low-liquid ammonia recycle percolation process. *Applied Biochemistry and Biotechnology* 133, 41–57.

Klein, I. (2007) Transportation logistics of biomass for industrial fuel and energy enterprises. *7th Annual Conference on Renewable Energy from Organics Recycling*, Indianapolis, Indiana.

Kumar, R., Mago, G., Balan, V. and Wyman, C.E. (2009) Physical and chemical characterizations of corn stover resulting from leading pretreatment technologies. *Bioresource Technology* 100, 3948–3962.

Kurakake, M., Kisaka, W., Ouchi, K. and Komaki, T. (2001) Pretreatment with ammonia water for enzymatic hydrolysis of corn husk, bagasse, and switchgrass. *Applied Biochemistry and Biotechnology* 90, 251–259.

Lamsal, B., Yoo, J., Brijwani, K. and Alavi, S. (2010) Extrusion as a thermo-mechanical pre-treatment for lignocellulosic ethanol. *Biomass and Bioenergy* 34, 1703–1710.

Laser, M., Schulman, D., Allen, S.G., Lichwa, J., Antal Jr, M.J. *et al.* (2002) A comparison of liquid hot water and steam pretreatments of sugar cane bagasse for bioconversion to ethanol. *Bioresource Technology* 81, 33–44.

Lee, D. and Yu, A.H.C. (1995) Evaluation of cellulase recycling strategies for the hydrolysis of lignocellulosic substrates. *Biotechnology and Bioengineering* 45, 328–336.

Mais, U., Esteghlalian, A., Saddler, J.N. and Mansfield, S.D. (2002) Enhancing the enzymatic hydrolysis of cellulosic materials using simultaneous ball milling. *Applied Biochemistry and Biotechnology* 99, 815–832.

Mani, S., Tabil, L.G. and Sokhansanj, S. (2004) Grinding performance and physical properties of wheat and barley straw, corn stover and switchgrass. *Biomass and Bioenergy* 27, 339–352.

Mani, S., Sokhansanj, S., Bi, X. and Turhollow, A. (2006) Economics of producing fuel pellets from biomass. *Applied Engineering in Agriculture* 22, 421–426.

McGinnis, G.D., Wilson, W.W. and Mullen, C.E. (1983) Biomass pretreated with water and high pressure oxygen: The WO process. *Industrial Engineering Chemical Product Research Development* 22, 352.

McMillan, J.D. (1994) Pretreatment of lignocellulosic biomass. In: Himmel, M.E., Baker, J.O. and Overend, R.P. (eds) *Enzymatic Conversion of Biomass for Fuels Production*. American Chemical Society, Washington, DC, pp. 292–324.

Millet, M.A., Baker, A.J. and Scatter, L.D. (1976) Physical and chemical pretreatments for enhancing cellulose saccharification. *Biotechology Bioengineering Symposium* 6, 125–153.

Miu, P.I., Womac, A.R., Igathinathane, C. and Sokhansanj, S. (2006) Analysis of biomass comminution and separation, processes in rotary equipment – a review. *ASABE paper no. 066169*. ASABE, St Joseph, Michigan.

Mohsenin, N. (1986) *Physical Properties of Plant and Animal Materials*, 2nd revised and updated edn. Gordon and Breach, Science Publishers, New York.

Mosier, N., Wyman, C., Dale, B., Elander, R., Lee, Y.Y. *et al.* (2005a) Features of promising technologies for pretreatment of lignocellulosic biomass. *Bioresource Technology* 96, 673–686.

Mosier, N., Hendrickson, R., Ho, N., Sedlak, M. and Ladisch, M.R. (2005b) Optimization of pH controlled liquid hot water pretreatment of corn stover. *Bioresource Technology* 96, 1986–1993.

Muthukumarappan, K. (2012) Novel integration of preprocessing and densification of different biomass feedstocks. Presented at: *The Advanced Biofuels in a Biorefinery Approach*, 28 February–1 March 2012, Copenhagen, Denmark.

National Research Council (2000) *Biobased Industrial Products – Priorities of Research and Commercialization*. National Academy Press, Washington, DC.

Nikolic, S., Mojovic, L., Rakin, M., Pejin, D. and Pejin, J. (2010) Ultrasound-assisted production of bioethanol by simultaneous saccharification and fermentation of corn meal. *Food Chemistry* 122, 216–222.

Park, J.-Y., Shiroma, R., Al-Haq, M.I., Zhang, Y., Ike, M. *et al.* (2010) A novel lime pretreatment for subsequent bioethanol production from rice straw – Calcium capturing by carbonation (CaCCO) process. *Bioresource Technology* 101, 6805–6811.

Puri, V.P. (1984) Effect of crystallinity and degree of polymerization of cellulose on enzymatic saccharification. *Biotechnology and Bioengineering* 26, 1219–1222.
Quesada, J., Rubio, M. and Gómez, D. (1999) Ozonation of lignin rich solid fractions from corn stalks. *Journal of Wood Chemistry and Technology* 19, 115–137.
Reshamwala, S., Shawky, B. and Dale, B. (1995) Ethanol production from enzymatic hydrolysates of AFEX-treated coastal bermudagrass and switchgrass. *Applied Biochemistry and Biotechnology* 51–52, 43–55.
Rolz, C. (1986) Ultrasound effect on enzymatic saccharification. *Biotechnology Letters* 8, 131–136.
Saha, B. and Bothast, R. (1997) Enzymes in lignocellulosic biomass conversion. In: *Fuels and Chemicals from Biomass. ACS Symposium Series* 666, 46–56.
Saha, B.C. and Cotta, M.A. (2007) Enzymatic hydrolysis and fermentation of lime pretreated wheat straw to ethanol. *Journal of Chemical Technology and Biotechnology* 82, 913–919.
Schell, D.J. and Harwood, C. (1994) Milling of lignocellulosic biomass: results of pilot-scale testing. *Applied Biochemistry and Biotechnology* 45/46, 159–168.
Sidiras, D.K. and Koukios, E.G. (1989) Acid saccharification of ball-milled straw. *Biomass* 19, 289–306.
Sierra, R., Holtzapple, M.T. and Granda, C.B. (2011) Long-term lime pretreatment of poplar wood. *AIChE Journal* 57, 1320–1328.
Silverstein, R.A., Chen, Y., Sharma-Shivappa, R.R., Boyette, M.D. and Osborne, J. (2007) A comparison of chemical pretreatment methods for improving saccharification of cotton stalks. *Bioresource Technology* 98, 3000–3011.
Sokhansanj, S. and Turhollow, A.F. (2004) Biomass densification–cubing operations and costs for corn stover. *Applied Engineering in Agriculture* 20, 495–499.
Sokhansanj, S., Cushman, J. and Wright, L. (2003) Collection and delivery of feedstock biomass for fuel and power production. *Agricultural Engineering International: The CIGR Journal of Science Research and Development*. Invited Overview Paper, Vol. V, February 2003.
Spinelli, R. and Hartsough, B. (2001) A survey of Italian chipping operation. *Biomass and Bioenergy* 21, 433–444.
Sugimoto, T., Magara, K., Hosoya, S., Oosawa, S., Shimoda, T. and Nishibori, K. (2009) Ozone pretreatment of lignocellulosic materials for ethanol production: Improvement of enzymatic susceptibility of softwood. *Holzforschung* 63, 537–543.
Sun, Y. and Cheng, J. (2002) Hydrolysis of lignocellulosic materials for ethanol production – a review. *Bioresource Technology* 83, 1–11.
Tabil, L., Sokhansanj, S. and Tyler, R.T. (1997) Performance of different binders during alfalfa pelleting. *Canadian Agricultural Engineering* 39, 17–23.
Tarkow, H. and Feist, W.C. (1969) A mechanism for improving the digestibility of lignocellulosic materials with dilute alkali and liquid ammonia. In: Gould, R.F. (ed.) *Cellulases and Their Applications.* American Chemical Society, Washington, DC, pp. 197–218.
Teymouri, F., Laureano-Pérez, L., Alizadeh, H. and Dale, B. (2004) Ammonia fiber explosion treatment of corn stover. *Applied Biochemistry and Biotechnology* 115, 951–963.
Tong-Qi Yuan, S.S., Feng, X. and Sun, R. (2011) Isolation and physico-chemical characterization of lignins from ultrasound irradiated fast-growing poplar. *BioResources* 6, 112–115.
van Walsum, G.P., Allen, S.G., Spencer, M.J., Laser, M.S., Antal, M.J. *et al.* (1996) Conversion of lignocellulosics pretreated with liquid hot water to ethanol. *Applied Biochemistry and Biotechnology* 57/58, 157–169.
Walch, E., Zemann, A., Schinner, F., Bonn, G. and Bobleter, O. (1992) Enzymatic saccharification of hemicellulose obtained from hydrothermally pretreated sugar cane bagasse and beech bark. *Bioresource Technology* 39, 173–177.
Wyman, C.E. (1994) Ethanol from lignocellulosic biomass: technology, economics and opportunities. *Bioresource Technology* 50, 3–50.
Wyman, C.E., Dale, B.E., Elander, R.T., Holtzapple, M., Ladisch, M.R. *et al.* (2005) Coordinated development of leading biomass pretreatment technologies. *Bioresource Technology* 96, 1959–1966.
Xu, J., Cheng, J.J., Sharma-Shivappa, R.R. and Burns, J.C. (2010) Sodium hydroxide pretreatment of switchgrass for ethanol production. *Energy and Fuels* 24, 2113–2119.
Yang, B. and Wyman, C.E. (2008) Pretreatment: the key to unlocking low-cost cellulosic ethanol. *Biofuels, Bioproducts and Biorefining* 2, 26–40.
Zheng, Y., Zhongli, P. and Ruihong, Z. (2009) Overview of biomass pretreatment for cellulosic ethanol production. *International Journal of Agricultural and Biological Engineering* 2, 51–68.

Appendix I Energy and Related Units

Basic Energy Units

1 calorie (thermochemical) = 4.184 J
1 calorie (15°C) = 4.1858 J
1 calorie (IT) = 4.1868 J
1 calorie (mean) = 4.1900 J
1 Btu = 251.9958 calories
1 Btu (thermochemical) = 1054.35 J
1 Btu (59°F) = 1054.80 J
1 Btu (IT) = 1055.06 J
1 Btu (mean) = 1055.87 J
1 kilowatt-hour (kWh) = 3.6×10^{6} J
1 kilowatt-hour (kWh) = 3412 Btu (IT)
1 therm = 100,000 Btu
1 electron-volt (eV) = 1.6022×10^{-19} J

Large-Scale Energy Units

1 quad = 109 MBtu = 1015 Btu
1 exajoule (EJ) = 1018 J
1 terawatt-year (TWyear) = 8.76×10^{12} kWh

Energy Units as Oil Equivalents

1 barrel of oil equivalent (BOE) = 5.80 MBtu
1 millions of barrels of oil per day (Mbd) = 2.12 quad $year^{-1}$
1 tonne of oil equivalent (TOE) = 41.868 GJ = 39.68 MBtu
1 megatonne of oil equivalent (Mtoe) = 4.1868×10^{16} J
1 gigatonne of oil equivalent (Gtoe) = 4.1868×10^{19} J = 41.868 EJ = 39.68 quad

Related Units

1 short ton (ton) = 2000 lb
1 metric tonne (t) = 1000 kg
1 ton = 0.907185 t
1 barrel (bbl) = 42 US gallons = 159.0 l
1 barrel of crude oil ~ 0.136 t
1 square mile = 640 acres = 2.590 km^2
1 hectare (ha) = 10^{-2} km^2 = 2.471 acres

©CAB International 2013. *Biofuel Crops: Production, Physiology and Genetics* (ed. B.P. Singh)

Appendix II Botanical Names

Common name	Botanical name
Abyssinian cabbage	*Brassica carinata*
Acacia	*Acacia* spp.
Baldcypress	*Taxodium distichum*
Bamboo	*Bambusa* spp.
Barley	*Hordeum vulgare*
Beech	*Fagus spp.*
Big bluestem	*Andropogon gerardii*
Birch	*Betula pendula*
Black mustard	*Brassica nigra*
Blue agave	*Agave tequilana*
Brown mustard	*Brassica juncea*
Cabbage	*Brassica oleracea*
Cabbage gum	*Eucalyptus amplifolia*
Camelina	*Camelina sativa*
Cassava	*Manihot esculenta*
Castor	*Ricinus communis*
Chickpea	*Cicer arietinum*
Chinese brake fern	*Pteris vittata*
Chinese kale	*Brassica alboglabra*
Clover	*Trifolium* spp.
Coconut	*Cocos nucifera*
Common reed	*Phragmites australis*
Cotton	*Gossypium hirsutum*
Crabgrasses	*Digitaria* spp.
Cultivated radish	*Raphanus sativus*
Eastern cottonwood/poplar	*Populus deltoides*
Ethiopian mustard	*Brassica carinata*
Eucalyptus/sugar gum	*Eucalyptus spp.*
Flax	*Linum usitatissimum*
Foxtail millet	*Setaria italica*

Continued

©CAB International 2013. *Biofuel Crops: Production, Physiology and Genetics* (ed. B.P. Singh)

Common name	Botanical name
Giant reed	*Arundo donax*
Grand eucalyptus	*Eucalyptus grandis*
Groundnut/peanut	*Arachis hypogaea*
Guayule	*Parthenium argentatum*
Guinea grass	*Panicum maximum*
Hemp	*Cannabis sativa*
Henequen	*Agave fourcroydes*
Hevea/rubber	*Hevea brasiliensis*
Indiangrass	*Sorghastrum nutans*
Jatropha	*Jatropha curcas*
Jew's Mallow	*Corchorus olitorius*
Jute	*Corchorus capsularis*
Kenaf	*Hibiscus cannabinus*
Lechuguilla	*Agave lechuguilla*
Loblolly pine	*Pinus taeda*
Lucerne (alfalfa)	*Medicago sativa*
Lupine	*Lupinus* spp.
Maize/corn	*Zea mays*
Miscanthus	*Miscanthus capensis*
Moso bamboo	*Phyllostachys pubescens*
Napiergrass/elephantgrass	*Pennisetum purpureum*
Nutgrass	*Cyperus rotundus*
Oak	*Quercus robur*
Oil palm	*Elaeis guineensis*
Pangola grass	*Digitaria decumbens*
Pearl acacia	*Acacia podalyriifolia*
Pearl millet	*Pennisetum glaucum*
Pigeon grass	*Setaria sphacelata*
Pongamia	*Pongamia pinnata*
Poppy	*Papaver somniferum*
Ramie	*Boehmeria nivea*
Rapeseed/canola/mustard	*Brassica napus*
Red clover	*Trifolium pratense*
Reed canarygrass	*Phalaris arundinacea*
Rice	*Oryza sativa*
Roselle	*Hibiscus sabdariffa*
Rubber	*Ficus elastica*
Safflower	*Carthamus tinctorius*
Shameplant	*Mimosa pudica*
Sisal	*Agave sisalana*
Sorghum/sweet sorghum	*Sorghum bicolor*
Southern pine/pine	*Pinus* spp.
Soybean	*Glyine max*
Spruce	*Picea* spp.
Sugarbeet	*Beta vulgaris*
Sugarcane	*Saccharum* spp.
Sugar gum	*Eucalyptus cladocalyx*
Sunflower	*Helianthus annuus*
Sunnhemp	*Crotalaria juncea*
Sweet potato	*Ipomoea batatas*
Switchgrass	*Panicum virgatum*
Tea	*Camellia sinensis*
Tobacco	*Nicotiana tabacum*
Tomato	*Solanum lycopersicum*

Continued

Common name	Botanical name
Tropical blue bamboo	*Bambusa atra*
Tung	*Aleurites fordii*
Turnip	*Brassica rapa* subsp. *sylvestris*
Wheat	*Triticum aestivum*
White wallrocket	*Diplotaxis erucoides*
Willow	*Salix* spp.

Index